AF478029

Aquaculture and Fisheries Biotechnology:
Genetic Approaches, 2nd Edition

Printed on FSC approved paper

Aquaculture and Fisheries Biotechnology: Genetic Approaches, 2nd Edition

Rex A. Dunham

Department of Fisheries and Allied Aquacultures
Auburn University
Alabama
USA

CABI is a trading name of CAB International

CABI
Nosworthy Way
Wallingford
Oxfordshire OX10 8DE
UK

Tel: +44 (0)1491 832111
Fax: +44 (0)1491 833508
E-mail: info@cabi.org
Website: www.cabi.org

CABI
745 Atlantic Avenue
8th Floor
Boston, MA 02111
USA

Tel: +1 (617) 682-9015

A catalogue record for this book is available from the British Library, London, UK.

Library of Congress Cataloging-in-Publication Data

Dunham, Rex A.
    Aquaculture and fisheries biotechnology: genetic approaches / Rex A. Dunham. – 2nd ed.
        p. cm.
    Includes bibliographical references and index.
    ISBN 978-1-84593-651-8 (alk. paper)
1. Fishes--Breeding. 2. Shellfish--Breeding. 3. Fishes--Molecular genetics.
4. Shellfish--Molecular genetics. I. Title.

    SH155.5.D86 2011
    639.3--dc22

                                    2011003482

ISBN-13: 978 1 84593 651 8

Commissioning editor: Rachel Cutts
Editorial assistant: Gwenan Spearing
Production editor: Simon Hill

Typeset by AMA DataSet, Preston, UK
Printed and bound by CPI Group (UK) Ltd, Croydon, CR0 4YY

First printed in 2011.
Transferred to print on demand in 2018.

Dedicated to

Petrcyle (Azset) – akong usa ka lawasnon, gihigugma ko ikaw hangtod sa hangtod
Aubrey, Nick and Julian; Christian, Nicole and Aaliyah; Amy; Gabrielle; Sean; and
Jeef – proof that genes work
Carol Jean – our Saint
Mom – enduring and loving
Scott, Earl and Dixie – educators all
Terry Abella – a true friend
The Dunham and Gealon families

In memory of my Dad
Richard Vincent (Dick) Dunham
December 1, 1919–August 11, 1971

Contents

1

History of Biotechnology, Genetics and Selective Breeding in Aquaculture and Fisheries

Aquaculture is an ancient form of farming dating back 2000 years or more in China and to the Roman Empire (Balon, 1995; Dunham *et al.*, 2001). However, only in the last few decades has aquaculture grown into a global practice resulting in tremendous worldwide production. Aquaculture production has enlarged dramatically since the early 1980s, and will become increasingly important as demand for fish products increases, world harvest by capture fisheries reaches a plateau or declines and human population numbers expand. Aquaculture now accounts for 50% of global fish consumption (FAO, 2009). The biomass of fish that can be produced per surface area is much greater than that for terrestrial animals, indicating that aquaculture could be the key for providing global food security.

Humans were hunter–gatherers prior to being farmers and fishermen before they were aquaculturalists. Although aquaculture is growing in importance and must expand to meet future demand for fish products, commercial harvest of natural populations has traditionally been of higher economic value than aquaculture and will continue to be of great importance. Even as aquaculture closes the gap or surpasses the value of commercial fisheries, the genetic management and conservation of natural fish stocks and gene pools will be of great

importance. Genetic variation is one key variable in the survival of various species. Also, natural populations are perhaps the best gene banks, a critical resource for genetic variation for current and future application in genetic improvement for farmed species and specialized sport-fish applications.

Recreational fishing is also of great importance in many countries. When the revenue from the fish, licences, fishing equipment, boats, travel, food and lodging is considered, recreational fishing is probably more than tenfold more important economically than aquaculture in the USA. Biotechnology is permanently linked not only to aquaculture, but also to commercial and recreational fisheries because of its potential positive and negative impacts on these resources.

Currently, the quantity of animal protein harvested from global aquatic sources via the capture of natural fish populations is at maximum sustainable yield. Many major fish stocks are showing precipitous declines in productivity due to overfishing and further increases are not anticipated under the current global conditions and environment. Wild fish stocks have been heavily fished or overfished, which has resulted in a noticeable levelling of fish landings at around 60 million t, with harvest from oceans unlikely to expand (Hardy, 1999). Almost two-thirds of marine stocks in the Pacific and Atlantic Oceans are being fully exploited or have already been overfished (Pauly *et al.*, 1998). The Food and Agriculture Organization of the United Nations (FAO) predicts a 36% increase in the world population, with only a 30% increase in production from aquaculture and fisheries. Hardy (1999) predicts a 55 million t shortage in demanded seafood products by 2025 resulting from levelled wild catch and increasing demand. By 2025, aquaculture will have to increase by 350% to cover the impending shortage (Hardy, 1999). In 1993, approximately 16 million t of aquacultured animal protein were produced, representing some 13% of the total aquatic animal protein harvested or produced (Tacon, 1996). This grew to 59.4 million t by 2004 (FAO, 2006) and was approaching 90 million t for 2009 (FAO, 2009). The growth of aquacultured animal protein increased at a rate of over 11.8% annually from 1985 to 1994, but slowed to 7.1% from 1995 to 2004, compared with the more modest growth of terrestrial meat production, which ranges from 0.7% (beef) to 5.2% (poultry).

With increased demand for aquacultured foods has come a need for more efficient production systems. Major improvements have been achieved through enhanced husbandry procedures, improved nutrition, enhanced disease diagnosis and therapies and the application of genetics to production traits. Although several aquaculture species have been greatly improved through the application of genetics, much greater improvements can be accomplished (Dunham *et al.*, 2001; Dunham, 2004). Genetics can greatly contribute to production efficiency, enhancing production and increasing sustainability. Resource utilization can be greatly improved and impediments to sustainability, such as slow growth of fish, inefficient feed conversion, heavy mortality from disease and the associated use of chemicals, loss of fish from low oxygen levels, inefficient harvest, poor reproduction, inefficient use of land space and processing loss, can all be diminished by utilizing genetically improved fish. Genetic enhancement of farmed fish has advanced to the point that it is now having an impact on aquaculture worldwide, but potential maximum improvement in overall performance is not close to being achieved. As space for aquaculture becomes more limiting, the necessity for more efficient production or increased production within the same amount of space will further increase the importance of genetic improvement of aquaculture species. Genetic research and its application have had a significant role in the development of aquaculture, and this role and impact will become increasingly important as aquaculture develops further.

Aquaculture genetics actually had its origin with the beginning of aquaculture in China and the Roman Empire more than 2000 years ago. Without realizing it, the first fish culturists changed gene frequencies and altered performance of the wild-caught fish, actually genetically enhancing the fish for fish-farming application by closing the life cycles and domesticating species such as the common carp, *Cyprinus carpio*. When the Chinese, Europeans and others observed mutations and phenotypic variation for colour, body conformation and finnage, and then selected for these phenotypes as well as for body size, genetics and selective breeding of fish and shellfish was born. Additionally, fish culturists and scientists who compared and evaluated closely related species for their suitability for aquaculture application over the past two millennia were also unknowingly conducting some of the first fish genetics research. Closely related species are reproductively isolated and have species status because of their genetic distance from one another; therefore, the comparison of different species is a genetic comparison (Dunham *et al.*, 2001). However, directed breeding and genetics programmes were probably not intense and strongly focused until the Japanese bred koi in the 1800s and the Chinese developed fancy goldfish.

Of course, fish biotechnology and molecular genetics research and development share the same beginnings as biotechnology and molecular genetics applied to other organisms when in 1665 Robert Hooke described cellular entities and developed the cell theory. Shortly thereafter, in 1667, Anton van Leeuwenhoek discovered that semen contained spermatozoa and theorized that they could fertilize eggs, although this was not substantiated for another 200 years. A series of discoveries during the 338 years that followed have led to the current state of biotechnology, molecular genetics and selective breeding.

The foundation for electrophoretic analysis was laid in 1816 when R. Pornet reported the effect of electric fields on charged particles, including proteins (Richardson *et al.*, 1986). Brown recognized the nucleus as a regular, constant cellular element within cells in 1831, and in the 1840s Carl Nageli had observed that the nucleus divided first in dividing cells but did not understand the significance of this observation. Although Charles Darwin (1859) was not the first to develop the theory of evolution based on selection of the fittest, this key concept was made believable in the 1860s by Charles Darwin and Alfred R. Wallace. Darwin's grandfather, Erasmus Darwin, was a proponent of evolution, and by the late 18th century Buffon and Lamarck had theorized that acquired characteristics were heritable. Buffon and Lamarck believed that the external environment brought about change, but Geoffroy Saint-Hilaire felt change was embryonic or germinal. Although their knowledge is not recorded in writing, obviously early fish breeders understood and utilized these basic concepts.

Of course, one of the most important keys for the emergence of the field of genetics occurred in 1866 when Gregor Mendel discovered the existence of genes and their transmission from generation to generation. Shortly thereafter, in 1869, Friedrich Miescher discovered deoxyribonucleic acid (DNA), although the full implications of this discovery were obviously not completely understood. In the 1870s, the German scientist Abbe developed the condenser and the oil-immersion lens, which enabled the description of chromosomes, and by 1879 Walter Flemming first observed the doubling of chromosomes.

Eduard and Benden (1880s) found that the nematode zygote received half its chromosomes from each parent, and in 1882 Flemming described the process and named it 'mitosis'.

In 1889, F. Galton laid the mathematical foundations for the study of quantitative variation and quantitative genetics. Galton was a cousin of Darwin, so that may have influenced him to examine genetics from a quantitative angle rather than the Mendelian qualitative approach. Ernst von Tschermak in Austria, Hugo DeVries in Holland and Carl Correns in Germany independently cited Mendel's research in 1900, and Mendel's work was then recognized and appreciated. About this time, Johanssen introduced the term 'gene'. Then, in the early 1900s, W. Johannsen, H. Nilsson-Ehle, E.M. East and R.A. Fisher tied the specific relationships of Mendelian genetics to biometrical approaches to develop the basis of quantitative genetics.

Sex chromosomes were discovered by C.E. McClung, E.B. Wilson and Nettie Stevens between 1901 and 1905 (Avers, 1980). In 1902–1903, Walter S. Sutton and Theodor Boveri linked Mendel's results with meiosis to explain Mendel's results, thereby connecting two independent disciplines, genetics and cytology, to develop the chromosomal theory of inheritance (Hartwell *et al.*, 2000). Correns, William Bateson and R.C. Punnett first discovered gene linkage – aberrations of Mendelian ratios – in 1905, but were unable to explain their results (Avers, 1980).

The relationship between genes and proteins was first suggested by Archibold E. Garrod in 1908. Also that year, G. Hardy and W. Weinberg independently developed some of the basic laws governing population genetics (Goodenough and Levine, 1974). Thomas Hunt Morgan and Calvin Bridges provided experimental proof of the chromosome theory in 1910. Thomas Hunt Morgan was one of the first to demonstrate the concept of linkage in *Drosophila*. Crossing over was first described by F.A. Janssens about 1909 and then verified in 1931 by Barbara McClintock and Harriet Creighton. Early in the 20th century, Thomas Hunt

Morgan and his student Alfred Sturtevant described single and multifactorial inheritance, chromosome mapping, gene linkage and recombination, sex linkage, mutagenesis and chromosome aberrations. Sturtevant described linear linkage in 1913.

During the 1920s and 1930s much progress was made in the field of population genetics. These efforts were led primarily by R.A. Fisher and Sewell Wright. Much of the research was related to selection, inbreeding and relatedness of individuals and populations, and also application and relevance to quantitative genetics and selective breeding. Also, Embody and Hayford (1925) conducted some of the first fish genetics research, a strain comparison of rainbow trout, *Oncorhynchus mykiss*, during this time period.

George Beadle and Edward Tatum advanced the hypothesis of one gene–one enzyme in 1941 and Avery, McCarty and MacLeod indicated that DNA was the physical material for heredity, bringing together Mendelian genetics, biochemistry and cytogenetics for the beginning of molecular genetics. By 1951 Barbara McClintock had identified movable control elements, but the understanding and appreciation of this concept would wait for many years. A major milestone was accomplished in 1953 when James Watson, Francis Crick and Maurice Wilkins discovered the molecular model for the chemical structure of DNA, the double-helical nature of DNA. Between 1961 and 1964, Marshall Nirenberg, Henry Matthaei, Severo Ochoa, H.G. Khorana and others deciphered the genetic code, and then Charles Yanofsky and Alan Garen followed with genetic evidence for the code (Goodenough and Levine, 1974; Avers, 1980).

Isozyme analysis was also developed in the 1950s. Oliver Smithies developed starch gel electrophoresis (Smithies, 1955), and Clement Markert and R.L. Hunter developed histochemical staining (Hunter and Markert, 1957) for the visualization of enzymes and isozymes (Richardson *et al.*, 1986; Whitmore, 1990). By the 1960s most of the key components for modern biotechnology were in place.

'Biotechnology' can be a confusing term. A Hungarian engineer defined biotechnology as all lines of work by which products are produced from raw materials with the aid of living things, in reference to an integrated process of using sugarbeets to produce pigs (Glick and Pasternak, 1998). However, the term was associated with industrial fermentation or ergonomics in 1961, when Carl Goran Heden's scientific journal led to biotechnology being defined as the industrial production of goods and services by processes using biological organisms, systems and processes. In the last two decades, biotechnology has often been associated with recombinant DNA technology, but in actuality recombinant DNA technology is a subdiscipline of biotechnology.

Fish genetics programmes first emerged in the 1900s after the basic principles of genetics and quantitative genetics had been established. However, there was not a substantial effort in fish genetics research and the application of genetic enhancement programmes until the 1960s because of the infancy and small scale of aquaculture, a lack of knowledge of fish genetics and a lack of appreciation of genetic principles by natural-resource managers regarding genetic enhancement, population genetics and conservation genetics. Slightly earlier, Ellis Prather conducted some of the first selection experiments with fish at Auburn University during the early 1940s, and appeared to have made significant improvement in the feed-conversion efficiency of largemouth bass, *Micropterus salmoides*, when fed minnows. Lauren Donaldson selected rainbow trout for increased growth in the 1950s, developing the Donaldson rainbow trout which appeared to have significantly altered the growth, body shape and fecundity of the rainbow trout. Unfortunately, many early fish culturists did not keep adequate genetic controls, and neither Prather nor Donaldson utilized genetic controls, making any genetic progress unverifiable.

Since the 1960s, fish genetics research and application of genetically improved fish and genetics principles have been gaining momentum with each passing decade.

In 1959, H. Swarup was one of the first to induce triploidy in fish – the three-spined stickleback, *Gasterosteus aculeatus*. Giora Wohlfarth and Rom Moav initiated a considerable amount of research on traditional selective breeding of common carp in the 1960s in Israel. This led to the development of the channel catfish traditional selective breeding efforts of Rex Dunham and R. Oneal Smitherman in the 1970s and 1980s in the USA after initial collaboration between the Israelis and Smitherman. Also in the 1970s, Trgve Gjedrem, Harold Kincaid and, later, William Hershberger initiated long-term selection programmes for various salmonids. The Auburn University catfish genetic enhancement programme and the Norwegian salmon breeding programme are now the longest ongoing fish genetic enhancement programmes, each 40 years old. This early work on selective breeding was the predecessor to later research on molecular genetics of aquaculture species. Also in the 1970s, Rafael Guerrero III and William Shelton developed sex-reversal technology for tilapia, which would later lead to the development and worldwide application of genetically male tilapia.

The next major technological breakthrough was the first isolation of restriction endonucleases by Werner Arber, Hamilton O. Smith and Daniel Nathans around 1970, which was the key discovery allowing the development of gene cloning (1978), genetic engineering (1978) and various restriction-fragment technologies (Glick and Pasternak, 1998). The discovery of reverse transcriptase by Howard Temin and David Baltimore was, of course, also key for the development of modern recombinant DNA technology. Then in 1973, Stanley Cohen and Herbert Boyer devised recombinant DNA technology (Cohen *et al.*, 1973). This type of research was further enhanced in 1975 with the development of procedures to rapidly obtain DNA sequences (Sanger *et al.*, 1977) and to visualize specific DNA fragments (Southern, 1975).

The 1980s saw more quantum leaps in molecular genetics biotechnology. Around 1980, Palmiter, Brinster and Wagner produced the first transgenic animals, mice, and Palmiter and Brinster demonstrated that the transgenesis could lead to greatly accelerated growth in the mice. Palmiter, Brinster and Wagner demonstrated the dramatic phenotypic alterations that could be realized through gene transfer. This provided the motivation and impetus for the dovolopmcnt of tcchnology for the generation of the first transgenic fish. In a year-and-a-half span from 1985 to 1987, Zhou first transferred genes into goldfish in China, followed by Ozato in Japan with medaka, Daniel Chourrout in France with rainbow trout and Rex Dunham in the USA with channel catfish.

In 1985, Jeffreys developed DNA fingerprinting technology (Jeffreys *et al.*, 1985), revolutionizing not only population genetic analysis and gene-mapping technology, but also forensic and criminal science. The current state of modern molecular genetics and genomics research would not have been possible without the revolutionary invention of the polymerase chain reaction (PCR) by Kary Mullis in 1985.

The new biotechnologies, such as sex reversal and breeding and polyploidy, began to have a major impact on aquaculture production in the late 1980s and early 1990s by not only improving growth rates but also allowing major improvement of flesh quality in species that exhibit sexual dimorphic and sexual maturation effects. Chourrout (1982) induced the first viable tetraploid from a normal diploid fish, rainbow trout; Standish Allen developed triploid technology for shellfish during the late 1980s; and Gary Thorgaard developed clonal lines of rainbow trout via androgenesis. The pioneering research on sex reversal and breeding technology by Shelton and Guerrero led to worldwide production of monosex Nile tilapia in the 1980s and 1990s, and Graham Mair took this technology one step further in the 1990s, leading to the development of YY populations of Nile tilapia and the production of genetically male tilapia (GMT) populations in many countries. Traditional breeding has already been utilized in concert with these new biotechnologies.

The 1990s brought continued rapid progress in molecular genetics and biotechnology. DNA marker and gene-mapping technology began to explode in the 1990s, and those advancements have accelerated in a remarkable manner during the past decade. Microsatellites were developed in 1989, radiation hybridization in 1990. random amplified polymorphic DNA (RAPD) and expressed sequence tag (EST) technologies in 1991, the amplified fragment length polymorphism (AFLP) technique in 1995 and single nucleotide polymorphism (SNP) procedures in 1998. Another major advance was the first nuclear cloning of a mammal, sheep, in 1997.

Environmental concerns about the application of biotechnology and genetic engineering emerged in the 1980s. Application of gene-transfer technology will not happen until genetic engineering is proved to be a safe technology. In the mid-1990s, Dunham conducted the first environmental-risk research with transgenic fish, channel catfish, demonstrating that in natural conditions the transgenics were slightly less fit than non-transgenic cohorts. Also in the mid-1990s, Du, Choy Hew, Garth Fletcher and Robert Devlin produced the first transgenic fish, salmon, exhibiting hyperlevels of growth – two- to sixfold and, in the case of Devlin's research, ten- to 30-fold increases in growth rate. Shortly thereafter, Norman Maclean produced transgenic tilapia with a two- to fourfold increase in growth rate.

Technological advances in DNA marker technologies and DNA microarray, gene chip and sequencing technologies have further accelerated the pace of aquaculture and aquaculture genomics. Genomic research has produced vast amounts of information towards an understanding of the genomic structures, organization, evolution and genes involved in the determination of important economic traits of aquatic organisms. Positional cloning of genes from aquatic species is no longer a dream. Zhanjiang Liu has led efforts to isolate and sequence more than 25,000 genes in catfish. Genomic sequences for fugu, tetraodon, medaka, zebrafish and stickleback are completed. Additionally, genomic sequences are near completion or well underway for Atlantic cod, Atlantic salmon, channel catfish, Nile tilapia and Pacific oyster.

In recent years, genomics technology has made remarkable advances. Entire genomes, 20,000–50,000 genes simultaneously, can now be observed for their expression in response to environmental variables with advanced microarrays. Next-generation sequencing will allow sequencing of entire genomes in weeks compared with the years previously required.

Commercial application of fish genetics and biotechnology are making advances as well. Another significant figure for fish genetics biotechnology has been Elliot Entis, CEO of AF Protein/AquaBounty Farms. His company is the first to attempt commercialization of transgenic fish, again salmon, in the developed world, beginning the process in the latter 1990s and on into the new millennium. Sam Lawrence of Eagle Aquaculture has led efforts to commercialize interspecific hybrid catfish in the USA, making this technology a reality. Several options for genetically enhanced tilapia are available in the Philippines with improved fish developed by selection, introgressive backcrossing as well as sex reversal and breeding.

In the early 1970s, research on the selective breeding of aquatic organisms gained momentum and long-term breeding programmes for salmon in Norway and catfish at Auburn University in the USA were established, which are now 40 years in duration. Selective breeding of tilapia in the Philippines began in the 1990s and now has been continuous for 21 generations. Since the early 1980s, research in aquaculture and fisheries genetic biotechnology has grown steadily, complementing the work in genetics and selective breeding, and the progress made in genomics and biotechnology since that time is almost unimaginable. Goro Yoshisaki has accomplished xenogenesis in fish, which could lead to a variety of interesting and novel genetic manipulations.

Currently, efforts are well established in the areas of traditional selective breeding, biotechnology and molecular genetics of aquatic organisms. Cultured fish are being improved for a multitude of traits, including growth rate, feed-conversion efficiency, disease resistance, tolerance of low water quality, cold tolerance, body shape, dress-out percentage, carcass quality, fish quality, fertility and reproduction and harvestability. For many years there has been a cry in the wilderness that aquaculture is impeded by the lack of genetically improved fish and the utilization of essentially wild fish. This is still true for some species and for new aquaculture species; however, for a few well-established aquatic species, large genetic gain has been realized, and there is evidence of up to tenfold improvement in some traits compared with poor-performing, unimproved wild strains by use of various combinations of traditional selective breeding and biotechnology. The development and utilization of genetically improved fish are widespread across the world in the 21st century. A variety of genetic techniques are being implemented commercially, including domestication, selection, intraspecific crossbreeding, interspecific hybridization, sex reversal and breeding and polyploidy, to improve aquacultured fish and shellfish. Genetically improved fish and shellfish from several different phylogenetic families are utilized. Genetic principles and biotechnology are also being utilized by fisheries managers and by researchers to enhance natural fisheries, to protect native populations and to genetically conserve natural resources. Genetically modified aquatic organisms are already having an impact on global food security in both developed and developing countries. However, in general, much more progress can and needs to be made. The combination of a variety of genetic improvement programmes − traditional, biotechnological and genetic engineering − is likely to result in the best genotypes for aquaculture and fisheries management.

2

Phenotypic Variation and Environmental Effects

One of the most ignored areas in aquatic genetics and biotechnological research is the effect of the environment and experimental procedure on genetic expression, the phenotype and phenotypic variation. Genetic potential cannot be realized without the proper environment. To conduct high-quality genetics and molecular genetics research the nuances of environmental effects must be understood and superior fish culture must be employed; otherwise the measurement of the genetic effects may not be accurate and may even be incorrect. Most scientific literature on molecular genetics and biotechnology describes detailed experimental procedures concerning the molecular aspects of the research, but often ignores the details concerning the fish culture, making it nearly impossible to properly evaluate the validity and value of the research. If experiments or genetic improvement programmes are not conducted properly, the environmental effects can mask the true genetic effects. This is a problem not only for molecular and biotechnological programmes, but for traditional selective breeding programmes as well. Expertise on culture and the control of environmental variation is as important as, if not more important than, the genetic aspects of genetics research and breeding programmes.

When evaluating genetic modification via traditional genetic approaches or molecular genetic and biotechnological approaches in aquatic organisms, it is critical not to forget the most important and basic equation for genetic improvement, which defines the components of the phenotype:

$$P = G + E + GE \qquad (2.1)$$

where P is the performance or phenotype (appearance or characteristics) of an individual, G is the genotype or genetic make-up of an individual, E is the environment of the individual and GE is the interaction between the genotype and the environment.

A farmer may be excellent at growing fish and may provide the best environmental conditions possible, but if the fish is genetically limited with poor genes, performance and production will be poor despite the superb fish culture. On the other hand, one may have the very best genotype of fish, but if poor environment and fish culture is provided, the fish's genetic potential will not be attained and performance and production will be poor. Genetics is not a silver bullet and both genetics and environment are important and critical.

The breeder or geneticist accomplishes genetic gain by utilizing the variation of phenotypes of individuals in a population or by introducing new genotypes to genetically improve the performance of individuals and populations. Phenotypic variation, V_P, is a function of the following:

$$V_P = V_G + V_E + V_{GE} \qquad (2.2)$$

where V_G is genetic variation, V_E is environmental variation and V_{GE} is variation from genotype–environment interactions. Variation in the phenotypes of aquatic organisms must exist or be introduced if genetic improvement is to be made.

Obviously, the goal of genetic modification is to utilize the component of genetic variation or to artificially introduce genetic variation to improve performance or the phenotype. To effectively utilize the components of genetic variation, make valid genetic comparisons, study gene expression, conduct quantitative trait loci (QTLs) analysis, utilize marker-assisted selection (MAS) or develop transgenic fish, among other activities, environmental variation must be controlled to allow accurate determination of the genetic value of an individual. Environment must be carefully considered in the design of tests to evaluate rate of growth and other traits in populations of aquatic organisms.

Environmental variation can be affected by culture techniques. Aquatic environments naturally have a large amount of environmental variation, which can be much more difficult to control than in terrestrial environments. For example, waste products in the aquatic environment are difficult to remove and oxygen levels can fluctuate dramatically, whereas in the terrestrial environment waste can be readily removed and oxygen levels are constant. Good or superior fish culture must be practised to control environmental variation so that accurate and meaningful genetic data or genetic enhancement is obtained. For these reasons, aquaculture genetics research is actually technically more difficult than general aquaculture research. Publications dealing with aquatic biotechnology often omit details concerning the fish husbandry, which in actuality is quite important for the critical evaluation of the data since the environmental conditions and environmental variability can have such a large effect on phenotypic expression.

There are numerous environmental effects and variations that can affect the phenotype and need to be controlled or corrected. These include age, mortality, stocking density, temperature, water quality, maternal effects, compensatory gain, competition, magnification effects, skewness, size effects and procedures such as communal stocking.

Stocking Density and Mortality

Obviously, stocking density affects growth rate and other performance factors. Differential mortality has the same consequence as having different stocking densities. The most severe error or problem with data analysis, of course, occurs when differences in stocking or survival would alter the true rank of the genotypes. In some cases, rank is not altered, but the true difference between genotypes may be underestimated. For instance, in the case of inbreeding, the inbreeding depresses both growth rate and survival. The inbreeding depression of survival lowers the density of the inbred replicates, which promotes their rate of growth. The depression in survival causes underestimation of the depression of growth.

Age, Temperature and Water Quality

Differences in spawning time and ultimately age of the experimental fish or shellfish result in additional environmental variation, potentially masking genetic effects. In some cases, minor age differences can result in major environmental effects.

One dramatic example is the effect of a 1-day difference in age on growth and survival of different genotypes of common carp (Wohlfarth and Moav, 1970; Fig. 2.1). The environment was a communal earthen pond and the common carp fry relied on zooplankton generated from fertilization for food; thus environmental conditions were somewhat severe. The two genetic groups were gold and blue-grey, two colour mutants. When both groups were spawned on the same day, survival was equal, and the gold common carp grew 30% faster than

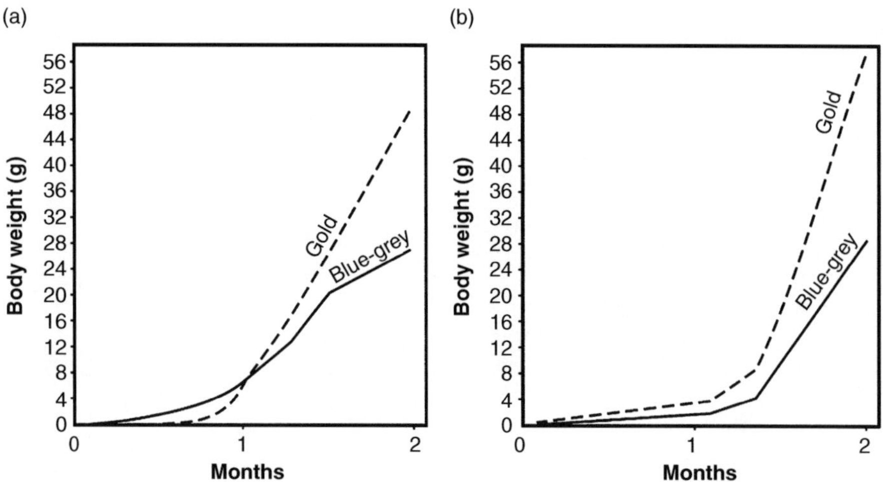

Fig. 2.1. The environmental effect of a 1-day age advantage on the growth of different genotypes of common carp, *Cyprinus carpio*. (a) Pond no. 1: blue-grey spawned on first day and gold on second. When spawned 1 day before the gold genotype, blue-grey, the inferior genotype, is larger for 1 month. Then the faster-growing genotype, gold, catches up and surpasses the size of blue-grey. (b) Pond no. 2: gold spawned on first day and blue-grey on second. When spawned 1 day before blue-grey, the gold genotype has the largest body weight throughout the entire experiment. (Adapted from Wohlfarth and Moav, 1970.)

the blue-grey individuals. When the blue-grey genotype was spawned earlier and had a 1-day advantage, it grew faster than the gold genotype for the first 30 days and blue-grey common carp had a 20 times greater survival than the gold individuals. However, after 60 days, the gold common carp were able to overcome the environmental disadvantage for growth, surpassed the blue-grey common carp in body weight and were 50% larger. With regard to survival, when the gold genotype was given a 1-day advantage, it was also able to translate this into the same 20 times greater survival that the blue-grey common carp were able to generate when they had the advantage. With this 1-day advantage, the gold common carp had a growth advantage of 50% higher body weight than the blue-grey individuals.

In the case of warm-blooded organisms, the age effects are much more easily corrected, simply by growing the test animals for a standard length of time. However, this is usually not a good solution for cold-blooded organisms, such as fish and shellfish, as large age differences can potentially subject them

to varying temperature regimes, which, of course, alter metabolism and performance. Correction by evaluating the experimental genotypes for a standard number of temperature days also has its shortcomings, as a single temperature day at optimum temperature may have a much greater effect on the phenotype than several days at a suboptimal temperature. In some cases, one alternative would be to carefully monitor environmental conditions and measure the aquatic organism at a common temperature for a standard number of days, even though the date of that time period may not be the same.

Obviously, water-quality differences can introduce additional environmental variance. Again, effects of pollution are greater for cold-blooded aquatic organisms, as they are more intricately linked to and living in their waste products compared with warm-blooded terrestrial animals.

Temperature and water quality affect not only production traits, such as growth and disease resistance, but also developmental traits. Temperature affects the mean for meristic traits. Temperature and water-quality

Fig. 2.2. Tailless and semi-tailless trait of channel catfish, *Ictalurus punctatus* (Dunham *et al.*, 1991). This is an example of an environmentally induced deformity without a genetic basis. (Photograph by R.O. Smitherman.)

degradation can induce congenital deformities that have no measurable genetic basis (Dunham *et al.*, 1991; Fig. 2.2). Temperature can affect sex ratio in a variety of fish, including channel catfish and Nile tilapia.

Biology and Physiology

Natural biological and physiological processes can also be responsible for environmental effects. As a fish grows, its gill surface to body volume ratio becomes smaller. Usually, larger fish have less tolerance of low dissolved oxygen than smaller fish. In a genetic evaluation, a smaller fish has an inherent advantage compared with a larger fish in a challenge test to evaluate low-oxygen tolerance. Therefore, the relationship between size and low-oxygen tolerance needs to be determined and used to standardize the genetic data to a common size for the experimental fish.

Usually a heavier, longer fish has a larger, longer head. Morphological measurements need to be standardized. One technique would be to standardize morphology by determining ratios. However, relative body shape changes as a fish grows. The body grows faster than the head in a young catfish. Therefore, the head size to total length ratio is naturally lower in a large fish. However,

as the fish grows and nears sexual maturity, the relationship changes and the head begins to grow faster than the body. Again, the relationship between body size and morphometric measurements needs to be established and used to correct data in genetic evaluations for valid genetic comparisons. The dress-out percentage varies with size in a similar manner to morphometric measurements and needs correction in a similar manner.

Size can also be related to disease resistance. Some pathogens attack smaller fish or sometimes preferentially attack larger fish because they are immunologically or physiologically more vulnerable.

Exposure to disease and stress can have long-lasting effects on the performance of fish. Permanent cell damage or alteration of metabolism may mask genetic potential. Another issue regarding genetic evaluation of disease resistance is prior exposure to pathogens before evaluation. Do prior exposure and the generation of antibodies give those fish an environmental advantage? Are results obtained with totally naïve fish realistic?

Meristic and morphological measurements are highly influenced by environmental conditions. Meristic traits are especially affected by temperature. Making genetic comparisons of fish for meristic and morphological traits when the fish are hatched at different geographic locations is dubious because of the strong influence of the hatching environment on these traits. Genetic comparisons should be made on fish hatched at the same location with the same water parameters and at the same time.

Maternal Effects

The maternal effect is a component of environmental variation. Maternal effects are the impacts made by the size, age and condition of the female upon quality of the eggs and upon the growth and viability characteristics of the embryo after fertilization (Kirpichnikov, 1981). Falconer and Mackay (1996) defined the maternal effect as the environmental influence that is attributed from the mother to the phenotypes of her offspring. Environmental paternal effects on the variability of the early life stages are usually insignificant (Heath *et al.*, 1999) since males have much smaller gametes than females (Chambers and Leggett, 1996). Maternal effects are of great importance in mammals since embryonic development is within the uterine environment. In fish, the maternal effects are potentially important during early life stages.

Many maternal effects for aquatic organisms are related to egg size. It is well documented that the size of the eggs is related to the size of the female in many species of fish, and this directly influences the fitness of the embryos and larvae. Egg size decreases as channel catfish, *Ictalurus punctatus*, females grow older and larger (r=0.57; Broussard, 1979). However, there is no correlation (r=0.04; Dunham *et al.*, 1983b and r=0.21; Bondari *et al.*, 1985) between female weight or total number of eggs in a spawn and egg size in females of the same year class. Disparity in egg sizes in pink salmon (Heath *et al.*, 1999), yellowtail flounder, *Pleuronectis ferrugineus* (Benoit and Pepin, 1999), channel catfish (Reagan, 1979), brown trout, *Salmo trutta*, and rainbow trout (Blanc and Chevassus, 1979) is a result of maternal effects. The size of the female is usually directly proportional to the size of the egg, as is the case for the black porgy, *Acanthopagnus schlegel* (Huang *et al.*, 1999). In

some cases, maternal effect does not exist for reproductive traits, such as egg hatchability in coho salmon, *Oncorhynchus kisutch* (Sato, 1980).

Female fish indirectly affect the survival of their progeny. In the case of Nile tilapia, *Oreochromis niloticus*, larger, older females produced larger eggs (Siraj *et al.*, 1983). The larger eggs also had higher hatchability and fry survival. Egg size and its correlation with fry size would be the first opportunity for an individual fish to gain a competitive environmental advantage. In Baltic cod, *Gadus morhua*, there is a high correlation between female size and egg diameter (Vallin and Nissling, 2000). Larger females (second-year spawning or repeat spawner) produced larger eggs. The larger eggs had neutral egg buoyancy at lower salinity, leading to greater larval viability.

Heath *et al.* (1999) found a high positive correlation between both egg size and larval body size and size of the female in chinook salmon, *Oncorhyncus tshawytscha*. The larvae hatched from larger chinook salmon eggs were more viable and initially grew faster. The maternal effect on larval body size decreased after 45 days post fertilization. The paternal influence on phenotype was dominant by day 116 post fertilization. The relationship between egg size and fry size disappeared by day 130. As usual, the maternal effect was temporary. Females that produced large eggs had lower fecundity than females that produced small eggs.

A positive correlation was observed between total length of female haddock and egg diameter and dry weight (Hislop, 1988). Hislop suggested that larger larvae were more viable under unfavourable environmental conditions. A positive correlation exists between egg size and early fry survival of brown trout (Einum and Fleming, 1999). Factors that led to greater starvation tolerance of larvae and ultimately higher survival were the amount of oil globules and longer incubation periods (Huang *et al.*, 1999) that were associated with larger eggs and females. Additionally, it appears that egg size directly correlates to fitness, which somehow affects later survival (Einum and Fleming, 2000a,b).

Nagler *et al.* (2000) examined embryo survival using single pair mating in rainbow trout. Survival rate of the embryos was highly correlated with female parent in all five developmental stages – second cleavage, embryonic keel formation, retinal pigmentation, hatch and swim-up stage – examined, while the paternal influence was insignificant. Such effects on embryo survival are expected as maternal mRNA and yolk contents influence translation of embryonic protein and nutrient supply for the embryo before feeding is initiated. Herbinger *et al.* (1995) also found a positive correlation between dam and offspring survival. A maternal effect was found for egg size, fry size through 90 days post hatch, swimming stamina and swimming stamina after starvation in brown trout, with all of these traits being positively correlated with large females and large eggs (Ojanguren *et al.*, 1996).

Coefficients of variation range from 6% to 7% for eggs in individual spawns of channel catfish (Dunham *et al.*, 1983b). The coefficients of variation were small, but some eggs were more than 20% larger than the mean. The small variation in egg size within individual populations may explain why skewness coefficients are small for channel catfish fingerling populations as no individual would have a large initial advantage due to egg size. Additionally, no correlation exists between egg size and per cent hatch (Broussard, 1979; Bondari *et al.*, 1985). The maternal effect accounts for 82.5% of the variance for egg diameter in yellowtail flounder (Benoit and Pepin, 1999).

The interrelationships of female size, egg size and fecundity are of great importance for both aquaculture and the management of fisheries resources. Egg size may be evolving in two directions. Large eggs have a higher probability for survival under poor environmental conditions, while under good environmental conditions the egg size is not as critical for fry survival. In the latter case, natural selection is in the direction of small eggs and small females (Einum and Fleming, 1999).

Relative fecundity and egg size are usually reciprocal. Heath *et al.* (1999) suggested that large females produce large numbers of eggs and therefore evolution is directed towards increased egg size. Conversely, small females produce a limited number of total eggs and therefore they might increase the relative fecundity by reducing the egg size to maximize the number of offspring. Natural selection occurs very early and can cause dramatic density-independent mortality in eggs and in early larval stages (Solemdal, 1997). The maternal effects may disappear at faster rates under culture conditions due to favourable environmental parameters under culture conditions (Einum and Fleming, 1999).

In the case of channel catfish, female brood-fish weight is positively correlated (r=0.72–0.83; Broussard, 1979; Bondari *et al.*, 1985) to the number of eggs per spawn. Some reports indicate that large egg masses of channel catfish have a lower per cent hatch than small egg masses (Bondari and Joyce, 1980; Jensen, J. *et al.*, 1983), but others indicate that there is no correlation between number of eggs per spawn and per cent hatch (Broussard, 1979; Bondari *et al.*, 1985). In some hatchery conditions, large egg masses have a low per cent hatch. This might suggest that under natural conditions there would be selection against large egg masses. Managers of commercial hatcheries overcome this potential maternal effect by splitting large egg masses into two or more smaller masses to allow better aeration and increased hatch. Although larger females produce more eggs than smaller females, the relative fecundity, eggs/kg body weight, is much higher for younger, smaller females than for older, larger females in both channel catfish and Nile tilapia.

The maternal effect for fecundity may actually be much more complex. Marshall *et al.* (1999) found that liver weight of females, as well as total lipid energy, was highly correlated to total number of eggs produced in the Pacific cod, *Gadus macrocephalus*.

The maternal effect due to egg size is usually temporary and short, and the durability of the maternal effect is related to the normal water-temperature regime for spawning of the fish and its developmental rate. The variance due to maternal effects is lost quickly after initiation of exogenous feeding. The longevity of the maternal effect is smaller in fish that are grown in higher water temperatures and that have more rapid developmental rates. For instance, *O. niloticus* has rapid embryonic development, spawns at high temperature and its maternal effect on fry size lasts for only 20 days (Siraj *et al.*, 1983), suggesting that maternal effects should not bias selection beyond this point. Palada-de Vera and Eknath (1993) also concluded that the initial size of fingerling *O. niloticus* did not affect growth; however, minimal replication was used in this experiment.

Channel catfish spawn at slightly lower temperatures, corresponding to slightly slower embryonic development, resulting in a longer maternal effect of egg size on fry and fingerling size. Egg diameter influenced the body weight of young channel catfish through 30 days of age (Reagan and Conley, 1977). Egg weight was not correlated to size of advanced fry in other studies (Broussard, 1979; Bondari *et al.*, 1985). Weights of advanced fry of several different strains were correlated to weights at 30 and 60 days (Broussard, 1979). Weights of 60-day-old fry from these different genetic groups were not correlated with fingerling weights.

Rainbow trout spawn at cold temperatures (below 12°C) and embryonic development is slow, and the maternal effect for egg size on fry size is correspondingly longer, 154 days before dissipating. Larger, older rainbow trout females produce larger eggs, which result in larger fingerlings through 75 days of growth, compared with younger smaller females (Gall, 1974). Maternal effect on growth and survival decreased with age and was not important for 1-year-old rainbow trout (Kanis *et al.*, 1976; Springate and Bromage, 1985; Springate *et al.*, 1985; Bromage *et al.*, 1992). A maternal effect also exists in salmon for survival to first feeding. Heath *et al.* (1999) did not detect maternal effects in 180-day-old fry of chinook salmon. The maternal effects influence only the early stages of the fish. The maternal effects are reduced with time, while the expression of the offspring's genome increases in importance (Heath *et al.*, 1999).

The nutrients that are contained in the fish eggs obviously affect egg quality and the viability of the larvae during early life stages, and this is another aspect of the maternal effect. The nutrition provided for female brood fish influences the nutrient concentration of the eggs. Controlling the environmental variance, including nutritional history, of experimental brood stock, particularly females, is an important component in designing and executing genetics experiments. Nagler *et al.* (2000) suggested that survival of rainbow trout from fertilization to swim-up can be determined from the egg quality. Blom and Dabrowski (1996) determined that rainbow trout females fed a diet with high ascorbic acid concentration produced more viable fry. However, the dietary ascorbic acid intake of the offspring was more important than the initial amount of ascorbic acid supplied in the eggs. The ascorbic acid levels of the fry fell rapidly in the first 14 days after first feeding, implying that nutritional maternal effects are also short term.

Disease resistance can also be influenced by maternal effects. The dam heritability for resistance to channel catfish virus disease is substantial in channel catfish, whereas the sire heritability is zero, indicating a considerable long-lasting maternal effect until the fish were at least 5–10 g. Maternal effects have been observed for salinity tolerance (Shikano and Fujio, 1997) and the specific gravity of the eggs and the otolith size (Solemdal, 1997) in guppies, *Poecilia reticulata*. In both rainbow and brown trout, alevins hatched from larger eggs had a greater number of pyloric caeca (Blanc and Chevassus, 1979).

Genetic maternal effects also exist for traits in shellfish. Wavy and smooth shell types are maternally inherited in Pacific oysters. The maternal effect can be used as a tool in fisheries management via fish-size regulations for commercial and sport fishing (Solemdal, 1997). The maternal effect can also be utilized in a beneficial manner in aquaculture, where the environmental conditions can be optimum, and smaller brooders can be used for more efficient fingerling production (Siraj *et al.*, 1983). Conversely, some marine species produce tiny eggs; therefore the live feed options are limited for first feeding of larvae. In that case, large females may be more desirable for brood stock.

Maternal effects can also have a genetic component. The selection for positive maternal effects may be complicated. Einum and Fleming (1999) stated that the direction of evolution for egg size is dependent on favourable environmental conditions. Females from large eggs would theoretically grow to large sizes rapidly and easily and therefore produce small eggs. However, if there were a genetic component, one should be able to manipulate or control the environment to conduct appropriate breeding to accomplish genetic improvement. Falconer and Mackay (1996) suggest that detailed knowledge of the maternal effect is necessary if selection for a maternal effect is to be successful.

The Minnesota strain of channel catfish produces large eggs and fry, not because of environment but because of genetics. Data analysis of Tave *et al.* (1990) indicates there is a maternal heterosis for body weight in *O. niloticus*. Dunham and Smitherman (1987) report maternal effects on combining ability in strains of channel catfish and species of catfish for producing heterosis in intraspecific crossbreeds and interspecific hybrids, respectively. Genetically based maternal effects can potentially be manipulated to enhance early and long-term performance in fish.

Correction of Growth Data

It is difficult to raise all genetic groups to exactly the same size to initiate genetic comparisons of growth. Obviously, initial size has a significant effect on final size (Dunham *et al.*, 1982b). The ideal comparison, of course, is when newly hatched fry of the same age are mixed together in the same experimental unit and grown communally. In this case, the environmental effects, other than maternal effect, should be equal for all individuals. However, it is difficult, if not impossible, to mark the fry to distinguish groups or individuals, except via DNA fingerprinting, which could be expensive.

There are many methods for measuring growth rate and correcting for initial size differences. Gain can be calculated, but larger individuals have an inherent advantage since they are initially larger and can consume more food and gain greater weight. Similarly, average daily gain is biased in favour of the individuals that have the largest starting weight. Per cent gain and instantaneous growth rate can be calculated. However, both of these corrections are biased in favour of the genetic groups with the small initial size as the relative growth rate of smaller fish is inherently faster, regardless of genotype. Also per cent gain in body weight has a bias for smaller fish since a smaller fish naturally grows at a faster rate and can double its weight much faster than a larger fish.

Some scientists have corrected initial size differences by grading the genetic groups to reach a common initial size to begin the experiment. Obviously, the grading is the same as conducting a selection, which alters the allele frequencies in the population and invalidates the comparison. Thus, the genetic groups and their genetic make-up have actually changed, and what is being compared is different than originally planned.

Regression can be a less biased method to correct for initial size differences – initial size, X, on final size, Y. This effect, b, the regression coefficient, is calculated by:

$$b = \frac{\text{final size difference (g)}}{\text{initial size difference (g)}} \qquad (2.3)$$

or:

$$b = \frac{(X - X_{mean})(Y - Y_{mean})}{(X - Y)^2} \qquad (2.4)$$

One technique is to calculate the regression coefficient from the data within the current experiment. However, this regression coefficient is confounded by genetic effects and does not correct only for environmentally induced differences. In this case, genotypes that are larger at the beginning of the experiment because of previous genetic effects or genetic advantage demonstrated during the actual experiment are penalized when the final weights are corrected.

An alternative technique is to take a single genetic group and split it into subsamples. The subsamples are intentionally grown to varying sizes. Then a growth experiment is conducted and the effect of initial size on final size calculated based on these fish. This regression coefficient should only be a result of environmental effects since all sub-samples had the same genetic make-up. Then this regression coefficient is applied in other experiments where different geno-types are being evaluated. Remarkably, the magnitude of this regression coefficient is quite similar – 3 – for experiments among numerous species, including salmonids, common carp, channel catfish, tilapia and largemouth bass, when the starting weights (15–30 g), final weights (250–500 g) and ini-tial differences (less than 50%) are similar in the experiments.

However, Dunham *et al.* (1982b) dem-onstrated that the absolute initial size, the absolute final size and the magnitude of the initial size difference can cause the regression coefficient for the effect of initial size on final size to vary tenfold in channel catfish. Regression coefficients b_{yx}=2.6–16.6 have been measured for the effect of initial size differences on final size of catfish in genetic experiments (Dunham *et al.*, 1982b; Table 2.1). The magnitude of the regression coefficient depends upon initial sizes, ini-tial size differences and final sizes of the experimental fish, and increases as these parameters increase. Size differences in common carp were also magnified as the fish grew, and final size differences were greater when initial differences were greater (Moav and Wohlfarth, 1974b).

Another alternative is to conduct mul-tiple rearing until all genetic groups reach the same weight, and then the actual com-parison is initiated. Multiple rearing is a technique to correct for initial size differ-ences (Moav and Wohlfarth, 1973). Groups of fingerlings are treated differently by adjusting feed and stocking rate so that all groups reach a predetermined size simulta-neously. However, depending upon the spe-cies, the artificial induction of skewness is a possible complication. If compensatory gain exists, multiple rearing is invalidated.

Skewness and Feeding Practices

Feeding practices affect the amount of envi-ronmental variation and the size distribution of the population. Skewness, an undesirable lack of symmetry in the frequency distribu-tion of a population, is often found in com-mon carp populations (Wohlfarth, 1977), as well as other species, and has both genetic and environmental components and origin. Skewness values of 1.0 are considered mod-erate, those greater than 2.0 are considered large.

Skewness coefficients for the weight of channel catfish can be affected by feeding rate, food particle size and other conditions resulting in competition for food (Moav and Wohlfarth, 1973). Small initial differences in size of fry that are caused by genetic or environmental advantages can be magnified

Table 2.1. Initial weights, final weights and the corresponding regression coefficient for channel catfish. (From Dunham *et al.* 1982b.)

Initial weight, X_1 (g)	Initial weight, X_2 (g)	Final weight, Y_1 (g)	Final weight, Y_2 (g)	b
1	1.88	12	15.6	4.1
2	3.5	20	23.9	2.6
2	4.0	36	54	9.0
30	12	100	28	4.0
30	12	200	74	7.0
30	12	300	98	11.2
30	12	400	146	14.1
30	12	500	282	12.1
30	12	600	301	16.6
32	18	500	445	3.9

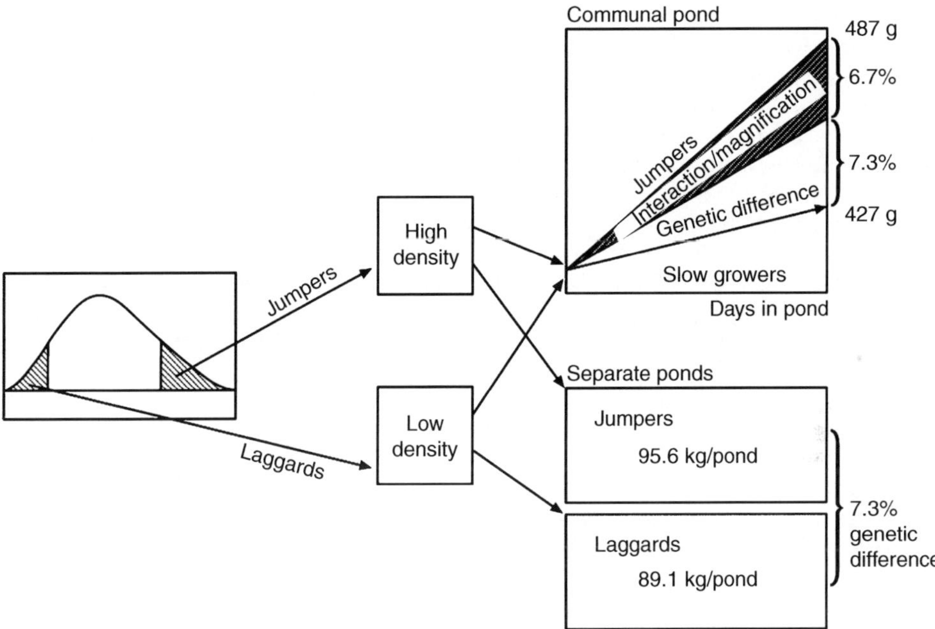

Fig. 2.3. The magnification effect in common carp, *Cyprinus carpio*. Jumpers and laggards are selected from a single population. They are then grown in separate ponds, high density and low density, respectively, until they reach the same size by manipulating feeding rate and stocking rate. Then they are grown in separate or communal ponds. The difference in growth in the separate ponds represents true genetic differences in growth potential. The larger growth difference in the communal ponds is a result of the true genetic difference from the separate evaluation and the magnification effect, an environmental effect, from competition for food. (Adapted from Wohlfarth, 1977.)

through competition – the magnification effect – allowing a subpopulation of larger individuals – shooters or jumpers – to gain exaggerated size advantages over their cohorts (Fig. 2.3).

Nakamura and Kasahara (1955, 1956, 1957, 1961) conducted a series of classical experiments that demonstrated the cause of skewness and the factors affecting skewness in common carp. Eggs and sac-fry demonstrated normal distributions for size. However, shortly after first feeding, skewness emerged. By growing the fish in individual containers, the populations remained normally distributed in the absence of competition. Decreasing particle size decreased skewness. Increasing feeding rates and increasing feeding frequency both decreased skewness. If the jumpers were removed from the population so as to regenerate a normal distribution, new jumpers emerged to occupy the vacant niche. The introduction of large artificial jumpers, such as goldfish of a larger size, prevented the emergence of skewness and the common carp population remained normally distributed.

Skewness is a result of competition for food. Individuals with slightly larger body size and consequently larger mouth size are able to magnify these initial size differences into extreme advantages. Wohlfarth (1977) later eloquently illustrated that this magnification effect had both environmental and genetic components.

Population distributions for body weight of channel catfish grown in cages were unexpectedly skewed (Konikoff and Lewis, 1974). Body weights of populations that were fed sinking pellets, which are more difficult to utilize for fish in cages than floating pellets, were skewed, whereas populations that were fed more accessible floating feed were normally distributed.

Food particle size affects uniformity of growth and skewness of population distribution in channel catfish fry grown in hapas (fine-mesh net cages) and troughs (McGinty, 1980). In these types of environment, the fry were largely dependent on the artificial food.

Treatments investigated included no feed, pellets and meal. In hapas, the fry fed pellets were more uniform in size and the population distribution had less skewness than in the other two treatments. This was a result of similar phenomena observed for the food size for fish in cages. Pellets were held within the hapas, allowing fry more time to consume the feed. Meal passed through the hapas quickly and allowed only a short time for feeding. This caused intensive competition for food and resulted in large amounts of variation, like the unfed treatment. The population fed meal had a positive skewness coefficient, indicating that more aggressive fish were monopolizing the meal as it passed through the hapa. The unfed population had a negative skewness coefficient, possibly an indication of small starving individuals lagging behind in growth. This treatment had the lowest survival.

Channel catfish fry fed pellets in troughs had lower survival, increased variation in size and a more positively skewed population distribution than fry fed meal. The meal, which was retained in the trough, was more accessible to the whole population than pellets and resulted in more uniform growth and higher survival.

Different species react differently to feed treatments resulting in extremes that can go beyond skewness. When some centrarchids were fed supplemental feed, only a portion of the population learned to feed resulting in a bimodal distribution.

Population skewness should be minimized in genetics and molecular genetics experiments, since the largest individual may result or partially result from environmental factors rather than genetic factors. Skewness coefficients for channel catfish fry grown in hapas and tanks were low to moderate compared with the large skewness coefficients found in common carp (Wohlfarth, 1977). Since skewness is a result of competition for food, species with relatively small mouths – small mouth size to body size ratios – appear to be more prone to skewness. Skewness for length was minimal in channel catfish fingerling populations stocked at 250,000/ha or less in earthen ponds and grown to 30 g (McGinty, 1980; Brooks *et al.*, 1982). Skewness can be reduced by increasing feeding rate, decreasing particle size and decreasing stocking rate (McGinty, 1980).

Compensatory Gain

Compensatory gain is a process in which animals that have been 'stunted' by an impoverished environment grow rapidly and catch up with animals of similar age once the environment is optimal. This phenomenon is common in mammals and has been documented in humans (Graham and Adrianzen, 1972) and cattle (Horton and Holmes, 1978). Some researchers report compensatory gain for fish (Rueda *et al.*, 1998; Soether and Jobling, 1999) but others suggest that the compensatory gain is limited in fish (Gaylord and Gatlin, 2000). This phenomenon is of great interest as it has several applications in aquaculture. Some farmers believe that stunted fingerlings exhibit extraordinary growth and compensatory gain and prefer them for stocking in grow-out ponds. If compensatory gain were to exist in fish, it would negate the validity of multiple rearing to produce fish of similar size for initiation of growth comparisons. The existence of compensatory gain is controversial in fish.

Wang *et al.* (2000) examined the existence of compensatory growth in hybrid tilapia (*Oreochromis mossambicus*×*O. niloticus*) reared in seawater. The fingerlings were deprived of feed for 1, 2 and 4 weeks, and no group exhibited full compensatory gain during the 4 weeks of re-feeding. The size of the fish was almost perfectly correlated with length of feed deprivation. Tilapia raised in seawater do not compensate for feed deprivation. Feed and nutrient digestibility and protein and energy retention efficiency were calculated, and no significant differences were found among the groups. Higher feed intakes and higher specific growth rates were observed for the deprived groups. The higher the deprivation period, the higher the feed intake and the specific growth rate upon re-feeding. This is, however, an expected result and not compensatory gain, because the final weights were smaller with increasing deprivation and specific growth rates and feed intake are naturally higher in smaller fish. Food deprivation altered body composition. Tilapia on the restricted feeding regimes tended to have higher moisture but lower protein, lipid, ash and energy levels compared with the full-fed controls, with, again, the greater the period of starvation, the greater the difference. However, the final values might be due to either the food deprivation or could be related solely to size since the deprived tilapia were small. It clearly shows that, given this period of time, they were unable to compensate with regard to body composition.

Similar experiments with African catfish, *Heterobaranchus longifilis* (Luquet *et al.*, 1995), and Alaska yellowfin sole, *Pleuronectis asper* (Paul *et al.*, 1995), gave virtually identical results to those found for the hybrid tilapia. In the case of the African catfish, again the conclusion was that compensatory growth occurred, although, again, it appears to be a case of the stunted fish performing at the level expected for smaller fish. However, Luquet *et al.* (1995) did indicate that no hyperphagy occurred after 2 weeks, once full feeding was restored.

Jobling and Koskela (1996, 1997) believed they observed compensatory growth in rainbow trout and that it was more pronounced when food was completely withheld than in partial restrictive feeding regimes. The compensatory gain was associated with hyperphagia. After being restrictively fed, immature rainbow trout had an increase in food intake and relative accumulation of visceral adipose tissues, carcass and hepatic liver content, carcass water content and liver wet weight when fed again (Farbridge *et al.*, 1992), but, again, this could be related to their smaller size compared with controls.

Dobson and Holmes (1984) also report compensatory gain for rainbow trout. This is another case where the fish culture methods are not reported in detail. True compensatory growth occurs when the fish grow at extraordinary rates compared with fish of the same size. Careful examination of the data of Dobson and Holmes (1984) shows that true compensatory growth did not occur. Weight gains of starved fish over the same period of time appear quite similar to those of fed controls.

Tiemeier (1957) indicated that compensatory gain occurs in channel catfish fingerlings in recreational ponds. Several other researchers report compensatory gain in channel catfish in high-density ponds (Kim and Lovell, 1995) and tanks (Gaylord and Gatlin, 2000, 2001a,b; Gaylord *et al.*, 2001), but Sneed (1968) concluded that it does not occur. Channel catfish fingerlings did not exhibit compensatory gain in ponds (Dunham *et al.*, 1982b). No growth differences were exhibited between stunted and normal fish after 26 days. The conflicting reports may be the result of data interpretation rather than differences in results.

The relative growth rate of small catfish compared with large catfish can give the impression that the small fish exhibit compensatory growth. The metabolism of smaller fish gives them the ability to grow at a relatively faster rate. The fish used by Tiemeier (1957) had been stunted for 1–2 years. Examination of the growth curves for his experiment shows that, although the stunted catfish were growing at a faster rate than other members of their age class, the stunted fish were growing at the same rate as similar-sized catfish of younger year classes. This suggests that growth rate is a function of size rather than age in subadult catfish.

Lovell (1979) had previously reported that missing a single day of feeding per week significantly reduced production. If compensatory gain exists, missing a single day of feeding should not decrease production. If the methods and results of Kim and Lovell (1995) are closely examined, there are several inconsistencies and explanations for the apparent compensatory gain. As in other experiments, the fish that had feed restriction the longest never completely compensated for the food deprivation. Again, details are missing from the methods, but the feeding technique presented was not true satiation feeding. Therefore, the control may have been somewhat restricted, allowing the short-term deprived fish to catch up. Mortality was not significantly different among groups, but the timing of mortality and the size of the moribund fish could have given certain treatments advantages or disadvantages. Specific growth rate was equal or higher for the control during every period except one compared with the short-term-restricted channel catfish. Even though the controls were twice the size of the short-term-restricted fish, during the first 3 weeks after the restricted fish were returned to full feed, their specific growth rates were almost identical. In the case of the long-term-restricted fish, when they were returned to full feed, their specific growth rates were the same as the control when the control was of similar size. This is not consistent with compensatory gain. Possible effects of variation in dissolved oxygen level were not analysed. Potential effects of varying temperature days were not analysed. When on full feed, the feed conversion ratios were equivalent, which is again inconsistent with compensatory gain.

Data analysis is misleading in the case of the evaluation of compensatory gain for channel catfish in tanks (Gaylord and Gatlin, 2000). If the data are converted to mean weights and the per cent gain is examined when the full-fed and 4-week-starved fish are of the same size, the per cent gains are virtually identical. Again, in this case the restricted fish never caught up and their final weight was less than that of the full-fed fish. Feed consumption rates are higher and feed conversion is more efficient for the restricted fish, but this is an expected result since they were smaller throughout the experiment. The data are very convincing for the conclusion that compensatory gain does not exist. All of these experiments demonstrate that compensatory gain was not exhibited by channel catfish fingerlings; therefore the multiple-rearing technique is valid for genetics experiments involving fingerling channel catfish.

Schaperclaus (1933) states that compensatory gain does not exist in common carp. Deprivation of protein stunted growth in common carp, but no compensatory growth was observed after re-feeding (Schwarz *et al.*, 1985). Restricted feeding could possibly result in additional problems, such as alterations in disease resistance. Lovell (1996) and Okwoche and Lovell (1997) reported that resistance to the bacterium *Edwardsiella ictaluri* was better in 1-year-old channel catfish that were fed compared with unfed channel catfish, but the opposite was true in 2-year-old fish. However, again, management of the environment is important in these evaluations because these results were possibly confounded by varying oxygen and ammonia levels in the experimental aquaria.

Several researchers have reported compensatory gain in fish. However, if the methods and results are carefully examined, it appears that the fish are growing and performing normally based upon their size; and until recently (Chatakondi and Yant, 2001) no data have convincingly shown that compensatory gain exists in fish. Chatakondi and Yant (2001) provide the first convincing evidence that compensatory gain exists in catfish. They starved channel catfish fingerlings (2.6 g) for 1–3 days and then fed them to satiation as long as they consumed more food than fish fed daily. Control fish were fed for 70 consecutive days and reached 25 g. Food deprivation and reinstatement equated to eight to 19 feeding cycles during the 70-day period. Chatakondi and Yant (2001) concluded that not only did compensatory gain occur, but also that the food-deprived fish (only the fish periodically feed-deprived for 3-day periods) had greater growth than controls. However, initial size differences account for all of the final size differences, so food-deprived fish did exhibit compensatory gain but did not have increased growth compared with controls. The 1-, 2- and 3-day-deprived fish consumed 33, 75 and 75% more food, respectively, than controls on the days when they were fed. The primary benefit from the food deprivation was that the food-deprived fish had 8–20% better feed-conversion efficiency. Previous experiments indicated that long-term food deprivation cannot be compensated for in comparison with continually fed controls. However, in this case, periodic short-term food deprivation can apparently be compensated for by a combination of hyperphagia and better feed-conversion efficiency.

Another result was that the resistance to the bacterial pathogen *E. ictaluri*, causative agent of enteric septicaemia of catfish (ESC), was better – 30–33% survival – for all food-deprived treatments compared with the fish fed every day – 0% survival (Chatakondi and Yant, 2001). One of the most effective treatments for ESC is to take the fish off feed. Apparently, the cyclical feeding placed the fish in at least a partial physiological condition to better resist this disease.

Communal Stocking/Evaluation

Communal stocking was developed by Israeli scientists (Moav and Wohlfarth, 1973) to overcome shortages of experimental units. In communal ponds, different genetic groups of aquatic organisms are stocked together for assessing differences among the groups. Communal experiments, stocking all genotypes in each replicate/experimental unit, are more efficient than experiments where the replicates are in separate ponds, cages, aquaria or tanks. Many more groups can be tested in fewer ponds or experimental units. The environment is identical for all groups, decreasing the error caused by between-pond variation in separate replicates. The results of communal experiments must be highly correlated with separate experiments for communal stocking to yield useful information (Moav and Wohlfarth, 1973). Communal stocking not only reduces the environmental component of variation but also reduces facility requirements.

Communal stocking is valid only when certain genetic, environmental and physiological criteria are met. Relative rankings of the genetic groups must be the same in both communal and separate evaluation. If the relative rankings are not the same, genotype–environment (GE) interactions are indicated and communal stocking is not valid. Results of both communal and separate evaluations

can be affected by initial size differences of the tested groups, necessitating elucidation of the effects of initial size on final size. The correlation between gain in communal ponds and separate ponds for different strains, cross-breeds, species and hybrids of catfish in six experiments was 0.89–0.97 and averaged 0.93 (Dunham *et al.*, 1982b). The mean correlation between replicate communal ponds was 0.91. These high correlations were obtained within several experiments in which various genetic crosses, ages of fish and stocking rates were involved. Rank of gain in communal ponds was the same as rank of gain in separate ponds. This allows the use of communal ponds for catfish genetics research. Communal stocking is valid for catfish genetics research (Dunham *et al.*, 1982b) since the correlations of the rankings between communal and separate trials were high, r=0.89–0.97. *Ad libitum* feeding further reduced competition among groups and the best agreements between communal and separate trials were obtained when this feeding regime was utilized. Communal stocking has been demonstrated to be a valid technique for genetic evaluations in several species, including Nile tilapia (Fig. 2.4).

Although communal stocking of channel catfish is valid, the effect of environmentally induced initial size differences on final size must still be considered. Although communal evaluation can accurately rank genotypes, caution must still be exercised in interpreting results and making recommendations prior to commercial application of genetically enhanced aquatic organisms. Potential GE interactions relative to the magnitude of the differences between genotypes in comparison with separate evaluation could occur even though the rank of the genotypes would be the same. This could be a potential cause of yield gap, the difference in production in the research environment and that in the farm environment. Commercial industry will be growing the improved genotype in a situation analogous to separate evaluation. In the case where the genetically enhanced aquatic organism was compared with other genotypes and controls in the communal environment and the magnification effect occurred, the genetic gain expected and

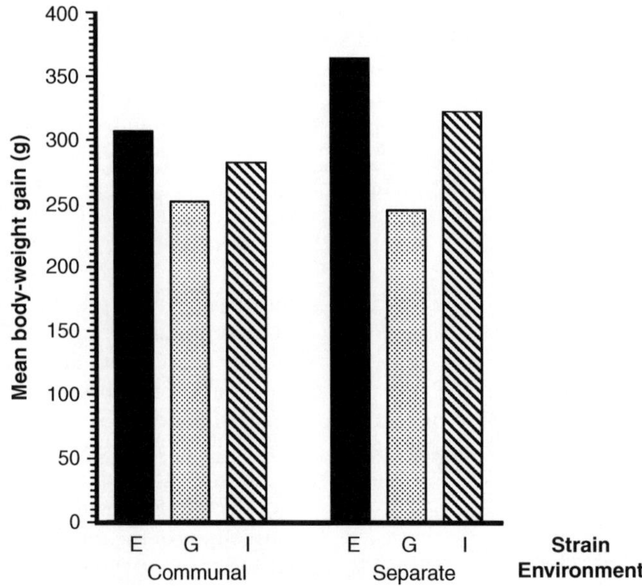

Fig. 2.4. Mean body-weight gain of three strains, Egypt (E), Ghana (G) and Ivory Coast (I), of Nile tilapia, *Oreochromis niloticus*, in communal and separate evaluation. The rankings in the two evaluations are identical, illustrating the validity of the communal stocking technique. (Adapted from Khater, 1985.)

indicated to private industry would be exaggerated compared with what they would obtain in the communal evaluation, and they should be tested once in a commercial setting. Therefore, if biotechnologically enhanced fish and shellfish are either developed or evaluated with communal evaluation, they should be tested once under separate evaluation to determine any magnification effects and GE interaction prior to release to farms.

3

Basic Genetics, Qualitative Traits and Selection for Qualitative Traits

Genetics is a management tool that can be used for genetic enhancement or maintenance in aquaculture and fisheries. Effective programmes have goals and plans, and this is also true for genetic enhancement programmes. Goals are the important traits of economic importance that we want to improve and the extent to which we would like or think feasible to improve them. Genetic enhancement programmes are then the plans that we use to accomplish these goals and objectives.

One assumes that a large number of individuals is needed to obtain genetic diversity. This is not always true. Sometimes a few genetically variable individuals may possess more genetic variability than a large number of essentially homozygous individuals. The number of genetically different gametes one individual can produce is related to the number of heterozygous loci that individual possesses. Thus the number of different gametes an individual can produce is 2^N, where N is the number of heterozygous loci. Table 3.1 shows that even with a few loci polymorphic an individual can produce an enormous assortment of genetically different gametes.

Variation in phenotypes for qualitative traits can be discrete, all or none type expression, black or white. For example, an individual is either albinistic or normal coloured. These traits are usually a result of gene expression from a single or only a few loci. Coloration (Fig. 3.1) and deformities are examples of qualitative traits. Qualitative traits such as changes in colour, finnage, scale pattern or deformities can be desirable or detrimental in aquaculture. Obviously, qualitative traits are important and the primary basis for the ornamental aquaculture industry.

The other type of variation is continuous. Production traits such as body weight are usually continuous, have a large number of phenotypes that blend together into a continuum and are usually controlled by a large number of loci.

Dominant and Additive Gene Expression

There are two basic types of gene action: dominance (recessive) and additive. In the case of dominance, only one copy of the dominant allele in a diploid organism is needed for expression of the associated trait. In the case of the recessive allele, two copies of the allele are needed for the recessive phenotype to be observed. In a completely dominant system, a large unit of phenotypic change occurs when going from the homozygous recessive genotype to the heterozygous genotype and no unit of change when going to the homozygous

Table 3.1. Number of different genetic combinations of gametes that can be produced by a single individual as related to the number of heterozygous loci for the individual.

Number of heterozygous loci, N	Number of possible gametes, 2^N
1	2
2	4
3	8
4	16
5	32
10	1,024
15	7,768
20	1,048,576
25	33,554,432
30	$1,074 \times 10^{12}$
50	$1,126 \times 10^{18}$
100	$11,267 \times 10^{33}$

Fig. 3.1. Coloration is an example of a qualitative trait.

dominant genotype. For additive gene action, the alleles, of course, act in an additive fashion with equal units of change when comparing the different genotypes; like mathematical addition, the stronger allele will make a greater contribution to the phenotype.

Different types of dominance exist: complete, incomplete, overdominance and co-dominance. As discussed earlier, in the case of complete dominance, the trait is fully expressed in the heterozygous and homozygous dominant genotypes (Fig. 3.2). Examples of complete dominance are found in Tave (2003: Table 1). The heterozygous genotype allows a major, but not complete unit of change in the phenotype for incomplete dominance. The homozygous dominant genotype is necessary to make the complete, maximum shift in the phenotype. Examples of incomplete dominance are found in Tave (2003: Table 2). In the case of overdominance, the phenotype of the heterozygous genotype is outside that of the two homozygous genotypes. A comparison of the expected means of F_1 relative to the two parental types for complete dominance, incomplete dominance, overdominance and additive gene expression is presented in Fig. 3.3. The phenotype associated with each allele is observed in the case of co-dominance.

Complete dominance

aa – red

     ] large unit of change

Aa – black

     ] no unit of change

AA – black

Incomplete dominance

bb – white

     ] large unit of change

Bb – dark grey

     ] small unit of change

BB – black

Overdominance

ss – malaria prone
Ss – healthy (best)
SS – sickle cell

Additive gene expression

rr – white

     ] large equal units of change

Rr – pink

     ] large equal units of change

RR – red

Fig. 3.2. Illustration of the concepts of complete dominance, incomplete dominance, overdominance and additive gene expression.

Explanation of Epistasis

The alleles at one locus can affect the expression of alleles at a second locus. This type of gene interaction is termed epistasis. Epistasis is the interaction of genes at different loci.

An understanding of the dihybrid cross is the basis for understanding epistasis. A monohybrid cross is one where the two mated individuals are heterozygous for one trait (Aa × Aa). When the trait is controlled by dominance, the monohybrid cross yields a 1:2:1 genotypic ratio and a 3:1 phenotypic ratio. A dihybrid cross is between two individuals heterozygous for two traits controlled by dominance (AaBb × AaBb). The dihybrid cross yields a 1:2:1:2:4:2:1:2:1 genotypic ratio and a 9:3:3:1 phenotypic ratio if both loci have dominance expression, independent assortment, no linkage and no epistasis (Fig. 3.4).

When a dihybrid cross yields a modification of the 9:3:3:1 phenotypic ratio, this is an indicator of epistasis. For example, if a 9:7 ratio is generated, the homozygous recessive genotype, aa (three of 16 individuals), or the homozygous recessive genotype, bb (three of 16 individuals), prevent

Overdominance	Complete dominance	Incomplete dominance	Midparent value additive gene expression	Incomplete dominance	Complete dominance	Overdominance
↓	↓	↓	↓	↓	↓	↓
F_1	F_1	F_1	F_1	F_1	F_1	F_1
	∧				∧	
	P_1				P_2	

Fig. 3.3. Illustration of the means of two parental types, P_1 and P_2, and their F_1 in the cases of complete dominance, incomplete dominance, overdominance and additive gene expression. For the three types of dominance, the mean of the F_1 can be in the direction of P_1 or P_2.

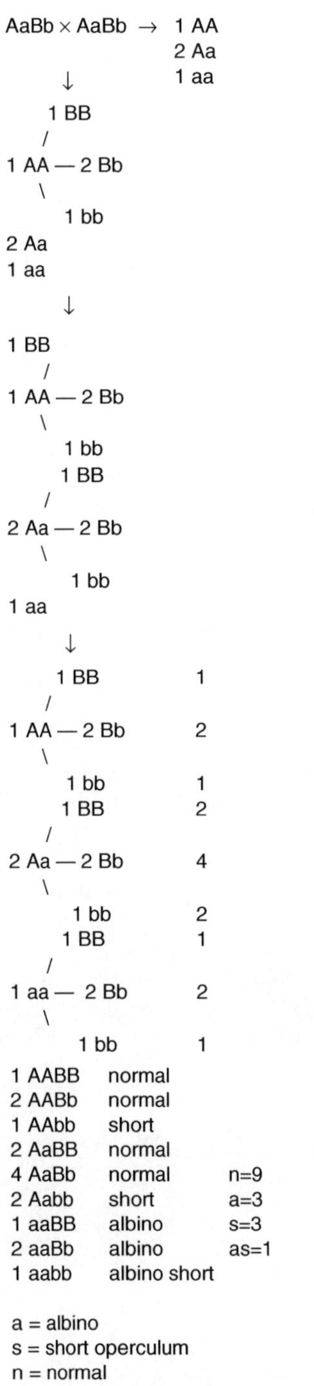

1 AABB	normal	
2 AABb	normal	
1 AAbb	short	
2 AaBB	normal	
4 AaBb	normal	n=9
2 Aabb	short	a=3
1 aaBB	albino	s=3
2 aaBb	albino	as=1
1 aabb	albino short	

a = albino
s = short operculum
n = normal

Fig. 3.4. Illustration of a dihybrid cross. The dihybrid cross yields a 1:2:1:2:4:2:1:2:1 genotypic ratio and a 9:3:3:1 phenotypic ratio if both loci have dominance expression, independent assortment, no linkage and no epistasis.

A –	B –	9 normal
aa	B –	3 albino
A –	bb	3 albino
aa	bb	1 albino

9 normal: 7 albinos
<u>9 : 3 : 3 : 1</u>

                aa   bb
    →   X   →   X

Coloration as a result of a series of biochemical reactions

Fig. 3.5. Illustration of complementary recessive epistasis. Colour as a series of biochemical reactions with a mutant homozygous recessive genotype at either of two steps resulting in albinism.

or mask the expression of the dominant allele. Thus, gene interaction at two loci has occurred. The double homozygous recessive, aabb, also produces the mutant genotype (one of 16 individuals), thus the 9:7 phenotypic ratio. This is an example of recessive epistasis; specifically, complementary recessive epistasis. The homozygous recessive genotypes are complementing each other to mask expression of the dominant alleles at the alternative loci (Fig. 3.5). We can realistically visualize this as a series of reactions in a biochemical pathway. For instance, coloration, deposition of melanin, is a series of biochemical processes. Blocking any of the steps in the progression would interrupt the deposition of melanin at the final step (controlled by the final locus in the operation).

A similar example of recessive epistasis can also be illustrated as a series of biochemical reactions in a pathway. Again, this can be thought of as a progression of biochemical steps. Blocking the process with a homozygous recessive genotype early in the pathway has maximum phenotypic alteration; for example, albinism in rainbow trout (Fig. 3.6). If the reaction passes that step, the phenotype changes, green coloration, but if a homozygous recessive disrupts the catalysis at the next step, green is the final product. If there is at least one dominant allele at

Rainbow trout coloration

B –	A –	black	9 normal
Bb	A –	red	3 green
B –	aa	white	3 albino
bb	aa	white	1 albino

9:3:<u>3:1</u>
9:3:4

	aa	or	bb
$\rightarrow$	X	$\rightarrow$	X
	albino		green

Progression of steps in a biochemical pathway

Fig. 3.6. Illustration of recessive epistasis resulting in three phenotypes.

A –	B –	9	black
A –	bb –	3	black
aa	B –	3	red
aa	bb	1	white

<u>9:3</u>:3:1
12:3:1

Fig. 3.7. Phenotypic ratios found in a dihybrid cross when single dominant epistasis exists for the two loci.

A –	B –	9	
aa	B –	3	> 15
A –	bb	3	
aa	bb	1	

<u>9:3:3:1</u>
15 : 1

Fig. 3.8. Duplicate dominant interaction. Two potential pathways exist, but only one needs to work to produce a normal phenotype.

each locus, all steps of the pathway are completed resulting in the normal dark phenotype.

There are also dominant forms of epistasis. If a dihybrid cross produces a 12:3:1 ratio, this is indicative of single dominant epistasis. In this case, there is a single dominant allele that has the most controlling influence at both loci. Thus, the phenotype belonging to the dominant allele A, black in the example in Fig. 3.7, is always expressed regardless of the genotype at the second locus B. The expression of the second locus (red) occurs only when the genotype at locus A is homozygous recessive. The double homozygous genotype results in a third phenotype, white.

Duplicate dominant epistasis may be a genetic mechanism that we will eventually learn is involved in critical pathways and for critical traits. Basic and vital systems of the body may have genetic reinforcement, alternative or backup genetic loci to ensure that if a mutation or an environmental problem occurs, the critical function is preserved. If any of the two loci has at least one dominant allele present, the normal phenotype will be expressed and abnormal phenotypes will be seen only when we have the double homozygous recessive genotype, aabb. This resulting phenotypic ratio is 15:1 in a dihybrid cross (Fig. 3.8). This can be

visualized as a biochemical pathway where there are two alternative routes to reach the end point. If one pathway is blocked, the second can be used. Urine production and glycolosis have alternative pathways that can lead to the final product. Perhaps epistasis exists in these systems to safeguard function.

Several types of gene action have now been discussed and, as explained earlier, phenotypic variation is affected by a combination of genetic factors, environmental factors and their interaction. Several types of gene expression exist. Thus, genetic variation can also be partitioned into additional components:

$$V_G = V_A + V_D + V_{AA} + V_{DD} + V_{AD} + V_{MH} \qquad (3.1)$$

where V_A is additive genetic variation, V_D is dominance genetic variation, V_{AA} is the variation from epistasis between additive genes, V_{DD} is the variation from epistasis between dominant genes, V_{AD} is the variation from epistasis between additive and dominant genes and V_{MH} is the variation due to maternal heterosis. The maternal effect might have a genetic component, maternal heterosis, that can be selected for.

Epistasis is the basis of some colour types in fish. For instance, it is the explanation for some red and black colour variants in tilapia. Scale pattern in common carp is also influenced by epistatic gene action (Tave, 2003). Examples of epistasis in fish are found in Tave (2003).

Pleiotropy, Lethality, Penetrance and Expressivity

Pleiotropy is when one gene affects more than one trait. Again, the alleles and loci that affect scale pattern in common carp are an excellent example of this phenomenon. These genes not only affect scale pattern, but also growth, survival, other meristic traits, tolerance of low oxygen, haemoglobin per cent, haematocrit and the ability to generate fins. Also, colour mutations usually have pleiotropic effects. Colour genes causing albinism in catfish and red colour in tilapia adversely affect other traits related to survival and reproduction (Tables 3.2 and 3.3).

Some mutations can also have semi-lethal or lethal effects, resulting in reduced viability or death. Again, certain alleles affecting scale pattern in common carp have semi-lethal or lethal effects, as does the allele responsible for saddleback in tilapia.

Penetrance is the ability of an allele to produce the product, trait or phenotype that it encodes. For instance, the allele that produces Huntington's chorea has 100% penetrance. If an individual possesses that allele, it is 100% certain to develop the disease. In contrast, not all individuals that possess genes making them susceptible to diabetes develop diabetes. Expressivity is the differential expression of genes. Individuals may have the same alleles, but the severity of the resulting trait or disease and/or the age of the individual when the trait appears may vary from individual to individual.

Table 3.2. Pleiotropy in albino channel catfish, *Ictalurus punctatus*.

	Albino	Normal
Egg weight (mg)	22	27
Hatch (%)	48	61
Early fry mortality (%)	30	22
Fry survival in ponds (%)	5	75
Fingerling survival (%)	80	95

Table 3.3. Pleiotropic effect of colour on mean body weight in tilapia.

Phenotype	Body weight (g)
R	26.7
RB	24.6
PRB	18.6
PR	16.3
P	13.3
PB	11.7

R, red; P, pink; B, black.
Body weight of R>PR>P.

Scale pattern in common carp illustrates several genetic phenomena: epistasis, pleiotropy and lethality. The scaled genotype is *S-nn*, mirror (scattered scales) is *ssnn*, line (on lateral line and/or back) is *S-Nn* and leather (almost nude) is *ssNn*. The genotypes *S-NN* and *ssNN* die so the *N* gene is a semi-lethal allele, reducing viability in the heterozygous state and causing death when homozygous. The *S* locus genotype epistatically affects the phenotype expressed at the *N* locus. The *N* allele affects expression at the *S* locus. Both *s* and *N* alleles have adverse pleiotropic effects on several traits, such as growth, survival, tolerance of poor water quality, ability to regenerate fins, gill-raker number, haemoglobin per cent and haematocrit.

Deformities

When deformities are observed, fish culturists often assume that the deformities have a genetic basis and are likely increasing because of inbreeding in the population. However, these assumptions are usually false. The majority of deformities observed are usually environmentally induced and are often related to low egg quality or poor water quality in the hatchery. Tailless (Fig. 3.9) and stumpbody (Fig. 3.10) traits in ictalurid catfish and tilapia have no genetic basis, and are apparently environmentally induced.

However, some deformities like saddleback in tilapia (an indented back, Fig. 3.11), which is caused by missing dorsal bones that support the dorsal flesh, does have a genetic basis. The saddleback gene, *S*, is not a sex-linked gene, is dominant and semilethal. The *SS* genotype is lethal. *Ss* heterozygotes are saddleback with greatly reduced survival, only about one-third survive, and *ss* is normal phenotype. If the saddleback gene is introgressed into other species the inheritance becomes much more complicated and epistasis is involved. This is a common phenomenon, and when mutations with simple inheritance are placed in an interspecific genome, epistasis often occurs.

Deformities can be valuable in the ornamental trade, but are usually undesirable in the food fish industries. Long fins on koi and fancy goldfish such as orandas and

bubble eyes are a result of heritable deformities. If these qualitative traits are a result of dominant gene action, they can be easily eliminated as all homozygous dominant and heterozygous (carrier individuals) phenotypes are obvious and those individuals can be immediately selected against, resulting in a population with none of the dominant detrimental allele. An example of this is the saddleback mutation in tilapia (Tave, 2003). On the other hand, it is extremely difficult to eliminate a deleterious recessive allele from a population as heterozygous carriers cannot be identified by simple visual observation. If the trait is of high economic importance or damage, the heterozygous carriers could be eliminated by mating them with individuals of known genotype and examining the phenotypic ratios in the progeny (progeny testing). Since fish are highly fecund and produce large numbers of progeny, progeny testing could eliminate the deleterious allele from the population in a single generation.

Simple Selection Procedures

A dominant allele is very easy to eliminate because every individual fish, each one that has a dominant allele, whether homozygous or heterozygous, expresses its phenotype. Therefore, it is easy to identify and cull individuals with the dominant allele. All such individuals, as is the case for saddleback tilapia, can be eliminated in a single generation, fixing (100%) the population for the wild-type recessive allele and the normal phenotype. This is the same as selecting for a recessive allele. For example, if you want an all-albino population, only albinos are chosen as brood stock: they are all homozygous and immediately a true breeding all-albino population is established.

If the goal is to eliminate a recessive allele from a population, the task is more difficult. Heterozygous individuals with the normal phenotype are carriers, and cannot be recognized and genotyped visually. In this case, selection is for a dominant allele or against the recessive allele.

Fig. 3.9. Environmentally induced tailless trait in tilapia.

Fig. 3.10. Environmentally induced stumpbody trait in channel catfish.

The effectiveness of the selection against the recessive allele can be predicted or determined after a certain number of generations by using the formula:

$$(1-q)_n = (1-q)_0 /[1 + N(1-q)_0] \qquad (3.2)$$

where $(1-q)_n$ is the allele frequency of the recessive allele after N generations of selection and $(1-q)_0$ is the initial gene frequency of the recessive allele.

If your population had 4% albinos and your boss gave you the assignment to transform the population to 1% albinos in ten generations or be fired, is your assignment reasonable and feasible? If a population has 4% albinos, the genotypic frequency aa = 0.04. What is our gene frequency? Allele frequencies must add to 1, so $(1 = p^2 + pq + q^2)$ where p and q are the gene frequencies of the two alleles in question. Thus 0.04 equals q^2 and thus $q = 0.2$. If ten generations and an initial gene frequency of 0.2 are inserted into our formula, our final gene frequency is 0.067. Ten generations of selection reduced our allele frequency from 0.20 to 0.067, a threefold reduction. Albinism is still present because of the heterozygous carriers. The genotypic frequency of aa is 0.067^2, which is 0.005 or 0.5% of albinos. When the selection was initiated ten generations earlier there were 4% albinos and now there is 0.5%, an eightfold reduction in the phenotype, and we meet the boss's expectations.

The equation can be factored to allow prediction of the number of generations needed to change an initial allele frequency to a desired final allele frequency. The resulting equation is:

$$N = [1/(1-q)_n] - [1/(1-q)_0] \qquad (3.3)$$

The result will usually be a high number of generations, which reflects how difficult it is to eliminate recessive genes from a population due to heterozygous carriers. Thus, if our population has 4% albinos and the acceptable proportion is a tenth of a per cent, the initial genotypic frequency is 0.04, the initial gene frequency is 0.20, the desired genotypic frequency is 0.001 and the square root of the genotypic frequency

Fig. 3.11. Saddleback trait in tilapia, a dominant semi-lethal trait.

for albinism is 0.01, the desired albinism gene frequency. Inserting these numbers into the formula gives $N = 95$ generations to reduce the albinism from 0.04 down to 0.001. Even after 95 generations of selection, the albinism allele and albinos are not completely eliminated. The solutions to these formulas are mathematical averages with associated standard deviations. Therefore, it is possible to eliminate albinism through selection, if we are lucky, but it is not probable. Also, there is the low probability that the genotypic frequency will be higher than the expected 0.001 after 95 generations of selection.

Progeny Testing

It may be very important to eliminate a deleterious trait or deleterious recessive allele or the boss may insist that the trait be eliminated from his farm. If the resources are available, progeny testing will allow elimination of targeted traits and alleles. Progeny testing is the estimation of the genetic value or make-up of an individual through measurement of the performance, appearance or other characters of a group of that individual's progeny.

At this point in time, progeny testing has not been necessary or utilized to eliminate deleterious traits in aquatic organisms. However, in terrestrial livestock progeny testing for qualitative traits can be quite important. The beef cattle industry once used a single bull – it appeared to be quite a valuable bull – that had not been properly progeny tested to impregnate cows industry-wide via artificial insemination. However, he was heterozygous, a carrier of the snorter dwarf allele. Individuals homozygous for this gene have respiratory problems, are dwarfs and are sterile, and often those calves will eventually die. After this bull was used, a large number of calves with snorter dwarfism were born causing a large economic loss, and increasing the allele frequency of this deleterious recessive gene in the cattle population because of inadequate progeny testing.

Two basic progeny testing systems can be used: heterozygous tester and homozygous tester systems. When utilizing a heterozygous tester, the animal with the unknown genotype and the normal phenotype (AA or Aa) is mated with a known heterozygote Aa to attempt to identify Aa genotypes and cull them from the population. The probability of not detecting the Aa carrier is $(3/4)^n$, where n is the number of progeny evaluated for the tested individual. The monohybrid cross, which is the same as the tester mated with the heterozygous tester, Aa × Aa, generates a 3:1 phenotypic ratio. Every individual progeny tested has a 75% chance of being normal so the carrier will not be detected. As the number of progeny n increases, obviously the higher the probability of identifying the Aa individual as each progeny is an independent event, and the probability of obtaining normal individuals repeatedly is multiplicative.

The alternative system uses a homozygous tester, which is a superior system. The unknown animal (Aa or AA) is mated with a known homozygous individual (aa). This mating will generate 1/2 normal individuals and 1/2 abnormal individuals. The probability of not detecting or making a mistake in identifying the genotype of the carrier is $(1/2)^n$, where n is the number of progeny that are tested. The probability of making a mistake is smaller with the homozygous tester as compared with the heterozygous tester.

If the boss will accept 6% mistakes, the probability of not detecting a carrier in a progeny testing programme of 0.06, how many progeny must be identified? In the case of the homozygous tester, our probability of not detecting or making a mistake is $(1/2)^n$ or $(1/2)^4$ is 0.06. To reach this level of effectiveness or this level of accuracy we must test four progeny from each individual. If we must use a heterozygous tester, that number of progeny evaluated more than doubles. To ensure an error rate of only 6%, we have to evaluate ten progeny for every individual. Again these are mathematical probabilities, so if we are fortunate we will detect the carriers with fewer progeny; or if unlucky, 6% of the time the predicted needed number of tested progeny will be insufficient.

Why not always use the homozygous tester if this system is more efficient? Sometimes heterozygous testers must be used because homozygous individuals are either not viable or have reduced reproduction capacity.

The progeny from the testing cannot be used for brood stock. They can be used for production purposes, but cannot be bred as the testing process is going to intentionally generate heterozygous individuals as a by-product.

Our discussion on the probability of not detecting heterozygotes is almost irrelevant in fish. Progeny testing is potentially very powerful in fish since they produce large numbers of progeny. With four to ten individuals the probability of not detecting a heterozygote is only 6%. Tilapia is considered a fish that has very low fecundity, yet a small young individual can still produce 100–200 progeny. If we progeny test 50–100 fingerlings, the probability of no detection is essentially zero. If economically important and with appropriate resources, deleterious alleles at any single locus trait can be eliminated in aquatic organisms in a single generation.

4

Strain Evaluation, Domestication and Strain Selection

Use of established, best-performing domestic strains is the first step in a genetic improvement programme and the mechanism to make the most rapid initial progress in genetic improvement. A strain is a population of fish within a single species that has the same history, the same origin and has at least one trait or a suite of traits that make it unique or make it different from other strains. The trait could be growth rate, disease resistance, colour, gill-raker number, any combination of traits or a unique DNA marker or isozyme marker that distinguishes that strain from other strains.

Strains of fish exhibit large amounts of variability for many different traits. Channel catfish, *Ictalurus punctatus*, strains differ in growth, disease resistance, body conformation, dressing percentage, vulnerability to angling and seining, age of maturity, time of spawning, fecundity and egg size (Dunham and Smitherman, 1984; Smitherman and Dunham, 1985). Rainbow trout, *Oncorhynchus mykiss*, strains exhibit similar variability (Kincaid, 1981). Okamoto *et al.* (1993) reported that a strain of rainbow trout resistant to infectious pancreatic necrosis (IPN) virus showed 4.3% mortality compared with 96.1% in a highly sensitive strain. Barramundi, *Lates calcarifer*, vary for upper temperature tolerance, loss of swimming equilibrium and disassociated caudal fin (Newton *et al.*, 2010).

In the case of Nile tilapia, *Oreochromis niloticus*, Egypt, Ghana and Ivory Coast strains were compared. There were differences in growth rates with Egypt and Ivory Coast growing the fastest and Ghana extremely slowly. However, the reproductive capacity of Ghana far exceeded the other two strains. There were also colour differences, with Egyptian strains having rosy markings on their cheek, under their chin and on their fins.

Breeders should take advantage of this tremendous amount of variability. Strain variation is virtually guaranteed for any species for any trait. Thus, the first step, if a very practical step-by-step genetic improvement programme is to be executed, is to conduct strain evaluation and selection. If a new farm or a new hatchery is initiated and known strains already exist with good performance, why start the new farm with an inferior fish? Utilize the best fish already available, and if possible a domestic strain. In a stepwise, practical genetic improvement programme, this is the simplest and easiest thing to do. If information is not available regarding the best-performing strain, choose four or five popular strains, evaluate them and decide which one to continue to work with.

Natural Selection in Strain Development

Natural selection and geographic origin lead to the formation of wild strains. Egypt strain of Nile tilapia is by far the most cold-tolerant strain and originates in the extreme north of the natural geographic range of Nile tilapia where it is subjected to colder temperatures than equatorial Nile tilapia (Fig. 4.1). There is a strong correlation between latitude of geographic origin and cold tolerance. Ghana originates the closest to the equator and is slightly less cold tolerant than the Ivory Coast strain which is slightly further from the equator.

For most strains of channel catfish, there is little difference in egg size and the time of spawning. But there are two outliers, Minnesota strain from the St. Louis River in Minnesota, which drains into Lake Superior and is the coldest of the Great Lakes, and Rio Grande strain from the Rio Grande River in Texas, the southernmost and hottest extent of the natural range of the channel catfish. Each has appeared to have undergone selection to adapt to its extreme environment. The Minnesota strain from the St. Louis River spawns at colder temperatures, 21°C, whereas most other strains start spawning at 24°C. Minnesota strain produces larger eggs than other strains of channel catfish resulting in larger fry. These adaptations make sense for Minnesota strain, since they are subjected to a very short growing season. By spawning earlier and producing larger fry, compensation is made for a short growing season and early winter to allow fingerlings to reach a size to better survive during winter.

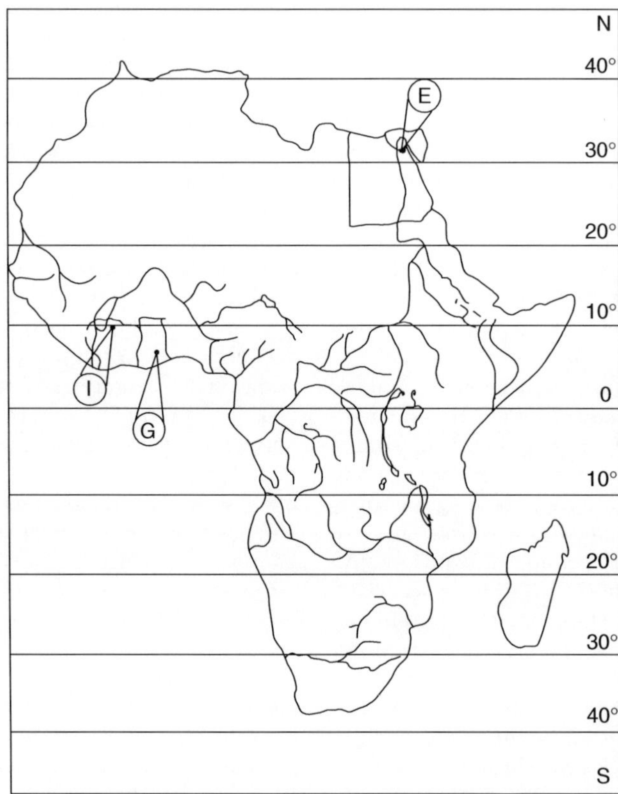

Fig. 4.1. Geographical origins of Egypt (E), Ghana (G) and Ivory Coast (I) strains of Nile tilapia, *Oreochromis niloticus*, result in strain variation in cold tolerance that is highly correlated with distance from the equator.

Rio Grande strain resides in the extreme opposite environment, and responds in an opposite direction. It is a late spawner with smaller eggs and greater fecundity compared with typical strains of channel catfish.

Domestic and Wild Strains

Domestic strains of fish usually have better performance in aquaculture settings than wild strains of fish. When wild fish are moved to aquaculture or hatchery environments, they are exposed to a new set of selective pressures that will change gene frequencies and results in domestication in the new environment (Dunham, 1996). In the aquaculture setting, fish must contend with a totally different environment. They are at high density, must learn to eat artificial feed, are subjected to more diseases and poorer water quality, and are relatively sheltered from predation. In the natural environment, fish are at very low density, seldom experience low oxygen, diseases are usually minimal, predators are a constant and they must forage for their own food. Radically different environments and this radically different selective forces are in place in the aquaculture and natural environments.

Domestication occurs even without directed selection by the fish culturist. Domestication effects can be observed in some fish within as few as one to two generations after removal from the wild. In channel catfish (*Ictalurus punctatus*) an increased growth rate of 3 to 6% per generation was observed due to domestication selection, and the oldest domesticated strain of channel catfish (99 years), the Kansas strain, has the fastest growth rate of all strains of channel catfish.

In general, wild strains are much less vulnerable to angling and ultimately have greater survival than domestic strains. A classical example of this phenomenon is found in rainbow trout. Government hatcheries used to not be cognizant of genetics and were judged by how many fingerlings they produced, not their suitability for the receiving environment. If one is producing fish in a hatchery environment, and the goal is to produce as many as possible, fish that are responding to the hatchery environment will best achieve the objective. These are not likely the best fish for the natural environment. If domestic and wild trout are observed at a trout hatchery, radical differences in behaviour are seen. When domestic trout are approached, they rush to the surface, follow humans and beg to be fed. In the case of wild strains, they dive for the corner of the tank and try to escape approaching humans. This has dramatic implications for survival of the fish depending on the environment to which they are released. Domestication makes the fish less fit to survive in nature. Often domesticated fish are eliminated from the natural environment in a few weeks due to competition, starvation and predation.

Domesticated common carp in Hungary exhibited better growth and resistance to *Aeromonas hydrophila* than wild strains (Ødegård *et al.*, 2010). Although most domesticated strains usually perform better in the aquaculture environment than wild strains, there have been exceptions, such as wild Nile tilapia, *Oreochromis niloticus*, and rohu, *Labeo rohita*, that appeared to grow better or the same (Nielsen *et al.*, 2010) in the aquaculture environment (Dunham *et al.*, 2001). However, the explanation for this anomaly appears related to a lack of maintenance of genetic quality and genetic degradation in the domesticated strains compared with these wild fish (Nielsen *et al.*, 2010; Dunham, 2011: this book). Poor performance of some domestic tilapia is related to poor founding (parental) lines, random genetic drift, inbreeding and introgression with slower-growing species, such as *Oreochromis mossambicus*, and slower-growing strains such as Nile tilapia from Ghana.

Domestication of farmed shrimp was relatively slow compared with that of finfish because of the use of wild brood stock and postlarvae, a lack of understanding of shrimp reproductive biology for domestication of the species, endemic disease challenges, laws restricting movements of shrimp and disease-free certification, and the relatively recent nature of shrimp aquaculture (Dunham *et al.*, 2001). As is the case with fish, domesticated shrimp are more cost-effective than wild brood stock (Preston *et al.*, 1999) and the reproductive performance of domesticated *Penaeus monodon* is similar to that of wild brood stock (Crocos and Preston, 1999).

Strain variation within a species can cause one to make incorrect generalizations about species. If one strain is chosen to define the characteristics of the species, mistaken generalizations could be made. For instance, the statement that channel catfish grow faster than blue catfish is essentially correct; however there is an overlap of performance. If a variety of strains of the two different species are examined, the fastest-growing channel catfish strains grow faster than the fastest-growing blue catfish strains. But there are some strains of blue catfish that will grow at the same rate or faster than slow-growing channel catfish strains. If we base our conclusion regarding the two species on only the latter observation, the exception rather than the rule has been identified.

Also, at one time it was believed that *O. mossambicus* was an exceptionally cold-sensitive species. However some reports stated that the cold tolerance of *O. mossambicus* is quite good. Of course, there is a wide geographic range for *O. mossambicus*. The strain that was stated as being cold tolerant was from South Africa. Of course, South Africa is far from the equator where this strain was subjected to a colder environment. Again, two laboratories are reaching two different conclusions for a species because of strain differences.

Strain Effects on Other Breeding Programmes

Strain variation is also important, since strain affects other genetic enhancement approaches, such as intraspecific cross-breeding, interspecific hybridization, sex control and genetic engineering. Thus, identifying high-performance strains positively affects these other genetic enhancement programmes.

Key Summary Points

Strain evaluation and selection is a key first step in a practical breeding programme. Strain variation can be found for all traits in all species. Strong environmental differences lead to the development of strains. Domestic strains perform better than wild strains in the aquaculture environment, and wild strains perform better than domestic strains in the natural environment. Identification of high-performance strains positively affects other genetic enhancement programmes.

5

Population Size, Inbreeding, Random Genetic Drift and Maintenance of Genetic Quality

An important aspect of genetic biotechnology is not only improvement of performance, but also the maintenance of genetic and stock quality by avoiding phenomena such as inbreeding. Inbreeding is the mating of individuals more closely related to one another than the average of the population. Sewell Wright defined inbreeding as the correlation of uniting gametes. Essentially, inbreeding is the mating of relatives. Inbreeding does not change gene frequency, but changes genotypic frequency. Therefore, inbreeding increases homozygosity, leading to inbreeding depression, decreased performance due to inbreeding. The effects of inbreeding depression include reduced rate of growth, viability, survival and reproductive performance and increased biochemical disorders and deformities from lethal and sub-lethal recessive alleles, as well as bilateral asymmetry. The genetic basis for inbreeding is dominance and epistasis.

Inbreeding occurs when there is selective or non-random mating, thus it can occur in large populations as well as small populations. If you have a small population, the probability of relatives mating increases. Familial lines emerge within the population.

Inbreeding is not an absolute. It measures the average change in homozygosity, and is a mathematical average. The inbreeding coefficient, F, ranges from 0 to 1. F predicts the proportion of loci that have become homozygous as a result of inbreeding. F also indicates the percentage of genes that are alike by descent. When determining F, if we do not know the parentage of an individual, the assumption is made that F equals 0. Obviously, this is an assumption, a starting point and may not reflect reality. F of progeny immediately is reduced to 0 by mating unrelated individuals, even if two inbred individuals are mated as long as they are unrelated. Whatever inbreeding depression was present is immediately corrected, and performance returns to the baseline of the original population prior to inbreeding.

Contrary to what seems obvious, inbred populations may actually have increased phenotypic variability. One purpose of heterozygosity is to enable individuals and populations to adapt and compensate for environmental variability. Genetic variability may decrease, but the inability of the organism to cope with environmental and micro-environmental changes may actually inflate the component of environmental variation, leading to an absolute increase in total phenotypic variability.

Potential Uses of Inbreeding

Is inbreeding good or bad? Usually inbreeding is detrimental and should be avoided. However, inbreeding can be good if we increase the number of individuals homozygous for some good genes. Individual or mass selection is actually a mild inbreeding programme.

One potential use of inbreeding is to develop lines for producing crossbreeds that will demonstrate heterosis. This approach was used to develop the 'hybrid' crossbred corn used today, but is highly inefficient. In the case of corn, hundreds and thousands of inbred lines were developed and evaluated to find the few that exhibited heterosis and performance above that of the initial population. Additionally, this resulted in severe inbreeding depression for seed production. Thus, the crosses could not be directly used in commercial corn production. This dictated the use of four inbred lines that had good combining ability to make a four-way hybrid. First, two sets of two lines were crossed to restore reproduction so that adequate seed could be produced. Next these two crosses were mated to produce a four-way hybrid exhibiting heterosis for production. Similarly, hundreds of inbred gynogenetic common carp lines were developed and crossed. Only a few exhibited potential for enhanced growth.

In the case of mice, when two inbred lines are crossed, they exhibit heterosis for growth. However, the level of growth is the same as that of the original population from which they were developed. Thus, no genetic enhancement has been accomplished compared with the base population. What has actually happened is negation of the inbreeding depression. If selection is coupled with the development of the inbred line, the probability of obtaining true genetic enhancement in the cross is increased.

Inbreeding can be useful to prepare for progeny testing. Progeny testing is most efficient if homozygous testers are utilized. The homozygous testers sometimes may be developed through inbreeding.

Line breeding is another use that has been used in the livestock industry. An outstanding individual may be mated to his daughters and granddaughters to increase the frequency of this outstanding individual's genes in the population. This approach has not been tested in fish yet.

Inbreeding Depression

The detrimental effects of inbreeding are well documented in aquatic organisms and can result in decreases of 30% or greater in growth production, survival and reproduction (Kincaid, 1976a,b, 1983a; Dunham, 1996) once the inbreeding coefficient reaches 0.125 to 0.375, most commonly 0.250. Inbreeding depression will affect some families more severely than others. This is an expected result because the heterozygosity, the genetic background, from one family to another will be different. Inbreeding depression for inbred families of rainbow trout ranged from 0 to 41% for percentage of the population recovered, from 6 to 60% for body weight, from 16 to 48% for total biomass, from 12 to 15% for number of eggs, from 0 to 53% for hatchability and from 5 to 9% only for fry survival.

Inbreeding can occur in both cultured and natural populations. When compared with much larger populations in Ukraine, the Hungarian meadow viper, *Vipera ursinii rakosiensis*, had low major histocompatibility complex (MHC) variability, which presumably reduced the immune response, and greater genetic uniformity (Újvári *et al.*, 2002). These small populations also had birth deformities, chromosomal abnormalities and low juvenile survival, suggesting that the Hungarian vipers are exhibiting inbreeding depression.

Two inbreeding studies in channel catfish gave contrasting results. One population was initiated by crossing six strains together, and is presumably more heterozygous than the second population originating from a single strain. At $F = 0.25$, the more heterozygous population exhibited no inbreeding depression, but the other population had inbreeding depression. When the more heterozygous population reached $F = 0.375$, it exhibited significant inbreeding depression. The more variable background of the 'heterozygous' population gave it a one generation buffer compared with the other population before the detrimental effects of inbreeding emerged. If a population has large amounts of heterozygosity, it apparently can tolerate more inbreeding.

This buffering, threshold effect or canalization is further illustrated by the relationship between homozygosity and inbreeding depression, which is not linear. Threshold effects are observed and epistasis is likely involved in establishing these threshold levels, and epistasis is also part of the explanation for the genetic basis of inbreeding. Initially as homozygosity increases, no inbreeding depression is observed. Once a critical point for homozygosity is reached, a threshold level, inbreeding depression occurs. As homozygosity increases the detrimental effects accelerate, so again we do not see linearity. As homozygosity continues to increase, another threshold is reached where increasing homozygosity does not cause additional depression in performance – a depression plateau has been reached. Duplicate dominant epistasis systems are a possible explanation for these threshold effects and plateaus.

This concept is illustrated by the following experiment. In the case of rainbow trout, isozymes were used to measure absolute homozygosity and its increase. Isozyme variability is then actually measuring the rate of inbreeding, and this rate of inbreeding was correlated to bilateral asymmetry. A 21% increase in homozygosity, as measured by isozymes, resulted in a 21% increase in bilateral asymmetry. When isozyme homozygosity increased to 44%, a doubling in homozygosity, bilateral asymmetry increased to 69%, a non-linear increase. A 3.5-fold increase in bilateral asymmetry occurred with a doubling in homozygosity.

Inbreeding depression is probably underestimated when measured in experiments. Lines depressed for one trait tend to be depressed for several traits, or if depression is not observed for one trait it is likely not observed for other traits either. Thus, weaker individuals and lines tend to die. At the end of the experiment when means are computed some bias exists because the worst-performing individuals did not make it to the end of the experiment and for many non-survival traits, such as growth rate, these mortalities are not included in the means. Then inbreeding depression is underestimated. In multiple generation experiments, weaker lines are dying and are eliminated over time, and are not impacting multi-generational means. Additionally, low surviving lines are now at lower densities because of differential mortality, thus their growth depression is further underestimated because they now have an environmental growth advantage.

GE interactions may occur for inbreeding depression, and laboratory studies, where organisms are pampered, may not reflect the inbreeding depression observed in more natural conditions. Inbred house mouse, *Mus domesticus*, males sired only one-fifth as many surviving offspring as outbred males because of their poor competitive ability and survivorship in large, semi-natural enclosures, but this reduced fitness is not observed in the laboratory where competition for mates does not exist, or in inbred females, which do not have to compete for mates in either environment (Meagher *et al.*, 2000). The reduced fitness of inbred males in the competitive environment was almost five times larger than what has previously been seen in the laboratory.

Correction and Avoidance of Inbreeding

If there is evidence that inbreeding depression exists, how is it corrected? As discussed earlier, inbreeding depression is easily correctible. When an inbred individual is mated to an unrelated individual the inbreeding coefficient of the progeny returns to 0, the effects of inbreeding on the performance of the progeny are also eliminated and inbreeding depression disappears. This was demonstrated in channel catfish (Bondari and Dunham, 1987). In most aquaculture businesses inbreeding will not become a problem because brood-stock populations are relatively large. If 50 breeding pairs are randomly mated per generation, the accumulated inbreeding should not result in inbreeding depression for 25–50 generations. Formulas for calculating inbreeding and determining the impact of random genetic drift are thoroughly discussed in Tave (1993).

How is inbreeding avoided? The objective is to prevent the inbreeding coefficient, F, of a population from reaching 0.25 or 25%. Utilization of adequate population size is one mechanism. If we have a large population size, the chance of relatives mating is minimal. Keeping pedigrees of fish to ensure that relatives do not mate is another option to avoid inbreeding. Utilization of molecular or biochemical genetic markers is another option. The latter two options require special facilities, much record keeping, skill and/or expense.

A simple technique is rotational line breeding. If we have three or more brood-stock ponds and they are generating the fingerlings for future brood stock, male brood-stock replacements are chosen from pond #1 and moved to pond #2. Brood-stock male replacements from pond #2 are stocked into pond #3, and brood-stock replacement males from pond #3 go into pond #1. Female brood-stock replacements from #1 go into #3, from #2 go to #1 and from #3 go to #2. With this system brothers and sisters, fathers and daughters, sons and mothers never mate as they have been physically separated. Inbreeding is not eliminated, but it is greatly slowed and reduced.

Avoidance of inbreeding and utilization of breeding schemes to avoid inbreeding are critical for the maintenance of genetic variance in both developed and developing countries; it may be just as important not to lose production from lack of genetic maintenance as to gain production from genetic enhancement. This is especially problematic when farming species with high fecundity such as Indian and Chinese carps, where only a few brood stock can easily meet fry-production needs.

Regardless of whether or not a commercial company or farm develops or obtains its own genetically enhanced aquatic organism or does not have an improved stock, it is important to maintain stock quality and prevent the current stock from regressing in genetic quality. Even when genetic improvement is not the goal, the avoidance of inbreeding depression and random genetic drift revolves primarily around population size.

A fish farmer can easily calculate the accumulation of inbreeding per generation as follows:

$$F/\text{generation} = 1/8 \times (\text{number of males that successfully breed}) + 1/8 \times (\text{number of females that successfully breed})$$

$$(5.1)$$

This indicates the change in F for each individual generation and must be calculated for each generation. The values are then added from all generations to determine the accumulated inbreeding to the present. Generally, if 50 pairs of fish randomly mate each generation, an F of 0.25 will not accumulate for 50 generations. Most large channel catfish farms, *Clarias* farms and tilapia farms maintain larger brood populations than 50 pairs, and inbreeding depression is probably not a concern in these cases. Of course, even with an adequate population size, inbreeding could still occur if progeny from these 50 pairs is not mixed or become segregated and care is not taken to randomly choose replacement brood stock that would represent all spawns from the 50 matings. Also, mass or individual selection programmes are actually mild inbreeding programmes, as mating is not random and, even with large population sizes, inbreeding could theoretically occur.

Avoidance of inbreeding and maintenance of genetic quality can be especially critical for genetic approaches to biotechnology. The nature of these biotechnologies makes it tempting to initiate populations with inadequate or small population numbers.

This could lead to inbreeding or drift, which might partially or completely negate or counter the improvements made through the biotechnology. Gene transfer, gynogenesis, cloning, sex reversal and breeding would all be potentially initiated with small population numbers without extra effort. Of course, these genetic changes could be bred into a population and genetic variation and population-size parameters addressed through traditional means. However, this could delay commercialization by a generation while the new genotype is being spread into a larger population base. However, another strategy would be to initiate commercialization with the narrower genetic base and correct the problem over time with conventional breeding strategies.

In theory, it might be a good idea to deliberately inbreed small, captive populations to eliminate inbreeding depression as inbreeding exposes deleterious recessive alleles in the homozygous state to selection, allowing their removal and rapid purging from the population (Doyle, 2003). This would eliminate future inbreeding depression even in small populations. Frankham *et al.* (2001) developed two populations of fruit flies by inbreeding (brother × sister mating) for 12 generations. One population was outbred until the inbreeding started and was presumably genetically variable, while the second type had been inbred for 20 generations and then hybridized just prior to the final inbreeding experiment. Thus the second population was also genetically variable, because of the hybridization, but purged of deleterious alleles because of the prior inbreeding. The purging was not successful, and there was a small but non-significant difference between the extinction rates at an inbreeding coefficient of 0.93 in the non-purged (0.74±0.03) and purged (0.69±0.03) treatments. This is consistent with other evidence indicating that the effects of purging are often small and purging using rapid inbreeding in a very small population is not a reliable method to eliminate the deleterious effect of inbreeding (Doyle, 2003). This is not a surprising result as in Chapter 3 we demonstrated that it is extremely difficult

to eradicate a homozygous recessive allele through selection.

Results of purging in the laboratory may differ from those in more natural conditions. The native silversword, *Argyroxiphium sandwicense sandwicense*, plants on the Hawaiian island of Mauna Kea were almost eliminated by cows and then were reintroduced in a recovery programme (Doyle, 2003). The population crash did not reduce the number of microsatellite alleles or heterozygosity. However, a population bottleneck resulted from the reintroduction, reducing the observed number of alleles, effective number of alleles and expected heterozygosity, though not the observed heterozygosity (Friar *et al.*, 2000). The data are consistent with other studies indicating that small populations, including those resulting from severe reductions, are not necessarily devoid of genetic variability and distortion can be caused when a population is founded or supplemented by a few individuals (Friar *et al.*, 2000).

Theoretically, the mixing of sperm from multiple males, followed by fertilization of eggs, should increase genetic diversity. However, in reality, this is not the case. When salmon milt was mixed, sperm from one male were more competitive and dominated the fertilization, so fertilization was essentially the same as using a single male. Ten microsatellite loci were analysed for the veliger larva of abalone; three loci were generally sufficient to assign parentage (Selvamani *et al.*, 2001). Most fertilizations were attributable to just one of the males in a pool of multiple males' sperm. Fertilizing with a mixed pool of sperm is not a good procedure for increasing genetic diversity, because almost always one male has faster, more competitive sperm, which fertilize the majority, if not all, of the eggs. The practical solution is to divide the eggs from single females into multiple lots, and then fertilize each lot with sperm from different single males to eliminate competition of the sperm and to maximize genetic diversity.

Similar results were obtained with Pacific oysters. Boudry *et al.* (2002) used a single, highly variable, microsatellite locus to trace the sources, showing large variation in parental contributions at various developmental stages of Pacific oysters, which led to a strong reduction of effective population sizes in experimental populations. Segregation distortions fluctuated with time, and variance in reproductive success was attributed to gamete quality, sperm–egg interaction and differential viability among genotypes. Sperm competition increased reproductive variance and decreased effective population size, as observed in abalone and salmon.

Waldbieser and Wolters (1999) studied reproductive behaviour in channel catfish using microsatellite markers. Multiple spawning by males was found in seven of the eight ponds examined, and 47% of the males fertilized two to six egg masses over 1–8 weeks, although most were 2–3 weeks apart. Four different spawns were identified as two full-sib families and were probably due to interrupted spawning, with the pair reinitiating mating at a later time and/or different location. The fish did not randomly spawn. A subset of males would dominate spawning for a couple of weeks and then they would be replaced by a new dominant subset.

To minimize the accumulation of inbreeding, each brood fish should contribute equally to the next generation, but this is often impossible to accomplish (Doyle, 2003). Sonesson and Meuwissen (2001) developed a scheme of overlapping generations to minimize inbreeding and evaluated it with simulation. Selection of older breeders from the age-class mixture reduced the rate of accumulation of co-ancestry and inbreeding, was practical when mature mortality rates are reasonably low and there is good control over family size and, theoretically, was even superior to pedigree analysis for reducing the accumulation of inbreeding. However, surely a combination of this overlapping-generation technique and pedigree maintenance would be even more effective.

Standard management practices recommend equalizing family sizes from one-on-one mating of males and females to minimize inbreeding and random genetic drift (Doyle, 2003), but theoretically this procedure can also reduce the intensity of natural selection (actually domestication) for fitness as mutational load will accumulate. However, computer simulations indicate that equalization of family sizes does not produce a high short-term threat to small conserved populations, up to about 20 generations, and the more efficient preservation of genetic variability outweighs the disadvantages of the procedure.

Indicators and Markers for Inbreeding

Bilateral asymmetry – unbalanced numbers for meristic traits on the right and left halves of the body – has been shown to be an indirect indicator of homozygosity in fish (Leary *et al.*, 1983, 1984, 1985a,b,c). Since the two halves of an embryo should be genetically identical, the random asymmetry can be interpreted as failure of the genotype to adequately regulate the development of the phenotype (Doyle, 2003). Thus another purpose of heterozygosity is to ensure that development is normal and symmetric. It is assumed that this bilateral asymmetry may also be an indicator of inbreeding depression and reduced fitness. Bilateral asymmetry of abdominal bristles in 32 fruit-fly lines was not correlated to two measures of fitness: productivity (a combined measure of fecundity and egg-to-adult survivorship) and competitive male mating success (Bourget, 2000). This indicates that, although bilateral asymmetry is usually not desirable, it may not always be a reliable indicator of reduced fitness.

Bryden and Heath (2000) found that bilateral asymmetry for the meristic and quantitative traits in chinook salmon is not heritable, and they conclude that bilateral asymmetry estimates in chinook salmon will not be confounded by appreciable additive genetic contributions and thus can be reliably used as an environmental and genetic stress indicator. This should also indicate that bilateral asymmetry has a dominance genetic component.

Allozymes do not always measure changes in inbreeding accurately and this was the case in guppy, *Poecilia reticulata* (Shikano *et al.*, 2001a,b). Three closed lines of guppies ($n = 10$) were propagated for six generations, achieving inbreeding levels of 0.19, 0.32 and 0.41 and salinity tolerances were 82.5, 71.7 and 67.6%, respectively, of their original values. Salinity tolerance was reduced linearly by 8.4% per 10% increase in inbreeding.

To monitor parentage to maintain genetic quality, the number of markers necessary to accomplish the goal needs to be known. The analysis of Bernatchez and Duchesne (2000) indicates that, if sufficient allelic diversity exists, a relatively low number of loci are required to achieve high allocation success to full-sib families even for relatively large numbers of possible parents, but there is no significant gain in increasing allelic diversity beyond approximately six to ten alleles per locus in population-assignment studies.

Microsatellites are an excellent tool for checking pedigrees and lineages. Because of high mutation rates, it is generally believed that microsatellites are powerful only for examining short-term lineage. However, Cunningham *et al.* (2001) examined Thoroughbred horse pedigrees extending back about 300 years with 12 microsatellite loci, and agreement between relatedness calculated from pedigree records and estimated from the microsatellites was extremely high.

Microsatellite analysis indicates that Atlantic salmon are polygynous and polygamous mating behaviour is the norm (Garant *et al.*, 2001). The number of offspring, the major component of fitness, was correlated with the number of mates, but not with body size. This seems almost contradictory to hatchery results, where artificial mixing of sperm does not result in a strong genetic contribution of multiple males. Perhaps, under natural conditions, the stimulation of natural mating results in more even maturation of sperm among the males.

Amos *et al.* (2001) examined neutral markers in three long-lived vertebrates – long-finned pilot whale, grey seal and wandering albatross – revealing negative relationships between parental similarity and genetic estimates of reproductive success. The negative correlation between parental relatedness and fitness was not merely the result of inbreeding depression, caused by the mating of close relatives, as the correlation extended to low levels of relatedness, where conventional inbreeding depression was unlikely. There appears to be a positive advantage to finding a dissimilar mate (Doyle, 2003), and this result was similar to that for Atlantic salmon, which may choose mates to maximize MHC diversity.

Experimental Measurement of Inbreeding Depression

To measure inbreeding depression the appropriate genetic control must be implemented. Two options exist: a randomly bred control and an outbred half-sib control. The randomly bred control may not exactly match the beginning genotype of the inbred families so this is a possible source of some experimental error. In the case of the outbred half-sib control, for each and every inbred family, they are also mated with an unrelated individual to produce the control. Each inbred family is matched with its own control, reducing error of measurement of inbreeding depression. Inbreeding depression is measured for each family, and each family will respond variably as each has its own genetic background and a different level of actual absolute starting homozygosity. Inbreeding depression for the population is the mean inbreeding depression of all of the families in the evaluation. Even in a small research population, each family has an internal control maximizing accuracy. Theoretically, the internal, outbred control should be the superior design. However, two inbreeding studies of rainbow trout on separate continents measured virtually identical inbreeding depression, one using a random control and one an internal outbred control. This may be a coincidence or perhaps with experiments of sufficient size error rate is minimal with either control.

Multiple methods exist for the development of inbred lines. The traditional way of making inbred lines in animals is, of course, to mate relatives such as brother × sister matings. One generation of brother and sister mating results in an inbreeding coefficient of 0.25. Two generations of brother and sister mating results in an inbreeding coefficient of 0.375.

With fish, there are other options to accomplish inbreeding because an individual can essentially be mated with itself through either gynogenesis or androgenesis. In the case of gynogenesis, all-female inheritance, all genetic material comes from the mother. Androgenesis is all-male inheritance; all genetic material comes from the father except, of course, for mitochondrial DNA since that is inherited through the egg.

A meiotic gynogen has essentially the same increase in homozygosity as with brother × sister mating, $F = 0.25$, as indicated by isozyme analysis. Mitotic gynogens and androgens are 100% homozygous. In these cases, F equals 1.0.

Random Genetic Drift and Bottlenecks

Random genetic drift is closely related to founder effect. Allele frequencies change or drift in random fashion because of low population numbers. Random genetic drift results in losses of alleles or genes that are present at low frequencies in the population. This occurs when the breeding population size is low; the low frequency of alleles are lost by random chance by the laws of probability. In the case of random genetic drift, allele frequencies are changing. In contrast, allele frequencies stay constant for inbreeding, but genotypic frequency changes.

For management purposes the concept of effective population size is important. From a genetics standpoint, this value accounts for genetic population size based on potential unequal contribution of the sexes in reproduction. Also from a genetic standpoint, individuals that are counted for population size are only those that breed and contribute to the next generation. Effective population size, N_e, equals (4 × number of males × number of females) divided by (number of males + number of females). This formula is important in conservation genetics and managing natural populations because of the goal of trying to maintain the status quo and genetically conserve populations. So this is a tool that we can use to adjust our population sizes etc. to make sure certain alleles remain present for a certain period of time or we can use this as a tool to show, in the case where we cannot adjust the population numbers, what is going to happen to our allele frequencies over time (Table 5.1).

Some controversy exists concerning the effectiveness and benefits of supplemental stocking programmes. Hedrick *et al.* (2000a,b) evaluated the effectiveness of a supplementation programme for winter-run chinook salmon in the Sacramento River in California. In this well-designed programme, supplementation stabilized and increased the effective population size. Returning spawners represented a broad sample of parents and not fish from only a few families.

Ryman and Laikre (1991) demonstrated that the effective population number of an endangered, wild population would actually be decreased by supplementing it with hatchery stock based on the first generation of stock supplementation; however, Wang and Ryman (2001) show, both analytically

Table 5.1. Effective population size (N_e) needed to keep an allele at its initial frequency (f) for one to 100 generations (G) with a probability of failure at $P = 0.05$ or $P = 0.01$. (Adapted from Tave, 1993.)

	N_e			
	$f = 0.05$		$f = 0.01$	
G	$P = 0.05$	$P = 0.01$	$P = 0.05$	$P = 0.01$
1	30	45	150	230
10	52	68	263	344
25	61	77	308	390
50	68	83	343	444
100	74	90	377	458

and by simulation, that the reduction in effective number (both drift and inbreeding) is quickly reversed in later generations as long as the census size of the population – the actual number of animals, including those breeding in the hatchery and the wild – is increased by supplementation. Selecting either wild or hatchery brood stock accomplished the increase in effective population size. This experiment also illustrates the point that erroneous conclusions can be reached in short-term experiments of insufficient length.

Poon and Otto (2000) developed a model to determine the balance of detrimental and beneficial mutation, and the modelling exercise indicated that the loss of fitness caused by random fixation of deleterious mutations is directly proportional to the dimensionality and inversely proportional to the effective population size. Poon and Otto (2000) conclude that the reciprocal relationship between the loss of fitness and effective population size implies that the fixation of deleterious mutations is unlikely to cause extinction when there is a broad scope for compensatory mutations, except in very small populations, and pleiotropy plays a large role in determining the extinction risk of small populations. This appears logical as, when one gene affects more than one trait, its importance and impact must be substantial for fitness. Compensatory mutations must be considered to increase the accuracy of predictions concerning the genetics of small populations (Doyle, 2003).

Doyle *et al.* (2001) describe a method to increase the genetic diversity of a bottlenecked brood stock without bringing in new breeders. The procedure is an extension of minimal kinship selection, which has been used to preserve diversity in brood stocks where complete pedigree records exist, such as in zoos. Unfortunately, pedigrees are rarely available in fish hatcheries; therefore, instead of using pedigrees to calculate kinships, Doyle *et al.* (2001) used Ritland's genetic relatedness estimator, calculated from microsatellites, to estimate the mean relatedness of each potential breeder to the whole population. A subset of breeders was then selected to maximize the number of founder lineages, in order to carry the fewest redundant copies of ancestral genes. Microsatellite data from a hatchery population of red seabream for which pedigrees were independently available were used to validate the method (Doyle *et al.*, 2001), resulting in higher standard measures of marker diversity in the selected subset of breeders than in randomly chosen subsets. In the next generation, a partial reversal of the effects occurred from genetic erosion and drift. The procedure differs from MAS in that DNA marker data were used to identify rare pedigrees or extended families, rather than to identify rare chromosome segments carrying QTLs, and the particular application emphasized the recovery of the genetic diversity lost when a hatchery is founded with a small and non-representative sample of an ancestral wild population (Doyle *et al.*, 2001).

Inbreeding and thus selection could change genetic covariances. Phillips *et al.* (2001) produced a large number of small populations of *Drosophila* to generate random drift, resulting in no change in average variance–covariance matrix (**G**) and overall structure among traits of the inbred lines relative to the outbred controls. However, there was a great deal of variation among inbred lines around this expectation, even changes in the sign of genetic correlations. Since any given line can be quite different from the outbred control, it is likely that in nature unreplicated drift will lead to changes in the **G** matrix; the shape of **G** is malleable under genetic drift and the evolutionary response of any particular population will probably depend on the specifics of its evolutionary history (Doyle, 2003). Thus, the establishment of a small aquaculture population from a larger population may result in genetic changes that are qualitatively and quantitatively different from those in the source brood stock.

Key Summary Points

The probability of useful outcomes of inbreeding is quite low. Crossing of inbred lines rarely results in true genetic improvement, only an increase in performance back to that of the base population, negating the effects of inbreeding depression. When the inbreeding coefficient reaches approximately 0.25, significant inbreeding depression usually occurs in aquatic organisms. Inbreeding can be avoided by using adequate population sizes, keeping pedigrees or rotating individuals away from their families in brood replacement ponds. Inbreeding can easily be corrected by one generation of crossbreeding.

6

Gynogenesis, Androgenesis, Cloned Populations and Nuclear Transplantation

Gynogenesis – all-maternal inheritance – and androgenesis – all-paternal inheritance – can be used to produce rapid inbreeding, clonal populations or monosex populations. Gynogenesis and androgenesis can also be used to elucidate sex-determining mechanisms in fish. These techniques can allow the rapid generation of inbred lines since they are selfing techniques. If the female is the homogametic sex, all the gynogens will be XX and female. If the female is the heterogametic sex, the gynogens will be ZZ, WZ and WW and both sexes will be found in the progeny. In the WZ system, the maleness gene or chromosome is Z and the femaleness gene or chromosome is W. If the male is the homogametic sex when androgens are produced, the androgens will be 100% ZZ and all male. If the male is the heterogametic sex, XX and YY androgens will result in equal proportions. Unlike mammals, YY individuals are viable in fish (Yamamoto, 1975; Parsons and Thorgaard, 1985) and should produce all-male XY progeny when mated with normal XX females.

Induction of Gynogenesis and Androgenesis

Gynogenesis is accomplished by activating cell division with irradiated sperm (Chourrout, 1986a,b; Thorgaard, 1986) and then restoring diploidy to the developing zygote. Irradiation breaks or destroys the DNA in the sperm, so there is no paternal contribution to the zygote, but the sperm is still motile and can penetrate the egg and activate cell division. Either ultraviolet (UV) irradiation or gamma irradiation can inactivate the nuclear genome of the male or female parent. UV irradiation has advantages and is more effective than gamma irradiation. If the dosage is too low, gamma irradiation can produce supernumerary chromosome fragments from the donor sperm, which can be integrated in the host, replicate independently for many cell divisions like a chromosome and be expressed in the zygote (Chourrout, 1984, 1986b). However, gamma irradiation has the advantage of being more penetrating than UV rays, and UV rays are effective only when the milt is diluted and spread in a thin layer (Chourrout, 1986b). Chemical mutagens, such as dimethylsulfate, can also inactivate large volumes of sperm (Tsoi, 1969; Chourrout, 1986b); however, this procedure also produces supernumerary chromosome fragments (Chourrout, 1986b; Fig. 6.1).

To ensure that genetically viable sperm do not fertilize the egg and produce normal diploids, sperm from related species (heterologous sperm) can be irradiated and used

Fig. 6.1. Karyotype containing supernumerary chromosome fragments produced by gamma irradiation of sperm. (From Thorgaard *et al.*, 1985.)

to activate cell division (Stanley and Jones, 1976; Chourrout and Quillet, 1982; Chourrout and Itskovich, 1983; Chourrout, 1986a,b) to allow verification that the embryos are gynogenetic. If a normal diploid or triploid hybrid is produced, either it is recognizable morphologically or it dies, depending on the viability of the hybrid.

After activation of the egg with irradiated sperm, the second polar body will be extruded, resulting in haploid embryos that eventually die if no treatment is used (Chourrout, 1986b). One alternative for producing diploid gynogens is to block the extrusion of the second polar body; then the diploid gynogen has two sets of chromosomes, both of maternal origin. This type of gynogen is referred to as a meiotic gynogen, or meiogen, since it was produced by blocking the second meiotic division. Treatments that result in high rates of triploid induction should also be the most effective for production of diploid meiogens since both are produced by retaining the second polar body. Retention of the second polar body is accomplished with temperature shocks or pressure treatments (Chourrout, 1980; Thorgaard *et al.*, 1981; Chourrout and Quillet, 1982; Benfey and Sutterlin, 1984a; Lou and Purdom, 1984). The treatment is applied shortly after sperm penetration prior to extrusion of the second polar body. The most effective time for these shocks varies among species (Dunham, 1990a; see Chapter 10).

Gynogens produced by blockage of polar-body extrusion are inbred since all genetic information is maternal. Meiotic gynogens are not totally homozygous since crossing over and recombination during oogenesis result in different gene combinations in the ovum and the second polar body (Thompson and Purdom, 1986). The chromosome sets in the ova and the second polar body are not identical. The rate of inbreeding or increase in homozygosity through meiotic gynogenesis is roughly equivalent to one generation of full-sib mating.

Diploid gynogens can also be generated by blocking first cleavage (Thompson and Purdom, 1986). Karyokinesis is allowed, resulting in the doubling of chromosome number to the diploid state, and, if the first cell division is blocked, a single diploid cell results. Gynogens produced by this technique – mitotic gynogens or mitogynotes – are 100% homozygous since a single set of chromosomes is duplicated (Thorgaard et al., 1981; Thompson and Purdom, 1986).

The first cell division can be blocked chemically (Sriramula, 1962; Squire, 1968; Valenti, 1975; Smith and Lemoine, 1979), with temperature shocks or with hydrostatic pressure (Chourrout, 1980; Thorgaard et al., 1981; Chourrout and Quillet, 1982; Benfey and Sutterlin, 1984a; Lou and Purdom, 1984). The timing of these treatments and those that are most effective are the same as for the induction of tetraploidy since both techniques rely on blocking the first mitotic division. Production of gynogens via blockage of first cleavage is more difficult than production of gynogens by retention of the second polar body (Thorgaard, 1986), just as triploids are also easier to produce than tetraploids (Wolters et al., 1981a; Bidwell et al., 1985; Thorgaard, 1986; Curtis et al., 1987; Rezk, 1988; Bury, 1989). This indicates that the second polar body is easier to manipulate than the first cell division or that embryos are more sensitive to environmental shocks near the first cell division compared with shortly after fertilization. Alternatively, gynogens produced by blocking the first mitotic division would be more likely to die during embryonic development because of a higher frequency of deleterious recessive genotypes found in these individuals, as they are all 100% homozygous (Scheerer et al., 1986).

Androgenesis is accomplished by irradiating eggs and then doubling the paternal genome, as described for mitotic gynogenesis (Thorgaard, 1986). Androgens are more difficult to produce than either type of gynogen (Scheerer et al., 1986). Diploidy can be induced in androgens only at first cell division, a difficult developmental stage for manipulating the embryo. Since androgens are 100% homozygous, like mitogynotes, a large percentage possesses homozygous, deleterious recessive genotypes, which may lead to death (Scheerer et al., 1986). The irradiation of eggs could, theoretically, destroy mitochondrial DNA (mtDNA) or have damaging effects on the cytoplasm which may adversely affect the development and viability of androgens in comparison with mitogynotes whose mtDNA and cytoplasm have not been irradiated. The function of loci found in the mitochondrial genome is primarily energy metabolism (Chapman et al., 1982; Avise et al., 1987; Moritz et al., 1987). mtDNA is almost solely maternal in origin, and paternal leakage is negligible (Chapman et al., 1982; Avise et al., 1987; Moritz et al., 1987).

If YY individuals are sub-viable, there will be a preponderance of female androgens, and the loss of YY males would be another explanation for the difficulty in producing androgens. May and Grewe (1993) examined the effects of gamma irradiation on nuclear DNA and mtDNA of brown and brook trout eggs fertilized with either brook trout or splake (*Salvelinus namaycush* × *Salvelinus fontinalis*) sperm. Only paternal allozymes were observed in embryos, confirming the inactivation of the nuclear genome in the eggs, and these embryos contained exclusively maternal mtDNA. May and Grewe (1993) suggest that mtDNAs are more resistant to gamma irradiation and cobalt-60 inactivation than nuclear DNAs, based on the structure or numerical superiority of the maternal nuclear DNA. These homozygous androgens have the nuclear genome of the paternal parent (androgenetic) and the mitochondrial genome of the maternal parent. Apparently, mtDNA is protected from irradiation by its additional double membrane, its small circular genome or its large copy number (May and Grewe, 1993; Brown and Thorgaard, 2002). Restriction fragment length polymorphism (RFLP), AFLP and sequencing analysis did not detect any point mutations from the irradiation treatments (May and Grewe, 1993; Brown and Thorgaard, 2002).

This also has implications in that it may be difficult to induce beneficial mutations via irradiation. Lee *et al.* (2000) studied factors that affect irradiation levels and cause target doses and actual absorbed doses to be different from those expected when irradiating eggs or sperm. They found high levels of vertical and horizontal variation in dose rates inside a small (but typical) container (14.5 cm high, 16.7 cm wide and 2.8 l in volume). Rotation or lack of rotation affected the variation in the dose actually applied. Rotated containers had vertical variation ranging between 1 and 21% and horizontal variation between 10 and 19%. As expected, variation was much higher when containers were not rotated, with vertical variation ranging between 6 and 28% and horizontal variation between 20 and 72% (Lee *et al.*, 2000). Rotated containers had much less variation in irradiation level when comparing air and water as surrounding medium. Dose rate in air was as much as 42% higher than in water when the container was rotated and as much as 218% higher in air when the container did not rotate. Samples irradiated in water probably receive dosages lower than desired. Lee *et al.* (2000) suggest that this type of variation and error complicates comparisons among studies and organisms, and that dosiometry studies should be conducted to verify actual experimental parameters.

Performance of Gynogens and Androgens

Homozygous gynogens of ayu show increased variation for size and for meristic traits (numerical traits such as anal-fin ray number). Rainbow trout mitogynotes exhibit greater amounts of bilateral asymmetry – unequal counts for meristic traits on the right and left sides of the body – than controls, isogenic crossbreeds and normal crossbreeds. Bilateral asymmetry has been previously documented in inbred rainbow trout and is associated with a reduction in biochemical genetic variation. Isogenic crossbreeds – crossbreeds resulting from mating two clonal lines (Fig. 6.2) – had similar bilateral asymmetry as outbred crossbreeds of common carp (Komen *et al.*, 1993). The coefficient of variation for size was much larger for the common carp clones compared with outcrossed populations. Homozygous gynogens of common carp exhibit growth depression compared with crossbreeds (Komen *et al.*, 1993). Growth reduction was slightly more when compared with outcrossed crossbreeds than when compared with isogenic gynogenetic crossbreeds.

Reproduction

Komen *et al.* (1993) compared the gonad development and fertility of common carp produced by either full-sib mating ($F = 0.25$), gynogenesis by retention of the second polar body ($F = 0.25$) or gynogenesis by endomitosis ($F = 1.0$) from a common female. Meiogenetic offspring were all female, but 50% of the mitogynotes were homozygous for a recessive mutant sex-determining gene, resulting in 50% males and intersexes. Another mutation affecting pigmentation in eggs was also detected in the mitogynote offspring but not in meiogenetic offspring. The variation in body weight, gonad weight and egg size in each group increased with increased inbreeding. Full-sib inbreds and meiogens ($F = 0.25$ for both) were comparable in mean gonadal development, but gonads from homozygous gynogenetic carp were often retarded. Ovulation response also decreased with increasing levels of inbreeding (Komen *et al.*, 1993).

Gynogenetic and polyploid fish often have abnormal ovarian development (Krisfalusi *et al.*, 2000). Piferrer *et al.* (1994) compared the gonadal morphology of normal and sex-reversed juvenile triploid, gynogenetic diploid and normal diploid coho salmon.

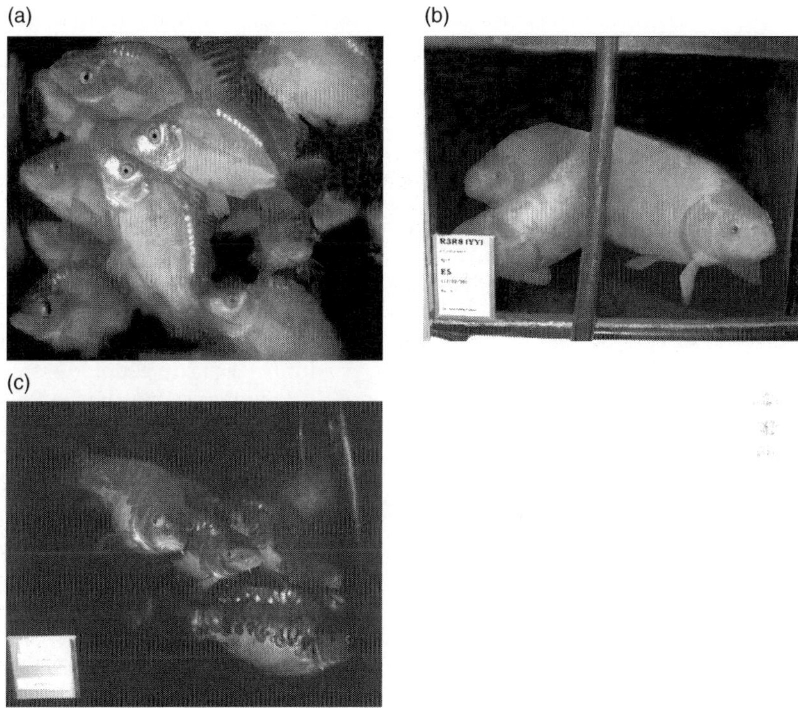

Fig. 6.2. Three photographs representing three homozygous cloned populations of common carp, *Cyprinus carpio*, maintained at Wageningen University, The Netherlands. Note the almost identical, scale pattern and body shape of the clones within each aquarium. (Photographs by Rex Dunham.)

Gynogens were 100% females, and 34% had reduced ovaries with areas devoid of oocytes. About 30% of gynogenetic rainbow trout females also had aberrant ovarian development (Krisfalusi *et al.*, 2000). Gynogenetic fish may be more prone to developmental abnormalities because of either increased homozygosity or incomplete inactivation of the paternal chromatin. In 90% of the gynogen rainbow trout examined, chromosomal fragments were positively correlated with aberrant ovarian development. The atypical gonadal morphology of the gynogens was similar to the ovarian morphology of triploid rainbow trout (Krisfalusi *et al.*, 2000). Disruption of the normal diploid chromosomal complement appears to alter germ-cell development in gynogenetic female rainbow trout and probably coho salmon due to the unbalanced nature of the genome. Contrary to what has been observed in cyprinids, male germ cells were not observed in the ovaries of the

gynogenetic coho salmon, perhaps because of differing mechanisms of sex determination for salmonids and cyprinids (Piferrer *et al.*, 1994). However, this again could be related to paternal chromatin and the incomplete destruction of the paternal genome, which might be affected by species or temperature.

Monosex Populations

In many species of fish, the female is the homogametic sex (Chen, 1969; Chourrout and Quillet, 1982; Goudie *et al.*, 1983, 1985; Wohlfarth and Hulata, 1983; Shelton, 1986; Shah, 1988). Theoretically, when this is the case, gynogenesis can be used to produce monosex female populations. Gynogenetic, all-female grass carp (Shelton, 1986), silver carp, *Hypophthalmichthys molitrix* (Mirza and Shelton, 1988), channel catfish (Goudie *et al.*, 1985) and rainbow trout (Chourrout and Quillet, 1982) have been produced. However, it is not feasible to produce commercial quantities of monosex gynogenetic fish because of the high mortality of the manipulated eggs (Stanley, 1976a,b; Donaldson and Hunter, 1982; Shelton, 1986). Hatching rates of eggs activated and treated to produce gynogens and survival of fry are usually less than 1%. If the male is the homogametic species, androgenesis would result in monosex male populations. Again, the low survival of eggs treated to produce androgens prevents large-scale production of androgens (Scheerer *et al.*, 1986). Gynogenesis and androgenesis can be combined with sex-reversal technology to produce commercial quantities of monosex fish as discussed in Chapter 12.

Cloned Populations

Clonal populations of fish can be generated in a non-specific manner, utilizing gynogenesis and sex reversal. Mitotic gynogens are produced by blocking the first cell division, and these individuals are 100% homozygous. However, they are not clones of their mother as their genotype is different from their dam because of independent assortment and recombination during meiosis and because of the absolute homozygosity induced from the gynogenesis. Additionally, each mitogynote is different from its siblings, and each is variably homozygous for a variety of allelic combinations. Further, the breeder has no control over which genotype survives and is generated by the gynogenesis. Clonal populations can be produced from these mitogynotes by performing a second generation of mitotic gynogenesis on the first-generation mitotic gynogen. Presumably, more than one offspring would be produced. Then a portion of the fry would be sex reversed, both sexes would exist, although genetically identical, and large numbers of identical fry would be produced via natural mating. Future male brood stock would again be perpetuated via sex reversal. Theoretically, clonal populations could be generated in a similar way from androgens. The primary difficulty in

this breeding scheme is that the mitogynotes and androgens are maximally inbred and 100% homozygous, often resulting in reproductive problems.

Phenotypic variability

Cloned populations have been produced in zebra fish, ayu, common carp and rainbow trout (Komen *et al.*, 1993). Theoretically, clones – individuals with identical genotypes for their entire genome – should have identical, non-variable performance. However, individuals with extreme homozygosity appear to lose significant ability to respond to the environment in a consistent, stable manner, and micro-environmental differences affect performance among individuals (Komen *et al.*, 1993). Thus, as the component of genetic variation decreases, actually becoming zero, the component of environmental variation increases and at a more rapid rate than the genetic component decreases, resulting in populations with extreme phenotypic variation. Homozygous gynogens of ayu show increased variation for size and for meristic traits.

Rainbow trout clones exhibit greater amounts of bilateral asymmetry compared with controls, isogenic crossbreeds and normal crossbreeds. Bilateral asymmetry has been previously documented in inbred rainbow trout and is associated with a reduction in biochemical genetic variation. Isogenic crossbreeds had a similar bilateral asymmetry to outbred crossbreeds (Komen *et al.*, 1993). Almost identical results have been obtained for clonal rainbow trout. Clonal lines of rainbow trout were produced by androgenesis, followed by another generation of androgenesis or gynogenesis, followed by retention of the second polar body (Young *et al.*, 1995). Clonal populations were more asymmetrical and showed higher variance in total meristic counts than either isogenic crossbreeds or outbred crossbreeds, indicating reduced fitness in the clonal populations and their ability to react to micro-environmental variation. There was no difference in asymmetry between outbred crossbreeds and isogenic crossbreeds. Fluctuating asymmetry of spotting pattern was also investigated as a potential measure of developmental stability. Asymmetry of spotting pattern was not different in homozygous clones compared with isogenic crossbreeds, but the spotting pattern had a low coefficient of variation within lines and may be useful for identifying isogenic lines and monitoring genetic divergence between lines.

The increased developmental stability and decreased variance of meristic traits in the isogenic crossbred rainbow trout indicate that they were less influenced by environmental conditions and may make better research organisms than the homozygous clones. Vandeputte *et al.* (2002) used a homozygous clone outcrossed to normal common carp as a powerful internal scaled control for the evaluation of mirror carp strains.

Regeneration of genetic variation in salmonids

Clonal individuals of rainbow trout can produce progeny with genetic variation. Salmonids have autotetraploid ancestry and tetrasomic inheritance (Brown and Thorgaard, 2002). The entire salmonid genome was duplicated (Allendorf and Thorgaard, 1984), and then the chromosome number was reduced via Robertsonian rearrangement (Thorgaard, 1976). In secondary tetrasomy, homologous regions on separate chromosomes can cross over during gametogenesis to produce unique genotypes (Brown and Thorgaard, 2002), causing problems in the production of clonal lines (Allendorf and Danzmann, 1997). Brown and Thorgaard (2002) were able to detect AFLP variation in clonal populations of androgenic rainbow trout. Of course, theoretically, the AFLP sequences do not have phenotypic importance, but this demonstrates that genetic variation for protein-encoding DNA sequences could also occur, resulting in genetic variation and phenotypic variation in what are supposed to be clonal lines.

Growth

Homozygous gynogens of common carp exhibit depression in growth compared with crossbreeds (Komen *et al.*, 1993). Growth reduction was slightly more when compared with outcrossed crossbreeds than with isogenic gynogenetic crossbreeds. Again, the coefficient of variation for size was much larger for the common carp clones compared with outcrossed populations. Müller-Belecke and Hörstgen-Schwark (2000) established six different homozygous clonal lines of Nile tilapia, which was verified by analysis of the allozymes of adenosine deaminase and DNA fingerprints. Up to first feeding, clones had a reduced survival rate (4.0%) compared with normal heterozygous fish (43.9%). Reproductive traits of the second-generation clones varied between and within clonal lines. The observed body weight of all-female clonal groups (45.5 g) kept under three different density classes was slightly less than that of corresponding all-female heterozygous control groups (50.2 g). In contrast to what has been observed for clonal populations of other species of fish, Nile tilapia homozygous clones had more uniform growth (coefficient of variation 23.1%) than in all-female heterozygous controls (coefficient of variation 34.1%).

Disease resistance

Kobayashi *et al.* (1994) produced clonal lines from different strains of amago salmon, *Oncorhynchus rhodurus*. As expected, the clones immunologically accepted operculum allografts from inbred sisters. Survival was similar among clones produced by dams from the same strain, but survival was different among clones using maternal parents from different strains. Han *et al.* (1991) produced clonal ayu, *Plecoglossus altivelis*, that accepted graft tissue transplanted from intraclonal siblings, but the graft tissue was rejected when transplanted into different clonal hosts.

Strain and genetic background affect the viability of clonal lines. Non-specific immune responses, serum lysozyme activity and phagocytosis were different among inbred clones of Nile tilapia (Sarder *et al.*, 2001). Resistance to *Aeromonas hydrophila* infection was positively correlated to non-specific immune responses. When a clonal line susceptible to *A. hydrophila* was crossed with a resistant clonal line, the resulting progeny demonstrated intermediate disease resistance, indicative of additive genetic variation rather than dominance gene action for this trait.

Four gynogenetic clonal lines of common carp were selected as high or low responders on the basis of their primary serum antibody production to dinitrophenyl - keyhole limpet haemocyanin (DNP-KLH) (Bandin *et al.*, 1997). Number of antibody-secreting cells and not serum antibody titres were the primary difference in progeny of high and low responders. High-responder clonal common carp lines produced 1.5- to twofold higher numbers of specific antibody-secreting cells than low-responder clonal lines, after both primary and secondary immunization. Individual variation within the inbred lines was reduced compared with that in outbred common carp.

Haematocrit, haemoglobin, erythrocyte count and mean corpuscular volume were measured for stressed and non-stressed clonal lines of ayu and also verified by DNA fingerprinting (del Valle *et al.*, 1996). Differences among clonal lines were observed for haematocrit and mean corpuscular volume under the no-stress environment. Heritability values were moderate for the no-stress measurements (mean of 0.24) and very low or zero for the stressed groups' traits (except one, 0.48).

Clonal hybrids

Salmonids are genetically unusual in many ways, and it may be possible to develop interspecific clonal hybrid lines. Clonal full-sib progeny groups of Atlantic salmon × brown trout hybrids were produced by gyno-genesis (Galbreath *et al.*, 1997). One hybrid female had a large portion of the oocytes undergo a premeiotic chromosome doubling, or possibly a complete suppression of meiosis, resulting in ovulation of diploid eggs, each possessing one full set of both Atlantic salmon and brown trout chromosomes identical to those in the maternal somatic cells. This opens the possibility of developing lines of clonal salmonid hybrids for experimentation.

Nuclear Transplantation – Nucleocytoplasmic Hybrids

A unique approach to genetic modification in fish is nuclear transplantation. Of course, this procedure is the key to the present-day controversial research regarding cloning of individuals. Intraspecific nuclear transplantation – transplantation of nuclei into enucleated eggs of the same species – has been accomplished in a variety of organisms, including amphibians (Briggs and King, 1952, 1953; Briggs, 1979; DiBerardino, 1987, 1997), mice (Illmensee and Hoppe, 1981; McGrath and Solter, 1984a,b; Kono *et al.*, 1991; Cheong *et al.*, 1993; Tsunoda and Kato, 1993), swine (Prather *et al.*, 1989), cattle (Prather *et al.*, 1989; Bondioli *et al.*, 1990; Sims and First, 1994), goats (Du *et al.*, 1995) and sheep (Willadsen, 1986; Smith and Wilmut, 1989; Campbell *et al.*, 1994). Bondioli *et al.* (1990) produced eight cloned male calves from the nuclei of a single bovine embryo. Then, in 1997, Wilmut *et al.* (1997) produced the famous cloned lamb, Dolly, from adult cells – cultured adult mammary epithelium cells. Although these are major scientific advances, none of these individuals are true 100% clones because donor eggs were used which have different cytoplasm and mtDNA from those of the individual donating nuclei.

Intraspecific clones of individual fish have been produced (Lee *et al.*, 2002). Fertile transgenic zebra fish were obtained by nuclear transfer using embryonic fibroblast cells from long-term cultures. The donor nuclei were modified with retroviral insertions expressing green fluorescent protein (GFP) and were transplanted into manually enucleated eggs. A typical 2% survival rate resulted in 11 adult transgenic zebra fish expressing GFP. These P_1 nuclear-transplant transgenics produced fertile, diploid F_1/F_2 progeny which also expressed GFP in a pattern identical to that of the founder fish. Slowly dividing nuclei from cultured cells can be reprogrammed to support rapid embryonic development, which should lead to targeted genetic manipulation in zebra fish.

Prior to this, fish research revolved around an interspecific approach. The nucleus from one species is transplanted into the enucleated zygote of a second host species, resulting in a hybrid with nuclear DNA from one species and cytoplasm and mtDNA from a second species. This technique is tedious and few embryos survive the procedure. This procedure has been accomplished in carps in China. The resulting embryos, with nuclei and nuclear DNA from the donor species and cytoplasm and mtDNA from the host species, have been termed nucleocytoplasmic hybrids (Yan, 1998). Few details have been published, but fish with altered performance have been produced.

Nucleocytoplasmic hybrids have been produced between different species, varieties, genera and subfamilies of fish, and resulted in individuals that produced viable offspring (Yan, 1998). This research demonstrated that most traits are controlled by nuclear DNA, a few are controlled by the cytoplasm/mtDNA and others by a combination of the nuclear and mtDNA genomes (Yan, 1998). The same conclusions were obtained from interspecific nucleocytoplasmic hybrids of frogs, *Rana* (Kawamura and Nishioka, 1963a,b). Interspecific nucleocytoplasmic hybrids of mammals have failed to develop past the blastocyst or morula stages (Mei *et al.*, 1993; Yan, 1998), again illustrating the greater genetic plasticity of fish compared with mammals.

In the case of fish, there are some restrictions on the type of egg that can be used as the host or recipient. Nile tilapia and Chinese bitterling, *Rhoedus sinensis*, have ellipsoid eggs that are rich in yolk and have only a single layer of chorion. After the enveloping chorion is removed, these eggs lose their elasticity and become flat,

frequently resulting in abnormal development (Yan, 1998). Species that have oil droplets in their eggs, causing buoyancy, are also difficult to handle as host cytoplasm. Eggs from females that are spawning for the second year appear to be more amenable to manipulation.

A stereomicroscope is utilized to accomplish the nuclear transplantation. Eggs are kept at optimum spawning temperature (Yan, 1998). Embryos are incubated in standard amphibian Holtfreter's solution. NaCl (0.35 g), KCl (0.005 g) and $CaCl_2$ (0.01 g) are mixed together in 100 ml of distilled water and then sterilized by boiling. This solution is cooled to 50°C and then $NaHCO_3$ (0.02 g) is added to adjust the pH to about 7. A Holtfreter's disassociation solution is utilized to separate fish blastula cells. This solution is made by adding NaCl (0.35 g), KCl (0.005 g) and ethylenediaminetetraacetic acid (EDTA) (55.8 mg) together in 100 ml of distilled water, and then sterilized by boiling. This solution is cooled to 50°C and then $NaHCO_3$ (0.02 g) is added to adjust the pH to about 7.

Fish eggs are placed on the bottom of a Petri dish, which is covered with a layer of 1.2% agar to prevent friction between the egg and the glass bottom. After activation of the egg, the cytoplasm coalesces and moves towards the animal pole, forming a blastodisc cap. Nuclear transplantation must be accomplished prior to the complete formation of the blastodisc (Yan, 1998). Optimum results are obtained if the operation is performed between the time of first cleavage and metaphase of the second meiotic division; otherwise abnormal cleavage and development of the nuclear-transplant eggs will probably occur.

The blastodisc is located just under the egg surface against the first polar body. The egg should be gently rotated until the blastodisc can be located. The egg nucleus is removed with a fine glass needle. Attempts to do this by irradiation or lasers have proved difficult to date. The host egg is now ready to receive the donor nucleus. The donor cells/nuclei are taken from the inner side of the blastoderm of a mid-blastula. Again, the blastoderm is separated with a fine glass needle and a hair loop. After the isolated blastoderm is placed in the disassociation solution, the cells separate in 1–2 min and must be removed before more than 2 min have elapsed. Durations longer than this will damage the cells and they will not develop. The cells should be moved directly from the Holtfreter's disassociation solution to the regular Holtfreter's. There is an inconsistency in the protocol of Yan (1998), as it indicates that the isolated blastula cells should be held in a calcium-free medium and yet Holtfreter's contains calcium. The cells need to be kept separate as they can reassociate, divide continuously and form a cell mass.

The donor egg is prepared by removing the chorion with forceps and the egg is rotated to locate the blastodisc, which is just below the polar body. The polar body has the appearance of a small ball. The egg nucleus is removed with a sharp glass needle by quickly taking a small slice, which includes the entire egg nucleus and a small portion of the cytoplasm. The damaged site will immediately heal in the Holtfreter's.

Nuclear transplantation is accomplished in Holtfreter's, and both the donor cells and recipient egg are placed in a Petri dish with the bottom coated with a layer of agar to prevent adhesion of the eggs and cells to the glass. Donor cells are sucked into a micropipette. The diameter of the opening of the pipette is slightly smaller than the donor cell (10 µm) to break the cell and free the nucleus. The donor blastula cell/nucleus is then injected into the enucleated recipient egg. If the nuclear-transplant cell gastrulates, it will normally complete development. Generally, 1–3% of the manipulated embryos survive to adulthood.

Tung and Yan (1985) accomplished intervariety nuclear transplantation. They transplanted the nucleus from a crucian carp, which is the wild type *Carassius auratus*, into the cytoplasm of domestic goldfish, *C. auratus*. These two varieties have vastly different morphology and behaviour. The nucleocytoplasmic hybrids had the coloration, the number of scales on the lateral line and the caudal fin to body length ratio of the donor nucleus species, wild crucian carp. No traits were exactly like the donor of the enucleated egg, domestic goldfish. The vertebrate number was intermediate, and the isozyme patterns were not exactly like the donor and host, indicating that these traits were affected by a combination of nuclear genes, mitochondrial genes and/ or cytoplasmic interactions. Three of the embryos grew to maturity, were females and produced normal eggs.

Intergeneric nucleocytoplasmic hybrids were also produced (Tung, 1980; Yan *et al.*, 1985). These nucleocytoplasmic hybrids had a pair of lateral barbels, pharyngeal teeth shape and number, morphology of pigment cells, colour, scale morphology, swim-bladder shape, haemoglobin patterns, lactate dehydrogenase (LDH) allozyme pattern, malate dehydrogenase allozyme pattern and DNA reassociation kinetics that were the same as those of the donor species of the nucleus, common carp. Male and female F_1 nucleocytoplasmic hybrids were fertile and were mated to produce F_2 and F_3 generations. Many of these common carp donor traits were then inherited by F_2- and F_3- generation nucleocytoplasmic hybrids (Yan, 1998), as expected. In the case of the lateral line scale count and the serum protein patterns, means were intermediate of the two parental types, but closer to those of the donor species, the common carp. Vertebrate number was nearer that of the enucleated host egg species, goldfish. Perhaps development rate and such meristic characters as vertebrate number are sometimes dictated by intrinsic factors of the cytoplasm or interactions among the cytoplasm, mtDNA and nuclear genome.

F_3-generation nucleocytoplasmic hybrids were compared with their parental donor species, red common carp, in ponds (Yan, 1998). The F_3 nucleocytoplasmic hybrids grew 22% faster than the red common carp and had 3.8% higher protein and 5.6% lower fat in the muscle than the red common carp. Reciprocal nucleocytoplasmic hybrids were produced by transferring the nucleus of the crucian carp to the enucleated eggs of common carp (Yan *et al.*, 1984a). As expected, reciprocal results were obtained. These fish had no lateral barbels and had pharyngeal teeth, vertebrate number and scale counts on the lateral line very similar to those of the donor species of the nucleus, crucian carp/goldfish.

Overcoming even greater genetic distances, intersubfamily nucleocytoplasmic hybrids were produced (Yan *et al.*, 1984b, 1985). Nuclei of blastula cells from the grass carp, *Ctenopharyngodon idella*, subfamily Leucinae, were transferred to enucleated eggs of the blunt-snout bream, *Megalobrama amblycephala*, subfamily Abramidinae. Overall morphology and appearance, head length/total length, body depth/total length, body width/total length, anal-fin ray number, gill-raker number, swim-bladder shape, number of pharyngeal teeth, teeth patterns and number of vertebrae were the same as those of the nuclear donor species and different from those of the host cytoplasm species. In some cases, the nucleocytoplasmic hybrids had extra anal-fin rays like the host cytoplasm species, blunt-snout bream. The karyotype and karyotype morphology were also the same as the nuclear donor species, grass carp. LDH isozyme patterns of the nucleocytoplasmic hybrids and the nuclear donor species were the same. However, the serum protein pattern and the immunoprecipitation of serum proteins were different for the nucleocytoplasmic hybrids compared with both the donor and host species, indicative of some type of cytoplasmic or mtDNA influence.

Reciprocal interfamily nucleocytoplasmic hybrids of goldfish, family Cyprinidae, and large-scale loach, *Paramisgurnus dabryas*, family Cobitidae, failed to produce adult fish (Yan *et al.*, 1990), as did hybrids between zebra fish donors and large-scale loach hosts (Yan *et al.*, 1993). This confirms and illustrates the importance of the interactions of the nuclear genome, the nucleus, the cytoplasm and the mtDNA. The genetic distance of families appears to be too great to allow the complete development of nucleocytoplasmic hybrids.

Interorder nucleocytoplasmic hybrids between Nile tilapia donors and large-scale loach hosts sometimes produced larvae, but never resulted in adult fish (Yan *et al.*, 1990, 1991). Evidence was generated – karyological and some morphological – indicating a functioning nuclear genome from tilapia, but there was some definite host influence on body shape. Apparently, one of the primary reasons for larval death was the incompatibilities in the eggs' developmental rate and the timing of gene expression of the nuclear donor genome.

Key Summary Points

Gynogenesis and androgenesis are both forms of intensive inbreeding. Therefore, their usefulness for genetic improvement is limited because these technologies are difficult to execute in some species, lead to inbreeding depression, and the performance of their crossbreeds is usually no better than the original population from which they were derived. However, gynogens and androgens can be quite useful for mapping and QTL research, and can be utilized as a powerful first step for elucidating sex determination and developing genetically monosex populations.

7

Intraspecific Crossbreeding

Both long-term and short-term genetic enhancement programmes are potential options for improvement of aquatic organisms. Intraspecific crossbreeding, interspecific hybridization and polyploidy are short-term programmes. Outbreeding or crossbreeding is the mating of unrelated individuals. The genetic basis of crossbreeding is dominance, epistasis and overdominance, similar to that of inbreeding. Most aspects of crossbreeding and inbreeding are opposite. Whereas inbreeding decreases growth, reproduction, vigour and survival and increases deformities, biochemical disorders and bilateral asymmetry, crossbreeding often has the opposite effects.

Heterosis

In the case of crossbreeding, we are usually referring to the mating of two or more strains, lines, races or breeds. The goal of using crossbreeding and hybridization is to obtain heterosis. The classical definition of heterosis is the mean of the reciprocal F_1 hybrids or crossbreeds minus the mean of the two parents divided by the mean of the two parents multiplied by 100. Obviously, heterosis can be positive, negative or zero. From a more practical standpoint, the problem with this definition is that in commercial industry and in some research situations the primary interest is making genetic improvement. This particular definition or value has a couple of weaknesses with regard to informativeness. In industry or in a genetic improvement programme, the goal is to find an F_1 crossbreed that performs better than the best parent. The heterosis from the classical definition includes the means of the poor-performing parent and the good-performing parent, and the good-performing reciprocal cross and the poor-performing reciprocal cross. Thus, one of the reciprocal crosses could have an outstanding mean, but because of poor performance of the reciprocal, a low or negative heterosis could be obtained. The face value conclusion is that these two strains do not combine well and are not good for crossbreeding. Another possibility is that neither reciprocal performs well compared with the best parent, but the inferior parent has such a low mean that positive heterosis is obtained. The face value conclusion is that these two strains combine well and are good for crossbreeding. From the standpoint of genetic improvement, both conclusions based on the classical calculation of heterosis are incorrect.

A more practical, informative definition of heterosis is the mean of the best reciprocal minus the mean of the best parent divided by the mean of the best parent multiplied by 100. This is a relevant calculation as our genetic enhancement goal is to identify and use a crossbreed that performs better than the best existing parent. This is true progress.

Sometimes in animal breeding, the term hybrid vigour is used interchangeably with heterosis. Technically, this can be confusing because hybrid vigour has a different meaning from a population genetics perspective. In that case, hybrid vigour is used in reference to fitness, and specifically, reproductive fitness. Hybrid vigour is present when the F_1 hybrid produces more progeny than either of the parental types.

Generally, values for heterosis tend to be larger when traits are measured in the younger life stages and decrease with age. In channel catfish, Marion strain females crossed with Kansas strain males result in greater heterosis for growth rate as fingerlings than as food fish, supporting the premise that in general heterosis is more dramatic at young ages. This is partially true; however, a large part of the explanation for this observation is that a normal biological effect is being observed associated with the fish's natural growth curve. The crossbreed is growing faster than the parent early in the experiment at a young age, developing a large relative or percentage difference in body weight. Relative growth rate slows as a fish increases in size. The genetically superior genotype reaches a size where growth rate begins to slow because of biological limits faster than the inferior genotype. At this time, the slower-growing genetic group is still in the faster growth stage. From a relative or percentage difference standpoint the inferior genotype closes the gap between the two, but the absolute size difference remains or increases. The genetically inferior never catches up as it eventually reaches the size where it is slowing down in growth as well. Percentage difference decreases with time. This phenomenon of decreasing percentage difference in size between the fast- and slow-growing genotype with time occurs in other genetic-enhancement programmes including selection and genetic engineering.

Crossbreeding Programmes

A variety of crossbreeding programmes exist. One is a simple two-breed cross. The next is a backcross, the mating between the F_1 and one of its parents. Often, an F_1 female is backcrossed to the male parental line. Obviously, it is important to identify which is the male parental line and which is the female parental line. There is scientific literature for which this critical piece of information is omitted.

The purpose of using the female as the F_1 in the backcross is to obtain maternal heterosis for the reproductive and early life-history traits. The weakness of the standard backcross is that although fry survival, egg and fry size, hatch and reproductive output may have been improved because of the maternal heterosis, the array of gene combinations, many of which are the parental types of the parent to which the F_1 was backcrossed, reduces heterosis that was sought for the other production traits.

A solution is to utilize a three-breed cross. The objective is to identify three strains, lines or breeds that combine well. Again, often an F_1 female is used to obtain maternal heterosis for the reproductive-related traits, and then a male from a third line is used which combines well with the first two lines to obtain heterosis for the main production traits.

Recurrent selection is selection for individuals in one parental line that combine well with the alternative parent line, or in other words, selection for increased heterosis. If the selection is conducted in both parental lines or species, it is reciprocal recurrent selection. This is a form of progeny testing.

Effects of Strains and Domestication on Combining Ability

Combining ability or nicking ability varies among strains. Nicking ability or combining ability is how well strains, breeds or lines combine together to produce heterosis. Two strains that combine well would have a lot of heterosis.

Domestication has a strong influence on the success of crossbreeding programmes. Domestic × domestic crosses are more likely to result in heterosis than wild × domestic and wild × wild crosses for several traits. Domestic × domestic channel catfish were more likely to exhibit heterotic growth rates than domestic × wild crossbreeds (Dunham and Smitherman, 1983a) as was the case in rainbow trout (Gall, 1969; Gall and Gross, 1978; Kincaid, 1981; Ayles and Baker, 1983). Bryden *et al.* (2004) examined the performance of reciprocal crossbreeds between a domestic and wild strain of chinook salmon, *Oncorhynchus tshawytscha*, for growth survival, saltwater growth, saltwater tolerance, stress response and recovery, and fecundity. No significant heterosis was observed. Crossbreeds between wild and domestic Atlantic salmon were intermediate in performance for body weight, condition factor and sexual maturation (Glover *et al.*, 2009). When wild strains of European sea bass, *Dicentrarchus labrax*, were crossed, large strain differences were obtained for survival, growth, shape, sex ratio, muscular fat content, visceral yield and spinal deformities, but not fillet yield (Vandeputte *et al.*, 2009). There was no heterosis among these wild strains and no G × E interactions. This helps illustrate the fact that heterosis is less likely to be obtained from wild strains than for domestic strains.

Apparently, domestication also affects the success of crossbreeding in crustaceans as well. When two wild strains of giant freshwater prawn, *Macrobrachium rosenbergii*, were crossed with a domestic strain (Thanh *et al.*, 2009, 2010), large additive strain effects were observed for growth, but no heterosis. Wild strains were involved and little heterosis observed. The domestic strain was the fastest growing of the strains. The most rapidly growing prawns were some of the crossbreeds.

There can be maternal effects on combining ability as well. In channel catfish strains, Auburn strain females have good combining ability but Auburn strain males do not. Of 50 interspecific ictalurid catfish hybrids that have been evaluated, the two best-performing ones have channel catfish as the female parent.

Reciprocals

Reciprocal crossbreeds usually do not grow at the same rate or have the same means for other traits. A strain/species female × B strain/species male does not equal B strain/species female × A strain/species male, contrary to what might be predicted. The nuclear genotypes/genomes should be the same. The reciprocal crossbreeds of the giant freshwater prawn did not have the same performance (Thanh *et al.*, 2009). Additionally, in the case of channel catfish, reciprocal crosses did not perform the same (Dunham *et al.*, 2001).

Three potential logical explanations for the difference in the reciprocals exist. The mitochondrial genome is inherited solely through the mother (except for rare paternal leakage). Thus, A × B and B × A are not equivalent genotypically. Additionally, there are interactions between the mitochondrial DNA and the nuclear DNA, potentially increasing the difference between the reciprocal. Obviously, the cytoplasm is inherited only from the mother. There may be cytoplasmic effects, nuclear cytoplasmic interactions or mitochondrial cytoplasmic interactions that have yet to be discovered. Differential methylation is another potential difference between reciprocals. In some organisms, there can be different systems of methylation. Methyl groups can be attached to a gene depending on whether they were inherited from the female versus the male. The nucleotides are the same but a different methyl group can be attached to the allele depending on whether it descended from the maternal line or paternal line. Methylation affects gene expression. The same allele could be inherited from the mother and father, not express and perform in the same way.

Body Weight

Intraspecific crossbreeding often increases growth rate of aquacultured species, but as expected, heterosis is not obtained in every case. Heterotic growth in excess of both parent strains occurred in 55% and 22% of channel catfish and rainbow trout crossbreeds evaluated, respectively (Dunham and Smitherman, 1983a; Dunham, 1996). Crossbred channel catfish sometimes grow 10–30% faster than the largest parental strain. Chum salmon crossbreeds have not shown increased growth rates (Dunham, 1996).

Common carp crossbreeds also expressed heterosis in a low percentage of the crosses examined (Moav *et al.*, 1964; Moav and Wohlfarth, 1974a; Nagy *et al.*, 1984); however, those exhibiting heterosis are the basis for carp industries in Israel, Vietnam, China and Hungary. The crossing of common carp lines in Hungary (Bakos, 1979; Bakos and Gorda, 1995) is a good example of the relative success of crossbreeding and its application. During the last 35 years more than 140 crossing combinations have been tested. Based on the experimental results, three crossbreeds were chosen for commercial application, whose growth rate and other qualitative features gave about 20% better performance than the parental lines and other control strains.

After the dissemination of the Hungarian common carp crossbreeds, about 80% of the farm production of common carp has come from the Szarvas crossbreeds. In Hungary the production of gynogenetic female lines and gynogenetic sex-reversed inbred male lines of common carp strains having the best combining ability was an important part of the crossbreeding programme. A higher heterosis was expected from the crossing of inbred lines but the growth rate of F_1 crossbreeds was only 10% higher than that of controls (Bakos and Gorda, 1995).

In Europe, the Czech Republic also had a national common carp breeding programme. The genetic diversity and uniformity of different common carp strains were identified for the lines and individuals for use in crossbreeding. As a result of the first top crossings, the crossbreeds South Bohemian × Northern mirror carp and the Hungarian 15 × Northern mirror carp gave outstanding results for growth rate (Linhart *et al.*, 1998).

In Israel, more than 20 years of experimental work demonstrated that the crossing of common carp sometimes resulted in positive heterosis. The synthetic strain DOR-70 is a crossbred line of Croatian Nasice and the Taiwanese big belly carp. This cross with fast growth became one of the most popular carps of the Israeli commercial fish production (Wohlfarth, 1993).

In Vietnam, eight local varieties of common carp and the foreign strains, Hungary, Ukraine, Indonesia and Czech, are maintained. Significant heterosis was observed in F_1 generations of crossbreeds. The double crosses among Vietnamese, Hungarian and Indonesian strains have been used as the starting material for the carp selective breeding and crossbreeding programme in Vietnam (Thien and Trong, 1995). The Vietnamese × Hungarian common carp crossbreed is very popular in both the northern and southern regions of Vietnam because of its fast growth and high survival rate under different production conditions.

In the case of the silver barb, *Barbodes gonionotus* (Bentsen *et al.*, 1996; Hussain and Islam, 1999), three strains – Bangladesh, Thailand and Indonesian – were crossed in all combinations. Females from six crosses had 23% higher growth rate than the parent strains, and even higher growth rates (35% improvement) were found in the three crosses using the Thailand strain as either the sire or dam (Hussain and Islam, 1999).

In the Vietnamese breeding programme, six different silver barb strains were used for crossing and a synthetic basic population has been developed (Bentsen *et al.*, 1996). Heterosis for growth was also found in some of these crosses.

Heterosis for growth rate has been observed in Nile tilapia (Circa *et al.*, 1994; Bentsen *et al.*, 1998; Neira *et al.*, 2009). About 30% of the crosses show heterosis for growth rate. The growth improvement was about 11%.

Crossbreeding can also improve performance in crustaceans and resulted in heterosis for growth rate, but not survival, in Chinese shrimp, *Fennropenaeus chinensis* (Tian *et al.*, 2006).

Survival and Disease Resistance

Crossbreeding more consistently improves survival traits and reproductive traits. Crossbreeding is more consistent for improving disease resistance than for improving growth rate. Crossbred F_1 fish have improved disease resistance in about 50–70% of the crosses examined. Combining abilities vary from strain to strain for survival traits such as resistance to bacterial diseases and parasitic diseases and also tolerance of poor water quality such as low oxygen, high ammonia and high nitrite. In general, crossbreeding often improves survival traits such as fry survival. In the case of Nile tilapia about 90% of crossbreeds had heterotic fry survival, with a mean of about 90% for the F_1 and about 60% for the parent lines. Crossbreeding of the Thai walking catfish, *Clarias macrocephalus*, improves resistance to *Aeromonas hydrophila* infections (Prarom, 1990).

Kirpichnikov (1981) selected a cold-resistant synthetic carp strain for fish farms in the cold zones of north Russia using the local and the Siberian wild carps from the River Amur for multiple crossing. As a result of this successful breeding programme a new strain, the Ropsa carp, was developed.

Wild strains of common carp were susceptible to koi herpes virus/carp interstitial nephritis and gill necrosis virus, whereas domestic strains tend not to be vulnerable (Shapira *et al.*, 2005). Two domestic strains, two domestic × wild crossbreeds and one domestic × domestic crossbreed were compared for viral resistance. In the laboratory, the most resistant genotype was one of the domestic × wild crossbreeds and one of the pure strains was the least resistant. The remaining genetic groups were intermediate in viral resistance. When the challenges were repeated in ponds, the results were the same except the other domestic × wild crossbreed had excellent resistance in ponds, although its performance had been intermediate in the laboratory.

Reproduction

Crossbreeding often results in increased reproductive performance. Spawning percentage of 3-year-old F_1 crossbred channel catfish brood stock was higher than that for 3-year-old pure strain brood stock. Apparently, heterosis exists for spawning ability in channel catfish. The exact same brood fish were used and the experiment repeated when the fish were 4 years old. No difference in spawning rate was found, so apparently the heterosis was for early sexual maturity. Again this is consistent with heterosis being stronger in younger life stages.

Crossbreeding for Improving Multiple Traits

The presence of heterosis is usually variable from one trait to another for a single crossbreed. However, in the case of channel catfish, the Marion female × Kansas male shows heterosis for several traits including growth, early sexual maturity, resistances to different diseases and water quality. Neira *et al.* (2009) found heterosis for body shape, fillet yield and visceral body fat percentage in Nile tilapia.

Effects of strain crossing of the Thai walking catfish, *C. macrocephalus*, on growth and resistance to *A. hydrophila* were studied (Prarom, 1990). Local strains of *C. macrocephalus* from Songkla (south, S) and Saraburi (central, C) were collected and semi-diallele crossing was conducted. Results showed that progenies from crossing of female from the south and male from the central grew better than other group (SC > CC, SS, CS). Heterosis was observed for body weight and length at the age of 2, 3 and 3.5 months. No heterosis was observed in the reciprocal cross. Although disease resistance was not significantly different among the crosses and there was no significant heterosis, progenies from strain crossing had observed resistance to *A. hydrophila* better than parent strains. There was no correlation between the following traits: growth and disease resistance, growth and survival rate, and survival rate and disease resistance. Crossbreeding improved phagocytosis activity in African walking catfish, *Clarias gariepinus*, but did not enhance body weight, total length, specific immune response to *A. hydrophila*, phagocytic index or male reproductive performance (Wachirachaikarn *et al.*, 2009).

Using Inbred Lines for Crossbreeding

One practice is to develop inbred lines to use in crossbreeding programmes to obtain heterosis. The existing data indicate that it is unlikely to obtain true genetic gain using this strategy. Hedgecock *et al.* (1995) and Hedgecock and Davis (2007) determined that the crossbreeding of inbred lines of Pacific oysters, *Crassostrea gigas*, often resulted in heterosis for growth and survival. Additionally, crossing of inbred lines of Pacific oysters reduces summer mortality (Suhrbier *et al.*, 2005). Similarly, the crossing of two inbred lines of cockle, *Fulvia mutica*, resulted in increased shell length and whole body weight, but intermediate survival (Fujiwara *et al.*, 2005). Likely, this is not true genetic enhancement. Bondari and Dunham (1987) demonstrated that inbreeding reduces growth in channel catfish. Crosses of the inbred lines

did indeed grow faster than the parental inbred lines. However, the performance of the crossbreeds only negated the inbreeding depression and was equivalent to that of the original population, so no true genetic gain was obtained as performance returned to the original baseline. Production of gynogenetic female lines and gynogenetic sex-reversed inbred male lines from common carp with the best combining ability was an important part of the Hungarian crossbreeding programmes. A higher heterosis was expected from crossing inbred lines, but the growth rate of F_1 crossbreeds was only 10% higher than that of controls (Bakos and Gorda, 1995).

Application of Crossbreeding Programmes

Application of crossbreeding requires greater effort than what is usually used in the aquaculture farm situation. A minimum of two brood-stock ponds are needed to perpetuate the two parental lines and at least one more pond is needed for the F_1 production fish (Fig. 7.1). Contamination of the parental lines must be avoided. Ideally, two ponds for each breeding purpose should be used in the event of some water quality or disease disaster in one of the ponds.

Strain incompatibilities for mating ability can sometimes exist, slightly impeding seed production from intrastrain crosses. These strain mating incompatibilities reduced fry output in channel catfish and

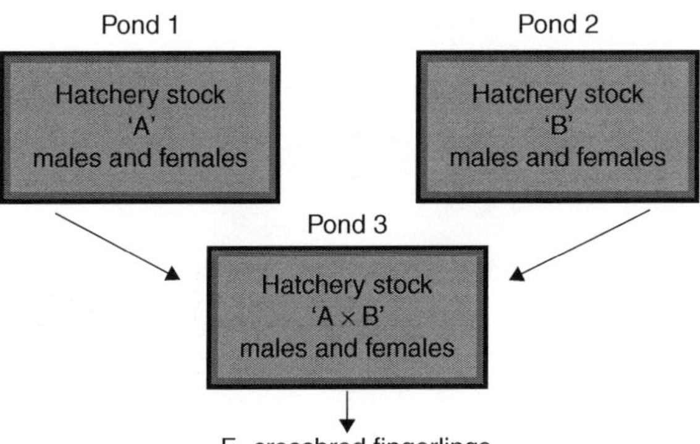

Fig. 7.1. The minimum ponds needed to apply intraspecific crossbreeding.

Nile tilapia and this appears to be more strongly influenced by the female than the male (Smitherman *et al.*, 1984, 1988).

Key Summary Points

Intraspecific crossbreeding improves growth rate in about 25% of the crosses tested, and in general that improvement can range from 10 to 50%. Domestic × domestic crosses are more likely to yield heterosis than those involving wild strains. Reciprocal crossbreeds do not have the same performance. Crossbreeding has a high probability of improving survival traits and disease resistance, with about 70% of tested crosses showing genetic improvement. Crossbreeding also has a high probability of improving various reproductive traits. In a limited number of examples, crossbreeding can improve multiple traits in a single crossbreed. Strain incompatibilities are possible; when they occur, fry production may be less efficient than for the parental strain matings.

8

Interspecific Hybridization

Interspecific hybridization is the mating between two different species. In general, interspecific hybridization programmes have a very low probability of success. The first obstacle that must be overcome is reproductive isolating mechanisms. Species remain separate because of reproductive isolating mechanisms, often making even artificial hybridization difficult.

The first reproductive isolating mechanism is geographical. Obviously, this is not a problem in the aquaculture or artificial reproduction situation. The second mechanism is mechanical, basically a lock-and-key mechanism: the copulatory organs do not fit. This applies primarily to mammals, but some fish species copulate as well. The third reproductive isolating mechanism is behavioural. Courtship displays, other mating behaviours and pheromonal cues do not match, preventing the two species from mating. The fourth one is temporal: the timing of reproduction is different for the two species. For instance, striped bass and white bass spawn at different times in different water temperatures. The fifth reproductive isolating mechanism is cellular. The gametes, sperm and egg are not compatible. The sperm cannot penetrate the eggs. The sixth one is gamete incompatibility. Even though the sperm is able to penetrate the egg, the two gametes are not genetically compatible

and embryonic development cannot be completed. The seventh mechanism is F_1 hybrid sterility. In this case, the two species can mate and successfully produce progeny, but the two genomes remain separate because the hybrids are sterile. The eighth reproductive isolating mechanism is F_2 breakdown. The F_1 individuals are fertile and can produce progeny. However, in F_2, F_3, F_n or backcross generations, reproduction breaks down, again keeping the two genomes separated. These are major reproductive isolating mechanisms that are responsible for maintaining different species.

An interesting example of a behavioural reproductive isolating mechanism was observed in the 1960s with centrarchid sunfish at the Illinois Natural History Survey. Some species of sunfish would naturally hybridize if they had no choice of mates and others would not. Bluegill females would not hybridize with redear sunfish males. One year, in a turbid pond, female bluegill produced large numbers of hybrid fingerlings with redear males. The bluegill has a black opercular flap and the redear has an opercular flap with a large, distinctive, red edge. Perhaps a visual cue, the distinctive red ear flap, was acting as the reproductive isolating mechanism. This hypothesis was tested by stocking bluegill females in a clear pond. Then the red fringe on the opercular flap of the redear males was removed and the males introduced into the ponds. Again large-scale hybridization occurred, so the red ear tab was allowing the bluegill female to distinguish bluegill and redear males, two closely related species.

In principal and genetic basis, interspecific hybridization is similar to intraspecific crossbreeding. This has been a popular breeding programme as over the years fisheries biologists have repeatedly tried to combine the best traits of more than one species, mostly with little success (Argue and Dunham, 1999). Interspecific hybridization rarely results in heterosis. Interspecific hybridization has been used with numerous species of fish to increase growth rate, manipulate sex ratios, produce sterile animals, improve flesh quality, increase disease resistance, improve environmental tolerance, and improve a variety of other traits to make fish farming more profitable.

Although interspecific hybridization rarely results in an F_1 suitable for aquaculture application, there are a few important exceptions to this rule, resulting in fish that can be quite valuable. The channel catfish (*Ictalurus punctatus*) female × blue catfish (*Ictalurus furcatus*) male is the only hybrid of nearly 50 North American catfish hybrids examined that exhibits superiority for growth rate, growth uniformity, disease resistance, tolerance of low oxygen levels, dressing percentage and harvestability (Smitherman and Dunham, 1985). This is by far the best genotype for ictalurid farming. However, mating blocks between the two species have prevented their commercial utilization. Like reciprocal catfish crossbreeds, characteristics of reciprocal catfish hybrids differ. Paternal predominance, the hybrid possessing traits more like the male parent than its reciprocal, was observed in blue × channel hybrids (Fig. 8.1) (Dunham *et al.*, 1982a).

Although they do not show heterosis for such a broad spectrum of traits, crosses of the silver carp (*Hypophthalmichthys molitrix*) and bighead carp (*Aristichthys nobilis*), black crappie (*Pomoxis nigromaculatus*) and white crappie (*Pomoxis annularis*) (Hooe *et al.*, 1994) and African catfish hybrids (*Clarias gariepinus*, *Heterobranchus longifilis* and *Heterobranchus bisorsalis*) (Salami *et al.*, 1993; Nwadukwe, 1995) all exhibit faster growth than parent species.

In the case of shellfish, various hybrids between the Thai oysters (*Crassostrea*

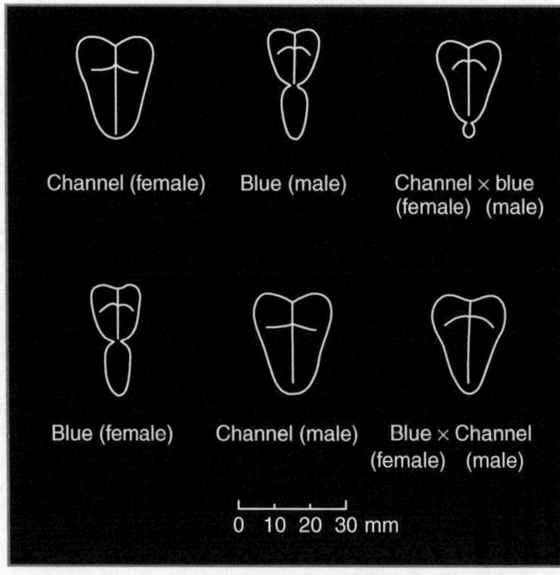

Fig. 8.1. Paternal predominance for swimbladder trait in hybrids between channel catfish, *Ictalurus punctatus*, and blue catfish, *Ictalurus furcatus*. (Adapted from Dunham *et al.*, 1982a.)

belcheri, *Crassostrea lugubris* and *Saccostrea cucullata*) were compared (Charoensit, 1995), but no heterosis was observed. Interspecific hybridization of *C. belcheri* × *C. lugubris* was successful only up to spat stage. Growth rates of hybrids and their reciprocal were significantly lower than those of the parents. Intergenetic hybridization was successful only when female *C. lugubris* crossed with male *S. cucullata*. Growth rates of the hybrid were significantly higher than that of *S. cucullata*, but not significantly different from that of *C. lugubris*. Shell morphology of the hybrid was intermediate between the two parental types. Diploid chromosome numbers of all three oyster species were the same at 20. Two types of chromosome, metacentric chromosomes and submetacentric chromosomes, were observed. Ratios between metacentric and submetacentric chromosomes were 16:4 for *C. belcheri* and *C. lugubris*, and 14:6 for *S. cucullata*. Hybrid oyster could be produced by crossing female *S. cucullata* and male *C. lugubris*. However, there was not much effect on growth rate. Shell morphology and karyotype could be used as methods to identify hybrids and their parental types.

In contrast to the results with Thai oysters, heterotic pearl production has been achieved in China using interspecific hybridization (Li *et al.*, 2009). The hybrid between the freshwater pearl mussels, *Hyriopsis schlegel* female and *Hyriopsis cumingii* male, increased pearl size by 23%, pearl output by 32% and the frequency of large pearls 3.7-fold.

A hybrid does not have to exhibit heterosis or overdominance for single traits to be commercially valuable. The overall genotype or breeding value for two or more traits may be superior or more valuable than the two parent species. In other words, the hybrid may be a good 'compromise' fish. The composite performance may make the F_1 the culture genotype of choice.

One of the best examples of applying interspecific hybridization is the hybrid between the Thai walking catfish (*Clarias macrocephalus*) female and the African walking catfish (*Clarias gariepinus*) male, the primary catfish cultured in Thai catfish

farming. There is no heterosis for any individual traits. The African walking catfish grows much faster than the Thai walking catfish and the hybrid has intermediate growth (Nwadukwe, 1995). Thai people prefer yellow coloured flesh in walking catfish. The faster-growing African walking catfish has a pink or reddish flesh that is unacceptable. Social factors can be important in genetics and in establishing genetic goals. The hybrid has the yellow flesh like its female parent, so it is acceptable for marketing. Walking catfish are marketed live. It is important to harvest the hybrid before it grows too big, otherwise customers will reject it thinking that it is an African walking catfish.

Another example of a hybrid that exhibits no heterosis for individual traits but has a valuable overall genotype is the hybrid, the sunshine bass, a cross between white bass (*Morone chrysops*) and striped bass (*Morone saxatilis*). Compared with either parent species, it grows faster, has good osmoregulation, high thermal tolerance, resistance to stress and certain diseases, high survival under intense culture, the ability to use soybean protein in feed, high handling tolerance and angling vulnerability (Smith, 1988; Colombo *et al.*, 1998). The striped bass grows much larger than the white bass, and the hybrid is intermediate in size. The white bass is extremely easy to catch by angling. In fact, some fishermen will use a double rig and catch two white bass on one cast, whereas the striped bass is very difficult to catch. Again, the hybrid is intermediate for angling vulnerability. The hybrid has heterotic survival. Striped bass are heat intolerant, white bass heat tolerant and the hybrid is intermediate. When the whole suite of traits is examined, the overall performance of the hybrid is better for both aquaculture and recreation.

Similarly, the rohu (*Labeo rohita*) × catla (*Catla catla*) hybrid grows almost as fast as pure catla, but has the small head of the rohu considered desirable in Indian aquaculture (Reddy, 2000). *C. catla* × *Labeo fimbriatus* (fringe-lipped peninsula carp) hybrids have the small heads of *L. fimbriatus*, plus the deep body and growth rate of *C. catla* (Basavaraju *et al.*, 1995). Dressing percentage is also improved in this hybrid

(Basavaraju *et al.*, 1995). Improved overall performance in experimental aquaculture conditions has also resulted from crosses of common carp with rohu, mrigal (*Cirrhinus cirrhosus*) and catla (Khan *et al.*, 1990); tambaqui (*Colossoma macropomum*) with pacu (*Piaractus brachypoma* and *Piaractus mesopotamicus*) (Senhorini *et al.*, 1988); green sunfish (*Lepomis cyanellus*) with bluegill (*Lepomis macrochirus*) (Tidwell *et al.*, 1992; Will *et al.*, 1994); and gilthead seabream (*Sparus aurata*) with red seabream (*Pagrus major*) (Knibb *et al.*, 1998a).

Another potential benefit of interspecific hybridization is that some species combinations result in progeny with skewed sex ratios or monosex progeny. Monosex populations of fish are desirable when growth differences exist between the sexes, sex-specific products such as caviar are wanted, reproduction needs to be controlled or when other exploitable sexual dimorphism exists. Monosex populations help reduce unwanted reproduction that would result from mixed-sex populations such as overpopulation and stunting of tilapia in production ponds.

Hybridization in tilapias or centrarchids often results in near monosex hybrids (Fig. 8.2). Hybridization between Nile tilapia, *Oreochromis niloticus*, and blue tilapia, *Oreochromis aureus*, results in predominantly male offspring (Hulata *et al.*, 1983). Tilapia matings that produce mainly male

offspring include Nile × wami (*Oreochromis urolepis honorum*) or greenhead (*Oreochromis macrochir*), and red (*Oreochromis mossambicus*) × wami (Wohlfarth, 1994). Conversely, the hybridization between striped bass and yellow bass (*Morone mississippiensis*) produced 100% female individuals (Wolters and DeMay, 1996).

Tilapia hybrids are also potentially quite valuable for other reasons. Tilapia hybrids are much easier to seine than their parent species, a significant advantage overlooked in many culture systems. In one study, 2% of the parent species were captured while 50% of the hybrids were caught. Some tilapia aquaculture in China is done in a polyculture system. Total harvest is not usually done going into the winter season because of the other species present. If the tilapias are not seined out of the ponds going into the winter season, a lot of the tilapia production is lost because of winter mortality. In this system, seinability should be a priority trait. This is also an overlooked trait in the American catfish industry. In multibatch systems such as this there will be losses to diseases, predation, decreased feed-conversion efficiency and oversized fish, if the fish are not seined and removed efficiently. A major advantage of interspecific hybrids is that, in general, they tend to be much easier to catch either by hook-and-line or seining.

In the case of hybridization of sunfish for recreational application, large size,

```
O. mossambicus      ♀ X  O. urolepsis honorum ♂
O. mossambicus      ♀ X  O. aureus ♂
O. niloticus        ♀ X  O. urolepsis honorum ♂
O. niloticus        ♀ X  O. aureus ♂

        XX              X        ZZ  = 100%  XZ  = 100% male


O. urolepsis honorum ♀ X  O. mossambicus ♂
O. urolepsis honorum ♀ X  O. niloticus ♂
O. aureus            ♀ X  O. niloticus ♂
O. aureus            ♀ X  O. niloticus ♂

        WZ              X        XY

   F₁ = 25% ♀     25%  WX  Female
      = 75% ♂     25%  WY  Male
                  25%  ZX  Male
                  25%  ZY  Male
```

Fig. 8.2. Sex-determining genotypes and sex ratios in reciprocal interspecific hybrids of tilapia.

minimizing reproduction and angling vulnerability are the three goals. The problem with the recreational fishery for hybrid sunfish is, of course, that heterosis is only observed in the F_1. Once the hybrids start reproducing, there is a huge assortment of genotypes leading to poor performance.

As is the case with tilapia, hybrid sunfish sometimes have altered sex ratios. In some cases they are fertile and sometimes sterile. The most consistent observation is catchability, as most interspecific centrarchid sunfish hybrids have very high angling vulnerability. The Illinois Natural History Survey has a 7 ha drainable lake that can be utilized for fishing experiments. The lake was stocked with 5000 sunfish hybrids and opened to public fishing. In 3 days, 82% of the hybrids were captured; after 1 week, close to 100% of the hybrids had been cleared from this lake. In a similar experiment, 4690 hybrids were stocked. In 15 days, 88% of the fish were caught, and at the end of the summer when the lake was drained, only 74 hybrids remained.

The meanmouth bass is an interspecific hybrid between largemouth bass females and smallmouth bass males evaluated by the Illinois Natural History Survey in the late 1970s. Anecdotally, they appeared to be very aggressive. Hybrids would fiercely protect their nest from researchers while largemouth bass or smallmouth bass would swim away when there was human disturbance. However, this is one example of interspecific hybridization where there was not heterosis for angling vulnerability, but incomplete dominance. The growth rate of the meanmouth was typical for hybrids, intermediate to its parents. These hybrids were aggressive sexually, and produced 2.5 times more nests than the parent species, perhaps creating a potential ecological risk.

Hybridization between species often results in offspring that are sterile or with diminished reproductive capacity due to problems with gonad development and chromosome pairing. Theoretically, just as is the case for fertile monosex hybrids, the production of sterile hybrids can reduce unwanted reproduction or improve growth rate by

energy diversion from gametogenesis or reduction in sexual behaviour. Karyotype analysis is believed to be a general predictor of potential hybrid fertility. Hybrids of Indian major carps are generally fertile because they share similar chromosome numbers (2N = 50). However, when they are mated with common carp (2N = 102), the hybrids have what is equivalent to a 3N chromosome number and are sterile (Khan *et al.*, 1990; Reddy, 2000). A natural triploidy also occurs when crossing between grass carp, *Ctenopharyngodon idellus*, and bighead carp, *Hypophthalmichthys nobilis*. Grass carp are commonly utilized for aquatic macrophyte control in the USA, but there is concern about their establishment in the natural environment resulting in potential impact on desirable vegetation in the ecosystem. The grass carp × bighead carp is not a viable option for weed control as although this triploid hybrid has reduced fertility, some progeny maintain diploidy and could be fertile (Allen and Wattendorf, 1987). However, there are exceptions to the fertility rule as crosses of some different sturgeon species with different chromosome numbers and most tilapia hybridizations produce fertile F_1 offspring (Steffens *et al.*, 1990).

Several salmonid hybrids have also been evaluated. As expected, salmonid hybrids were more viable when made within genera than between genera (Chevassus, 1979). Salmonid hybrids do not express heterosis for growth rate. Hybridization is a good programme to improve disease resistance such as coho salmon (*Oncorhynchus kisutch*) hybrids, which are considered resistant to several salmonid viruses (Dorson *et al.*, 1991). Viral resistance in the hybrids was improved, but overall viability was poor. The induction of triploidy can increase the hatchability of these potentially important hybrids (Parsons *et al.*, 1986). If interspecific hybrids hatch, their subsequent survival and disease resistance are often better than the parent species. Triploid hybrids from rainbow trout (*Oncorhynchus mykiss*) and char (*Salvelinus* spp.) were resistant to several salmonid viruses, but grew more slowly than their diploid counterparts. Rainbow trout and coho salmon triploid hybrids had increased resistance to infectious haematopoietic necrosis

(IHN) virus, but these hybrids grew more slowly than diploids (Dorson *et al.*, 1991). Hybrid triploids of Atlantic salmon (*Salmo salar*) × brown trout (*Salmo trutta*) showed survival and growth rates comparable to those of Atlantic salmon monospecific diploids. Triploid Pacific salmon hybrids have also shown earlier seawater acclimatization (Seeb *et al.*, 1993).

Hybrids may have increased environmental tolerances when one parental species has a wide tolerance such as euryhaline species, a specific tolerance such as cold tolerance, or because of increased heterozygosity (Nelson and Hedgecock, 1980; Noy *et al.* 1987). Hybrids between red tilapia, *O. mossambicus* (high salinity tolerance), and Nile tilapia, *O. niloticus* (low salinity tolerance), display enhanced salinity tolerance (Lim *et al.*, 1993). Florida red strain hybrids (*O. mossambicus* × *O. urolepis honorum*) can reproduce in salinities as high as 19 ppt (Ernst *et al.*, 1991), which is not necessarily a good trait when considering the potential environmental impact. Nile tilapia × blue tilapia hybrids also show enhanced salinity tolerance (Lahav and Lahav, 1990; Wohlfarth, 1994). Urmaza and Aguilar (2005) also compared the growth of hybrids between Nile tilapia and red tilapia and their parents in saline environments. Heterosis for body weight was observed in several cases, but heterosis for survival was found in only one hybrid.

Hybridization among marine species, and among marine and freshwater spawning species, has shown little promise for developing improved fish for aquaculture application. Reciprocal hybrids between gilthead seabream, *S. aurata*, and red seabream, *P. major*, developed vestigial gonads at 2–3 years and were sterile (Knibb *et al.*, 1998a), and no growth or survival superiority was observed compared with the parent species until sexual maturity. Hybridization between European sea bass (*Dicentrarchus labrax*) females and striped bass (*M. saxatilis*) resulted in viable fry. Surprisingly 28% were triploids, only triploid fry survived to 6 months of age, and at 8 months the survivors showed poor growth compared with diploid *D. labrax*. Such hybrids would only

be of commercial value where reproductive confinement is needed for ecological reasons and highly desirable flesh quality was obtained. Several hybrids in the family Sparidae have been produced in the Mediterranean with the cross between *P. major* and common dentex, *Dentex dentex*, being fast growing (Colombo *et al.*, 1998).

Just as was the case for intraspecific reciprocal F_1 crossbreeds, reciprocal F_1 interspecific hybrids usually show different phenotypes and performance. Reciprocal hybrids of *O. niloticus* (N) × *O. mossambicus* (M) demonstrate different salinity tolerances (Thanakijkorn, 1997). Genetic maternal effects were evident as the hybrid with the *O. niloticus* mother had a higher survival rate after salinity challenges at 20 ppt than pure *O. niloticus*, but lower survival rates than those of the reciprocal hybrid and *O. mossambicus*. Survival percentages after direct transfer to 20 ppt salinity were 71.1, 35.6, 99.4 and 100.0% for the hybrid N × M, *O. niloticus*, M × N and *O. mossambicus*, respectively. Survival percentages after acclimatized transfer were 58.89 ± 8.31, 73.33 ± 2.72, 98.89 ± 1.57 and 100.00 ± 0.00%, respectively.

At 30 ppt salinity, a direct transfer killed all tilapia with *O. niloticus* maternal ancestry, but the reciprocal hybrid (M × N) and *O. mossambicus* had similar survival percentages at 30 ppt (97.8 and 99.4%, respectively). An acclimatized transfer at 30 ppt resulted in survival percentages of 34.5 and 43.3% for *O. niloticus* and the hybrid N × M, respectively, which were significantly lower than those of the hybrid M × N and *O. mossambicus* (98.3 and 99.4%, respectively).

Growth rates of N × M hybrids were comparable to those of Nile tilapia, while growth rates of the M × N hybrids and *O. mossambicus* were comparable to, but lower than, those of the first two groups, an additional example of maternal genetic effects. There were no differences in carcass percentages of the four groups.

Backcrosses were also evaluated. Based on the direct transfer results, backcross hybrid MN × N showed the highest salinity tolerance (comparable to that of *O. mossambicus*), but no significant differences in salinity tolerances were found in the remaining backcross (N × NM, NM × N, N × MN) or pure *O. niloticus*; thus some type of maternal effect from the maternal nuclear genome, cytoplasm or mitochondrial genome continued to be transmitted to the backcross generation. Carcass yield of the backcross hybrids, however, tended to be higher than those of the parent species. Differences in some morphological traits were observed among the hybrids and the parental species.

Backcrosses and F_2 Hybrids

Most backcrosses have poor performance or performance that is inferior to the F_1 hybrid. There are rare examples where a backcross will have the same heterotic performance of the F_1, and this is a result of epistasis.

Usually, backcrossing is used to introgress a particular trait, gene or set of genes from one species into another. For instance, it may be desirable to make a backcross that is essentially a channel catfish, but introgressed with the genes from blue catfish for catchability, resistance to ESC and dress-out percentage. A synthetic breed of sturgeon was made through seven generations of backcrossing. One of the problems with interspecific backcrossing programmes is reproductive problems, especially in the early generations. In the case of the Russian sturgeon, after seven generations the gene combinations in the backcross were selected and found to break out of the reproductive problem and exhibit adequate reproductive rates.

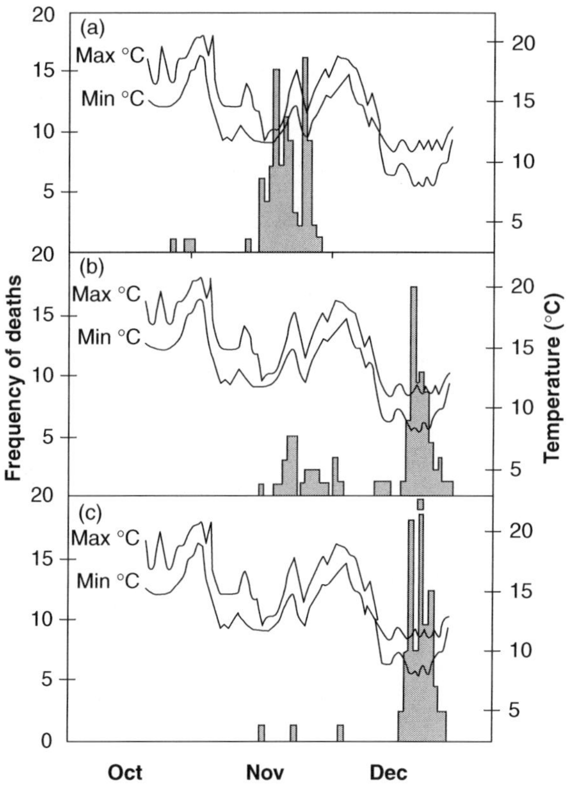

Fig. 8.3. Mortality curves of red *Oreochromis mossambicus* (a), *Oreochromis aureus* (b), and their F_1 backcross (c) under ambient conditions in autumn, Alabama, USA. (Adapted from Behrends and Smitherman, 1984).

Table 8.1. Survival time, mean lethal lower temperature (MLLT) and range of MLLT of red *Oreochromis mossambicus*, *Oreochromis aureus* and their F_1 backcross under ambient conditions in autumn, Alabama, USA.

Genetic group	Survival time (days)	MLLT (°C)	Range of MLLT (°C)
O. aureus	60.8	11.6	8.0–16.5
O. mossambicus	40.1	13.8	11.9–19.0
F_1 backcross	67.8	10.0	8.0–17.0

The red coloration of *O. mossambicus*-based red tilapia has been successfully backcrossed into faster-growing more cold-tolerant blue tilapia, *O. aureus* (Fig. 8.3, Table 8.1). The red tilapia grows very slowly and has poor cold tolerance. Half of the F_1 hybrids are red. Red F_1 males were backcrossed to blue tilapia females. The F_1 backcross had better cold tolerance and survived for a longer period of time than either of the parent species. Epistasis is the explanation for this result. The mean lower lethal temperature of the backcross is about 1.5°C lower than that of the blue tilapia.

Red Nile tilapia and red blue tilapia have been developed. *O. niloticus* is less cold tolerant than *O. aureus*. Not surprisingly, the red Nile tilapia backcross has cold tolerance very similar to Nile tilapia, and the red blue tilapia backcross has cold tolerance very similar to blue tilapia. Naturally, the genes of the parent species are dictating the characteristics of the backcross hybrids.

Reciprocal backcross hybrids are possible of course. F_1, F_2 and F_1 channel catfish × blue catfish backcross hybrids have been compared. The greater the percentage of blue catfish genes in the $F_{\times hybrid}$, the more blue catfish-like performance observed. The greater the percentage of channel catfish genes in the $F_{\times hybrid}$, the more channel catfish-like performance observed. However, once again, the early generations of backcrossing were actually more difficult to produce than F_1 hybrids.

There is at least one example where an F_2 hybrid was valuable and applied in commercial aquaculture. Cold tolerance is an important trait for tilapia in Israel. At one time, virtually the entire Israeli tilapia industry was based on the F_2 hybrid between *O. niloticus* females and *O. aureus* males. The F_1 hybrids are monosex males, grow fast and have increased cold tolerance and seinability. However, F_1 hybrid fry production is difficult and inefficient. The F_1 individuals are fertile and produce large numbers of F_2 hybrid. The F_2 loses some of the growth, cold tolerance and seinability of the F_1, and is 75% male. However, the overall performance is still superior to the original parent species and seed production is adequate. Methyltestosterone is used then to sex reverse the 25% females to maleness.

9

Selection and Correlated Responses to Selection

Selection is often used to improve quantitative or production traits. Quantitative traits exhibit continuous variation, a range of phenotypic expression controlled by many loci. Examples of quantitative traits in animals include milk production, body weight, feed conversion, egg production and disease resistance. In contrast to crossbreeding and hybridization, selection is a long-term programme with continual improvement over a multitude of generations. However, there are short-term products, as each generation of selection should result in an additional step of genetic improvement and each generation of fish should be superior to the previous generation.

Heritability

One of the keys to selection is heritability. There are two types of heritability: broad sense and narrow sense. Broad-sense heritability is the total genetic variation divided by total phenotypic variation. Narrow-sense heritability is defined as the additive genetic variation divided by the total phenotypic variation. It is the more useful of the two.

Theoretically, heritability can range from 0.0 to 1.0 because there is the extremely rare possibility that all of phenotypic variation is due to additive genetic variation. The other extreme is that none of the phenotypic variation is due to additive genetic variation. Heritability can be used as a predictor. In general, when heritability is less than 0.15, it is considered a low heritability. Heritability between 0.15 and 0.3 would be considered a moderate heritability. If heritability is 0.35 or higher, it is a high heritability. If heritability is significant, moderately low or higher, enough additive genetic variation exists that selection should be successful. When heritability is zero or low, crossbreeding may be more effective than selection to improve the trait. Heritability is useful from the standpoint that it can be used as a predictive tool for which genetic improvement programme should be pursued.

Heritability varies from population to population or from strain to strain, and also varies from generation to generation. This is logical because different strains and populations have different genetic make-up and potentially different levels of additive genetic variation. Heritability will change from generation to generation because allele frequencies are altered via the intentional selection. Therefore, the amount of additive genetic variation is changing because of selection for the best alleles. Additive genetic variation could increase in some generations because of recombination events and other genetic mechanisms that result in increased diversity.

Broad-sense heritability (V_G/V_P) is generated through analysis of variance from a series of full-sib families. The weakness of broad-sense heritability is that it contains all components of genetic variation: additive, dominance, epistasis, maternal heterosis. Additive genetic variation is needed for selection to be successful. Narrow-sense heritability measures the relative proportion of additive genetic variation:

$$h^2 = V_A / V_P \qquad (9.1)$$

Or

$$h^2 = V_A / V_E + V_A + V_D + V_{AA} \\ + V_{DD} + V_{AD} + V_{MH} + V_{GE} \qquad (9.2)$$

Broad-sense heritability is not specific enough. It identifies total genetic variation, but not the specific types. If the broad-sense heritability is 0, it is informative as it tells you there is no genetic basis for the trait and additive genetic variation is 0; thus selection will not be successful. However if broad-sense heritability is 0.5, additive genetic variation could be a maximum of 0.5, but it could be any value below 0.5 including 0.0.

There are different ways to generate narrow-sense heritability. From this point forward, heritability will mean narrow-sense heritability as broad-sense heritabilities are seldom calculated and heritability is usually synonymous with narrow sense in the scientific literature. One of the more common methods is a full-sib–half-sib analysis. Regression of parent on offspring is an alternative analysis to determine heritability. Both of these techniques are quick and results are generated within a single generation.

Full-sib–half-sib analysis results in sire heritability (h_s^2), dam heritability (h_D^2) and sire + dam heritability (h_{SD}^2). The dam heritability comes from the dam component of variation (σ_D):

$$h_D^2 = \frac{4\sigma_D^2}{\sigma_S^2 + \sigma_D^2 + \sigma_{SD}^2 + \sigma_W^2} \qquad (9.3)$$

The numerator is multiplied by 4. The dam component of variance contains all of the maternal effects. Maternal effects are usually not due to additive genetic variation. If there is a significant amount of maternal effects in the dam variance, the error in the heritability is multiplied fourfold, inflating heritability (Table 9.1).

For sire heritability the numerator is multiplied by 2:

$$h_S^2 = \frac{2\sigma_S^2}{\sigma_S^2 + \sigma_D^2 + \sigma_{SD}^2 + \sigma_W^2} \qquad (9.4)$$

The sire component of variation (σ_S) contains dominance. If dominance is significant the heritability will be inflated as much as twofold for h_s^2. For sire + dam heritability again the numerator is multiplied by 2, so potential bias is possible once more:

$$h_{SD}^2 = \frac{2(\sigma_S^2 + \sigma_D^2)}{\sigma_S^2 + \sigma_D^2 + \sigma_{SD}^2 + \sigma_W^2} \qquad (9.5)$$

If the sire and dam heritabilities are the same, this is indicative that the h^2 has been measured accurately. If one is higher than the other, the conservative or smaller value is likely the most accurate. In the above, σ_W is the within or error component of variation.

The third method is the realized heritability estimate. Realized heritability is generated from an actual selection experiment. The parent generation is evaluated and the selection response of the progeny generation evaluated to determine the realized heritability.

Table 9.1. Sire and dam heritabilities for length and weight in Nile tilapia, *Oreochromis niloticus*, illustrating the effect of maternal effects on heritability and the temporary nature of maternal effects in Nile tilapia.

	h_s^2	h_D^2
45-day length	0.100 ± 0.190	0.545 ± 0.231
45-day weight	0.045 ± 0.145	0.349 ± 0.192
90-day length	0.065 ± 0.056	−0.021 ± 0.067
90-day weight	0.036 ± 0.062	0.041 ± 0.085

Realized heritability equals response to selection divided by selection differential, and is sometimes termed realized heritability estimate. Realized heritability takes two generations of experimentation, thus it is a slower technique than full-sib–half-sib analysis and parent on offspring regression. Heritability estimates for body weight from a full-sib–half-sib analysis tend to give higher values than an actual realized heritability in fish.

The parent population has a normal distribution, and a mean. There is a minimum value for which individuals possessing this minimum value or higher are selected. This minimum value is the point of truncation. The selected individuals also have a mean. The difference between the mean of the original parent population, or randomly bred population, and the mean of the selected parents is the selection differential (SD or S). The difference between the means of the progeny from the selected parents and the progeny of the control parents is the genetic gain (ΔG), or the response to selection (R) or the progress (P). Thus:

$$h^2 = \frac{\text{(mean of selected progeny} - \text{mean of control progeny)}}{\text{(mean of selected parent} - \text{mean of control parent)}} \qquad (9.6)$$

or

$$h^2 = R/SD \qquad (9.7)$$

Some formulas use intensity of selection (i), which shows how strongly a population is selected. The height of the point of truncation is called 'Z'. The area under the curve to the right of the point is called 'P'. Selection intensity equals Z divided by P. This area is difficult to calculate, but tables exist that indicate the value of i based on what percentage of the population has been selected.

Response to selection or progress can be projected. The response or progress equals $i\sigma_{ph}h^2$. Thus, if you know the heritability for your trait, the phenotypic variation (standard deviation, σ_{ph}) and intensity, response to selection is predicted. Thus, $i\sigma_{ph}$ equals selection differential. Selection differential × heritability predicts response to selection.

In the practical or commercial situation, selection differential or intensity will be dictated by the number of brood fish needed to make production quotas. The selection intensity chooses us rather than we choose the selection intensity, unless we have extra resources to devote to achieve desired selection intensities.

During several generations of selection, response is usually not linear and in some generations can even be negative. Eventually, a selection plateau may be reached and progress ceases. However, there are mechanisms to break out of the plateau and start making progress again.

The rate at which a selection plateau is reached is related to selection intensity. A strong selection intensity results in the most rapid gain, but theoretically the selection plateau will be reached more rapidly and at a lower level than for less intense selection. If we reduce our selection intensity the progress or rate of gain is slower, but theoretically the selection plateau will be reached more slowly and at a higher point. Selecting at a higher intensity, fewer or a smaller percentage of individuals are chosen, thus by chance there is a greater probability of accidentally culling individuals possessing good and unique alleles or gene combinations. Therefore, additive genetic variation may decrease at a more rapid rate, potentially causing a selection plateau at a lower level. By reducing selection intensity a smaller percentage of individuals are selected, reducing the rate of gain and reducing the likelihood of culling individuals with unique, superior alleles, resulting in a higher selection plateau. Given enough resources – facilities, money, time and labour – two lines could be developed: one for short-term maximum gain for immediate commercial use and a second for slower gain and maximum long-term genetic gain for future application.

Even when selection plateaus have been reached, continued selection can be important. Heritability may appear to be zero but hidden additive genetic variation may exist. If selection pressure is released at this point, performance of the population may decrease and some of the progress may be lost. This has been observed in poultry. Epistasis may be masking additive genetic variation, or potential genetic variation may be tied up in linkage. Recombination events may eventually generate some new genotypes, resulting in additional additive genetic variation.

Maintaining selection pressure may eventually allow the selection plateau to be overcome. Some theorize that additive genetic variation is inexhaustible, which seems illogical, as there must be limits and heritability can be zero. However, there have been some experiments with mice, Tribolium beetles and maize where genetic gains were being made even after 100+ generations of selection. Theoretically, with time mutations will add to genetic variation; however, this is surely a very slow process.

One method to break out of a selection plateau quickly is to introduce new genetic variation with crossbreeding. One option is top-crossing. A few individuals from a population that is superior to the individuals in the plateaued population are introduced. If affordable, the more introduced individuals the better. If the new or introduced blood is lower performing, performance will likely be intermediate to the original and introduced lines once they are crossed. The decrease in performance should be temporary, selection response should reinitiate and eventually the selection plateau will be surpassed. Ideally, the introduced line will be equal to or better than the original line or some immediate gains will be made from heterosis or epistatic gene action.

Selection can be made in either the positive direction or the negative direction. If selection is made in both directions it is called bidirectional selection. Then a realized heritability is generated for the up selection, and another one for the down selection. Often the two values will be averaged and the average reported as the heritability for the trait. However, selecting for larger or smaller values is actually conducting selection for two separate traits. Sometimes equal responses in both directions will be obtained, a symmetrical response. But often the selection response will be unequal, an asymmetrical response to selection. In the case of an asymmetric response with the down direction giving the higher h^2, the average h^2 of the positive and negative direction for the trait will overestimate the selection response that would be obtained if selection were conducted to increase the value for the organism.

Another reason why selection in the up and down direction should be considered two different traits is that usually it is much easier to make things smaller than it is to make things larger. Often there are biological limitations to making certain traits increasingly larger. The selection for increased egg size in chickens has a limitation. If the egg becomes too large, it will not be able to pass through the oviduct or cloaca.

Multi-generation selection experiments allow calculation of a cumulative realized heritability. The nature of this calculation inflates the denominator relative to the numerator, so cumulative realized heritability will often be smaller than heritability in individual generations:

$$h^2 = R/(SD_1 + SD_2 + ... + SD_n) \qquad (9.8)$$

where R is the cumulative response to selection and SD_1, SD_2, ..., SD_n are the selection differentials for generations 1, 2, ..., n.

The standard error of cumulative realized heritability can be calculated as:

$$\sigma_{h^2} = \sigma_R/(SD_1 + SD_2 + SD_3) \qquad (9.9)$$

where σ_R is the standard error of the response to selection. σ_R is calculated from the equation:

$$\sigma^2 = V_p[(th^2/N_e) + (1/m)]R \qquad (9.10)$$

where V_p is phenotypic variance, t is the number of generations, N_e is the effective population number and m is the number of individuals measured (Falconer, 1989).

Heritability is not the absolute predictor of potential selection response. High heritability does not guarantee biologically or economically significant selection response. Despite very low heritability, selection response may still be biologically or economically significant. Phenotypic variability is an important component of selection response as indicated in the equations above. A trait with very high heritability but with extremely low phenotypic variability may result in hardly any progress from selection from an economic or biological perspective as phenotypic variation is inadequate. Essentially, progress equals intensity of selection × variation of phenotype × heritability. Something cannot be made from nothing. Conversely, a trait with a very low heritability could be significantly improved with selection if a large quantity of phenotypic variation exists. Although the proportion of the total phenotypic variation due to additive genetic variation is low, the absolute amount of additive genetic variation could be quite high.

Genetic Controls

All selection and genetic experiments must have a genetic control so that progress and genetic improvement can be measured. A genetic benchmark is necessary. Several options exist.

The original parent population can be re-spawned. One advantage of this approach is that if a high percentage of these fish can be re-spawned, the original parent population is re-created and the original genetic make-up is regenerated, which appears to be a very strong control. The potential weakness is that the control brooders will be older and larger than the brooders used for the select lines. If maternal effects are important, maternal effects are introduced into the evaluation. The other disadvantage is that in a long-term experiment with multiple generations of selection, the control is eventually going to die out. If mortality is not random over time compared with the original mean of the trait, the selection differential may change significantly without being noticed, reducing the accuracy of measurement of heritability.

The second option is a random control. A random sample of the fingerlings from the control population is chosen for future brood stock and randomly bred in each generation of experimentation. In the first generation of experimentation, the random sample should be taken from the original population before the select individuals are chosen for the select line. If the selection is conducted first, the population mean is shifted and the control actually becomes a negative selection, underestimating the selection differential and inflating the calculation of the selection response.

Other potential sources of error exist. The random sample may not actually be perfectly random, so the mean of the sample might be different from the original population. If we measure the value for each individual chosen to make the control population, this problem is solved as we can now calculate the selection differential exactly. Not all fish will survive or actually spawn, which may result in the mean of the fish that produce the control progeny being different from the mean of the control brood stock used to calculate the selection differential, another source of error. This may be an unsolvable problem if the fish are unmarked and there is group spawning, as genetic contributors are unknown. If the fish are marked and individually mated, the true mean of the control parents can be calculated, resulting in a more accurate measurement of heritability. The same source of survival and mating error can result in the select line, compounding error further in the selection differential unless the select individuals are also marked and individually mated.

Sometimes the size of the fish in the parent generation and the progeny generation will be radically different at the conclusion of the experiment, for environmental reasons or poor experimental execution. Under these circumstances the response to selection can actually be larger than the selection differential, which is, of course, dubious from a genetic standpoint. Conversely, the selection differential could be inflated causing underestimation of the realized heritability.

One way to address this problem is to use a standardized selection response and a standardized selection differential. Instead of reporting and calculating the selection differential and the response to selection in absolute units of the measured trait, they are calculated in units of standard deviation. This will work as a correction factor, as the coefficient of variation tends to remain fairly constant as the mean of a trait increases. Therefore, if the response to selection is 100 g and the standard deviation is 25 g, the response to selection is 2.5 standard deviations. If the selection differential is 50 g but the standard deviation is 10 g, the selection differential is 5.0 standard deviations. Heritability is then 2.5 standard deviations divided by 5.0 standard deviations, which is 0.5, rather than 100 g divided by 50 g which gives us the unrealistic result of 2.0. Another alternative is to standardize the response to selection and selection differential by reporting the difference between the selects and controls as a percentage of the control.

Another technique that has been implemented is utilizing the select line from the previous generation as the genetic control in an attempt to measure genetic gain generation by generation. This may be adequate to measure single-generation response. The weakness of this approach is that the control, the benchmark, is changing every generation. There is likely error every time the control changes and thus error is accumulating over time, reducing the accuracy of measuring the multi-generation gain. The motivation and temptation to use this approach is more likely in commercial settings because of the financial motivation not to waste culture space on inferior animals, the randomly bred control.

A further type of control is a standard crossbreed. This becomes the initial benchmark. This should be a fairly strong control as some error in the random selection of the parent lines may be corrected for by over-dominance and heterozygosity in the F_1 crossbred control, as heterozygosity can increase standardized performance and reduce phenotypic variability.

A seemingly perfect solution is to use a 100% homozygous clone as the multi-generational control. Every individual in a cloned population is genetically the same, so the component of genetic variability is zero. These clones are 100% homozygous, which leads to two problems. Often clones have severe reproductive problems as many homozygous deleterious recessive genotypes are generated. The second problem is that 100% homozygous individuals are less adaptable to micro-environmental differences, thus these individuals are actually phenotypically more variable than normal populations as the component of environmental variation becomes very large.

The ideal genetic control is an F_1 crossbreed between two different clones. Each individual now has heterozygosity but there is no genetic variation: each individual is genetically identical. These individuals/populations can adjust to micro-environmental variation and have relatively standard performance. The population distribution will have some variation, but less than a normal outbred population. They will also be monosex male or monosex female. One hundred years later, our control is still exactly the same. As spawning time can sometimes vary in an experiment or multiple experiments initiated in a season, the same control can be generated repeatedly throughout the season to correct for age difference among the tested genetic groups.

Practical Aspects of Selection

The population that is being selected for increased size may be in several ponds or tanks. The selection should be conducted pond by pond, and the individuals should not be pooled prior to selection. If the fish are pooled first, all of the selected individuals – the largest individuals – may come from a single pond owing to environmental effects such as differential mortality, age effects or other environmental factors such as water quality associated with that single pond. If this occurs, selection will be less efficient as some superior individuals will not be recognized and the number of families selected will be reduced, leading to faster rates of inbreeding.

Sexual dimorphism, males and females growing at different rates, or other sexual dimorphic traits can complicate selection.

In the case of Nile tilapia and channel catfish, males grow faster than females while the opposite is true in salmonids, common carp, silver barb and largemouth bass. If selection is conducted at an age or size before the fish are sexable, all or the majority of the individuals may be a single sex. Obviously, this has implications for reproduction. The options are to wait until the fish are a sexable size, to use a sexing technique such as dyes or microscopy to sex the immature fish, to retain some extra fish at random to ensure that enough brood stock of both sexes will be available or to conduct the selection in only one sex. If the selection is in only one of the sexes, the selection intensity is reduced by half and the rate of genetic gain is reduced by half.

Individual and Family Selection

The main two basic types of selection are individual (mass) and family selection. In the case of individual selection, individuals are selected based on their own performance. In the case of family selection, individuals are selected based on the mean of their family, not their individual mean. Combined selection utilizes both family and individual selection. The best individuals in the best families are selected. Theoretically, combined selection would yield the most rapid rate of and largest genetic gain.

The advantage of the mass selection or individual selection is simplicity. The best performers are chosen without tracking pedigrees. Minimal facilities are required, as well as minimal information and data collection. No marking or separate facilities are required for individual families, which reduces time, labour, facilities and expenses. The primary disadvantages are the possibility of inbreeding since the relationship of the selected individuals is not known, and the fact that relatives tend to have more similar performance. Theoretically, this problem would be worse in small populations, but it could occur in large populations also, and the inbreeding depression could negate some of the selection progress.

One advantage of family selection is that it is based on more genetic information, so the selection should be more accurate. Records and pedigrees are maintained, and individuals/families are marked or kept separate so that inbreeding can be avoided. Other advantages of family selection are that it allows selection for traits that are not easily measured from individuals, allows one to select more intensely for sex-limited traits and allows selection for lethal traits. For instance, individual selection for dressing percentage is not possible since after the measurement of the trait the individual is no longer capable of breeding. In the case of sex-limited traits, males cannot be individually selected for fecundity or egg size. If information is available on mothers, sisters or nieces, males can be selected for fecundity based on the mean of their female relatives.

Also, when additive genetic variation is small or difficult to measure, selection response can sometimes be obtained with family selection when mass selection will yield no response.

Environmental Shock and Manipulation-Creative Selection Techniques

Flies usually have six bristle hairs, and it is likely important to have this number just as it is important for us to have two hands and ten fingers. Flies have a major locus for bristle hair number, but apparently have several modifying loci, as a canalization method to ensure the proper number of bristle hairs. If there is a mutation at the major locus, the modifying loci act as backup. By cold shocking the fly eggs, phenotypic variation and abnormal bristle hais numbers were observed. This alteration of embryonic development revealed additive genetic variation masked by epistasis and other mechanisms. Selection for increased bristle hair number was conducted on the flies that developed from cold-shocked eggs. Genetic gain was accomplished and the elevated bristle hair number was maintained even after the shock treatments stopped.

Another novel selection technique was attempted to increase eggs laid per day for chickens. A population had reached a selection plateau of 1.9 eggs per day for a 24 h day. The chickens were then exposed to a 22 h day, 11 h of daylight and 11 h of darkness. The chickens still laid 1.9 eggs per 'day', but for a 22 h day. Thus the number of eggs laid per 24 h (2.07) increased by 10%. Additional additive genetic variation was revealed and eggs laid per day responded to selection under the 22 h regime. The environmental manipulation tricked the chickens and also uncovered hidden additive genetic variation.

Selection in Aquaculture

Research on selection in fish for relevant aquaculture traits began in the 1920s (Embody and Hayford, 1925), but very little selection was conducted prior to 1970. Unfortunately, during this time period, several potentially high-impact experiments did not include adequate genetic controls to prove genetic gain. From 1970 to the present, research on selection and traditional selective breeding has continued to grow rapidly (Dunham, 1996; Dunham *et al.*, 2001) despite the excitement about and increased funding in the area of molecular genetics and genomics. A few hundred heritability estimates have been obtained for several traits of cultured fish and shellfish. Even in developing countries like Thailand, a large amount of research has been conducted on quantitative genetics of aquatic organisms (Table 9.2).

The selection programmes that have been attempted for cultured fish have mainly been successful (Table 9.3). In general, the response to selection for growth rate in aquatic species is very good compared with that for terrestrial farm animals and these programmes have been highly successful. Fish and shellfish often have higher genetic variance than farmed land animals; Gjedrem (1997) indicates that the genetic variation for growth rate is 7–10% in farmed terrestrial animals and 20–35% in fish and shellfish. The estimated gains vary from 2 to 20% per generation for aquatic organisms, which is much higher than that usually obtained in

farm animals. Fecundity is also higher in aquaculture species compared with warm-blooded agriculture animals, allowing for higher selection intensity for aquaculture production improvement.

Selection for increased body weight has a high probability of success in the vast majority of aquatic organisms and in the vast majority of strains within a species. Selection is a consistent technology to improve body weight of aquacultured fish and usually, but does not always, improve body weight (Dunham, 1996). In some cases growth has been doubled after several generations of selection. In some species, such as common carp and Nile tilapia, the success of selection is inconsistent. This is not unexpected as genetic background and quantities of additive genetic variation can vary from one population to another. Depending on the quantity of additive genetic variation and the influences of other contributors of total phenotypic variation, different selection programmes, family or mass selection, could yield different results.

Body weight of channel catfish has been improved by 12–20% with one or two generations of mass selection (Bondari, 1983; Dunham and Smitherman, 1983a; Smitherman and Dunham, 1985; Dunham and Brummet, 1999). Selection has been successful in all eight lines of channel catfish that were evaluated, and the fastest-growing strains yielded the fastest-growing select lines (Smitherman and Dunham, 1985). The best select line grew twice as fast as average strains of channel catfish (Burch, 1986). After three generations of selection, growth rate of channel catfish was improved by 20–30% when grown in ponds (Rezk, 1993) and four generations of selection in Kansas strain resulted in a 55% improvement in growth rate (Padi, 1995).

In channel catfish and other species, the response to selection for body weight is much higher in the first couple of generations compared with later generations. One explanation is that additive genetic variation is reduced after a couple of generations. Although theory says that additive genetic variation is needed for selection response, obviously dominance-based loci

Table 9.2. Heritabilities for aquacultured species in Thailand. (Adapted from Dunham *et al.*, 2001.)

Species	Trait	h^2	Reference
Pangasius sutchi	TL at 99 days	$h_D^2 = 0.052 \pm 0.171$	Leeprasert (1987)
	TL at 126 days	$h_D^2 = 0.195 \pm 0.227$	
	TL at 182 days	$h_D^2 = 0.173 \pm 0.204$	
	TL at 240 days	$h_D^2 = 0.062 \pm 0.440$	
	BW at 99 days	$h_D^2 = 0.056 \pm 0.273$	
	BW at 126 days	$h_D^2 = 0.139 \pm 0.199$	
	BW at 182 days	$h_D^2 = 0.122 \pm 0.213$	
	BW at 240 days	$h_D^2 = 0.126 \pm 0.424$	
Labeo rohita	TL at 118 days	$h_D^2 = 0.075 \pm 0.155$	Supsumrarn (1987)
	TL at 202 days	$h_D^2 = 0.046 \pm 0.074$	
	TL at 285 days	$h_D^2 = 0.893 \pm 0.094$	
	BW at 118 days	$h_D^2 = 0.102 \pm 0.107$	
	BW at 202 days	$h_D^2 = 0.024 \pm 0.044$	
	BW at 285 days	$h_D^2 = 0.093 \pm 0.088$	
Clarias macrocephalus	TL at 9 months	$h_R^2 = 0.39$	Jarimopas *et al.* (1989)
	BW at 9 months	$h_R^2 = 0.84$	
C. macrocephalus	TL at 210 days	$h_R^2 = 0.23$	Chamnankuruwet (1996)
	BW at 210 days	$h_R^2 = 0.31$	
C. macrocephalus	Resistance to *Aeromonas hydrophila*	$h_S^2 = 2.34 \pm 0.07$ $h_D^2 = 0.15 \pm 0.22$	Rasatapana (1996)
	BW at 53 days	$h_S^2 = 2.01 \pm 0.95$ $h_D^2 = 0.31 \pm 0.51$	
Puntius gonionotus	TL at 111 days	$h_D^2 = 0.012 \pm 0.055$	Jitpiromsri (1990)
	TL at 170 days	$h_D^2 = 0.044 \pm 0.246$	
	TL at 231 days	$h_D^2 = 0.168 \pm 0.541$	
	TL at 276 days	$h_D^2 = 0.202 \pm 0.332$	
	BW at 111 days	$h_D^2 = -0.217 \pm 0.090$	
	BW at 170 days	$h_D^2 = -0.067 \pm 0.322$	
	BW at 231 days	$h_D^2 = 0.055 \pm 0.626$	
	BW at 276 days	$h_D^2 = 0.291 \pm 0.517$	
Hypophthalmichthys nobilis	TL at 124 days	$h_D^2 = 0.019 \pm 0.159$	Nukwan (1987)
	TL at 208 days	$h_D^2 = 0.078 \pm 0.122$	
	TL at 292 days	$h_D^2 = 0.038 \pm 0.109$	
	TL at 362 days	$h_D^2 = -0.014 \pm 0.039$	

(Continued)

Table 9.2. *Continued*

Species	Trait	h^2	Reference
	BW at 124 days	$h_D^2 = 0.077 \pm 0.186$	
	BW at 208 days	$h_D^2 = 0.069 \pm 0.128$	
	BW at 292 days	$h_D^2 = 0.048 \pm 0.156$	
	BW at 362 days	$h_D^2 = 0.004 \pm 0.043$	
Saccostrea cucullata	LW at 15 months	$h_R^2 = 0.184$ (50 oysters/net)	Thavornyutikarn (1994)
	LW at 15 months	$h_R^2 = 0.148$ (150 oysters/net)	

TL, total length; BW, body weight; LW, live weight; h_D^2, heritability estimated from dam variance; h_R^2, realized heritability; h_S^2, heritability estimated from sire variance.

Table 9.3. Examples of response to selection in growth rate. (Adapted from Dunham *et al.*, 2001.)

Species	Mean body weight (g)	Gain per generation (%)	No. of generations	Reference
Coho salmon	250	10.1	4	Hershberger *et al.* (1990)
Rainbow trout	3.3	10.0	3	Kincaid *et al.* (1977)
	4000	13.0	2	Gjerde (1986)
Atlantic salmon	4500	14.4	1	Gjerde (1986)
	6300	14	6	Gjerde and Korsvoll (1999)
Channel catfish	450	14	4	Padi (1995)
	67	20	1	Bondari (1983)
Tilapia	100	15	5	Rye and Eknath (1999)
Rohu carp	400	17	2	Mahapatra *et al.* (2000)
Shrimp	20	4.4	1	Fjalestad *et al.* (1997)
	15	10.7	1	Hetzel *et al.* (2000)
Oysters	42	9–12	1	Toro *et al.* (1996)
	36	9	2	Nell *et al.* (1999)
Clam	33	9	1	Hadley *et al.* (1991)

also respond to selection. Thus some of this high response in the first couple of generations could be due to epistatis and dominance, and likely they become near fixation after a couple of generations contributing to the slowing of rate genetic gain.

Strain effects are important in selection programmes. The data on selection for body weight in channel catfish indicate that if selection is conducted for a poor-performing strain, a better fish will result. Selection upon a good-performing strain will result in the best fish (Fig. 9.1).

Measurement of selection response for body weight is likely underestimated. Inbreeding depression probably affects the select line more than the randomly bred control. Also, domestication in randomly bred channel catfish increases body weight by 2–6% per generation. The selected line probably does not benefit in the same way, resulting in measurement of a smaller selection response than what actually occurs.

Selection for increased body weight has also been successful in other families of catfish. Estimates of selection response for

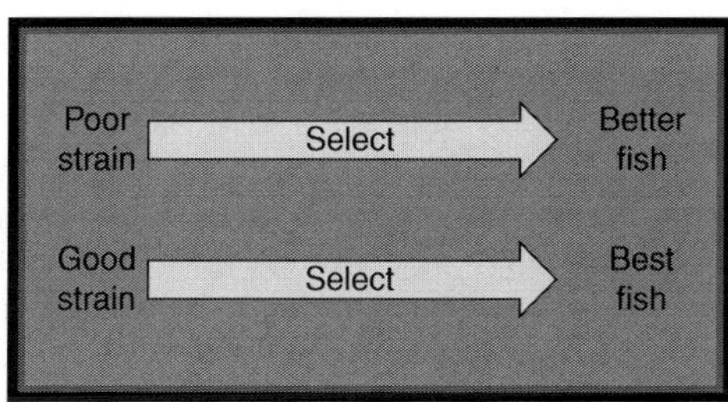

Fig. 9.1. Strain effect on final product.

body weight and total length were determined for *Clarias macrocephalus* (Jarimopas *et al.*, 1989). Four generations of individual selection increased body weight by 50.5 g and total length by 0.88 cm.

Six generations of selection increased body weight by 30% in rainbow trout (Kincaid, 1983b), an increase of 7% was achieved within a single generation (Gjedrem, 1979) in Atlantic salmon and an increased growth rate of 50% was achieved with ten generations in coho salmon (Hershberger *et al.*, 1990). Gjerde and Korsvoll (1999) reported a realized selection differential in Atlantic salmon of 83.9% during six generations, or 14% increase in growth rate per generation, and a reduction of 12.5 units in sexual maturity or 8% per generation. Gjerde (1986) estimated a response to selection of 13 and 14.4% per generation in rainbow trout and Atlantic salmon, respectively. The genetically improved salmon show improved feed efficiency; the increased growth rate is mainly due to a much higher feed intake per kilogram of body weight.

Heritability, additive genetic variation and selection response for body weight can vary among strains. Tilapia and common carp are two of the more prominent examples of this phenomenon. Body weight of common carp initially appeared unresponsive to selection as five generations of selection for increased body weight resulted in no genetic gain, and five generations of family selection resulted in modest gains of about 5–10% (Moav and Wohlfarth, 1974a). However, different strains of common carp may possess varying amounts of additive genetic

variation. Smisek (1979) estimated heritabilities for body weight of 0.15–0.49 in a Czechoslovakian strain of common carp. Vietnamese common carp had a heritability of 0.3 for growth rate and six generations of selection increased body weight by 5% per generation (Tran and Nguyen, 1993). In a selection experiment to increase growth rate in rohu carp at the Central Institute of Freshwater Aquaculture, India, the response was 34.5% for the first two generations (Gupta and Acosta, 2001).

Mass selection has improved body weight in *Oreochromis mossambicus*, *Oreochromis shiranus*, red tilapia and *Oreochromis aureus* (Dunham *et al.*, 2001). Mass selection has improved body weight in an inbred line of *O. mossambicus* (Chan, 1971) and increased body weight by 15% in *O. aureus* (Bondari *et al.*, 1983). Sanchez and Ponce (1988) also improved body weight in *O. aureus* through five generations of selection, achieving 17 g per generation. Two generations of selection for body weight in *O. shiranus* resulted in a 13% increase in body weight (Maluwa and Gjerde, 2007).

Selection for increased body weight in red tilapia has been variable. Selection for increased body weight in red tilapia was not successful (Behrends *et al.*, 1988); however, growth rate was improved 16% by five generations of selection in Thai red tilapia (Jarimopas, 1990). Great variability to the selection response has been observed in Nile tilapia (Hulata *et al.*, 1986; Huang and Liao, 1990; Khaw *et al.*, 2008), from no response in some strains to 1–7% gain per generation in others and as much as 11% per generation in the Philippines.

Two generations of within-family selection for increased body weight for Philippine and Thai strains of Nile tilapia resulted in a 4.5% increase in body weight (Abella *et al.*, 1990; Uraiwan, 1990) and one generation of selection increased length by 2.3% in Indonesian farm stocks (Brzeski and Doyle, 1995). Eight generations of within-family selection increased body weight by 18% in Nile tilapia when grown to about 30 g in tanks (Bolivar *et al.*, 1994). Selection for increased body weight in Nile tilapia was not successful in three populations (Hulata *et al.*, 1986; Teichert-Coddington and Smitherman, 1988; Huang and Liao, 1990).

The lack of response in some strains may reflect a narrow genetic base in the founder stock or sole use of mass selection in cases where additive genetic variation was low. Some selection programmes for Nile tilapia were moderately successful; 14% body weight increase over two generations for a synthetic Egyptian strain (Rezk *et al.*, 2009). Selection for increased growth in a GIFT strain of Nile tilapia gave a much different result, a growth improvement of 77 to 123%, and demonstrated that a combination of family, mass and index selection can double the growth rate of Nile tilapia in a low-input environment in developing countries. The genetic gain was superior to that reported for crossbreeding of Nile tilapia (Tave *et al.*, 1990; Bentsen *et al.*, 1998). During the first two generations of selection, Charo-Karisa *et al.* (2006) found similar responses to selection in Nile tilapia grown in low-input environments as was found in the first two generations in the GIFT population. Production trials and socio-economic surveys in five Asian countries revealed that the cost of production per unit of fish produced is 20–30% lower for the GIFT strain than for other Nile strains in current use. Price elasticity data indicate that yield increases by using GIFT strain would benefit mostly the vast group of poor consumers.

The 11% genetic gain per generation in GIFT tilapia was better than that obtained in most other species of fish, which typically averages 5–7% per generation as demonstrated for salmonids and common carp following approximately five to ten generations of selection (Gjedrem, 1979; Kincaid, 1983a; Hershberger *et al.*, 1990; Tran and Nguyen, 1993). However, other exceptional examples exist such as that for channel catfish (Padi, 1995), which had an increased body weight of 14% per generation over four generations, and the 13–14% increase per generation observed by Gjerde (1986) for salmon.

Four generations of selection for body weight in blunt-nose bream resulted in 19% improvement in China. Two generations of selection for body weight of silver barb in Bangladesh increased body weight by 15% (Gupta and Acosta, 2001).

Response to selection can differ depending on the direction of selection. Body weight of common carp in Israel was not improved over five generations, but could be decreased in the same strain selected for small body size (Moav and Wohlfarth, 1976). Virtually identical results were also reported for Nile tilapia (Teichert-Coddington and Smitherman, 1988). In general, it is easier to select to make traits smaller rather than larger, which, of course, would rarely have aquaculture significance. There are exceptions to the above observations above as common carp in the Czech Republic responded to selection for increased body weight, but not for decreased body weight (Vandeputte *et al.*, 2008).

Selection for body weight has also been successful in marine species such as gilthead seabream, *Sparus aurata* (Knibb *et al.*, 1997a). The first step in the breeding programme to improve growth rate in gilthead seabream was strain evaluation (Knibb *et al.*, 1997a). Then family selection was attempted but difficulties were encountered in efficiently producing single-pair offspring groups (full and half sibs), which led to the conclusion that family mating designs were inappropriate for the group-spawning *S. aurata* (Gorshkov *et al.*, 1997). Progeny testing, where single males were stocked with groups of females, which more closely simulate the natural group-spawning behaviour of the species, yielded genetically related groups more successfully. The small number of families obtained (four each for full and half sibs) did not allow generation of reliable estimates of heritability for growth, yet large (14%–29%) sire components of the offspring weight variance were evident (Knibb *et al.*, 1997a,b). Mass selection proved more effective and resulted in significant heritability estimates for growth (Knibb *et al.*, 1997b, 1998a,b).

Selection improved growth of the shrimp, *Penaeus japonicas* (Preston *et al.*, 1999; Hetzel *et al.*, 2000). Improvement in the growth, survival and total yields were obtained in two selected lines (10–15% increase in mean yields) (Preston and Crocos, 1999). For *Penaeus vannamei*, Fjalestad *et al.* (1997) estimated a response in one generation of selection of 4.4% for growth rate and 12.4% for survival after challenge test against Taura syndrome. Hetzel *et al.* (2000) obtained a selection response of 10.7% in the first generation of selection in *P. japanocus*. Commercial companies have intimated good success for increasing growth in shrimp using selection.

Selection is an effective genetic enhancement programme to improve growth rates in shellfish. In Australia, one generation of mass selection for growth rate in Pacific oysters (Ward *et al.*, 2000) increased growth by 8%. Haley *et al.* (1975) reported that mass selection of adult oysters gave a strong response to selection for growth rate in *Crassostrea virginica*. They concluded that because of large environmental variability, a combination of family and mass selection would be required to achieve maximum response. Newkirk (1980) obtained a 10–20% gain in growth rate of oysters after one generation of selection. Nell *et al.* (1999) observed a genetic gain of 9% increased growth rate in Sydney rock oysters (*Saccostrea commercialis*); similar results were obtained in the Chilean oyster (*Ostrea chilensis*) (Toro *et al.*, 1996). A genetic gain of 9% per generation of selection for growth rate has been estimated for the hard-shell clam or quahaug (*Mercenaria mercenaria*) (Hadley *et al.*, 1991). Rate of genetic improvement appears similar to that of finfish.

Selection has been effective for improving disease resistance, but not as consistently as selection for body weight. Strain variation in selection response is more prevalent for disease resistance than for body weight, and often no selection response is found for some strains while others will exhibit significant enhancement of disease resistance as a result of selection. In the case of salmonids, selection for disease resistance has been particularly successful (Embody and Hayford, 1925). As early as 1925, Embody and Hayford selected surviving brook trout from a population with endemic furunculosis and increased the survival rate from 2% in the initial population to 69% after three generations of selection. Ehlinger (1977) obtained reduced mortality

rates due to furunculosis in brown trout and brook trout after selection. Okamoto *et al.* (1993) reported on a strain of rainbow trout resistant to IPN virus with average mortality of 4.3% compared with 96.1% in a highly sensitive strain. One generation of selection increased resistance to *Flavobacterium psychrophilum* (bacterial coldwater disease) in rainbow trout by an absolute 32% and a relative 105% (Leeds *et al.*, 2009). Kirpichnikov *et al.* (1993) reported a successful selection for increased resistance to dropsy (spring viraemia of carp) in common carp. Three stocks were involved and the response to mass selection was moderate. Schaperclause (1962) reported a positive response to selection for survival in common carp, which showed an average mortality rate of 11.5% in 65 ponds stocked with progeny of selected fish versus 57% in 76 ponds stocked with progeny of non-selected fish.

Fjalestad *et al.* (1997) estimated a response for one generation of selection of 4.4% for growth rate and 12.4% for survival when exposed to Taura syndrome virus in *P. vannamei*. More dramatically, Gitterle *et al.* (2009) improved resistance to Taura syndrome virus by an absolute 30% and a relative 100% with two generations of selection in *P. vannamei*. Additionally, growth and pond survival were being improved by 5–6% per generation.

Body conformation can be changed dramatically via selection (Fig. 9.2). Whereas body weight of common carp was unresponsive to selection, body conformation of common carp can be dramatically altered. Heritability for body depth was high, 0.40–0.80 (Ankorion *et al.*, 1992), and deep-bodied lines have been developed. Recently, a significant heritability was found for deformities in the Atlantic cod, *Gadus morhua* (Praebel *et al.*, 2009). The implication is that deformity rate could be reduced through selection.

Heritabilities for fat percentage in catfish and trout are about 0.50, indicating selection should work to decrease fat content. However, heritabilities for dress-out percentage have been zero to date. Selection should be successful for altering body composition, but may not improve carcass yields.

Fig. 9.2. Potential change in body depth in common carp, *Cyprinus carpio*, through selection.

Reproductive traits tend to have high selection responses. Neira *et al.* (2006) were able to shift spawning date in coho salmon, *Oncorhyncus kisutch*, by about 14 days with four generations of selection. Age at sexual maturity also has high realized heritabilities in salmonids.

Long-term selection appears feasible in fish. Ponzoni *et al.* (2009) evaluated the eighth to 14th generations of GIFT Nile tilapia selected for increased body weight. The response per generation was 13% even after this length of time.

Meristic traits tend to have very high heritability. However, phenotypic variation is also important for a biologically significant selection response. The phenotypic variation for most meristic traits is near zero, so it will be difficult to make significant changes in meristic traits, if it were desirable. There is likely little incentive to change meristic traits except for academic purposes.

Four generations of bidirectional selection for dorsal-fin ray number was conducted in rainbow trout. The starting point was about 7 dorsal-fin rays per fish. A relatively symmetrical response was obtained.

The average dorsal-fin ray number decreased by about 0.3 or 5% in three generations and increased by 0.3 in four generations.

National Breeding Programmes

Today there are few national breeding programmes in fish and shellfish which aim at producing improved stocks for the industry in their countries. In 1975 a national breeding programme for Atlantic salmon and rainbow trout was started in Norway and as of 2001 it was supplying about 75% of the industry with improved eyed eggs. In Canada, a similar breeding programme for Atlantic salmon is in operation at the Atlantic Salmon Federation, St. Andrews (Friers, 1993). In 1993 the Philippines National Tilapia Breeding Programme (PNTBP) was started with brood stock from the GIFT project. The operation was conducted by the GIFT Foundation until about 1999 and then a private company took over. The National Bureau of Fisheries and Aquatic Research also maintains a branch of this breeding programme. In Israel (Wohlfarth, 1983) and Hungary

(Bakos, 1979) crossbreeding programmes with common carp exist. Vietnam has national selective breeding programmes. Additionally, some private companies in several countries including the USA have their own breeding programmes. Several of these national breeding programmes utilize family selection; however, private breeding companies rarely have selection programmes that utilize family selection.

Unintentional Selection – Effect of Everyday Farm/Hatchery Practices

How fish are fed and handled, and how a hatchery or farm is operated, has genetic implications because – consciously or unconsciously – the culturist is performing selection. In the 1950s to 1970s, Auburn University was not cognizant of these implications. Auburn strains of Nile tilapia and channel catfish are among the most difficult to capture by seining. Nile tilapia and channel catfish replacement brood stock were likely fish that were left over from experiments, the ones most difficult to catch. Unintentional selection for fish that were difficult to harvest was probably conducted for several years.

Similar unintentional negative selections have occurred in the commercial environment as well. For example, in 1984 I was interviewing a farmer for a survey that I was conducting on the ancestry of cultured catfish in the USA. I asked one farmer, 'Where do you get your fish?' He replied, 'I got them from Auburn about 1960.' This man was a small farmer, and marketed his fish through fee fishing, hook-and-line. He commented, 'You know in the beginning my fish were real easy to catch.' Now 25 years had transpired and he continued, 'It's much more difficult for customers to catch the fish. They still eat and grow real good but they're harder to catch.' Then I asked, 'So where did you get your brood-stock replacements?' He replied, 'Well, we would fish a pond for a couple of years...' Then his voice slowed, and I could hear the wheels turning in his head as he realized that for 25 years he was selecting for fish that are difficult to catch

by hook-and-line, but hook-and-line vulnerability was what he needed to market his fish. He continued: 'We would fish a pond for 2 years and then we would drain the pond and whatever was left over, we would use for brood stock.' Small culture practices can have significant genetic implications.

Correlated Responses to Selection and Indirect Selection

All traits are genetically correlated either positively or negatively, or they are not correlated. When selection is conducted upon one trait, positive, negative or no correlated responses to selection can occur for other traits depending on the nature of genetic correlations among the traits. The correlated response is any phenotypic change in a second trait, when selection is conducted for a trait of interest. However two traits might be phenotypically correlated, but not have a high genetic correlation.

In the case of a positive genetic correlation, this is the ideal situation because if we select for first trait the second trait will also be improved. If there is no genetic correlation, selection for the first trait will have no effect on the second trait. Of course, the worst-case scenario is if we have a negative genetic correlation resulting in decreased performance for the second trait when selection is conducted for the primary trait of interest. Thus it is important to monitor other important economic traits when selection is conducted for a trait of interest to ensure that there are no adverse effects on such traits. The monitoring is also important in mass selection programmes as there is the potential for long-term decrease in performance in some traits because of the accumulation of inbreeding.

Genetic correlations are also important as they may lead to alternative and sometimes better options for selection such as indirect selection. Indirect selection is valuable when traits are difficult to measure, lethal or sex-limited, so it has some of the same advantages as family selection. In the case of indirect selection, one trait is selected for with the intent of improving a second trait. Essentially, selection is being conducted seeking a positive correlated response.

The correlated response of a second trait Y when selecting for trait X can be predicted as follows:

$$CR_y = i \cdot h_x \cdot h_y \cdot r_A \cdot \sigma_y \qquad (9.11)$$

where CR_y is the correlated response of trait Y, i is the selection intensity, h_x and h_y are the square roots of the heritabilities of trait X and trait Y, r_A is the additive genetic correlation and σ_y is the phenotypic standard deviation of trait Y.

This equation can be factored to allow prediction of the relative effectiveness of direct and indirect selection for a trait. The formula is:

$$CR_x / R_x = r_A \cdot (i_y / i_x) \cdot (h_y / h_x) \qquad (9.12)$$

where R_x is the direct response of trait X. If this ratio is larger than one, indirect selection for trait X should be more effective than direct selection for trait X. If the value is one, then both methods are equally effective; and if it is less than one, then direct selection is the superior technique. The ratio of the selection intensities is a weighting factor as it may be possible to select for one trait more intensely than another, affecting the efficacy of indirect versus direct selection. If the selection intensities for the two traits are the same, this part of the equation cancels becoming one. As the genetic correlation, selection intensity for the second trait and heritability of the second trait increases, the likelihood that indirect selection for the first trait will be more effective also increases.

If there is a negative genetic correlation between two traits, three options exist: (i) focus on selecting for the more important trait; (ii) use multiple trait selection with selection indices; or (iii) use a different and superior genetic-enhancement programme for one of the traits.

Generally, body weight at different ages, total length at different ages, and body weight and total length have high genetic correlations. Although selection for body weight has generally been associated with positively correlated responses such as increased survival and disease resistance, in some cases long-term selection results in decreased bacterial resistance due to either changes in genetic correlations or inbreeding depression. Increased fecundity, fry survival and disease resistance were correlated to selection for increased body weight in channel catfish after one generation of selection for body weight (Dunham and Smitherman, 1983b; Smitherman and Dunham, 1985). No effect was observed for spawning date, spawning rate, hatchability of eggs, or survival of sac-fry. Three and four more generations of selection resulted in no change in body composition or seinability (Rezk, 1993; Padi, 1995). Progeny from select channel catfish had greater feed consumption, more efficient feed conversion and greater disease resistance than controls (Dunham, 1981; Al-Ahmad, 1983). A positive genetic correlation between body weights at different ages in channel catfish is evident from selection experiments. Select-line progeny grew faster during the fingerling stage than control populations in all strains examined (Dunham and Smitherman, 1983b). Two of the three select lines grew more rapidly during winter, and all select lines grew slightly faster than controls during the second season (to food-size) of growth. Feed consumption had a greater effect on body weight than feed conversion (Al-Ahmad, 1983). In all breeding programmes, strain selection, crossbreeding, hybridization, selection or genetic engineering, increased growth rate is due to both feed consumption and feed-conversion efficiency.

Atlantic salmon experience a positive correlated response in feed conversion when selected for growth rate (Thodesen, 1999). Wild salmon had a 17% higher intake of energy and protein/kg growth compared with fish from the fourth generation selected for growth rate. The wild fish had 8% lower retention of both energy and protein. Quinton *et al.* (2007) and Burch (1986) also reported a strong genetic correlation between growth rate and feed conversion in European whitefish (*Coregonus lavaretus*) and channel catfish, respectively, and the nature of the heritabilities and genetic correlations indicate that indirect selection for feed conversion by selecting for growth rate would more effectively improve feed conversion than direct selection for feed-conversion efficiency.

The relationship between body weight and carcass traits is not consistent from one species to another. Navarro *et al.* (2009a) examined the heritabilities and genetic correlations among body weight, visceral fat, muscular fat, muscular moisture and muscular ash in gilthead seabream, *S. auratus*. The nature of the heritabilities and genetic correlations would allow development of a selection index to improve growth, fat content, texture and carcass yield simultaneously. Dress-out and fillet percentages also had positive heritabilities (Navarro *et al.*, 2009b). However, body weight and fillet percentage had a negative genetic correlation, indicating that it might be difficult to simultaneously select for both traits. In the case of sea bass, *Dicentrarchus labrax*, body weight, viscera percentage, visceral fat percentage, fillet percentage and head weight percentage all had significant heritability (Saillant *et al.*, 2009). Body weight had a positive correlation with each of these traits except a negative correlation to head weight percentage, indicating that selection for increased body weight would also increase fillet percentage, but result in fish with a higher visceral fat percentage. Body weight, fat percentage, relative head length, relative body height, relative body width, processed body percentage and fillet yields had moderate to high heritabilities in common carp in the Czech Republic (Kocour *et al.*, 2007a,b). Body weight was highly correlated with fat percentage. Relative head length had strong negative correlation with fat percentage, dress-out percentage and fillet percentage. Thus, indirect selection for reduced relative head length should increase fillet percentage and dress-out percentage, but also increase fat percentage. Selection for increased body weight would also result in a fattier common carp.

Surprisingly, the relationship between body shape and dressing percentage is not strong in channel catfish. However, there is a strong relationship between body density and dressing percentage. Therefore, indirect selection can be conducted for dressing percentage by measuring body density without killing the animal.

A positive correlated response to selection occurs for dress-out percentage when channel catfish are selected for increased body weight. The lines selected for increased body weight had a 1.0–1.5% increase in dress-out percentage. This appears small; however, if 300 million kg are processed, 5 million kg more product is produced to sell with negligible additional input, raising profit by US$20 million.

When mice are selected for body weight, increased fat percentage and obesity often result. In the case of channel catfish, three generations of selection for body weight had negligible impact on body composition. However, muscle lipid content in rainbow trout responded to bidirectional selection (Quillet *et al.*, 2005). Selection for muscle lipid content did not impact dress-out percentage or fillet percentage.

A variety of genetic relationships exist among growth and survival traits. In the case of salmonids, negative genetic correlations and correlated responses exist between resistances for some pathogens. Selection for growth had minimal effects on survival for tilapia. In seven large-scale genetic experiments, genetic correlations between growth rate and survival or survival for specific diseases have been estimated. Although the genetic correlation is not high, average close to 0.30 (Table 9.4), selection for rapid growth will increase survival rate as a correlated response.

In general, selection for increased disease resistance in salmonids has been successful. However, in some cases, there is a negative genetic correlation for resistance between two diseases. The genetic correlation between furunculosis and gill disease is negative in sockeye salmon.

Selecting for resistance to bacterial cold-water disease in rainbow trout did not affect body weight or thermal growth coefficients (Leeds *et al.*, 2009). Resistance to furunculosis, infectious salmon anaemia and infectious pancreatic necrosis all had relatively high heritabilities in Atlantic salmon

Table 9.4. Genetic correlation between growth rate and survival.

| Survival | Genetic correlation | Species | Number | | Reference |
			Sire	Dam	
Survival, fingerling	0.30	Brook trout	32	32	Robison and Luempert (1984)
Coldwater vibriosis	0.18	Atlantic salmon	53	329	Standal and Gjerde (1987)
Survival, fingerlings	0.37	Atlantic salmon	187	1404	Rye *et al.* (1990)
Survival, fingerlings	0.23	Rainbow trout	213	1062	Rye *et al.* (1990)
Furunculosis, fingerlings	0.30	Atlantic salmon	25	50	Gjedrem *et al.* (1991)
Survival, fingerlings	0.31	Atlantic salmon	100	298	Jonasson (1993)
Survival, fry	0.38	Shrimp	26	52	Suarez *et al.* (1999)

(Kjoeglum *et al.*, 2008). Additionally, the genetic correlations among these traits were all zero, which should allow simultaneous selection for all three traits without any negative correlated responses to selection.

Heritability for upper thermal tolerance is significant in rainbow trout (Perry *et al.*, 2005). Genetic correlation between this trait and body weight was essentially zero, thus no correlated responses to selection would be expected when selecting for either of these traits on the corresponding trait. Melanin deposits of Atlantic salmon were negatively correlated with pericarditis (Olesen *et al.*, 2009) and pericarditis was not correlated with body weight; however, pericardial fat was correlated with body weight. Shell closing strength has a high heritability in Japanese pearl oysters, *Pinctada fucata* (Ishikawa *et al.*, 2009). This trait is correlated with high summer survival as the ability to close the shell tightly appears to be a major survival trait.

A negative correlated response to selection occurs for tolerance of low oxygen when body weight is selected in channel catfish again for three generations. Similar data exist for rainbow trout. Selected lines of catfish growing about 30% faster than controls became stressed in half the time as controls when exposed to critically low levels of oxygen. Perhaps these select lines are growing faster partially because of a higher metabolic rate. If there is a higher metabolic rate, oxygen may be consumed more rapidly. And although they did not measure the oxygen tolerance, this was the case in rainbow trout (Kinghorn, 1983).

Selection for increased growth rate in red swamp crawfish was not successful. However, age of sexual maturity was negatively correlated with growth rate. Even though they were not growing faster, the crawfish eventually reached larger sizes because sexual maturity was delayed. If the harvest cycle could be delayed, size and production were larger (Lutz and Wolters, 1989).

Late sexual maturity has a significant heritability in Nile tilapia. Therefore selection can delay sexual maturity, which is important for Nile tilapia culture as this will help delay reproduction in production ponds. Unfortunately, late sexual maturity is negatively correlated with growth rate. Selection for late sexual maturity in Nile tilapia will decrease the growth rate.

Multiple Trait Selection and Selection Indices

More than one trait can be selected for simultaneously. To date that is not a common practice in aquaculture and fisheries. The best way to accomplish multiple trait selection is through selection indices. A selection index can also be used to more efficiently select for a single trait. This is done by developing an index for selection based on records of relatives or by developing an index based on highly correlated traits. Utilization of breeding values for many traits is called multivariate analysis, and if the index is for a single trait, univariate analysis.

Other multiple-trait selection techniques such as tandem selection and independent culling levels exist, but they are less efficient and do not utilize weighting, thus it is easier to lose valuable alleles if they do not reside in the most desirable individuals. Advantages of using a selection index are that it indentifies individuals with valuable heritable traits allowing us to increase selection power and can allow for multiple trait selection.

In a selection index, each trait is weighted. This weighting factor is a regression coefficient. To calculate the regression coefficient for a multiple-trait selection index, four items are needed.

1. The economic value of each trait. The economic component will reflect how much influence a trait has on profit margins; some genes are more important than others due to their economic value.
2. The heritability for each trait.
3. The genetic correlations among the traits.
4. The phenotypic correlation of the traits.

Each individual will have an index or breeding value:

$$I = b_1 X_1 + b_2 X_2 + ... + b_n X_n \qquad (9.13)$$

where b_1, b_2, ..., b_n is the regression coefficient for each trait, X_1, X_2, ..., X_n.

The breeding value or index value (I) is assigned for each individual in a population. The selection of the individual is based on the breeding value, a composite genetic value. The breeding value is valid only for the population of the individual, and is not valid across populations. The average members of the population have a value of 1.0.

Another definition of breeding value is the value of an individual compared with the mean value of its progeny, since genes are passed on with reproduction and not genotypes. The average effect of the parents in the progeny determines the mean value of the progeny. The breeding value is valid only in the population that is being studied, not across populations. A breeding value is equal to the sum of the average effects of the genes it carries, the summation being made over the pair of alleles at each locus and over all loci (Falconer and Mackay, 1996).

Selection indices are mathematically and statistically complex. Matrix algebra is used to obtain the regression coefficients and equations. The best program that was developed and currently used for this task is BLUP (best linear unbiased predictor), to develop these indices with least bias. Predecessors to BLUP were BLP (best linear predictor) and then BLUE (best linear unbiased estimator). The breeding value used in index selection is a composite of several characteristics based on economic value. BLUP is used to evaluate the fixed effects while simultaneously evaluating the breeding value for the overall economic value.

The more records we have the more accurate our assessment of an individual's value is. Although multiple trait selection is not commonly used in aquaculture and fisheries, this breeding programme is potentially advantageous in aquatic organisms since they have large population sizes and progeny groups.

Key Summary Points

A genetic gain of 10–15% per generation is possible and growth rate can be doubled in six generations through selection; however, 6–7% improvement per generation is an average. The turnover rate for the crop is accelerated, and maintenance requirements reduced. Survival increases because the fish are cultured a shorter time. Retention of energy and protein increases as growth rate increases. Feed resources are better utilized. Production costs are strongly reduced. Selection has a high probability of improving body weight: about 90% of selection programmes have been successful.

Selection for disease resistance and survival traits is also highly successful, with about 60–75% of the selection programmes evaluated being successful. Reproductive traits, fat percentage and body conformations usually have high heritabilities.

Traits can have negative genetic correlations and therefore negative correlated responses to selection can occur for secondary traits when selection is conducted for the primary trait. It is important to monitor all economically important traits when selection is practised.

10

Polyploidy and Xenogenesis

Polyploidy has been thoroughly studied in fish and shellfish. The polyploid state refers to individuals with extra sets of chromosomes. The normal and most common chromosome complement is two sets (diploid). Triploidy refers to individuals with three sets of chromosomes and tetraploidy refers to individuals with four sets. Hexaploids have six sets, and aneuploids have at least a diploid set with one or more additional chromosomes, but not a full complement, to the set. Polyploidy is lethal in mammals and birds (Chourrout *et al.*, 1986a), but has led to the development of many useful, improved plant varieties, including domestic wheat (Strickberger, 1985). Triploid fish are viable (Thorgaard *et al.*, 1981; Wolters *et al.*, 1981a,b; Chourrout, 1984; Cassani and Caton, 1986a) and are usually sterile, which is a result of lack of gonadal development (Allen and Stanley, 1981a,b; Wolters *et al.*, 1982b; Purdom, 1983; Casani and Caton, 1986a; Chourrout *et al.*, 1986a).

Culture of triploid fish can be advantageous for several reasons. The potential of increased growth (Chourrout *et al.*, 1986a), increased carcass yield, increased survival and increased flesh quality are the main culture advantages (Bye and Lincoln, 1986; Thorgaard, 1986; Hussain *et al.*, 1995; Dunham, 1996). At the onset of sexual maturity, reduced or inhibited gonadal development may allow energy normally used in reproductive processes to be used for growth of somatic tissue (Thorgaard and Gall, 1979; Lincoln, 1981; Wolters *et al.*, 1982b). The sterility of triploids would be desirable for species such as tilapia, where excess reproduction may occur in production ponds (Shelton and Jensen, 1979). Use of sterile triploids can prevent the permanent establishment of exotic species in otherwise restricted geographical locations (Shelton and Jensen, 1979). Induction of triploidy in interspecific hybrids can prevent the backcrossing of hybrids with parental species (Curtis *et al.*, 1987), and also allows viability in some unviable diploid hybrids (Allen and Stanley, 1981b). Other potential uses include supplemental stocking of natural populations without compromising the genetic integrity of the resident population, disruption of reproduction in nuisance species and sterilization of transgenic fish, all mechanisms for reducing environmental risk and applying genetic conservation. The performance of triploid fish is dependent on the species and age of the fish as well as the experimental conditions. GE interactions are common for triploid fish.

Triploid production has great potential to enhance performance in fish and shellfish; however, many problems exist. The first is that triploids can sometimes be fertile, defeating the advantages of sterility. Polyploidy can decrease performance for some traits. For many species, polyploid production may not be economically feasible.

Polyploid Induction in Fish

Triploidy is induced by allowing normal fertilization and then forcing retention of the second polar body (Chourrout, 1980, 1984; Lou and Purdom, 1984). The second polar body is retained by applying temperature (hot or cold), hydrostatic pressure, anaesthetics or chemical shocks shortly after fertilization (Thorgaard *et al.*, 1981; Wolters *et al.*, 1981a; Chourrout and Itskovich, 1983; Benfey and Sutterlin, 1984a; Chourrout, 1984; Cassani and Caton, 1986a; Curtis *et al.*, 1987; Johnstone *et al.*, 1989; Table 10.1). Ueda *et al.* (1988) were able to induce triploidy in rainbow trout, *Oncorhynchus mykiss*, by applying high pH and high calcium to either sperm or eggs. In some cases this caused fusion or adhesion of sperm, resulting in dispermy, and apparently could also suppress expulsion of the second polar body. Among failed procedures is the application of diethyl ether. Diethyl ether alone or in combination with hydrostatic pressure did not result in triploid induction in rainbow trout (Lou and Purdom, 1984). Nitrous oxide application induced 80% triploidy in Atlantic salmon, *Salmo salar* (Johnstone *et al.*, 1989). Freon was moderately effective, halothane and ethane induced less than 10% triploidy, and cyclopropane was ineffective for inducing triploidy in Atlantic salmon (Johnstone *et al.*, 1989). Hydrostatic pressure (Fig. 10.1) produces more consistent results, survival of treated eggs and per cent triploidy than temperature shocks and other treatments (Cassani and Caton, 1986a; Bury, 1989).

When hydrostatic pressure is applied to the loach, *Misgurnus fossilis*, the polar body starts to extrude but then merges again with the egg cytoplasm and the meiotic spindle is destroyed, resulting in two female

pronuclei (Betina *et al.*, 1985). The male pronucleus then fuses with the female pronuclei, resulting in the triploid zygote. The pronuclei are smaller and located closer to the surface of the egg than in the untreated diploid zygotes. At the beginning of first cleavage (anaphase, furrow formation), the blastodisc in the control eggs is thicker than in the triploid zygotes and the nuclear transformations are accelerated in the triploid eggs.

The success of treatments to induce polyploidy depends on the time of initiation of the shock, the magnitude of the shock and its duration. The best time for initiation of the shock varies widely among different species but is related to the rate of development and, specifically for triploidy, the timing of the second meiotic division and, for tetraploidy, the timing of the first mitotic division. Naturally, within a species, the timing of these cell-division events is based on temperature, so results can vary depending on temperature. However, by standardizing for temperature shifts, consistent results can be obtained. One method is to apply the shocks at a certain accumulated number of temperature degree-minutes (Palti *et al.*, 1997). Cherfas *et al.* (1990) applied the shocks based on τ_0, the percentage of time until a division event occurs, which is, of course, temperature dependent.

Often the hatch of embryos that have been induced for triploidy is lower than that of controls. In turbot, *Scophthalmus maximus* (Piferrer *et al.*, 2000), and sea bass, *Dicentrarchus labrax*, the lower hatch is due to handling and treatment of the eggs and not due to the state of triploidy.

The quality of the gametes can affect the efficiency of polyploidy and gynogenesis in rainbow trout (Palti *et al.*, 1997). When viability of bighead carp (*Hypophthalmichthys nobilis*) eggs was above 59%, high rates of triploidy were produced, but when the viability was less than 40%, no triploids were produced (Aldridge *et al.*, 1990). Delaying fertilization of ovulated grass carp (*Ctenopharyngodon idella*) eggs did not affect the rate of triploid induction (Cassani and Caton, 1986a).

Table 10.1. Techniques for production of gynogen, androgen, triploid and tetraploid fish and shellfish.

Species	Technique	Shock intensity Temperature	Initiation time after fertilization	Duration	% Polyploidy	Reference(s)
FISH						
Triploidy						
Atlantic salmon (*Salmo salar*)	heat	28°C	50 min	10 min		Refstie (1984)
	heat	30°C	50 min	5 min		Refstie (1984)
	heat	32°C	20 min	5 min	95	Sutterlin *et al.* (1987)
Coho salmon (*Oncorhyncus kisutch*)	heat	28–30°C	10 min	10 min	58–84	Utter *et al.* (1983)
Chinook salmon (*Oncorhyncus tshawytscha*)	heat	28–30°C	10 min	10 min	58–84	Utter *et al.* (1983)
Pink salmon (*Oncorhyncus gorbuscha*)	heat	28–30°C	10 min	10 min	58–84	Utter *et al.* (1983)
Rainbow trout (*Oncorhyncus mykiss*)	heat	26°C	1 min	10 min	100	Solar *et al.* (1984)
	hydrostatic		40 min	10 min	80–90	Lou and Purdom (1984)
	hydrostatic	7000 psi	40 min	4 min	100	Chourrout (1984)
gynogen	heat	26°C	95–508°C·min	20 min		Palti *et al.* (1997)
gynogen	hydrostatic	9000 psi	273–693°C·min	3 min		Palti *et al.* (1997)
	nitrous oxide/ pressure	11 atm	0 min	30 min	90	Shelton *et al.* (1986)
Brook trout (*Salvelinus fontinalis*)	heat	28°C	10 min	10 min	98–100	Galbreath and Samples (2000)
Yellow tail flounder (*Pleuronectes ferrugineus*)	hydrostatic	7000 psi	5 min	10 min	93–100	Manning and Crim (2000)
Turbot (*Scophthalmus maximus*)	cold	0°C	5 min	20 min	90	Piferrer *et al.* (2000)
Gilthead seabream (*Sparus aurata*)	heat	37°C	3 min	2.5 min	100	Gorshkov *et al.* (1998)
Common carp (*Cyprinus carpio*)	cold				91.7	Ueno (1984)
	cold	0–2°C	$0.3–0.7\tau_0$	pb II, 45–60 min		Cherfas *et al.* (1990)
	heat	40°C	4 min	1–1.5 min	80–100	Hollebecq *et al.* (1988)
Grass carp (*Ctenopharyngodon idella*)	hydrostatic	5.6×10^3 kPa	4 min		95	McCarter (1988)
	hydrostatic	7000–8000 psi	4 min	1–2 min	98	Cassani and Caton (1986a)
	heat	42°C	4 min	1 min	67 (0–100)	Cassani and Caton (1986a)

(Continued)

Table 10.1. *Continued*

Species	Technique	Shock intensity Temperature	Initiation time after fertilization	Duration	% Polyploidy	Reference(s)
Bighead carp (*Hypophthalmichthys nobilis*)	hydrostatic	500 atm	4 min	1.5 min	80–100	Aldridge *et al.* (1990)
Rohu (*Labeo rohita*)	heat	42°C	7 min	1–2 min	12	Reddy *et al.* (1990)
Silver barb (*Puntius gonionotus*)	cold	2°C	0.5 min	10 min	72.5	Koedprang and Na-Nakorn (2000)
Blue tilapia (*Oreochromis aureus*)	heat	39.5°C	3 min	3.5–4 min	100	Don and Avtalion (1986)
African catfish (*Clarias gariepinus*)	cold	5°C	3 min	40 min	80–100	Henken *et al.* (1987), Richter *et al.* (1987)
Thai walking catfish (*Clarias macrocephalus*)	cold	5°C	2 min	30–60 min		Vejaratpimol and Pewnim (1990)
	cold	7°C	0 min	25 min	80	Na-Nakorn and Legrand (1992)
Loach (*Misgurnus anguillicaudatus*)	cold	1°C	5 min	30–40 min	100	Chao *et al.* (1986)
Loach (*Misgurnus fossilis*)	hydrostatic	400–500 kg/cm^2	6 min		60–86	Betina *et al.* (1985)
Ayu (*Plecoglossus altivelis*)	cold	0–0.5°C	6 min	30–60 min	100	Taniguchi *et al.* (1987)
Zebra fish (*Danio rerio*)	hydrostatic	6140–6240 psi				Gestl *et al.* (1997)
	heat	41°C	5 min	4 min	100	Kavumpurath and Pandian (1990)
Tetraploidy						
Rainbow trout	hydrostatic	7000 psi	5 h 50 min	4 min	100	Chourrout (1984)
gynogen	heat	29°C/31.5°C	981–2616°C·min	10/5 min		Palti *et al.* (1997)
gynogen	hydrostatic	9000 psi	2730–3696°C·min			Palti *et al.* (1997)
Bighead carp	hydrostatic	500 atm	36 min	1.5 min		Aldridge *et al.* (1990)
Rohu	heat	39°C		2 min	70	Reddy *et al.* (1990)
	cold	10–15°C		10 min	30–55	Reddy *et al.* (1990)
SHELLFISH						
Triploidy						
Atlantic Bay scallop L-super (*Argopecten irradians*)	CB	0.1 mg/l	10 min	10 min	94	Tabarini (1984a,b)

Species						Reference
Scallop (*Chlamys nobilis*)	CB	0.5 mg/l			71.4	Komaru and Wada (1989)
Geoduck clam (*Panope abrupta*)	6-DMAP	600 mol/l			>95	Vadopalas and Davis (1998)
Manila clam (*Ruditapes philippinarum*)	CB	1 mg/l	20–35 min		75.8	Dufy and Diter (1990)
Pacific oyster (*Crassostrea gigas*)	6-DMAP	450 µmol/l	30% of zygotes show pb I	10 min	94	Tian *et al.* (1999a,b)
	CB	1 mg/l	30 min	15 min	72	Downing *et al.* (1985)
American oyster (*Crassostrea virginica*)	CB	0.5 mg/l (25°C)	25 min	15 min		Shatkin and Allen (1989)
European flat oyster (*Ostrea edulis*)	CB	1 mg/l	30–35 min or 90–100 min	20 min	69	Gendreau and Grizel (1990)
Pearl oyster (*Pinctada albina*)	heat/caffeine	31.5/13 mmol/l			53.8	Durand *et al.* (1990)
Japanese pearl oyster (*Pinctada fucata*)	CB	0.5 mg/l	20–50 min		100	Wada *et al.* (1989)
Mussel (*Mylitus edulis*)	heat	32°C	20 min	10 min	97.4	Yamamoto and Sugawara (1988)
Shrimp (*Penaeus chinensis*)	heat	28–32°C	8–20 min	8–16 min	39–75	Fu-hua *et al.* (1999)
Pacific abalone (*Haliotis discus hannai*)	6-DMAP	100–150 µmol/l	pb II	15 min	46–54	Zhang, G. *et al.* (1998)
Tetraploidy						
Chinese mitten-handed crab (*Eriocheir sinesis*)	CB	0.5–2.0 mg/l	10–20 min	10–20 min	14–58	Chen *et al.* (1997a,b,c)
Scallop (*Chlamys farreri*)	CB	0.5 µg/ml	10 min and 1 h (pb I and 1st mitosis)	10 min and 10 min	40 and 21	Yang *et al.* (1997)
Pacific oyster	CB		15–30 min after pb I		91	Stephens and Downing (1988)
Chinese mitten-handed crab	CB	1.5 mg/l	9.2–9.5 h	18 min	57–58	Chen *et al.* (1997a,b,c)
European flat oyster	CB	1 mg/l	5–25 min or 260–2800 min	20 min	40–53	Gendreau and Grizel (1990)
Pacific abalone	CB	0.8 mg/l	8 min	20 min	36	Zhang *et al.* (2000b)

CB, cytochalasin B; 6-DMAP, 6-dimethylaminopurine; pb I, polar body I; pb II, polar body II.

Fig. 10.1. Hydrostatic pressure chamber for inducing polyploidy. It is important to utilize a design that allows the bleeding off of all air to prevent the chamber from exploding and becoming a projectile. (Photograph by Rex Dunham.)

Strain or family effects may have a bearing on polyploid induction efficiency. The strain of rainbow trout may react differently for different temperature shocks to induce triploidy (Anders, 1990). This is not surprising because one might expect genetic differences in rates of embryonic development that might affect optimal parameters for ploidy manipulation. Additionally, Blanc *et al.* (1987) found sire effects on performance and ploidy induction for triploid and tetraploid rainbow trout. A heat shock of 30°C was required to produce 100% triploidy in a domestic strain of rainbow trout, whereas 28°C was required for the same result in a wild strain (Solar and Donaldson, 1985).

Triploidy can occur naturally in untreated individuals, and this has been documented in pink salmon (Utter *et al.*, 1983). There are species that are naturally triploid or tetraploid rather than diploid (Dunham *et al.*, 1980). Polyploidy is one mechanism of speciation. Diploid and tetraploid races can exist within the same species (Saitoh *et al.*, 1984), and different species of the same genera can be either diploid or tetraploid (Dunham *et al.*, 1980). Although diploid, salmonids have tetraploid ancestry and sometimes still show vestiges of tetrasomic inheritance. Some species of fish reproduce and exist as gynogens and triploids (Monaco *et al.*, 1984).

Polyploid Induction in Shellfish

Polyploid induction in shellfish is different from that in fish as both polar bodies I and II can be present after fertilization (Tian *et al.*, 1999a,b). Therefore, tetraploids and aneuploids can be produced, as well as triploids, during triploid induction for Pacific oysters, *Crassostrea gigas* (Tian *et al.*, 1999a,b,c). Temperature and salinity can affect the rate of induction of polyploidy in shellfish and invertebrates, specifically the geoduck clam, *Panope abrupta* (Vadopalas and Davis, 1998). The timing of the meiotic divisions and extrusion of the first and second polar bodies are different for fish and shellfish. The timing of application of cytochalasin B (CB) affects ploidy induction and segregation of the chromosomes in Pacific oysters (Stephens and Downing, 1988; Guo *et al.*, 1989). Fertilized eggs are 2N, 3N or 4N if the treatment affects polar body I, 3N if the treatment affects polar body II, and 5N if the treatment affects both polar bodies I and II.

However, Stephens and Downing (1988) produced at least some of the three ploidy levels, 2N, 3N and 4N, when attempting to block either polar body I or II. If polar body I is blocked, 60% of the polyploids are aneuploid, but if polar body II is blocked, the number of aneuploids is no different from that in diploid controls. An unusual configuration of the maternal chromosomes occurred following the blockage of polar body I, which possibly leads to the high proportion of aneuploids and also abnormal tetraploids. The 20 maternal dyads formed three division planes oriented in a tripolar configuration (Guo *et al.*, 1989). Temperature affects the rate of triploid induction in Pacific oysters, and Downing (1988a) recommended that the optimum temperature for producing triploids (88%) was 25°C for 30–45 min after insemination when using CB. However, the concept of τ_0 could probably be applied to shellfish, and the application of the shocks or treatments applied based on developmental time (temperature minutes) rather than absolute time associated with a single temperature. CB also adversely affects survival for the first 48 h after treatment.

Sperm quality affects rate of triploid induction in Pacific and American oysters and their hybrids (Downing, 1989b), with poor sperm quality leading to very low triploid induction rates. The application of CB to induce the polyploidy results in the loss of some embryos, and usually the survival of monospecific embryos is higher than that of their interspecific diploids and triploids. Triploid induction for eastern oysters, *Crassostrea virginica*, was 100% at 20 ppt, but was reduced to 84% at 30 ppt, whereas triploid induction for Pacific oysters was 100% at both salinities.

Wada *et al.* (1989) compared CB, heat shock and cold shock for producing triploid Japanese pearl oysters, *Pinctada fucata martensii*. Cold shock produced triploids, but CB was the most effective treatment. CB was also more effective in producing triploidy than hydrostatic pressure in scallops, *Chlamys nobilis* (Komaru and Wada, 1989).

Pressure shocks, thermal shocks, CB and 6-dimethylaminopurine (6-DMAP) have been used to induce triploidy in gastropods (Zhang, G. *et al.*, 1998). Treatments of 300 µM of 6-DMAP for 15 min for blocking polar body II yielded the highest percentage of triploids, but none of the veligers survived. Treatments of 100–150 µM for 15 min to block polar body II gave 46–54% triploids and allowed 90–95% survival of the trocophores. Both heat and cold shocks can be effective for producing triploid mussels, *Mytilus edulis* (Yamamoto and Sugawara, 1988).

CB was slightly more effective than 6-DMAP for producing tetraploidy in Pacific abalone, *Haliotis discus hannai* (Zhang *et al.*, 2000b). However, Zhang *et al.* (2000a) concluded that 6-DMAP was more efficient than CB for producing triploid Pacific oysters.

Tetraploid Pacific oysters were produced by using polyethylene glycol (PEG) treatment to fuse blastomeres (Guo *et al.*, 1988). This resulted in 1–4% tetraploids. Viable tetraploid *C. gigas* have been produced by inhibiting the first polar body of eggs from triploids that had been fertilized with sperm from diploids (Eudeline and Allen, 2000), but repeatability for producing high yields of tetraploids is inconsistent. Varying the duration of the treatment to inhibit polar body I of triploid eggs had definite effects for optimizing tetraploid production. Short treatments 15–35 min after fertilization (approximately half the period of meiosis I in triploid eggs) yielded individuals with tetraploid and heptaploid cells. Longer treatments 7–43 min after fertilization (about three-quarters of the period of meiosis I in triploid eggs) yielded embryos with 100% heptaploid cells. Tetraploid induction was most consistent when treatments were applied to eggs from individual triploid females rather than from multiple pooled females (Eudeline and Allen, 2000). Eggs from individual triploids were fertilized and CB (0.5 mg/l) was added after 10 min. A subsample of the fertilized eggs was untreated. When 50% of the untreated eggs exhibited polar body I extrusion, the CB treatment was terminated. Percentage tetraploidy ranged from 13 to 92% after 8 days, with a mean of 55%. Seven of eight replicates went through metamorphosis and settlement, and at settlement the percentage of tetraploids ranged from 7 to 96%, with a mean of 45%, allowing the establishment of tetraploid brood stock for the natural production of triploids.

Application of this technique – the crossing of triploid females with diploid males to produce tetraploids – was also attempted in the pearl oyster, *Pinctada martensii* (He *et al.*, 2000a,b) and eastern oyster (Supan *et al.*, 2000b). Ploidy in the embryos ranged from 2N to 5N. At 1 year, the majority of the survivors were diploid, triploid and aneuploid but 2% viable tetraploids remained. The number of ripe female triploid eastern oysters was extremely low, making production of tetraploid individuals from CB-treated 2N × 3N crosses extremely difficult, but it has been accomplished.

Triploid Cells

Triploids have an extra set of chromosomes and therefore a larger nucleus (Fankhauser, 1945). The cytoplasm/nucleus ratio is constant; thus the larger polyploid nucleus causes an increase in volume of cytoplasm and total cell size (Fankhauser, 1945; Swarup, 1959b). Theoretically, triploids would reach a larger size than diploids because of their larger cell size. However, the stickleback and the ayu (Aliah *et al.*, 1990) adjust to this increased size by reducing cell number, and triploid sticklebacks are the same size as diploids. Conversely, triploid European sea bass had a similar number of cells to diploids (Felip *et al.*, 2001a,b,c). The cell size increases in all tissues for polyploids. There was a uniform increase in cell size for erythrocytes, leucocytes, brain cells and retinal cells in triploid coho salmon and Atlantic salmon (Small and Benfey, 1987).

The increase in DNA content per cell could slow the growth of triploids by decreasing the mitotic rate (Fankhauser, 1945) or the metabolic rate of the cell (Szarski, 1970). Gene dosage compensation could also reduce the growth rate of triploids (Myers, 1985). Polyploid cells reduce overall protein production proportional to cell size such that the overall concentration of proteins remains the same as in non-fish species (Lucchesi and Rawls, 1973).

Ploidy Determination

Karyotyping, the actual visualization and enumeration of chromosomes, is, of course, the most accurate method for determining ploidy level. This technique is tedious and slow and does not allow evaluation of samples *en masse*. Additionally, karyotyping may not always detect mosaic ploidy types because of the difficulty in sampling large numbers of cells. Several other techniques can be utilized, including flow cytometry, cell-size measurement with a Coulter Counter Channelyzer or blood smears, silver staining of nucleolar organizing regions (NORs) and, in interspecific hybrids or intraspecific crossbreeds with fixed differences, isozyme analysis.

Flow cytometry allows rapid analysis, but the equipment is extremely expensive. The Coulter Counter Channelyzer also allows rapid analysis and the machinery is relatively expensive, although much less expensive than for flow cytometry. Flow cytometry unequivocally identified diploid and triploid coho and chinook salmon, whereas 11% of the samples analysed with a Coulter Counter Channelyzer were inconclusive (Johnson *et al.*, 1984). However, Benfey *et al.* (1984) found that the Coulter Counter Channelyzer was comparable to flow cytometry for speed and accuracy. They also determined that measuring erythrocyte dimensions from blood smears was a valid technique to determine ploidy level. The Coulter Counter Channelyzer is a highly effective technique for measuring ploidy *en masse* for grass carp, and a three-person team is capable of evaluating up to 2400 fish in 8 h (Wattendorf, 1986).

One procedure for flow cytometry utilizes fixation of the erythrocytes with formalin. Burns *et al.* (1986) found that this formalin fixation lowers the amount of fluorescence, which could lead to problems in data interpretation.

Analysis of NORs is another technique for ascertaining ploidy level because, in many species, as the ploidy increases, the number of NORs increases (Phillips *et al.*, 1986). This is a highly accurate technique for salmonids, rainbow trout, chinook salmon and coho salmon, which have one chromosome with a NOR per haploid genome, as do most fish species. Any tissue can be used since dividing cells, rather than chromosomes, are needed for the analysis, and samples can be taken from fish as small as 7–8 mm in length without sacrificing the animal. Analysis of NORs is the least expensive technique and is more rapid and technically easier than karyotyping, but again does not allow measurement of a massive number of samples. This technique tends to be much more accurate in younger fish that have a high rate of cell division, as the NORs are usually only found during mitosis. The technique can be much less accurate in older fish, which have a much slower rate of cell division.

Blood-smear techniques for identification of triploid walleye (*Stizostedion vitreum*) × sauger (*Stizostedion canadense*) hybrids were 94% accurate, but it took 1 h per fish to process the samples (Garcia-Abiado *et al.*, 1999). However, Kucharczyk *et al.* (1997a,b, 1999) were able to distinguish 100% of haploid, diploid and triploid bream, *Abramis brama*, and northern pike, *Esox lucius*, without sacrificing the fish utilizing NORs, but needed to examine 40 cells per fish to obtain that accuracy. When this technique is applied to turbot, *Scophthalmus maximus*, it is 97% accurate (Piferrer *et al.*, 2000).

Mahmoud Rezk (unpublished results) developed another technique for ploidy determination, resulting in karyotypes and NORs being visualized on the same slide (Fig. 10.2). This is also a relatively slow procedure, but increases accuracy and the information generated.

When the correct marker alleles are available for dimeric isozymes, diploids display a 1:2:1 banding pattern and triploids, as demonstrated in brown trout (Crozier and Moffett, 1989a), have a 4:4:1 staining ratio. In the case of brown trout × brook trout hybrids, triploids could be distinguished from diploids by examining the relative contribution of the maternal alleles, which should be double that of the paternal alleles for most, but not all, isozyme loci (Scheerer and Thorgaard, 1987). Sugama *et al.* (1988) were also able to utilize isozyme markers to evaluate the success of triploid induction in red seabream, *Pagrus major*, black seabream, *Acanthopagrus achlegeli*, and their hybrid. The triploid induction rate was between 87 and 97%.

Triploid Fish Performance

Growth

Triploid fish may grow faster (Valenti, 1975; Purdom, 1976; Thorgaard and Gall, 1979; Wolters *et al.*, 1981a; Krasznai and Marian, 1986), at the same rate (Nile tilapia: Don and Avtalion, 1986; Richter *et al.*, 1986; Dunham, 1990a; Hussain *et al.*, 1995) or slower (channel catfish and rainbow trout: Refstie, 1981;

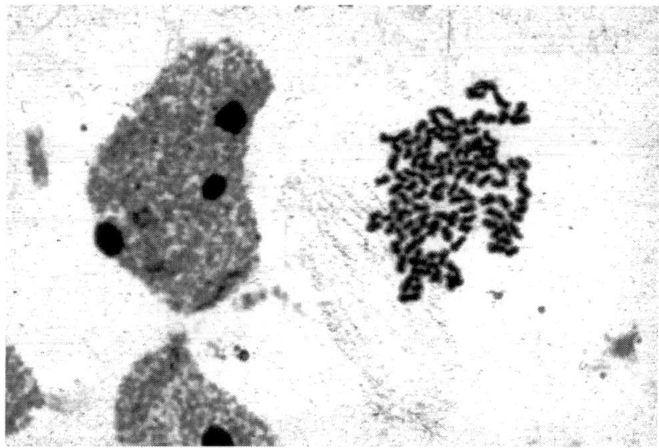

Fig. 10.2. Karyotype and nucleolar organizer regions of striped bass, *Morone saxatilis*, stained on a single slide. (Photograph by Mahmoud Rezk.)

Wright *et al.*, 1982; Chourrout *et al.*, 1986a; Krasznai and Marian, 1986; Shah and Beardmore, 1986; Wolters *et al.*, 1991) than diploid fish. Triploids rarely grow faster than diploids during early stages of culture prior to maturation effects (Swarup, 1959a,b; Purdom, 1973, 1983; Allen and Stanley, 1981b; Wolters *et al.*, 1982b; Chourrout, 1984; Chourrout *et al.*, 1986a; Don and Avtalion, 1986; Johnson *et al.*, 1986; Taniguchi *et al.*, 1986; Penman *et al.*, 1987; Richter *et al.*, 1987). Generally, diploids grow faster than triploids until the onset of sexual maturity and then the triploid grows faster and converts feed more efficiently (Purdom, 1976; Thorgaard and Gall, 1979; Wolters *et al.*, 1981a; Dunham, 1990b, 1996).

In some cases, the growth of triploids is inferior to that of diploids (Gervai *et al.*, 1980; Wright *et al.*, 1982; Utter *et al.*, 1983; Chourrout *et al.*, 1986a). Triploid walking catfish, *Clarias macrocephalus*, grew more slowly than diploids (Na-Nakorn and Legrand, 1992). Triploid white bass (*Morone chrysops*) female × striped bass (*Morone saxatilis*) male hybrids grew more slowly than diploid hybrids – 868 g versus 1153 g through 15 months of age when evaluated in earthen ponds (Kerby *et al.*, 2002). Diploid male and female common carp grew faster than corresponding male and female triploid common carp up to 100–400 g; however, gutted weights were equivalent for the two ploidy levels (Basavaraju *et al.*, 2002). First-year growth of triploid all-female rainbow trout (Solar *et al.*, 1984) and especially mixed-sex rainbow trout (Tabata *et al.*, 1999) and loach, *Misgurnus anguillicaudatus* (Suzuki *et al.*, 1985), was slower than that of diploids. After sexual maturation, triploid mixed-sex and especially female triploid rainbow trout grew faster than males (Tabata *et al.*, 1999). Triploid European sea bass, *D. labrax*, grew more slowly than diploids, but after sexual maturity they grew faster than diploids, resulting in adult fish of similar size (Felip *et al.*, 2001b).

One explanation for the examples of slower growth in triploids is that the larger cell size of triploids may be compensated for by a decrease in total cell number (Muntzung, 1936). This probably results from a reduced mitotic rate in triploids (Astaurov, 1940). Another potential cause for slowed growth in juvenile triploids may be the effects of the triploid induction treatment (Myers, 1986). Kafiani *et al.* (1969) and Newport and Kirschner (1982) have shown that initiation of the mid-blastula transition is dependent on the nucleus/cytoplasm ratio. The triploid zygote may initiate this process earlier than normal. The period between activation of the genome and gastrulation, however, remains unchanged and organogenesis will probably begin with a reduced number of precursor cells (Chulitskava, 1970; Kobayakawa and Kubota, 1981). This may cause initial aberrant embryonic development, which may ultimately affect growth in juvenile fish.

No difference in growth is another potential outcome. There was no difference in the growth of triploid and diploid land-locked Atlantic salmon, although triploids were longer and had a lower condition factor than diploids (Benfey and Sutterlin, 1984b,c). However, Quillet and Gaignon (1990) observed slower growth by triploid Atlantic salmon through 2.5 years of age. Growth and survival of triploid Atlantic salmon were the same as those of diploids in both communal and separate evaluation. Coho salmon triploids had growth similar to that of diploids. The first-year growth of triploid and diploid silver barb, *Puntius gonionotus* (Koedprang and Na-Nakorn, 2000), pre- and post-smolt coho salmon in seawater (Withler *et al.*, 1998), rainbow trout (Solar and Donaldson, 1985; Kim *et al.*, 1986; Oliva-Teles and Kaushik, 1990a), zebra fish, *Brachydanio rerio* (Kavumpurath and Pandian, 1990), African catfish, *Clarias gariepinus* (Henken *et al.*, 1987; Richter *et al.*, 1987), and common carp in Israel (Cherfas *et al.*, 1994a) was the same. Although triploids generally do not grow faster than diploids, triploid rainbow trout regenerated fins more often and more rapidly than diploids (Alonso *et al.*, 2000).

In some cases, triploids do have enhanced growth compared with their normal diploid siblings (Valenti, 1975; Wolters *et al.*, 1982b; Taniguchi *et al.*, 1986). This increased growth rate can be a result of lack of sexual development, since the growth rate of fish slows as they approach sexual maturity, or theoretically due to increased cell size. Channel catfish triploids become larger than diploids at 8–9 months of age (90 g) when grown in tanks (Wolters *et al.*, 1982b; Chrisman *et al.*, 1983). This is slightly after the time when sexual dimorphism in body weight is first detected in channel catfish. Triploid coho salmon grew 14.5% faster than diploids to about 20 g (Utter *et al.*, 1983); however, Johnson *et al.* (1986) found no differences in growth and condition factor from 18 to 30 months between diploid and triploid coho salmon after smoltification when the fish were grown in either fresh or salt water. Triploid Chinese catfish, *Clarias fuscus* (Qin *et al.*, 1998), and European catfish, *Silurus glanis* (Krasznai and Marian, 1986), grew faster than diploids during the first year of growth.

When the onset of sexual maturity and gonad development slows the growth of diploids, triploids can surpass the diploids in size (Purdom, 1976; Thorgaard and Gall, 1979; Chourrout *et al.*, 1986a). Upon reaching 2 years of age, the growth of diploid rainbow trout slows and triploid rainbow trout become larger than diploids (Bye and Lincoln, 1986). After the onset of sexual maturity, triploid channel catfish (Wolters *et al.*, 1982b), rainbow trout (Thorgaard and Gall, 1979) and plaice × flounder hybrids, *Pleuronectes platessa* crossed with *Platichthys flesus* (Purdom, 1976), not only exhibit faster growth but also better feed-conversion efficiency than diploids. During the second year of life, as sexual maturation approached, triploid common carp grew faster than diploids (Bieniarz *et al.*, 1997).

The effect of triploidy on the growth of interspecific hybrids can vary from that of parental species. Triploid channel catfish × blue catfish hybrids grew at the same rate as diploids in ponds and in tanks, and channel white catfish, *Ameirus catus*, triploid hybrids at the same rate as diploids in ponds (Lilyestrom *et al.*, 1999). However, diploid channel × blue hybrids had higher condition factors when grown in tanks.

Triploid Atlantic salmon female × brown trout (*Salmo trutta*) females grew at the same rate as Atlantic salmon in saltwater net pens (Galbreath and Thorgaard, 1997). Gilthead seabream (*Sparus aurata*) eggs × *P. major* triploid hybrids (Gorshkova *et al.*, 1995) had no growth superiority compared with either parent up to age 2 and 3 years.

The first-year growth of triploid hybrids between female rainbow trout and male brook trout (*Salvelinus fontinalis*), Arctic charr (*Salvelinus alpinus*) or lake charr (*Salvelinus namaycush*) was inferior to that of diploid and triploid rainbow trout made with the same dams in a farming environment (Blanc *et al.*, 2000). After 3 years, the growth differences remained but were small. Sexual maturation resulted in males of all genotypes being smaller than females.

H. nobilis triploid hybrids grew faster and had fewer abnormalities compared with diploid hybrids (Beck *et al.*, 1984; Cassani *et al.*, 1984). Triploid salmonid hybrids exhibited growth similar to (Quillet *et al.*, 1987) or slower than (Parsons *et al.*, 1986) that of diploid hybrids. As is often the case with intraspecific triploids, interspecific salmonid triploids grew faster than controls once the maturation period was reached (Quillet *et al.*, 1987).

Family and strain effects for the performance of triploids exist. Diploid Arctic charr, *S. alpinus*, grew faster than triploids (Chiasson *et al.* 2009). However both ploidy level and family affected growth, and family predicted the performance of triploids. Triploids from fast-growing families grew more rapidly than diploids from slow-growing families. Growth performance of sibling triploid and diploid oysters was correlated, but not their ability to reproduce (Normand *et al.*, 2009).

Within and between families, variance in growth was higher for triploid Atlantic salmon in seawater compared with diploids (Friars *et al.*, 2001). Withler *et al.* (1998) found an interaction between family and ploidy level for growth of coho salmon. Blanc and Vallee (1999) found a strong correlation between family performance for diploid and triploid brown trout and rainbow trout. However, hybridization seems to break down this relationship. There was no correlation between the performance of triploid hybrids of female rainbow trout and brown or brook trout males and the growth of maternal and paternal species family controls (Blanc and Vallee, 1999). Blanc and Poisson (1988) found that diploid rainbow trout × Arctic charr were not viable, but their triploid was. Individual breeders affected survival at different developmental stages as well as the 90-day weight of the alevins. Paternal effects were important before the eyed stage, while maternal effects predominated at hatching time and during the early life of the alevins (Blanc and Poisson, 1988).

Guo *et al.* (1989) found that the growth of triploid rainbow trout was affected by maternal strain. Triploid Nile tilapia grew 66–90% faster than diploids and had decreased sexual dimorphism for body weight; yet other studies indicate no growth advantage of triploid Nile tilapia. Strain effects or GE interactions are possible explanations for these discrepancies. Use of specific strains, crosses or families may improve growth or other traits of triploid fish.

Triploidy could affect growth of different tissues disproportionately. However, the sizes of the retina, optic tectum and trunk kidney of triploid and diploid ayu, *Plecoglossus altivelis*, were the same (Aliah *et al.*, 1990).

Some growth differences between triploids and diploids might be explained by differences in muscle development. Johnston *et al.* (1999) examined muscle growth and development in mixed-sex and all-female diploid and triploid Atlantic salmon. Rate of somite formation at 6°C was the same for diploids and triploids. The rostral-to-caudal development of myotubes and acetylcholinesterase staining at the myosepta were slightly more advanced in triploids than in diploids. However, family differences were larger than the differences between triploids and diploids. Satellite cells are used in post-embryonic growth, and they were more abundant in diploid than in triploid smolts. Cells expressing the myogenic regulatory factor myf-6, which indicates that the satellite cells are committed to differentiation, were about 20% more abundant in diploids. The

rate of muscle-fibre recruitment in seawater stages was double for all-female diploid Atlantic salmon compared with all-female triploid Atlantic salmon.

Metabolic utilization of endogenous reserves, ammonia excretion, oxygen consumption and protein, energy and lipid utilization were no different for diploid and triploid rainbow trout embryos (Oliva-Teles and Kaushik, 1987). Apparent digestibility coefficients of protein and energy, oxygen uptake, ammonia excretion and digestive and metabolic utilization of diets were no different in juvenile diploid and triploid rainbow trout (Oliva-Teles and Kaushik, 1990a). Ammonia and urea excretion and oxygen consumption were no different from fertilization to complete absorption of the yolk-sac for diploid and triploid rainbow trout (Oliva-Teles and Kaushik, 1990b). Similarly, diploid and triploid grass carp sequestered and excreted similar amounts of phosphorus when fed sago pond weed, *Potamogeton pectinatus* (Chapman *et al.*, 1987).

It has been suggested that the mechanical induction of triploidy could have enduring adverse effects on the survival and growth of triploids. However, Blanc *et al.* (1987) found that triploids produced from shocks had the same growth and performance as triploids naturally produced by mating diploid rainbow trout females with tetraploid males. In contrast, triploid rainbow trout produced by crossing female diploids with tetraploid males grew at the same rate as diploids and about 5% better than triploids produced by cold shock (Chourrout *et al.*, 1986a). This slight difference in growth could be a result of increased heterozygosity in the naturally produced triploids compared with the mechanically produced triploids, rather than any effects from the induction procedure.

Feed conversion and consumption

The differences or lack of differences in growth between triploids and diploids could be related to feed-conversion efficiency and feed consumption. Triploid channel catfish convert feed more efficiently than diploids in the tank environment (Wolters *et al.*, 1982b; Chrisman *et al.*, 1983), and in this example the triploids also grew faster.

However, triploid channel catfish × blue catfish hybrids (Lilyestrom *et al.*, 1999) and African catfish, *C. gariepinus* (Henken *et al.*, 1987), had the same feed conversion as diploids in tanks. Age class 0 and 1 diploid and triploid rainbow trout had the same feed conversion (Oliva-Teles and Kaushik, 1990a). The feed conversion of diploid Thai walking catfish was better than that of triploids in the first month, but diploid and triploid fish had similar means in the second and fourth months of growth (Na-Nakorn and Legrand, 1992). In all of these examples, the triploids did not have better feed-conversion efficiency than diploids and the growth of the triploids was also no better than that of the diploids.

To prevent the establishment of an exotic species in the USA, triploid rather than diploid grass carp are utilized for weed control. Unfortunately, triploid grass carp are less efficient plant consumers than diploids (Thomas *et al.*, 1986), but despite this, their application is still an effective management tool.

Morphology, meristics and identification

Traits related to external appearance, including pigmentation, which is darker in triploid channel catfish than in diploid channel catfish, differ between diploids and triploids. Triploid grass carp × bighead carp had fewer scales on the lateral line, a traverse series below the lateral line and a relatively longer gut compared with diploids. The morphology (27 measurements) and meristics (six scale counts and five fin formulae) of diploid and triploid grass carp are different, but not to the extent that they can be separated on this basis with 100% accuracy (Bonar *et al.*, 1988). Analysis of covariance and discriminant analysis indicated a classification accuracy of 65–85%.

Family differences may affect these differences in morphology and meristics in grass carp. When examining batches of triploid grass carp with a Coulter Counter Channelyzer at the initiation of an evaluation, technicians could not distinguish diploids and triploids externally, but after a large number of individuals were examined, some characteristic would be observed that distinguished the two genotypes (R. Wattendrof, Florida Fish and Game Commission, 1986, personal communication). However, when a subsequent batch was examined, the distinguishing factor would not be present. After a substantial number of individuals had been examined in the second batch, a unique distinguishing characteristic for differentiating the ploidy levels externally would become apparent specifically for the second group.

Triploid channel white catfish were more similar morphologically to their maternal parent than their paternal parent (Lilyestrom *et al.*, 1999), which is not expected since interspecific hybrids of ictalurid catfish exhibit paternal predominance (Dunham *et al.*, 1982a). Perhaps the double dose of the maternal chromosome complement superseded or overwhelmed the genetic mechanisms for paternal predominance. However, triploid channel × blue catfish did have appearances more similar to the male than the female parent (Lilyestrom *et al.*, 1999), consistent with the paternal predominance observed in diploid channel × blue catfish hybrids (Dunham *et al.*, 1982a).

Tiwary *et al.* (2000) examined several body proportions in triploid Indian catfish, *Heteropneustes fossilis*, to distinguish the two genotypes, triploid and diploid. Several significant differences were found for various morphometric ratios between the two ploidy levels; however, only the ratio between standard length and body depth was a precise indicator to separate triploids from diploids.

Carcass traits and flesh quality

A potential benefit of polyploidy is positive changes in carcass traits (Dunham, 1990a). Reduced gonadal development leads to less waste in processing. Chrisman *et al.* (1983) reported an average of 6% higher dress-out percentage in tank-raised triploid channel catfish at 3 years of age compared with diploids. The increased carcass yield results from the lack of gonadal development in triploids, but these fish were well past the size and age when catfish are usually marketed. If dress-out percentage had been evaluated at younger ages and smaller sizes, the result might have been different. Hybrid channel × blue catfish triploids had no dress-out advantage when grown in tanks (Lilyestrom *et al.*, 1999). However, triploid channel catfish × blue catfish and channel × white catfish hybrids consistently had a slightly lower dress-out percentage than diploids when grown in ponds. Head weight and visceral percentages did not differ between triploid and diploid channel × blue and channel × white hybrid catfish. Triploid Atlantic salmon × brown trout hybrids had the same dress-out percentage as Atlantic salmon in saltwater net pens (Galbreath and Thorgaard, 1997).

Female triploid hybrids between female rainbow trout hybridized with male brook trout (*S. fontinalis*), Arctic charr (*S. alpinus*) or lake charr (*S. namaycush*) had a higher dress-out percentage than female triploid rainbow trout made with the same dams in a farming environment because of lower visceral losses (Blanc *et al.*, 2000). Similar results were obtained for triploid common carp in India. Triploid populations had a higher dress-out percentage particularly because of the lack of gonadal development in the females (Basavaraju *et al.*, 2002). This would be an advantage in most countries, but not necessarily in India, where the ovaries and eggs are considered edible; in some countries they are a delicacy. Although diploid male and female common carp grew faster than the corresponding sexes of triploids, the gutted weights for diploids and triploids were the same. In reality, then, triploid production is more efficient. Triploid African catfish also yielded more product than diploids because of lower visceral percentage at high feeding levels, but not at low feeding levels (Henken *et al.*, 1987). Triploid and diploid Thai walking catfish had the same carcass percentages (Lakhaanantakun, 1992).

The flesh quality of triploid rainbow trout females is improved relative to diploid females because post-maturational changes are prevented (Bye and Lincoln, 1986). However, proximate body composition was no different in juvenile diploid and triploid rainbow trout (Oliva-Teles and Kaushik, 1990a). Hussain *et al.* (1995) did not detect differences between triploid and diploid Nile tilapia for biochemical composition.

Triploid ayu had higher rates of fat accumulation in the muscle compared with diploids when fed on both low- and high-fat diets (Watanabe *et al.*, 1988). After maturation, fat in the muscle decreased rapidly for all treatments except that triploids fed a high-fat diet continued to accumulate fat in the muscle. Similarly, triploid African catfish had less protein, more fat and more energy deposited per gram of growth compared with diploids (Henken *et al.*, 1987). Female triploid tench had large visceral fat deposits and males had fat deposits along the connective-tissue capsule (Flajshans, 1997).

With regard to fatty acid composition, triploid common carp had higher monoenes 16 : 1 and 18 : 1, but lower polyenes, 22 : 6, than diploids (Lee, E.-H. *et al.*, 1989b). The neutral lipid triglyceride was higher in the triploids and in the diploids during the non-spawning season in the belly and dorsal muscles compared with the diploids during the spawning season, but the glycolipid and the phospholipid phosphatidylcholine were lower for triploids and diploids in the non-spawning season compared with diploids in the spawning season. The total lipids of triploid common carp and diploids out of spawning season were higher than in diploids during the spawning season. The sterility of the triploid apparently prevents them from mobilizing fat stores for reproduction.

The moisture, crude protein, lipid, mineral content, fatty acid composition, free amino acid composition and histology of triploid and diploid amago salmon, *Oncorhynchus masou ishikawae*, were evaluated (Saito *et al.*, 1997). In October, the lipid content in the triploid was two to three times greater than in the diploids and the moisture content of the triploid was low. The cross-sectional area of the muscle fibres decreased in the triploid between March and June as compared with the diploid, but decreased at the same rate as in the diploids in September. The two genotypes did not differ for any of the other traits evaluated for the proximate analysis.

Fat affects flavour, as do compounds such as free amino acids, nucleotides, total creatinine, betaine and trimethylamine oxide (TMAO). These flavour compounds, excluding fat, were highest in diploid common carp outside the spawning season, intermediate for triploids and lowest for diploids during the spawning season (Lee, E.-H. *et al.*, 1989a). Total mineral content was highest for the triploids. Flavour evaluations were not conducted but, based on this analysis, might be highest for diploids outside the spawning season, intermediate for triploids and lowest for diploids during the spawning season.

Flesh colour is another important carcass trait for salmonids. Triploidy did not induce flesh-colour changes of economic importance in rainbow trout, and both genotypes showed a similar ability to fix canthaxanthin (Choubert and Blanc, 1985). When 12 diploid and triploid families were fed canthaxanthin, triploids did not differ from diploids in average dominant wavelength and excitation purity; however, triploids were inferior for luminosity after canthaxanthin ingestion. When canthaxanthin was fed to maturing females for 37 days, the canthaxanthin retention was three times higher for diploids during the first 4 weeks, but later the canthaxanthin deposit in the flesh of triploids was higher than that of diploids (Choubert and Blanc, 1989). The canthaxanthin concentration in the muscle and ovary combined for the diploids was not different from the total in the muscle of the triploids, as apparently the triploids diverted deposition to the flesh as their ovaries are underdeveloped. Four weeks after spawning the canthaxanthin level reached the same level in the muscle of both diploids and triploids. Additionally, when fed astaxanthin for 39 days to enhance redness in the flesh, triploids had a lower chroma than diploids, but no differences were found for lightness and hue angle (Choubert *et al.*, 1997).

Survival

The relative survival of triploids is variable. Atlantic salmon female × brown trout males had the same survival as Atlantic salmon in saltwater net pens (Galbreath and Thorgaard, 1997). Triploid and diploid blue tilapia, *Oreochromis aureus*, had the same survival in tanks (Byamungu *et al.*, 2001). Triploid channel × blue and channel × white catfish hybrids (Lilyestrom *et al.*, 1999), white bass × striped bass hybrids (Kerby *et al.*, 2002) and common carp (Bieniarz *et al.*, 1997) had survival similar to diploids in ponds. Surprisingly, long-term survival of triploid common carp – survival during the third year – was lower for triploids than for diploids (Bieniarz *et al.*, 1997). Triploid walking catfish had lower survival than diploid controls (Na-Nakorn and Legrand, 1992).

Early survival of rainbow trout through 60–105 days was lower for triploids than for diploids (Solar *et al.*, 1984; Solar and Donaldson, 1985; Tabata *et al.*, 1999), primarily around the time of hatching (Guo *et al.*, 1990); however, as rainbow trout approach sexual maturity during the second year of growth, the mortality of diploids, especially males (Tabata *et al.*, 1999), is higher (Guo *et al.*, 1989; Tabata *et al.*, 1999). Juvenile triploid rainbow trout have poorer survival than diploids when subjected to harsh environmental conditions (Quillet *et al.*, 1987).

The lower early survival of triploid rainbow trout may be related to differences in embryonic development. Diploids and triploids hatched at the same time (Happe *et al.*, 1988), but triploid embryos developed more rapidly than diploids, perhaps causing some problem between development time and hatch. Absorption of water by yolk reserves was similar for both ploidy levels, but at swim-up triploids were 5% smaller than diploids. Starvation resulted in an increase in water content and a decrease in wet weight. These changes were greater for triploids than for diploids, and these differences in response to starvation could be related to differences in mortality rates.

Through 2.5 years of age, triploid Atlantic salmon had lower survival than diploids (Quillet and Gaignon, 1990). Triploid Atlantic salmon often have higher on-farm mortality compared with diploids (Sadler *et al.*, 2000a,b). Sadler *et al.* (2000a) measured the stress response in terms of levels of plasma cortisol, glucose, lactate, haematocrit, erythrocyte count, mean cell volume, blood haemoglobin concentration, mean cell haemoglobin, mean cell haemoglobin concentration (MCHC), total protein and erythrocyte ATP. The magnitude of the stress response was similar for diploid and triploid smolts. Triploids had higher mean cell haemoglobin, but a lower blood haemoglobin concentration than diploids. The isohaemoglobin components were the same. The triploids had enlarged erythrocytes, which exhibited reduced shear dependence on blood viscosity at constant haematocrit and should not contribute to greater peripheral vascular resistance. Sadler *et al.* (2000b) concluded that the higher farm mortality of triploid Atlantic salmon is not due to failure in respiratory homeostasis when stressed.

Similar results were obtained for rainbow trout and brook trout (Benfey and Biron, 2000). Plasma cortisol, glucose, chloride, haematocrit levels, haemoglobin concentration, total blood-cell concentrations (erythrocytes and leucocytes) and differential leucocyte concentrations and their relative proportions were measured in diploid and triploid rainbow and brook trout before and after acute handling stress. Resting blood-cell concentrations were lower in triploids than in diploids. The triploids exhibited a typical acute stress response that was no different from that of diploids. Additionally, the critical thermal maxima for diploid and triploid brook trout at 1 year (25 g) and 2 years (668 g) were not different (Benfey *et al.*, 1997).

O'Keefe *et al.* (2000) did find a difference in one stress response between diploid and triploid Atlantic salmon. Erythrocytes do not normally divide in the bloodstream; however, O'Keefe *et al.* (2000) hypothesized that mitosis of erythrocytes in the plasma could be effective for coping with stress. Triploids did produce higher percentages of mitotic erythrocytes in response to handling stress and transportation. Perhaps this assists them in coping with stress or, alternatively, it may be an indication that they are experiencing a higher level of stress and are responding in kind.

The blood of triploid brook trout contains high numbers of immature erythrocytes, which may indicate a premature release of mitotically dividing cells into the peripheral circulation in response to stress (Atkins *et al.*, 2000). Exercise could reduce stress and enhance growth. Exercise did reduce the incidence of dividing erythrocytes in triploid brook trout, but did not affect the growth of diploids or triploids (Atkins *et al.*, 2000).

Swimming ability should be related to survival in natural conditions, as this trait would be key for both capturing prey and avoiding or escaping predators. Triploid coho salmon had similar haematocrit, lower haemoglobin content and the same sustained swimming ability as diploids (Small and Randall, 1989).

Svobodova *et al.* (1998) hypothesized that the blood characteristics of triploid tench would both reduce oxygenation capacity of the blood and lower non-specific immunity compared with diploids. They base this hypothesis on their findings that triploids had lower erythrocyte count, lower haemoglobin content, higher median corpuscular volume, higher median corpuscular haemoglobin, lower median corpuscular haemoglobin concentration, observed lower haematocrit, observed lower leucocyte count and a lower total protein in the plasma compared with diploids. Similar measurements were made for triploid rainbow trout, but in this case triploids had greater erythrocyte volume and haemoglobin concentration, lower erythrocyte count and no difference in the other parameters compared with diploids (Ranzani-Paiva *et al.*, 1998).

Tolerance of low oxygen

The data on the tolerance of low oxygen of triploids are contradictory. Stillwell and Benfey (1999) indicated that triploids have impaired performance when oxygen is limited, and this should be related to blood oxygen-transport capacity. Blood haemoglobin concentration should be a measure of the ability of the blood to transport and supply oxygen. Triploid fish of various species have higher, lower or equivalent blood haemoglobin concentrations. Blood haemoglobin concentrations in brook trout triploids and diploids were equal, and Stillwell and Benfey (1999) suggested that differences between triploids and diploids may be artefacts of intermittent endocrinological and physiological fluctuations. They also hypothesized that the blood of triploids and diploids should be equally effective in biological oxygen requirements. Benfey and Sutterlin (1983, 1984b,d) report triploid Atlantic salmon equally effective in biological oxygen requirements, and had similar oxygen-carrying capacity when stressed (Sadler et al., 2000b). But this does not necessarily mean that the triploids are able to deliver that oxygen into the tissues at the same rate as diploids. Triploid brook trout consume 20% less oxygen than diploids during exertion, despite the fact that they have the same quantity of haemoglobin, the same opercular movement rate and the same number of tail beats per minute as diploids. Triploid amago salmon, *O. masou macrostomus*, had higher mean erythrocyte volume, lower erythrocyte counts, but the same oxygen update as diploids (Nakamura et al., 1989).

Hybrid triploid channel × blue and channel × white catfish had decreased tolerance of low dissolved oxygen compared with diploids (Lilyestrom et al., 1999). Some data indicate that triploid salmonids, like triploid hybrid ictalurids, may have a lower affinity for oxygen. This was measured directly for some salmonids.

Several potential explanations exist for the decreased oxygen tolerance of triploids. Ueno (1984) reports that the erythrocytes and nuclei of triploid common carp had changed to long ellipsoids, which had grown more in the major axis than the minor axis and had surface areas 1.44 and 1.40 times greater, respectively, compared with those of diploids. The increased cell volume could affect how rapidly oxygen or metabolites are exchanged in and out of the cells and into the circulatory system. Additionally, the triploid common carp compensated for the larger erythrocyte size by reducing the number of cells to 60% that of the diploids (Ueno, 1984). The reduced erythrocyte number could also adversely affect the ability of the triploids to transport oxygen.

The affinity of triploid Atlantic salmon blood for oxygen was the same as that of diploids; the haemoglobin–oxygen loading ratio – Hufner's constant – was only 77% of that for diploids (Graham et al., 1985). The reduced haemoglobin–oxygen loading ratio in combination with the lower haemoglobin concentration resulted in the triploids having a maximum blood oxygen content only 68% that for diploids. This reduced ability to transport and bind oxygen may not harm the resting triploid but may hinder its ability to obtain oxygen during exertion or stress.

Triploid rainbow trout clearly have decreased aerobic swimming capacity (Virtanen et al., 1990). The blood haematocrit value and plasma lactate concentration increased and MCHC decreased in exercise, indicating swelling of erythrocytes and accumulation of anaerobic wastes, and these changes were either not present in diploids or more dramatic in triploids. These are indicators that the triploids have lower aerobic capacity than diploids. Additionally, plasma insulin decreased more in triploids than in diploids during exercise and plasma glucagon and glucagon-like peptide (GLP) levels decreased for diploids but increased for triploids during exercise.

Another possible explanation for reduced low-oxygen tolerance of triploids could be related to gill structure and deformities. Sadler et al. (2001) examined skeletal deformity throughout the development for all-female triploids, mixed-sex triploids, all-female diploids and mixed-sex diploids from the Tasmanian strain of Atlantic salmon. Incidence of skeletal deformities was higher in triploid populations. Lower-jaw deformity was observed in 2% of triploid fry, 7% of triploid freshwater smolt, 14% of triploid saltwater smolt and 1% of diploid freshwater smolt. Short opercula were found in 22% of triploids and 16.6% of diploids. Up to 60% of triploids and 4% of diploids suffered from the absence of primary gill filaments during freshwater development prior to transfer to salt water. Gill surface area was reduced in both normal triploids and triploids afflicted with gill-filament deformity compared with diploids. The reduction of gill surface area, coupled with the reduced or deformed gill filaments and opercula, probably affects the capacity for metabolic gas exchange under exercise or suboptimal environmental conditions.

Disease resistance

Disease resistance has been little studied in triploid fish. Physiological studies indicate that triploids should have inferior disease resistance. However, diploid and triploid Thai walking catfish had similar resistance to *Aeromonas hydrophila* (Lakhaanantakun, 1992). Triploid ayu, *P. altivelis*, had resistance to *Vibrio anguillarum* similar to that of diploids (Inada *et al.*, 1990).

Differences in haematology between triploids and diploids could lead to differences in disease resistance. Svobodova *et al.* (2001) found no differences in total or differential leucocyte counts between diploid and triploid 3-year-old tench, in which lymphocytes dominated (>90%) the total leucocyte count. When the fish were 4 years old, there were no differences in total leucocyte counts or in differential lymphocyte, monocyte and myelocyte counts, but triploids had lower metamyelocyte counts and higher neutrophil granulocyte counts.

The immune system of triploids may be inferior to that of diploids. Sibling diploid and triploid Atlantic salmon injected intraperitoneally with lipopolysaccharide were evaluated for the alternative complement pathway activity (measured by the titre of haemolytic activity against rabbit erythrocytes) (Langston *et al.*, 2001). Serum iron concentrations decreased to very low levels by day 2 post-injection in the diploid fish and by day 3 in the triploid fish. The longer time taken for the triploids to recover complement activity and the slower onset of the hypoferraemic response suggest that triploids may be at a disadvantage compared with their diploid siblings in their defence against bacterial invasions.

Reproduction

Triploid fish are usually and essentially sterile. Generally, triploid females have minimal production of sex hormones. However, triploid males, although almost always completely sterile, have sex-hormone profiles that mimic those of diploid males. Grass carp and some salmonid triploid males may exhibit sexual behaviour and mate with females without fertilizing the eggs (J. Casani, Lee County Hyacinth Control District, Fort Mvers, Florida, 1995, personal communication). Experimentally, there have been extremely rare occasions where triploid males produced small numbers of viable progeny.

Triploid loach were completely sterile (Suzuki *et al.*, 1985). Most of the 1-year-old triploid male and female common carp had undeveloped gonads and were sterile (Cherfas *et al.*, 1994a). Long-term sterility was demonstrated in white bass × striped bass hybrids, which had reduced and dysfunctional gonads at 5 years of age (Kerby *et al.*, 2002). Triploid female Atlantic salmon were sterile (Benfey and Sutterlin, 1984c; Refstie, 1984). The gonadosomatic index (GSI) was reduced by 7.7% compared with diploids (Benfey and Sutterlin, 1984c; Cotter *et al.*, 2000b; Murphy *et al.*, 2000) or had the external appearance of undeveloped gonads (Benfey and Sutterlin, 1984c), but the ovaries actually had a small number of oocytes – one to 12 compared with several hundred for diploid females. Female silver barb did not undergo vitellogenesis (Koedprang and Na-Nakorn, 2000), and female triploid European sea bass had lower hepatosomatic indices, possibly indicating a lack of oestradiol-mediated hepatic synthesis of vitellogenin (Felip *et al.*, 2001a).

Male triploid Atlantic salmon (Refstie, 1984) and rainbow trout (Benfey *et al.*, 1986; Tabata *et al.*, 1999) developed secondary sexual characteristics, but had abnormal, reduced gonads (Refstie, 1984; Benfey *et al.*, 1986). However, Murphy *et al.* (2000) indicated that Atlantic salmon male triploid grisle had testicular development similar to diploids when grown in sea cages. Triploid rainbow trout had a lower spermatocrit, exhibited post-spawning mortality and their sperm were aneuploid (Benfey *et al.*, 1986).

The GSI of male and female triploid African catfish (Henken *et al.*, 1987), catfish, *H. fossilis* (Tiwary *et al.*, 2000), and European catfish (Krasznai and Marian, 1986) was lower than that for diploids. Richter *et al.* (1987) further indicate that triploid male African catfish had testes containing cysts, with primary spermatocytes that were arrested in prophase 1 of meiosis. Triploid male *H. fossilis* also had greatly reduced numbers of germ cells in their seminiferous tubules (Tiwary *et al.*, 2000). Triploid tench, *Tinca tinca*, had islets of germ cells and spermatids (Flajshans, 1997). The GSI of triploid Atlantic salmon males was 52% of that of diploid males (Benfey and Sutterlin, 1984c), testes were well developed but contained few spermatids and no spermatozoa, and no triploid males reached spermiation. The GSI of male and female triploid silver barb was reduced by about 50% compared with controls (Koedprang and Na-Nakorn, 2000), and spermatogenesis and oogenesis were retarded in triploids. Triploid males had all stages of spermatogenic cells in silver barb (Koedprang and Na-Nakorn, 2000) and goldfish (Yamaha and Onozato, 1985), including a few spermatozoa. However, functional sperm from triploid males results in aneuploid individuals that are unviable (Arai, 2001); specifically in zebra fish (Kavumpurath and Pandian, 1990). Triploid and diploid European sea bass males had similar testicular development; however, triploid males never spermiated (Felip *et al.*, 2000).

Pre-adult triploid channel × blue hybrid catfish had GSIs equivalent to those of diploids; however, diploids had normal gonadal development and histology, whereas triploid hybrids had abnormal gonadal development (Lilyestrom *et al.*, 1999). The triploid males had abnormal gonadal histology, and no sperm were present in the seminiferous tubules. Females had one only ovary 36% of the time.

Gonad development in bluegill (*Lepomis macrochirus*) × green sunfish (*Lepomis cyanellus*) hybrid triploid males was more advanced than that for triploid females (Wills *et al.*, 2000), but males failed to yield any milt when hand-stripped. Some spermatozoa were completely differentiated, having tails, but all cells were at least 3N and most were 6N, indicating that, despite completion of spermatogenesis, meiosis was interrupted at the first reduction division. Similar results have been obtained with grass carp (Allen *et al.*, 1986b). Triploid male grass carp produced sperm of abnormal shape and variable size that were 1.5, 3 and 6N.

The detrimental effect of triploidy on gonadal development is even more severe for females. Male and female triploid Atlantic salmon × brown trout hybrids and Atlantic salmon males had a lower GSI than Atlantic salmon females when grown in saltwater net pens (Galbreath and Thorgaard, 1997). Female triploid rainbow trout (Nakamura *et al.*, 1987; Sumpter *et al.*, 1991b; Krisfalusi *et al.*, 2000) and European sea bass (Felip *et al.*, 2000) had little or no gonadal development. Triploid female rainbow trout have string-like ovaries lacking developing oocytes, and their odd set of chromosomes apparently disrupts oogenesis (Krisfalusi *et al.*, 2000). Their ovaries were composed of numerous cysts containing small oocytes in varying stages of degeneration and oogonia and no evidence of vitellogenesis (Nakamura *et al.*, 1987). Triploid African catfish had ovaries containing primarily oogonia and some oocytes arrested in the previtellogenic stage (Richter *et al.*, 1987). Ovaries of triploid European sea bass contained primarily oogonia, and those of tench were of low maturity (Flajshans, 1997). Triploid goldfish had predominantly degenerated oocytes of the chromatin–nucleolus stage and had no oocytes at the perinucleolus stage (Yamaha and Onozato, 1985). Gonads of female bluegill × male green sunfish triploid hybrids were immature compared with those of diploid hybrids (Wills *et al.*, 2000), and they primarily possessed oogonia and immature oocytes. Most of the ovarian cells were triploid, possibly indicating that the second reduction division does not occur in the triploids. Triploid female *H. fossilis* only had atresic follicles in the ovaries (Tiwary *et al.*, 2000). Triploid rainbow trout females do not develop secondary sexual characteristics and have the appearance of juveniles (Tabata *et al.*, 1999). Female triploid brook trout had no signs of maturation at the time of ovulation of diploid females, and maintained a silvery skin colour and no protruding vent, typical of immature fish (Smith and Benfey, 1999). These triploid females had lowered GSI and reduced diameter.

Given sufficient time and age, triploid females may be able to develop mature oocytes. Smith and Benfey (2000) examined three age classes of adult triploid and diploid female brook trout. Triploid females always had lower means for reproductive traits. However, triploid females began to produce some mature oocytes when 4 years of age, while diploids began producing mature oocytes at 2 years of age. Similarly, triploid tilapia, *Oreochromis mossambicus*, females had small ovaries containing degenerating oocytes, abnormal oocyte and yolk development and few developing oocytes (Pandian and Varadaraj, 1988a). At 3 months, diploid females readily spawned, but triploids did not produce any mature oocytes. By 10 months of age triploid females did contain some matured oocytes.

Triploidy greatly reduces the production of sex hormones in females. Triploid European sea bass (Felip *et al.*, 2000), rainbow trout (Nakamura *et al.*, 1987) and brook trout (Smith and Benfey, 1999) females have reduced levels of oestradiol and testosterone compared with diploids. Triploid rainbow trout females also had reduced levels of $17\alpha,20\beta$-dihydroxy-4-pregnen-3-one (Nakamura *et al.*, 1987). Sex-steroid levels were lower in female triploid Atlantic salmon grisle grown in sea cages compared with diploids at the time of maturation (Cotter *et al.*, 2000b). The decrease in endogenous levels of oestradiol does not explain the failure of ovarian development in female triploid rainbow trout (Krisfalusi, 1999), and treatment of triploid rainbow trout females (Krisfalusi, 1999) and brook trout females (Smith and Benfey, 2000) with oestradiol does not stimulate oocyte development but will result in the development of secondary sex characteristics in triploid brook trout females. However, the data of Benfey *et al.* (1989, 1990) suggest that in some species, such as coho salmon, lack of oestradiol does have a role in the lack of full maturation of post-meiotic oocytes in triploids. Injections of 17-β-oestradiol increased vitellogenin levels, hepatosomatic levels and pituitary gonadotrophin. Vitellogenin levels were the same in diploid and triploid coho salmon.

Sex-hormone levels are not much affected by triploid induction in males. Testosterone levels are similar for diploid and triploid European sea bass males (Felip *et al.*, 2000). Gonadal steroid and gonadotrophin hormone profiles were similar for diploid and triploid Atlantic salmon male grisle when grown in sea cages (Cotter *et al.*, 2000b).

The probability is extremely low but it is possible for triploid males to produce progeny; however, even if progeny were produced, this would not be likely to result in a reproducing population, as the progeny would probably be sterile. Triploid rosy bitterling, *Rhodeus ocellatus ocellatus*, develop nuptial coloration regardless of their spermiation ability (Kawamura *et al.*, 1999). Nearly 80% of the spermatozoa from triploid males were abnormal, having malformation of the head and mitochondrion, excessive formation of the head, mitochondrion and flagellum, or no flagellum. Spermatazoa with multiple flagella were common and they often had a saccate-like organ (Kawamura *et al.*, 1999). The motile lifespan of spermatozoa from triploids was the same as that in diploids, but they were not able to advance like spermatozoa from diploid males. They spun rather than advancing in a forward movement. Sperm density of the triploids was less than 2% of that of diploids, and normal spermatids and spermatozoa were rare inside the testes. One embryo developed from the mixing of sperm from triploid males with 1500 eggs from diploid females. The resulting embryo had a ploidy level of 2.5N (Kawamura *et al.*, 1999). If this embryo survived to adulthood, it is likely that the line would die: if it was a female, it would be sterile and if it was a male, it would have greatly reduced fertility if not sterility.

However, there may be one exception to the fertility rule, as triploid male grass carp can produce small numbers of viable diploid progeny when mated with diploid females (van Eenennaam *et al.*, 1990). When diploid females were artificially inseminated with milt from triploid males, the fertilization rate was one-half to one-third of that for diploid matings. Diploid × triploid embryos that hatched had smaller bodies and yolk-sacs and utilized their yolk more slowly than diploid × diploid embryos. Most hatched diploid × triploid embryos had notochord deformities and died before first feeding. However, 0.1–0.2% of the embryos survived to 5 months and were diploid (van Eenennaam *et al.*, 1990).

This result is a little surprising since Allen *et al.* (1986b) did not detect any 1N sperm in triploid grass carp. Again, individual, family or strain variation could be an explanation for these variable results from one experiment to another. Alternatively, aneuploid embryos may have lost chromosomes during development and reverted to the diploid state. Chouinard and Boulanger (1988) were also able to artificially backcross a triploid rainbow trout × brook trout hybrid with brook trout and produce viable progeny.

The ramification is that there must not be any diploid females in the system in order to be 100% certain that a breeding population cannot be established or that there is absolutely no possibility of a genetic impact. If only triploids are in the system, the reproductive limitations of the females

in concert with the males' vastly reduced fertility should ensure that no breeding populations are established. If diploid fish are already present in the system, an effective alternative would be to use monosex female triploid populations since the triploid females have not been known to produce any progeny.

One complication regarding these experiments demonstrating the fertility of triploid males is that artificial insemination was utilized. The possibility exists that male triploid grass carp or other species might not be capable of producing progeny naturally. Videos taken by Dr Cassani in Florida show that triploid male grass carp exhibit normal courtship behaviour in spawning tanks but, compared with diploids, have no visible ejaculate when the female ovulates. Therefore, these grass carp males may not be able to produce progeny naturally, but if artificially hand-stripped might be capable of producing progeny.

Embryonic development

Divergent embryonic development might explain some of the differences between diploids and triploids. Rainbow trout diploids, diploid gynogenetics, triploids from heat shock or from 2N × 4N mating, diploid rainbow trout coho salmon and triploid hybrids were compared for hatching time (Quillet *et al.*, 1988). Gynogenesis did not change hatching time but induced variability for hatching time. The higher the level of ploidy, the more rapid the development time; however, the variability in hatching time was also higher for the polyploids in comparison with the diploid controls. Triploid hybrids also had more rapid embryonic development than diploid hybrids, but had a more uniform hatch time than diploid hybrids.

Sex ratio

If the female is homogametic when triploidy is induced, all the progeny receive two X chromosomes from the dam and either an

X or a Y chromosome from the sire. XXX individuals are female and XXY are male, and the sex ratio should be 1:1. If the male is homogametic, triploid progeny should have the genotype WWZ, WZZ or ZZZ, but the sex of these potential genotypes has not been determined.

Triploid *O. aureus*, a species where the female is heterogametic, WZ, were effectively sterile and were 80% female (Byamungu *et al.*, 2001). This may indicate that WWZ and WZZ genotypes are female; otherwise there should have been a majority of males. This is consistent with the femaleness chromosome being dominant in fish WZ sex-determining systems. Triploid tilapia, *O. mossambicus*, produced both male and female individuals (Pandian and Varadaraj, 1988a,b). Triploid channel × blue catfish had a 1:1 sex ratio (Lilyestrom *et al.*, 1999) as expected since an XY system determines the sex in both parent species.

Hybrid viability

Triploidy may also benefit hybridization efforts. Triploid induction can also allow production of otherwise non-viable or sub-viable diploid hybrids (Allen and Stanley, 1981b; Chevassus *et al.*, 1983; Chourrout and Itskovich, 1983; Scheerer and Thorgaard, 1983; Parsons *et al.*, 1986; Seeb *et al.*, 1986, 1988; Quillet *et al.*, 1987; Scheerer and Thorgaard, 1987), probably because of the presence of a balanced maternal chromosome set in triploids that is not present in diploid hybrids.

Diploid Nile tilapia females × *Tilapia rendalli* male hybrid embryos experience near 100% mortality (Chourrout and Itskovich, 1983). However, this hybrid combination is viable when triploidy is induced. Several salmonid hybrids are non-viable in the diploid state but are viable when triploidy is induced (Chevassus *et al.*, 1983; Parsons *et al.*, 1986; Seeb *et al.*, 1986; Quillet *et al.*, 1987; Scheerer and Thorgaard, 1987). This has allowed production of otherwise unviable rainbow trout × coho salmon triploid hybrids with increased resistance to IHN virus (Parsons *et al.*, 1986)

and osmoregulatory ability (Seeb *et al.*, 1986). Interspecific triploid and diploid hybrids were produced between female Atlantic salmon, brown trout and rainbow trout and male Arctic charr, bull trout (*Salvelinus confluentus*), lake trout and brook trout (Shah *et al.*, 1999). Triploids were more viable than diploids. Only one diploid hybrid, Atlantic salmon × Arctic charr, had good viability at swim-up stage. Reciprocal diploid and triploid hybrids between masu salmon, *Oncorhynchus masou*, and brook trout did not differ in viability (Arai, 1988). Also, a small number of triploid brook trout × kokanee salmon (*Oncorhynchus nerka)* survived to first feeding (Arai, 1988), and 62% of pink salmon (*Oncorhynchus gorbuscha*) × Japanese charr (*Salvelinus leucomaenis*) triploids survived to eyed stage (Yamano *et al.*, 1988), while all diploids of these two hybrids died during embryonic development. Reciprocal triploids between brook, brown and rainbow trout had greater hatching and fry survival compared with diploids (Dobosz and Goryczko, 1988).

Survival of the tiger trout, a brown trout female crossed with a brook trout male, is raised from 5 to 34% when triploidy is induced (Scheerer *et al.*, 1986). This may allow utilization of this attractive sport fish, which is sterile and exhibits fast growth (Scheerer and Thorgaard, 1987). The rainbow trout × coho salmon triploid had decreased growth, but had increased resistance to IHN virus. This same triploid hybrid had total resistance to viral haemorrhagic septicaemia (VHS) virus, while diploid and triploid rainbow trout were sensitive to this virus (Dorson and Chevassus, 1985). Survival of triploid hybrids between female rainbow trout and male brook trout (*S. fontinalis*), Arctic charr (*S. alpinus*) or lake charr (*S. namaycush*) was inferior to diploid and triploid rainbow trout made with the same dams in a farming environment (Blanc *et al.*, 2000), and the Arctic charr hybrid was the weakest. Two peaks of mortality occurred for these triploid hybrids, one during the embryonic and larval stages and the other due to sexual maturation of the males. Similar results were obtained for triploid rainbow trout and triploid hybrids between rainbow trout females and males of brown trout, brook trout and coho salmon (Quillet *et al.*, 1987, 1988). Triploids had lower survival, especially early survival, compared with parent species, and growth was intermediate to that of the parents. The triploids had faster growth once the parent species began to mature sexually. In seawater, triploid females had better growth and survival during the maturation period (Quillet *et al.*, 1987), except that brook trout and their hybrids were not evaluated in that study.

Not all triploid hybrids exhibit greatly increased viability, as brook trout × rainbow trout hybrids produced by Chouinard and Boulanger (1988) exhibited only slightly increased viability. However, it did allow them to eventually produce backcrosses utilizing the triploids.

The triploid female grass carp × bighead carp male exhibited heterosis for temperature tolerance (Bettoli *et al.*, 1985). The preferred temperature and the critical thermal maximum were higher for the triploid hybrid than for both parent species, both experimentally and as observed in the field. Wu *et al.* (1997) supply a possible explanation – enzymatic gene dosage – for the increased viability of triploid hybrids in comparison with diploid hybrids. They state that three kinds of abnormal isozymic expression – paternal allele inhibition, delay in parental gene activation and preferential gene expression of the maternal allele – can occur in hybrids from distantly related paternal species. Since there are incompatibilities and every allele controls only one-half of the gene product, the hybrid embryos act like haploids and die during embryonic development from double-haploid syndrome. There may also exist incompatibilities in temporal and spatial gene expression between the two species, not only for isozyme loci but for other loci as well, which could also force the hybrid embryos to essentially function and attempt to develop in an analogous manner to haploids. Since the triploid hybrids have two maternal genomes, the insufficient dosages, the disturbances of spatial and temporal expression of paternal and maternal alleles

are compensated for or inhibited by the double set of maternal alleles. Thus normal metabolism and development, hatching and growth occur for the triploid hybrids from the two distantly related species.

Heterozygosity in triploids

Triploids should have an increase in heterozygosity since they have two chromosome sets from the mother and one from the father. Leary *et al.* (1984, 1985a,b,c) measured isozyme heterozygosity at 42 loci in triploid rainbow trout and found an increased heterozygosity of 30% compared with diploids. This led to an increase in developmental stability as fluctuating asymmetry decreased by 14%. However, Leary *et al.* (1985c) predicted even greater developmental stability in these triploids, based on the relationship between enzyme heterozygosity and asymmetry in random mating populations of rainbow trout. Heterozygosity in triploid ayu was 60% higher than for diploids (Taniguchi *et al.*, 1987).

Behaviour

Triploidy alters behaviour and overall appears to result in a calmer, less aggressive fish. Triploid ayu have lower sensitivity to sound and light than diploids (Aliah *et al.*, 1990). Triploid grass carp feed and behave similarly to diploids (Allen and Wattendorf, 1987). Triploids appear to be more lethargic and not as aggressive at feeding when observed in tanks or ponds.

Invertebrate Triploid Performance

Recently, a considerable amount of research has focused on the production of triploid shellfish and various invertebrates (Dunham *et al.*, 2001). The use of triploid oysters can enhance oyster culture, primarily by delaying sexual maturity and increasing flesh quality and secondarily by allowing some growth improvement. Potential applications of triploidy for invertebrates are the same as for fish, including genetic conservation of native gene pools when stocking sterile conspecifics, utilization of exotic species without their establishment, increasing the viability of interspecific hybrids and sterilization of transgenics.

Growth

In contrast to fish, triploidy increases the growth rate of shellfish in the vast majority of examples. Triploid Pacific oysters have a 20–25% growth enhancement. Triploids grew linearly during the reproductive season, while diploids grew little until spawning and then lost 64% of their body weight (Allen and Downing, 1986). After spawning, the growth of triploids and diploids was equivalent. In China, triploid Pacific oysters had 14, 8, 35 and 73% greater shell length, shell height, body weight and wet meat weight compared with diploids (Zeng *et al.*, 1999b).

Stanley *et al.* (1984) compared the growth of triploid eastern oysters, *C. virginica*, produced by blocking meiosis I or II. Triploids produced during meiosis I grew faster than diploids through 3 years of age, whereas triploids created during meiosis II grew at the same rate as diploids. Stanley *et al.* (1984) attributed the greater growth of meiosis I triploids to the fact that they had higher heterozygosity than meiosis II triploids. However, increased growth should have been observed for meiosis II triploids because of the lack of sexual development alone. In Alabama and on the Atlantic coast, triploid eastern oysters grew more rapidly than diploids (Anderson and Rouse, 1998; Guo *et al.*, 2000). Eastern oyster triploids exhibited increased growth compared with diploids in the majority of growth comparisons evaluated. Triploid *Crassostrea ariakensis* grew faster than diploid eastern oysters in a variety of salinities (Calvo *et al.*, 2000).

Triploidy retards reproduction and increases growth rate in Japanese pearl oysters. Heterozygosity was not an explanation for the increased growth. Soft-shell clams had twice the heterozygosity as diploids, but had no growth improvement. Consistently, heterozygosity has not had an effect on the growth of triploid shellfish. As expected, the growth advantage of triploid Japanese pearl oysters was less than that of diploids at young ages – 5–15% – than at adult ages – 20–90%.

Triploid Sydney rock oysters, *Saccostrea commercialis*, grew faster than diploids at 6–24 months of age, resulting in a 41–90% increase in body weight (Hand *et al.*, 1999; Dunham *et al.*, 2001). The triploids have a higher dry meat weight and condition index. After 2–2.5 years of growth on commercial farms in Australia, the Sydney rock oyster triploids were 31% heavier and had 9% higher shell height than diploids (Hand *et al.*, 1998a). Similar to results with fish, the growth advantage of the triploids was size specific rather than age specific. The growth advantage does not become obvious until the oysters are larger than 5–10 g or have a shell height of more than 30–40 mm. Triploidy significantly decreased the amount of time needed for the Sydney rock oysters to reach minimum market size.

The size-specific growth advantage is also found in triploid scallops, *Pecten fumatus* (Heasman *et al.*, 1998), and triploid scallops, *C. nobilis* (Komaru and Wada, 1989). Triploid *P. fumatus* are larger and have larger muscle tissue once they reach the juvenile stage. Through 9 months of age, there was no size difference between triploid and diploid scallops, *C. nobilis*, but by 14 months of age the shell width and meat weight of triploids were higher. Yearling triploid Atlantic Bay scallops exhibited 36% greater body weight, shell inflation and 73% greater adductor-muscle weight (Tabarini, 1984a,b) compared with diploids. Shell height and length were unaffected.

Triploidy does not always result in faster growth of shellfish. Growth and survival were not different for the blacklip abalone, *Haliotis rubra* (Liu *et al.*, 2006, 2009), up to 30 months of age; however the triploids had a more elongated shell and greater foot muscles than diploids. Triploids had higher feed consumption than diploids, but diploids had superior feed-conversion efficiency. Diploids also grew faster than diploids for greenlip abalone, *Haliotis laevigata*. However, the triploid abalone yielded up to 30% greater meat weight compared with the same-length diploid abalone during the spring–summer maturation periods at 36 and 48 months. Diploid abalone produced equivalent meat weights to triploid abalone between the maturation periods, 42 months. Fatty acid composition of the meat was the same for triploids and diploids.

After 3 years, triploid hard-shell clams, *Mercenaria mercenaria*, had lower dry tissue weight and shell parameters compared with diploids (Hidu *et al.*, 1988). Aneuploid Pacific oysters – 2N+1, 2N+2, 2N+3, 3N–2, 3N–1 – grew more slowly than diploids (Guo *et al.*, 2000). Trisomics grew more slowly than diploids in most families and in the remaining few families there was no growth difference.

The fast growth of triploids has been hypothesized to be a result of three factors: sterility, cell size and heterozygosity. Wang *et al.* (2002) compared diploid, triploid induced and triploid mated (2N × 4N) individuals. Heterozygosity was strongly correlated with growth when comparing the three groups; however, within groups or among individuals the correlation between heterozygosity and growth was weak or no different from zero. Heterozygosity has an influence on triploid growth but obviously does not explain all of the growth differences among diploids and various triploid genotypes.

Hawkins *et al.* (2000) obtained similar results with Pacific oysters. Microsatellite and allozyme variation, feeding rate, absorption efficiency, net energy balance and growth efficiency were measured in meiosis I triploids, meiosis II triploids and diploid siblings. Improved physiological performance in triploids was associated with increased allelic variation, rather than with the quantitative dosage effects of ploidy status, leading Hawkins *et al.* (2000) to suggest that it may be preferable to induce triploidy by blocking meiosis I, rather than meiosis II as has traditionally been undertaken during commercial breeding, because

genetic variation was highest in individuals triploidized at meiosis I.

Energy storage and bioenergetics

Mason *et al.* (1988) examined energy budgets to ascertain the cause of growth differences between triploid and diploid clams, *Mya arenaria*. They concluded that the blockage of gametogenesis resulted in the shunting and reallocation of energy to somatic-tissue growth.

Ripe yearling diploid Pacific oysters have a negative energy balance, while triploids remain in a state of positive energy balance during the time of peak reproductive condition (Davis, 1988a,c). The reduced metabolic costs and nitrogen excretion of the triploids contributes to the differences in the energy available for tissue production. Lower O/N ratios in diploids possibly indicate that germinal-tissue growth combined with warmer water may contribute to stress and a negative energy balance during this time of year. Triploid Pacific oysters spent 26% more energy for growth and 13% more energy for assimilation than diploids (Zhou *et al.*, 2000). In contrast to the results of Davis (1988a,c), Zhou *et al.* (2000) found that the triploids were excreting more nitrogen and twice the ammonia compared with diploids. Oxygen consumption was the same, but triploids consumed 97% less energy for respiration (Zhou *et al.*, 2000). Dry gonad weight was five times greater for diploids. In contrast, ammonia excretion rates were no different for triploid and diploid scallops, *Chlamys farreri* (Liu *et al.*, 2000).

Survival and disease resistance

In general, triploid shellfish appear to have higher mortality rates than diploids. Salinity affected embryonic survival for diploid Pacific oysters and even more severely for triploids, but did not affect the survival of diploid and triploid American oysters (Downing, 1989b). Mortality rates of triploid

Pacific oysters were 2.5 times those of diploids when exposed to elevated temperature and low dissolved oxygen (Cheney *et al.*, 1998). With regard to summer mortality, performance of triploid Pacific oysters was much more erratic than that of diploids (Suhrbier *et al.* 2005). Triploid greenlip abalone, *H. laevigata*, had heavy mortality compared with diploids in several life stages (Dunstan *et al.*, 2007). Survival of triploid scallops, *P. fumatus*, was lower than that of diploids immediately following treatment for triploid induction and during early larval rearing (Heasman *et al.*, 1998). Glycogen levels of triploid Pacific oysters were initially higher than those for diploids but, after 130 days of starvation, survival of diploids was higher than that for triploids (Davis, 1988c).

However, other examples exist where triploids – Sydney rock oyster – have better survival than diploids. Triploid Sydney rock oysters had better survival on farms in Australia than diploids (Hand *et al.*, 1998b). Winter mortality was the same between triploid and diploid Sydney rock oysters in commercial trials. Survival of triploid *C. ariakensis* was higher than that of eastern oysters in Chesapeake Bay (Calvo *et al.*, 2000). Guo *et al.* (2000) were able to produce trisomic Pacific oysters. In most families the percentage of trisomics decreased from 50% at the two-cell stage to 5–25% when the oysters were 1 year old, but in some families the trisomics had high survival and remained at a frequency of 40–61%.

When exposed to the parasite *Mikrocytos roughleyi*, triploid Sydney rock oysters had lower mortality (12%) than diploids (35%) (Hand *et al.*, 1998b). Triploid *C. ariakensis* had greater resistance to *Perkinsus marinus* and *Haplosporidium nelsoni* (MSX) in Chesapeake Bay (Calvo *et al.*, 2000) than native oysters. Triploid *C. gigas* was more resistant than triploid *C. virginica* to *P. marinus* and MSX (Calvo *et al.*, 1999a).

Reproduction

Gonadal development in triploid Pacific oysters is one-seventh that of diploids. In another evaluation, triploid male Pacific oysters had gonads 50% as large as diploids and female triploids had gonads 25% as large as diploids (Allen and Downing, 1986). Gonadal development was also retarded in Atlantic Bay scallops, *Argopecten irradians* (Tabarini, 1984a,b; Allen and Downing, 1985). Triploid eastern oysters have only slightly reduced gonadal development compared with diploids (Allen and Downing, 1985).

The gonads of triploid soft-shell clams are greatly reduced (Allen and Downing, 1985), but a small fraction develop full gonads (Allen, 1987). Triploid males did not mature and had undeveloped gonads and females had some oocytes but abnormal maturation during the reproductive season of their second year (Allen *et al.*, 1986a).

Gametogenesis proceeds past the spermatocyte stages in both triploid Pacific and eastern oysters in about 50% of the individuals, and triploid Pacific oysters produce numerous spermatocytes (Allen, 1987; Allen and Downing, 1990). During their third year, triploid male eastern oysters had macroscopically visible gonads with no follicular inhibition (Lee, 1988). Primary spermatocytes were present, but spermatids and spermatozoa were absent. By late season triploid males had well-developed follicles filled with primary spermatocytes (Lee, 1988), and diploids were fully ripe and many had spawned. Pacific oyster triploid males have a reduction in gametes (Davis, 1988a). Triploid scallops, *C. nobilis*, had spermatocytes and developing oocytes; however, there were no spermatozoa or mature oocytes (Komaru and Wada, 1989).

During their third year, triploid eastern oyster females exhibited ovogonial proliferation, but very few primary ovocytes developed and these female oysters had underdeveloped follicles (Lee, 1988). In some cases, triploid female Pacific oysters have virtually no gametes (Davis, 1988a). Although oocyte development was severely retarded, some females produced large numbers of eggs (Allen and Downing, 1990). Gonads of diploid catarina scallop, *Argopecten ventricosus*, had higher concentrations of proteins, carbohydrates, lipids and acylglycerols than those of triploids, indicative of the fertility of the diploids and the sterility of the triploids (Ruiz-Verdugo *et al.*, 2001b).

During vitellogenesis, oocytes of diploid Pacific oysters were oval-shaped with well-developed organelles and numerous yolk granules evenly distributed in the cytoplasm (Zeng *et al.*, 1999c). Microvilli encircled the oocytes. Most oocytes of triploids were oblong or irregular in shape, with a small number of organelles and fewer yolk granules. Granules in some oocytes were deformed. No microvilli were observed outside oocytes of triploid Pacific oysters. Development of triploid oocytes was blocked during vitellogenesis.

The dynamics of polar body release are important for creating polyploid shellfish (Eudeline *et al.*, 2000). The timing of 50% first polar body (polar body I) release in eggs of triploid Pacific oyster is important for efficient production of tetraploid individuals. Polar body I release is generally slower in triploid eggs than in diploid eggs at 26°C. Lowering the temperature (from 26 to 19°C) slowed development in diploid eggs, but nearly stopped development in triploid eggs (Eudeline *et al.*, 2000) At any temperature, the variability in 50% polar body I release was much higher for triploid eggs than for diploid eggs.

Both within- and between-female variation occurred for release of polar body I (Eudeline *et al.*, 2000). Greater synchronization of polar body I release in triploid females was not achieved through varying the amount of time eggs remain in seawater between the time they are stripped and when they are fertilized (or time of hydration), increasing the time of hydration or using serotonin. Because of this variability, utilization of triploid eggs from a single female at a time should result in greater production of tetraploid embryos of Pacific oysters than treating eggs *en masse* from more than one female.

Diploid female Pacific oyster × male triploids produce a larger number of aneuploids – 80–95% – and fewer triploids than the reciprocal cross – 16–20% aneuploids and 20–53% triploids (Guo *et al.*, 2000). Fertile trisomic oysters were identified and mated with diploids to produce viable progeny. High frequencies (50%) of trisomic progeny were produced, but in most families the trisomics were subviable.

Reversion to diploidy

Triploidy may not be a fail-safe technique to genetically sterilize oysters, as triploids are able to revert a portion of their cells back to the diploid state. Either triploids had lower survival than diploids or some reverted to diploids. Some Pacific oysters grown in the Chesapeake Bay also reverted from triploidy to diploidy (R. Mann, Virginia Institute of Marine Science, 1996, personal communication). Allen (2000) also found reversion in *C. ariakensis*. A large proportion of diploid/triploid mosaics were detected in adult Sydney rock oysters (Hand *et al.*, 1999).

Sex ratio

Triploid soft-shell clams were 77% female and 16% female-like (perhaps this 16% was intersex) (Allen *et al.*, 1986a), and 7% were undifferentiated (Allen, 1987). The proportion of hermaphrodites is considerably higher in triploid Pacific and eastern oysters than in diploids (Allen, 1987; Allen and Downing, 1990). The sex ratio in both triploid and diploid Pacific oysters was 1:1 (Allen and Downing, 1990).

However, Pacific oysters are protandric dioecious: young oysters mature as males and then change to females (Guo and Allen, 1998). Guo and Allen (1998) propose that the primary sex determination is a single locus where the XY genotype results in true males and XX individuals can change sex. At 1 year of age, diploids were 23% female, 75% male and 2% hermaphrodites and, by 2 years of age, the percentage of females had increased to 46%. Triploids had a higher percentage of females at 1 year of age, 46% for CB-induced triploids and 91% for triploids from the mating of a 2N female and a 4N male. These last two genotypes would have a higher mean number of X or female-ness chromosomes, so perhaps there is some dosage effect for the hormone with regard to time to feminization.

Flesh quality

Meat quality is preserved in triploid shellfish during reproduction (Allen, 1987). Diploid Pacific oysters decreased glycogen content in their flesh by 72% prior to spawning and then glycogen increased after spawning (Allen, 1988b). Female triploids matured less than males and mobilized less glycogen than males. Male triploids matured half as much as diploids and twice as much as triploid females. In triploids, glycogen levels decreased by only 8% prior to spawning but continued to decline for an additional 8 weeks (Allen, 1988b). Numerous studies document that glycogen levels of triploid shellfish of many different genotypes and crosses remain higher in triploids than in diploids during gametogenesis (Downing, 1988a,b,c). This higher glycogen content has been provided as an explanation for the superior flavour of triploids (Allen and Downing, 1991); however, glycogen is flavourless (Maguire *et al.*, 1994b). Nell (2002) suggests that the firmer texture of the triploid compared with gravid diploids may be the real taste or consumer preference.

The adductor muscles of triploid catarina scallop were larger than those of diploids, but no differences were observed for biochemical composition (Ruiz-Verdugo *et al.*, 2001b). Triploid Sydney rock oysters are readily accepted by processors in Australia during the cool months, but these triploids develop a discoloration on the gonads during the warmer months which may adversely affect marketability (Hand *et al.*, 1998b). These oysters were especially prone to localized discoloration of the gonad (Hand and Nell, 1999). This brown patchiness is distinctive from the grey gonad patchiness seen in diploids following partial spawning (Nell, 2002). This problem exists in more than one species as 6% of triploids, but not diploids, developed brown patches on the meat in the summer in Tasmania (Maguire

et al., 1994a). Triploid Pacific oysters (Allen and Downing, 1991; Maguire *et al.*, 1994b) and triploid Sydney rock oysters have higher consumer acceptance than diploids, based on sensory evaluations by taste panels.

Triploid shrimp

Triploidy is the only technique that can guarantee shrimp populations are skewed towards the faster-growing sex, the female (Sellars *et al.*, 2009). Triploidy is also used to prevent the theft of elite stocks/germplasm. In the case of *Penaeus* (*Fenneropenaeus*) *chinensis*, triploids had a reduced number of haemocytes (Xiang *et al.*, 2006). This may be a key for explaining the trend of reduced tolerance of low oxygen in finfish as well as having implications for crustaceans. The triploid shrimp grew faster during sexual maturation, but not before that time.

Tetraploids

Mechanical production of triploids can be tedious and difficult (Allen and Stanley, 1981a,b; Wolters *et al.*, 1982a,b; Purdom, 1983; Cassani and Caton, 1986a; Chourrout *et al.*, 1986a), high egg mortality can occur and the reproductive biology of some species limits or prevents triploid induction. Survival and triploid induction rate can vary from laboratory to laboratory or hatchery to hatchery, even when the same protocol is attempted (Wolters *et al.*, 1981a,b; Bidwell *et al.*, 1985; Cassani and Caton, 1986a; Rezk, 1988; Bury, 1989). A possible alternative is the natural production of triploids by crossing tetraploids with diploids (Chourrout *et al.*, 1986a).

Tetraploids have a balanced set of chromosomes, which can result in viability and fertility. Theoretically, the progeny of matings between tetraploids and diploids should be 100% triploid. If the tetraploids are fertile, replacement of 4N brood stock is accomplished by mating tetraploids with each other to produce the next generation of tetraploids.

Naturally produced triploids may have advantages compared with those generated mechanically (Chourrout *et al.*, 1986a). The possibility exists that the stress and mechanical damage the embryo might experience during treatments to retain the second polar body could have long-lasting effects on performance (Myers, 1985, 1986; Chourrout *et al.*, 1986a; Scheerer *et al.*, 1986) and this stress may be a partial explanation for the lack of improved performance by triploids prior to maturation effects on diploids. Myers (1985) also theorized that naturally produced triploids would overcome the dosage compensation and aberrant embryonic development that depress the performance of triploids. Triploids produced by mating tetraploids with diploids have potential for greater heterozygosity than those produced mechanically (Allen and Stanley, 1981b) and could have improved performance through dominance or overdominance effects.

Tetraploid induction

Allowing karyokinesis while blocking cytokinesis produces tetraploids. Similar techniques – temperature shocks and hydrostatic pressure – are used to induce both triploidy and tetraploidy (Bidwell *et al.*, 1985; Chourrout *et al.*, 1986a; Rezk, 1988; Bury, 1989), but of course later in embryonic development for tetraploids. Again, pressure treatments appear to be more consistent than temperature shocks for producing both triploidy and tetraploidy (Rezk, 1988; Bury, 1989). Mosaic individuals having 3N, 4N and 5N cells sometimes result (Bidwell *et al.*, 1985), and catfish mosaics usually have compressed bodies or caudal deformities.

Multiple peaks can be observed with regard to optimum time during embryonic development in which tetraploidy can be induced. Myers *et al.* (1986) found that, when they tried to produce tetraploid chinook salmon, coho salmon, rainbow trout and coho salmon × Atlantic salmon, two periods of tetraploid induction occurred, corresponding to 55–75% and 100–110% of the interval to first cleavage.

Tetraploid catfish (Bidwell *et al.*, 1985; Rezk, 1988) and rainbow trout (Chourrout,

1982; Chourrout *et al.*, 1986a; Blanc *et al.*, 1987) have been produced. These tetraploids were sub-viable and had half the survival rate of diploids and triploids. The growth rate of tetraploid rainbow trout was 25% slower than that of diploids (Chourrout *et al.*, 1986a). Attempts were made to produce tetraploid tilapia (Myers, 1985, 1986). Tetraploid tilapia embryos were detected, but these individuals all died prior to swim-up stage. The number of allotetraploids (interspecific hybrid tetraploids) detected was higher than that of autotetraploids (pure-species tetraploid) for Nile tilapia and *O. mossambicus.* Tetraploid grass carp survived no longer than 50 days (Cassani *et al.*, 1990).

Fertility and performance

Although tetraploid rainbow trout are subviable, tetraploid males matured sexually and produced sperm (Chourrout *et al.*, 1986a). When tetraploid males were mated with diploid females, the fertility of the 2N × 4N matings was variable (0 to 97%) and lower (40% of control) than matings between diploids. Second-generation tetraploids were 95% male (Chourrout *et al.*, 1986a). These tetraploid rainbow trout had very poor growth and survival (Fig. 10.3).

Eggs of certain diploid females were more likely to be fertilized by tetraploid males than those of others (Chourrout *et al.*, 1986a). Females that had large micropyle openings for their eggs had higher rates of fertilization with 2N sperm from 4N males than females with small micropyle openings (Fig. 10.4). The diploid sperm (cell) is twice the size of haploid sperm, and apparently has difficulty penetrating the micropyle of diploid eggs with small-diameter micropyles. If tetraploid females are fertile, the problem of compatibility between diploid micropyle diameter and tetraploid sperm diameter should be alleviated since, theoretically, the smaller sperm from diploids should not have trouble penetrating the micropyle of a 2N egg.

The triploid progeny produced by mating diploid rainbow trout females with

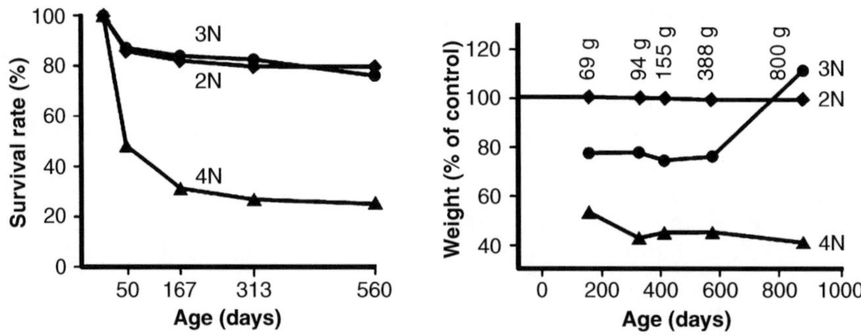

Fig. 10.3. Relative survival and growth of diploid, triploid and tetraploid rainbow trout, *Oncorhynchus mykiss*. (Adapted from Chourrout *et al.*, 1986a.)

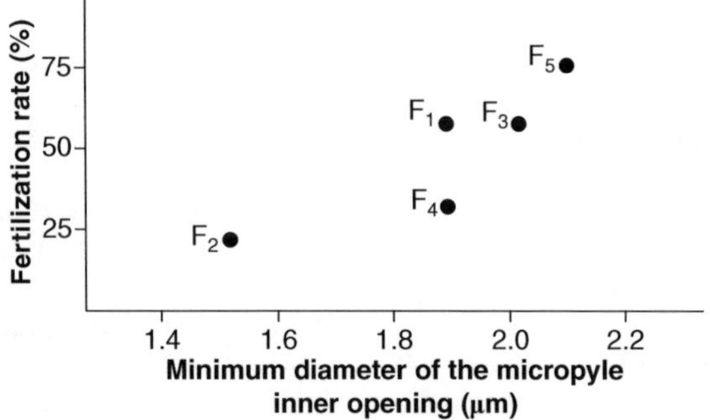

Fig. 10.4. The relationship between the minimum diameter of the micropyle inner opening and fertilization rate for diploid rainbow trout, *Oncorhynchus mykiss*. Females fertilized with sperm from tetraploid males. (Adapted from Chourrout *et al.*, 1986a.)

tetraploid males had survival and growth rates similar to mechanically (heat shock) produced triploids, which are slightly depressed compared with diploids (Blanc *et al.*, 1987). Apparently, heat shock does not produce any long-term damage to performance of triploids produced by this method, and the expected higher heterozygosity of the naturally produced triploids does not benefit growth or survival.

The percentage of triploids produced by mating 2N rainbow trout females and 4N rainbow trout males was 97% (Chourrout *et al.*, 1986a). The remaining individuals consisted of diploids and aneuploids – individuals with unbalanced chromosome sets. Apparently, the meiotic process of tetraploid rainbow trout males does not always allow even division of chromosomes

to produce 2N sperm or some chromosomal material was lost during embryonic development. If 100% triploid rainbow trout populations are required, all individuals would need to be screened for ploidy level to allow culling of non-triploid individuals.

Yamaki *et al.* (1999) were able to produce diploid–triploid mosaic amago salmon, *O. masou ishikawai*. Some of these females produced both diploid and triploid progeny. A mosaic male produced only diploid progeny. The haploid sperm may have been more motile, faster and competitive, or similar to the results of Chourrout *et al.* (1986a), perhaps the 1.5N sperm had more difficulty penetrating the micropyles of the haploid eggs.

Metabolic differences may explain growth and survival differences between tetraploids and diploids. The uptake and metabolic utilization of acetate and glutamic acid did not differ among tetraploid, triploid and diploid rainbow trout fry; however, oxidation of glucose was lower and oxidation of leucine was higher for tetraploids than for the other two genotypes (Fauconneau *et al.*, 1989).

Hexaploid Fish

Hexaploids have three pairs of chromosomes. Arai *et al.* (1999) produced hexaploid loach, *M. anguillicaudatus*, by inhibiting the extrusion of the second polar body when a natural pair of tetraploids was mated. The sex ratio of the hexaploids was 1:1, and all-female tetraploid gynogens were produced from these hexaploid fish, suggesting a male heterogametic sex-determination system. The progeny of 6N × 6N, 6N × 4N, 6N × 2N, 2N × 6N and 4N × 6N matings produced hexaploid, pentaploid, tetraploid, tetraploid and pentaploid progeny, respectively. The hexaploid loach produced viable triploid sperm and eggs and produced second-generation progeny of many ploidy levels when mated with loach of varying ploidy levels.

Tetraploid Shellfish

Similar to tetraploid fish, tetraploid shellfish appear to be sub-viable. Tetraploidy was induced (40–64%) in the Manila clam, *Ruditapes philippinarum* (Dufy and Diter, 1990), Sydney rock oyster (Nell *et al.*, 1998) and the European flat oyster, *Ostrea edulis* (Gendreau and Grizel, 1990), but no tetraploids were detected in 4-month-old spat. Tetraploid and pentaploid Pacific oysters all died within 72 h post-fertilization (Cooper and Guo, 1989) or almost all died within 12 days (Stephens and Downing, 1988). All triploid and tetraploid clam (*C. farreri*) embryos died, leaving only diploid juveniles (Yang, H. *et al.*, 1999). Guo and Allen (1998) were able to produce

viable tetraploid Pacific oysters. A male tetraploid was mated with a diploid female to produce triploid progeny.

Limitations and Constraints

One limitation to the application of triploid induction is identification of protocols and procedures that consistently or always yield 100% triploids. A second limitation is that, since newly fertilized embryos are needed for manipulation, the species of interest must be one that has reproductive traits conducive to large-scale hand-stripping of eggs and artificial fertilization. Production of commercial quantities of triploids will be relatively easy for some species, but difficult or near impossible for others. Triploid induction of tilapia is an example that would have great applicability in aquaculture if the biology of the species were only more compatible for triploid induction. Brämick *et al.* (1995) suggest that triploidy would suppress reproduction and significantly increase yields from pond culture. While it is fairly easy to induce triploid tilapia under laboratory conditions and in field trials, the low number of eggs per batch and asynchronous spawning of female tilapias currently prevent commercial production of triploid tilapia. A solution to the problems of mechanical triploid induction and the necessary artificial spawning to obtain adequate numbers of gametes is to produce tetraploids and mate them to naturally produced all-triploid populations. This has been demonstrated with rainbow trout, although only 96% triploids were produced. Fortunately, the remaining progeny were aneuploids, which are also sterile or unviable. This breeding scheme will not be widely used because induction of tetraploidy has proved extremely difficult for almost every species evaluated. Additionally, when tetraploid induction is successful, tetraploids are weak and have low survival, which may be problematic for producing commercial-sized brood populations. Another potential problem is that tetraploid males have sperm twice the size of those of diploid males. The tetraploid's sperm can be too large for the ova's micropyle, the portal of entrance for the sperm, reducing fertility in 2N female × 4N male matings. Theoretically, the reciprocal cross should not possess this problem.

Even when sterile triploid fish are produced, adverse ecological impacts are possible through competition for food, space and mates. Triploid female salmon and grass carp do not undergo sexual development and their sex-hormone levels are significantly lower than in normal diploid females. Triploid males (salmon and grass carp) have sex-hormone cycles and sexual development similar to those of diploid males, although the triploids are sterile. These males also exhibit sexual behaviour and can court and mate diploid females, although all progeny are usually aneuploid and die.

Obviously, if sufficiently large numbers of triploid males were in a breeding population, they could disrupt spawning and potentially cause an adverse effect on recruitment. Not only do male triploid fish exhibit sexual behaviour, but male triploid brook trout apparently emigrate in response to sexual maturation (Warrillow *et al.*, 1997), potentially and naturally placing them in a spawning habitat. A positive application might be to use triploid males to control reproduction of nuisance species. The complication of this strategy is that the introduction of large numbers of nuisance-species triploids could in the short term aggravate the adverse ecological effects, although long-term benefit may be derived. Another constraint of triploidy is that for some species unequivocal sterility cannot be guaranteed. Triploid males can sometimes produce viable diploid progeny. Oysters can revert from triploidy to at least partial diploidy. Even trisomic oysters can exhibit fertility.

Fisheries Management Applications

There are several ways in which triploidy could be used for sport-fish management. Since female triploid brook trout do not emigrate, Warrillow *et al.* (1997) suggest that stocking of triploid females could reduce autumn emigration and reduce the loss of catchable brook trout from Adirondack lakes with outlets and little spawning habitat. This benefit would have to be weighed against the likely higher mortality of the triploids.

Triploids could also be used to increase angling opportunities while not jeopardizing genetic conservation. It is impossible to have a hatchery for every stream and to propagate a specific genetic population individually for each stream. There may be cases where anglers may clamour for increased stocking to enhance a fishery, but the only way to meet demand would to be to bring in conspecifics from an outside source, thereby compromising the genetic integrity of the original native population. The utilization of sterile triploids to stock the system would allow angler demand to be met while preserving the genetic integrity of the population already occupying the stream.

To evaluate this concept, triploid rainbow trout were compared with diploid rainbow trout after stocking 18 Idaho streams (Dillon *et al.*, 2000). Return to anglers was identical for both genotypes and mean time to harvest was the same. Estimated cost for producing the catchable-sized triploids was 15% higher than for diploids. These results demonstrated that this management strategy could allow the simultaneous satisfaction of consumptive demand for fishing and conservation of native gene pools.

Environmental Protection

Use of triploids has been proposed to prevent domestic fish and transgenic fish from having a genetic impact on native gene pools. Cotter *et al.* (2000a) evaluated triploid technology for minimizing the interaction between domestic and wild Atlantic salmon. Triploid and diploid Atlantic salmon were released from hatcheries and sea cages. The return of triploids to the coast and to fresh water was reduced compared with diploids. No salmon from the cage releases returned to their home stream and return to the streams was minimal. The return of a small number of hormonally deficient triploid females to fresh water indicates that reproduction is not the sole trigger for the homing migration. Apparently, reduced survival of triploids relative to diploids in the natural marine environment was part, if not the major part, of the explanation for the lack of triploids returning to the spawning grounds.

Brook trout in the Adirondack lakes have a high incidence of emigration corresponding to maturation and spawning. Warrillow *et al.* (1997) found that only mature triploid males, diploid males and diploid females emigrated. No triploid females emigrated, triploid males had a reduced rate of maturation and more diploids emigrated than triploids. This example also illustrates the possible success of triploidy for protecting native gene pools from outside sources of fish.

The reduced survival of triploids in the natural environment, the reduced migration success of the triploids and their infertility indicate that application of triploids would greatly reduce the genetic and ecological impact of escaped farmed fish on native populations.

Another potential application of triploidy is to prevent the backcrossing and introgression of hybrids into the parent species. For example, an important sport fish with potential for application in aquaculture is the hybrid between striped and white bass. There are concerns and efforts to limit the use of this hybrid in sport-fish management and aquaculture because these diploid hybrids are fertile and have the potential to backcross with the parent species or produce F_2 progeny (Bayless, 1972). Individuals that are probably F_2 progeny have been detected in the wild (Avise and Van Den Avyle, 1984).

Triploid striped bass × white bass hybrids have been produced with hydrostatic pressure (Curtis *et al.*, 1987) and, since *Morone* bass have high fecundity, there is a good possibility of commercial production of triploid hybrids. The viability, growth, behaviour, angling vulnerability and fertility of these triploid *Morone* hybrids must be evaluated to ascertain their

potential for sport-fish management or aquaculture when it is desirable to use triploids rather than diploids.

Razak *et al.* (1999) evaluated the potential of utilizing triploidy to sterilize transgenic Nile tilapia containing growth-hormone genes. Gonadal development of the triploid males and females was retarded and they were apparently sterile. However, there was an adverse effect on growth enhancement from the triploid induction. Diploid transgenic Nile tilapia grew 3–4.4 times faster than diploid controls. Triploid transgenics grew 1.3–1.8 times faster than the diploid controls. Triploid growth-hormone transgenics did grow faster than normal diploids, but the benefits of transgenesis from growth-hormone gene transfer were partially negated by the triploidy.

Xenogenesis

A xenogenic organism is comprised of elements typically foreign to its species. Xenogenesis is a method of reproduction in which successive generations differ from each other.

The first step for fish xenogenesis is to produce sterile triploid host embryos. Then diploid donor stem cells are introduced into triploid host embryos. If the stem cells are transplanted into a sterile host at the blastula stage or around the time of hatching, the stem cells have the potential to migrate to the genital ridge, colonize and form gametes (both ova or sperm) of the donor species. Lastly, male and female host are bred, resulting in pure diploid donor progeny.

Xenogenesis has been accomplished in fish using stem cells from testicular tissue. Testes contain six major types of cell: primordial germ cells (PGCs), spermatogonia A, committed spermatogonia B, spermatids, mature sperm cells and somatic cells. An isolated germ cell within a spermatocyst is a type A spermatogonium, and these cells possess stem-cell potential (Schulz and Miura, 2002). These cells divide during spermatogenesis producing isogenic germ cells committed to meiosis. Okutsu *et al.*

(2006a) proved that these spermatogonia A were stem cell-like in fish when they transplanted them into developing rainbow trout (*O. mykiss*) embryos and these embryos produced sperm or eggs derived from the transplanted cells. Similar results were previously found for *Drosophila* male germline stem cells, which can regenerate by spermatogonial de-differentiation (Brawley and Matunis, 2004).

Both of these cell types, PGCs and spermatogonia A, have been transplanted from a donor species to a related host species (Takeuchi *et al.*, 2003; Okutsu *et al.*, 2006a, 2007; Saito *et al.*, 2008) with the recipient species producing sperm and eggs (originating from testicular PGCs or spermatogonial stem cells (SSCs)) of the target species (Okutsu *et al.*, 2006b). This procedure can be successful utilizing fresh, cultured or cryopreserved donor cells, opening up the possibility of many potential applications. In the case of the salmonid xenogens, testicular development was normal (Okutsu *et al.*, 2007). A 30–70% success rate was achieved among injected host embryos. The xenogenic individuals did have reduced fecundity (possible age effects), but the F_1 offspring were normal in genotype, performance and appearance.

Culture and transformation of spermatogonial stem cells

Shikina *et al.* (2008) developed a technique to isolate and culture type A spermatogonia from the testes of immature rainbow trout. The cultured spermatogonia were maintained for more than 1 month, and were able to colonize recipient gonads and proliferate upon transplantation into embryos near hatching. This could be a vital link to making cell-mediated gene transfer possible in fish, as knockout mice have been produced from spermatogonial cell lines (Kanatsu-Shinohara *et al.*, 2006), opening many possibilities for breeding systems and genomic studies in fish.

Alternative methods also have the potential to allow xenogenesis and autogenesis via PGC transplantation. Saito *et al.* (2008) transplanted a single PGC from pearl danio, *Danio albolineatus*, into the blastula of a zebra fish whose native PGC production had been knocked out by an antisense morpholinos oligonucleotide against dead end. The donated PGC formed a single testis that produced pearl danio sperm. Xenogenic pearl danio males were sex-reversed to femaleness and mated with untreated males to produce normal, fertile pearl danio offspring. Similarly, the zebra fish host was able to develop goldfish, *Carassius auratus*, and loach, *M. anguillicaudatus*, testis that produced donor sperm from the injection of a single donor PGC. Normally, a few dozen PGCs are needed to form gonads containing germ cells (Saito *et al.*, 2006). This study showed that one PGC and perhaps a single SSC are capable of producing a single testis.

Applications of xenogenesis

Xenogenesis could be put to the following uses.

1. To develop an improved system for making hybrid fish.
2. To couple an embryonic stem-cell transplantation system with homologous recombination to create a tool for gene knockout.
3. To gain a better understanding of the communication and interaction between genes and their tissue/cellular environment. We will have genes in hametes programmed to produce gene products at young or older age residing in tissues and an organism programmed to mature at a younger or older age than the gametes. Which will dictate maturity: the gonads and gametes, or the tissues and body, or both?
4. To develop technology to clone genetically superior male fish and multiply them into large populations (PGCs and SSCs have not undergone meiosis and are isogenic).
5. To develop a cryopreservation technology for the entire diploid and mtDNA genome of a fish individual.
6. To create a mechanism to regenerate extinct lines, strains and species of fish in the future, in the event of an epizootic or cataclysmic event, if we have frozen testes.

7.   To bank important natural populations of fish by inexpensively cryopreserving their testes, thus protecting our natural resources.
8.   To develop a method to improve spawning of a difficult-to-spawn species.
9.   To greatly accelerate all genetic/reproductive research.

Key Summary Points

Very rarely does the induction of triploidy enhance growth in fish, but it is very effective for enhancing growth in shellfish. However, in fish species such as salmonids, oysters and common carp, which are sometimes harvested and marketed at an age and size when sexual-maturation effects can slow growth and decrease flesh quality and carcass yield, the triploid genotype can be superior for growth, flesh quality and carcass yield. However, the tolerance to low dissolved oxygen is reduced in triploid fish compared with diploids. Triploid fish are generally sterile and females have a greatly reduced production of sex hormones, but triploid males can develop secondary sexual characteristics, exhibit spawning behaviour and induce females to expel eggs even though they are unable to fertilize them. Triploid shellfish are much more likely to exhibit low levels of reproduction. Triploidy can restore viability to unviable interspecific hybrids. Triploid oysters can sometimes revert some of their cells back to the diploid state. Polyploidy has good potential for applications in genetic conservation and environmental protection.

11

Sex Reversal and Breeding

Monosex or sterile populations of fish are desirable in aquaculture for a variety of reasons. The male grows faster in some species and the female faster in other species (Fig. 11.1). In this case, monosex culture of the faster-growing sex can increase production, and sexual dimorphism for growth occurs in most cultured fish (Dunham, 1990a).

Some species of fish mature at small sizes and young ages prior to the desired time of harvest. This can decrease production because unwanted reproduction results in crowding of the fish and higher densities than intended in the culture pond as well as wasted energy from the sexual activity of the stocked fish. Sex differences may also exist for flesh quality and carcass yield. Sexual maturity is also closely linked with both carcass yield and growth rate: as the fish become sexually mature, growth rate slows and carcass yield decreases. Although not well documented, sexual dimorphism could also exist for other economic traits, such as disease resistance or tolerance of poor water quality.

In some countries, the introduction of potentially valuable exotic fish species for aquaculture is met with resistance or is not allowed because of potential adverse ecological impacts and conservation concerns. Controversy also exists for the utilization of fertile, diploid hybrids in aquaculture or in sport-fish management, which may escape, backcross with and genetically contaminate parental species. The utilization of monosex or sterile populations of fish is a solution or partial solution for the problems associated with sexual differences, sexual maturation and unwanted reproduction.

Several techniques are available for producing monosex or sterile populations of fish for aquaculture. Manual sexing, sterilization, hybridization, gynogenesis, androgenesis, polyploidy, sex reversal, and sex reversal combined with breeding are options, all with advantages and disadvantages for producing the desired populations. Monosex populations can be produced by direct hormonal sex reversal. Phenotypic sex can be altered by administration of sex hormones, oestrogens or androgens, at the critical period of sex determination to produce skewed or all-female and skewed or all-male populations, respectively. The development of fish makes them conducive to the manipulation of their sex. Although the male or female genotype is established at fertilization, phenotypic sex determination occurs later in development. The artificial elevation of the appropriate sex hormone is sufficient to overcome the natural hormone or gene product during the period of sexual differentiation and to dictate the sex of the

Fig. 11.1.   Sexual dimorphism for size in Nile tilapia, *Oreochromis niloticus*. The male is the larger individual. (Photograph by R.O. Smitherman.)

individual. The period of sexual determination is related to size rather than age. Sex reversal is another manipulation for controlling reproduction in fish and ultimately improving the rate of growth.

Sexual Dimorphism

Sexual dimorphism for commercial traits of aquacultural fish, game fish or ornamental species is the key to the potential benefit of producing monosex populations. One sex grows faster than the other in most fish species. Channel catfish and tilapia males grow faster than females (Beaver *et al.*, 1966; Stone, 1981; Brooks *et al.*, 1982), whereas the opposite relationship exists for grass carp, *Ctenopharyngodon idella* (Hickling, 1967), rainbow trout and other salmonids (Johnstone *et al.*, 1978; Bye and Lincoln, 1986; Gall, 1986) and cyprinids, where females grow faster than males.

Sexual dimorphism can develop at young ages and small sizes. Two examples are channel catfish, which become sexually dimorphic for body weight by 6 months of age, when the fish are approximately 12 cm

in length (Brooks *et al.*, 1982); and tilapia, which not only become sexually dimorphic in size at 2 or 3 months, but also become sexually mature (Shelton *et al.*, 1978). The timing and degree of sexual dimorphism may vary among strains within a species. Although male channel catfish usually grow faster than females, the amount of sexual dimorphism differs among strains, and the Marion strain exhibits no sexual dimorphism for size (Benchakan, 1979; Dunham and Smitherman, 1984).

Faster growth of one sex is presumably a result of a combination of genetic and hormonal factors, although competition and suppression of one sex and magnification of initial size differences are alternative explanations for sexually dimorphic growth. Most measurements of growth have been made when both sexes are in competition in the same environment. If competition were eliminated and the two sexes grown in separate environments, sexual dimorphism for growth rate might decrease, although Stone (1981) and Hanson *et al.* (1983) demonstrated that, when male and female Nile tilapia are grown in separate and communal environments, growth differences are similar and competition apparently has no large effect. Male Nile tilapia grew 2.5 and 2.2 times faster than females when grown in cages mixed and separate, respectively (Stone, 1981), so there is about a 12% magnification of the sexual dimorphic growth from competition.

Manual separation of the sexes (Chimits, 1957; Mires, 1969; Lovshin and Da Silva, 1976) requires the least amount of technology for monosex culture, but it is extremely wasteful, tedious and inefficient. Although this technique has been applied effectively, especially in developing countries (Lovshin and Da Silva, 1976; Popma, 1987), it has numerous disadvantages. The first is that the fingerlings must be grown to a size large enough to visually determine sex before manual separation, resulting in half the fingerling production being wasted since the slower-growing sex is culled unless the slower-growing sex is utilized in separate ponds, when prevention of reproduction is the primary consideration. Manual separation of sexes is extremely labour-intensive and has the potential for mistakes (Lovshin and Da Silva, 1976; Popma, 1987).

An alternative method to hand-sexing is mechanical separation by size grading, which can separate most of the fish by sex. Some of the same disadvantages exist, as half the initial fingerling production is wasted. If elimination of reproduction in early-maturing species in production ponds is the primary goal, grading will probably result in too many mistakes to adequately separate the sexes to eliminate reproduction.

Timing of sexual maturity and carcass traits are often sexually dimorphic. Sexual maturity, growth rate, maximum attainable size, size at sexual maturity, degree of sexual dimorphism and carcass traits are all interrelated, but different interrelationships exist among various species (Alm, 1959; Smitherman and Dunham, 1985; Tveranger, 1985; Gall, 1986; Dunham and Smitherman, 1987; Dunham, 1990a). In the case of rainbow trout and Atlantic salmon, males mature at younger ages than females (Johnstone *et al.*, 1978; Bye and Lincoln, 1986;

Gall, 1986), which prevents a portion of these males from reaching harvestable size and reduces the flesh quality in males by the time they reach typical marketable sizes. Aggravating this problem further, some Atlantic salmon males – streak males – exhibit precocious sexual maturity and reproduction. The most rapidly growing salmonid males may reach maturity early, but the slower-growing, later-maturing males will eventually surpass the early-maturing males in size (Tveranger, 1985; Gall, 1986). Age at sexual maturity has a high heritability in salmonids (Naevdal, 1983; Gjerde, 1984; Gjerde and Gjedrem, 1984), and a high genetic correlation exists between growth rate and age of sexual maturity (Thorpe *et al.*, 1983; Gjerde and Gjedrem, 1984). Similar to salmonids, rapidly growing male mosquito fish, *Gambusia affinis*, also exhibit early sexual maturity (Busack and Gall, 1983).

These interrelationships are different and even the opposite in other species, so no overall principle exists. Early- and late-maturing platyfish, *Xiphophorus maculatus*, males grow at the same rate until maturity (Kallman and Borkoski, 1978). Early- and late-maturing platyfish males differ at a sex-linked maturation locus, and late-maturing males reach twice the mature size of early-maturing males (McKenzie *et al.*, 1983). The fastest-growing strain of channel catfish, Kansas, matures a whole year later, on average, than slow-growing strains of channel catfish (Dunham and Smitherman, 1984, 1987). The slower-growing strains of channel catfish, such as Rio Grande, exhibit more sexual dimorphism for body weight than the faster-growing strains (Benchakan, 1979; Dunham and Smitherman, 1984, 1987). This opens the possibility that selection for increased body weight could affect age and size at sexual maturity and might also reduce sexual dimorphism in size.

Examination of the sexual maturity, sexual dimorphism and growth-rate interrelationships among species of catfish, rather than among strains within a species of catfish, reveals interrelationships parallel, in some aspects, to those for salmonids.

White catfish, *Ameirus catus*, have rapid early growth, early sexual maturity, small mature size and much sexual dimorphism for body weight (Brooks *et al.*, 1982; Dunham and Smitherman, 1984). Blue catfish, *Ictalurus furcatus*, have slow early growth, late sexual maturity, large mature size (maximum for ictalurids) and little sexual dimorphism for body weight, while channel catfish, *Ictalurus punctatus*, are intermediate for all of these traits. Among ictalurid catfish, fast early growth is correlated with early sexual maturity and large sexual dimorphism for body weight and is inversely related to mature size and maximum attainable weight. The same interrelationships exist among species of centrarchid sunfishes (Childers, 1967).

Some, but not all carcass traits of catfish are sexually dimorphic. Carcass yield in male and female channel catfish is similar as the males develop larger heads and females have increased visceral percentages, resulting in equal reduction of carcass yield (El-Ibiary *et al.*, 1976; Dunham *et al.*, 1985). The flesh texture of sexually mature 4-year-old male channel catfish is much tougher than that of females when the fish are processed during the spawning season, similar to the sexual dimorphism in flesh quality seen in salmonids. This may be related to the weight loss and increased muscularization exhibited by males prior to spawning compared with the weight gain, water retention and relative lack of increased muscularization exhibited by females during the same time (Jensen, J. *et al.*, 1983).

Sexual maturation in both sexes of rainbow trout results in deterioration of flesh quality and alterations in skin and muscle pigmentation, reducing market acceptability (Johnstone *et al.*, 1978; Bye and Lincoln, 1986). Rainbow trout are usually marketed at 2 years of age or less, so the primary carcass problems are related to early-maturing fish. The degree of this problem varies depending on market demand, as the US market historically requires a smaller fish while the European market demands a larger fish, aggravating the problem. Early maturation is sexually dimorphic in rainbow trout, with males more likely to mature early – jacks – resulting in marketing problems (Johnstone *et al.*, 1978; Bye and Lincoln, 1986).

Chemical and Mechanical Sterilization

Sterilization is an alternative to prevent unwanted reproduction in ponds or to prevent the establishment of escaped, useful introduced or exotic species, domestic and transgenic aquatic organisms or hybrids in the natural environment. Sterilization might also promote growth, change the behaviour of the fish, alter body composition or improve carcass yield. Sterilization can potentially be accomplished through surgery, immunology or radiation or with chemicals. Sterilization can also be accomplished genetically through hybridization, ploidy manipulation or transgenesis, which will be discussed later. Unfortunately, none of these non-genetic technologies have shown much promise until recently. The obvious disadvantages of surgical sterilization include large labour expenditures and the possibility of mortality, which make it impractical on a large scale. Although tilapia, grass carp, salmonids and catfish all have high survival after gonadectomy (Brown and Richards, 1979; Akhtar, 1984; Underwood *et al.*, 1986; Bart, 1988; Bart and Dunham, 1990), the labour expenditure does not allow the use of this technique for large-scale commercial production. Surgical sterilization might be feasible for applications when only a few, valuable fish need sterilization. Stocking of small numbers of sterile grass carp for weed control is a possible example.

Surgical sterilization has not been successful in fish and does not allow permanent sterilization because fish have the ability to regenerate gonadal tissue. Both testes and ovaries of grass carp fully regenerate and function even after the surgery is conducted on sexually mature fish (Underwood *et al.*, 1986). Regeneration occurred despite removal of the entire gonad, including the surrounding mesentery. Sonneman (1971) and Brown and Richards (1979) felt that regeneration came from tissue close to the removed gonad, but the data on grass carp refute this hypothesis. Clippinger and Osborne (1984) also found that regeneration occurred when only one-third of the gonad was removed in immature grass carp, and complete regeneration was more frequent in mature grass carp males (91%) than in mature females (38%) (Underwood *et al.*, 1986). Testicular regeneration occurs in both adult blue catfish and channel catfish, although it is slow and may take several years for complete regeneration. However, the ability of male and female fish with regenerated gonads to ejaculate or ovulate on their own has not been determined, although ictalurid catfish with regenerated testes produce viable sperm (Bart, 1994). The gonads of tilapia also have the ability to regenerate. Ovariectomized Nile tilapia females regenerated ovaries in 119 days (Akhtar, 1984) and, despite being fed methyltestosterone, two-thirds of the females regenerated ovarian tissue.

Partial removal of gonads can also cause compensation by the remaining tissue; Peters (1957) observed that, after removal of one ovary or part of both ovaries, the remaining tissue regenerated to the point of producing the normal mass of tissues and eggs. Partial removal of the testes in rainbow trout caused the remaining testes to grow at an accelerated rate (Robertson, 1958). However, cauterization may have potential applicability, as cauterization of the seminal vesicles of the catfish, *Heteropneustes fossilis*, during the prespawning period caused complete regression of the seminal vesicle and, after cauterization, only one fish showed any sign of regeneration (Sundararaj and Naygar, 1969).

Regenerative ability may be species or technique specific, as Donaldson and Hunter (1982) indicated that surgical sterilization was somewhat successful in salmonids (Robertson, 1961; McBride *et al.*, 1963) and that relatively rapid gonadectomy techniques have been developed for fish greater than 25 cm total length (Brown and Richards, 1979). Again, labour and the necessity of complete testicular removal to prevent development of secondary sexual

characteristics were cited as major problems (Brown and Richards, 1979; Donaldson and Hunter, 1982). Additional problems were found when sterilizing sockeye salmon, *Oncorhynchus nerka*, which regenerated testicular tissue in the posterior region adjacent to the spermatic duct, resulting in spermatogenesis after gonadectomy, and females regenerated ovarian tissue and developed a limited number of eggs (Robertson, 1961).

Immunological sterilization of fish has received limited attention. Laird *et al.* (1978, 1981) injected juvenile Atlantic salmon with a homogenate of homologous gonadal tissue with Freund's adjuvant and obtained some inhibition of gonadal development.

Sterilization of fish by irradiation, X-rays and gamma rays has also been relatively unsuccessful. Sterilization was achieved in some cases (rainbow trout, tilapia), but similarly to surgical sterilization, the effects were temporary (Kobayashi and Mogami, 1958; Al-Daham, 1970). Some experiments have demonstrated promise for inhibiting gonadal development of salmonids with irradiation. Isotopes depressed the primary sex cells of Atlantic salmon (Migalovskii, 1971), and gamma rays depressed sexual maturation in rainbow trout (Tashiro, 1972) and gonad development in chinook salmon (*Oncorhynchus tshawytscha*) fry (Bonham and Donaldson, 1972). X-rays reduced ovarian size in rainbow trout fingerlings (Kobayashi and Mogami, 1958) and suppressed gametogenesis for up to 7 months, sterilizing some pink salmon (*Oncorhynchus gorbuscha*) males (Pursov, 1975). Irradiation of adult rainbow trout did not inhibit sexual maturation (Foster *et al.*, 1949), while irradiation of young chinook salmon can alter sexual maturation and inhibit anadromous migration in at least some individuals (Hershberger *et al.*, 1978).

Chemosterilization has shown some potential, although, as with the other sterilization procedures, the effects have been temporary (Stanley, 1979). Chemosterilants can interfere with mating behaviour or with the production, release or utilization of sex steroids. Chemosterilization had some success in blocking mating competitiveness in sea lampreys, *Petromyzon marinus* (Hanson and Manion, 1978). Overdoses of hormones and anabolic steroids can lead to sterilization (Hirose and Hibiya, 1968; Yamazaki, 1976). Hormone antagonists, such as methallibure, have shown the most potential (Shelton and Jensen, 1979; Donaldson and Hunter, 1982). This compound acts as a chemical hypophysectomizing agent by interrupting the production and release of gonadotrophins (Paget *et al.*, 1961; Pandey and Leatherland, 1970). Methallibure may act by blocking production or release of gonadotrophins at one or more sites (Bell *et al.*, 1962; Malven *et al.*, 1971). Implantation of methallibure in the hypothalamus of rats prevented ovulation, indicating hypothalamic control of gonadotrophin release from the pituitary (Malven *et al.*, 1971). Methallibure reduced plasma but not pituitary gonadotrophin in goldfish, *Carassius auratus*, indicating that methallibure acted by inhibiting the gonadotrophin-releasing factor (Breton and Jalabert, 1973; Merch *et al.*, 1975). Methallibure has prevented spermatogenesis and testicular steroidogenesis in sea perch, *Cymatogaster aggregata* (Wiebe, 1968), blocked gonadal differentiation in guppies, *Poecilia reticulata* (Pandey and Leatherland, 1970), suppressed spermatogenesis in guppies (Pandey and Leatherland, 1970) and caused degeneration of cells in testes of adult tilapia (Dadzie, 1972). This drug also reduced gonad weight and maturation of gonads in both sexes of goldfish, sea perch and stickleback, *Gasterosteus aculeatus* (Hoar *et al.*, 1967). Methallibure, a dithiocarbamoylhydrazine derivative (ICI 33,828), inhibited sexual development in males more than females in pink salmon (Donaldson, 1973; Flynn, 1973).

Cyproterone acetate is a progestogen with powerful antiandrogenic properties that reduce gonadal development and spermatogenesis in mammals (Shelton and Jensen, 1979). This compound blocked the uptake of testosterone in rainbow trout (Schreck, 1973). Cyproterone acetate was not effective in producing sex reversal in Japanese medaka, *Oryzias latipes* (Smith, 1976),

or blue tilapia, *Oreochromis aureus* (Hopkins, 1977), but administration of cyproterone acetate delayed spawning and reduced mating and total fry production in blue tilapia. Methallibure also had a similar effect of delaying spawning and reducing reproduction in blue tilapia, but also reduced body weight by 60% after 6 weeks of treatment (Shelton and Jensen, 1979). However, there were no long-term effects on growth rate after termination of the methallibure treatment.

Trenbolone acetate (TBA), which has been approved for use in cattle, has shown promise for masculinizing and sterilizing channel catfish (Davis, K.B. *et al.*, 2000). Channel catfish fry were fed with 0, 50 or 100 mg TBA/kg feed for 60 days. Fingerlings were identified as males. At 3 years of age the gonadal weight, GSI and plasma testosterone concentration were higher in control fish than in treated fish. These TBA-treated male channel catfish readily spawned, but all spawns were infertile. Histological examination of the gonads indicated that TBA interferes with normal gonadal development of both testis and ovaries at the concentrations used, but did not functionally masculinize channel catfish.

Another new means of chemical sterilization that has potential is administration of γ-aminobutyric acid (GABA). *In vivo* administration of GABA-A agonists to embryonic mice decreased migration of gonadotrophin-releasing hormone (GnRH) neurones out of the nasal placode, and antagonism of GABA-A receptors resulted in disorganized distribution of GnRH neurones within the forebrain (Bless *et al.*, 2000). GABA directly acts on GABA-type receptors, becoming a migratory stop signal during GnRH neuronal development in mammals. This temporary migrational pause is necessary for proper organization of the GnRH neurones in the forebrain.

Overall, the effects of chemosterilants appear to be temporary to date, although blue tilapia fry treated with methallibure for 6 weeks after hatch had reduced fry production when they were 1 year old (Shelton and Jensen, 1979). However, Hoar *et al.* (1967)

found that, within 1 week after cessation of methallibure treatment, spermatogenesis resumed. Mechanical and chemical sterilization techniques have shown promise, but have not been adequately developed for practical application and 100% effectiveness. Additionally, food-safety issues and government approval of these treatments will need to be addressed before there is any future application. Some of these methods warrant further long-term evaluation to prove efficacy and improve efficiency. If they were effective, the impact of utilization on exotic species, hybrids, transgenics and domestic fish could be enormous and have major implications for the preservation of biodiversity and genetic biodiversity.

Hormonal Sex Reversal

Monosex populations can sometimes be produced through hormonal sex reversal (Quillet *et al.*, 1987). The development and sex-determination mechanisms of fish make them conducive to the manipulation of their sex. Although the male or female genotype is established at fertilization, phenotypic sex determination occurs later in development and the timing of sex determination varies among species. Production of monosex populations by direct hormonal treatment requires elucidation of the labile period of sexual differentiation during which the fish are susceptible to hormonal masculinization or feminization.

Phenotypic sex is determined prior to hatch in some salmonid species (Goetz *et al.*, 1979), during the first 3–4 weeks after hatch in channel catfish (Goudie *et al.*, 1983) and Nile tilapia (Shelton *et al.*, 1978), and even as late as fingerling stage (85–200 mm) for grass carp (Boney, 1982; Jensen and Shelton, 1983; Boney *et al.*, 1984). Sex determination is dependent on size and developmental stage rather than on age, and this is critical for planning the initiation and cessation of the hormone treatment.

The phenotypic sex can be altered by administration of oestrogens or androgens during the critical period of sex determination to produce skewed, all-female or all-male populations. The dosage of artificial hormone is sufficient to overcome the natural hormone or gene product and dictate the sex of the individual.

Several androgens have been used to produce monosex male populations (Yamazaki, 1983; Dunham, 1990a). Most of these androgens are synthetic derivatives of testosterone. One of the most efficacious and widely used is 17-methyltestosterone (Dunham, 1990a). Other compounds used include 19-norethynyltestosterone and 11-ketotestosterone (Yamazaki, 1983). Thyroid hormones have also been used to alter sex or growth rate (Howerton *et al.*, 1988; Reddy and Lam, 1992).

Several oestrogenic compounds have been used to produce monosex female populations. The hormone 3-oestradiol has been one of the most efficacious compounds, and oestrone and ethynyloestradiol can also be used for feminization (Yamazaki, 1983; Dunham, 1990a).

Sex reversal can be accomplished by administering the exogenous hormones by bath (Donaldson and Hunter, 1982; Yamazaki, 1983), in feed (Shelton *et al.*, 1978) or through implants (Boney, 1982; Boney *et al.*, 1984), depending on the developmental and culture characteristics of the species. Coho salmon have been sex-reversed to femaleness by bathing the embryos in 17β-oestradiol (25 µg/l), followed by oral administration of 17β-oestradiol (10 mg/kg) to the fry (Goetz *et al.*, 1979). Immersion of alevin of masu salmon, *Oncorhynchus masou*, in a bath with 17β-oestradiol (0.5–5 µg/l) produced 100% females (Nakamura, 1981). Tilapias have been sex-reversed to all-maleness by feeding 10–60 mg 17α-methyltestosterone/ kg feed for 21–35 days post swim-up (Shelton *et al.*, 1978; Hanson *et al.*, 1983; Popma, 1987). Grass carp were sex-reversed to all-maleness with silastic implants of 17α-methyltestosterone in the abdominal cavity (Boney, 1982; Boney *et al.*, 1984). The androgenic hormone was released from the implant over a 60-day interval encompassing the period of sex determination for grass carp (Shelton, 1986).

In the majority of species, sex reversal has been accomplished by feeding the young fish hormone-treated feed. This feed is prepared by dissolving the hormone in ethanol and spraying it on the feed (Shelton *et al.*, 1978). The feed is air-dried and is then ready for use. All-female populations of rainbow trout are desirable because of their late sexual maturity, relative faster growth and superior flesh quality compared with males. Successful sex reversal of salmonids with oestrogens has been inconsistent (Ashby, 1957; Jalabert *et al.*, 1975; Simpson, 1975; Simpson *et al.*, 1976; Johnstone *et al.*, 1978, 1979a,b; Donaldson and Hunter, 1982), and one problem has been the incomplete sex reversal of genotypic males, resulting in hermaphroditic individuals with ovotestes (Jalabert *et al.*, 1975).

Sex reversal of genotypic male blue tilapia to phenotypic femaleness can be similarly difficult (Hopkins, 1977; Liu, 1977; Sanico, 1977; Hopkins *et al.*, 1979; Jensen and Shelton, 1979; Shelton and Jensen, 1979; Meriwether and Shelton, 1981). Again, a large percentage of individuals with ovotestes resulted when oestrogens were administered to genetic males (Meriwether and Shelton, 1981). Apparently, for many species it is easier to change genetic females to phenotypic males than to change genetic males to phenotypic females.

Channel catfish are an exception to the trend in that it is problematic to change genetic females to functional phenotypic males (Goudie *et al.*, 1983). All-female populations can easily be produced with a variety of oestrogens, including 17β-oestradiol, and all attempts to produce all-male populations of channel catfish with androgens failed (Goudie *et al.*, 1985; Davis, K.B. *et al.*, 2000). In fact, administration of testosterone to channel catfish fry results in populations skewed towards femaleness – paradoxical feminization. Apparently, biofeedback systems of channel catfish react to the elevated levels of androgen by digesting and converting the excess androgen to oestrogen-like compounds, ultimately elevating oestrogen levels and sex-reversing genetic males to phenotypic females. Compounds exist that inhibit or block the enzymes that convert

testosterone to oestrogen. Initial attempts at feeding these inhibitors simultaneously with testosterone still resulted in populations skewed towards femaleness (Goudie *et al.*, 1985). As is the case with Nile tilapia, 100% female populations of channel catfish are not desirable for commercial application since males grow faster than females.

Sex reversal to all-maleness in tilapia, primarily Nile tilapia, is now routine throughout most of the world in both industrialized and developing countries. Feeding swim-up fry 10–60 mg 17α-methyltestosterone/kg feed for 21–35 days results in populations with 95–100% males (Clemens and Inslee, 1968; Guerrero, 1974, 1975; Rodriguez-Guerrero, 1979; Hanson *et al.*, 1983; Obi and Shelton, 1983; Shepperd, 1984; El-Gamal, 1987; Popma, 1987; Jo *et al.*, 1988; Tiwary *et al.*, 1999). Sex reversal to all-maleness using methyltestosterone is widely done in Israel, where it is commonly applied in conjunction with interspecific hybridization of blue tilapia and Nile tilapia to ensure 100% all-male hybrids.

Several factors influence the effectiveness of sex reversal in tilapia and other fish, including species of fish (Yamazaki, 1983), genetics (Shepperd, 1984; El-Gamal, 1987), type of hormone (Shepperd, 1984), dosage of hormone (Guerrero, 1974; Jo *et al.*, 1988), duration of treatment (Popma, 1987) and timing of treatment (Popma, 1987). Genetic effects are evident as sex reversal to all-maleness has been less successful for red tilapia than for Nile tilapia, and this may be directly related to the fact that the red tilapia populations had sex ratios skewed towards femaleness (Shepperd, 1984; El-Gamal, 1987). The increased percentage of non-sex-reversed females is probably related to the increased proportion of females in the initial population.

Originally, researchers believed that the sex-reversal procedure needed to be implemented in the hatchery environment in water relatively devoid of plankton. Theoretically, if plankton were available, the hormone-treated feed might not be a sufficient proportion of the diet for effective sex reversal, effectively reducing dosage rates.

However, sex-reversal rates in environments with and without phytoplankton gave the same high rates of sex reversal (Lopez-Macias, 1980; Buddle, 1984; Chambers, 1984).

Timing and duration of hormone treatment are extremely important. The critical time for initiation of hormone treatment of Nile tilapia was originally established at 12 mm for efficient sex reversal (Shelton *et al.*, 1978; Popma, 1987). However, Popma (1987) later learned that tilapia fry as large as 14–15 mm can be effectively sex-reversed, and that the duration of the treatment was also important, not in terms of absolute time but in terms of the growth of the fry. Treatments of 28 days' duration yielded higher percentages of males than 21-day treatments. A large percentage of females was found in populations where some of the fry were 12 mm or less when treatment was terminated; however, most of these females were sex-reversed if the treatment was not terminated until all the fry reached 13–14 mm. Size is important for the timing of sex determination in some fish (Dutta, 1979; Popma, 1987).

Grass carp are difficult to train to accept artificial feed, limiting options to accomplish sex reversal. However, sex reversal of grass carp from genetic females to phenotypic males is effective utilizing silastic implants of methyltestosterone. Sex reversal is accomplished when the implants are placed in the fish at 85 mm and the hormone is released until the fish reach about 200 mm. These sex-reversed males (genetic female, phenotypic male) produce viable sperm capable of fertilizing normal eggs and producing all-female progeny (Boney *et al.*, 1984).

Anabolic effect

Feeding male hormone to fish can result in an anabolic, a catabolic or no effect on the subsequent growth of the fry after cessation of treatment (Hanson *et al.*, 1983). Even though the hormone-treated feed was only delivered during the initial 3–4 weeks of life, the feeding of methyltestosterone-treated feed to tilapia fry affected their growth rate to 200 g or larger (Hanson *et al.*, 1983). Coho salmon treated with methyltestosterone also grew faster than untreated controls (Fagerlund and McBride, 1975), and both treated coho salmon and tilapia were more efficient at converting feed than were the untreated controls (Fagerlund and McBride, 1975; Hanson *et al.*, 1983; El-Gamal, 1987). Feed conversion was increased in sex-reversed blue tilapia, Nile tilapia and their respective red backcrosses compared with control populations.

The type of male hormone utilized for sex reversal determines the extent of anabolic effect and whether or not it occurs (Hanson *et al.*, 1983; Dunham, 1990a). Sex reversal with ethynyltestosterone depressed the rate of growth by 5% in Nile tilapia and blue tilapia relative to controls (Anderson and Smitherman, 1978), whereas sex reversal with 17-methyltestosterone increased the growth rate of Nile tilapia by 30% relative to controls (Hanson *et al.*, 1983). Adrenosterone increased initial growth of tilapia fry, but by the time the fish reached 3 g there was no difference in size between treated and control fish (Katz *et al.*, 1976).

The treatment level and duration of treatment also have an effect on the extent of the anabolic response. Coho salmon (Fagerlund and McBride, 1975) and goldfish (Yamazaki, 1976) that were fed 1 mg methyltestosterone/kg feed grew faster than controls, but those fed 10 mg methyltestosterone/kg feed did not exhibit accelerated growth, and goldfish fed 30 mg methyltestosterone/kg feed had depressed growth. The anabolic response of Nile tilapia and blue tilapia increased with the level of 17-methyltestosterone and with duration (Rodriguez-Guerrero, 1979; Tiwary *et al.*, 1999); however, no synergism between level and duration existed for Nile tilapia (Tiwary *et al.*, 1999). Female blue tilapia expressed larger anabolic responses than males (Jo *et al.*, 1988). In general, it appears that increasing levels of androgens have the potential to increase growth, but, if the peak application rate is reached, no further growth enhancement is obtained or growth can actually be depressed.

Sex hormones may have different effects on older fish. The anabolic effect decreased with increased levels of 17-ethynyltestosterone and increased duration of feeding in fingerling tilapia, and different size classes gave different responses (Rothbard *et al.*, 1988). Responses in the hatchery environment differed from those in field tests. Large doses of androgens can also result in impaired gamete production and sterility in both salmon adults and embryos; however, the sterility can lead to extended lifespan for the fish (Donaldson and Hunter, 1982).

One explanation for the anabolic effect is that the conversion of females to males results in increased mean body weight. However, the hormone is primarily responsible for the anabolic effect, not the conversion of the fry from female to male (Hanson *et al.*, 1983). Genotype also has an effect, either by allowing enhanced effects of the artificial hormone or by additional benefit of increased gene expression and the production of natural hormones. Hanson *et al.* (1983) compared the growth of untreated Nile tilapia males (all XY, hand-sexed), sex-reversed males (all XX) and treated sex-reversed males (mixed XX and XY, presumably 1:1) both sex-reversed with 17α-methyltestosterone. Sex-reversed males (all XX) were slightly larger than untreated normal males, despite the fact that they had the female genotype, indicating that the administration of the hormone and its effects were more important than the genotype of the fish. The heaviest group was the treated sex-reversed males (mixed XX and XY), indicating that genotype has an important effect also, as treated XY males grew faster than treated XX males and untreated XY males.

Additional data indicate that genetic effects influence the anabolic response, as the anabolic effect is more dramatic in Nile tilapia than in blue tilapia (El-Gamal, 1987). The anabolic effect was also larger in back-crossed red × Nile tilapia than in back-crossed red × blue tilapia, providing further evidence for the genetic basis for variable anabolic effects.

In contrast to androgens, oestrogens may cause catabolic effects (Dunham, 1990a). Several examples exist illustrating that oestrogen-treated salmon embryos and fry exhibit depressed growth and sometimes high mortality (Donaldson and Hunter, 1982).

Health issues

In some countries restrictions exist on the sale of hormone-treated aquatic organisms unless it is proved that there are no risks to human health from consuming these animals. This presents a problem for marketing fish that have been treated with sex hormones in countries with these regulations. Of course, natural sex-hormone levels will be higher in the juvenile or adult fish when marketed than in the treated fry and the hormone fed to the fry will have been long metabolized; however, regulators do not have an appreciation for this. Radiolabelled methyltestosterone fed to rainbow trout and *Oreochromis mossambicus* fry essentially disappears within 100 h after consumption (Johnstone *et al.*, 1983), and similar results have been obtained for blue tilapia (Goudie, 1984; Goudie *et al.*, 1986).

Sex Reversal and Breeding

Sex reversal is not always 100% effective and can be difficult or impossible for some species. Additionally, marketing of hormone-treated fish is allowed in some countries but is illegal in others, such as the EU countries and India.

One method of circumventing some of these problems is by combining sex reversal with breeding. When sex reversal is combined with genetic manipulation, populations can be produced that naturally and genetically produce monosex progeny when mated. Theoretically, this can allow natural production of monosex populations by mating sex-reversed and untreated individuals of like homogametic genotype. Alternatively, if the desired sex is heterogametic, progeny having 100% of this genotype can be produced by mating one sex artificially homogametic for a genotype that does not naturally occur with the other sex that is naturally homogametic for the alternative allele. For example, in some species, sex reversal is used as the first step for eventually producing YY individuals, which, if viable and fertile, could be mated to XX females to produce 100% XY male populations.

Production of monosex populations via sex reversal and breeding requires an understanding of the sex-determining mechanism of the desired species. Fish express a wide variety of sex-determination mechanisms (Dunham *et al.*, 2001). Many commercially cultured species (carps, salmonids, catfishes) exhibit an XX/XY sex-determination mechanism, but some marine species and shellfish are sequential hermaphrodites (gilthead seabream, groupers, oysters), while others exhibit mixed XX/XY genetic, polygenic and temperature control over sex determination, such as Nile tilapia (Dunham, 1990a; Mair *et al.*, 1991a; Dunham *et al.*, 2001), hirame, *Paralichthys olivaceus* (Yamamoto, 1999), and loach (Nomura *et al.*, 1998). The sex-determination mechanism of the European sea bass, again a marine species, is still not clear, although temperature does appear to be important (Dunham *et al.*, 2001). Different mechanisms may be found in closely related species. The Nile tilapia has an XX/XY system, while the blue tilapia has a WZ/ZZ system (Dunham, 1990a; Mair *et al.*, 1991b), and evidence for similar diversity in sex determination exists for centrarchids (Childers, 1967) and ictalurids (Dunham and Smitherman, 1984). Androgenesis (Bongers *et al.*, 1994; Myers *et al.*, 1995), gynogenesis, analysis of chromosomes in meiosis (synaptonemal complexes) (Carrasco *et al.*, 1999) and a variety of molecular-genetics techniques have been used to assist in elucidating sex-determining systems.

Various breeding strategies utilizing sex reversal and breeding, progeny testing, gynogenesis and androgenesis can lead to the development of all-male or female or near 100% all-male or female populations that are homozygous for the female genotype, XX, or the YY genotype, which has been given the misnomer super-male. The objectives of sex reversal and breeding are similar to those of polyploidy: to control reproduction or prevent the establishment of exotic species and to enhance sexually dimorphic traits, such as flesh quality and growth. XX populations have been successfully developed for salmonids, carps and tilapias and YY populations have been established for Nile tilapia, salmonids and, marginally, for channel catfish.

All-female XX systems

One approach is to combine gynogenesis and sex reversal to produce monosex populations. The problem of low yield of gynogens can also be overcome by this approach. When the female is the homogametic sex, the gynogenetic progeny are all XX. If these fry are sex-reversed to phenotypic maleness and if these sex-reversed XX males are fertile, they can then be mated to normal XX females and the progeny from this pairing should be 100% XX females. Because of the natural mating, the yield of XX embryos should be high. When male brood-stock replacements are needed, some of the monosex XX fry are sex-reversed to males to produce the future brood stock.

This system of combining gynogenesis and sex reversal for producing all-female populations has been accomplished for grass carp (Boney *et al.*, 1984; Shelton, 1986). First, meiogenetic grass carp (genotype XX) were produced and then they were sex-reversed to phenotypic males. When the gynogenetic grass carp reached 75 mm, time-released methyltestosterone silastic implants were surgically placed in their abdomens and fertile sex-reversed XX male grass carp resulted. When mated with normal XX females, all-female progeny were produced (Boney, 1982; Boney *et al.*, 1984). Grass carp sexually differentiate between 85 and 200 mm (Shelton and Jensen, 1979; Shelton, 1986).

If this technology had developed prior to the introduction of grass carp to the USA, there would have been no need for sterilization of grass carp via triploidy to prevent the establishment of this exotic species. When this type of system results in 100% males or 100% females, the fish can be introduced and applied anywhere outside their natural geographical range without fear of their establishment if that species is not already present. Replacements of the phenotypically reversed sex must be generated by hormone application to the fry. If both sexes were to escape, one generation of progeny could be produced, but, when the original escapees die, only one sex remains in the system and the species dies out. Ecological impact is temporary unless irreversible damage occurs during the short time that the exotic species is present, such as extinction of a native competing species. Interspecific hybridization with a native species is another potential ecological/genetic risk, which is not eliminated with the sex-reversal system.

This system of producing all-female populations appears feasible for species where the female is homogametic and can be sex-reversed; however, not all species possess these criteria. The disadvantages of the gynogenetic system are that in some species gynogenesis is difficult or impossible, the low survival of the gynogens could make generation of commercial-scale populations difficult and the gynogenetic approach could greatly increase inbreeding unless substantial effort is implemented to utilize a large number of progenitors. The inbreeding and random genetic drift can be prevented or corrected by initiating the population with several gynogenetic families or by crossbreeding, respectively.

An alternative to the approach utilizing gynogenesis and sex reversal is using sex reversal with progeny testing and breeding to develop monosex female populations (Fig. 11.2). In contrast, one advantage of the gynogenetic system is that it eliminates the progeny testing necessary to identify and eliminate normal XY males. Of course, both systems are useful only when monosex female populations are desirable or acceptable. A similar process combining androgenesis and sex reversal could produce all-male populations or all-female populations or both, depending on whether the males are the homogametic or heterogametic sex and whether males can be sex-reversed to be functional females.

A classical example of producing monosex populations through sex reversal, progeny testing and breeding to produce all-female populations was accomplished with Nile tilapia (Calhoun, 1981; Calhoun and Shelton, 1983). In this case, females are XX and males are XY. Fry were fed 17α-methyltestosterone to sex reverse the group to all males, half with the genotype XX and half with the genotype XY. The treated fish were then progeny-tested by mating them with normal XX females. XX males should produce 100% female progeny and were kept as founder brood stock. XY males should produce progeny that are approximately 1 M:1 F and these males were culled. Thus, male and female brood stocks were all XX and produced nearly all-female progeny. When needed, male brood-stock replacements were generated by sex reversing the all-female progeny.

This system was tested on a commercial scale for Nile tilapia and was essentially successful (Calhoun and Shelton, 1983). Of course, all-female populations are not desirable for Nile tilapia because the females grow 40% slower than males. However, the research demonstrated that this system has potential application for species

Take a sample of fry and sex reverse to all-male, resulting in XX and XY males

↓

Progeny test with normal XX females: XY × XX yields 1:1 sex ratio, XX × XX gives all-female XX

↓

Sex reverse a sample of XX fry to males for future brood stock

Fig. 11.2. Scheme for producing all-female XX populations of fish.

with similar sex determination and when monosex female populations are desirable.

A small percentage of males were produced in the commercial-scale test of sex reversal and breeding to produce all-female Nile tilapia. Genetic variation for sex determination probably explains these unexpected results, although temperatures experienced during development can affect phenotypic sex in fish.

These systems for producing monosex progeny that are affected by genetic variation at modifying loci might be made more efficient through selection. Individuals or families that produce the desired sex ratios would be kept as brood stock and selected, and those producing aberrant sex ratios would be culled from the population.

XX all-female populations have been successfully developed for salmonids, carps and tilapias. All-female populations are desirable in salmonids because of the more rapid growth of females, the early sexual maturity and associated slow growth of males and the poor flesh quality of males, especially early-maturing males. Production of all-female populations through sex reversal alone was inconsistent, and the use of sex hormones on food fish had legal implications. Sex reversal and breeding offered a good genetic alternative to solve these problems.

Sex reversal and breeding have been effective for producing monosex female populations of salmonids. A combination of sex reversal and breeding to produce all-female XX populations is the basis for the majority of the rainbow trout industry in the UK (Bye and Lincoln, 1986; Fig. 11.3) and the USA (Dunham *et al.*, 2001) and the entire chinook salmon industry in Canada. All-female populations are desirable because males display early maturity at a small size and have poorer flesh quality.

Female rainbow trout are the homogametic genotype, XX. Fry were sex-reversed with methyltestosterone and sex-reversed XX males identified by progeny testing (Bye and Lincoln, 1986). These fish mated with normal XX females and produced 100% female progeny. The primary hindrance to

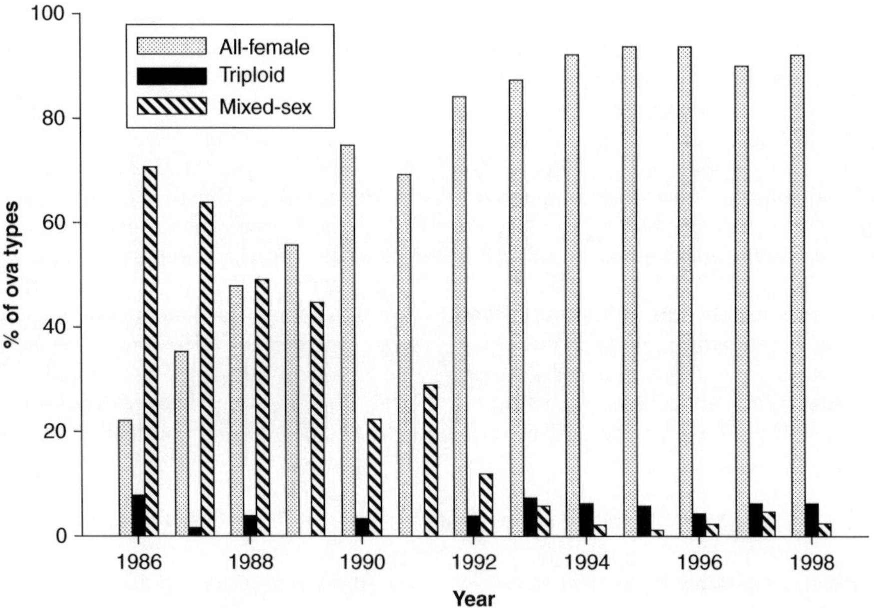

Fig. 11.3. Percentage of all-female, triploid and mixed-sex rainbow trout, *Oncorhynchus mykiss*, utilized in Scotland from 1986 to 1998, illustrating the increasing and almost exclusive adoption of the all-female production technology. (Courtesy of David Penman.)

this procedure occurs when generating the XX male brood stock (Bye and Lincoln, 1986). The sperm ducts of these XX males do not develop properly and become closed during sex reversal. However, viable sperm are produced by these fish, the testes can be removed and milt released through maceration of the testes, circumventing the problem of the closed sperm ducts. Decreasing the level of methyltestosterone for sex-reversing the fish allows opening of the sperm ducts but decreases sex-reversal rates. Optimum treatments for opening sperm ducts without decreasing sex reversal are needed.

Monosex female populations of both chinook salmon and coho salmon crossed with chinook salmon (Hunter *et al.*, 1983) have been generated. Sex-reversed males in this study did not have the problem of closed sperm ducts. Sixty-two per cent of progeny groups from sex-reversed males contained between 1 and 8% males, and 38% were all-female progeny. This suggests possible autosomal influence or polygenic inheritance for sex in chinook salmon as found in tilapia and other fish species (Hunter *et al.*, 1983). Since a large proportion of the families produced the expected 100% female progeny, selection might again be used to develop lines that consistently produced the all-female spawns.

Common carp is another species where all-female populations would be desirable because of sexually dimorphic growth and the fact that in some developing countries the eggs are considered a delicacy. At the Dor station (Agricultural Research Organization (ARO)) in Israel, all-female common carp populations (Cherfas *et al.*, 1996) have been produced by sex-reversing XX gynogenetic females to males (Gomelsky *et al.*, 1994) and using these XX males for breeding. All-female seed was released to commercial farms and resulted in 10–15% yield improvement over existing commercial stocks.

Gynogenesis and sex reversal were successfully induced in striped bass, *Morone saxatilis*, in an attempt to obtain brood stocks producing monosex populations to avoid limitations on transfers of this exotic species (Gomelsky *et al.*, 1998, 1999). If inheritance is as expected, 100% XX all-female populations

should result, which could not become reproductively viable populations upon any potential escape from captivity. The development of monosex female silver barbs, *Barbodes gonionotus*, is another example of carrying out basic research on the sex-determination system of a cultured species and then adapting these findings to produce genetically monosex XX females on a commercial scale (Dunham *et al.*, 2001). This was accomplished over the relatively short period of 8 years, and primarily on research stations in developing countries in Thailand and Bangladesh. Pongthana *et al.* (1995) demonstrated that gynogenetic silver barb were all females, that it was possible to hormonally masculinize such gynogenetic fish (Pongthana *et al.*, 1999) (Fig. 11.4) and that most such neomales produced all- or nearly all-female progeny. Monosex female batches produced higher yields in pond culture than mixed-sex batches by demonstrating both better growth and higher survival rates than the mixed-sex fish. The higher survival was probably a result of them growing fast enough to reach a size that allowed them to utilize a larger and more abundant prey source (N. Pongthana, National Aquaculture Genetics Research Institute, Department of Fisheries, Thailand, 1999, personal communication).

All-male ZZ systems

A similar scheme of sex reversal and breeding for species such as blue tilapia where the male is homogametic, ZZ, would theoretically allow the production of all-male progeny (Hopkins *et al.*, 1979; Jensen and Shelton, 1979; Shelton, 1987) (Fig. 11.5). Fry are fed oestrogen, resulting in ZZ sex-reversed females and WZ normal females. Treated females are progeny-tested by mating them with normal ZZ males. ZZ sex-reversed females should produce 100% all-male ZZ progeny when crossed with normal ZZ males. WZ females produce a normal 1:1 sex ratio and are culled. After the progeny testing and culling, a brood population is generated where the males and females have the same genotype, ZZ. When the fish are spawned, all-male progeny

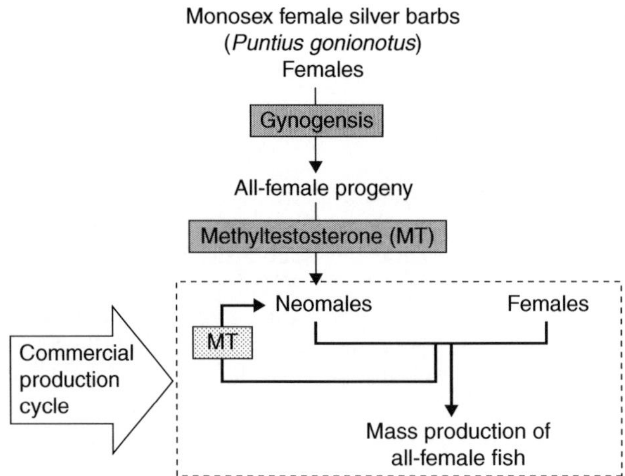

Fig. 11.4. Strategy for production of all-female silver barb, *Puntius gonionotus*. (Courtesy of David Penman.)

Take a sample of fry and sex reverse to all-female, resulting in WZ and ZZ females
↓
Progeny test with ZZ males: ZZ × ZZ gives 100% ZZ males
↓
Sex reverse a few progeny to females for future brood stock

Fig. 11.5. Scheme for producing all-male ZZ populations of fish.

result and, when brood replacements are needed, a portion of the fry is fed oestrogen to generate replacement ZZ females.

A functionally sex-reversed ZZ female was identified and mated to a normal ZZ male (Simpson, 1975; Shelton *et al.*, 1978; Jensen and Shelton, 1979) and the resulting progeny were 100% male, as expected. However, this system is not yet feasible on a commercial scale for blue tilapia because of the small numbers of functionally sex-reversed ZZ individuals that were generated. Most of the oestrogen-treated ZZ males were partially sex-reversed and developed abnormal genitalia and/or ovotestes. These fish did not spawn and exhibited abnormal sexual behaviour.

Although the functionally sex-reversed ZZ female spawned, this fish either did not exhibit or weakly exhibited normal sexual behaviour or perhaps did not have normal pheromone production (Meriwether and Shelton, 1981). When the ZZ female was in competition with normal WZ females, the male blue tilapia always chose to mate with the normal WZ females rather than the phenotypic ZZ female. However, when the normal WZ females were removed, the ZZ female spawned three times in succession with the male. The sex-reversal and breeding programme for production of all-male blue tilapia will not be commercially viable until more effective hormone treatments are developed for functionally sex-reversing male fry.

Theoretically, we could also produce all-female WW populations in fish with WZ sex determination (Fig. 11.6). However, this has not yet been attempted.

All-male YY systems

The most successful sex-reversal and breeding programme with the greatest economic impact in tilapia has been the

Sex reverse to maleness, resulting in WZ and ZZ males
↓
Progeny test with WZ females: WZ × WZ gives
1 WW female : 2 WZ females: 1 ZZ male
↓
Progeny test with ZZ males: WZ female yields 1:1 sex ratio, WW female results in WZ females
↓
Sex reverse sample of progeny from WZ female mated with WZ male to
maleness, resulting in 1 WW male: 2 WZ males :1 ZZ male
↓
Progeny test with WZ females: ZZ males yield 1:1 sex ratio, WZ males yield 3:1 females,
WW males yield all females
↓
Now mate WW males with WW females, results in all-female WW progeny
↓
Sex reverse a sample to maleness for future brood stock

Fig. 11.6. Scheme for producing all-female WW populations of fish.

production of all-male XY populations of Nile tilapia from YY males. First tilapia fry are sex-reversed to all females, resulting in sex-reversed XY and normal XX females, which are then progeny-tested with normal XY males (Fig. 11.7). Spawns with 75% males (25% YY males, 50% XY males) and 25% XX females result from the XY × XY mating, which allows identification of the XY females. Male progeny from this mating are then grown for a second generation of progeny testing. When the YY males are test-crossed with normal XX females, 100% XY male progeny result. The repeated, tedious progeny testing is not an efficient system, so one more genetic step can be applied to develop a more effective technology.

Mating the YY males with XY females would also result in 100% males, both XY and YY genotypes. If a subset of these fry is sex-reversed to femaleness, YY and XY females are produced. When these fish are mated to normal XY males, the YY females produce only male offspring and XY females produce 25% female fry. At this point, YY males and YY females have both been identified. All fingerlings produced from YY × YY matings are YY and male, making mass production of YY males technically feasible. Replacement YY females are easily produced by sex reversal without the three generations of progeny testing, and progeny testing is no longer needed.

This system of YY production has been demonstrated for Nile tilapia and can increase production of tilapia by 50%. YY male genotypes of Nile tilapia have viability and fertility equal to normal XY males. Many, but not all, YY males produce all-male progeny, once again illustrating the probable polygenic nature of sex determination in Nile tilapia. All-male progeny, XY, known as genetically male tilapia (GMT), produced from YY males are now mass-produced on a commercial scale. The YY male technology provides a robust and reliable solution to the problem of early sexual maturation, unwanted reproduction and overpopulation in tilapia culture (Beardmore *et al.*, 2001; Dunham *et al.*, 2001).

Genetics of Sex Determination

The precise mechanism by which sex is determined in tilapia, specifically Nile tilapia, is still not fully understood. Early hypotheses were based on the sex ratios observed in hybrid crosses of different species (Chen, 1969). However, so far, no theory based on hybrid sex ratios successfully explains all observed sex ratios. The theory of autosomal influence (Hammerman and Avtalion, 1979) was developed to explain aberrant sex ratios. Sex ratios can be highly variable in hybrids between Nile tilapia, blue tilapia and others.

Sex reverse to all-female, resulting in XX females and XY females
↓
Progeny test with XY males: XX yields 1:1 sex ratio, XY × XY gives 1 XX: 2 XY: 1YY = 3 males : 1 female
↓
Take the male progeny XY and YY and progeny test with XY females from mother's, generation:
XY males yield sex ratio of 3:1, YY males yield a genotypic ratio of 1 XY: 1 YY – all male
↓
Take sample of these fry and sex reverse to female, resulting in XY and YY females
↓
Progeny test with XY males: XY females yield 3:1 sex ratio, YY females yield all males
↓
Mate YY males with YY females to generate all YY progeny
↓
Sex reverse a sample to females for future brood stock

Fig. 11.7. Scheme for producing all-male YY populations of fish.

Genetic contamination from other species or polygenic inheritance of sex, the same explanations as for failure to produce all-male hybrid tilapia, are also possible explanations for the aberrant sex ratios in all-female XX Nile tilapia populations (Calhoun and Shelton, 1983) and in all-male YY populations of Nile tilapia (Mair *et al.*, 1997). There were families in the commercial-scale test for production of monosex female Nile tilapia that did produce the expected 100% female progeny (Calhoun and Shelton, 1983). When full-sibling males and females were mated, 100% females were produced, further evidence for genetic variation of sex determination within Nile tilapia. Nile tilapia has a predominantly monofactorial mechanism of sex determination with heterogametic XY males and homogametic XX females (Calhoun and Shelton, 1983; Shelton *et al.*, 1983; Mair *et al.*, 1991a; Trombka and Avtalion, 1993), with some additional polygenic effect. This also indicates the potential for using selection to produce populations that give the desired 100% monosex result when polygenic inheritance occurs.

Before data were generated suggesting the existence of polygenic inheritance or autosomal loci influencing sex, 1:1 sex ratios were assumed for individual spawns of Nile tilapia because 1:1 sex ratios were observed for the overall population. Calhoun and Shelton (1983) examined sex ratios within individual spawns in the Ivory Coast strain of Nile tilapia and sex ratios were normally distributed. Numerous spawns significantly deviated from 1:1 sex ratios and, in fact, sex ratio was normally distributed among individual spawns. Some females produced as high as 90% male progeny and others as high as 70% female progeny. When those females were re-spawned with other males, similar sex ratios were observed in the progeny.

The female component of variation was responsible for 13-fold more variation in progeny sex ratios than the male component of variation (Calhoun, 1981), suggesting the possibility that the modifying loci may actually lie on the X chromosome rather than the autosomes. This would result in X chromosomes of varying strength, if there was genetic variation at these modifying loci. The female would have more influence on the sex ratio of the progeny than the male since females have two X chromosomes and the male has only one. The female also has more influence on sex ratio than males in blue tilapia (Shelton *et al.*, 1983).

Cytoplasmic effects or interactions with the nuclear genome could also explain or contribute to the stronger influence of the female on the sex ratio. Cytoplasm of the embryo and mtDNA found in the cytoplasm originate from the female only. Although genes influencing sex have not previously been found on mtDNA (Chapman *et al.*, 1982; Avise *et al.*, 1987; Moritz *et al.*, 1987), if they existed in tilapia, they would be inherited only through the female. Interactions, epistasis or pleiotropy with or in the nuclear genome could also influence sex determination.

Additionally, sex-ratio distribution can also vary among strains of Nile tilapia (Shelton *et al.*, 1983; Mair *et al.*, 1991a). This might be expected because alleles and allele frequencies vary for polygenic traits among strains. The percentage of males in the Chitralada strain of Nile tilapia was 50.5%; however, progeny sex ratios ranged from 15.5 to 100% male and over 53% of the pairings produced sex ratios significantly different from 1:1 (Tuan *et al.*, 1999). Both maternal and paternal effects on progeny sex ratio were evident. Capili (1995) demonstrated that the rare females in progenies of YY males conform to the expected XY genotype in progeny tests with known genotypes in the same strain. Capili (1995) also noted the existence of both paternal and maternal effects on the occurrence of rare females in the progeny of YY males. Sex ratios of progeny from two YY males were also highly variable, ranging from 36 to 100% male, with a mean of 80.6%. These high levels of heterogeneity for sex ratio are additional evidence for polyfactorial sex determination in this species. Lester *et al.* (1989) observed considerably more heterogeneity in the sex ratios of families collected from a mix of strains, some of which were known to be introgressed with *O. mossambicus* (Macaranas *et al.*, 1986). The authors interpreted these high levels of heterogeneity for sex ratio as evidence for polyfactorial sex determination in this species, as other researchers have concluded, especially when introgressed with other species. When species introgress, inheritance appears to become more complicated, with increased epistatic effects (Tave *et al.*, 1983; Halstrom, 1984), which would be consistent with the high levels of heterogeneity observed by Lester *et al.* (1989).

YY male genotypes of Nile tilapia sired a mean of 95.6% males (Mair *et al.*, 1997; Beardmore *et al.*, 2001) when mated with XX females. The overall sex ratio from YY males mated with XX females can range from 95% male to 100% male. Again, genetic variation for sex determination probably explains these unexpected results, although temperatures experienced during development can affect phenotypic sex in fish.

Scott *et al.* (1989) observed no females in 285 progeny of a single YY male crossed to ten separate females. Similarly, Varadaraj and Pandian (1989) observed no females among the progeny of eight YY females in *O. mossambicus*. Hussain *et al.* (1994) hypothesized the existence of an autosomal sex-modifying locus (with alleles *SR* and *sr*) epistatic to the gonosomal locus and which induces female to male sex reversal when *sr* is homozygous (Dunham *et al.*, 2001). This hypothesis explained the occurrence of varying proportions of males in heterozygous and homozygous meiotic and mitotic gynogenetic progeny (Mair *et al.*, 1991a; Hussain *et al.*, 1994). However, this hypothesis still does not account for some of the aberrant sex ratios observed in the progeny of hormonally sex-reversed fish, indicating additional autosomal influences and environmental influences (Mair *et al.*, 1990; Trombka and Avtalion, 1993). When sex ratio varied in XY GMT progenies, there was no apparent trend in the occurrence of these aberrant sex ratios that would indicate the segregation of a single autosomal sex-modifying locus, as postulated for Nile tilapia by Hussain *et al.* (1994) and demonstrated for blue tilapia (Mair *et al.*, 1991b). These results substantiate the hypothesis of predominantly monofactorial sex determination, and the occurrence of occasional females from the action of several autosomal sex-modifying genes.

Many systems of sex determination may be working synergistically or antagonistically in tilapia. Sarder *et al.* (1999) found males in gynogenetic clonal lines of Nile tilapia. This line appeared to be homozygous for an allele or combination of alleles at an autosomal locus or loci resulting in the sex reversal of females to maleness but with limited penetrance. Guan *et al.* (2000) identified two sex-determining genes in tilapia that appear to be different versions of the 'doublesex' gene first identified in *Drosophila*. Drosophila has only one doublesex gene, and produces male and female products by RNA splicing after transcription. One of these genes is already known in vertebrates (including zebra fish), and is expressed mainly in the testis. Tilapia have a variant of this same gene which is expressed in the ovary and is structurally somewhat like the female version of the *Drosophila* doublesex gene – again, a partial explanation of why female tilapia may have a greater influence on sex ratios than males. The expression of the two tilapia doublesex genes is mutually exclusive in any particular fish, and therefore these genes are probably under the control of additional genes and environmental factors. Guan *et al.* (2000) suggest that there are many targets for genetic intervention in the teleost sex-determining system.

Sexual differentiation in *Drosophila* is controlled by a short cascade of regulatory genes (Doyle, 2003), and perhaps similar systems exist in tilapia and other aquatic organisms. Their expression pattern determines all aspects of maleness and femaleness, including complex behaviours displayed by males and females. Doublesex is expressed near to the end of the cascade. Mutational and transgenic manipulation of specific genes in the cascade produced *Drosophila* that are genetic (XX) females but develop and behave as males. This genetic sex reversal can be blocked by other gene manipulations in the cascade. If the blocking gene is coupled with the appropriate promoter, the blocking action can be stopped by applying a heat shock during development. The result is that the transformed female flies develop as males, display vigorous male courtship and yet still produce female pheromones, thus attracting males and possibly continually stimulating themselves with their own female pheromones.

Chromosome analysis can now provide hints concerning sex determination in tilapia. In tilapias the heterogametic sex chromosomes cannot be distinguished by their appearance in karyotypes. When homologous chromosomes were observed with electron microscopy and were tightly paired, two unpaired regions were observed in the nuclei of female *O. aureus*, the heterogametic sex (Campos-Ramos *et al.*, 2001). The two regions were on different chromosomes, one of which has not yet been identified in the karyotype. However, there were no unpaired regions in *O. aureus* males. One of the unpaired regions was closely related to the sex-determining region of male *O. niloticus*, which is also the heterogametic sex. However, male Nile tilapia have only one such region. It appears that *O. aureus* may have two pairs of sex chromosomes, which would contribute to the aberrant sex ratios sometimes observed in tilapia.

Effects of temperature

Further complicating the inheritance of sex in tilapia and other fish are the effects of temperature on sexual differentiation. Putative all-female progeny from androgen sex-reversed males (XX) crossed with normal females in Nile tilapia yield higher proportions of unexpected males in progeny reared at high temperatures (36°C) during the period of sex differentiation (Baroiller *et al.*, 1995a,b). Similar results occur in different strains of Nile tilapia (Dunham *et al.*, 2001). However, the temperature effect on sex determination still does not explain the occurrence of a small percentage of males in the expected all-female progeny reared at ambient temperatures.

Abucay *et al.* (1999) exposed different sex genotypes (putative all-female, all-male and all-YY males) to varying temperatures and salinities (putative all-female progeny only) for a minimum period of 21 days after first feeding. The majority of putative all-female progeny exposed to high temperature – 36.5 ± 0.4°C – produced higher percentages of males compared with putative all-female controls reared at ambient temperature – 27.9 ± 1.4°C. Conversely, at high temperature, some of the XY all-male and YY male progenies had a lower percentage of males compared with controls. Sex differentiation in YY males was more labile than in normal XY males. Alternatively, this could be an effect of inbreeding. Low temperature – 25.8 ± 0.2°C – and salinity varying from 11.3 to 26.7 ppt did not affect sex ratios.

Similarly, constant high temperatures had a strong masculinizing effect (M:F sex ratios of 7.33–19.00:1.00 at 35°C versus 0.75–0.82:1.00 in controls reared at 27°C) in

O. aureus (Baras *et al.*, 2000). Fluctuating temperatures (day at 35°C, night at 27°C, and vice versa) produced less masculinization, but still produced sex ratios skewed towards maleness (M:F sex ratios of 2.33–11.50:1.00).

Temperature also affects sex ratio in normal diploid loach, *Misgurnus anguillicaudatus*, which was 1:1 when the fish were reared at 20°C, but skewed towards maleness when the fish were reared at 25 and 30°C for 214 days from day 6 after hatching (Nomura *et al.*, 1998). Gynogenetic diploids were all genetic males and females at 20°C, but gynogenetic males and intersexes were observed in fish reared at high temperatures (25 and 30°C) for 220 or 240 days from day 11 after hatching and in fish reared at 28°C for 1 month from day 11 after hatching.

Genetics of temperature effects on sex determination

The effect of temperature on sex ratio apparently has a genetic basis in Nile tilapia. D'Cotta *et al.* (2001) examined genetically all-female progenies (sired by XX phenotypic males) and all-male progenies (sired by YY phenotypic males) that developed at 27 and 35°C. The complementary DNA (cDNA) transcript, *MM20C*, was differentially expressed. This gene has minimal expression at normal temperature but is strongly expressed by both sexes at the higher masculinizing temperature, especially by the genetic males. Apparently, *MM20C* is a gene that stimulates testicular development of tilapia and is up-regulated with elevated temperature.

In conclusion, sex determination in tilapia is predominantly monofactorial, with an underlying mechanism of male heterogamy playing the major role, but which is influenced by several genetic and environmental factors.

YY system in channel catfish

The YY system of production for Nile tilapia was almost completed for channel catfish. All-male progeny would be beneficial for catfish culture since they grow 10–30% faster than females, depending on strain of catfish (Benchakan, 1979; Dunham and Smitherman, 1984, 1987; Smitherman and Dunham, 1985). However, there are strains of channel catfish that have no sexual dimorphism for growth rate and all-male populations would not be beneficial when culturing these strains unless other sexually dimorphic traits exist and favour males, such as flesh quality. Sex reversal and breeding allowed the production of YY channel catfish males that can be mated to normal XX females to produce all-male XY progeny. Channel catfish were sex-reversed to femaleness with β-oestradiol (Goudie *et al.*, 1983). XY phenotypic females were fertile and identified through progeny testing by mating them to normal XY males (Goudie *et al.*, 1985).

If the YY genotype is lethal, as it is in mammals, the sex ratio of the progeny should be 2 M:1 F. If the YY genotype is viable, the sex ratio of the progeny should be 3 M:1 F. The sex ratio of progeny from the mating of XY female and XY male channel catfish was 2.8 M:1 F, indicating that most, if not all, YY individuals were viable. YY males have now been demonstrated as viable in salmonids, Nile tilapia, goldfish and channel catfish (Donaldson and Hunter, 1982; Goudie *et al.*, 1985; Dunham, 1990a; Dunham *et al.*, 2001). When YY channel catfish males were mated with normal XX females, 100% male progeny were produced. The channel catfish YY system has stalled, however, because YY females have severe reproductive problems, and large-scale progeny testing is not economically feasible to identify YY males.

The type of sex-reversal and breeding programme appears to have an effect on success and the elucidation of hidden genetic variability. Goudie *et al.* (1995) utilized an alternative approach to produce all-male channel catfish progeny. Gynogenetic progeny were produced from XY females, which should have resulted in YY males and XX females. Sex ratios of offspring from matings of gynogenetically derived YY male channel catfish with normal XX females produced the expected 100% male progeny in only seven of 18 males tested. Since YY males previously developed from sex reversal and progeny testing of XY × XY matings gave the expected 100% male progeny, aberrant sex ratios (<100% males) were surprising and suggest some disturbance in the meiotic process or that instability in the sex-determination system occurs as a result of induced gynogenesis of XY females (Goudie *et al.*, 1995). Alternatively, temperature may have been the cause of the unexpected female progeny, or the polygenic nature of sex determination in channel catfish does exist but was not previously detected because of allele frequency differences in different families or strains of channel catfish.

Constraints and Sex Markers

Progeny testing to identify sufficient YY channel catfish males for commercial production would be tedious and require a large commitment of facilities as the generation interval is long (4 years), requiring 8 years to develop the YY males, two-thirds of the tested males must be discarded and most catfish farms are large, requiring thousands of brood stock. Either YY females must be generated and be fertile, or DNA markers are needed to reduce progeny testing constraint and time.

The constraint of progeny testing can also be alleviated if DNA or protein markers are found for sex-determining chromosomes. Segments of the Y chromosome may have distinctive DNA sequences that differ from those found in the X chromosome. Once the marker is found and utilized, genotype for sex can be assayed and determined regardless of phenotype, eliminating the need for progeny testing, which requires facilities, extra effort and increased record keeping. Although morphologically distinctive sex chromosomes have not been found in channel catfish, distinctive DNA sequences associated with the maleness chromosome may exist, and such sequences are currently being sought in channel catfish. Sex-determining markers have been isolated for coho salmon, but not for other species. Of course, a limitation for this strategy is availability of expertise and funding to conduct the DNA or protein screening. If these DNA markers are found, rapid evaluation of fingerling YY males would easily allow large-scale commercial production of all-male channel catfish populations. To date, the YY system has stalled and not become commercially feasible for channel catfish because YY females do not reproduce or have severe reproductive problems, making large-scale production of YY males too difficult.

There is a general lack of sex-specific markers to aid in breeding programmes aimed at producing monosex fish. Some markers may have application across a genus, such as *Oncorhynchus* (Nakayama *et al.*, 1999), but it appears likely that other sex-specific markers may be species specific.

Some of the problems of the sex-reversal and breeding programmes are that not all fish species are responsive to sex hormones for sex reversal and the effort needed to determine proper dosages and length of treatment time (proper initiation and termination times). Additionally, space may be limiting for progeny testing for some farms and for research stations. The multiple generations of progeny testing required to develop a genetically homogeneous monosex population are the primary restraint on this technology for species with long generation intervals. However, once a research institution has developed an XX, YY, WW or ZZ population, progeny testing is no longer required, assuming these genotypes can be functionally sex-reversed to produce both sexes, removing the major impediment to application. However, farmers in both developed and developing countries would need to be supplied with brood-stock replacements, since the opposite sex will eventually die out, or they will need to be taught how to sex reverse brood replacements. This latter strategy is feasible in both developed and developing countries, especially if hormone-treated feed is provided by government agencies or feed mills. Another strategy is to sell only heterozygous monosex offspring to protect proprietary brood stock.

Key Summary Points

Sex reversal and breeding to produce monosex female populations is an excellent method to increase growth and production of some species of fish. For some species, monosex females also have superior flesh quality and fry survival. Sex reversal and breeding to produce monosex male populations is effective for increasing growth and production in other species, especially tilapia. Monosexing can be an effective method to limit or control reproduction.

12
Biochemical and Molecular Markers

Markers are necessary to study genomes, conduct gene-linkage mapping, locate genes on chromosomes, isolate genes, determine gene expression, study the biochemical and molecular mechanisms of performance, conduct population-genetics analysis and apply MAS. Knowledge of gene locations can be utilized along with physical mapping to clone useful genes by positional cloning. However, before positional cloning of useful genes is possible, thousands of molecular markers must be identified for any aquatic species of interest (Liu and Dunham, 1998a).

Technology has advanced to the point that a tremendous array of biochemical and molecular markers are available to study the genetics of fish and aquatic invertebrates. One of the earliest and most tedious analyses was blood typing, but now this technique is seldom utilized. Traditional markers included isozymes (Liu *et al.*, 1992), RFLP markers (Miller and Tanksley, 1990) and mtDNA analysis (Curtis *et al.*, 1987). Several powerful DNA markers have been developed, including RAPD (Welsh and McClelland, 1990; Williams *et al.*, 1990; Liu *et al.*, 1998a), microsatellites or simple sequence repeats (SSRs) (Hughes and Queller, 1993; Queller *et al.*, 1993; Liu *et al.*, 1999e,f; Tan *et al.*, 1999), AFLP (Vos *et al.*, 1995; Liu *et al.*, 1998c), ESTs (Liu

et al., 1999a; Ju, 2001) and SNPs (Kocabas *et al.*, 2002a). The new DNA technologies have allowed the construction of gene maps in a matter of months, rather than the years that were the case for the construction of gene maps with conventional molecular markers, such as RFLP (Liu *et al.*, 2003).

Isozymes and Enzymes

Isozymes are multiple molecular forms of individual enzymes. These multiple forms can be alleles of one another at a single locus – allozymes – or can be products of different loci where there are multiple copies of genes making the same enzyme or enzyme subunits Temporal differences in isozyme expression exist, which can be utilized in the study of developmental genetics – spatial or tissue-specific expression as well as allelic variation. Isozyme and enzyme analyses are technically easy, but are limited in both the numbers of loci available and polymorphism. For example, isozyme variation is low in Nile tilapia (Abdelhamid, 1988; Rognon *et al.*, 1996; Agnese *et al.*, 1997).

However, one major advantage is that genetic variation is being measured, which is directly related to protein products that actually affect performance. For example,

Hallerman *et al.* (1986) demonstrated that isozyme variation is associated with growth rate in channel catfish. Isozyme variation has also been linked with disease resistance, temperature tolerance, developmental speed and salinity tolerance in fish (Dunham, 1996).

Additionally, when 1+-year-old smolts of five hatchery strains of Atlantic salmon were released into a Danish river in 1996, three of the strains went to sea almost immediately, but two strains waited for more than 2 weeks before migrating (Nielsen *et al.*, 2001). Differences in the temporal expression of gill enzyme development were highly correlated with migration pattern, and early-migrating strains reached high enzyme activity earlier than late-migrating strains. The strains with delayed enzyme development and migration exhibited a delayed regression of seawater tolerance compared with the early strains.

Northern and southern populations of the minnow, *Fundulus heteroclitus*, have different levels of expression of the LDH-B gene, *Ldh-B* (Schulte *et al.*, 2000). The northern strains, such as Newfoundland fish, have superior expression at lower temperatures, while the southern strains, such as Florida fish, have superior expression at higher temperatures. Deletion studies have been carried out to identify the approximate location within the regulatory sequence where the adaptive changes in the transcript occurred. A difference of only 1 base pair (bp) in the regulatory sequence accounted for the adaptive difference in *Ldh-B* expression between the northern and southern populations.

Another major advantage is that isozymes are inherited in a co-dominant fashion. This makes heterozygotes and homozygotes readily distinguishable, thus strengthening applications for gene mapping, population-genetics studies and determining parentage.

Isozymes can be separated in an electric field passed through a matrix, such as starch, cellulose acetate or polyacrylamide, based on their size, shape and charge, since most frequently different isozyme forms of the same enzyme vary in one or more of these parameters. Slices of the matrix are incubated in a specific histochemical stain to visualize the desired enzyme. Most staining procedures result in the deposition of dye at the site of enzyme activity, but a few stains involve a reverse process in which only the site of activity remains unstained, such as is the case for superoxide dismutase (SOD). Upon termination of the staining, the intensity of zones of staining reflects the proportions in which the gene products are present provided that the staining is terminated before overexposure occurs. The resulting zymogram is interpreted genetically.

Most gene products migrate towards the anode, and varying the pH of the buffer can affect the mobility of gene products and, in rare cases, the direction of migration. Various enzymes/isozymes can be encoded by single- or multiple-locus systems. In all single-locus systems, the homozygous genotype yields a single zone of activity (band) on a starch-gel slice. Heterozygous genotypes yield multiple bands. The phenotype of heterozygotes will include the superimposition of the homozygous patterns in a diallelic situation, plus additional zones attributed to allozyme and isozyme heteromers of multimeric enzymes.

Enzymes are made of single or multiple protein subunits – series of polypeptide chains. Enzymes can be monomers, dimers, trimers, tetramers, hexamers and octamers made of one, two, three, four, six and eight subunits, respectively. Allozymes encode a single subunit, and these subunits bind in the cell or tissue to form the entire protein/enzyme. The dimeric system is the most common. The trimeric, hexameric and octameric systems are rare. Of 100 enzymes frequently examined in human-genetics studies, there were 28% monomers, 43% dimers, 4% trimers, 24% tetramers and 1% octamers (Harris and Hopkinson, 1976).

In the case of single-locus monomeric systems, the homozygous genotype, both alleles make the same subunit, resulting in a single isozyme/band. When these loci produce different allelic products in the heterozygous condition, each allele produces a different subunit but the subunit represents the entire isozyme, so there is expression of two products in the heterozygote at a 1:1 ratio.

Trimeric proteins are rare, and the trimeric enzyme system most utilized for fish is purine-nucleoside phosphorylase, which is usually a multilocus system (Whitmore, 1990). A heterozygote for a single-locus trimeric system would produce a 1:3:3:1 banding ratio, with the heterotrimers being more abundant than the two homotrimers.

Of course, in the case of a single-locus tetrameric system, the homozygotes will express a single band on starch gels because the gene products or subunits that combine to form the whole protein/enzyme are identical entities. In the case of the heterozygote, five subunit combinations are expected in a ratio of 1:4:6:4:1 reflecting random combinations of AAAA, AAAB = AABA = ABAA = BAAA, AABB = ABAB = BAAB = BABA = BBAA = ABBA, ABBB = BABB = BBAB = BBBA, and BBBB (Whitmore, 1990). The three heterotetramers should have electrophoretic mobilities spacing them equivalently between the homotetramers. If limited separation of the homotetramers occurs, the heterotetramers may not be discrete and will appear as an elongated smear of activity on the gel slice (Whitmore, 1990).

In the case of single-locus dimeric systems, the homozygotes express a single band on starch gels because the gene products – subunits that combine to form the whole protein/enzyme – are identical. The resulting bands are homodimers. Heterozygotes express two allelic subunits, and these subunits generated by the homologous chromosomes usually bind together in random fashion to produce the enzyme proper (Whitmore, 1990). Heterozygotes express both parental homodimers and a heterodimer from the two allelic subunits in a 1:2:1 ratio, with the heterodimer being the most intense band (approximately twice the intensity of the homodimers). The heterodimer should be intermediate to the two homodimers in electrophoretic mobility.

Multilocus systems are more complex. The gene products or subunits from different loci may combine if they are expressed in the same tissue and cell type, thereby forming homomers within loci, heteromers within loci and heteromers between loci (Whitmore, 1990). Alternatively, expression can be strictly tissue specific, or expression can be temporally or spatially separated within a tissue, preventing formation of heteromers between loci. Since they are under different regulatory control, the quantity of product produced by each nuclear locus may not be equivalent, as is the case for heterozygotes at a single locus. Thus, ratios of banding intensity in multilocus systems can be, but are not necessarily, a predictable symmetrical series (Whitmore, 1990). Interactions of products in a multilocus system will not usually obscure the predictable ratios of activity of products within a given heterozygous locus, except in unusual cases where duplicate loci have allelic variation in the same tissue and when distinct loci have alleles with the same mobility being expressed in the same tissue.

Mitochondrial and supernatant cytoplasmic loci are assembled independently; therefore there is no possibility of heteromer formation of multimeric enzymes between these two types of locus (Whitmore, 1990). However, some subunits of mitochondrial enzymes, such as nicotinamide adenine dinucleotide (NADH) dehydrogenase and cytochrome C oxidase, are coded in the nucleus and transported to the mitochondria after transcription and translation.

Isozymes are a strong tool for studying evolutionary genetics. Knowledge of evolutionary genetics is necessary in some cases to fully understand and interpret isozyme data.

Genes may be duplicated via the tandem duplication of a single gene or a set of neighbouring genes (Whitmore, 1990). Tandem duplications are rare (Ferris and Whitt, 1977; Buth, 1979; Crabtree and Buth, 1981); until they have diverged enough to express different allelic products, tandem duplications are difficult to detect (Whitmore, 1990).

Alternatively, genes can be duplicated and speciation will occur following polyploidization events of ancestral fish; this has occurred in lineages of at least six orders (Buth, 1983). Species within the families Salmonidae and Catostomidae are derived from tetraploid ancestors. Salmonids are presumed to be of autotetraploid – intraspecific polyploid event – origin (Schultz, 1980) and catostomids of allotetraploid – interspecific hybridization and polyploid event – origin (Ferris and Whitt, 1977). As is the case with tandem duplicates, the salmonid genes duplicated by the polyploid event were initially identical and under the same regulatory control (Whitmore, 1990). In cases where these conditions have been maintained, equivalent gene products are produced by isoloci, making genetic interpretations of salmonid zymograms more complex due to this form of genetic control.

Duplicated loci can eventually diverge, both in their allelic composition and in regulatory aspects of their tissue expression, which would result in distinct loci with tissue-specific expression. As long as at least one of these loci maintains its original metabolic function, the other is free to evolve and acquire new functions through mutation and be selectively maintained (Ohno, 1970). This results in the evolution of multilocus systems. Another possibility is the silencing of a duplicated locus via mutation or functional diploidization, if no selective advantage is afforded the diverged duplicate locus (Ferris and Whitt, 1977, 1978a). Silenced loci might be able to retetraploidize and resume their former function if a favourable regulatory mutation occurs, and this may have occurred for glucose-phosphate isomerase expression in *Moxostoma lachneri* (Buth, 1982).

Another phenomenon that can complicate the interpretation of zymograms is null alleles that do not produce or encode protein products (subunits) or produce much reduced amounts of the subunits, which can yield skewed ratios of activity or unexpected proportions of certain genotypes. Of course, null alleles act as recessive alleles as they cannot be detected in the heterozygote individuals. Progeny testing is one of the best ways of detecting null alleles. Null alleles are rare but have been detected in carp (Engel *et al.*, 1973; Lim and Bailey, 1977), rainbow trout (Allendorf *et al.*, 1984) and oysters (Gaffney, 2002).

Heteropolymer restriction can also complicate interpretation of zymograms. In some cases, random association of subunits of multimeric enzymes does not occur and formation of heteromers is restricted (Whitmore, 1990). There is bias in the assembly of the multimer, with similar allelic products being more likely to combine. For example, the creatine kinase (CK) product predominating in skeletal muscle (*Ck-A* locus) often has restriction of intralocus heterodimer assembly in fishes. The heterodimeric combination is not formed, yielding heterozygotes exhibiting two zones of activity or bands – the two CK parental forms – on starch gels, as would be expected for heterozygotes for monomeric product (Ferris and Whitt, 1978b).

The restriction of heteromer assembly among products of different loci is more common (Whitmore, 1990). Interlocus restriction of assembly occurs in the LDH system. In most fishes, both *Ldh-A* and *Ldh-B* products are usually present in skeletal muscle and interact to form three interlocus heterotetramers and two homotetramers in the double-homozygous genotype. However, in several fish, such as darters, Etheostomatini, the assembly is restricted to the homotetramers (AAAA and BBBB) and the symmetrical heterotetramer (AABB – in some or all of its six possible assemblies) (Buth *et al.*, 1980). In *Etheostoma fonticola* (Buth *et al.*, 1980) and *Gyrinocheilus aymonieri* (Rainboth *et al.*, 1986), all interlocus heterotetramers are restricted and not assembled, limiting expression to the homotetramers.

Vertebrate haemoglobins can also be restricted in their assembly (Whitmore, 1990). Two loci (α and β) are contributing tetrameric subunits to the haemoglobin molecule, but assembly is restricted to form only the symmetrical heterotetramer, $\alpha_2\beta_2$ (Ingram, 1963).

Restriction Fragment Length Polymorphism

RFLP was once widely used and is still very useful, having been used to construct genetic maps of many species (Vaiman *et al.*, 1996; Smith, E.J. *et al.*, 1997; Smith, T.P. *et al.*, 1997; Yang and Womack, 1997). Restriction endonuclease enzymes are used in this method to directly cut the DNA at restriction sites. Base substitution at the restriction sites, insertions, deletions or DNA-fragment rearrangements at or between the restriction sites cause the polymorphism. The resulting products are then separated on agarose gel (Fig. 12.1), transferred to a membrane and hybridized with labelled probes to produce DNA fingerprints. The advantages of RFLP include co-dominant inheritance and easy interpretation and scoring. This technique is now less frequently used because it is time-consuming and requires tedious Southern blotting. Additionally, probe development is required for RFLP analysis, polymorphism is generally low and sequence information is needed if using PCR. This technique is too slow and tedious to generate large numbers of markers.

Mitochondrial DNA

The analysis of mtDNA variation is an alternative for studying population genetics in fish (Capili and Skibinski, 1996; Agnese *et al.*, 1997). For species such as striped bass where isozyme variation is minimal, significant mtDNA variation was observed (Wirgin *et al.*, 1989). The mutation rate of mtDNA is about an order of magnitude higher than that of the nuclear genome, and the control region is particularly hypervariable, thus allowing studies on recent evolution.

Since the mitochondrion is the major site of cellular respiratory metabolism and a possible source of maternal effects, genetic improvement programmes should be

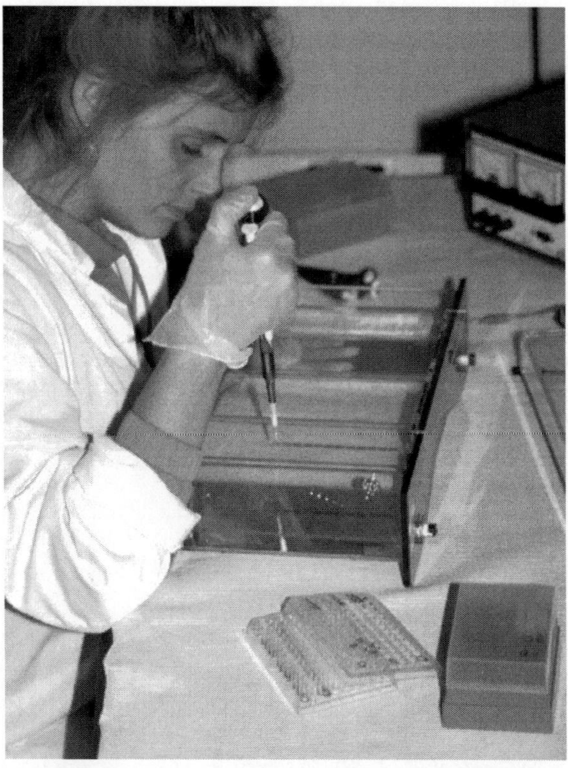

Fig. 12.1. Loading DNA onto a gel.

concerned with mtDNA as well as nuclear DNA. MtDNA analysis often revealed genetic differences among populations of fish that were homogeneous for isozyme variability. Three types of polymorphism can be detected for mtDNA in fish: (i) length polymorphisms; (ii) restriction-site polymorphisms caused by base-pair additions, deletions or both; and (iii) heteroplasmy.

MtDNA heteroplasmy is the existence of more than one form (genotype or haplotype) of mtDNA in an individual. Natural mtDNA heteroplasmy has been observed in bowfin, *Amia calva* (Bermingham *et al.*, 1986), American shad, *Alosa sapidissima* (Bentzen *et al.*, 1988), striped bass (Wirgin *et al.*, 1989), white sturgeon, *Acipenser transmontanus* (Buroker *et al.*, 1990), dwarf cisco, *Coregonus artedii* (Shields *et al.*, 1990), red drum, *Sciaenops ocellatus* (Gold and Richardson, 1990), Atlantic cod, *Gadus morhua* (Arnason and Rand, 1992), and anchovy, *Engraulis encrasicolus* (Magoulas and Zouros, 1993).

Heteroplasmy could result from one of two mechanisms. On rare occasions, mtDNA can be inserted and inherited from the male parent – paternal leakage. A second mechanism would be a mutation in the mtDNA genome, with some type of selective force or random process resulting in an increased population of the mutated mtDNA until it was detectable. It is likely that

heteroplasmy is more frequent than what is detected because the secondary haplotype could be at frequencies too low to detect.

MtDNA does encode genes and could affect the performance of aquatic organisms. Sequence and restriction analysis readily detects variation in mitochondrial genes. Genetic variation exists for the *ND5/6* gene of mtDNA in different strains of Nile tilapia, *Oreochromis niloticus* (L. Sifa, Shanghai Fisheries University, 1999, personal communication). The indices of haplotype diversity and nucleotide diversity of *O. niloticus* were 0.69 ± 0.10 and 0.03 ± 0.10, respectively.

Randomly Amplified Polymorphic DNA

RAPD markers are polymorphic DNA sequences separated by gel electrophoresis after PCR, using one or a pair of short (8–12 bp) random oligonucleotide primers (Liu and Dunham, 1998a; Liu *et al.*, 1998a, 1999b). Polymorphisms are a result of base changes in the primer-binding sites or of sequence-length changes caused by insertions, deletions or rearrangements. RAPD is very powerful in detecting large numbers of polymorphisms because oligonucleotide primers scan the whole genome for perfect and subperfect binding sites in a PCR reaction (Fig. 12.2). When two binding sites are

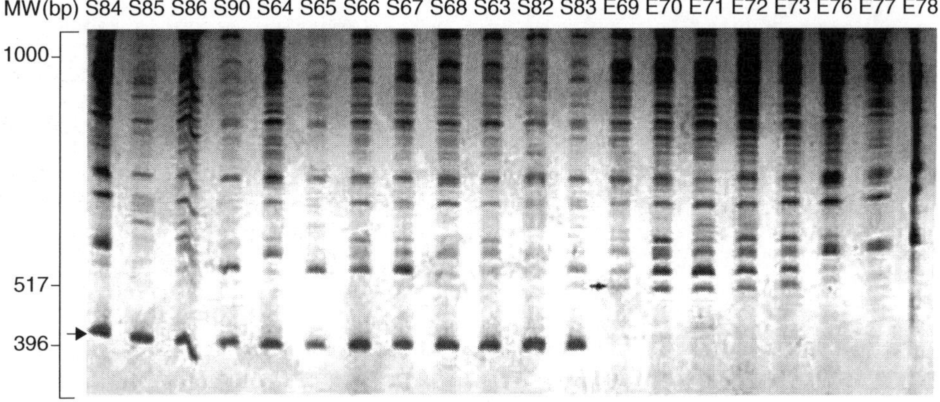

Fig. 12.2. Fixed RAPD differences in Ozark hellbender populations found by evaluating a small number of primers. Population S has a band at approximately 400 bp that is absent in population E. (Adapted from Kucuktas *et al.*, 2002.)

close enough (3000 bp or less), an RAPD band is produced on the gel. Each RAPD primer usually amplifies several bands, some of which are polymorphic in even closely related populations, which can be either tremendously advantageous or disadvantageous. This will be discussed in more detail later.

RAPD markers are expressed and scored as dominant alleles. The amplified DNA product is scored based on size and presence. A polymorphism occurs when a band is present in one parental type but absent in the other. Even if a homologous fragment exists in the other parent, exhibited as a band with a different size, it would be scored as a distinct marker, although it actually represents the same locus or the same general location of the DNA sequence. Technically, RAPDs are not genes or alleles as they do not code for gene products. A potential disadvantage for RAPD analysis is that these dominant banding patterns fail to distinguish between heterozygous and homozygous individuals. Of course, inheritance of the markers could be verified by progeny testing, but this is not simple because of the large number of bands. Potentially, sequence-tagged site (STS) markers could be developed from the RAPD markers by cloning and sequencing of the RAPD markers, and the STS markers would be co-dominant, increasing the power of the analysis (Liu *et al.*, 1999a, 2001).

RAPD markers are particularly useful for efficient, economic, non-radioactive DNA fingerprinting of genotypes for the determination of genetic relationships and rapid construction of genetic-linkage maps (Grattapaglia and Sederoff, 1994; Johnson *et al.*, 1994; Liu and Dunham, 1998a,b). RAPD does not require any known probes or sequence information necessary for RFLP or microsatellite analysis. RAPDs are highly polymorphic and the technique is simple and fast. RAPD markers meet the requirements of a good marker system: the generation of large numbers of polymorphic markers, simplicity, economical, reproducible and normal Mendelian inheritance.

The primary drawback with RAPD analysis is the potential for reduced reproducibility (Hedrik, 1992; Riedy *et al.*, 1992; Scott *et al.*, 1992) because of the use of short random primers (usually ten nucleotides long), which necessitate lower annealing temperatures for PCR. Consequently, such short primers can bind to both their perfectly homologous binding sites and sites that are not completely homologous – suboptimal (non-specific) regions – especially during the first few cycles. This creates the risk of non-specific amplification and possible variation in results among experiments between different aliquots of the same sample at different dilution, and especially among laboratories. However, reproducibility was excellent for RAPD analysis of ictalurid catfish within the size range of 400–1500 bp (Liu *et al.*, 1998a, 1999b). Amplified products larger than 2 kilobase pairs (kbp) and smaller than 200 bp showed lower reproducibility. The bands were reproducible in number and over time. Higher concentrations of DNA template and primers led to amplification of more bands, making scoring more difficult. However, quantifying DNA before RAPD analysis and using the same concentration of primers gave consistent and reproducible results. The concentration and purity of genomic DNA templates used for PCR are the major factors for obtaining reproducible results. Genomic DNA should be prepared using a constant procedure, and the quantity of DNA should be determined before starting RAPD. Primer concentration should be kept constant to obtain reproducible results. Another weakness is that the testing of a large number of primers is required to generate a large number of markers.

RAPD has been used in the guppy, *Poecilia reticulata* (Foo *et al.*, 1995), tiger barb, *Barbus tetrazona* (Dinesh *et al.*, 1993), and medaka, *Oryzias latipes* (Kubota *et al.*, 1992), to study genetic variation. Only low levels of RAPD polymorphism existed among strains of channel catfish and strains of blue catfish (Liu *et al.*, 1998a). Fewer than 5–10% of the bands appeared to be polymorphic in some strains within each species. Fixed differences in RAPD genotypes were found for hellbender (salamander) (Kucuktas *et al.*, 2002) and for striped bass (R.A. Dunham,

H. Kucuktas and Z. Liu, unpublished results). It appears that the likelihood of finding strain-specific markers is greater via RAPD analysis than via isozyme analysis.

As expected and as it is for all types of genetic marker, interspecific RAPD variation was much greater than intraspecific RAPD variation for channel catfish and blue catfish (Liu *et al.*, 1998a, 1999b). More than 40% of the bands were polymorphic when comparing the two species. Each random RAPD primer amplified 5.3 bands from channel catfish and 5.5 bands from blue catfish. About 47.3% of the amplified bands from channel and blue catfish with random RAPD primers were species specific.

Liu *et al.* (1998a) examined the inheritance of RAPD sequences in channel catfish × blue catfish hybrids. All polymorphic paternal and maternal bands were amplified, and the sum of all RAPD bands from both parents existed in RAPD profiles of F_1 hybrids, indicating full penetrance, the dominant nature of RAPD markers and a dominant Mendelian pattern of inheritance in the F_1 progeny (Fig. 12.3). All polymorphic paternal and maternal bands were amplified. In all cases, channel catfish RAPDs segregated in F_2 and reciprocal F_1 blue backcrosses and blue catfish RAPDs segregated in F_2 and reciprocal F_1 channel backcrosses in expected ratios, confirming

Segregation of RAPD markers

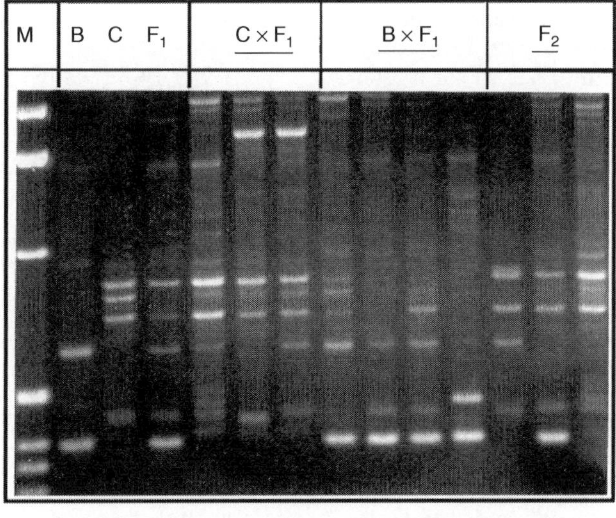

Primer A20

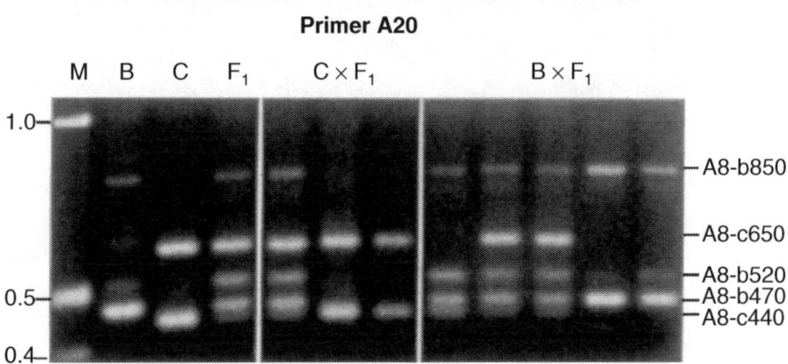

Fig. 12.3. Mendelian transmission of RAPD markers in backcross channel catfish (*Ictalurus punctatus*) × blue catfish (*Ictalurus furcatus*) hybrids. (Adapted from Liu *et al.*, 1998a.)

the dominant nature of the RAPD markers and the Mendelian inheritance.

Reciprocal F_1 hybrids (channel catfish female × blue catfish male and blue catfish female × channel catfish male) were also evaluated (Liu *et al.*, 1998a) because of paternal predominance (Chappell, 1979; Dunham *et al.*, 1982a) being prevalent for many of the phenotypic traits of the reciprocal hybrids. However, all RAPD bands were transmitted into the F_1 hybrids, regardless of the sex of the parents, and there was no apparent relationship between RAPD markers and paternal predominance. In channel × blue hybrids, by adjusting experimental conditions, almost any RAPD primer can be made to yield data. However, certain analyses, such as gene mapping, require testing of large numbers of individuals for each primer, and it is impractical and too time-consuming to elucidate and implement all of these conditions or to utilize low-yielding primers (Liu *et al.*, 1998a, 1999b). RAPD primers for gene-mapping analysis need to fulfil certain stringent criteria. An RAPD primer for gene mapping should generate a reasonable number of RAPD bands, detect high levels of polymorphism, be easy to use without special conditions and have high reproducibility. The number of bands amplified is important because, if too few bands are generated, a large number of primers and gel runs would be required to produce enough data points. If too many bands are amplified, it may make scoring and analysis difficult or impossible. High levels of polymorphism can reduce the numbers of primers and gel runs required to give the same numbers of markers.

RAPD markers must also pass some stringent tests to be useful for gene-mapping analysis. A useful RAPD marker for gene mapping needs to be polymorphic and segregational, highly reproducible, prominent in band intensity and well separated from the other bands for ease of scoring. Liu *et al.* (1998a) found that 52% of the RAPD primers tested for the blue catfish × channel catfish system were good to excellent primers, generating seven or eight RAPD markers per primer. In contrast, RAPD was inefficient in producing intraspecific polymorphic markers in catfish and would not be a good choice of molecular marker for intraspecific gene-mapping strategies.

A RAPD primer's usefulness for application across multiple species may be dictated by genetic distance. The primers evaluated by Liu *et al.* (1998a) worked well for both channel catfish and blue catfish, two highly related species. However, selected primers that were used for gene mapping in zebra fish (Johnson *et al.*, 1994) performed worse than random primers in the channel catfish × blue catfish evaluation, indicating that primers appropriate for one family of fish are not the most appropriate for another family. Similarly, high-GC primers (Kubelik and Szabo, 1995) were less useful than random RAPD primers in ictalurid catfish (Liu *et al.*, 1998a).

Amplified Fragment Length Polymorphism

AFLP (Vos *et al.*, 1995; Liu and Dunham, 1998a) combines the strengths of RFLP and RAPD. Genomic DNA is digested with two restriction enzymes *Eco*RI and *Mse*I, suitable adaptors are ligated to the fragments and the ligated DNA fragments are selectively amplified with different primer combinations (Vos *et al.*, 1995; Liu *et al.*, 1998c); then the products are resolved by gel electrophoresis. AFLPs are highly polymorphic and the technique is simple and fast. The molecular bases of AFLP polymorphism are base substitutions at the restriction sites, insertion or deletion between the two restriction sites, base substitution at the pre-selection and selection bases, and chromosomal rearrangements. The advantages of AFLP include its PCR-based approach, requiring a small amount of DNA; no requirement for any known sequence information or probes; and the specific amplification of a subpopulation of the restriction fragments because the long PCR primers in the procedure allow high annealing temperature and high repeatability. Perhaps the greatest advantage of AFLP analysis is that it is capable of producing large numbers of polymorphic bands in a single analysis at a relatively low cost per marker (Vos *et al.*, 1995; Liu and Dunham, 1998a). The generation of hundreds up to thousands of bands with limited numbers of primer combinations makes AFLP a very efficient and economical system for genetic analysis. There can be more than 100 loci per primer combination and over 4000 primer combinations to evaluate, and more than one restriction enzyme can be used, making it possible to generate tremendous numbers of markers. AFLP bands are widespread and evenly distributed, giving near genome-wide coverage. AFLP is highly reliable because it combines the advantage of RFLP and RAPD, and is devoid of the disadvantages of the slow speed and low levels of polymorphism of RFLP and the low reproducibility of RAPD. The AFLP procedure allows the genetic analysis of closely related populations. The disadvantages of AFLP are that these markers are dominantly inherited and that more technically demanding and specialized, expensive equipment such as DNA sequencers is required.

In some ways, these new DNA technologies can be almost too powerful. If enough markers are utilized, probably every population or sample can eventually be distinguished from any other population or sample. Criteria need to be established to match the new technologies to define when populations are actually differentiated. This will not be an easy concept to develop.

Liu *et al.* (1998c, 1999d) indicate that, among the polymorphic DNA fragments, two subsets can be distinguished: the presence or absence of bands and band-intensity polymorphisms. Presence or absence polymorphisms result from the gain or loss of restriction sites, insertions, deletions or reversions between restriction sites, or from the fact that selective nucleotides of the primer used in the AFLP procedure and the sequences adjacent to the restriction site are not complementary.

The intensity of the polymorphism is very difficult to score. In reality, even if band intensities could be quantified, the intensities of the amplified band are not really polymorphisms. If the samples are obtained and analysed in exactly the same way, the DNA at the location of a specific AFLP marker should be amplified equally among all individuals possessing that sequence, or allele. If band intensities could actually be quantified, there would be two explanations for varying band intensities. One case would be the difference between homozygous individuals, which would have two copies of the marker, and heterozygous individuals, which would have one copy of the marker. The cumulative effects of different chromosomal locations generating AFLP fragments of exactly the same size could also generate bands of different intensities. This would indeed be real genetic variation but would not be specific to a single chromosomal location, or locus. Differences in band intensity could also be the result of a combination of both these phenomena, making interpretation potentially difficult or impossible.

In the case of catfish (Liu *et al.*, 1998c), the polymorphic paternal and maternal bands showed dominant Mendelian pattern of inheritance in the F_1 interspecific progeny of channel and blue catfish. Little segregation of polymorphic AFLP markers was observed in F_1 individuals, thus indicating that the majority of the polymorphic loci were homozygous in the parental species. The majority of the AFLP markers must represent species-specific markers. The markers segregating in F_1 individuals represented allelic variation within a species. Those AFLP markers were heterozygous in the paternal or maternal parent. Since low levels of polymorphisms were detected within either channel catfish or blue catfish, most AFLP loci associated with interspecific variation were homozygous within the species. This also indicates that multiple chromosomal locations for the same-sized bands utilizing the same primers must be a rarity; otherwise the transmission of the AFLP markers to the F_1 hybrids would have been more complicated. That leaves no logical genetic mechanism that would generate bands of different intensities in ictalurid catfish. This may be true for other species of fish as well. The results also indicated that chromosomes of channel catfish and blue catfish paired properly and their AFLP markers followed Mendelian inheritance.

The application of AFLP markers in genetic linkage and QTL mapping (Meksem *et al.*, 1995; Cho *et al.*, 1996; Mackill *et al.*, 1996; Otsen *et al.*, 1996; Liu *et al.*, 2003) and analyses of genetic resources (Folkertsma *et al.*, 1996; Travis *et al.*, 1996; Keim *et al.*, 1997a,b) has greatly sped research in these areas. The efficient, rapid, economical development of AFLP markers linked to disease resistance genes has allowed application of these markers in MAS programmes (Meksem *et al.*, 1995).

Liu *et al.* (1998c, 1999b) examined the characteristics of the production of high-quality AFLP markers. AT-rich selection bases were more associated with a lower quality of AFLP fingerprinting. With the exception of two primer combinations (Liu *et al.*, 1998c), all of the lower-quality primer combinations were from primer combinations with primers M-CAT or M-CTT. This may indicate that the genomes of channel catfish and blue catfish are AT-rich. The AT-richness would create a greater number of amplified bands when the selective bases are AT-rich, especially at positions nearest the 3′ end. AT-rich primer combinations also resulted in greater numbers of total amplified bands in barley using AFLP (Qi and Lindhout, 1997). The weak intensities of these bands may be caused by low efficiency of primers with AT 3′ ends in PCR reactions. However, if the AT-richness does cause amplification of a greater number of bands, the intensities should be similar among all amplified bands, with the exception of bands amplified from highly repetitive elements. That was not the case for channel and blue catfish. Alternatively, these weak bands could be from non-specific priming at the mismatched sites. Most of the poor primer combinations had T as the 3′-terminal base in the primer (Liu *et al.*, 1998c). Kwok *et al.* (1990) indicate that primers with 3′-terminal T mismatches can be efficiently utilized by *Taq* polymerase when the nucleotide concentration is high. Amplification of a T mismatched with a C, G or T may be initiated frequently, although at a lower efficiency than with a matched base. The catfish results of Liu *et al.* (1998c, 1999d) are consistent with this explanation. If this is correct, T should be avoided from the terminal selective bases for the design of AFLP primer kits.

The primers and selection bases can affect the variation of AFLP profiles. For channel catfish and blue catfish, *Eco*RI primers had a large effect on the total number of amplified bands – 49–267 bands (Liu *et al.*, 1998c). Eight *Eco*RI primers produced 69–161 total bands per primer combination when combined with the eight *Mse*I primers, and the eight *Mse*I primers produced 93–145 total amplified bands when combined with the eight *Eco*RI primers.

The terminal selective bases have large effects on both the total number of amplified bands and their reproducibility (Liu *et al.*, 1998c, 1999d). Terminal T exhibited the lowest levels of selectivity, producing the largest numbers of amplified bands and the highest levels of background bands in ictalurid catfish. Among eight *Eco*RI primers, E-ACG produced the least number of total amplified bands. Eight *Mse*I primers produced a mean of 69 bands per primer combination. E-ACT produced the largest total number of amplified bands, with a mean of 161 bands per primer combination, when used with the eight *Mse*I primers. Among eight *Mse*I primers, M-CTC generated the lowest number of amplified bands, with a mean of 93 bands per primer combination, and M-CTT produced the largest number of total amplified bands, with a mean of 145 bands per primer combination. The terminal-T selection base generated large numbers of amplified bands compared with other selective bases.

The position of selective bases, rather than the AT-richness, has the greatest effect on AFLP fingerprinting patterns. Although the percentage of G/C and A/T bases in the selection bases of *Eco*RI primers used by Liu *et al.* (1998c, 1999d) when analysing channel catfish and blue catfish was the same (50%) as in the *Mse*I primers (50%), their positions are different. Two of the eight *Eco*RI primers tested harboured A/T at their 3 terminus, while four of the eight *Mse*I primers had a terminus of A/T bases.

Similarly, at the second-most 3 position, *Eco*RI primers had two A/T bases out of eight. In contrast, *Mse*I primers had eight A/T bases out of eight. Combining the second- and the third-most terminal bases, *Eco*RI primers had four of 16 being A/T bases and *Mse*I primers had 12 of 16 being A/T bases. This may explain the larger variation of total number of amplified bands generated from *Eco*RI primers (variation from high G/C to high A/T at the terminal bases) than that from *Mse*I primers (variation from medium A/T to high A/T at the terminal bases). The polymorphic rates of the AFLP bands were inversely correlated to the total numbers of amplified bands in blue and channel catfish (Liu *et al.*, 1999d). Primer combinations that produced large numbers of amplified bands had lower percentages of the bands being polymorphic. Primer combinations that produced small numbers of amplified bands generated higher rates of polymorphism. Eight *Eco*RI primers produced a mean of 33–62% polymorphic bands. The eight *Mse*I primers resulted in a mean of 30–50% polymorphic bands. Among the *Eco*RI primer combinations, primer combinations that produced the smallest numbers of total amplified bands ($\bar{X}$ = 69) had the highest mean rate of polymorphism – 62%. Similarly, among the *Mse*I primer combinations, primer combinations that produced the smallest number of total bands ($\bar{X}$ = 96) exhibited the highest mean polymorphism – 49%. In contrast, the primer combinations that produced the largest mean number of amplified bands – 161 – among the *Eco*RI primer combinations had the lowest mean polymorphic rate (33%) among the *Eco*RI primer combinations. Similarly, the primer combinations with *Mse*I primer 8 produced the highest mean total of amplified bands – 145 per primer combination – but their mean polymorphic rate (30%) was also the lowest among the *Mse*I primer combinations. Again, the total numbers of amplified bands appeared to be related to the terminal selective bases. Primers with the terminal T produced large numbers of amplified bands, which was correlated with a lower percentage of polymorphic bands than primers with G, C or A terminal bases.

High reproducibility is required for good DNA markers. Liu *et al.* (1998c, 1999d) tested the reproducibility of AFLP markers in channel and blue catfish by using DNA templates from different individual fish isolated at different times. High levels of reproducibility were observed as all individuals tested over time always exhibited identical banding patterns.

Different numbers of bands may be observed as a function of different radioisotopes (Liu, 1999d). Generally, the same AFLP profiles were obtained with either ^{32}P or ^{33}P for ictalurid catfishes. However, ^{33}P allowed detection of more AFLP bands than ^{32}P. This difference is probably related to the low-energy radioisotope, ^{33}P, allowing longer exposure than the high-energy radioisotope, ^{32}P. For instance, with 1 week's exposure to ^{33}P it was possible to obtain reasonably clean autoradiograms, while 1 week's exposure to ^{32}P generated overexposed autoradiograms.

Longer exposure allowed some weaker bands to be detected, which otherwise were not visible or were just too weak. While the detection of more bands assisted robust analysis of many loci simultaneously, too many bands could make analysis extremely difficult, especially for adjacent markers all segregating. Another factor for the reproducibility of AFLP bands is the size of amplified products. Generally, amplified products with large sizes displayed on the top of sequencing gels have lower reproducibility. Often the large fragments were not efficiently amplified to generate strong bands. AFLP bands of 50–500 bp exhibited the highest reproducibility. Large variation in the number of bands amplified was observed when different primer combinations were used. This is not a reproducibility issue, but is relevant for primer selection for genetic-linkage analysis.

Similarly to RAPD analysis, intraspecific polymorphism for AFLP can be relatively low. Intraspecific polymorphisms in ictalurid catfishes were generally less than 10% of all bands (Liu *et al.*, 1999d). However, over 50% of AFLP bands were polymorphic between channel catfish and blue catfish, and were suitable for use in mapping analysis using the interspecific hybrid system. As expected and similarly to the case with RAPD markers, there was no relationship between AFLP genetic variation and paternal predominance in channel × blue catfish hybrids (Liu *et al.*, 1998c). Even though strong paternal predominance was observed for the morphological and meristic traits of interspecific F_1 hybrids, transmission of AFLP markers to F_1 was normal, and the two reciprocal F_1 hybrids of channel catfish × blue catfish had identical AFLP profiles. Most AFLP markers were inherited in expected Mendelian ratios in F_2 or backcross hybrids (Liu *et al.*, 1998c). AFLP can generate large numbers of DNA bands without any previous knowledge of a fish genome. This technology offers a robust analysis of genomes, allowing approximately 100 catfish genomic fragments to be displayed in a single analysis. The high levels of interspecific polymorphism between the channel catfish and blue catfish suggest that AFLP analysis should be highly useful for generating a large number of markers for genetic-linkage analysis, using the interspecific hybrid system to produce mapping populations. Exactly as in the case for RAPD molecular markers, the low level of intraspecific polymorphism makes AFLP an inefficient marker system for mapping analysis using an intraspecific mating design. More intraspecific polymorphism was detected with AFLP analysis than with RAPD analysis for channel catfish and blue catfish (Liu *et al.*, 1998a,c).

Segregation of channel catfish and blue catfish AFLP markers followed expected ratios except that there was strong segregation distortion for two blue catfish markers in the F_2 and backcross hybrids selected for increased growth rates (Liu *et al.*, 1998c). These two markers were either not detected or were detected at much lower frequencies than expected. Since the F_2 and backcross hybrids were selected for increased growth rate, these markers and the corresponding markers in channel catfish may be linked to growth loci. The segregation could be caused by selection pressure for increased growth or possibly by natural selection. If there were genes or genomic sequences in blue catfish that had a negative effect on growth rate in comparison with the channel catfish sequences, selection for increased growth rate should select against these genes or genomic sequences. Since both markers were absent from the individuals analysed, the two markers may be linked or they could be independently linked to two separate growth-encoding loci. Alternatively, the segregation distortion may be due to natural selection for survival. If some genes or genomic sequences have a negative effect on survival, they would be selected against and detected at a reduced frequency.

Microsatellites

Microsatellites are simple-sequence, tandem repeats of 1–6 bp. Dinucleotides are the most common. The molecular basis of microsatellites is the difference in number of repeats (Liu and Dunham, 1998a; Liu *et al.*, 1999e; Tan *et al.*, 1999). Microsatellites containing one type of repeat are called simple microsatellites; those with more than one type of repeat are called composite microsatellites. Microsatellite markers are ideal molecular markers because they are highly polymorphic, evenly distributed in genomes and co-dominantly inherited. In general, larger microsatellites are the most polymorphic. Approximately, 1–4% of the genome consists of microsatellites, and one microsatellite occurs about every 10 kb.

Microsatellites are highly useful among various types of DNA markers because their high rate of polymorphism and co-dominant inheritance allow precise genetic analyses, increase mapping accuracy, maximize the genetic information generated and allow lineages of individuals or families (individual spawns) to be accurately traced (Waldbieser and Wolters, 1999). High levels of polymorphism also indicate that microsatellite markers may be highly useful for population-genetics analysis and strain identification. Microsatellite loci are short in size, facilitating genotyping via PCR. Their disadvantage is that microsatellite analysis requires great effort, time and expense in library construction, screening, sequencing and PCR primer analysis, and they may have non-specific bands (Liu and Dunham, 1998a). Characterization of large numbers of microsatellites for the construction of genetic maps with high resolution is a tedious and strenuous task.

The majority of microsatellite loci can be amplified in both channel catfish and blue catfish (Fig. 12.4), suggesting evolutionary conservation between these two catfish species (Liu *et al.*, 1999e), and microsatellite markers can be used as co-dominant markers in the interspecific hybrid system for genetic-linkage analysis. Liu *et al.* (1999e) confirmed that microsatellite markers are inherited as co-dominant markers in catfish. Microsatellites were highly polymorphic in both channel catfish and blue catfish. The microsatellites were developed from channel catfish. Only 10% of microsatellite loci could not be amplified in blue catfish. In the case of these loci, the divergence may be quite dramatic so that primers could not bind at all, or it could be minor so that primer binding might require lower temperatures. One microsatellite locus, *Ip351*, was amplified at 40°C to produce allelic fragments, which were not amplified from blue catfish at 50°C or higher temperatures, indicating conservation of the locus with minor base substitution at the premier binding sites.

If large percentages of primers can amplify across species borders, high levels of genomic conservation are indicated. The high levels of genomic conservation of channel catfish and blue catfish were consistent with the indistinguishable karyotypes of the two species (LeGrande *et al.*, 1984).

Conservation of microsatellite loci among closely related species is expected, although successful amplification of these loci across species boundaries depends on the conservation of primer sequences (Liu *et al.*, 1999e; Tan *et al.*, 1999). In some genera, families, orders or phyla, microsatellites are deficient and primers from one species will often not work in related species. Sequence variations at the primer region can significantly affect the success of PCR amplification. Base changes at the 3′ end of the primer-binding sites are more critical than at the 5′ end (Liu *et al.*, 1999e). Caution should be exercised when interpreting the existence of null alleles, because they can be obtained from amplifications from only one allele when one or both primers fail to bind to the alternative allele, thus skewing the data towards a higher frequency of homozygosity (Menotti-Raymond and O'Brien, 1995).

Genomic conservation of microsatellite loci has also been compared among channel catfish, blue catfish, white catfish (*Ameiurus catus*) and flathead catfish (*Pylodictus olivaris*) (Liu *et al.*, 1999e; Tan *et al.*, 1999), all in the family Ictaluridae. The microsatellite loci were highly conserved in all genera tested from Ictaluridae. All channel catfish primers tested successfully amplified genomic DNA from flathead catfish, and 86% of the channel catfish primers successfully amplified the genomic DNA from white catfish. Southern blot analysis confirmed allelic amplification. If the amplification is allelic, all the amplified bands from all species should harbour microsatellite sequences

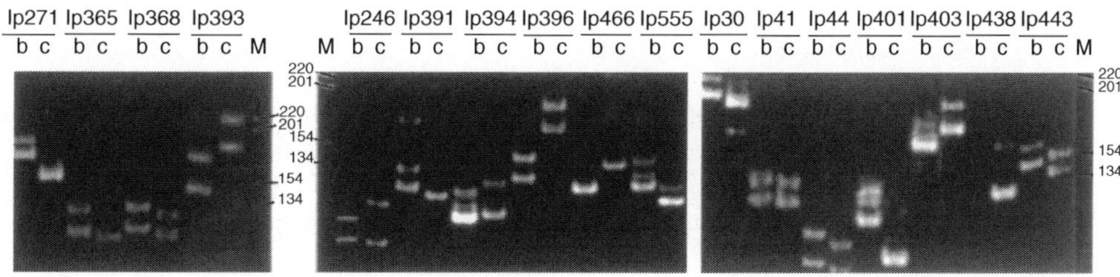

Fig. 12.4. Amplification of blue catfish, *Ictalurus furcatus*, microsatellites with channel catfish, *Ictalurus punctatus*, primers. (Adapted from Liu *et al.*, 1999e.)

and thus should hybridize to the microsatellite probes. All amplified bands from channel catfish, flathead catfish and white catfish hybridized to the $(CA)_{15}$ probe, confirming allele-specific amplification.

Ictalurid catfish have similar genomic organization. Microsatellites appear to evolve free of selection pressure, and thus evolve towards high evolutionary divergence because they are non-coding sequences (Liu *et al.*, 1999e). This may be the general case; however, some microsatellites are found within coding regions and in this case might be subject to strong selective pressures (Liu *et al.*, 1999f). The discovery of high conservation of microsatellite marker loci across wide taxonomic borders in fish may indicate either that such sequences evolve under unknown selection pressure or that sequence evolution in aquatic animals is slower than in land animals (Liu *et al.*, 1999e).

Various classes of microsatellite exist and can be found at varying frequencies. In eastern oysters, dinucleotide motifs within microsatellites were dominated by AG, trinucleotide microsatellites had all possible motifs in equal frequencies and tetranucleotides were more prevalent than trinucleotides and were strongly associated with specific repetitive sequences, which was not the case for other classes of microsatellite (Gaffney, 2002).

Expressed Sequence Tags

ESTs are short, single-pass cDNA sequences reverse-transcribed from mRNAs and generated from randomly selected cDNA-library clones (Adams *et al.*, 1991; Liu *et al.*, 1999a). The single-pass sequencing is at both the upstream and the downstream segments of cDNAs. The basis of EST analysis is that specific mRNAs and transcript quantities vary in different tissues, in different developmental stages or when the organism faces different environmental conditions. Characterization of ESTs is a relatively easy and rapid way for identification of new genes in various organisms (Tilghman, 1996).

Because of the relative ease of EST analysis, the EST database is the fastest-growing division of GenBank (Karsi, 2001). More than 415,000 human ESTs have been characterized (Wolfsberg and Landsman, 1997). Extensive EST analysis is not only an efficient way to identify genes, but is also powerful for the analysis of their expression (Karsi, 2001). ESTs indicate when, where and how strongly genes are expressed, and each EST represents a gene, so they can be used in functional genomics analysis. ESTs are particularly useful for the development of cDNA microarrays, which allow the study of differentially expressed genes in a systematic way. EST analysis is useful for comparative genomics by determining orthologous counterparts of genes through evolution. Profiling of expression provides a rapid means of examining gene expression or differential gene expression in specific tissue types, in biological pathways, under specific physiological conditions, during specific developmental stages or in response to various environmental challenges. ESTs are also efficient molecular markers for genomic mapping (Hudson *et al.*, 1995; Schuler *et al.*, 1996), and microsatellites can be found within ESTs, making ESTs even more useful for gene-mapping research. However, only 4.6% of ESTs characterized from the skin of channel catfish contained microsatellites (Karsi, 2001; Karsi *et al.*, 2002a). Disadvantages of EST analysis are that a large amount of preparatory work is required, the cDNA library may not contain transcripts of low abundance and normalization of the cDNA may be necessary.

EST analysis greatly speeds the pace of genomics research. Implementing a transcriptomic approach, all of the 47 ribosomal protein genes in the 60S ribosome (Patterson *et al.*, 2003) and all of the 32 ribosomal protein genes in the 40S ribosome of channel catfish (Karsi *et al.*, 2002b) were characterized in a few months, which would have taken years of analysis, requiring much greater resources to accomplish, prior to the development of EST procedures. Such systematic EST analyses allowed identification of alternative spliced transcripts as well as alternatively polyadenylated transcripts, demonstrating the greater power and value of EST analysis compared with that given solely by surveys of genes and their expression (Karsi, 2001).

Microsatellites have been discovered in ESTs generated from a variety of tissues in channel catfish (Karsi *et al.*, 2002a; Kocabas *et al.*, 2002a). The percentage of ribosomal protein genes (ESTs) containing microsatellites was 4.6% (Karsi, 2001). Ribosomal protein S16 of channel catfish contained a compound microsatellite composed of CT and CA repeats. The microsatellite locus for this protein had both types of simple sequence repeat. They were short and both were expanding. The CT repeat had genotypes ranging from eight to 11 repeats, while the CA repeat was found to have genotypes with four or five repeats. This polymorphic microsatellite should allow the genomic mapping of the *S16* gene (Karsi, 2001).

Microsatellites of channel catfish are also found in muscle-specific genes. The catfish myostatin gene possesses multiple microsatellites (Kocabas *et al.*, 2002a), making this gene highly polymorphic and potentially useful for gene mapping and probably MAS. These microsatellites were found in numerous regions of the gene, including the upstream untranslated region (UTR), exons and introns. The upstream UTR contained highly repetitive sequences, including several hundred base pairs of simple sequence repeats with a simplified consensus of TGGTAG (Kocabas *et al.*, 2002a). One (CAG) repeat was found within the first exon encoding a polyglutamine tract. Three microsatellite sequences existed in the first intron with $(GTTT)_7$, $(TG)_{11}$ and $(TA)_{38}$ repeats. One microsatellite repeat of $(GAA)_7$ existed in the second intron. In the 3-UTR, one AC repeat was present.

Single Nucleotide Polymorphism

SNP is caused by base variation among individuals at any site of the genome (Kocabas, 2001). An SNP is a single base variation. This single base variation can be determined by DNA sequencing, primer extension typing, the designing of allele-specific oligo and gene-chip technology. An SNP can be in the form of a transition or transversion from one base to another, but can also be a single-base indel, deletion or insertion. Transitions change a purine to a purine or a pyrimidine to a pyrimidine. Transversions change a purine to a pyrimidine or the reciprocal. Transversions have the probability of happening twice as often as transitions, but transitions are found just as frequently as transversions, perhaps because transitions are less likely to change the amino acid and thus are neutral, being less likely to be selected against. SNPs are inherited in a co-dominant fashion.

SNPs are obviously found in both non-coding and coding regions. SNP replacement polymorphisms change the amino acid, and SNP synonymous polymorphisms change the codon but not the amino acid. SNP regulatory polymorphisms can occur which alter gene regulation. The channel catfish myostatin gene not only contained microsatellites, but also had many SNPs (Kocabas *et al.*, 2002a). Many of these SNPs were neutral and did not change amino acid sequences, but five SNPs caused changes in amino acid sequences.

The advantages of SNP are that SNP sites are abundant throughout the entire genome (3×10^7 different sites have a SNP in humans), they are highly polymorphic and the SNP analysis is the only system that identifies every single difference or polymorphism among individuals. Microsatellites have more alleles per 'locus'; however, SNPs are at a higher density. SNPs are found anywhere from every 76 to every 2000 bp in various organisms (Liu, 2007). Channel catfish has a SNP every 76 bp.

Originally, SNP analysis had several disadvantages, including the need for sequence information, the necessity of probes and hybridization, high expense and difficult genotyping. However, with the development of TaqMan® technology, gene-chip technology, pyrosequencing and MALDI-TOF (matrix-assisted laser desorption ionization–time of flight mass spectrometry), large-scale SNP analysis is now becoming affordable and will soon be even more affordable than other genetic markers, becoming the marker of choice for many applications. SNP analysis will result in the ability to perform whole genome selection, the ultimate MAS, when traits and their corresponding SNP variation are fully defined.

Relative Costs of Different Markers

In some cases, costs may dictate what type of markers may be used for a genetic application. RFLP, RAPD, AFLP, microsatellites and SNPs were compared for their relative costs (Table 12.1). The primary advantage of RFLP and RAPD is the very low start-up costs. The quality and power of the data sets are lower for RAPD and RFLP. Costs of AFLP, microsatellite and SNP analysis are similar. However, if extremely large numbers of SNP markers are developed, the cost per sample becomes quite small compared with other marker systems.

Isozyme costs are about $16,000/160 samples for 40 loci. Therefore, for small sample numbers isozyme analysis is going to be more cost-effective than DNA markers, and the biological meaningfulness is going to be greater. However, the ability to find fixed differences between populations will be vastly inferior for isozymes compared with RAPD and AFLP and inferior to microsatellites for tracing lineages. Once sample sizes reach 1000 or more, the DNA analyses are more cost-effective than isozymes, with the same advantages and disadvantages for biological information generation as for the smaller sample sizes.

Relative Effectiveness of Markers

The various markers have different strengths and weaknesses. Yuan *et al.* (2000) compared RFLP, RAPD, SSR and AFLP for their

Table 12.1. Relative costs (US$) of RFLP, RAPD, AFLP, microsatellites and SNPs at variable numbers of individuals sampled, markers tested and technicians required. The laboratory, basic equipment, water, glassware and other miscellaneous items were assumed to already exist. Several technologies, such as DNA sequencing, allele-specific oligo and gene chip, can generate SNP genotypes. DNA sequencing and gene chip are expensive, so costs were estimated for allele-specific oligo, which is a widely used technique. Start-up costs include labour, chemicals, sequencing, etc. to develop markers. (Costs were determined with assistance from Huseyin Kucuktas and Eric Peatman.)

Fish/markers/technicians	RFLP	RAPD	AFLP	Microsatellites	SNPs[a]
Start-up costs	2,000	2,000	10,000	15,000	10,000
30/3/1	135	135			
30/10/1			600	750	750
30/50/1			3,000	3,750	3,750
100/3/1	445	445			
100/10/1			1,980	2,475	2,475
100/50/1			9,900	12,500	12,500
1,000/3/1	4,450	4,450			
1,000/10/2			19,800	25,000	25,000
1,000/50/2			99,000	125,000	125,000
10,000/3/4	44,500	44,500			
10,000/10/4			198,000	250,000	250,000
10,000/50/4			999,000	1,250,000	1,250,000
10,000/10,000					10,000,000

Estimated cost/marker/fish: RFLP = $1.50; RAPD = $1.50; AFLP = $2.00; microsatellites = $2.50; SNP = $2.50.
[a]The cost of SNPs varies significantly due to different platform costs. There are different platforms for different applications. For high numbers of fish, low marker numbers (as above) one would probably use multiplexed TaqMan® probes and costs would be as listed (including genomic DNA extraction). But if you ran 10,000 fish on 10,000 SNPs your costs would be $0.10/genotype or less. Also start-up costs are hard to partition. You can spend $10,000 and develop 100,000 SNPs, but your proportional costs for developing small numbers of SNPs are much higher per SNP.

effectiveness in delineating inbred lines of maize (Doyle, 2003). The information content, as measured by expected heterozygosity and mean number of alleles polymorphic, was best for SSR analysis. AFLP resulted in the lowest polymorphic values. The genetic-similarity trees were highly correlated except for those generated by RAPD. Yuan *et al.* (2000) concluded that AFLP was the most efficient system because of the large number of bands generated and that both AFLP and SSR could replace RFLP for maize population-genetics analysis.

Desvignes *et al.* (2001) examined allozyme and microsatellite gene frequencies in domestic carp strains in France and the Czech Republic. French strains had lower heterozygosity but a higher number of alleles. The two marker systems generated data giving similar results and conclusions, but the microsatellites better discriminated strains within and between countries.

Allendorf and Seeb (2000) examined 21 allozyme polymorphisms, 15 nuclear DNA polymorphisms and mtDNA variation in four Alaskan populations of *Oncorhynchus nerka*. Concordance was obtained among markers in the amount of genetic variation within and among populations, with the striking exception of one allozyme locus (*sAH*), which exhibited more than three times the amount of among-population differentiation as other loci. Allendorf and Seeb (2000) concluded that information should be gathered on many loci for population comparisons, but that the type of locus or marker system is of secondary importance.

Microsatellite markers underestimate genetic divergence among populations when gene flow is low (Balloux *et al.*, 2000). Microsatellites are extremely powerful but discriminate less well at the species level than at the subpopulation level (Doyle, 2003). Balloux *et al.* (2000) examined genetic divergence among two chromosome races of the common shrew in which gene flow is reduced by a number of factors, including male hybrid sterility. The genetic divergence of these two races of shrew estimated from mtDNA, proteins and karyotypes was much larger than that estimated from microsatellites, with the exception of one microsatellite located on the male (Y) chromosome. Balloux *et al.* (2000) utilized computer simulations to show that the discrepancy arises mainly from the high mutation rate of microsatellite markers for F statistics and from deviations from a single-step mutation model for R statistics.

Gerber *et al.* (2000) compared dominant markers, such as AFLPs, with co-dominant multiallelic markers, such as microsatellites, for reconstructing parentage. Both sets of markers produced high exclusion probabilities; dominant markers with dominant allele frequencies in the range 0.1–0.4 were more informative (Doyle, 2003). Not unexpectedly, dominant markers were less efficient than co-dominant markers for reconstructing parentage, but can still be used with good confidence when loci are deliberately selected according to their allele frequencies.

Controversy and uncertainty surround the issue of whether or not a population at risk of extinction is actually worth saving, given that resources are limited and priorities must be set. Genetic uniqueness and genetic diversity are often cited as the appropriate criteria for expending effort to save populations (Doyle, 2003). The genetic diversity and uniqueness of neutral markers may have no correlation with the quantitative genetic variation, which is the actual basis of adaptation and short-term adaptation. McKay *et al.* (2001) found that small, peripheral populations of the rare plant sapphire rockcress are genetically adapted to local microclimates, and that local adaptation occurs despite the absence of divergence at almost all marker loci and very small effective population size, as indicated by extremely low levels of allozyme and DNA sequence polymorphism. This empirical evidence proves that setting conservation priorities based exclusively on molecular-marker diversity may lead to the loss of locally adapted populations.

Comparison of living and fossil samples of marine snails revealed that the northern populations have come to be dominated by snails with a new, probably adaptive, thick-shelled morphology (Hellberg *et al.*, 2001) and have become morphologically more diverse despite being relatively uniform at the mtDNA-marker loci. Similar conclusions can be drawn from data on freshwater fishes, in which post-Pleistocene colonization of new habitats has also led to evolutionary divergences (Hellberg *et al.*, 2001; Doyle, 2003). When given the opportunity, quantitative traits can diverge at more rapid rates than neutral markers.

Freeland and Boag (1999) found that the differentiation of Darwin's ground-finch species based on morphological data is not reflected in either mtDNA- or nuclear DNA-sequence phylogenies, and inferred that genealogies based on mitochondrial and nuclear markers are not even concordant with each other. This is obviously another example illustrating that DNA markers do not always correlate with important quantitative and qualitative traits, which Lewontin (1985) predicted based on the theory that the two measures should be very poorly correlated.

Similarly, estimates of molecular and quantitative genetic variation were essentially uncorrelated in natural populations of *Daphnia* (Pfrender *et al.*, 2000). Although molecular markers provided little information on the level of genetic variation for quantitative traits within populations, they may be valid indicators of population subdivision for such characters, as molecular measures of population subdivision gave conservatively low estimates of the degree of genetic subdivision at the level of quantitative traits (Pfrender *et al.*, 2000; Doyle, 2003).

Reed and Frankham (2001) examined 71 sets of data where quantitative variation and marker variation were both estimated. The mean correlation between molecular and quantitative measures of genetic variation was weak ($r = 0.22$), and there was no significant correlation between molecular variation and life-history traits ($r = 0.11$) or heritability ($r = 0.08$). DNA markers may have only a very limited ability to predict quantitative genetic variability (Doyle, 2003).

Individuals that are more heterozygous at allozyme loci are often more fit, by several measures of fitness, than individuals in the same random-mating population that are less heterozygous (Doyle, 2003). Several possible explanations exist for this observation, including that allozyme heterozygosity may itself be beneficial, which implies that allozymes are not selectively neutral; allozymes may be neutral markers for chromosome segments carrying unknown genes that enhance fitness when heterozygous (associative overdominance); or marker homozygosity may be an indicator of inbreeding depression.

Thelen and Allendorf (2001) analysed ten allozyme and ten microsatellite loci in a hatchery population of rainbow trout. Allozyme heterozygosity correlated positively with condition factor, but microsatellite, non-coding DNA heterozygosity did not. The observed relationship between heterozygosity at allozyme loci and condition factor in rainbow trout appears to be the direct effect of the allozymes themselves, rather than associative overdominance or linkage to unidentified selected genes. The results indicate that allozymes and microsatellites are differentially affected by natural selection, and that allozymes are selected whereas microsatellites are not.

Additionally, isozymes/enzymes may experience quite different rates of evolution in different species. The neutrality theory predicts that the rate of molecular evolution will be constant over time (Ayala, 2000). However, the variance of the rate of evolution is generally larger than expected according to the neutrality theory. Several modifications of the theory have been proposed to account for the 'overdispersion' of the molecular clock, such as effects due to generation time, population size, slightly deleterious mutations, repair mechanisms and more. An extensive examination of two proteins, glycerol-3-phosphate dehydrogenase (GPDH) and SOD, indicates that none of these modifications can simultaneously account for the disparate patterns observed in the two proteins (Ayala, 2000). GPDH evolves very slowly in *Drosophila* species, but several times faster in mammals, other animals, plants and fungi, whereas SOD evolves very fast in *Drosophila* species and also in mammals, but much more slowly in other animals and still more slowly when plants and fungi are compared with one another or with animals. Sometimes generalizations cannot be made even with a single marker.

This series of experiments indicates that genetic markers need to be matched properly with the objective of the research and the type of information and analysis being pursued. For some applications, the type of marker is not so important. For others, some markers are more powerful than others. In some cases, the type of genetic variation detected and the genetic conclusions can vary depending on the type of marker utilized. One unfortunate result is that the lack of marker genetic variation does not eliminate the possible existence of environmentally and evolutionarily important quantitative genetic variation in the very same individuals, and this quantitative genetic variation can exist without any discernible biochemical- or molecular-marker variability.

13

Population Genetics and Interactions of Hatchery and Wild Fish

Genetic Variation, Population Structure and Biodiversity

Genetics is in reality a relatively new field. For centuries, humans moved fish among countries and stocked conspecifics in new watersheds without any knowledge or concern regarding genetic principles, impact or consequences. During the first 70 years of the 20th century, fish movement and stocking were rampant. Beginning in the 1970s, conservation genetics has become recognized and is a burgeoning issue, as are biodiversity and genetic biodiversity. In general, individual countries and natural-resource agencies now take a much more conservative approach to stocking programmes, genetic conservation and biodiversity. However, many decisions and policies are made and implemented without data on population genetics and the genetic interactions of fish populations. There is a need for much more research in this area.

When data are available, lack of accurate stocking histories complicates data interpretation. Another void is a lack of data demonstrating the relationship between performance and biochemical and molecular markers, which was introduced in the previous chapter. These are very difficult data to generate, as it is not easy to replicate the natural environment, rivers, reservoirs, lakes and oceans in a realistic manner. Geneticists sometimes give natural-resource managers conflicting advice regarding the desirability of increased or decreased genetic variation and the policies and mechanisms to achieve various goals.

The first question that needs to be considered is the importance of genetic variation in natural populations. Is it better to have more genetic variation or less genetic variation? These are difficult questions to answer, and the answer may be different depending on the individual circumstances. Do population structures dictate the need for the quantity and type of genetic variation? In the circumstances leading to different population structures, have the selective pressures led to the optimum genotypes in a particular environment or have limitations on gene flow in that environment limited the development of the optimum genetic structure of a population?

Theoretically, genetic variation is beneficial and important. Genetic variation is important for the long-term survival of a species. Genetic variation can ensure the fitness of a species or population by giving the species or population the ability to adapt to changing environments.

Obviously, a lack of genetic variation or too much homozygosity can be detrimental to an individual's or a population's survival traits and fitness. The cheetah is a prime example of the potential detrimental effects of excess homozygosity. This highly homozygous species has severe reproductive problems. Homozygosity has also been correlated with bilateral asymmetry (fluctuating asymmetry) – unbalanced meristic counts on the right and left halves of the body – in fish. Additionally, highly or totally homozygous individuals and populations actually exhibit greater phenotypic variation than outbred controls because they are more greatly affected by environmental or micro-environmental change and have reduced homeostatic ability compared with more heterozygous individuals and populations. Inbreeding in small, natural populations increases extinction rate (Doyle, 2003).

Inbreeding depression resulting from increased homozygosity is well documented in fish (Dunham *et al.*, 2001). Field crops have been endangered when they did not have the genetic variation to respond to new pathogens or plagues. Clearly, the existence of genetic variation is important to the long-term survival and fitness of a species. Many natural populations respond to different forms of selection, such as directional, bidirectional, cyclical and stabilizing selection, which help to ensure the maintenance of the genetic variability and/or fitness of population.

Levels of homozygosity and inbreeding can be important not only in domestic or aquaculture populations, but in wild populations as well. Inbreeding does adversely affect reproductive success in wild deer (Slate *et al.*, 2000). Microsatellite heterozygosity was utilized as an indicator of individual inbreeding coefficients among unmanaged deer on the island of Rhum, Scotland. Heterozygosity was correlated with lifetime breeding success (total offspring) in both males and females.

The majority of inbreeding experiments on fish (Dunham *et al.*, 2001) and other organisms have been done in aquaculture and laboratory-type environments. Some have hypothesized that inbreeding

depression would be more severe and affect fitness more adversely in the harsher natural environment compared with the laboratory environment or aquaculture environment where animals are well taken care of. However, the fitness of mosquito populations declined to the same extent in natural tree holes as under favourable laboratory conditions for mosquitoes (Armbruster *et al.*, 2000).

Depending on population structure, inbreeding can be prevented in natural populations via migration. However, migration rates may need to be larger than previously expected to prevent inbreeding. A simulation by Vucetich and Waite (2001) indicates that in real populations the number of immigrants needed to prevent inbreeding is actually much greater than one individual per generation, which is the theoretical requirement in idealized Fisher–Wright populations (Doyle, 2003). In random-mating populations, where reproductive variance follows a Poisson distribution, one immigrant per generation will theoretically prevent inbreeding if the numerical population size is larger than about 20. However, variation in mating success caused by spawning frequency, fecundity and mortality differences in real populations increases reproductive variance and causes the effective population size, N_e, to be considerably less than the numerical (census) size. The reproductive variance of immigrants is highly variable as well, exacerbating the problem. Therefore, more than one immigrant individual is needed to prevent inbreeding and the situation becomes worse the greater the discrepancy between actual and effective population sizes. The required number of immigrants increases with the census size of population; this is not the case in idealized, theoretical populations, in which the census and effective sizes are equal (Doyle, 2003).

Migration can counteract the deleterious effects of inbreeding. Experimental populations of mustard, *Brassica campestris*, were maintained at a census population size of $N = 5$ for five generations, with three levels of migration (0, 20 and 50%) per generation, and the result was that several measures of fitness were lower in populations that had experienced no in-migration (Newman and Tallmon, 2001). Similarly, in a natural population of warbler on an island off the west coast of Canada, 98% of the population died in the winter of 1989, with the resulting inbred birds suffering higher mortalities (Keller *et al.*, 2001). All measures of genetic diversity dropped dramatically, and then quickly recovered to pre-crash values due to the immigration of only one animal per year. This immigration made the bottleneck undetectable. The increased mortality of inbred individuals during the crash also eliminated many deleterious recessive alleles.

Small populations of Scottish Atlantic salmon had similar genetic diversity, heterozygosity, as large ones and had not lost genetic diversity over a 50-year period (Consuegra *et al.*, 2005). High levels of asymmetric gene flow, migration, probably maintained genetic diversity in these peripheral populations based on tagging studies.

However, immigration can also have negative effects. A single immigrant warbler immediately introduced deleterious alleles back into the population (Doyle, 2003) and, because inbreeding was high, the overall level of inbreeding and inbreeding depression (Keller *et al.*, 2001) increased rapidly to their high pre-crash levels. The purging of deleterious alleles in the bottleneck by one set of immigrations was counteracted by immigration from a single individual. Wright called this phenomenon immigration load (Doyle, 2003).

Data from houseflies support this population phenomenon of recovery from bottlenecks and purging of deleterious alleles. Housefly lines were inbred either rapidly or severely, followed by population expansion or by chronic low population size over a long period (Reed, 2001), both resulting in the same inbreeding. As expected, inbred populations have consistently lower fitness than outbred populations across a range of environments. However, the bottlenecked populations had lower inbreeding depression for a given level of inbreeding in all environments than populations kept at a constant small size. Populations initiated from a small number of founders are able to recover fitness and survive novel environmental challenges, provided that habitat is available for rapid population growth (Doyle, 2003).

However, this is not the usual case as inbreeding usually reduces fitness without purging deleterious recessive alleles. Meta-analysis indicated that results from one study to another are quite variable, however the usual outcome is that strong inbreeding for a generation or two usually does not reduce subsequent inbreeding (Leberg and Firmin, 2008). Serial bottlenecks in mosquitofish caused a strong decline in population growth and increased rates of extinction. Attempting to purge recessive alleles through bottlenecks has the risk of extinction in some lines and we have inadvertently observed this for inbred channel catfish.

Doyle (2003) also explores how a bottleneck – a brief period of very low numbers – affects the ability of a population to meet new selective challenges and evolve new adaptations. In the most extreme case, with one pair of breeders passing through the bottleneck, only a maximum of four alleles per locus will be available for population regrowth and evolutionary adaptation, and some of those alleles may be identical. Possibly the effects of individual genes simply add together, genetic variance will decrease when genes are lost by drift and the rate at which the population can evolve will decrease because of the bottleneck. Generally, only additive genetic variance allows an evolutionary response to selection. Theoretically, when some alleles are lost by drift from a non-additive genetic system, the remaining genes will sometimes make a contribution to additive genetic variance (Doyle, 2003), and in this case the capacity to respond to selection will be enhanced by the bottleneck.

R.A. Fisher theorized that fitness traits such as fecundity and survival would have low heritability or additive genetic variation since they would be subject to strong natural selection over time causing additive genetic variation to disappear (Doyle, 2008), and indeed wild populations generally show low heritability for fitness traits. This is misleading as usually the absolute quantity of additive genetic variation is not reduced, but factors in the denominator such as environmental variation, variation due to GE interactions and other genetic variances caused by dominance, epistasis and maternal heterosis are high, causing the ratio h^2 to be low. However, consistent with Fisher's expectations, Teplitsky *et al.* (2009) found that life-history/fitness traits in New Zealand gulls did have low absolute additive genetic variance compared with morphological traits. In contrast, experiments with butterflies indicated that additive variance will increase in traits closely related to fitness, such as fecundity and survival. The traits related to fitness have an especially high proportion of dominance variance, complicating the interpretation of the data related to the theories presented here.

Quantitative traits that are under weak selection pressure will lose additive genetic variance in the bottleneck, as do neutral marker alleles (Doyle, 2003). Additive genetic variance of wing size (supposedly a neutral fitness trait) decreased during bottlenecks, while additive variance and heritability of egg-hatching rate – a fitness component – increased in the butterfly *Bicyclus anynana* (Saccheri *et al.*, 2001). The change of wing-size variance, but not hatching rate, followed random-sampling expectation exactly. Dominant genes can be selected for and therefore, especially in early generations of experiments, this response could increase the narrow-sense heritability as reflected by the increased response to selection from the dominant alleles.

Bottlenecks do not always have deleterious effects. Atlantic salmon in the River Eo, Spain did not lose genetic diversity as measured by allele-frequency distributions and heterozygosity excess during a prolonged stock collapse (Ribeiro *et al.*, 2008). There may be genetic mechanisms that help maintain genetic diversity during these events.

Although genetic variation is usually considered desirable, cases may exist where a lack of genetic variation may enhance an organism's short-term fitness when a population is highly adapted for a particular environment. Theoretically, the introduction of inferior genotypes could reduce a population's fitness, and some conservation geneticists have coined the term outbreeding depression for this population phenomenon, although it is not well documented. Outbreeding depression is usually related to temporary relaxation of selection pressure. For some critical developmental events and biochemical pathways, canalization and epistasis negate potentially detrimental genetic variation.

Outbreeding depression also refers to decreased performance of F_1 and subsequent generations when two different populations, strains or lines mate, or a loss of fitness in the F_2 generations and beyond. This phenomenon of outbreeding depression or F_2/F_3 breakdown does not occur often (Doyle, 2008). In fact, meta-analysis indicates that crossbreeding of wild populations is usually beneficial (Doyle, 2008). McClelland and Naish (2007) found that 576 comparisons of parents and F_1 hybrids, and 94 comparisons of F_2 hybrids, demonstrated that outbreeding is generally positive. The weakness of the analysis was that there were not many studies of F_2-generation hybrids in natural or stressful environments.

When outbreeding depression does occur, it may be a result of incompatabilities between the nuclear and mitochondrial genomes (Ellison and Burton, 2008). Compatability between the mtDNA genome and the nuclear genome is potentially important since they both encode different polypeptide subunits of respiratory and other electron-transport enzymes that need to join to form these critical proteins. F_3 crossbreeds of the wild marine copepod, *Tigriopus californicus*, had low fitness (Ellison and Burton, 2008). When maternal backcrosses were made fitness was completely restored. When paternal backcrosses were made fitness was not restored. This is evidence that a matching full-haploid nuclear genome and an mtDNA genome from the same population were required for maximum fitness.

Although crossbreeding can be used to increase genetic diversity, there is one other case where crossbreeding can lead to outbreeding depression and decreased fitness in wild populations. Genetically distinct stocks of pink salmon, *Oncorhynchus gorbuscha*, can have differing locally adapted rates of embryonic development (Wang *et al.*, 2007). Pink salmon crossbreeds required more degree-days to develop in both the F_1 and backcross F_2 generations, leading to decreased survival rates.

However, we need to remember that outbreeding or crossbreeding can be beneficial for rescuing small, unfit populations that are under duress because of inbreeding or genetic drift. Crossbreeding increased fitness for more than one generation both in the presence and absence of competition in the rare plant, the buttercup (Willi *et al.*, 2007).

Organisms near the periphery of a recent range or habitat expansion are often genetically less variable than those at the heart of the geographical range (Doyle, 2003). For example, mtDNA variation is low in the current northernmost populations of marine snails, which have recently been recolonized from glacial refugia located further south (Hellberg *et al.*, 2001). These homozygous populations at the edges of geographical ranges or in suboptimum habitats may be a result of founder effects and drift or they may be a result of selection. They could be strengthened, weakened or not affected by the introduction of new genetic variation, but this is little studied and needs to be.

Various forms of stabilizing selection may lead to wild homozygous lines, and outbreeding depression may be a natural phenomenon required for the long-term fitness of a population or a natural product of a particular genetic structure. In some cases, selection acts to develop lines that then mate to produce fit offspring, although these offspring do not have hybrid vigour in terms of ecological or reproductive fitness. Landry *et al.* (2001) found that wild Atlantic salmon chose mates with genotypes different from their own, maximizing the heterozygosity of offspring for the MHC. Microsatellite allele

and MHC data indicated that enhancing the diversity of the peptide-binding region of MHC appeared to be the mating objective, not solely the avoidance of inbreeding. Such an apparent genetic structure and process are counter to information indicating that salmon mating was highly random (Doyle, 2003). This stabilizing selection is a major influence on MHC gene-frequency distributions only at the local population level, such as within rivers, but over larger geographical distances migration and random drift were the dominant evolutionary process at the MHC locus, as indicated by the similar geographical pattern of MHC allele frequencies and neutral microsatellite variation (Landry and Bernatchez, 2001). This same balancing selection exists at the MHC locus in the endangered chinook salmon of the Sacramento River, has apparently maintained MHC diversity for millions of years in these fish and continues to counteract potential random loss of diversity via genetic drift caused by the recent, local population.

Mounting evidence indicates that this balancing selection to maintain high levels of genetic diversity via overdominance for disease resistance is common in salmon and has been documented in 31 populations of a third species, sockeye salmon (Miller *et al.*, 2001). Again, balancing selection took place locally, within sockeye salmon populations. However, directional selection also occurred at the MHC locus in several of the sockeye populations, illustrating that different forms of selection can be prevalent in different populations.

In the case of Atlantic salmon, genetic distances between populations as measured at the MHC locus correlated well with genetic distances measured at neutral microsatellite loci and also with geographical distance (Landry and Bernatchez, 2001). Additionally, divergence of the Atlantic salmon populations was essentially a random process. Conversely, the apparent heterogeneity in selection at MHC loci in sockeye salmon resulted in strong genetic differentiation between geographically proximate populations with and without detectable levels of balancing selection, in stark contrast to observations at neutral loci (Miller *et al.*, 2001; Doyle, 2003). Miller *et al.* (2001) conclude that, based on the distribution of MHC class II diversity throughout the Fraser drainage, conservation of sockeye salmon must be conducted on the basis of individual lake systems.

Fontaine and Dodson (1999) established the relatedness of salmon fry (in their first summer of life) and parr (in their second and third summers of life) captured in adjacent territories by examining microsatellites, and found that fish collected near each other were not full sibs, which possibly has implications on how to collect brood stock for genetic conservation, assuming that a similar distribution of individuals is found for adults. However, the distribution of individuals appears to be different for other species. Pouyaud *et al.* (1999) found that in mouth-brooding black-chin tilapia, *Sarotherodon melanotheron*, related individuals tended to aggregate in open water environments and that mating occurred preferentially within small groups of kin, based on heterozygote deficiencies and similarity indices at four microsatellite loci. However, this inbreeding did not take place in riverine populations. If similar breeding structures exist in other tilapias, inbreeding may be higher in aquaculture populations than expected.

Various population structures exist, such as panmixia, sympatry, disjunct and stepping-stone (May and Krueger, 1990). When all of the fish constitute a single reproductive unit, panmixia exists and mating is random. Whether a panmictic (a single stock) exists, or several discrete, non-interbreeding stocks, dictates the genetic management strategy.

The conclusion that individuals within a body of water represent one panmictic population can be considerably strengthened if a study uses genetic data in combination with fish movement data relative to spawning areas and observations of reproductive behaviour (May and Krueger, 1990). Based on allozyme data, most panmictic populations are marine species, such as the milkfish, *Chanos chanos* (Winans, 1980), or the southern African anchovy, *Engraulis capensis* (Grant, 1985). Low levels or lack of genetic differentiation may be observed in freshwater species such as northern pike, *Esox lucius* (Seeb *et al.*, 1987), even though geographical isolation through lake and drainage boundaries prevents panmixia.

Populations that are genetically differentiated but apparently have free access to spawn with each other because they live in the same body of water are sympatric. Reproductive isolation among sympatric populations is not due to geographical boundaries, such as waterfalls or lake shorelines, but instead is due to processes such as olfactory homing to natal areas, assortative mating, behavioural selection of different spawning substrates or physiologically based differences in the timing of spawning (May and Krueger, 1990). Populations of Pacific salmon, such as the coho salmon, mix during part of their life cycle and then subsequently assort to natal waters prior to spawning, representing one form of sympatry (Wehrhahn and Powell, 1987). Temporal rather than spatial reproductive isolation – such as is the case with pink salmon, *O. gorbuscha*, which have a strict 2-year life cycle, resulting in genetic differentiation between odd- and even-year populations that use the same spawning stream (Aspinwall, 1974; Beacham *et al.*, 1985) – represents another form of sympatry. Lake Superior is large enough to have discrete spawning areas for lake trout, and genetically differentiated populations appear to occupy the same body of water for their entire life cycle (Dehring *et al.*, 1981; Goodier, 1981; Krueger *et al.*, 1989).

Genetically differentiated sympatric populations may exhibit differences in quantitative life-history traits that are important to fishery management (May and Krueger, 1990). Brown trout individuals sampled from a single body of water often demonstrate genetic differentiation between life-history categories, suggesting that the traits were population specific. Morphologically dissimilar brown trout in Lake Brunnersjoma, Sweden, have fixed gene differences, indicating two isolated sympatric populations (Allendorf *et al.*, 1976; Ryman *et al.*, 1979), and sympatric populations of brown trout have different feeding habits (Ferguson and Mason, 1981) and migration traits (Krieg and Guyomard, 1985; Krueger and May, 1987).

Discrete or disjunct populations, such as those that live in separate ponds or lakes with no outlet or in headwater streams with inaccessible barriers to upstream migration, have no possibility of reproductive contact between them, and these isolated breeding units tend to diverge genetically with time (May and Krueger, 1990). The extent of differentiation of these populations from their nearest neighbour will be directly proportional to the time of their separation and will be influenced by effective population size, selection, mating pattern, migration and mutation rates. Allozyme differentiation for disjunct populations has been measured in populations of largemouth bass (Philipp *et al.*, 1981, 1983a, 1985; Norgren *et al.*, 1986), bluegill, *Lepomis macrochirus* (Felley and Avise, 1980), and brook trout (Stoneking *et al.*, 1981; Dunham *et al.*, 2002e).

Stepping-stone is a population structure where localized breeding populations are adjacent to one another such as in many tributary streams of a major river system, and the populations maintain reproductive isolation by homing to their hatching location but occasionally stray to neighbouring streams, leading to gene flow among populations (May and Krueger, 1990). Genetic differentiation among populations is then directly proportional to geographical or stream distance and the intensity or frequency of homing to natal areas. This population structure will often lead to genetic similarity among stocks within a region or area and increasing levels of differentiation as geographical distances increase between sources, as observed in Alaskan chinook salmon (Gharrett *et al.*, 1987) and sea lamprey, *Petromyzon marinus*, in the Great Lakes (Krueger and Spangler, 1981; Wright *et al.*, 1985).

Continuous geographical changes in allele frequency within a population are called clines (Richardson *et al.*, 1986); the change in allele frequency along a cline may be gradual or there may be steps in allele frequency. Clines can be produced by selection or drift, and can be counteracted by migration and stabilizing selection.

In regard to identifying significant evolutionary units, what is the relationship between biochemical and molecular markers and meaningful performance or quantitative traits? Should genetic conservation be based on markers or on performance? Obviously, classification by performance would be ideal, but it is extremely difficult to accurately measure true differences in performance of wild populations because of problems with replication, environmental variation and GE interactions. Very few data comparing markers and performance are available for wild aquatic organisms.

However, some data are being generated to provide some answers to these problems relevant to genetic management. Merilä and Crnokrak (2001) analysed 18 independent studies of the degree of differentiation in neutral marker loci and genes coding quantitative traits with standardized and equivalent measures of genetic differentiation, F_{ST} and Q_{ST}, respectively. Quantitative-trait divergence among populations was almost always larger than neutral-marker divergence. Natural selection rather than random genetic drift drove the populations towards different means for quantitative traits, physiology and behaviour (Merilä and Crnokrak, 2001): that is, mean differences in appearance are due to selection, not drift. Merilä and Crnokrak (2001) conclude that selection pressures, and hence optimal phenotypes, in populations of the same species are unlikely to be often similar; natural selection promotes unique local adaptations and the unique quantitative features of local populations are primarily due to selection, not drift.

This also indicates that genetic conservation based on neutral DNA markers may be too conservative and may miss ecologically important genetic differences among populations. Both fitness and non-fitness traits have significant additive variation (Doyle, 2003). This is illustrated in the European flounder, *Platichthys flesus* L., that had strong selection and divergence at the Hsc70 heat shock locus for adjacent populations that were either benthic spawners (which live inshore) or pelagic dwellers and spawners (Hemmer-Hansen *et al.*, 2007).

Nine microsatellite loci did not show much divergence based on the small spatial differences of the fish. If genetic conservation was based on these neutral markers, a tremendously important difference among the populations based on their spawning strategy and heat shock locus would be missed.

Effects of Geography and Environment on Population Variation

The amount and type of genetic variation and population structure in a fish population vary depending on geographical location, environmental conditions, selective forces, stocking of conspecifics, hybridization and interactions between domestic and wild populations. An understanding of these factors is essential and must be accounted for to properly determine the effects of stocking – intentional and accidental – of native conspecifics, domesticated fish, hybrids or transgenic fish on genetic variation in wild populations. Geographical location, environment and climate have major effects on population genetic variation (May and Krueger, 1990). Generally, the greater the geographical separation of breeding areas, the more genetically differentiated will be the populations of fish that use them (Wright, 1943).

Location within the geographical range affects genetic variation, and there appears to be a relationship between periphery of range, climate extremes and genetic variation. The southern subspecies of largemouth bass exhibits more isozyme variation than the northern subspecies (Philipp *et al.*, 1983a, 1985). The same trend is observed in northern and southern populations of striped bass (Dunham *et al.*, 1989) and the same relationship has been observed in mummichog, *Fundulus heteroclitus* (D. Powers, Stanford University, 1990, personal communication).

The opposite trend was seen for brook trout, where isozyme variation was higher in northern strains than in the southern strain (Dunham *et al.*, 2002d). As was the case for largemouth bass, mixed or intergrade populations had intermediate levels of genetic variability.

Possibly, the real explanation for these clinal trends is not actually the north–south change in latitude but the change in temperature as it relates to thermal limits. As the populations near climates that approach a thermal limit for the species, genetic variation becomes restricted. This may be due to founder effects, random genetic drift or selection associated with the harsher environment. In the case of the largemouth bass, isozyme variation nears zero in the more northerly populations living near the end of the geographical range of the species and possible limit of the temperature regime for their survival. In the case of the brook trout, the most southern populations are near the southernmost extent of the geographical range and are at the uppermost thermal limit of the species, and they too have low levels of isozyme variation.

Physiological and ecological differences among Florida largemouth bass, *Micropterus salmoides floridanus*, northern largemouth bass, *Micropterus salmoides salmoides*, and their hybrids have been documented that are probably related to natural selection at different environmental temperatures (Philipp *et al.*, 1985; Figs 13.1 and 13.2). A number of studies have shown a difference in their response to various temperature regimes (Fields *et al.*, 1987; Carmichael *et al.*, 1988). Other studies have shown differences in timing of spawning, growth rate, reproductive success and survival of the two subspecies that are probably related to temperature and selection (Isley *et al.*, 1987; Maceina *et al.*, 1988; Philipp and Whitt, 1991).

Such selective pressures related to temperatures are well documented in other species. Strains of Nile tilapia that originated from geographical locations furthest from the equator (Khater, 1985) and populations domesticated furthest from the equator (Li *et al.*, 2002) have the best cold tolerance. The Minnesota strain of channel catfish, originating from the St. Louis River, which empties into the coldest of the Great Lakes, Superior, spawns at colder temperatures and produces larger eggs than other

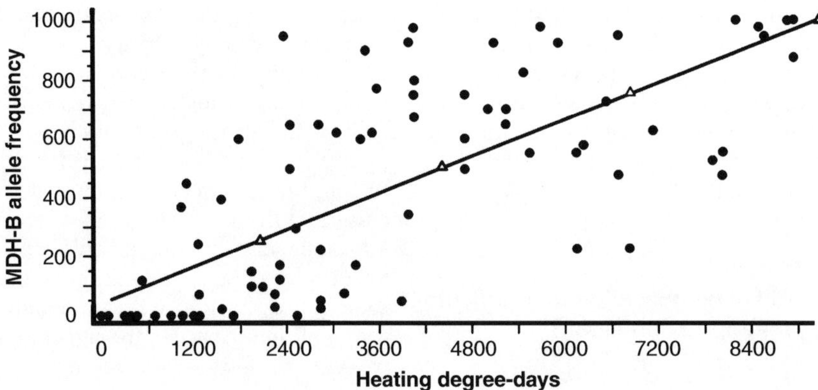

Fig. 13.1. Change in malate dehydrogenase (MDH)-B allele frequency in largemouth bass, *Micropterus salmoides*, with heating degree-days. (Adapted from Philipp *et al.*, 1985.)

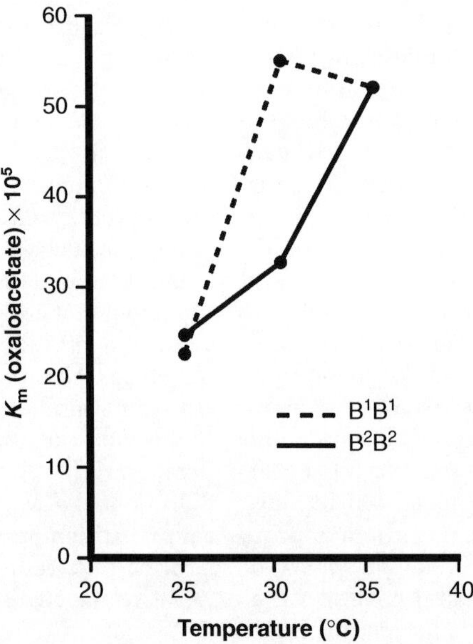

Fig. 13.2. The effect of temperature on the K_m rate constant for malate dehydrogenase allozymes in largemouth bass, *Micropterus salmoides*. (Adapted from Hines *et al.*, 1983.)

strains of channel catfish. The Rio Grande strain of channel catfish, from the hottest and southernmost extent of the geographical range for channel catfish, spawns later at higher temperatures and has smaller eggs and greater fecundity than other strains of channel catfish.

In the case of shoal bass, *Micropterus* sp. cf. *Micropterus coosae*, and redeye bass, *M. coosae*, the more northerly populations of both species in Georgia exhibited the greatest genetic variation as measured by number of alleles per locus, percentage loci polymorphic and mean heterozygosity compared with more southern Georgia populations and with the South Carolina redeye bass, although a notable exception was the high variability of the Ocmulgee River redeye bass, one of the most southerly populations sampled. The higher genetic variability in the middle of the range of redeye bass may be a result of this being a zone of intergradation. These are more specialized species that have very restricted geographical ranges. In this case, the borders of the geographical range do not define thermal maxima or minima. The greater level of genetic variation may be the expected result in the centre of the geographical range, which could historically have had the greatest gene flow.

Allele and genotype frequencies are also influenced by selective factors. Certain isozyme alleles are selectively advantageous for traits such as growth (Dunham and Smitherman, 1984; Hallerman *et al.*, 1986) or temperature tolerances (Philipp *et al.*, 1985). Pollution could also selectively influence genetic variation by increasing variation through mutation or disruptive selection, by decreasing variation through differential mortality of some genotypes or by reducing population sizes, which would result in random genetic drift.

The fish populations in Weiss Lake, a polluted lake in Alabama, exhibit interesting patterns of genetic variation. Largemouth bass and spotted bass, *Micropterus punctulatus*, populations in this lake exhibit large amounts of genetic variation relative to other bass populations in Alabama. However, black crappie, *Pomoxis nigromaculatus*, and white crappie, *Pomoxis annularis*, in Weiss Lake have little isozyme variation compared with other south-eastern crappie populations. These extremes are difficult to explain, and may represent different responses to pollution or may be a result of totally different factors. Brown bullheads, *Ictalurus nebulosus*, exhibit decreased genetic variation in polluted waters, possibly due to selective reductions in population size (Murdoch and Hebert, 1994).

Environmental heterogeneity is a major factor in maintaining and structuring genetic variation in natural populations. The patchwork effect of freshwater habitats could have important consequences for the genetic structure and evolution of a species.

Geographical location can also affect genetic variation because of its relationship to humans and exploitation. Populations of brook trout living in nine eastern Canadian lakes had lower allozyme heterozygosities than trout in the adjacent streams, and a positive correlation existed between the magnitude of the lake–stream difference and the distance of the lakes from the nearest all-season road (Jones *et al.*, 2001). Jones *et al.* (2001) hypothesize a negative causal relationship between angling mortality and heterozygosity in the lake, with bottlenecks and random genetic drift being the cause of the increased homozygosity. They suggest that angling delays recovery from natural population crashes and reduces long-term effective population number, and suggest that managers should prevent human-induced mortality at any indication of a large natural mortality event to allow populations to rapidly increase in size following a decline. However, there is an alternative explanation. Increased homozygosity might be an expected result in populations even when not in decline. If heterozygous individuals are faster-growing and more aggressive, they will be differentially selected by angling, resulting in increased homozygosity in the populations.

Geography, in terms of marine, fresh or combinations of both environments, affects genetic variation in fish. DeWoody and Avise (2000) reviewed microsatellite data from thousands of individuals from about 80 species, and found that freshwater fish displayed levels of population genetic variation similar to those of non-piscine animals. Marine fish populations exhibited higher heterozygosities and had nearly three times the number of alleles per locus. Anadromous fish were intermediate to marine and freshwater fish for these parameters. These results were consistent with earlier results using allozymes (DeWoody and Avise, 2000; Doyle, 2003), indicating that allozymes and microsatellites have congruence with regard to measuring overall genetic variation. DeWoody and Avise (2000) suggest that the consistent difference between freshwater and marine within-population genetic diversity is due to characteristic differences in evolutionarily effective population size (Doyle, 2003).

Factors Affecting the Establishment of New Genotypes in Established Natural Populations

Several factors might affect the establishment of new genotypes, either conspecific wild or domestic populations, by accidental or intentional release and their opportunity or ability to interact with and influence wild populations. These include size of fish, number of fish stocked, number of repeat stockings or releases, timing of stocking or release, selective value of the new genotype and other environmental variables. These variables have not been completely evaluated, but a growing database illustrates their importance and function (Kulzer *et al.*, 1985; Philipp *et al.*, 1985; Norgren *et al.*, 1986; Isley *et al.*, 1987; Maceina *et al.*, 1988; Dunham *et al.*, 1992b). These studies indicate that it is difficult to establish a new genotype, even wild rather than domesticated, in an established natural population (Dunham *et al.*, 1992b).

One example illustrating this point is that of the massive stockings of Florida largemouth bass into established northern or native largemouth bass, *M. salmoides salmoides*, populations (Kulzer *et al.*, 1985; Norgren *et al.*, 1986; Maceina *et al.*, 1988; Dunham *et al.*, 1992b). Genes from Florida largemouth bass were established in these populations at varying levels and, in some cases, the stocking of the Florida largemouth bass was unsuccessful, resulting in no genetic changes in the populations. Key factors in the establishment of Florida alleles were total numbers of fish stocked, number of years since initial stocking, number of repeat stockings, elevation, age of the lake, and water clarity.

Size of fingerling or subadult fish stocked has an effect that is not well defined. Kulzer *et al.* (1985) did not find a correlation between size of fingerling or subadult largemouth bass stocked and the success of the introduction in largemouth bass. However, the results were confounded by lack of replication and a multitude of additional variables. Studies on the success of stocking large and small subadult trout have had contradictory results (Pycha and King, 1967; Plosila, 1977). The large subadults had greater survival when introduced in one study and lower survival in another. For years, the state of Alabama in the USA has heavily stocked small Florida largemouth bass fingerlings into specific reservoirs. In these cases, 10 years of stocking have not changed the diagnostic isozyme allele frequencies in these reservoirs, indicating no genetic impact. In response to this result, the state of Alabama has altered its stocking programme and is attempting to stock larger fingerlings to gain a genetic impact. In contrast, the state of Arkansas has been quite successful in shifting allele frequencies towards the Florida genotype via stocking in Arkansas reservoirs. The factors resulting in these differing results and impacts need to be determined.

A similar example and result were found in the stocking brook trout and striped bass. Hatchery brook trout stocks from the north were extensively stocked in southern waters from the 1930s to the 1970s. Despite this stocking, 22–56% of the populations remain pure southern in the states of Georgia, North Carolina and Tennessee (an undetermined number, perhaps 25%, never received stocking) (Dunham *et al.*, 2002e). Forty to fifty per cent of the current populations are a hybrid mix. Historical stocking records are not clear and complete, but many of these hybrid populations were probably established by the stocking of hatchery populations that were already F_1 hybrids into environments already devoid of brook trout. In summary, it appears that where the native genotype was still present and established, there was little genetic impact from the stockings. Additionally, in mixed populations, there appears to be natural selection occurring that favoured the original southern genotype.

Historically, massive stockings of northern strains of striped bass also took place in the Deep South of the USA. Preliminary evidence based on mtDNA genotypes suggests that these stockings did not have a genetic impact either and that the prevalent genotype in the South is the native Gulf genotype.

The main conclusion of these series of studies is that it is difficult to establish new genotypes in established wild populations.

In the event that conspecifics of native fish populations were able to successfully crossbreed, short-term outbreeding depression could result based on results obtained for Swiss weeds. Keller *et al.* (2000) crossed local (Swiss) weeds of three species with weeds of the same species obtained from other parts of Western Europe, and grew the F_1 crossbreed and F_2 backcross progenies in Switzerland. Outbreeding depression occurred, as indicated by smaller plants and lower survival in some F_1 and many, but not all, of the F_2 backcrosses (Doyle, 2003). Seed mass showed positive benefits from crossbreeding in one species in the F_1, but depression occurred in the F_2 backcross, an expected result as backcrosses generally have intermediate performance. Temporary outbreeding depression might be an expected result because of independent assortment. However, long-term benefits might accrue from selection on the new genotypes generated from the introgression and this needs study.

Individual heterozygosity, a result of outbreeding, is associated with increased fitness, especially under harsh conditions (Doyle, 2003). Outbreeding is associated with the survival of young male bats in Great Britain (Rossiter *et al.*, 2001). Heterozygosity itself was not associated with fitness, but the influence of mean d^2, a measure of total outbreeding due to a wide multilocus effect rather than single loci, representing heterosis as opposed to solely negating inbreeding depression, had a positive impact on fitness (Rossiter *et al.*, 2001; Doyle, 2003). Mean d^2 may also be correlated with immunocompetence, which influences mortality.

Interspecific Hybridization

Ecological concerns also pertain to the application of interspecific hybridization. Fertile hybrids have the potential to backcross and genetically contaminate the parent species in the wild if they are accidentally or intentionally released. The long-term ecological, evolutionary and genetic consequences are unknown. Hybridization is, of course, one mode of evolution and speciation. If backcrossing were to occur, theoretically a species' fitness could be improved, unaffected or weakened. If backcross individuals were detrimental, they might be eliminated by selection.

Argue and Dunham (1999) extensively reviewed interspecific hybridization in fish. In general, reproductive isolating mechanisms appear to be quite effective in fish. Many F_1 hybrids are sterile. Even if F_1 hybrids are fertile, F_2 breakdown may occur or the F_1 hybrids or F_2 hybrids or backcrosses may be less competitive reproductively compared with the parent species. F_2 breakdown can occur in ictalurid and tilapia hybrids. Attempts to produce F_3 generations of mixed-species hybrids have resulted in fish that have severe reproductive problems. Hundreds of cases of hybridization were reviewed and permanent introgression into one of the parent species almost never occurred.

Hybridization is, of course, a natural phenomenon. Hybridization occurs frequently between wild populations of black crappie and white crappie. Isozyme analysis indicates that F_1 hybrids, F_2 hybrids and backcrosses have reduced reproduction compared with the original parents and therefore the parental species remain the dominant genotype. Striped bass × white bass hybrids have backcrossed with parental species in natural settings. However, this occurrence is rare and F_2 individuals are rare.

In the case of various species of Darwin's ground finch in the Galapagos Islands, differentiation of the species based on morphological data is not reflected in either mtDNA- or nuclear DNA-sequence phylogenies, and genealogies based on mitochondrial and nuclear markers are not concordant with each other (Freeland and Boag, 1999). Freeland and Boag (1999) interpret the absence of species-specific DNA sequence lineages as evidence for ongoing hybridization involving all six species of *Geospiza*. Interspecific hybrids have an advantage in some years, so there is no selection against hybridization, but there is strong, ongoing selection for morphological traits such as the size of the beak in relation to the size of the seeds available for food (Doyle, 2003). Apparently, in this case, even interspecific hybridization did not lead to outbreeding depression and interspecific hybridization is being utilized for strong evolutionary selection.

Natural hybridization may be a mechanism for species to exchange genetic resources with the goal and possible outcome of strengthening fitness. Shoal bass had alleles commonly found in redeye bass at very low frequencies and, in some cases, redeye bass had alleles common for shoal bass at high frequencies (R.A. Dunham, K. Norgren and H. Kucuktas, unpublished results). This relationship is not surprising since shoal bass were more homozygous than redeye bass and therefore less likely to possess alternative alleles common in redeye bass. Either hybridization events had occurred between the two species several generations ago or the two species share alleles from a common ancestor. No genotypes were observed that would be expected for an F_1 hybrid, indicating no recent hybridization. Virtually all polymorphic loci for both species were at Hardy–Weinberg equilibrium, which indicates random breeding, and no selection, mutation or migration. Any hybridization would have had to occur several generations ago with genotype frequencies now stable and at equilibrium.

Hybridization appears to be most damaging, if we assume change is bad, when species of limited geographical distribution must compete with stockings of related species or hybrids. The rare and less dominant species can be genetically compromised through hybridization and backcrossing. Hybridization of *Micropterus*, black bass, in Texas, USA, is an example of this phenomenon.

The risk of genetic impact by interspecific hybrids on related species in the natural environment is probably low. However, current scientific opinion is to view hybrids and their application in a conservative manner. These views could lead to a restrictive policy limiting the application of hybrids. However, there appears to be less scientific concern regarding hybrids that naturally occur in nature or that are already widely utilized (Hallerman *et al.*, 1998). Natural-resource agencies, particularly in the USA, are much more aware of the potential adverse ecological effects of hybrids, although strong management pressure exists to utilize hybrids already considered beneficial to a fishery.

In some cases, there may be legitimate concern. The impact of artificially high population numbers from aquaculture and of domestication may lead to behaviours, genotypes or environmental situations more conducive to interspecific hybridization.

High frequencies of Atlantic salmon × brown trout hybrids have been detected in rivers near intensive salmon-farming sites in Norway and Scotland, which may be indicative of a breakdown or partial breakdown of reproductive isolating mechanisms between these two species (Matthews *et al.*, 2000). Fish from several rivers, located both near and far from salmon farms, were analysed. All hybrids were Atlantic salmon female × brown trout male, even though careful sampling was conducted to avoid biasing the sample towards spawning sites dominated by salmon (Doyle, 2003). Hybrid parr representing 1.0% of the reproduction were recorded from one of the rivers distant from farms, but were present at frequencies ranging from 0.7 to 3.1% in seven out of ten systems located within 38 km of salmon farms. Hindar *et al.* (1998) also indicate that in western Scotland hybridization between escaped farmed female Atlantic salmon and brown trout occurs about an order of magnitude more frequently compared with their wild counterparts, and that the rate of hybridization between Atlantic salmon and brown trout in Norway is increasing relative to pre-aquaculture levels.

Interactions Between Domestic and Wild Fish

There is concern that domesticated fish used in aquaculture, such as domestic strains, selected lines and intraspecific crossbreeds, could have a genetic impact on natural populations when or if these fish escape or are intentionally released. A potential philosophical conflict exists between enhancers, such as aquaculturists, and preservationists, such as ecologists and conservation geneticists. Fish in the natural environment have little contact with humans, are at relatively low densities, have relatively low exposure to diseases, live in relatively good water quality, are almost never subjected to low oxygen levels, must catch and feed on prey items and must avoid predators. In contrast, the associated selective pressures in the aquaculture or hatchery environment are very different. In this environment, fish are in frequent contact with humans, are at tremendous densities, have frequent exposure to pathogens, are often exposed to poor water quality and low oxygen levels, are provided with artificial diets and are relatively protected from predators.

When fish are removed from the natural environment and placed in the culture environment, random genetic drift and domestication effects (new and greatly different selective forces act upon fish in the domestic environment compared with the natural environment) alter gene frequencies and reduce genetic variation as measured by isozyme analysis and DNA markers. Domestication reduces genetic variability in fish (Allendorf and Utter, 1979; Allendorf and Phelps, 1980; Ryman and Stahl, 1980; Stahl, 1983; Dunham and Smitherman, 1984; Hallerman *et al.*, 1986; Koljonen, 1989) through both selective processes and random genetic drift. The majority of this research was demonstrated with salmonids. The same trend has been observed in limited studies of channel catfish. One sample of six wild fish was compared with 15 populations of domesticated channel catfish (Dunham and Smitherman, 1984; Hallerman *et al.*, 1986). The small sample of wild fish had more genetic variation than any of the domesticated populations and also possessed two new alleles. The analysis of five microsatellite loci also indicated that hatchery sea trout, *Salmo trutta*, had fewer alleles

but the same heterozygosity as wild sea trout in Poland (Was and Wenne, 2002). Finnish populations of land-locked Arctic charr had high population differentiation even among nearby lakes, based on microsatellites (Primmer *et al.*, 1999). The hatchery stocks were similar to wild populations for allele number and mean heterozygosity; however, hatchery populations demonstrated higher levels of single- and multilocus genotypic disequilibrium. One lake population was completely dependent on stocking, had increased egg and alevin mortality, had increased disease susceptibility and exhibited particularly low levels of genetic variation; however, some abundant unstocked natural populations demonstrated similar low levels of microsatellite variation.

Atlantic salmon in Norway and Ireland have reduced genetic variability, and founder effects and subsequent selection have a stronger effect on the genetic differentiation among domestic strains than geographical origin (Norris *et al.*, 1999). Domestic populations of Atlantic salmon in north-west Ireland had a reduction in mean heterozygosity (0.281 ± 0.057) at three minisatellite loci compared with wild Atlantic salmon (0.532 ± 0.063) (Clifford *et al.*, 1997, 1998).

The relative amounts of genetic variation of wild and domestic populations of fish need further definition, although wild populations are expected to continue to have more genetic variation. Understanding of this genetic variation is important not only for risk assessment, but also for evaluation of future genetic resources for utilization in biotechnology.

Domesticated populations with reduced genetic variability are propagated in large numbers, sometimes reaching population numbers much greater than those found in natural populations. Purposeful or accidental (as a result of flooding or escape during harvest) introduction of the domestic fish may then allow mixing of the domestic and natural populations. The population genetics of such mixing of domestic and wild populations of fish is not well understood. If the domesticated fish do not survive or reproduce or if their progeny do not survive, no effect on gene frequencies of the wild population will be observed. If large numbers of domesticated fish survive and reproduce relative to wild fish, genetic variation might be reduced since much genetic variation may have been lost during the domestication process (Allendorf and Utter, 1979; Allendorf and Phelps, 1980; Ryman and Stahl, 1980; Stahl, 1983; Dunham and Smitherman, 1984; Hallerman *et al.*, 1986; Koljonen, 1989). If both domestic and wild fish reproduce or multiple domestic stocks of different origin are introduced, genetic variation of the fish in the affected natural environment may increase.

Initial studies also indicate that wild fish generally outcompete domestic strains of fish in the natural environment (Rawson, 1941; Greene, 1952; Miller, 1952; Smith, 1957; Vincent, 1960; Anderson, 1962; Buettner, 1962). Again, almost all of these observations were on salmonids (Flick and Webster, 1962, 1964; Mason *et al.*, 1967; Pycha and King, 1967; Moyle, 1969; Cordon and Nicola, 1970; Kempinger and Churchill, 1970; Fraser, 1972, 1974, 1980, 1981; Flick and Webster, 1976; Reisenbichler and McIntyre, 1977; Van Velson, 1978; Reimers, 1979; Hynes *et al.*, 1981; Maclean *et al.*, 1981; Ryman and Stahl, 1981; Webster and Flick, 1981; Lerder *et al.*, 1984; Krieg and Guyomard, 1985; Petrosky, 1985; Johnsen and Ugedal, 1986; Seelbach and Whelan, 1988), which are coldwater fish, and were localized experiments or observations. More recent results (Dunham *et al.*, 2002e) examining the interaction of hatchery and wild brook trout provide circumstantial evidence corroborating these earlier observations. Despite a long history of systematic restocking and the geographical proximity of two unstocked tributary streams, Atlantic salmon in the Penobscot River in Maine and the two tributaries are genetically differentiated, based on microsatellite loci, from each other (Spidle *et al.*, 2001). In the River Shannon system in Ireland, extensive stocking of Canadian Atlantic salmon following the construction of a hydroelectric dam had no genetic impact and no Canadian alleles were detected using minisatellites (Galvin *et al.*, 1996). Populations from the tributaries below the hydroelectric scheme appeared distinct from those above; however, the differentiation detected in the upper system was attributed primarily to genetic drift resulting from the poor escapement of adult salmon to some of the upper tributaries.

Hatchery-reared Atlantic salmon have immigrated into wild populations in the Baltic Sea (Vasemägi *et al.*, 2005) after being stocked into rivers for supplementation programmes. An 18-year evaluation indicated that the overall immigration rate into the non-supplemented wild population averaged a bit less than 10% but was as high as 25% between 1993 and 2000. The genetic relatedness between the wild and hatchery populations of Atlantic salmon appears to be increasing in the Baltic region.

Ruzzante *et al.* (2001) found that relatedness, but not inbreeding, appears to differ among locations within rivers for Danish brown trout based on microsatellite data. The presence of both native and hatchery stocks was detected at most locations, and the number of domestic individuals detected was correlated with the intensity of the stocking. However, a disproportionately large number of wild brown trout were found at locations where stocking with domestic fish no longer takes place, suggesting the limited long-term genetic success of stocking and further demonstrating the greater competitiveness of wild genotypes. The data of Hansen *et al.* (2001) with brown trout gave even more dramatic results as there was an almost complete absence of stocked, domesticated trout in samples of trout from Danish rivers.

Larger-scale examples with more species, including warm-water species, are needed to determine the interactions between wild and domestic fish and to confirm or refute the hypothesis that wild fish generally outcompete domestic fish in the natural environment. In the case of channel catfish, domestic fish have escaped into the natural environment routinely. AFLP analysis indicated that domestic populations were all related to one another, forming a single branch in the phylogenetic analysis, while wild populations were more related to one another than to domestic populations or were distinctive. There was no molecular genetic evidence for apparent impact of domestic catfish on wild populations.

Several factors may contribute to the domesticated fishes' potential decreased fitness and competitiveness in these examples. Possibly the loss of certain variation may make these fish less adaptable in the natural environment. Wild trout tend to be stronger than domesticated trout and exhibit superior swimming stamina (Dickson and Kramer, 1971; Woodward and Strange, 1987). Wild trout have a greater ability to raise blood parameters in response to stress than domesticated strains (Woodward and Strange, 1987). Lepage *et al.* (2000) obtained similar results with sea (brown) trout. The metabolic stress responses of wild and domesticated fish originating from the same river were measured by placing the fish in a new environment, alone as well as in combination with predators. This stress induced elevated plasma concentrations of glucose and cortisol and brain levels of cortisol, dopamine, serotonin and metabolites of dopamine and serotonin. The stress responses in the domestic brown trout were weaker that those of wild cohorts, and alterations in brain monoamine neurotransmission were part of this effect.

The behaviour of domesticated fish varies from that of wild fish. Domesticated trout have a lag phase when released into the natural environment, during which they do not know how to feed properly (Johnsen and Ugedal, 1986). During this 1- to 2-week period, analysis of stomachs indicated that the domesticated trout were consuming algae prior to learning how to catch appropriate prey items. Domesticated hatchery masu salmon fed higher in the water column than domestic sea-ranched and wild masu salmon (Reinhardt, 2001). Sea-ranched masu salmon were intermediate in the water column when feeding. Aggression was about the same in all three genotypes. Hatchery techniques are needed to develop fish that avoid the surface or to prevent selection for surface-seeking behaviour in order to increase the survival of post-release

ranched salmon. Alternatively, utilization of wild strains already possessing this trait should achieve the same objective.

Domestic fish are more aggressive, which may be related to their increased vulnerability to predation or harvest (Moyle, 1969; Dickson and Kramer, 1971; Fraser, 1974). The nervous behaviour and wariness of wild salmonids relative to domestic salmonids is well documented (Moyle, 1969; Fraser, 1974). Wild trout position themselves deep in hatchery tanks and domesticated trout orient themselves nearer the surface of the water (Moyle, 1969). Johnsson *et al.* (2001) found that 1-year-old wild Atlantic salmon had a stronger heart rate and flight response from a simulated predator than seventh-generation farmed salmon derived from the same founder wild population, but the differences were weaker or reversed in 2-year-old fish.

Wild trout are less vulnerable to angling, have greater survival traits, live longer and contribute much more to the total biomass produced in the natural environment (Miller, 1952; Smith, 1957; Mason *et al.*, 1967; Fraser, 1972, 1980, 1981; Flick and Webster, 1976; Webster and Flick, 1981). Mezzera and Largiadèr (2001) utilized microsatellite analysis to demonstrate that in angling a stocked hatchery strain of brown trout and its hatchery crossbreeds were selectively removed compared with wild brown trout. They suggest that angling might be used to reduce the genetic impact of supplementary breeding programmes or domestic escapees to help preserve wild gene pools (Doyle, 2003). Domesticated common carp are also more vulnerable to angling than wild common carp (Beukema, 1969). Wild carp also exhibit more wariness and greater seine or harvest avoidance (Hoffmann, 1934; Hulata *et al.*, 1974; Moav *et al.*, 1975; Wohlfarth *et al.*, 1975b, 1976; Suzuki *et al.*, 1976).

Wild fish can also be more reproductively competitive than domestic fish. Wild coho salmon males outcompeted captively reared males and controlled access to spawning females in 11 of 14 paired trials in laboratory stream channels and, in two cases where domestic satellite males were observed participating in spawning, DNA genotyping indicated that they did not sire any of the progeny (Berejikian *et al.*, 2001). This has additional practical management implications when the objective is to enhance the effective population number of a naturally spawning population with fish reared in captivity (Doyle, 2003).

Farmed domestic salmon were competitively and reproductively inferior to wild fish after deliberate release into the Imsa River in Norway, achieving less than one-third the breeding success of the native salmon (Fleming *et al.*, 2000). Most of the introgression was a result of native males mating with farmed females, as domestic males were relatively uncompetitive. There were also indications of selection against farm genotypes during early survival but not thereafter, leading to a lifetime reproductive success, adult to adult, of the domestic salmon that was 16% that of the native salmon. The productivity impact of the domesticated salmon was higher than the genetic impact as the productivity of the native population was reduced by more than 30% through resource competition and competitive displacement.

If fast growth rate is advantageous, why do aquatic organisms not evolve to reach larger and larger body sizes? One reason why domestic fish may not compete well with wild conspecifics is that fast growth rate might, in some cases, be detrimental to fitness. *Drosophila* lines selected for high feeding rates as larvae grew faster but had reduced lifespan, and those selected for a low larval feeding rate grew more slowly to adulthood, had lower mortalities and had enhanced expression of genes known to promote resistance to stress (Foley and Luckinbill, 2001). Aquatic organisms that grow too fast or have very large appetites might be selected against under natural conditions.

There are additional explanations for the reduced fitness of domestic genotypes compared with wild genotypes in the natural environment. In a culture environment, if selection is relaxed, some of the genetic gain will be lost when this selection pressure is removed. When aquatic organisms are removed from the natural environment, again a selective pressure is removed and gene frequencies can change. Therefore, reduced natural selection in the benign, domestic environment can theoretically permit unfavourable genes to accumulate in the population by drift and/or mutation pressure even though they are not selected (relaxed selection), which could result in a catastrophic loss of fitness when the organisms are exposed again to the full force of natural selection, such as in fisheries-stock enhancement (Doyle, 2003). These effects can occur in captive populations of any size since they are related to selective pressure, not population size. If the size is small for the captive population, inbreeding and the accumulation of deleterious homozygotes can further exacerbate the reduced fitness for natural settings, unless the population is selected for fitness traits (Doyle, 2003).

This theory is demonstrated in houseflies. Large (500) and small (50) housefly populations were maintained in an environment that eliminated selection on traits that are expressed after 21 days of age by killing all the flies at 21 days (Reed and Bryant, 2001). The rate of loss of later-life fitness components due to relaxed selection was equivalent to the rate of loss due to inbreeding in populations with an effective size of 50 individuals, illustrating how domestication could reduce fitness. Even if captive populations are kept large to avoid inbreeding, breeding them in benign environments where natural selection for fitness traits is absent may reduce the capability of these populations to exist in natural environments within a few generations (Doyle, 2003).

Another possible interaction between domestic and wild populations of fish is the establishment of sympatric, but reproductively isolated, populations. Although strains of fish do not usually have reproductive isolating mechanisms preventing them from interbreeding, occasionally behavioural mating blocks prevent or decrease the rate of interstrain matings. Marion channel catfish females preferentially mated with their own strain rather than with Kansas males (Smitherman *et al.*, 1984), and the Ghana strain of *Oreochromis niloticus* was more likely to mate with its own strain than with other strains (Smitherman *et al.*, 1988). The existence of reproductively isolated, sympatric populations of trout (Brown *et al.*, 1981; Lerder *et al.*, 1984), especially brown trout, *S. trutta*, is well documented. Some strains of domestic and wild rainbow trout are sympatric, but are reproductively or near to reproductively isolated. This occurs because of behavioural differences, including temporal or spatial differences in spawning (Smitherman *et al.*, 1988).

Reports have been widely publicized that domestic salmon have escaped in Norway and are observed on spawning grounds, as 20–30% of the Atlantic salmon spawning in local rivers have escaped from aquaculture operations (Saegrov *et al.*, 1997). Spawning does not guarantee genetic impact, but the first examples of genetic impact have now been demonstrated. The genetic effects of sea-ranched and domestic salmonids on local stocks are reviewed by Hindar *et al.* (1991) and Hindar (1999). Three effects have been identified: inter-breeding, competition and disease contamination by parasites and diseases for which the potential for impact may be higher on genetically naïve local strains (Thresher *et al.*, 2009). Impacts on native salmonid species and strains range from complete eradication to no effect. Hindar (1999) concludes that ecological effects on native populations are typically negative – reduced population size and lower survival rates – which he attributes to the effects of genetic pollution of locally adapted stocks. Because of the widespread nature of these potential impacts of sea-ranched fish, great effort is under way to preserve native strains for both Pacific salmonid stocks on the North American west coast and Atlantic salmon on the east coast (Thresher *et al.*, 2009).

Domestic populations of shrimp may be affecting genetic variation in wild populations. Wild populations of *Penaeus monodon* in the Philippines could be differentiated from each other and from cultured populations in the same general area on the basis of six polymorphic microsatellite loci (Xu *et al.*, 2001). There was a weak (non-significant) trend for the wild populations to have less genetic variation in areas where aquaculture was intense.

Atlantic salmon that escape from farms in north-west Ireland can interbreed with wild Atlantic salmon (Clifford *et al.*, 1998). Escaped domestic juvenile Atlantic salmon completed their life cycle and bred with each other and native fish upon their return to the river in north-west Ireland, based on mtDNA and minisatellite markers (Ferguson *et al.*, 1998). Escaped fish homed accurately, as adults, to the site of escape, the area adjacent to the hatchery outflow in the upstream part of the river. The proportion of juveniles of maternal farm parentage in two rivers averaged 7%, but reached a high of 70% in an individual sample, illustrating the importance of adequate sampling and possible clustering of different genotypes (Clifford *et al.*, 1998). Juveniles of domestic parentage survived to at least the 1+ summer stage. Only a small proportion of 29,000 adult farm salmon that escaped in spring 1992 appear to have bred successfully in the rivers studied (Clifford *et al.*, 1998).

The genetic impact of land-locked Atlantic salmon was greater in Canada, where salmon migrate in and out of four rivers that enter a large freshwater lake (Tessier *et al.*, 1997). Hatchery stocks had changed allele frequencies and losses of low-frequency alleles, but no reduction in heterozygosity, and some of the riverine populations were supplemented by hatchery stock originating in the same rivers. In one of the riverine populations, susceptibility to genetic drift and inbreeding doubled after only one generation of supplementation.

Genetic impact may vary depending on the type of wild population, its behaviour and its life history. The introgression of genes, microsatellites and mtDNA markers from hatchery brown trout used for stocking was much higher in the resident, non-migratory wild brown trout in Danish rivers than in the migratory sea trout (brown trout) that spawn in the river but also spend time in the ocean (Hansen *et al.*, 2000). In contrast to our previous examples with Atlantic salmon, domestic males had a greater genetic impact than domestic females; however, this may be due to natural selection rather than sexual selection. Hansen *et al.* (2000) conclude that stronger selection acts against stocked hatchery brown trout that become anadromous compared with hatchery brown trout that become resident, and, since most resident brown trout are male, this might also explain the apparent gene flow from hatchery to wild brown trout being male biased.

However, Thompson *et al.* (1998) provide evidence that domestic Atlantic salmon males can be competitive and breed with wild females at high rates, again in contrast to our earlier examples for Atlantic salmon, where primarily domestic females were mating with wild fish. Secondary males – either subdominant adults or, more often, parr, which mature in fresh water – can successfully fertilize Atlantic salmon ova that the primary male is trying to protect and fertilize. Eight wild and 11 ranched, domestic redds were sampled from the Burrishoole system, western Ireland, in two consecutive years. Multiple paternal genotypes were detected in 18 of the 19 samples and, of 1484 progeny analysed, 29% could not have been derived from the primary adult male genotype. However, the percentage of progeny sired by domestic redds was significantly higher (mean 42%). When only 1995 was considered, the average level of secondary male contribution to parentage was almost double for ranched redds compared with wild redds.

The performance of domestic Atlantic salmon in the natural environment is of great importance, as is the introgression of their genes into the wild populations. McGinnity *et al.* (1997) examined this question in a natural spawning tributary of the Burrishoole system in western Ireland, and found that survival of the progeny of farmed Atlantic salmon to the smolt stage was lower than that of wild salmon, with increased mortality being greatest in the period from the eyed egg to the first summer. Progeny of domestic Atlantic salmon grew fastest and competitively displaced the smaller native fish downstream. The offspring of domestic Atlantic salmon had a reduced incidence of

male parr maturity compared with native fish, and the native fish had a greater tendency to migrate as autumn presmolts. The growth and performance of domestic × wild crossbreeds were generally either intermediate or no different from the wild fish. The domestic and crossbred progeny can survive in the wild to the smolt stage, survive at sea and home to their river of origin, indicating the potential for genetic impact on natural populations. Anglers were pleased with the crossbred Atlantic salmon and domestic salmon and how they matured and homed at different ages, thereby improving fishing (A. Ferguson, Queen's University of Belfast, 1997, personal communication).

One hypothesis has been that use of hatchery salmon one or two generations removed from the same river and wild population to which they would be returned would minimize genetic impacts. However, Caroffino *et al.* (2008) found that although the fry from the fish that were domesticated for one or two generations had lower survival than wild fry, the recently domesticated fish were more prevalent at 2 years of age. This indicates that even this strategy could affect the rate of random genetic drift in a population. However, other studies have shown that genetic diversity was not affected after several generations of such a supplemental stocking even though the allelic richness of the wild population was higher than that of the hatchery fish (Doyle, 2008).

Additionally, the practice of supplemental stocking with fish from the same river even when domesticated for a single generation may have significant shortcomings that are different from and the opposite of that found by Caroffino *et al.* (2008). Araki *et al.* (2007) observed that the fitness of steelhead trout selected for one generation of hatchery rearing, measured as the number of adult offspring they produced in the wild which returned to the river, was reduced by nearly 50%. This has significant implications to be considered not only for the understanding of domestication and population genetics, but also for supplemental stocking strategies.

Does genetic variation and population structure in natural populations affect global food security, fisheries and aquaculture? Should we be concerned about the effect of domesticated aquaculture stocks on the genetics of natural populations? Aquaculturists, fisheries managers, conservationists and ecologists should work together to preserve genetic variation in natural populations. This is probably desirable to ensure the short- and long-term survival and health of native populations. Additionally, natural populations are the best form of gene banks, and preservation of these gene pools may enhance our ability to utilize these genetic resources in beneficial ways in the future. If genetic variation enhances a species' long-term survival, genetic variation also enhances global food security.

Integrated Management Strategy

Natural-resource agencies have a variety of mandates and agendas. One objective is to provide quality fishing and to serve the public. Another objective is to preserve and protect natural resources, including biodiversity and genetic biodiversity, possibly with a programme of genetic conservation. On occasion, these agendas and goals could be contradictory. In some cases, the goal may be to maintain the native, natural genetic status quo. In other cases, it may be desirable to intentionally mix or genetically enhance populations. Genetic variation can be important at an individual level or a population level, and some natural populations are already mixed or in highly altered environments. Genetic enhancement and modification may be desirable for some wild populations, and there may be specialized applications, such as artificial urban fishing environments, where genetically modified 'wild' fish may be of great benefit. A three-tiered management plan may be the answer, worldwide, to meet these potentially conflicting goals for natural populations. This plan may allow genetic enhancement of some native populations while also accomplishing wise conservation of our natural genetic resources, global food security and the establishment of natural live gene banks for species preservation and fitness enhancement, as well as future exploitation. Under this plan:

1. unique distinctive and representative populations would be identified, protected and preserved;

2. some areas and populations would be designated for genetic improvement to meet recreational goals (humans are part of the environment) or to restore or enhance genetically damaged populations; and

3. some areas would be designated for intentional mixing, deposition and propagation of many genotypes, to form large, diversified, living gene banks for future utilization in fisheries and aquaculture.

14
Genomics, Gene Mapping, Quantitative Trait Locus Mapping and Marker-assisted Selection

Basic Genomics

Genomics has several definitions (Liu, 2007; Gibson and Muse, 2009), including the study of genes and their functions, the study of the genome, the molecular characterization of all genes in a species, the comprehensive study of the genetic information of a cell or organism, the study of the structure and function of a large number of genes simultaneously, the study of the structure, content and evolution of genomes, and the science of the study of the genome. Correspondingly, the science of the study of the transcriptome is transcriptomics and the science of studying the proteome is proteomics. Gene mapping and QTL mapping are areas of study within genomics.

There are numerous areas of study within genomics. Two of the major disciplines are structural genomics – the study of the structure, organization and evolution of genomes or the elucidation of the tertiary structure of each class of protein found in cells – and functional genomics – the study of the expression and function of genomes.

Gibson and Muse (2009) indicate that there are eight core aims of genomics. These aims include the establishment of an integrated web-based database and research interface, assemblage of physical and genetic maps of the genome, generation and ordering of genomic and expressed gene sequences, identification and annotation of all genes encoded with the genome, characterization of DNA-sequence diversity, compilation of atlases of gene expression, compilation of functional data such as biochemical and phenotypic properties of genes, and generation of resources for comparison of genomes across species (comparative genomics).

Bioinformatics is a critical aspect of genomics. Increased computer power and advances in programming and databases have been key aspects of the tremendous progress that has been made in the aims and goals of genomics.

Total length of genomic DNA is the sum of all chromosomal DNA. Genome sizes can vary 100,000-fold among organisms. In general genome size is correlated with biological complexity, but – like everything in biology – many exceptions occur. Phenotypically, fish and crustaceans are quite diverse, and correspondingly their genome size is highly diverse. Mollusc genome size is more uniform than those of fish and crustaceans. It is difficult to realistically compare genome sizes among species as their ploidy level can vary between 2N and 6N. However, one way to accomplish this is to use standardization, the C-value, which is the size of haploid genomes in pg.

There are many parameters that define genomes that can be variable among species. Genomes/chromosomes can vary in GC content. Heterochromatin can have long stretches with highly repetitive DNA, transposons, insertions of mtDNA and duplications. Centromere and telomeres are not the same in this regard. The repetitive sequences include transposons, inactive mRNA that is derived from copies of cellular genes (pseudogenes), SSRs, microsatellites and short tandem repeats/variable-number tandem repeats (VNTRs), segmental duplications (as large as 330 kb) and blocks of interspersed repeats. Microsatellites are SSRs with repeat length up to 13 bp, in contrast to minisatellites which have longer repeat lengths.

One of the key developments in molecular genetics and genomics was the basic chain-termination method of sequencing developed by Frederick Sanger in 1974. Variations of this technique were the basis for sequencing for three decades. In Sanger or dideoxy sequencing, single-stranded molecules are synthesized, randomly terminated by the addition of a labelled dideoxy nucleotide (ddA, ddT, ddC, ddG) and then separated by electrophoresis. The sequence is visualized either by radioactivity, with manual sequencing one lane per base on a fixed gel, or each base is identified by computer fluorescently as it emerges from automated lanes. Most sequencing during the 1980s was done manually. The key to this technology is to generate molecules that are different by one nucleotide.

During the 1990s high-throughput sequencing with fluorescent labels became more prevalent, greatly improving sequencing output and research. Major advances included laser reading, improvement in chemistry and the use of capillary electrophoresis instead of slab-gel electrophoresis.

Recently, next-generation sequencing using massive pyrosequencing has become a reality with machines such as the 454 Life Sciences pyrosequencer and the Illumina Genome Analyzer. Now sequencing that took many years and millions of dollars can be done in a matter of weeks at a cost of tens of thousands of dollars. Sequencing

advances have resulted in the 1995 price of US$2 per base being lowered to 4 cents per base in large-scale sequencing projects.

Genetic Maps

Genetic maps are useful for understanding the genomic organization of a species, the comparative evolution of organisms and genetic mechanisms, to a lesser degree for gene isolation and ultimately for genetic improvement. Different types of gene map can be constructed including genetic maps, physical maps and cytological maps, which can be integrated.

A genetic map is a description of the relative order of genetic markers in linkage groups in which the distance between markers is expressed in units of recombination (Liu, 2007). The unit of recombination is the centimorgan (cM), which is the standard unit of genetic distance, an expression of the percentage of progeny in which a recombination event has occurred between two markers. Animal chromosomes average about 100 cM in length.

Several gene-mapping techniques exist. One option is to utilize recombinant lines that have polymorphism. Another option is radiation hybrid mapping, for which fragments of chromosomes are incorporated into a panel of hamster fibroblast cell cultures. Species-specific PCR is then used to see which loci are present in each line/fragment. A third technique is single-sperm typing. This is a PCR-based method that actually counts the number of crossovers between markers. A potential cautionary note is that recombination can be focused on hotspots. Although not commonly obtained in this manner, recombination rates can also be estimated from population data based on linkage disequilibrium.

A mapping function adjusts for the fact that the probability of crossovers leading to recombination does not increase linearly with physical distance. The Kosambi mapping function adjusts for interference, the presence of one crossover reducing the likelihood of another crossover in the same vicinity. In general centromeres and telomeres are less recombinant than euchromatin.

A physical map is an assembly of contiguous stretches of chromosomal DNA, contigs. A contig is a set of sequence fragments that have been ordered into a contiguous linear stretch on the basis of sequence overlaps at the fragment ends.

Two strategies exist for constructing physical maps. Alignment of the physical map can be generated from randomly isolated clones based on shared restriction fragment length profiles of these clones. The second strategy involves hybridization. A common probe is used for the terminal sequence of one clone as a probe to identify a set of overlapping clones that share the probe sequence. Contigs can then be extended by end sequencing the clones, leading to the identification of STSs. These can then be used as hybridization probes to either extend the chromosome walk or fill in gaps in the contig. The contigs are then utilized to build a scaffold, a set of contigs constituting a whole-genome sequence.

Cytogenetics is the study of the physical appearance of chromosomes. A cytological band is an area of the chromosome that stains differently from areas around it. Cytological maps are the banding patterns observed through a microscope on stained chromosome spreads. A cytological map is also a type of chromosome map whereby genes are located on the basis of cytological findings obtained with the aid of chromosome mutations.

Choice of Markers

Segregation analysis of polymorphic markers allows the assignment of isozymes, biochemical markers, DNA fragments and genes to chromosomes, ordering of these genetic markers and genes, establishment of genetic linkages, mapping of important genes and identification of modes of inheritance of production traits by demonstrating the linkage of major genes, either singly or as part of polygenic systems (Liu and Dunham, 1998a). Initially, most of this research was conducted with isozymes and easily scored qualitative traits; however, advances in DNA technology during the past decade have allowed rapid identification of very large numbers of markers for producing genetic-linkage maps – the mapping of genes on chromosomes. Isozymes and other biochemical markers, such as genes associated with the immune system, have the probable advantages of being conserved across taxa and linked to quantitative traits, but have the disadvantage of being few in number. However, the advent of ESTs is rapidly overcoming this shortcoming of type I markers – actual genes – for gene mapping.

Assuming a recombination genome size to be 2000 cM, 200 evenly distributed genetic markers would be required for a map with a resolution of 10 cM (or a map for any locus being within 5 cM on average to the nearest marker) (Poompuang and Hallerman, 1997; Dunham and Liu, 2002). Theoretically, a genetic map with 1000 molecular markers can place any locus in close proximity to less than a million base pairs. Thus, several thousands of markers may be needed to construct genetic-linkage maps that can tightly localize specific genes for practical applications, such as gene isolation. Gene isolation as an application of gene mapping is less useful than in the past as other molecular and genomic techniques have become advanced and powerful.

Microsatellites and AFLP markers are the most reliable, efficient and abundant markers for detailed genetic-linkage mapping in catfish (Liu and Dunham, 1998a,b; Liu et al., 1998a,b,c, 1999a,b,c,d,e,f, 2003; Waldbieser et al., 2001) and perhaps other aquatic organisms. However, mapping with SNP markers is on the verge of being the most powerful and economic marker system. Fine linkage mapping depends on the availability of large numbers of ESTs and the anchoring of well-ordered contigs of bacterial artificial chromosome (BAC) clones to linkage maps.

Conservation of microsatellites across a broad range of taxa in aquatic organisms has potentially important applications and implications (Liu et al., 2003) for gene mapping, MAS, cloning of genes and evolutionary studies. High levels of genetic conservation allow comparative gene mapping, which is especially important considering the rare availability of type I markers (markers that encode genes) in fish (Z.J. Liu, unpublished results). Comparative gene mapping would facilitate rapid advancements in gene mapping of major aquaculture species.

Conservation of microsatellite loci across a broad range of species has been demonstrated among various taxa (Moore *et al.*, 1991; Deka *et al.*, 1994; FitzSimmons *et al.*, 1995; Fredholm and Wintero, 1995; Menotti-Raymond and O'Brien, 1995; Coote and Bruford, 1996; Rico *et al.*, 1996; Sun and Kirkpatrick, 1996; Zardoya *et al.*, 1996; de Gortari *et al.*, 1997; Surridge *et al.*, 1997; Liu *et al.*, 1998d, 1999e, 2003). In the majority of these studies, primers designed from microsatellite-flanking regions of one species were successfully evaluated in closely related species. For example, primers from humans were tested in other primates (Coote and Bruford, 1996), or primers from one species member of a family were evaluated for their conservation among other species in the family (Fredholm and Wintero, 1995; Menotti-Raymond and O'Brien, 1995).

Microsatellites are highly conserved among species of fish. Most primers for channel catfish, Ictaluridae, amplified microsatellites from Cichlidae (Liu *et al.*, 1999e). Homologous microsatellite loci have endured for approximately 300 million years in turtle (FitzSimmons *et al.*, 1995) and for about 470 million years in fish (Rico *et al.*, 1996). Microsatellite-flanking sequences of fish may evolve at a slower rate than those of mammals (Liu *et al.*, 2003) based on fish gene maps generated to date. The identification of homologous chromosome segments in siluriform, cyprinodontiform and salmoniform fishes supports the hypothesis (Morizot, 1994) that teleostean gene arrangements may have diverged more slowly from those of the vertebrate ancestor than those same gene arrangements in mammalian orders. Gene map locations in one teleost may be highly predictive of map locations in other fish (Dunham *et al.*, 1998).

Mapping Systems

Gene mapping can be accomplished by using either intraspecific or interspecific systems. Interspecific hybridization systems are powerful for the construction of genetic-linkage maps. Interspecific approaches to gene mapping have frequently been used, in recent years utilizing both microsatellites and AFLPs (Agresti *et al.*, 2000) and in the early fish gene maps generated with isozymes (Morizot and Siciliano, 1979; Pasdar *et al.*, 1984; Johnson and Wright, 1984; Johnson *et al.*, 1987), because of the power and high level of polymorphism found in interspecific systems. Liu *et al.* (2003) utilized a channel catfish × blue catfish hybrid system for gene mapping. The F_1 hybrids are fertile and F_2 hybrids or backcross progeny can be readily produced (Argue, 1996; Liu *et al.*, 1998a). The primary advantages of using this hybrid system for genetic-linkage analysis are the high level of polymorphism between channel catfish and blue catfish and the fact that few reference families are needed because single families can be generated that are heterozygous for virtually every genetic marker. Additionally, there are large performance differences between species for production traits, which could expedite QTL mapping.

Conservation of microsatellite loci between closely related species, as demonstrated by Liu *et al.* (1999e) for ictalurid catfish, allows construction of unified maps among those generated from intraspecific and interspecific mapping systems. One individual channel catfish was heterozygous for 24 of 31 microsatellite loci. Similar results were generated for blue catfish, white catfish and flathead catfish. These results indicate that small numbers of interspecific or intraspecific reference families would be sufficient for gene mapping of microsatellite loci, and microsatellite markers would be almost as numerous as RAPD and AFLP markers. The microsatellites would, in fact, be even more powerful because of their co-dominance and ability to identify heterozygotes.

Utilization of haploid gynogenesis is another powerful mapping strategy. Single-sperm typing is now possible but haploid gynogenesis has advantages compared with single-sperm typing (Lie *et al.*, 1994). Cell division in eggs is activated by irradiated sperm, resulting in haploid individuals, representing a single maternal meiotic event with no paternal genome contribution. The haploid gynogenesis strategy represents the female counterpart of sperm typing but has the advantages that no individual sorting is needed, repeated tests are possible on the same individual because of the large amount of cells and DNA, it is not entirely dependent on PCR, and markers that can be mapped are not restricted to non-coding nuclear DNA and include isozymes, mtDNA and ESTs (Lie *et al.*, 1994). Both single-sperm typing and haploid gynogenesis allow resolution of recombination rates below 0.5% and discrimination of loose linkages above 45%, up to no linkage, 50%. Computer simulation indicates that lethal genes, which may eliminate specific haplotypes and cause segregation distortion of markers linked with such genes, do not interfere with the recombination estimate (Lie *et al.*, 1994). Additionally, haploid gynogenesis indicates the location of putative lethal genes relative to the informative markers. This strategy is efficient in distinguishing between variants, and allowed Lie *et al.* (1994) to detect segregation distortion of microsatellites in Atlantic salmon, probably due to preselection of eggs or embryos resulting in differential mortality of certain genotypes.

Linkage Disequilibrium

When conducting linkage or population genetic analyses, linkage disequilibrium is sometimes observed. Linkage (or gametic) disequilibrium is when there is a lack of fit for observed two-locus gametic frequencies compared with those expected based on the product of the single-locus allelic frequencies (May and Krueger, 1990). The frequency of an A_1B_2 gamete (loci A and B) in the population should be equal to frequency of the A_1 allele times the frequency of the B_1 allele. Linkage disequilibrium can also be defined as non-random associations between two sites (loci) or as occurring when alleles at two or more loci do not segregate independently (Liu, 2007). The cause of this could be various Hardy–Weinberg factors or epistatic selection, and can include either coupling or repulsion of alleles.

Linkage disequilibrium should decay by $1r$ each generation for random mating, where r is the recombination rate between the two loci (May and Krueger, 1990). The value for r can vary from zero for complete linkage to 0.5 for no linkage; therefore, linkage disequilibrium will decay by one-half each generation for most pairs of loci.

Linkage disequilibrium can be caused by mixtures in the sample of two or more populations with different allelic frequencies, a founding population already in disequilibrium, selection for certain heterozygous genotypes or random genetic drift to high frequencies of particular chromosome types (May and Krueger, 1990). In most population studies, the linkage disequilibrium is caused by mixing or founder effects. If a population has a bottleneck with a low effective population size, linkage disequilibrium might be expected for several generations.

For most population studies, variance components of the linkage disequilibrium values will only help indicate recent mixing (zero to two generations) of two highly divergent intraspecific gene pools (May and Krueger, 1990). Forbes and Allendorf (1989) found linkage disequilibrium values for linkages that have not yet decayed through recombination for fixed alternate alleles for several mixed populations of two distinct subspecies of cutthroat trout, Westslope and Yellowstone, after five to 15 generations of interbreeding.

Isozyme Maps

Isozymes were the first biochemical or molecular markers utilized for gene mapping in fish. Pasdar *et al.* (1984) examined linkage relationships of nine enzyme loci – aconitase (ACON), esterase (EST), glucose-phosphate isomerase A and B (GPI), glycerate-2-dehydrogenase (G2DH), malic enzyme (ME), phosphoglycerate kinase (PGK), phosphoglucomutase (PGM) and superoxide dimutase (SOD) – in backcrosses of reciprocal F_1 hybrids between green sunfish (*Lepomis cyanellus*) and redear sunfish (*Lepomis microlophus*) to each of their two parental species. A three-point linkage map containing G2DH, PGK and SOD was obtained, with frequencies of recombination between G2DH and PGK and between PGK and SOD at 45.3 and 24.7%, respectively. The remaining six loci assorted independently.

Salmonid isozyme maps revealed the phenomenon of pseudolinkage. In the case of pseudolinkage, an unexpected excess of recombinant genotypes is observed in progeny transmission vectors compared with parental phases. Pseudolinkage occurs when the meiotic disjunction processes are regulated or altered, preventing the random assortment of chromatids following crossing over in meiosis I and meiosis II (Danzmann and Gharbi, 2007). During meiosis, the chromosomes of maternal origin are preferentially segregated to the first polar body, while paternal chromosomes segregate to the disintegrating second polar body.

Although partial tetraploidy is widespread in the teleost genome, salmonids present one of the most obvious and most studied problems in fish gene mapping because of their tetraploid ancestry. Johnson and Wright (1984) and Johnson *et al.* (1987) examined the joint segregation analyses of isozyme loci in males and females of seven species and three fertile species hybrids of trout, charr and salmon, and identified 15 linkage groups. Johnson and Wright (1984) were proponents of the interspecific approach to gene mapping, and concluded that, since linkage groups are highly conserved among species and hybrids, they can be combined to form a common linkage map, in this case one for salmonids. Pseudolinkage was observed and was explained as preferential multivalent pairing and disjunction of metacentric (centrally fused)

chromosome arms with homoeologous arms of other chromosomes in male salmonids, in contrast to bivalent pairing in females. Five pseudolinkage groups were detected and were highly conserved among species and hybrids. Conservation of pseudolinkages among salmonids must have been a result of major chromosomal fusions in a common tetraploid ancestor before the radiation of salmonid species.

Earlier, Davisson *et al.* (1973) and Lee and Wright (1981) had observed the pseudolinkage found in some salmonid males. Both cytological and linkage analyses indicated that spontaneous centric fusion and fission could account for the curious patterns of pseudolinkage of two LDH loci in males of brook trout and in the F_1, F_2 and backcross generations of lake trout × brook trout hybrids (Davisson *et al.*, 1973). Intraindividual polymorphisms for acrocentric and metacentric chromosomes in somatic and gonadal tissue of these fish are consistent with the proposed polyploid evolution in Salmonidae. Mitotic and meiotic analyses of the tetraploid-derivative species, brook trout, indicated that the process of diploidization was incomplete (Lee and Wright, 1981).

The diploid number was 2N = 84, with 16 metacentrics and 68 acrocentrics for both males and females from different sources, and no inter- or intraindividual Robertsonian polymorphism was present. Oocytes at pachytene had the expected 42 bivalents with eight metacentric and 34 acrocentric pairs. However, variable numbers of tetravalents, with a total of 35–40 bivalent plus tetravalent elements, were observed in metaphase I cells of males. Each tetravalent was composed of two acrocentric and two metacentric chromosomes. Variability in the number of tetravalents was found not only among different brook trout sources, but also among different cells of the same fish (Lee and Wright, 1981). Differential homoeologous pairing was proposed to account for the variable number of tetravalents and to explain the occurrence of pseudolinkage in some salmonid males.

Some of the linkage relationships in salmonids are due to duplicated loci resulting from the tetraploid ancestry. Hollister *et al.* (1984) observed variable genotypes for the duplicate loci encoding the enzyme peptidase D (PEPD) in lake trout, brook trout and their fertile hybrid (splake). Non-random assortment was observed among progeny of parents doubly heterozygous for the PEPD-1 and PEPD-2 loci, the duplicate loci encoding GPI and the locus sorbitol dehydrogenase (SDH). Linkage groups were PEPD-1 with GPI-1 and PEPD-2 with GPI-2 with SDH. The results fitted and were consistent with the earlier-determined chromosomal model involving preferential tetravalent pairing of homoeologous chromosomes – pseudolinkage.

Disney and Wright (1990) later observed extensive multivalent pairing in lake trout, which, along with data on hybrid splake (brook lake) trout, supported a meiotic model to explain pseudolinkage. Additionally, C-banding of mitotic and meiotic lake trout chromosomes revealed an intraindividual polymorphism for a Robertsonian fusion, and silver staining showed that the chromosomes with active NORs located proximal to a centromere were not involved in the fusion event.

Null alleles can also complicate linkage analysis in salmonids and in oysters. Unusual phenotypic distributions were observed at the muscle-specific, duplicate aspartate aminotransferase (AAT) locus in a wild population of brook trout, and analysis of these phenotypic distributions eliminated disparate gene frequencies, non-random association between the two loci and inbreeding as possible explanations (Stoneking *et al.*, 1981). Models incorporating a null allele and inheritance data from hatchery populations of brook trout fitted the data and confirmed a null-allele polymorphism. This AAT null allele, along with other null-allele polymorphisms in salmonids, is evidence that loss of duplicate gene expression is still occurring; however, there is no such evidence of ongoing loss of duplicate gene expression in the Catostomidae, another tetraploid-derivative lineage (Stoneking *et al.*, 1981).

Early gene-mapping research was also conducted with *Xiphophorus*, utilizing interspecific approaches. Morizot *et al.* (1977) obtained a three-point linkage group comprised of loci coding for adenosine deaminase (ADA), glucose-6-phosphate dehydrogenase (G6PDH) and 6-phosphogluconate dehydrogenase (6PGD) for *Xiphophorus* (Poeciliidae) by utilizing reciprocal backcross hybrids from crosses between either *Xiphophorus helleri guentheri* or *Xiphophorus helleri strigatus* and *Xiphophorus maculatus*. The alleles at this linkage group assorted independently from the alleles at isocitrate dehydrogenase (IDH)-1 and -2 and glyceraldehyde-3-phosphate dehydrogenase (GAPDH)-1, and the latter assorted independently from each other. The linkage group was conserved in all populations of both species of *Xiphophorus* examined. Data from *X. helleri guentheri* backcrosses indicate the linkage relationship, ADA-6%-G6PDH-24%-6PGD, and ADA-29%-6PGD (30% when corrected for double crossovers), but results from backcrosses of *X. helleri strigatus* gave different recombination frequencies for the same gene order. Possible explanations include differences due to an inversion or a sex effect on recombination. The linkage of 6PGD and G6PDH exists in species of at least three classes of vertebrates (Morizot *et al.*, 1977).

Recombination data from backcross hybrids among three species and four subspecies of *Xiphophorus* indicate four additional linked loci, linkage group (LG) II, (esterase) EST-2-0.43-EST-3-0.26-(retinal lactate dehydrogenase) LDH-1-0.19-(mannose phosphate isomerase) MPI (Morizot and Siciliano, 1979). Interference was detected in the EST-3 to MPI region, and LG II assorted independently from the six loci of LG I and from GAPDH-2 and IDH-2.

Next, Morizot *et al.* (1991) analysed 76 polymorphic isozyme loci in backcross hybrid individuals from intra- and interspecific crosses of the genus *Xiphophorus* (Poeciliidae), identifying 17 multipoint linkage groups containing 55 protein-coding loci and one sex-chromosome-linked pigment-pattern gene. Gene orders were determined for ten linkage groups, and total genome length was estimated to be 1800 cM.

Comparisons of the *Xiphophorus* linkage map with those of other fishes, amphibians and mammals suggested that fish gene maps are remarkably similar and probably retain many syntenic groups (Morizot *et al.*, 1991); this will be discussed in more detail later. Synteny is the conservation of gene order on a chromosome across species (two or more). Synteny is broken up by chromosome fusions, splits, inversions and reciprocal translocations. Homologs are highly conserved loci that come from a common ancestor. Orthologs are true homologs. Paralogs are similar genes that arise from duplication in one or both lineages prior to an evolutionary split or perhaps afterwards as well. Entire gene complexes can undergo multiple rounds of duplication, which complicates the ability to distinguish orthologs from paralogs.

Six isozyme linkage groups have been established for ictalurid catfish, using the channel × blue catfish interspecific hybrid system. Gene–centromere distances were estimated for six loci in gynogenetic channel catfish (Liu *et al.*, 1992) and for additional polymorphic loci in blue × channel triploid hybrids. At least 28 polymorphic isozyme loci were found and used to establish channel catfish multipoint LGs I–VI, comprising 18 loci (R.A. Dunham, B. Argue and D. Morizot, unpublished results). Eleven unlinked loci may bring the total of isozyme-marked chromosomes to 17 of the 29 chromosome pairs of channel and blue catfish. The extensive genetic variability within and between blue and channel catfish at approximately 70 isozyme loci (Dunham and Smitherman, 1984; Hallerman *et al.*, 1986; Carmichael *et al.*, 1992) could allow significant expansion of this isozyme-based gene map. Ictalurus LG I is comprised of loci coding for glutathione reductase and PGM (Morizot *et al.*, 1994). Three other loci assort independently from the LG I pair, providing markers for four chromosomes.

Three isozyme loci assigned to catfish linkage group LG II are also linked in LG II of poeciliid fishes, indicating the evolutionary conservation of both neutral and physiological markers. Comparison of the gene

maps of poeciliid and salmonid fishes substantiates that this result is not a rare event. Orthology is sometimes difficult to establish because poeciliids are diploid and salmonids are recently tetraploid derived (Johnson *et al.*, 1987). However, at least four cases of linkage-group conservation have been identified, and linkage-group divergence has yet to be observed (Morizot, 1990), again illustrating the strength of the linkage conservation in fish. Within poeciliids, LGs II of *Xiphophorus* and *Poeciliopsis* are homologous (Morizot *et al.*, 1989) and are homologues of *Poecilia* LG I (Narine *et al.*, 1992). Additionally, poeciliid and salmonid syntenies can be identified in centrarchid fishes and ranid frogs and are apparently even evolutionarily conserved in segments of human chromosomes 12, 15 and 19 (Morizot, 1990). Morizot (1990, 1994) and Morizot *et al.* (1991) proposed that the chromosomal arrangements of duplicated genes in fishes suggest retention of patterns produced by three rounds of tetraploidization.

In *Xiphophorus* poeciliids, carbonic anhydrase (CA1) (orthologous to catfish CAH-2) has been assigned to LG XXIV (the *X. maculatus* sex-chromosome linkage group), and GAPDH-2 (orthologous to catfish GAPDH-1) is linked to peptidase C (PEPC) (orthology with catfish dimeric peptidases is uncertain) in LG U3. Sex-chromosome linkage groups vary widely even within orders for fish.

Catfish LGs II and III are strong examples of evolutionary conservation of linkage groups in fishes and between fishes and mammals. The orthologues of the three catfish LG II isozyme loci are also linked in *Xiphophorus* LG II, and GPI and MPI orthologues are also linked in salmonid LG 13. Orthologues of MPI and α-mannosidase are syntenic on human chromosome 15 (O'Brien, 1993). Catfish LG III shows homology with other fish linkage groups. *Xiphophorus* LG IV contains pyruvate kinse (PK), GPI, PEPD and probably a cytosolic IDH locus, which is orthologous to the catfish arrangement of loci. Salmonid LG 3 also contains GPI and PEPD loci, as well as human chromosome 19.

The sole example of definitive divergence between catfish, poeciliid and salmonid linkage groups involves muscle IDH (mIDH). The location of mIDH in catfish LG III is different from that in poeciliids and salmonids, and the orthologies of purine nucleoside phosphorylase (PNP) and EST loci in catfish LG III with those of loci in other fishes are not yet absolutely determined. mIDH is linked to LDH-A, LDH-C and two esterase loci in *Poeciliopsis* LG I, which is homologous to *Xiphophorus* LG II. This linkage is conserved in salmonid LG 13, which is also partly homologous to *Xiphophorus* LG II. Loci for muscle PK, mIDH and MPI appear to be conserved on human chromosome 15, suggesting a symplesiomorphic vertebrate gene arrangement (Morizot, 1994). In contrast, the catfish position of mIDH in LG III is within a segment conserved in *Xiphophorus* LG IV, salmonid LG 3 and human chromosome 19. The apparent homology of catfish LG III with synteny of GPI, PEPD and mIDH on mouse chromosome 7 might suggest a polymorphic symplesiomorphy.

Morizot (1994) proposes that linkage of duplicates of GPI, IDH, PK and, by extension from salmonids, AAT and malate dehydrogenase (MDH) loci indicates that poeciliid LGs II and IV (catfish LGs II and III) are derived from ancestral chromosome duplications and translocations, which are more likely to occur between homoeologous chromosomes, perhaps because of mispairings in regions of residual homology. The translocations of mIDH on to catfish LG III and mouse chromosome 7, therefore, both represent a relatively frequent type of chromosomal rearrangement.

Since chromosome numbers vary widely among fish, some linkage-group divergence among fish gene maps should be expected, but the observed example differs from the expected pattern of linkage-group splitting to produce new linkage groups. The loci assessed for linkage in ictalurid catfish include markers assigned to multipoint linkage groups in poeciliid and/or salmonid fishes. These studies indicate that many, but not all, gene arrangements among fish gene maps can probably be predicted from the gene arrangements of other fish gene maps. While no mechanisms have been found that result in gene-arrangement conservation, it is possible that highly conserved areas of the genome may be under stringent selection, and that disruption of these arrangements may result in disturbances of gene regulation.

Gene sequencing has now begun to confirm these hypotheses and observations from gene-mapping studies. Ninety-five per cent of the *Fugu rubripes* genome has been sequenced and, as in the human genome, gene loci are not evenly distributed but are clustered into sparse and dense regions (Aparicio *et al.*, 2002). Some 'giant' genes had average coding-sequence sizes, but were spread over genomic lengths significantly larger than those of their human orthologues. Three-quarters of predicted human proteins match strongly to *Fugu*, but approximately one-quarter have highly diverged from *Fugu* or have no pufferfish homologues, illustrating the extent of protein/gene evolution in the 450 million years since teleosts and mammals diverged. Some *Fugu* genes, either specific to teleosts or lost in human lineages, are no longer found in humans (Gajewski and Voolstra, 2002; Gilligan *et al.*, 2002). In general, conserved linkages of chromosomal segments were preserved from the common vertebrate ancestor of *Fugu* and humans, but with considerable scrambling of gene order. Additionally, several conserved non-coding sequences in the promoter and intronic regions were identified that are probably involved in gene regulation (Gilligan *et al.*, 2002). More than 80% of the gene assembly of *Fugu* was multigene-sized scaffolds. Repetitive DNA accounted for less than one-sixth of the sequence, and gene loci occupied about one-third of the genome in this small 365 mega-base genome (Aparicio *et al.*, 2002).

The evolution of mitochondrial genomes is not necessarily parallel to that of the nuclear genome. The gene order, nucleotide composition and evolutionary rate of the mtDNA genome of *Fugu* correspond to those of other teleosts, suggesting that the evolution of this genome was not affected by the processes that led to the dramatic reduction of the size of the nuclear genome of *Fugu* (Elmerot *et al.*, 2002).

Divergences of catfish and *Xiphophorus* occur for gene arrangements where human and *Xiphophorus* gene arrangements are also dissimilar and where rearrangements are found in human or rodent lineages. These chromosomal segments provide a focal point for critical testing of homology of fish gene arrangements in three teleost orders, involving gene families with members located on human chromosomes 11, 12, 15 and 19.

Human chromosome 19 contains the most highly conserved genetic linkage in vertebrates, the GPI and PEPD loci. Synteny of these genes has been shown for humans, a variety of other mammals, amphibians, salmonid fishes and the poeciliid fishes *Xiphophorus* and *Poeciliopsis* (O'Brien, 1993). Not a single case of asynteny has been found among vertebrate gene maps.

A second strongly conserved DNA segment is homologous to arrangements on human chromosome 15 and includes PKM2 (PK isoenzyme M2) and MPI, which are conserved in poeciliids, salmonids and mammals. The only exception to the linkage of these two genes so far identified in mammals is in bovines, where PKM2 is located on chromosome 10 and MPI on chromosome 21. Linkage groups of fishes homologous to these segments of human chromosomes 15 and 19 appear to have originated by chromosome duplication, as evidenced by the linkage of GPI, IDH and PK loci in each linkage group (Morizot, 1994). The presence of duplicate genes for cytosolic and membrane-bound isozymes of α-mannosidase on human chromosomes 15 and 19 is another example of this pattern of linkage of gene duplicates (Lundin, 1993).

Surprisingly, the first definitive case of linkage-group divergence among teleost gene maps involved two linkage groups that are otherwise highly conserved throughout vertebrate evolution. Gene maps of mammals other than mouse and human contain numerous examples consistent with the hypothesis that most translocations occur between paralogous chromosomal segments.

For example, the most common location of nucleoside phosphorylase (NP) in mammalian gene maps appears to be in syntenic groups homologous to human chromosome 15 segments. Synteny of purine NP with creatine kinase B (CKB), PKM2 and usually MPI occurs in many mammalian species. However, NP and CKB are syntenic on chromosome 14 in humans, while in apes NP and CKB are asyntenic. Rabbits have diverged even further not only by having NP asyntenic with CKB, but also by relocating NP on to yet another paralogue, syntenic with peptidase B (PEPB), which is found on human chromosome 12, syntenic with LDH-B, an arrangement also exhibited by salmonid fishes. Mechanisms must exist that make certain chromosome segments across orders susceptible to rearrangements and translocation.

Linkage of LDH-A (human chromosome 11) to human chromosome 15 loci, mIDH, PK2 and MPI in poeciliid fishes, similar to synteny of mIDH and LDH-A in frogs, further illustrates this cycle of translocations among paralogous segments of human chromosomes 11, 12, 15 and 19, which are predicted to have arisen through a minimum of two rounds of tetraploidization (Morizot, 1990, 1994; Lundin, 1993). The presence of mIDH in poeciliids on a human chromosome 15 homologue and in catfish in a human chromosome 19-related syntenic group is logical, as exactly the same pattern is evident in mammals by the location of sorbitol dehydrogenase (SORD) on human chromosome 15 syntenic with GPI and PEPD (human chromosome 19) in African green monkeys.

For practical genetic improvement for aquacultural species, understanding why, how and in what directions linkage arrangements have diverged is less important than the predictiveness implied by slow divergence of teleost gene arrangements from those of the vertebrate ancestor. This should allow research with one species usually to be of practical application in another species. To provide the most conclusive estimates of rates of gene-map divergence in fishes, the segments where fish and mammalian gene maps have substantially diverged should be examined. The comparison of catfish gene arrangements of orthologues located on human chromosome 19 with those mapped in *Xiphophorus* illustrates the clearest test of teleost linkage-group conservation. Chromosome 19 in humans carries loci of GPI, PEPD, CKM (= CK-A), ERCC2, LIGI and CMT genes (another possible orthologue, PGK2, is listed in many human gene-map summaries, but in fact is a pseudogene; (D.C. Morizot, University of Texas, M.D. Anderson Hospital and Tumor Institute, 1994, personal communication). In *Xiphophorus*, GPI and PEPD loci are linked in LG IV (and in catfish LG III), but, in contrast to humans, ERCC2 and CK-A are assigned to LG XIV and LIGI to LG VI and CMT is polymorphic but is as yet unassigned to the gene map.

Loci for muscle PK, mIDH and MPI appear to be conserved even further on human chromosome 15, suggesting a symplesiomorphic vertebrate gene arrangement (Morizot, 1994). The catfish position of mIDH, in contrast, is within a segment conserved in *Xiphophorus* LG IV, salmonid LG 3 and human chromosome 19. The apparent homology of catfish LG III with synteny of GPI, PEPD and mIDH on mouse chromosome 7 might suggest a polymorphic symplesiomorphy; however, the following explanation proposed by Morizot (1994) seems more likely. Linkage of duplicates of GPI, IDH, PK and, by extension from salmonids, AAT and MDH loci indicates that poeciliid LGs II and IV (catfish LGs II and III) are derived from ancestral chromosome duplications, and translocations are more likely to occur between homoeologous chromosomes, perhaps because of mispairings in regions of residual homology. The translocations of mIDH on to catfish LG III and mouse chromosome 7, then, represent examples of a relatively frequent type of chromosomal rearrangement.

DNA Markers and Maps

The development of powerful DNA techniques, such as AFLP, RAPD and microsatellites, which generate large numbers of molecular markers, has allowed rapid progress in aquatic organism linkage mapping. Construction of framework genetic-linkage maps has progressed rapidly in aquaculture species such as catfish (Li *et al.*, 2000; Waldbieser *et al.*, 2001; Liu *et al.*, 2003), tilapia (Lee and Kocher, 1996; Kocher *et al.*, 1998), salmonids (Allendorf *et al.*, 1994; Moran *et al.*, 1997; Reed *et al.*, 1997; Hoyheim *et al.*, 1998; Young *et al.*, 1998; Rexroad *et al.*, 2002), shrimp (Li *et al.*, 2000; Alcivar-Warren *et al.*, 2002) and oysters (Guo and Allen, 1997; Gaffney, 1999; Hubert *et al.*, 2000), and especially for model species such as zebra fish and *Fugu* (Aparicio *et al.*, 2002). Radiation hybrid panels have been developed in tilapia (Kocher *et al.*, 2002), BAC libraries in catfish (G.C. Waldbieser, unpublished results) and normalized cDNA libraries constructed for EST analysis and functional genomics analysis (Lui and Feng, 2001).

Over 350 microsatellite markers and over 600 AFLP markers have been mapped (Li *et al.*, 2000; Waldbieser *et al.*, 2001; Liu *et al.*, 2003) for channel catfish (Fig. 14.1). Outbred populations of channel catfish contained an average of eight microsatellite alleles per locus and an average heterozygosity of 0.70 (Waldbieser *et al.*, 2001). A total of 293 microsatellite loci were polymorphic in one or two families, with an average of 171 informative meioses per locus, and these data were used to construct a genetic-linkage map of the channel catfish genome (N = 29). Nineteen type I loci, 243 type II loci and one EST were placed in 32 multipoint linkage groups covering 1958 cM (Waldbieser *et al.*, 2001). Nine additional type II loci were contained in three two-point linkage groups covering 24.5 cM, 22 type II loci were unlinked and multipoint linkage groups ranged from 11.9 to 110.5 cM, with an average intermarker distance of 8.7 cM.

In the case of the interspecific mapping strategy of Liu *et al.* (2003), a total of 607 polymorphic AFLP loci were produced, with the 64 *Eco*RI/*Mse*I primer combinations, using a (channel catfish female × blue catfish male) × blue catfish male backcross family. A total of 101 markers (16.6%) were

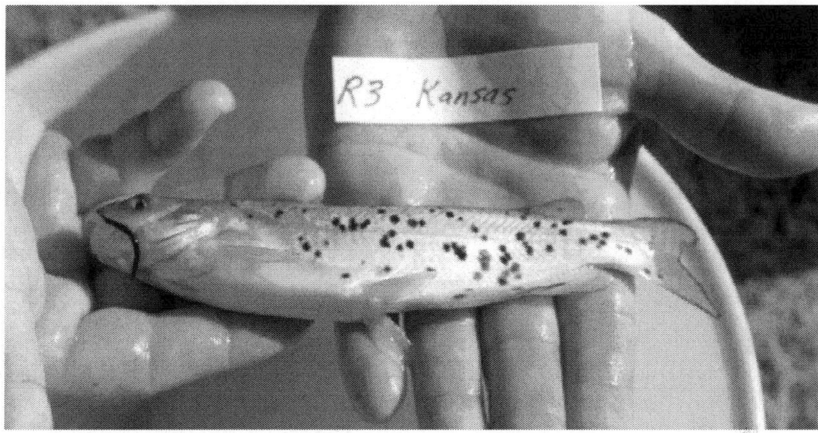

Fig. 14.1. Channel catfish, *Ictalurus punctatus*, the major aquaculture species in the USA. Currently about 1000 markers have been mapped, including isozyme, microsatellite and AFLP markers.

not used for the construction of the linkage map because they showed distortion – linkage disequilibrium – from the expected 1:1 ratio. Four hundred and forty-five markers were assigned to 40 linkage groups, with 29 markers unlinked, and 32 markers were excluded because of large map distances. The number of markers on the 40 linkage groups ranged from two to 33, compared with two to seven markers per linkage group for fish isozyme maps (Johnson *et al.*, 1987), 13 to 49 for various types of markers, but primarily AFLPs, in medaka, *Oryzias latipes* (Naruse *et al.*, 2000), and 73 to 201 ESTs for extensive zebra fish, *Brachydanio rerio*, maps (Hukriede *et al.*, 1999, 2001). There were 25 major linkage groups with five to 33 AFLP markers and 15 small linkage groups with two to four AFLP markers, with genomic coverage of this AFLP linkage map spanning 2511 cM Kosambi, compared with the 1958 cM channel catfish map based on microsatellites (Waldbieser *et al.*, 2001). The largest linkage group spanned 216.9 cM, with 19 AFLP markers, and the smallest 0.0 cM, with two AFLP markers, with the mean for the linkage map being one AFLP marker every 5.6 cM. However, the distances between any two given markers varied greatly, ranging from 0 to 51.2 cM. The microsatellite catfish map (Waldbieser *et al.*, 2001) was less variable in regard to

range of length of linkage groups, with multipoint linkage groups ranging from 11.9 to 110.5 cM, with an average intermarker distance of 8.7 cM, similar to the mean intermarker distances for the AFLP map. Eight of the AFLP linkage groups were longer than the longest microsatellite linkage group, indicating that in many cases the microsatellite strategy may be underestimating the size of the linkage group which may be related to marker distribution.

Karyotypes of channel and blue catfish are indistinguishable and both species have small chromosomes. However, some channel catfish chromosomes are larger than those of blue catfish, with length ranging from 1.5 to 3.5 μm and 1.0 to 2.0 μm for channel and blue catfish chromosomes, respectively (LeGrande *et al.*, 1984). In both species, the relative size of the individual chromosomes ranges from 2 to 10% of the total complement length. Based on chromosome length (LeGrande *et al.*, 1984) for channel catfish, the largest linkage group would be expected to be approximately 2.3 times larger than the smallest linkage group. Future fine-tuning of the catfish map and consolidation of the linkage groups should bring the variation in linkage length closer to this much tighter range, compared with the much more variable length of the microsatellite and AFLP catfish linkage maps.

Based on segregation ratios in 41 haploid embryos derived from a single *Oreochromis niloticus* female, a tilapia gene map was developed from 62 microsatellite and 112 AFLPs (Kocher *et al.*, 1998). Linkages were identified for 162 of the markers, with 95% of the microsatellites and 92% of the AFLPs linked on the map. The map covered 704 cM Kosambi in 30 linkage groups covering the 22 chromosomes of Nile tilapia and 24 of these linkage groups contained at least one microsatellite polymorphism. From the percentage of markers 15 or fewer cM apart, a total map length of 1000–1200 cM was estimated (Kocher *et al.*, 1998). High levels of interference were observed, consistent with measurements in other species of fish. The *Oreochromis* gene map now contains more than 500 microsatellites, indicating that the genome size is about 1 gigabase and the map length is 500 kb/cM (Kocher *et al.*, 2002).

Agresti *et al.* (2000) utilized the interspecific approach to develop tilapia gene maps. A synthetic stock (artificial centre of origin (ACO)) was produced by crossing five groups of fish, *O. niloticus* (wild type (On) and red (ROn) strains), *Oreochromis aureus* (Oa), *Oreochromis mossambicus* (Om) and *Sarotherodon galilaeus* (Sg). Three-way cross families (3WC) and four-way cross families (4WC) were produced, to introgress all four species into the ACO. An Om × (Oa × ROn) family was used to develop a gene map using microsatellite and AFLP DNA markers. The female (Om) parent had a total of 78 segregating markers (17 microsatellites, 61 AFLPs), and 62 markers (13 microsatellites, 49 AFLPs) were linked in 14 linkage groups covering a total of 514 cM. The F_1 hybrid male parent had a total of 229 segregating markers (62 microsatellites, 167 AFLPs), and 214 of these markers (60 microsatellites, 154 AFLPs) were linked in 24 linkage groups covering a total of 1632 cM.

Young *et al.* (1998) conducted gene mapping for rainbow trout by utilizing doubled haploids – androgens. The sex-determining locus was at a distal position on one of the chromosomes. AFLPs appeared to be primarily clustered near centromeres, VNTRs were frequently more telomeric and salmonid-specific small interspersed nuclear elements were intermediate in distribution compared with the other two marker types. This is another example of the choice of markers affecting the outcome of a genetic analysis.

Chromosome painting uses fluorescent *in situ* hybridization (FISH). In the first species the chromosomes are labelled with probes of different colours and the second species is hybridized with the same probes; single-colour chromosomes indicate complete correspondence between the two species and multicoloured chromosomes are indicative of breakage/fusion events. Paint probes of whole-arm chromosomes of rainbow trout were specific for single pairs of arms, suggesting that the majority of chromosomes from ancestral tetraploids have diploidized (Phillips *et al.*, 2000). Paint probes for the sex chromosomes of lake trout paint different autosomal chromosome pairs for rainbow trout and chinook salmon, indicating separate evolution of sex chromosomes in *Salvelinus* and *Oncorhynchus*. Similar fluorescent techniques have been used for gene mapping in salmonids. The Y chromosome in chinook salmon has been identified using FISH with a probe to a male-specific repetitive sequence isolated from this species. The probe lights up the distal end of the short arm of an acrocentric chromosome (Stein *et al.*, 2001). A large number of ribosomal RNA (rRNA) genes and simple repeat sequences have been mapped for a long list of molluscs using FISH (Guo *et al.*, 2007: Table 17.1).

Sakamoto *et al.* (2000) constructed a rainbow trout – a tetraploid-derivative species – gene map with 191 microsatellites, three RAPDs, seven expressed sequence marker polymorphisms (ESMPs) and seven allozyme markers. Twenty-nine linkage groups were identified, with potential arm displacement in the female map because of male-specific pseudolinkage arrangements. Synteny of duplicated microsatellite markers confirms some previously reported pseudolinkage arrangements based on allozyme markers. Fifteen centromeric regions (20 chromosome arms) were identified with a half-tetrad analysis using gynogenetic diploids (Sakamoto *et al.*, 2000). Female map length was about 1000 cM, but this is a substantial underestimate, as many geno-typed segments remained unassigned at the LOD score (logarithm (base 10) of odds) threshold of 3.0. Female:male map distances (ratio 3.25 F:1 M) were tremendously different. Females had an eightfold lower recombination rate in telometric regions compared with males, whereas proximal to the centromeres the female recombination rate was tenfold greater than that of males. Sakamoto (2000) proposed that quadrivalent formations, which were almost exclusive in males, are the cause of the sex-specific differences in recombination rates.

The initial linkage map of zebra fish, *B. rerio*, was constructed using haploid genetics and was saturated with 401 RAPD markers and 13 SSRs, spaced at an average interval of 5.8 cM (Postlethwait *et al.*, 1994). This strategy allowed rapid mapping of lethal and visible mutations. Johnson *et al.* (1996) then expanded the zebra fish map using half-tetrads. A total of 652 identified PCR-based markers were closely linked to each of the 25 centromeres of zebra fish.

Since then, radiation hybrid maps have been generated for zebra fish containing 4226 markers, including 459 genes and 3867 ESTs, covering more than 88% of the genome (Hukriede *et al.*, 1999, 2001). The map was 14,372 centirays, and average breakpoint frequency corresponded to 1 centiray = 118 kb. The concordance between radiation hybrid maps and meiotic maps was 96%. The distribution of the ESTs was very uniform, ranging from 73 to 201 ESTs per linkage group, indicating that there are no gene-rich or gene-poor chromosomes in zebra fish.

The zebra fish map now contains more than 1267 expressed sequences (Postlethwait *et al.*, 2002). Postlethwait *et al.*'s (2002) data indicate that in some cases the entire content of some human chromosomes has been conserved for the 450 million years since the lineages of zebra fish and humans diverged. Intrachromosomal rearrangements have been frequent, resulting in altered gene orders within conserved syntenies. The maps indicate that about 30% of

the zebra fish genes have been retained from a genome duplication that probably occurred before teleost radiation. Teleost and mammalian genomes have about the same number of chromosomes despite the teleost genome duplication, probably because of fissions of human chromosomes not because of fusions of fish chromosomes (Postlethwait *et al.*, 2000). Zebra fish chromosomes can be identified by AT-rich repetitive sequences at the centromere and GC-rich sequences adjacent to the centromeres (Phillips *et al.*, 2000).

Wada *et al.* (1995) found 227 informative RAPD markers in the Japanese medaka, *O. latipes*, segregating at 170 loci, including three pigment-pattern loci, five enzyme-coding loci and one male-determining factor. A gene map was constructed consisting of 28 linkage groups spanning about 2480 contiguous cM with an average of 323 kb/cM. Naruse *et al.* (2000) used a reference-typing DNA panel from 39 cell lines derived from backcross progeny to map 633 markers (488 AFLPs, 28 RAPDs, 34 interspersed repetitive sequences (IRSs), 75 ESTs, four STSs and four phenotype markers) for the medaka, *O. latipes*, of the order Beloniformes. The total map length of 24 linkage groups – the haploid number for medaka – was 1354.5 cM, and 13–49 markers were obtained for each linkage group. Conserved synteny between medaka and zebra fish was observed for two independent linkage groups; however, unlike zebra fish, the medaka linkage map exhibited obvious restriction of recombination on the linkage group containing the Y locus compared with the autosomal chromosomes. Different genomic phenomena can be seen among different species of fish.

Gene mapping in Pacific oysters (Hedgecock, 2002) and eastern oysters (Gaffney, 2002) is difficult because of distortion of Mendelian segregation ratios, and this might be aggravated by a high frequency of null alleles (Gaffney, 2002). Hedgecock (2002) found by utilizing microsatellite markers in F_2 crossbreeds that this distortion is caused by a very large number of deleterious recessive alleles/mutations carried by oysters. Wilson *et al.* (2002) used 673

AFLP polymorphic markers and some microsatellites that were free from segregation distortion to develop a low-density linkage map for black tiger shrimp, *Penaeus monodon*. A total of 116 markers segregated in more than one of the experimental families, resulting in 20 distinct linkage groups covering a total genetic distance of 1412 cM, although black tiger shrimp have more than 40 chromosome pairs.

The Major Histocompatibility Complex and Oncogenes

Genes other than isozymes and DNA markers have been mapped in fish, such as MHC class genes and tyrosinase genes, which, in the future, may serve as good starting points to search for related QTLs. McConnell *et al.* (1998) isolated two MHC class II B genes from an inbred *X. maculatus* strain, Jp 163 A, and mapped one of these genes, *DXB*, to LG III, linked to a malic enzyme locus, which is also syntenic with human and mouse MHC. A second type of class II B clone, a DAB-like gene, was orthologous to class II genes identified in other fishes, and was 63% identical to the *X. maculatus DXB* sequence in the conserved β_2-encoding exon. This second *DXB* gene is an unlinked duplicated locus not previously identified in teleosts, and was assigned to a new linkage group LG U24.

Interspecific hybrids of *Xiphophorus* have a number of simple oncogenes and tumour-suppressor genes, and these fish serve as malignant melanoma models (Morizot *et al.*, 1998; Gómez *et al.*, 2002). The gene map of *Xiphophorus* has about 100 genes assigned to at least 20 independently assorting linkage groups, in addition to more than 250 anonymous DNA sequence markers. Morizot *et al.* (1998) mapped the tumour-suppressor locus, DIFF, which is one of two genetic determinants of melanoma formation in the hybrid melanoma, the Gordon–Kosswig melanoma model. The other gene responsible for melanoma formation in this model is a sex-linked tyrosine kinase gene, *Xiphophorus* melanoma receptor kinase (*Xmrk*), which is related to the gene encoding epidermal growth factor receptor (EGFR). The cellular oncogene homologues of the non-receptor tyrosine kinase family orthologous to yes and fyn are overexpressed in malignant melanomas of *Xiphophorus* and may be involved in tumour progression (Morizot *et al.*, 1998). The *Xiphophorus* yes gene, *YES1*, belongs to LG VI, closest to the *EGFR* gene, and a *fyn* gene homologue to LG XV, linked to the gene for cytosolic α-galactosidase (Morizot *et al.*, 1998). The *EGFR*-related sequence (EGFRL1) previously assigned to *Xiphophorus* LG VI was determined to be the *EGFR* orthologue. Morizot *et al.* (1998) conclude that the presence of expressed duplicates of members of the tyrosine kinase gene family in teleost fishes may increase the potential number of targets in oncogenic cascades in fish tumour models.

Wellbrock *et al.* (2002) reviewed the genetics of melanoma in *Xiphophorus* and indicate that the primary event for tumour formation is the cell lineage-specific overexpression of a structurally altered tyrosine kinase receptor. This phenomenon is also found in many tumours of birds and mammals. Once expressed at high levels, the *Xiphophorus* melanoma-inducing receptor kinase *Xmrk* shows constitutive activation. Analyses of the different signalling cascades induced by the Xmrk receptor has led to the identification of the src-kinase Fyn, the mitogen-activated protein (MAP) kinases ERK1 and ERK2, the signal transducer and activator of transcription (STAT5) and the phosphatidylinositol-3 (PI3) kinase as its major downstream substrates (Wellbrock *et al.*, 2002).

Effects of Karyotypes, Clustering and Distortion

Theoretically, mapping results can be affected by the karyotype, DNA profile and chromosome characteristics of the species of interest. The mapping results of Liu *et al.* (2003) may be unique because of the karyotype of ictalurid catfish, just as the nature of the tetraploid ancestry of salmonids makes mapping of their genome unique because of phenomena such as pseudolinkage. The haploid chromosome number for channel catfish is 29. The number of linkage groups is expected to be equal to the chromosome number of 29, but initial linkage maps identify 35–40 linkage groups (Waldbieser *et al.*, 2001, Liu *et al.*, 2003). The ancestral ictalurid karyotype is proposed to be 2N = 58, which is the 2N for both channel and blue catfish, with a relatively high arm number (FN) (LeGrande, 1981), and which is probably the ancestral karyotype for all siluriforms. FN for the ancestral ictalurid was probably greater than 80, and that for both blue and channel catfish is estimated to be 90–92 (LeGrande, 1981; LeGrande *et al.*, 1984). This high arm number is characteristic of ictalurids and siluriforms in general. It may not be surprising that six to 11 extra linkage groups were indicated by the microsatellite and AFLP maps, respectively, because of the large arm number, 92, of channel catfish. Channel catfish have 17 pairs of metacentric/submetacentric chromosomes and some of these AFLP and microsatellite linkage groups may represent single arms of metacentric/submetacentric chromosomes.

AFLP primer combinations affected marker distribution and number of markers in channel catfish gene mapping (Liu *et al.*, 2003). Various numbers of markers were produced, depending on the primer combinations, with an average of 9.4 markers produced per primer combination. Several primer combinations produced over 20 AFLP markers; however, six primer combinations produced two or fewer AFLP markers. Additionally, the AFLP markers had an uneven distribution on the catfish gene map. A highly clustered distribution was observed for 133 AFLP markers, and these markers tended to be distributed at the end of several major linkage groups, perhaps centromeric, telomeric or both. Their distribution at the end of these linkage groups does not confirm a telomeric or centromeric location, as there were 11 more linkage groups than chromosome pairs, eventually requiring combining of some linkage groups. Therefore, many of the clustered AFLP markers could be at a position close

to centromeres, or alternatively near telomeres. Additionally, several very small linkage groups were obtained, and some of these might belong to the 12 pairs of chromosomes that are subtelocentric in channel and blue catfish.

Clustering may be a characteristic of AFLP markers, as this phenomenon has been observed in other gene maps (Liu *et al.*, 2003). Highly clustered AFLP markers were found in potato (Van Eck *et al.*, 1995), barley (Becker *et al.*, 1995; Powell *et al.*, 1997) and soybean (Keim *et al.*, 1997b) and clustered near centromere regions (Alonso-Blanco *et al.*, 1998) in *Arabidopsis thaliana*. In the case of fish, AFLPs appeared to be primarily clustered near centromeres in rainbow trout (Young *et al.*, 1998), as stated earlier in this chapter.

Different markers appear to have different distributions in fish, as VNTRs were frequently more telomeric and salmonid-specific small interspersed nuclear elements were intermediate in distribution compared with each other and AFLPs in the rainbow trout (Young *et al.*, 1998). In the case of zebra fish, the distribution of the ESTs was very uniform, ranging from 73 to 201 ESTs per linkage group, indicating that there are no gene-rich or gene-poor chromosomes in zebra fish (Hukriede *et al.*, 1999, 2001), and zebra fish chromosomes can be identified by AT-rich repetitive sequences at the centromere and GC-rich sequences adjacent to the centromeres (Phillips *et al.*, 2000). This may be related to the fact that certain AFLP primers tend to amplify regions rich in specific nucleotides in catfish (Liu *et al.*, 1999d), and this might influence clustering and distribution patterns. This may be true for other aquatic organisms as well, and marker type definitely has an influence on the type of coverage and marker-distribution pattern in the fish gene map. Preliminary results hint at consistent results among fish when using the same class of markers.

Several explanations have been put forward for the AFLP clustering in gene maps, such as the possibility that a small proportion of the clustered markers results from allelism between some AFLP bands, since AFLP are allelic markers; a reduced recombination rate around centromere regions and/or telomere regions; an actual enrichment of AFLP markers in these regions due to uneven distribution of restriction sites; the presence of highly repetitive elements within these genomic regions with great variation in both lengths and sequences among the repetitive elements; or a combination of these explanations (Alonso-Blanco *et al.*, 1998; Liu *et al.*, 2003). The pericentromeric regions contain mainly repeated sequences of unknown functions (Maluszynska and Heslop-Harrison, 1991; Fransz *et al.*, 1998) in *Arabidopsis*. High levels of marker clustering hinder the effectiveness of AFLP markers and therefore warrant study of the nature of the genomic sequences surrounding the regions of clustering markers.

Segregation distortion is common in fish gene maps, and was observed for 101 of 607 markers in ictalurid catfish (Liu *et al.*, 2003). The distorted markers appeared to be correlated with certain primer combinations. When primers *Mse*I-CAC and *Mse*I-CTG were used, large numbers of markers segregated in non-Mendelian ratios.

The distortion rate of 16% for catfish AFLP markers was lower than that found for other organisms when AFLPs were utilized to construct gene maps – 65% for clubroot (Voorrips *et al.*, 1997) and 54% for silkworm (Tan *et al.*, 2001). Several reasons may account for the observed marker distortion in channel catfish and other aquatic organisms. Channel catfish have highly abundant *Tc1*-like transposable elements (Liu *et al.*, 1999d), and fragments amplified from such elements or other types of repetitive element may not follow Mendelian segregation ratios. Additionally, competition among gametes for preferential fertilization (Lyttle, 1991), sampling in finite mapping populations or amplification of a single-sized fragment derived from several genomic regions (Faris *et al.*, 1998) may cause marker distortion. Segregation distortion of microsatellites in Atlantic salmon appeared to be related to preselection of eggs or embryos (Lie *et al.*, 1994).

These distortions in fish gene mapping are expected, as linkage disequilibrium is caused by mixing of highly differentiated genotypes – species certainly qualify – and selection for heterozygous genotypes (May and Krueger, 1990), which are purposely generated in mapping resource families. Perhaps linkage disequilibrium could also be caused by selection for homozygous genotypes. If selection for heterozygotes or homozygotes has occurred in these fish gene-mapping studies, the distorted markers that were unmapped because of distortion could, in fact, be important and of significance for QTL mapping, selective genotyping and MAS.

A great number of recent, higher-density linkage maps have been generated for a wide variety of aquatic organisms including salmonids, tilapia, ictalurid, clarrid, flounder, European sea bass, ayu, carp, yellowtail, shrimp, oyster, scallop, abalone, sea urchin and others (Danzmann and Gharbi, 2007).

Quantitative Trait Locus Mapping

A great acceleration in genomics and gene-mapping research for aquatic organisms has occurred recently. A large number of fish genes and regulatory sequences have been identified and isolated, the structure of the fish genome is much better understood and extensive gene maps for fish, oysters and shrimp have recently been generated. QTL mapping and MAS are becoming a reality. QTL markers for growth, feed-conversion efficiency, tolerance of bacterial disease, spawning time, embryonic developmental rates and cold tolerance have been identified in species such as channel catfish, rainbow trout and tilapias. A large number of molecular markers are needed to map QTLs and economic trait loci (ETLs) for MAS programmes and for cloning genes from various organisms for gene transfer and genetic engineering.

All types of biochemical and molecular marker are potentially useful for QTL mapping. Genes coding for isozymes and other actual genes have an inherent advantage because, of course, they are actual genes and thus have a good probability of being correlated with quantitative and qualitative traits of economic importance. Again, their disadvantage is that they are not numerous. ESTs are a solution to this problem, but much follow-up work is needed to identify polymorphism in the ESTs. The EST approach is productive for generating type I (expressed sequence) markers for gene mapping. Although many other types of molecular marker are currently available for use in genome mapping, ESTs represent actual genes. Because DNA sequences are more conserved in genes than in non-expressed sequences, comparative anchorage maps can be constructed using ESTs. Type I markers – known genes – which facilitate mapping of QTLs of aquaculture performance traits important to MAS (Poompuang and Hallerman, 1997) have been identified and are numerous (Liu *et al.*, 1999f, 2001; Kim *et al.*, 2000; Liu and Feng, 2001; Karsi *et al.*, 2002a).

In contrast, neutral DNA markers have the dual advantage for QTL analysis of being abundant and highly polymorphic. Microsatellites are particularly good candidates for QTL mapping and MAS as they are sometimes found within genes, they are highly polymorphic and they are inherited co-dominantly. Microsatellites have been identified within ESTs (Karsi, 2001; Kocabas *et al.*, 2002a), and these microsatellites can be potentially useful for genomic mapping if they are polymorphic. Microsatellites were found in 4.6% of skin ESTs in channel catfish (Karsi *et al.*, 2002a). Polymorphic microsatellites were found within dopamine receptor, profilin ribosomal protein S16, S100-calcium-binding protein A14, urokinase receptor and protein tyrosine phosphatase IF1 genes of channel catfish. High levels of polymorphism in the channel catfish myostatin gene should facilitate genomic mapping of this gene. Variation was found in microsatellites and SNPs in myostatin (Kocabas *et al.*, 2002a).

Theoretically, there are certain principles that can affect the reliability and applicability of QTL analysis. Care must be taken when searching for QTLs from mixtures of populations with different gene frequencies, as Hardy–Weinberg disequilibrium is not a powerful tool for detecting admixtures and, if populations are mixed, many false-positive associations can occur (Deng *et al.*, 2001). Additionally, correlations between a molecular marker and disease susceptibility or resistance in a single generation of evaluation of an aquaculture species may be false because aquaculture brood stocks usually consist of relatively few groups of siblings and/or admixed populations that have different marker gene frequencies, according to Deng (2002). He suggests that possible solutions to this QTL problem include controlled breeding and progeny testing or doubled-haploid analyses (Doyle, 2003).

Novel computerized QTL mapping may be on the horizon. Grupe *et al.* (2001) have developed an accelerated QTL mapping procedure – *in silico* mapping – which can be utilized if homozygous lines, phenotypic data on the lines and a marker database are available. A computational procedure predicts the chromosomal regions that regulate the phenotypic traits, using a database of SNPs. A linkage-prediction program scans a murine SNP database and, on the basis of known inbred strain phenotypes and genotypes, predicts the chromosomal regions most likely to contribute to complex traits (Doyle, 2003). This computational prediction method does not require the generation and analysis of experimental intercross progeny, but still correctly predicted the chromosomal regions identified by analysis of experimental intercross populations for multiple traits analysed. Of course, the experimental crosses had to be produced to verify that the program worked. A total of 19 of 26 experimentally verified QTL loci affecting ten traits were correctly identified. The computational algorithm is available free at www.mouseSNP.roche.com (Doyle, 2003).

Quantitative trait locus markers in agriculture

Many QTL markers have now been identified. Disease resistance-linked markers have been identified in plants and animals

(Michelmore *et al.*, 1991; Ballvora *et al.*, 1995; Meksem *et al.*, 1995; Knight *et al.*, 1999). Percentages of the phenotypic and genetic variation attributable to QTLs have been estimated. QTL alleles accounted for 27% of phenotypic variation and 46% of genotypic variation for crop yield and quality in barley, with a heritability ranging from 0.43 to 0.80 (Romagosa *et al.*, 1999; Zhu *et al.*, 1999). Fifty-nine QTLs explained 2.2–15.4% of trait variation in the height of maize plants (Austin and Lee, 1996). Multilocus analysis indicated that the cumulative action of all significant QTLs accounted for 43.8% of the total phenotype variation for cold tolerance of tomatoes (Foolad *et al.*, 1998). In maize, three loci, on chromosomes 1, 7 and 10, explained most of the variation of seedling tolerance for herbicide (Sari-Gorla *et al.*, 1997). QTLs accounted for 24–61% of the variation of adaptation differences between highland and lowland tropical maize (Jiang *et al.*, 1999). RFLP markers discriminated with 100% accuracy between the semi-double and double flower type of carnations (Scovel *et al.*, 1998). Isozymes can also act as QTLs, and there is a link between inherited morphological traits and four enzyme loci in oat (*Avena* spp.). QTLs do not always act in simple additive form. Nine clock genes – potential QTL genes – affect the circadian behaviour of mice, but few or none of the 14 circadian QTLs found were clock genes (Shimomura *et al.*, 2001). Two additional pairs of QTL loci that had strong effects on circadian behaviour together but not separately, indicating variation in behaviour, were determined not by single QTL genes acting additively, but by multiple genes acting interactively. This has implications for the design of MAS programmes.

Quantitative trait loci of aquatic organisms

In aquaculture species, initial efforts have begun for QTL mapping. QTL markers for growth, feed-conversion efficiency, tolerance of bacterial disease, spawning time, embryonic developmental rates and cold tolerance have been identified in channel catfish, rainbow trout and tilapias. Putative linked markers to the traits of feed-conversion efficiency and growth rate have been identified for channel catfish (Dunham, 2004). Chromosomal positions have been determined for some quantitative and performance traits in aquatic organisms.

A great amount of QTL research has been conducted for sex determination and sex linkage in fish because of the great interest in producing monosex populations of certain species. Sex-determining genes and their locations vary in fish. Sex-linked inheritance in fish was first reported by Aida (1921) in medaka. Chromosomal positions have been identified for single genes in *Xiphophorus* livebearers, which determine age and size at sexual maturity (Kallman and Borkoski, 1978). Waldbieser *et al.* (2001) examined 293 polymorphic microsatellite loci in channel catfish, and seven of these loci were closely linked to the sex-determining chromosome region. Sex-linked markers have been found for medaka (Matsuda and Nagahama, 2002).

May *et al.* (1989) backcrossed second-generation sparctics (*Salvelinus fontinalis* × *Salvelinus alpinus*) to *S. fontinalis*, allowing identification of a tight classical linkage of phenotypic sex, *Sex-1*, the primary sex-determining locus in salmonids, with *Ldh-1*, *Aat-5* and *Gpi-3*. The map order was centromere–*Ldh-1*–(*Aat-5* and *Gpi-3*), with the latter two loci being tightly linked. There was no association of phenotypic sex (presumably *Sex-1*) with these same three loci and other loci known to be linked to these loci from maps generated from splakes (*S. fontinalis* × *Salvelinus namaycush*) and cutbows (*Oncorhynchus mykiss* × *Salmo clarki*). The linkage of *Sex-1* with these loci is observed only in *S. alpinus*, indicative that *Sex-1* lies across the centromere from these three loci in *S. alpinus*, and represents a Robertsonian fusion not found in any of the other four species (May *et al.*, 1989). Young *et al.* (1998) conducted gene mapping for rainbow trout by utilizing doubled haploids – androgens. The sex-determining locus was at a distal position on one of the chromosomes.

Oreochromis QTLs for sex and colour have been mapped (Kocher *et al.*, 2002). MAS may be a mechanism to improve the efficiency of monosex male production in Nile tilapia. A microsatellite marker has been found on LG 23 that is associated with sex determination. In some cases a single QTL for sex determination has been found in tilapia (Moen *et al.*, 2004a; Lee *et al.*, 2005), and in other cases upwards to three unlinked loci (Cnaani *et al.*, 2004). QTLs for sex ratio in tilapia have also been located (Palti *et al.*, 2002; Shirak *et al.*, 2002). Lee *et al.* (2003, 2004) found epistasis for QTLs for sex determination for LG 1 of the XY locus and LG 3 of the WZ locus in tilapia. This finding may help to further elucidate the occurrence of sex ratios in tilapia that do not adhere to 1:1.

Strong relationships also exist between DNA markers and reproduction in fish. Fishback *et al.* (2000) examined the maturation of a domestic strain of rainbow trout for which the spawning season had been expanded from 2 weeks to 8 months through selection. The spawning time for the majority of the females, but not the males, could be predicted based on genotypes from 14 microsatellite loci. The distributions of the neutral microsatellites diverged along with the selective divergence of spawning date, and the changes in spawning date were correlated with changes in the frequencies of QTL markers. Sakamoto *et al.* (1999) also found spawning date to be polygenic in rainbow trout with 11 markers. This is consistent with the high heritability and strong selection response in rainbow trout for spawning date.

The correlation between growth-related traits and male age at maturation was also studied in rainbow trout; QTLs were identified for each and there may have been epistasis between growth and precocious male maturation (Martyniuk *et al.*, 2003). O'Malley *et al.* (2003) also found QTLs for growth rate and a reproductive trait, spawning date, and in this case found possible pleiotropic effects between the two traits. Two QTLs influencing time of hatch were found in rainbow trout (*O. mykiss*) (Robison *et al.*, 2001).

QTLs for growth and feed-conversion efficiency have been identified in fish and shellfish. Chromosomal positions have also been identified for single genes in *Xiphophorus* that control growth rate from birth (Kallman and Borkoski, 1978). Hallerman *et al.* (1986) observed that several isozyme loci were correlated with growth in channel catfish. In contrast, variation at 14 allozyme loci, as measured by expected heterozygosity, was not lost after two generations of selection for weight in oysters (English *et al.*, 2001), indicating a lack of correlation between growth and isozymes for this species. This may be a result of the vastly different reproductive strategy of oysters, discussed later. Putative linked markers to the traits of feed-conversion efficiency and growth rate have been identified for channel catfish (R.A. Dunham, Z. Liu, R. Yant and N. Chatakondi, unpublished results; Fig. 14.2). Tanck *et al.* (2001) utilized 11 microsatellites

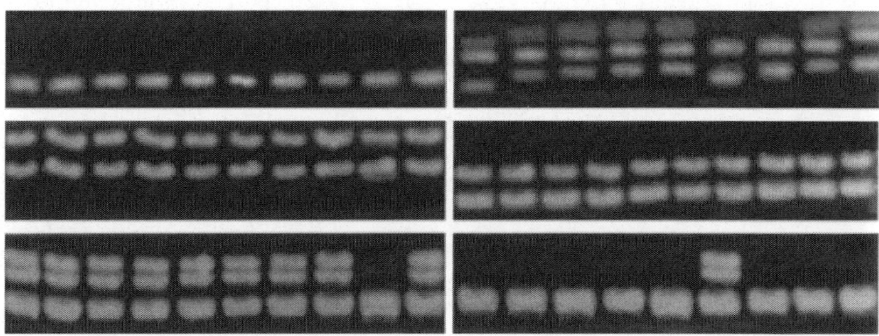

Fig. 14.2. Microsatellite markers linked to QTLs for feed-conversion efficiency in channel catfish, *Ictalurus punctatus*. Best-converting individuals (left) are compared with worst-converting individuals (right).

and found that they were correlated with mass and length in common carp.

Length polymorphisms have been detected in the growth-hormone gene (*GH- 2*) intron of pink salmon and could serve as QTLs, and a null allele at this microsatellite locus was responsible for an apparent deficit in heterozygotes in populations of pink salmon. QTLs were found for body mass and condition factor in rainbow trout (Martyniuk *et al.*, 2003), and for body weight in Asian seabass, *Lates calcarifer*. Reid *et al.* (2005) found QTLs, one major and one minor locus for growth rate, and one major locus and three minor loci for condition factor in rainbow trout. Reid *et al.* (2005) had the same findings in Arctic char, opening the possibility that genetic relationships for the same traits and QTLs may be similar in related species, which could lead to more rapid discovery and application in this technology. Cnaani *et al.* (2004) found three unlinked QTLs for growth rate in tilapia. An SNP for RuvB-like protein in giant tiger shrimp is associated with fast growth rate (Prasertlux *et al.*, 2010).

QTLs have also been identified for disease resistance in fish. In rainbow trout, a DNA marker linked to IHN virus has been identified and associated with resistance to this disease (Palti *et al.*, 1999, 2001), and this trait was polygenic. Two or three loci affect resistance to IHN with one having possible pleiotropic effects on embryo length (Robison *et al.*, 2001; Khoo *et al.*, 2004). Rodriguez *et al.* (2004) indicate that three separate linkage groups are involved. Resistance to *Ceratomyxa shasta* was also polygenic in rainbow trout (Nichols *et al.*, 2003). Ozaki *et al.* (2001) used 51 microsatellite markers to identify several chromosome regions containing putative QTL genes that affect resistance to IPN in rainbow trout. Two putative QTLs affecting disease resistance were detected on chromosomes A (IPN R/S-1) and C (IPN R/S-2), respectively, suggesting that this is a polygenic trait in rainbow trout. MHC polymorphism is linked to immunity to infectious diseases in rainbow trout (Palti *et al.*, 2001). QTLs for IPN virus have been obtained for rainbow trout (Okamoto *et al.*,

2002). A single QTL for killer cell-like activity was identified (Zimmerman *et al.*, 2004).

In Atlantic salmon, a multistage analysis for QTLs was performed for disease resistance traits, and two QTLs affecting resistance to infectious salmonid anaemia were found (Grimholt *et al.*, 2003; Moen *et al.*, 2004b). QTL markers have also been found for resistance to *Aeromonas salmonicida* in Atlantic salmon (Grimholt *et al.*, 2003). In trout and salmon, a candidate DNA marker linked to IHN disease resistance has also been identified, as has a marker for IPN disease resistance in Atlantic salmon. A single QTL found on LG 1 accounts for most of the variation in IPN resistance and a highly resistant line can be developed by selecting this marker.

Putative microsatellite markers are linked to resistance to the bacterium *Edwardsiella ictaluri* in channel catfish (R.A. Dunham, Z. Liu, R. Yant and N. Chatakondi, unpublished results). QTL maps have been developed that have multiple markers for bacterial disease resistance in Japanese flounder. Six loci spread across five linkage groups were found for immune response in tilapia (Cnaani *et al.* 2004), and a QTL has also been found for lysis level (Shirak *et al.*, 2006).

Tanck *et al.* (2001) utilized 11 microsatellites and found that they were correlated with stress-related plasma cortisol levels and basal plasma glucose levels in common carp. Cnaani *et al.* (2004) identified seven loci for stress response in five linkage groups for tilapia.

QTLs for fitness traits and survival have been identified in fish and shellfish. A microsatellite accounted for 7.5% of the variance in thermal tolerance in unselected populations of rainbow trout (Perry *et al.*, 2001). In a tilapia hybrid (*O. mossambicus* × *O. aureus*) several QTLs were found affecting fitness traits using 20 microsatellites and then a second experiment confirmed the findings using six microsatellites in one linkage group (Cnaani *et al.*, 2003).

In rainbow trout, *O. mykiss*, one major and two minor QTLs were found for high-temperature tolerance using backcrosses (Jackson *et al.*, 1998) and F_2 or crossbreeds

(Perry *et al.*, 2001; Cnaani *et al.*, 2003). In both Arctic char (*S. alpinus*) and rainbow trout two QTLs were found for upper-temperature tolerance (Somorjai *et al.*, 2003), matching the virtually identical results that Reid *et al.* (2005) found for growth rate and condition factor between Arctic char and rainbow trout. There appears to be genetic conservation for the number of major and minor loci affecting traits as would be expected.

Jackson *et al.* (1998) found two QTLs for upper thermal tolerance in rainbow trout, which had additive epistatic interaction. Danzmann *et al.* (1999) identified QTLs for heat tolerance in rainbow trout, and found that epistasis was affected by the genomic background of paternal alleles. One potential problem with QTL analysis is that the markers found in one strain may not be applicable in another strain.

Sun and Liang (2004) identified a locus associated with cold tolerance that will be used as a starting point for QTL identification in the common carp (*Cyprinus carpio* L.) at LG 5 as well as three additional markers. However, only one QTL for cold tolerance was discovered in tilapia (Cnaani *et al.*, 2003).

Negative QTLs can also exist. Oysters and elm trees rely on sexual recombination and high fecundities, with millions of highly varied offspring, to respond adaptively to environments that are highly heterogeneous in space and time, and usually exhibit strong heterosis in nature, with homozygous offspring at marker loci expiring quickly as a cohort of offspring ages (Doyle, 2003). This can result in unusual, non-Mendelian inheritance of allozyme and microsatellite marker loci. Heterosis and segregation distortion are due to linkage between the (neutral) markers and deleterious recessive alleles at nearby loci in Pacific oysters (Launey and Hedgecock, 2001). The heterosis in Pacific oysters was due to linkage, not to the intrinsically higher fitness of shellfish that are heterozygous for the markers, and the non-Mendelian inheritance of markers in the oyster was due to the selection against deleterious homozygotes at linked loci. Oysters carry a high load of

deleterious mutations, but the strategy of producing millions of recombinant offspring allows them to thrive (Doyle, 2003). Similar results have been obtained in *Drosophila* (Rice and Chippendale, 2001), which demonstrated that evolution proceeds in sexual organisms because recombination allows favourable genes to overcome deleterious genetic backgrounds. These factors affect the ability to identify QTLs and, in some cases, identification of negative QTLs to select against may be as valuable as identifying positive QTLs to select for.

QTLs have also been identified for morphology. Head length, head depth, head width, body depth, body width, caudal depth, caudal width, total length and body weight were strongly correlated in channel × blue backcross hybrid catfish (Hutson, 2008). Morphometric traits had minimal variation while body weight and total length had large components of variation. Forty-four linkage groups have been studied in these backcrosses using AFLP markers. A total of 12 of 44 linkage groups had at least one significant QTL for at least one of the nine traits. LG 19 was unique as it had multiple QTLs for every trait measured, except for caudal width for which no QTL was indentified on any linkage group. Additionally caudal depth was represented on the map by the fewest linkage groups, two. Approximately half of the markers measured were associated with positive effects on the traits and half had negative effects. LGs 5, 7, 9, 39 and 40 were significant for multiple traits and always had a trait negative effect. Total length was represented on the map by the most linkage groups and the most markers. The linkage relationships found among body weight, total length and the seven morphometric traits indicated that multiple-trait MAS to increase body weight, body depth, body width and caudal depth while decreasing head size might be difficult. Certain QTLs seemed more promising for accomplishing the goal, and focusing on MAS on these markers might yield positive results.

In the case of meristic traits in rainbow trout, a different QTL was associated with each trait, and as expected, based on more traditional studies, significant maternal and environmental effects occurred (Nichols *et al.*, 2004). Pyloric caecae number in rainbow trout has three major QTLs (Zimmerman *et al.*, 2005).

QTLs for flesh colour have been identified and were influenced by additive genetic variation (Araneda *et al.*, 2005). QTLs have also been observed for body and peritoneum colour in tilapia (Lee *et al.*, 2005).

Marker-assisted Selection

Theoretically, MAS has the potential to greatly accelerate genetic improvement in livestock, aquatic organisms and plants. Highly saturated genetic maps have been generated for rice, maize, wheat, tomato, cotton, soybean, cattle, pigs and sheep as well as aquatic organisms, providing the genetic framework and tools for developing MAS programmes. MAS potentially enables improvement in economically important traits and may provide a powerful alternative for improving traits that are difficult to breed for, such as carcass yield, disease resistance, feed-conversion efficiency and sex-limited traits, compared with traditional approaches, such as family and indirect selection. MAS may be particularly useful for low-heritability traits and those complicated by dominance effects. Theoretical calculations predict that MAS would increase the rate of genetic gain by 25–50% over traditional animal-breeding programmes (Weller, 1994). Traditional selective breeding in cattle, pigs and sheep results in a genetic progress per year of approximately 1% for several traits (Korver *et al.*, 1988). MAS may be less dramatic for fish than for livestock in comparison with selection for improving growth, since typical genetic gain for fish growth rates are 6–14% per generation (Dunham *et al.*, 2001), which is equivalent to 2–14% per year, with an average of about 3–4% per year.

To commercialize MAS technology, QTLs are located and their effects on the phenotype measured. Then the markers are evaluated in commercial populations. Lastly, the markers are combined with phenotypic and pedigree information in genetic evaluation for predicting the genetic merit of individuals within the population to allow actual MAS.

Marker-assisted selection in agriculture

Initial experiments with maize, tomatoes, barley, pigs, dairy cattle and aquatic organisms (Kashi *et al.*, 1990; Lande and Thompson, 1990; Meuwissen and Arendonk, 1992; Soller, 1994; Stromberg *et al.*, 1994; Miklas *et al.*, 1996; Spelman and Garrick, 1998) have all given positive results, indicating that the utilization of DNA and protein markers has the potential to speed genetic improvement in comparison with traditional selection. MAS programmes have been utilized successfully in various animal and plant systems, but only a few examples exist for fish. Since not many data exist for fish, the potential usefulness of MAS will be illustrated with examples from plants and livestock as well as fish.

MAS has been used successfully to improve several traits in plants. This approach has been very successful for improving disease resistance in plants, which translates into great improvements in yield. Yield was increased in wheat by about 87% by utilizing MAS to increase resistance to stripe rust fungus, using the *YrH52* resistance gene from wild wheat (Peng *et al.*, 2000). In contrast, traditional selection resulted in an 84% yield increase for wheat, but only after eight successive generations of selection (Wallace and Yan, 1998). A moderate resistance against barley mosaic bymovirus, which is transmitted by a soil-borne fungus and cannot be controlled by chemical methods (Ordon *et al.*, 1999), was obtained via MAS. Increasing pathogen resistance in tomato increased yield by 24% (Rast, 1975), reduced the diameter of lesions caused by bacteria from 18.5 mm in control plants to 2.5 mm in resistant plants (Chen *et al.*, 2000), and improved insect resistance by 47.4% in recombinant inbred lines (Groh *et al.*, 1998). Soybean resistance to cyst nematodes by conducting MAS with the *rhg1* gene was improved by 50% (Cregan *et al.*, 1999) and 54% (Concibido *et al.*, 1996). Traditional plant breeding for insect resistance in plants has had the impediment that resistance has normally been lost during the course of breeding programmes (Murray, 1991), perhaps as the insects adapted to overcome the phenotypic manifestations of the genetic enhancement of the plant. Time will be needed to determine the durability of genetic improvement from MAS.

Yield has also been improved in crops via MAS. The grain and malt quality of a maize hybrid was improved by about 60% using genetic maps that spanned 1413 cM and 1601 cM in the genome (Austin *et al.*, 2000). In spring turnip rape, 89.0% and 86.7% of F_2 individuals were homozygous for the marker alleles, fad high and SCAR-h17b, respectively, associated with high oleic acid, while the wild type was 63.4% for fad low and 65.1% for SCAR-H17c (Tanhuanpaa and Vilkki, 1999); a 2.5-fold variability was measured by traditional selection (Vlahakis and Hazebroek, 2000). MAS improved potato-tuber starch content by 21% and tuber yield by 63% (Schaefer-Pregl *et al.*, 1998). MAS in seedless grape increased dry matter by 78.7% (Lahogue *et al.*, 1998), and the markers accounted for 64.9% of the phenotype variation. MAS improved drought resistance by 43% and increased photosynthesis by 24% in small-grained cereals (Quarrie *et al.*, 1999). Lodging can reduce grain yield of wheat by 40%, but MAS improved waterlogging and the molecular markers accounted for 77% of the phenotypic variation in plant height and stiffness (Keller *et al.*, 1999), although traditional selection for adaptation of barley improved yield by 81% (Wallace and Yan, 1998). MAS increased osmotic adjustment ability up to 70% (Zhang *et al.*, 1999).

MAS has been effective in livestock. It increased beef production efficiency by 29% (Winkelman *et al.*, 1996; Keele *et al.*, 1999). Selecting for the SW790 marker on chromosome 8 increased ovulation rate and reproduction in pigs by 25% in F_2 progeny (Rathje *et al.*, 1997).

Initial data on MAS in livestock indicate that MAS is sometimes more productive than traditional selective breeding, but in some cases is equally effective or less effective than traditional selection. MAS and traditional selection were evaluated for several conformation and reproductive traits in dairy cattle (Rice, 1979; Schrooten *et al.*, 2000). The following genetic improvements were obtained: udder size, MAS genetic gain = 25%, traditional selection = 21%; chest width, MAS = 43%; gestation length, MAS = 57%; stature, MAS = 36%, traditional = 51%; body capacity and size, MAS = 78%, traditional = 27%; dairy character, MAS = 39%, traditional = 19%; angularity, MAS = 29%; fore-udder attachment, MAS = 64%, traditional = 21%; and front-teat placement, MAS = 25%, traditional = 31%. Short-term MAS in chicken increased antibody response by 50 and 62% in chickens with the FF and UU genotypes, respectively (Yonash *et al.*, 2000). Long-term, traditional selection of inbred chicken lines improved disease resistance about twofold (Owen and Axford, 1991).

Marker-assisted selection in fish

Only a few examples of actual MAS exist for fish. The growth rate of rainbow trout was increased by 26% based on selection for an mtDNA marker, but this method was strain specific because the relative performance of fish with the specific haplotype was consistent across males within strains but not across strains (Ferguson and Danzmann, 1999). In contrast, six generations of traditional selection increased body weight by 30% in rainbow trout (Kincaid, 1983b). MAS improved feed-conversion efficiency by 11% for aquaculture species, while traditional selection improved feed-conversion efficiency by 4.3% (Davis and Hetzel, 2000). In rainbow trout, 25% of progeny exhibited a high degree of upper-temperature tolerance after MAS for heat tolerance (Danzmann *et al.*, 1999). MAS has not been broadly applied in fish. However, MAS led to the development of a line of Japanese flounder, *Paralichthys olivaceus*, resistant to lymphocystis disease (Fuji *et al.*, 2007). These fish were widely applied on farms and demonstrated high levels of disease resistance and survival. It appears that MAS has the potential to accelerate genetic improvement of aquaculture species under certain conditions.

MAS programmes have been successfully evaluated in various animal and plant systems. Much theoretical research has been conducted which indicates that MAS has the potential to greatly accelerate genetic improvement in breeding programmes. Initial experiments with corn, tomatoes, barley, swine and dairy cattle all have given positive results, indicating that the utilization of DNA and protein markers has the potential to accelerate genetic improvement in various crops or terrestrial animals.

However, MAS is not always the most efficient or cost-effective method. These initial experiments indicate that when heritability for a trait is high, MAS does not provide any faster rate of genetic gain than traditional selection. However, when heritability is low, the rate of genetic gain obtained from MAS can be substantially higher than that for traditional selection. Theoretically, new schemes based on whole-genome selection may enhance rate of genetic gain even for traits with high heritability.

State of Aquaculture Genomics

Aquaculture genomics has generated an explosion of information during the past 15 years. Framework linkage maps have been generated with large numbers of markers, particularly type I markers of known genes, for a number of aquaculture species. Normalized cDNA libraries for EST analysis and functional analysis have been constructed. Radiation hybrid panels in tilapia and analysis of BAC libraries in catfish have greatly advanced the area of physical mapping of fish genomes. Sequencing of fish genomes is well advanced for some aquaculture species and is nearing completion in some cases. This progress in the past 15 years is quite remarkable.

15

Gene Expression

The central dogma is that DNA is transcribed into RNA which is translated into protein (Fig.15.1). Another original concept was that one gene produces one protein. Both of these concepts are essentially true, but exceptions are now known. For example, the enzyme reverse transcriptase can take an RNA template and produce cDNA. Additionally, a single gene can produce more than one protein in special cases that will be discussed below.

The total number of genes is approximately 25,000 in humans, a number much smaller than originally thought. Many species of fish have a similar gene number. For some aquaculture species such as catfish and salmon, the majority of the genes have been isolated and cloned. Catfish have about 30,000 genes of which currently around 50% are of known function. Bacteria have about 5000 genes. Increased biological complexity is associated with increased gene number although, as usual, there are exceptions to this rule.

The transcriptome is dynamic and complex. Information in the transcriptome can be amplified and diversified. Sometimes a DNA sequence actually encodes more than one gene product through various transcriptional and translational events that are a result of three possible major mechanisms. Each transcript encodes a single protein; however, the primary protein may be further processed to produce more than one active polypeptide via post-translational glycosylation, acetylation, phosphorylation or other post-translational mechanisms. Surprisingly, highly related gene products sometimes can have opposite biological functions and effects, which may actually be an efficient mode of operation.

Multiple gene products can result from alternative splicing. A single gene can be transcribed into heterogeneous nuclear mRNA (hnRNA) through splicing, resulting in the production of more than one mRNA molecule. This can result in a phenomenon where introns of one transcript may be exons of another. A second mechanism for production of alternative transcripts occurs when a single gene possesses more than one promoter, leading to the generation of related but distinct transcripts. Different transcripts can result from different polyadenylation sites. Some genes are non-protein coding and their products are not protein but a functioning RNA. These transcripts are not polyadenylated so they will not be observed in standard cDNA libraries. Sequence divergence is larger for transfer RNAs (tRNAs), ribosomal RNAs (rRNAs), microRNAs and other RNAs, thus genetic similarity of these transcripts is not always conducive to comparative genomics.

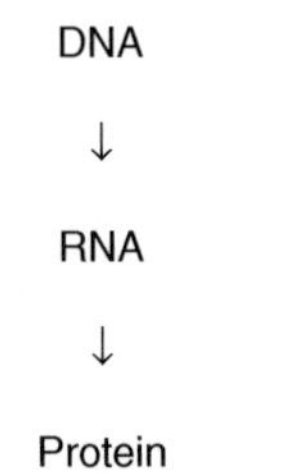

DNA

↓

RNA

↓

Protein

Fig. 15.1. Illustration of the central dogma.

Annotation, linking a gene's sequence to genetic data about function, expression, mutant phenotypes of the protein associated with that locus, as well as to comparative data from homologous proteins in other species, is critical for understanding gene expression/functional genomics data. Sometimes function cannot be determined and these genes of unknown function are known as orphans.

There are some cautionary notes regarding annotation, and great care must be taken for interpreting some functional genomics data. For instance, close matches in BLAST can be misleading because function can sometimes be quite different despite the close match. Annotation by molecular function can sometimes be inadequate as molecular function does not always predict biological function, and molecular function can sometimes remain the same but physiological function has changed. However, the usual case is conservation of gene function.

Many techniques exist to study gene expression, each, of course, with advantages and disadvantages. EST analysis is one option that is not used as much as in the past as it leads to more advanced microarray analysis. Microarray analysis, cDNA or oligonucleotide microarrays, is a powerful tool to study gene expression. A microarray is a collection of gene probes that have been spotted on to a glass slide or synthesized as oligonucleotides on a silicon chip or collection of beads. The microarrays are hybridized to fluorescently labelled cDNA from the treatments being examined, and the relative intensity of signals between samples provides relative quantification of each transcript. Serial analysis of gene expression (SAGE) measures absolutely all transcripts and is a highly accurate reflection of overall gene expression, but is more tedious and expensive. Next-generation sequencing is being adapted for powerful gene expression experimentation, but is not yet perfected. Some experiments require only single gene analysis and Northern blotting or quantitative PCR is adequate for this type of study.

Our ability to isolate and clone beneficial genes for biotechnological applications relies on an understanding of genomic structure and organization. Elucidation of gene function and tissue physiology through the identification of genes that are specifically expressed under a specific environment or physiological condition allows an understanding of molecular mechanisms controlling performance traits. The transcriptomic approach is an efficient technique for the systematic analysis of abundantly expressed genes (Karsi *et al.*, 2002b). The role and function of the majority of transcribed genes are still unknown. Functional genomics has advanced rapidly and the knowledge of gene expression responsible for growth, disease resistance and response to cold temperature and other traits has been greatly expanded for aquaculture species utilizing tools such as EST analysis and microarrays.

Gene Expression and Expressed Sequence Tags

Sequencing technology allows the production of vast numbers of ESTs representing a large percentage of the transcriptome – the overall transcriptional activity – of an organism. Determination of overall transcripts of tissues and organs not only produces large numbers of ESTs, but also generates expression profiles by using non-normalized cDNA libraries. Since gene expression is tissue specific, no cDNA library contains all genes. However, this problem is partially overcome by making multiple cDNA libraries from a series of tissue types.

EST cataloguing and profiling provides the basis for functional genomics research. ESTs – short, single-pass cDNA sequences – are generated from randomly selected library clones (Adams *et al.*, 1991; Karsi, 2001). Characterization of ESTs is a rapid, efficient, cost-effective method for the identification of new genes (Tilghman, 1996). The EST approach is particularly valuable for the characterization of genes with low levels of expression and for systems where biological samples are limited, such as with the pituitary gland, as well as for identifying genes in general. One attribute of EST analysis using non-normalized libraries is its ability to produce expression profiles. The frequency of cDNAs in a cDNA library is a reflection of mRNA abundance in the mRNA pool of the tissue from which the library was derived. For the purpose of EST cataloguing for the development of molecular tools and gene isolation, repeated sequencing of highly expressed genes is wasteful. Normalized cDNA libraries (Patanjali *et al.*, 1991; Sasaki *et al.*, 1994) are extremely important for the characterization of large numbers of unique ESTs, as they equalize the number of clones for each gene, allowing the detection of genes that have low expression levels and reducing wasteful redundant sequencing.

The EST approach is used extensively to analyse genes from various species (Franco *et al.*, 1995; Aliyeva *et al.*, 1996; Ju *et al.*, 2000; Cao *et al.*, 2001; Karsi *et al.*, 2002a,b), and large-scale EST analysis is an efficient method for the identification of genes and the determination of their expression profiles. This technology offers a rapid, informative, initial examination of genes expressed in specific tissue types under specific physiological conditions or during specific developmental stages and in specific biological pathways. ESTs are an efficient tool for genomic mapping (Hudson *et al.*, 1995; Schuler *et al.*, 1996).

ESTs will be useful for comparative genomics and functional genomics using cDNA microarray technology, for the determination of their orthologous counterparts through evolution, for mapping by PCR analysis using radiation hybrid panels and for identification of polymorphic markers in genes of known functions (type I markers), all of which will increase gene cloning (Kocabas *et al.*, 2002a,b). ESTs have greatly advanced the study of gene expression via microarray analysis (Peatman *et al.*, 2008).

Hybridization-based cDNA microarray technology allows simultaneous quantification of gene expression levels for thousands of genes among samples of different genotypes, developmental stages, tissues, physiological states, stresses and challenges, and various environmental conditions (Karsi *et al.*, 2002a,b). Microarray analysis can determine the temporal expression of suites of genes and examine which genes are turned on before, during and at different points of time after the initial gene induction, thus revealing information on gene interrelationships and interactions. A cDNA microarray is an orderly arrangement of cDNA inserts. This technology can monitor the entire transcriptome or a major fraction of the transcriptome on a single chip or membrane, allowing interactions among thousands of genes to be ascertained simultaneously. Old technologies were sometimes limited to only qualitative evaluation, determining which genes were inactive, active or very active, whereas microarray technology allows relative levels of expression among genes to be evaluated much more precisely.

The construction of cDNA microarrays is dependent on information gained from ESTs, allowing the evaluation of large sets of clones and sequences. The importance of EST analysis has its basis in the fact that mRNA quantities vary in different tissues, different organs and different developmental stages or when the organism faces different environmental conditions (Ju *et al.*, 2000). After sequence analysis, EST inserts can be amplified using PCR to produce cDNA inserts. A cDNA microarray is developed by arraying large numbers of unique ESTs on a gene filter or chip.

To conduct a cDNA microarray analysis, cDNA inserts are spotted and fixed on to specially treated nylon membranes or glass slides (Kim, 2000). Next, probes are made from mRNAs isolated from the control sample and the treatment sample, using reverse transcriptase. Usually, the two probes are labelled with different fluorescent dyes, and labelled probes are then hybridized to the arrayed substrates. The relative intensity of each fluorescent spot between the two probes quantifies the relative level of gene expression for each treatment or sample. Differential expression of specific genes can then be more thoroughly examined by conventional techniques, such as Northern blot or quantitative reverse transcriptase PCR (RT-PCR).

These analyses allow the detection of relative expression of groups of genes at the same point in time and under the same environmental conditions, resulting in information on function (Kim, 2000; Karsi, 2001). Data from a series of experiments can be combined to assign function to genes, with genes showing similar expression profiles across differing states and conditions probably functioning in common physiological or metabolic pathways. The identification of genes expressed in cells of a specific tissue is necessary to understand gene function and tissue physiology. Sequenced ESTs can be catalogued according to tissue specificity (Hishiki *et al.*, 2000) or biochemical pathways (Mekhedov *et al.*, 2000) or as a high-fidelity set of non-redundant transcripts (Boguski and Schuler, 1995). These can then be used for more extensive functional annotation and assessment and integrated with linkage and physical maps.

The characterization of a large number of ESTs from various organisms allows the assembly of EST sequences into tentative consensus sequences or gene-indexing databases, such as the UniGene (Boguski and Schuler, 1995), STACK (Burke *et al.*, 1998) and TIGR gene indexes (Quackenbush *et al.*, 2000). Such tentative consensus sequences can be used to assign genes to functional annotation, to link the transcripts to mapping and genomic sequence data, and to provide links between orthologous and paralogous genes (Quackenbush *et al.*, 2000; Karsi, 2001; Kocabas *et al.*, 2002a,b). More than one gene product can result from alternative splicing, as there are many more variant proteins than there are genes and alleles. EST analysis can assist in identifying and understanding the origin of these alternatively spliced variant proteins (Karsi, 2001; Karsi *et al.*, 2002a,b).

Gene expression as it relates to several performance traits will now be discussed. The presented data and the discussion are not comprehensive with regard to gene expression data for fish, which is expanding rapidly. The intention is to present a sampling of gene expression data that is important with respect to the phenotypic variation observed for these traits.

Growth

A considerable amount of research has been oriented towards the isolation of growth hormone (GH) genes, and many have already been applied for transgenesis. Pseudo-GH genes have also been isolated. A male-specific GH pseudogene, *GH-Y*, is found in two of three subspecies of *Oncorhynchus masou* – masu and biwa but not amago (Zhang *et al.*, 2001). About 95% of these males contain this marker and a few females. These abnormal female individuals appear to be the result of sex reversion, which may have an autosomal modifying locus basis.

The utilization of EST analysis has resulted in the identification of several growth- and muscle-related genes. ESTs from Pacific oyster gonads and early embryos typically included tubulin, actin, the mitochondrial genes cytochrome B and NADH dehydrogenase, and the general classifications of proteins, protein kinases, protein phosphatases, cell-cycle proteins and genes for DNA replication (Shimizu *et al.*, 2002). In zebra fish, precursor cells give rise to fast and slow muscles and these precursor cells are regulated by competing influences of two key signalling molecules, hedgehog and transforming growth factor-β (TGF-β) (Westerfield *et al.*, 2000).

Fibre recruitment to growth and the number and behaviour of satellite cells are affected in Atlantic salmon by genetic variation, ploidy level and temperature during early development (Johnston, 2000). Fibre density is important as high density is associated with firmer texture and the absence of gaping during filleting. Postembryonic growth of skeletal muscle is a result of myotomal cross-section fibre number in Atlantic salmon, which is 10,000, 180,000 and 1,000,000 for fry, smolts and fish in seawater for 1 year. Muscle growth in older fish is a result of hypertrophy alone, with fibres reaching a maximum diameter of 240 μm. Myostatin functions as a negative regulator of fibre number. Myogenic regulator factors from the β-helix loop helix family of transcription factors (MyoD, myf-5 and myogenin, myf-6) are related to the commitment and differentiation of muscle cells, respectively. Eighty per cent of the mononuclear cells of fast muscle in Atlantic salmon fry are actively dividing satellite cells that are differentiating and expressing one or more myogenic regulator factors.

The growth and development of fast- and slow-twitch muscles are also related to swimming ability. Swimming speed is associated with temperature acclimatization in goldfish and carp (Kobiyama *et al.*, 2002). Goldfish fast muscle has increased adenosine triphosphase (ATPase) activity in the myofibres after acclimatization to cold temperatures. Myosin increased actin-activated Mg^{2+} ATPase in carp after acclimatization. Twenty-nine isoforms of fast-muscle heavy-chain myosin exist in carp. Two different specific forms are up-regulated in carp when they are cold-acclimatized to 10°C or heat-acclimatized to 30°C.

Myostatin is a gene that inhibits muscle growth by acting as a negative regulator of muscle deposition, as confirmed from knockout experiments in mice which resulted in tremendous increases in skeletal muscle mass. Natural mutations of the myostatin gene have been found in two breeds of cattle, the Belgian Blue and the Piedmontese, which results in a double-muscling phenotype (Kocabas, 2001). Knockout of myostatin could increase muscling, growth, flesh quality, flavour and processing yields of fish. The catfish myostatin was encoded by a single-copy gene, making the knockout

less complicated and more feasible than for many other traits.

The myostatin gene is a member of the TGF-β superfamily (McPherron and Lee, 1996), although myostatin shares only low levels of identity with other well-known members of this superfamily which include the inhibins, the TGF-βs and the bone morphogenetic proteins (McPherron *et al.*, 1997). The deduced amino acid sequences of the channel catfish myostatin were highly conserved compared with a variety of other organisms (Kocabas *et al.*, 2002a) with regard to sequence identity, structure and organization of the gene. Genomic sequences are available only from chicken, bovines and seabream (GenBank accession numbers AF346599, AF320998 and AF258447, respectively), and all had three exons and two introns. Such high levels of conservation through evolution suggest intense selective constraints and the importance of the function of myostatin.

Myostatin is expressed by cells from a variety of tissues in mature sheep and fish (Sharma *et al.*, 1999; Kocabas, 2001; Roberts and Goetz, 2001; Rodgers *et al.*, 2001). The myostatin gene was expressed in various tissues and developmental stages at differential levels in channel catfish, suggesting complex regulation of this gene and perhaps additional roles for myostatin in addition to growth inhibition (Kocabas *et al.*, 2002a). Expression was strongest in the muscle, intermediate in brain, eye, intestine and trunk kidneys, but low in gill, spleen, heart and liver; the lowest expression was observed in the stomach, head kidney and swim bladder. Expression in the stomach was miniscule but detectable with intense PCR.

As expected, expression was variable during development and at various ages in channel catfish (Kocabas *et al.*, 2002a). Expression of a variety of muscle genes varies with the age of fish (Fig. 15.2). Myostatin was not detectable in unfertilized eggs; however, expression was detectable in 1-day-old embryos. Myostatin expression increased gradually as the embryos developed and hatched, reaching high levels of expression about 2 weeks after fertilization. Expression

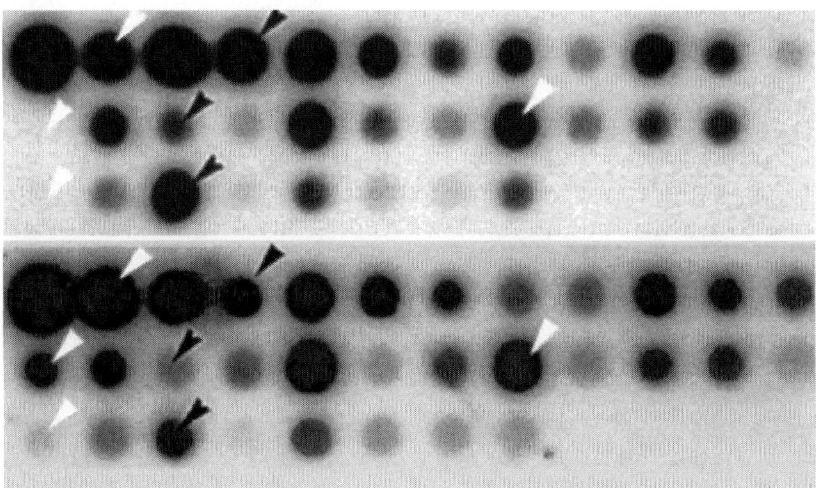

Fig. 15.2.   Dot-blot analysis of gene expression in muscle tissues at 9 months (young fish, upper) and 4 years (mature fish, lower) in channel catfish, *Ictalurus punctatus*. White arrows indicate genes that are up-regulated and black arrows indicate genes that are down-regulated in mature fish.

of myostatin was high before channel catfish reach 3 years of age, but levels of expression dropped three- to fivefold at the ages of 4–7 years. This was a surprising result, but the older fish may be losing the ability for rapid muscle growth, reducing the need for negative regulation by myostatin.

Biochemical analysis might also indicate what genes are important for examination of the growth process. The biochemical components in serum of four strains of *Oreochromis niloticus* were determined quantitatively, including serum potassium, sodium, chlorine, calcium, cholesterol, urea nitrogen, total protein, albumin, globulin, LDH, glutamic oxaloacetic transaminase (GOT), alkaline phosphatase (ALP) and μ-amylase (J.S. Li, M. Dey and R.A. Dunham, unpublished results). There were significant differences for strain and sex in most biochemical components.

The relationship between the biochemical components in serum and growth, as well as feeding habits, was determined. The contents of serum potassium, sodium, chlorine and calcium maintain the balance of electrolytes and osmotic pressure in the internal milieu. The appropriate ratio of sodium to the other electrolytes (potassium, calcium) is necessary to maintain the

normal neuromuscular response of the body. There was a difference between females and males in serum components, except for the activity of serum μ-amylase, but there was a difference among four strains of Nile tilapia in the activity of serum μ-amylase. The endoenzyme of starch is μ-amylase; it acts at the end of the starch chains and at the glucosidic bond in the inner part of the starch chains. The activity of serum μ-amylase is related to feeding habits (Lou *et al.*, 1995). Since sex did not affect the activity of serum μ-amylase, but strain differences did, the difference in feeding habit could be larger between strains than between the sexes.

Both sex and strain affected the content of serum cholesterol, and the content of serum cholesterol of females was higher than that of males. Differences also exist between females and males for serum total protein, albumin and globulin, related to the metabolic difference between females and males in synthesizing serum proteins in the liver. Strain also affected serum total protein and globulin. Serum globulin is related to the immune system.

GOT is one of two important transaminases that can catalyse the amino-transportation between glutamic acid and oxaloacetic acid. Serum GOT is probably related to the growth of fish, with higher activity of serum GOT promoting growth (Lou *et al.*, 1995). The serum GOT activity of the male is higher than that of the female *O. niloticus*, so it is one of the physiological factors that causes sexually dimorphic growth. Serum GOT was also highest in the fastest-growing strains of Nile tilapia.

Sex and strain also affected serum urea nitrogen. The fastest-growing strain of Nile tilapia had the highest serum urea, so growth may be related to amino acid metabolism.

LDH is a highly studied glycolytic enzyme, delivering the hydrogen of coenzyme I; it can turn lactic acid into pyruvic acid by dehydrogenation. ALP is a nonspecific enzyme that catalyses the hydrolysis of organic monophosphate esters (Qi, 1988). Again, differences in the activities of these two serum enzymes were detected between both sex and strain.

Ovulation and reproduction

When channel catfish were induced to ovulate with carp pituitary extract, expression of the gonadotrophin (GtH) β subunit was elevated by 149% and prolactin by 176%

(Karsi *et al.*, 1998). Levels of several peptide hormones were increased dramatically as a result of induced ovulation (Fig. 15.3). GtH β-I and β-II, GH and pro-opiomelanocortin (POMC) increased by 486, 393 and 345%, respectively. POMC accounted for about 20% of all transcriptional activity in the pituitaries after induced ovulation. GtH, GH and POMC may be important for final oocyte maturation and/or ovulation in channel catfish and other species of fish and shellfish. One application of this information could be the design of improved spawning reagents for spawning induction utilizing these hormones.

Induction of ovulation in catfish using currently available reagents, such as carp pituitary extract, is extremely inefficient, and this might be corrected by the development of alternative ovulation-inducing reagents using the homologous catfish GtHs. GH and POMC may play important roles in regulating final oocyte maturation or ovulation and might be used along with the GtHs to further enhance ovulation, synchronize ovulation and increase the quality and hatching potential of artificially handstripped gametes.

Almost two-thirds of the known gene products found in the pituitary cDNA library of channel catfish generated after

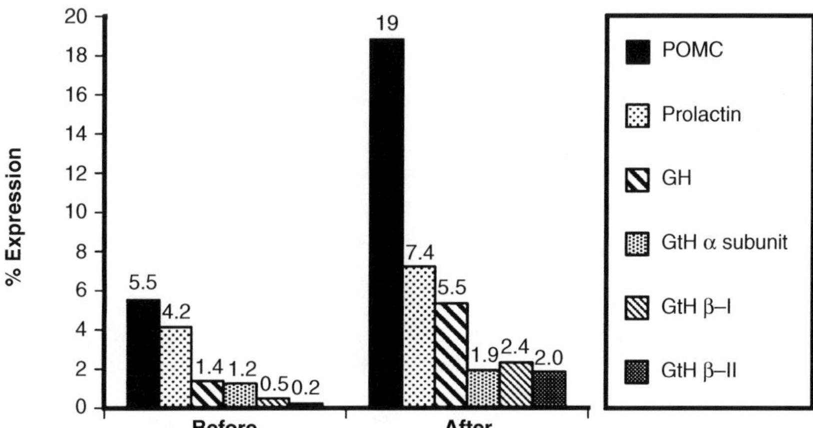

Fig. 15.3. Changes in hormone profiles after induced ovulation in channel catfish, *Ictalurus punctatus*. POMC, pro-opiomelanocortin; GH, growth hormone; GtH, gonadotrophin. (Adapted from Karsi *et al.*, 1998.)

induced ovulation were hormonal or cellular regulator-gene products (Karsi *et al.*, 1998). Hormones and other regulators are the major transcriptional products in the catfish pituitary just prior to spawning. Thirty known genes were identified and can be grouped into three major groups: hormones and other cell-cycle regulators, translational proteins and enzymes. Hormonal clones – POMC, GH, prolactin, GtHs and somatolactin – accounted for 54.2% of clones. Cellular regulators – RAP1A, cyclic adenosine monophosphate (cAMP)-responsive element modulator, cyclin 2b and guanosine triphosphate (GTP)-binding protein – represented another 8% of the clones.

The second major group of genes expressed in the catfish pituitary after hormone induction (18.6%) included those encoding translational-machinery proteins, such as large and small ribosomal proteins (Karsi *et al.*, 1998). This is an expected result for all tissues since the ribosomal proteins are, of course, necessary for all translational activity. Among these were seven clones of large ribosomal proteins representing six proteins, L7a, L10, L18, L30, L35 and L41 (two clones), three small ribosomal protein clones, ribosomal protein S11, S24 and S29, and ribosomal phosphoprotein P1.

The third group of genes represented by the ESTs included those for several enzymes: alcohol dehydrogenase, cytochrome c oxidase, vascular ATPase and acylphosphatase (Karsi *et al.*, 1998). Additionally, β-globin cDNA, MHC B protein and ictacalcin were detected.

The major transcriptional activities in the pituitary are related to hormone and cell regulation (Karsi *et al.*, 1998). The other gene products probably support the hormonal and regulatory functions of the pituitary. Since the pituitary cDNA library was generated from females induced to ovulate during the spawning season, the pituitary may be more polarized at this time than at other times of the year.

The exact mechanism of induction of ovulation is not completely understood. GtHs from the injected carp pituitary extract may directly elicit ovulation at the ovarian

follicle, since pituitary extract can induce oocyte maturation and ovulation *in vitro* (Kime and Dolben, 1985; Kime *et al.*, 1989), although the indirect effect from GnRHs may be able to increase endogenous gene expression of certain hormones such as GtH, resulting in ovulation (Liu *et al.*, 1997).

GnRH is the key hormone in fish reproduction. Information on GnRH genes is expanding rapidly. Medaka, eel (*Anguilla japonica*) and arowana (*Scleropages jardini*) all produce three forms of GnRH (Okubo *et al.*, 2002). In the medaka the three forms are expressed in the preoptic area, midbrain tegmentum and nucleus olfactoretinalis, which hypothetically would stimulate GtH secretion from the preoptic region and act as neurotransmitters in the other two regions. The medaka has at least two GnRH receptor genes.

GtH-II is believed to have a role in final oocyte maturation and has been termed maturational GtH (Liu *et al.*, 1997, 2001; Karsi *et al.*, 1998). The description of the role of the maturational GtH in fish reproductive physiology was derived from studies using immunoassays to measure pituitary and plasma levels of GtH during sexual maturation and the subsequent reproductive cycles (Hassin *et al.*, 1995). Hassin *et al.* (1995) indicate that many of the data obtained prior to 1986 are difficult to interpret since the cross-reactivity of the antisera used in these assays with GtH-I was not known. Additionally, most research of this kind was done with salmonids (Swanson, 1991), which respond strongly to photoperiod for their reproductive cycles, whereas temperature is a more important natural regulator of reproductive cycles of catfish and many other species, further complicating comparisons among species.

GtH β-I and GtH β-II were differentially regulated in channel catfish (Karsi *et al.*, 1998), and the increased gene expression from GtH β after induced ovulation suggests an important role for GtH β in the final oocyte maturation and ovulation of catfish, similar to what was found for final oocyte maturation and ovulation of salmonids and other fishes (Swanson, 1991; Prat *et al.*,

1996; Zohar, 1996). Perhaps the different reproductive active physiology and responses to photoperiod and temperature of salmonids and ictalurids do not complicate the interpretation of the GtH data.

GH expression was also up-regulated during induced ovulation of channel catfish (Karsi *et al.*, 1998). Although GH is a main regulator of growth, it has been used together with GtHs for the induction of ovulation in mammals (Artini *et al.*, 1996) and may act with GtH to augment the actions of follicle-stimulating hormone (FSH) and luteinizing hormone (LH) on oestradiol and progesterone production (Sharara and Giudice, 1997; Liu *et al.*, 2001). GH accelerates bovine oocyte maturation *in vitro* (Izadyar *et al.*, 1996). The great increase in GH expression during induced ovulation in channel catfish indicates that GH may have similar reproductive functions in mammals, chickens and fish, and may play important roles in the final oocyte maturation and ovulation in fish. Although similarities in GH up-regulation were observed among channel catfish, mammals and chickens in reproductive cycles, prolactin was enhanced only slightly in channel catfish in contrast to the warm-blooded animals.

POMC was the most abundant gene product in the pituitaries of ovulating channel catfish, indicating potentially significant roles of opioid peptides in the final oocyte maturation or for actual spawning (Karsi *et al.*, 1998). Alternatively, this could be a response to stress from the injections and handling of the fish. POMC is the common precursor of adrenocorticotrophin (ACTH), ACTH-related peptides, endorphin (Eipper and Mains, 1980) and melanotrophins (melanocyte-stimulating hormone (MSH)) (Nakanishi *et al.*, 1979). POMC is synthesized in the pars distalis (PD) in the pituitary as a precursor protein and is then cleaved to produce the biologically active mature peptide. Transcriptional activation and/or repression of POMC have been reported in several studies in late gestation (McMillen *et al.*, 1988; Yang *et al.*, 1991; Myers *et al.*, 1993; Matthews and Challis, 1995) and during active labour (Matthews *et al.*, 1994) in mammals. The opioid peptides inhibit GnRH-mediated LH release and interfere with the GtH stimulatory effect on gonadal sex-steroid production (Fabbri *et al.*, 1989); however, there are exceptions (Mahmound *et al.*, 1989). The opioid peptides do not have an unconditional inhibitory effect on GtH release (Kandeel and Swerdloff, 1997), as they are ineffective at the time of the pre-ovulatory GtH surge (Van Vugt *et al.*, 1983; Piva *et al.*, 1985) or they would inhibit ovulation. Opioids have a role in regulating the activity of the hypothalamus–pituitary–gonadal axis in the frog (*Rana esculenta*) (Facchinetti *et al.*, 1992, 1993) and in goldfish (Rosenblum and Peter, 1989). Their role in fish reproduction needs elucidation, but the current information suggests a role in the ovulatory process.

POMC also has a role in stress response in mammals (Hollt *et al.*, 1986; Wu and Childs, 1991; Larsen and Mau, 1994), but the elevation in POMC expression in mammals under stress is less than 100% compared with the three- to fourfold increases associated with reproduction. Although stress response is a possible explanation for the spike of POMC during catfish ovulation, POMC is probably up-regulated due to the dramatic increase in GtH-β expression during induced ovulation in channel catfish, as POMC increases due to reproduction were much greater than those induced by stress in mammals.

Diseases

Diseases cause tremendous economic losses in the aquaculture industry. Disease resistance may actually be the most important trait for genetic enhancement. The skin is the first layer of defence, and is involved in both physical and immune defences against pathogenic bacteria, viruses and parasites. Host responses against skin-inhabiting parasites and infectious diseases are often observed, but the mechanisms and genetics of defence reactions against diseases in fish skin are not completely understood (Karsi *et al.*, 2002a). Identification of key gene expression during response to attack by pathogens may allow identification of important genes for possible transgenic manipulation.

The fish spleen is comparable to that of mammals, and protects the fish from blood-borne antigens (Ellis, 1992; Zapata *et al.*, 1996). The spleen contains immune cells, such as T and B cells, that proliferate and secrete specific cell-surface molecules, enzymes and specific antibodies to either destroy the antigen or neutralize it to prevent further disease manifestations (Kaattari and Piganelli, 1996). Identification of functional genes in immune cells during disease processes may facilitate the understanding of mechanisms involved in disease resistance and disease defences, allowing future genetic improvement in aquatic organisms for disease resistance. Thus, the identification of genes in the spleen and skin and the characterization of their expression profiles are among the objectives of transcriptome analysis in aquatic organisms (Karsi *et al.*, 2002a; Kocabas *et al.*, 2002b).

Gene regulation in the skin of the channel catfish is highly polarized; the most abundantly expressed genes in channel catfish skin were those for calcium-binding proteins, indicated by high expression levels of ictacalcin – 4.8% of all cDNA clones – and S100-like calcium-binding protein (Karsi *et al.*, 2002a). This high level of expression surpassed that of the most abundantly expressed gene in the head kidney of channel catfish, β-actin (~2%). Other highly expressed genes were ribosomal proteins, cytoskeleton genes, kinases and phosphatases, receptors, mitochondrial genes, enzymes, immune-related genes, other proteases, translational factors, transcriptional factors and known ESTs with unknown function. Ictacalcin (4.8%), ribosomal protein S02 (2.0%), ribosomal protein L11 (1.6%), ribosomal protein L41 (1.6%), creatine kinase (1.2%), keratin type I (1.2%), ribosomal protein S20 (1.1%), ribosomal protein S09 (1.0%), dopamine receptor (0.9%) and ribosomal protein L35 (0.8%) were responsible for 16.2% of all gene expression in the skin. This expression profile in the channel catfish skin is more polarized than in its brain or head kidney

(Ju *et al.*, 2000; Cao *et al.*, 2001), partially because of the high levels of expression of translational ribosomal protein genes in the skin (Karsi *et al.*, 2002a; Patterson *et al.*, 2003).

The majority of EST clones in channel catfish could be identified by similarity comparisons with other organisms, indicating that intense EST analysis is an efficient way for gene annotation in less well-studied species (Karsi *et al.*, 2002a,b; Patterson *et al.*, 2003). Similar to the data generated from cDNA libraries from the brain and head kidney, 2.8% of skin clones were similar to known sequences of unknown function from model systems, such as *Homo sapiens*, *Mus musculus*, *Caenorhabditis elegans*, *Bos taurus*, *Macaca fascicularis* and *Arabidopsis thaliana*. Once a gene is characterized and its function known in any one of these species, comparative functional genomics will allow the explanation and understanding of these orthologous genes.

The percentage of known genes in the channel catfish spleen was comparable to that of the catfish brain, head kidney, skin and liver (Ju *et al.*, 2000; Cao *et al.*, 2001; Karsi *et al.*, 2002a; Kocabas, 2002b). The percentage of ESTs that can be characterized and identified is also a function of the EST length, which is important when attempting to compare data. Long ESTs probably produce sequences within the coding regions, while short ESTs often produce sequences in the UTRs. As sequence conservation is not prominent in UTRs, short sequences often lead to the production of unknown ESTs.

The most abundantly expressed genes in the channel catfish spleen were involved in the translational machinery, including 33 ribosomal proteins (Kocabas *et al.*, 2002b). The genes for haemoglobin β chain (2.6%) and α-globin (1.2%) were highly expressed in the spleen. In addition to the many structural genes, other genes highly expressed in the spleen were for the β_2-microglobulin precursor, the high-affinity immunoglobulin E (IgE) receptor, the immunoglobulin heavy chain, interferon-induced protein 1–8D, MHC class I α chain and C1q-related factor. Many of these abundantly expressed genes in the spleen were related to immunological functions, as expected for a disease defence organ such as the spleen.

Ribosomal genes also dominate expression in tissues of other species of fish. In the case of Japanese flounder, *Paralychthys olivaceus*, 45% of EST clones were known genes from the liver and leucocytes (Aoki *et al.*, 2002). Ribosomal protein L23 and gelatinase (1.34%), and apolipoprotein A-I (4.29%), were the most commonly expressed genes in the liver and leucocytes, respectively. As expected, concanavalin A (ConA)-treated and hirame rhabdovirus-infected leucocytes had differential expression, as these treatments should elicit an immune response and its associated gene expression. Hirono and Aoki (2002) further examined the immune-related genes in Japanese flounder. The cytokines interleukin 1β, tumour necrosis factor (TNF) and chemokine were isolated as well as the cytokine receptors TNF 1 and 2, interleukin 1R and 6R; chemokine receptors; toll-like receptors; cell-surface molecules such as T-cell receptors, IgM, IgD, CD3 and CD8; transcription factors such as IRF, CEBP and Stat3; complement components C3, 8 and 9; the antimicrobial proteins lysozyme, transferrin and Mx; and proteases and protease inhibitors. The immune and defence genes isolated were similar to those found for humans and mammals.

Additionally, *VHSV-G* gene elicited strong humoral and cellular immune responses probably protecting Japanese flounder during infection from VHS (Byon *et al.*, 2005). Non-specific immune up-regulation was observed from genes such as NK, Kupffer cell receptor, MIP1-α and Mx1 protein gene, and specific immune-related genes such as the CD20 receptor, CD8 α chain, CD40 and B lymphocyte cell-adhesion molecule.

Fish also possess antibacterial peptide genes, which are quite effective in swine and cecropus moths. The antimicrobial peptide-2 expressed in channel catfish liver was down-regulated in response to ESC infection (Peatman *et al.*, 2007). It seems odd that an antimicrobial peptide would be down-regulated during bacterial infection, but this might provide a clue to the effectiveness and lethality of ESC, which may have evolved to counteract native antibacterials. This peptide should be in increased in expression if it is to battle invading bacteria.

Japanese flounder have two lysozyme (muramidase) genes – c and g type (Hikima *et al.*, 2002) – and both are single-copy genes. Their gene products were able to lyse *Vibrio anguillarum* and *Pasteurella piscida*, but not the two major pathogens of Japanese flounder, *Edwardsiella tarda* and *Beta streptococcus*. *E. tarda* induced lysozyme production even though the lysozyme was not effective in lysing this bacterium. It seems logical that the most devastating pathogens for a species would be ones that have evolved to circumvent the natural immune and defence mechanisms. In this case the invader induces the counterattack, but has developed its own defence mechanism. Likewise, long-term selection for disease resistance may never achieve total resistance, as the pathogen counters with its own selection response and the variety of strains within a single pathogen species may allow circumvention of host defences and result in varying heritabilities dependent on the serotype used for the challenges. If this is true, gene transfer of constructs from distantly related species or different taxa may be more effective in genetically improving disease resistance than transgenically manipulating genes intraspecifically. Here we have two examples where the native antibacterial peptides had little effect on disease resistance, but transgenes from completely different taxa – insects, birds, mammals – and synthetics were very effective in imparting disease resistance.

The liver is an important organ for gene expression related to the immune response in fish. The acute-phase response (APR) is one important initial defence against disease caused by pathogens. The liver is a central site for the APR component of innate immunity (Peatman *et al.*, 2008) and production of acute-phase proteins (APPs). APPs have positive roles in mediating the complex inflammatory response and restoring homeostasis after infection or injury. A large number of cytokines, complement components, pathogen-recognition receptors (PRRs) and antimicrobial peptides from several aquaculture species have been found and studied (Dunham, 2009).

In the case of channel catfish, 35 of the 127 genes up-regulated in response to exposure to the bacterium *Edwardsiella ictaluri*, causative agent of ESC, were APPs (Peatman *et al.*, 2007), including coagulation factors, proteinase inhibitors, transport proteins and complement components. The highest up-regulated genes (>50-fold) were involved in iron homeostasis (intelectin, haemopexin, haptoglobin, ferritin and transferrin). Up-regulation also occurred for the majority of the complement cascade including the membrane attack complex components and complement inhibitors. PRRs and chemokines were also differentially expressed in the liver following infection; chemokines appear to have a key role in ESC resistance.

Similarly, in the closely related species, blue catfish (Peatman *et al.*, 2008), at least 20 of the 98 up-regulated genes represented APPs, including the APR complement activation and metal ion binding/transport categories, and were among the most highly up-regulated transcripts following ESC infection. An active complement response to infection was observed, with three forms of complement C3 up-regulated along with C4 and members of the membrane attack complex (C7 and C9). The complement regulatory protein factor H was strongly up-regulated (>14-fold). Another similarity to channel catfish was that genes involved in iron binding and transport were strongly induced following ESC infection, including intelectin, haptoglobin, haemopexin/warm-temperature-acclimation-related 65 kDa protein, caeruloplasmin and transferrin. Additional APPs up-regulated included pentraxin (serum amyloid P-like), fibrinogen and angiotensinogen. Immunosurveillance, immune signalling, and immune-cell activation genes were up-regulated. Other major categories of up-regulated genes included those involved in protein processing, localization and folding, and protein degradation.

When channel catfish and blue catfish are invaded by bacteria, up-regulation of genes in the liver is the norm, but very few are down-regulated. Channel catfish and blue catfish had 207 and 98 up-regulated genes, respectively, and each had five down-regulated genes in the liver after infection with the Gram-negative bacterium *E. ictaluri*. Three blue catfish genes down-regulated were selenoprotein P1b and selenoprotein H, which may possess antioxidant properties, plus a cell-cycle gene, anaphase-promoting complex subunit 13. Similarly, in channel catfish only a few genes including antimicrobial peptide-2 and thioredoxin-interacting protein (Bao *et al.*, 2006), which functions in the oxidative stress response in mammals, were significantly down-regulated following infection.

Similar to ictalurid catfish, Atlantic salmon had a much smaller number of transcripts down-regulated than up-regulated when challenged with *Aeromonas salmonicida* (Ewart *et al.*, 2005), apparently a characteristic result of transcriptomic analyses of bacterial infections. However, Atlantic salmon challenged with the bacterium, *Piscirickettsia salmonis*, had 71 transcripts up-regulated and 31 different transcripts down-regulated in macrophages, and 30 different transcripts up-regulated and 39 different transcripts down-regulated in haematopoietic kidney (Rise *et al.*, 2004), indicating that the relationships between the number and proportion of up- and down-regulated genes in response to attack by pathogens may vary by tissue, cell type and species of bacterial pathogen.

Fish have some unique, unidentified disease resistance genes yet to be fully described and understood. Several unidentified genes were up-regulated or down-regulated by viruses, bacterial endotoxins or mitogens in Japanese flounder leukocytes (Hirono and Aoki, 2002). Atlantic salmon challenged with *A. salmonicida* differentially expressed genes for humoral components of the innate immune system, genes not previously associated with infection, and a number of genes with no known homologues (Ewart *et al.*, 2005). Eighty unknown differentially expressed genes responded to bacterial exposure in channel catfish and 20 in blue catfish (Peatman *et al.*, 2007, 2008). The numerous unidentified genes that are differentially expressed in fish during infection could represent a subset of genes that respond to infectious agents that do not have this role in or could be absent from warm-blooded animals.

Some genes seem to be especially crucial for disease resistance in fish. As previously mentioned, iron homeostasis, both binding and transport, in channel catfish appears to be a critical immune response. These transcripts include intelectin, the most highly up-regulated gene observed (>85-fold), haptoglobin (>34-fold), haemopexin (>25-fold), caeruloplasmin (8.5-fold), transferrin (>7-fold) and ferritin (>2-fold). Intelectin was also the most highly up-regulated gene, an amazing 455-fold, in ESC-infected blue catfish. This may be particularly important as blue catfish are almost completely resistant to ESC, making intelectin a good candidate for genetic manipulation.

Consistent with the catfish data, intelectin was also up-regulated in carp (Reynders *et al.*, 2006) and rainbow trout (Gerwick *et al.*, 2007), but after exposure to cadmium or inflammatory stimulus, respectively, not bacterial exposure. In mammals, intelectin is believed to be involved in pathogen defence mechanisms by recognizing galactofuranose in carbohydrate chains of bacterial cell walls (Tsuji *et al.*, 2001) and by acting as a receptor for lactoferrin, an iron-sequestering homologue of transferrin (Suzuki *et al.*, 2001).

Chemokines are also key disease resistance genes in ictalurid catfish exposed to bacterial pathogens. Two blue catfish CC chemokines, SCYA106 (>10^5-fold) and SCYA113, were highly induced (Bao *et al.*, 2006; Peatman *et al.*, 2006). Based on comparative mammalian data, SCYA106 and SCYA113 may be regulators of dendritic-cell trafficking to secondary lymphoid organs. Up-regulation of CCL19-like genes was found after exposure to live bacterial vaccine or amoebic gill disease in Atlantic salmon (Martin *et al.*, 2006; Morrison *et al.*, 2006). Perhaps chemokines respond to both

bacteria and parasites. A catfish orthologue of CXCL14 chemokine also exhibited elevated expression in the liver after bacterial infection (Baoprasertkul *et al.*, 2005). In mammals, CXCL14 acts as a chemoattractant for activated monocytes, immature dendritic cells and natural killer (NK) cells (Starnes *et al.*, 2006).

Acquired immunity genes such as MHC genes should also play an important role in disease resistance. McConnell *et al.* (1998) isolated two MHC class II B genes from an inbred *Xiphophorus maculatus* strain, and these genes contained six exons and five introns. The encoded β_1 domain had three amino acids deleted and a cytoplasmic tail nine amino acids longer than what has been observed in other teleost class II β chains; thus it was more similar to human leucocyte antigen (HLA)-DRB, clawed frog Xela-F3 and nurse shark Gici-B (McConnell *et al.*, 1998). Key residues for disulfide bonds, glycosylation and interaction with α chains were conserved, and these same features are also present in the swordtail, *Xiphophorus helleri*.

The catfish bacterial challenges showed that acquired immunity genes also responded to *E. ictaluri* invasion (Peatman *et al.*, 2008), and a large number of genes with functions in protein modifications and degradation were up-regulated in the liver of catfish following infection (Peatman *et al.*, 2007). Members of these two groups of genes were likely related to the endoplasmic reticulum's (ER) unfolded protein response, which up-regulates chaperones and genes for protein degradation upon the accumulation of unfolded proteins during stress, or to the degradation and processing of antigens for the MHC class I molecule. Minimally, 15 unique transcripts were up-regulated in these two categories including chaperones, proteasome activators and proteasome subunits.

The up-regulation of two different genes encoding MHC class I components, α chains and β_2-microglobulin (B2M), indicated that active antigen processing and presentation were likely occurring in the blue catfish liver after exposure to *E. ictaluri*, an intracellular bacterium, as part of a cell-mediated immune response (Peatman *et al.*, 2007). Similar to the response in mammals, genes associated with the generation of peptides and peptide loading for the MHC class I molecules, PA28-α and -β, were up-regulated in blue catfish liver. Both PA28-α and -β proteasome activator subunits were up-regulated in blue catfish, suggesting a shift towards MHC class I antigen processing. Two ER chaperones, calreticulin and endoplasmin (GRP94), were also induced, providing further evidence of an active MHC class I-mediated response, as was tapasin (2.3-fold), which is also involved in MHC class I antigen loading. The coordinated up-regulation of MHC class I α chain, B2M and PA28-β was also observed in large yellow croaker (*Pseudosciana crocea*) following injection of polyinosinic:polycytidylic acid (poly I:C) (Liu *et al.* 2007).

One application of the transcriptomic studies is genetic engineering to improve disease resistance. Comparison of the blue catfish and channel catfish shows the complexity in choosing candidate genes for transgene manipulation, which will impact the schemes and designs needed for gene manipulation in the future as their gene expression in response to ESC was significantly different. Both species shared a wide spectrum of similarities in gene expression profiles after infection including an APR and strong induction of complement components and iron regulatory genes at day 3 after infection. However, a total of 58 genes were differentially expressed in blue catfish liver, but not in channel catfish liver at day 3 after infection. CC chemokine SCYA106, the most highly induced transcript in blue catfish, was not differentially expressed in channel catfish. Several MHC class I-related components, as well as several other immune-related genes, were up-regulated in blue catfish but not in channel catfish.

A few large expression differences between blue and channel catfish were observed at 24 h after infection, with little evidence of early induction of these genes in channel catfish. However, matrix metalloproteinase 13 (MMP-13) was up-regulated more than 20-fold in channel catfish at 24 h with only a slight elevation of expression relative to blue catfish. MMP-13 was also up-regulated during *E. tarda* infection in Japanese flounder (Matsuyama *et al.*, 2007).

Expression patterns in moribund fish might be considered when choosing candidate genes. Up-regulation was generally higher in moribund catfish. Expression of SCYA106 and lysosomal-associated membrane protein 3 (LAMP3) rose sharply in both dying blue catfish and channel catfish relative to controls, but showed greater up-regulation in blue catfish. MMP-13 expression in moribund channel catfish was much higher than that observed in moribund blue catfish.

Another major difference between blue catfish and channel catfish expression profiles was the induction of genes involved in MHC class I cascades in blue catfish. An earlier, or more efficient, MHC class I/cytotoxic T-cell response to the intracellular bacteria might explain some of the differences in resistance in the two species. MHC class I-related components including MHC class I α chain, B2M and proteasome activator PA28-α genes showed little up-regulation in channel catfish or blue catfish at 24 h but were strongly up-regulated in moribund fish of both species. MHC class I components were also up-regulated in channel catfish, but later than in blue catfish. CC chemokine SCYA106 expression was also barely induced in either species at 24 h, but strongly elevated in moribund fish.

Species differences in gene expression when challenged with ESC were also observed in the kidney as well as the liver. After ESC infection dramatic up-regulation up to 200-fold was found for CXCL10 chemokine in head kidney of channel catfish, which is susceptible to ESC, while CXCL10 showed a modest induction 24 h after infection in blue catfish, which is resistant to ESC. In the case of channel catfish, CXCL8 chemokine expression increased by about three- to fivefold at 24 h after infection, while the up-regulated expression was not detected in the blue catfish until 72 h after infection.

Gene expression in response to infection can vary among cell types and tissues.

In Atlantic salmon, ten antioxidant genes were up-regulated in infected macrophages, but not in infected haematopoietic kidney (Rise *et al.*, 2004). Transcripts of the adaptive immune responses such as T-cell-receptor α chain and CC chemokine receptor 7 were down-regulated in infected haematopoietic kidney, but not in infected macrophages, which may be indicative of infection-induced kidney tissue damage.

Several generalizations among fish and mammals have been observed with regard to gene expression for disease resistance. The blue catfish and channel catfish APR as measured 3 days after infection included many of the components of the typical mammalian APR and was also similar to the APR in salmonids and carp (Dunham, 2009). APPs such as haptoglobin, haemopexin, transferrin, caeruloplasmin, fibrinogen, angiotensinogen, pentraxin and several complement components accounted for a significant percentage of up-regulated transcripts in blue catfish (Peatman *et al.*, 2008), rainbow trout (Gerwick *et al.*, 2007) and zebra fish (Lin *et al.*, 2007), indicating the likely conservation of function of the vast majority of APPs between mammals and teleost fish, although zebra fish also exhibit significant differences compared with mammals.

Pentraxin, up-regulated 4.1-fold in blue catfish, initiates the complement cascade and possesses opsonizing activity in the snapper, *Pagrus auratus* (Cook *et al.*, 2003, 2005). The complement system of teleost fish exhibits conserved roles in sensing and clearing pathogens (Boshra *et al.*, 2006) as indicated by the presence of C3, the central component of the complement system, in multiple forms in fish, possibly serving as an expanded pathogen-recognition mechanism (Sunyer *et al.*, 1998). Three forms of C3 important for activation of the lectin and classical complement pathways, and two components of the membrane attack complex which carries out cell lysis, C7 and C9, were up-regulated in blue catfish liver. Complement factor H exhibited the highest up-regulation among complement-related factors (14.5-fold), which may inactivate C3b in the alternative complement pathway, suggesting that the host fish were attempting to modulate the complement response. Two lesser-known immune transcripts, LAMP-3 and galectin-9, were elevated both in blue catfish following ESC infection and in zebra fish following infection with *Mycobacterium marinum* (Meijer *et al.*, 2005). A catfish gene with highest similarity to lymphocyte antigen 6 complex, locus E (LY6E), the putative disease resistance gene for Marek's disease virus in chickens, was up-regulated.

Several genes that are not considered APPs were up-regulated in channel catfish, carp, trout and zebra fish, including microfibrillar-associated protein 4, Toll-like receptor 5, neurotoxin/differentially regulated trout protein SEC31/high-affinity copper uptake protein and SEC61. A major difference of ESC-induced gene expression in catfish is the absence of hepcidin, an iron-regulatory hormone, from the transcriptomic profile of catfish liver, although it is highly up-regulated in other teleost livers (Dunham, 2009). This could be useful for future transgenic application as hepicidin is involved in the pathway that leads to drastically decreased plasma iron levels during infection, a potential host defence mechanism to deny bacteria access to the critical metal. When plasma iron levels decrease, a feedback mechanism probably down-regulates hepcidin production in the liver.

Brain

Gene expression profiles in the channel catfish brain are similar to those in humans and mice (Adams *et al.*, 1991; Lee *et al.*, 2000). The expression profile in the catfish brain (Ju *et al.*, 2000, 2002) was much less polarized than those of the catfish pituitary (Karsi *et al.*, 1998) and muscle (Kim *et al.*, 2000), which might be expected since the brain is a complex organ with many different cell types.

Only 49.5% of clones sequenced from channel catfish brain cDNA library were known genes; the remainder were unknown (Ju, 2001). Considering the large numbers of genes already sequenced and known from the human brain, a large number of novel genes must be quite specific to teleost fish brains.

The variety of genes expressed in the channel catfish brain was tremendously diverse as might be expected (Ju *et al.*, 2000, 2002; Kocabas, 2001). The following classes of genes were expressed in the channel catfish brain: protein-translational machinery, such as ribosomal proteins, translational factors or tRNA genes; cellular structural genes, such as actins, tubulins, keratins and histones; enzymes, transcriptional factors, DNA-binding proteins, DNA-repair proteins; genes involved in the immune system; metal-binding proteins, ionic channels and genes involved in protein sorting and transportation; proto-oncogenes, tumour repressors and tumour-related proteins; hormones, receptors and regulatory proteins; developmental genes, such as clock genes and genes involved in tissue or organ differentiation; stress-induced genes, such as heat-shock proteins and cold-acclimatization proteins; genes in lipid metabolism; genes homologous to human mental disease-related genes; genes homologous to known sequences of unknown functions; mitochondrial genes; and others. Genes involved in protein translation accounted for the greatest proportion of expression in the brain (21.4%), as has been repeatedly found for other tissues and organs, followed by mitochondrial genes (6.2%), structural genes (3.1%), enzymes (2.7%), hormones and regulatory proteins (2.5%), immune-related proteins (2.1%) and transcriptional factors or genes involved in DNA binding or DNA repair (1.6%), and those for transport and translocation of small molecules accounted for 1.8% of expression. Many genes in this last category encoded voltage-gated ionic channels, metal-binding proteins such as metallothionein and calmodulin, and amino acid transporters. A total of 1.2% of the gene expression was from proto-oncogenes, tumour suppressors and tumour- or malignancy-related proteins, which may make channel catfish another valuable organism for studying cancer mechanisms. Three

fatty-acid-binding protein genes accounted for 1.1% of all expression in the brain (Ju, 2001).

Genes involved in mental processes were also expressed (Ju, 2001). Six genes shared high levels of similarity to known human genes involved in a number of human mental disorders, such as Huntington's disease, the atrophin-1 gene of dentatorubral and pallidoluylsian atrophy (DRPLA) disease (Khan *et al.*, 1996), the CGI-108 gene of clinical global disease and the small EDRK-rich factor 2 of spinal muscular atrophy (Scharf *et al.*, 1998), again indicating a possible role for fish research to contribute to mental health research. Four stress-induced genes were expressed in the channel catfish brain (Ju *et al.*, 2002): the heat-shock proteins hsp70 and hsp90; the ependymin gene, which has been demonstrated to be important for cold acclimatization in fish (Tang *et al.*, 1999); and the stress-inducible homologue of mouse (Blatch *et al.*, 1997).

Mitochondrial genes were among the most highly expressed genes, perhaps emphasizing the importance of respiration and energy transport in the brain (Ju *et al.*, 2000, 2002). Alternatively, there may be an abundance of mitochondrial transcripts because of the abundance of mitochondria, of which there are 200 to thousands of copies in individual cells; however, if this were the primary explanation, higher levels of expression of mtDNA genes should be found in other tissues. Of course, this large copy number could also be a mechanism for increasing levels of expression.

The mitochondrial tRNA-Val gene was the most highly expressed gene – 4.5% of clones – in the channel catfish brain (Ju, 2001). Several other mitochondrial genes were also expressed at high levels, including cytochrome c oxidase I (1.8%), cytochrome oxidase III (0.8%) and cytochrome b (0.6%).

Several nuclear genes were expressed at high levels in the channel catfish brain (Ju, 2001), such as ribosomal protein genes L41 (1.5%), L24 (0.8%), S27 (0.8%), L35 (0.7%), immunoglobulin heavy chain (1.1%) and fatty-acid-binding protein (0.6%). The high levels of expression of these genes may indicate high copy numbers for the loci encoding these proteins in the catfish genome, but more probably their promoters were either strongly or continually active (Ju *et al.*, 2002).

Genes encoding α- and β-globin were repeatedly sequenced – redundancy factor of 10.5 – but these globin genes were presumably from blood contained in the brain, despite great care in avoiding contamination with blood (Ju *et al.*, 2002). Globin genes are normally expressed at extremely high levels in the blood, and even minimal blood contamination may result in the false appearance of high expression of globin genes in the brain or other tissues.

Mitochondrial genes of channel catfish (Ju, 2001) had a high level of sequencing redundancy (4.63), a measure of relative expression level. Translational proteins, such as ribosomal proteins, were also highly expressed, with a redundancy factor of 3.89. Other categories of genes with a redundancy factor of more than 2.0 included lipid-binding proteins (3.75), genes involved in immune systems (2.50) and stress-induced proteins (2.25). The redundancy factor was lowest for development and differentiation-related genes (1.00), followed by genes homologous to known sequences of unknown functions (1.04), proto-oncogenes (1.07), enzyme genes (1.10), brain genes homologous to human disease-related genes (1.17) and transcriptional factors (1.27).

Cold tolerance

Environmental stressors, such as temperature change, have a large effect on fish metabolism, physiology, growth and disease resistance (Kocabas, 2001; Ju *et al.*, 2002). Low temperature can threaten the survival of cells by decreasing the fluidity of membranes, disassembling the cytoskeleton, slowing enzymatic processes, inhibiting secretory processes and decreasing metabolic rates (Tang *et al.*, 1999). Adverse effects at the cellular level can result in drastic reduction in feeding, growth and sometimes survival of warm-water and tropical fish.

Several genes that protect cells from damage during temperature change have been discovered. Cold acclimatization is attained by altering structure from induction of the desaturase gene for the desaturation of phospholipids, resulting in increased membrane fluidity, or by functional means involving the expression of metabolic isozymes, as well as the induction of protecting proteins, such as chaperons, to assist enzyme folding under cold stress (Tang *et al.*, 1999).

In the case of channel catfish (Ju *et al.*, 2002), gene induction was rapid – within 2 h of temperature shift. Exposure to cold induced the chaperons, hsp70 and hsp70/hsp90 organizing protein; transcription factors and genes involved in signal-transduction pathways, such as zinc-finger proteins, calmodulin kinase inhibitor, the nuclear autoantigen SG2NA, interferon regulatory factor 3 and inorganic pyrophosphatase; genes involved in lipid metabolism, such as TB2 and acyl-coenzyme A (CoA)-binding protein; and genes involved in the translational machinery, such as ribosomal proteins. Some of these genes were induced transiently and others continually. Down-regulated genes were primarily ribosomal protein genes, indicating reduced metabolic activity and need for translation of mRNA when channel catfish are held at low temperature for extended lengths of time. Channel catfish responded to low temperature by adjusting the expression of a large number of genes. The rapid induction of proteins involved in signal transduction and chaperons suggest that both *de novo* synthesis of cold-induced proteins and modification of existing proteins are required in the adaptation and tolerance of fish to low temperature (Ju *et al.*, 2002).

Osmoregulation

Osmoregulation is critical in fish. Fish cells accomplish this by storing osmolites, such as amino acids and their derivatives, sugar alcohols, urea and methylamines. One of the most important osmolites is taurine, 2-aminoethanesulfonic acid. Taurine transporter genes are up-regulated in carp epidermal cells *in vitro* in response to hyperosmotic conditions. The taurine transporter gene in *Oreochromis mossambicus*, a euryhaline species, was up-regulated in all organs examined – kidney, stomach, intestine, gill, eye, liver, fin and muscle – when the fish were challenged with saline water (Takeuchi and Toyohara, 2002). The response was time-dependent in all tissues except fin and muscles, which had an acute, delayed response. The Mozambique tilapia taurine transporter gene encodes 629 amino acids and has 12 putative membrane-spanning domains.

Genetic Imprinting and Paternal Predominance

Genetic imprinting based on parental origin of genes has been demonstrated for a variety of organisms (Swain *et al.*, 1987; Silva and White, 1988; El-Sherbini, 1990). Specific methylation of genes based on origin of gametes, followed by inactivation of gene expression, is the molecular mechanism for genetic imprinting.

Genetic imprinting or chromosome incompatibility could be explanations for the strong paternal predominance observed in catfish (channel catfish female × blue catfish male, blue catfish female × channel catfish male) (Dunham *et al.*, 1982a). This phenomenon has been well documented only in interspecific hybrids in ictalurids and in hybrids between the horse and donkey. The paternal predominance is greatly reduced in intraspecific crosses of channel catfish (El-Sherbini, 1990). Chromosomal incompatibility, followed by maternal chromosomal losses, in the interspecific hybrid could cause the paternal predominance; however, comparative karyology studies of the parental channel and blue catfish and their reciprocal hybrids did not reveal any differences in karyotypes or chromosomal losses (LeGrande *et al.*, 1984). If chromosomal breakage or losses occurred, changes in DNA fragments located in these regions should be detectable by AFLP analysis.

However, AFLP profiles of the two reciprocal hybrids were the same, confirming that chromosome breakage and loss probably do not account for paternal predominance (Liu *et al.*, 1998c).

Genetic imprinting is an alternative explanation for paternal predominance in catfish. If certain maternal genes are selectively methylated in an interspecific system, causing them to be inactivated or to be more weakly expressed than the same paternal genes in the F_1 channel × blue hybrid, phenotypic manifestation of paternal predominance could occur. Similar genetic mechanisms might also explain the paternal predominance in horse × donkey reciprocal hybrids. However, intraspecific gynogens and androgens of fish are viable, indicating that there may not be specific gender imprinting in fish. Alternatively, the lowered viability of gynogens and androgens might be partially explained by weakening of expression of some genes from imprinting.

Transposable Elements

Transposable elements can be classified as retrotransposons and DNA transposons based on their mode of transposition (Finnegan, 1989; Karsi, 2001). Retrotransposons are more abundant in vertebrates, while DNA transposons are common in bacteria and invertebrates (Henikoff, 1992). Retrotransposons move – 'jump' – via RNA intermediates and depend on reverse transcriptase, while DNA-mediated transposons move through DNA intermediates by cut-and-splice mechanisms, relying on transposases (Plasterk, 1996). Transposition of DNA transposons requires short terminal inverted repeats (IRs) in *cis* and transposase encoded in *trans* (Ivics *et al.*, 1997; Karsi, 2001). Examples of DNA transposons include the *P* elements of *Drosophila melanogaster*, the *mariner* element of *Drosophila mauritiana*, the *Ac* and *Spm* elements of maize and the *Tc1* elements from *C. elegans*. The *P* element has been developed and utilized as a powerful tool for genomic manipulations in *Drosophila*, such as insertional mutagenesis, transposon tagging, enhancer trapping and enhanced integration of transgenes into the germ line (Karsi, 2001). Unfortunately, *P* elements from *Drosophila* cannot be widely applied among various organisms because they are unable to transpose in non-host species. Theoretically, specific host factors are required for activity (Rio *et al.*, 1988; Handler *et al.*, 1993; Gibbs *et al.*, 1994).

The first vertebrate DNA transposon discovered was in a fish, channel catfish, utilizing electronic screening (Henikoff, 1992), and then in other vertebrates (Radice *et al.*, 1994; Smit and Riggs, 1996). *Tc1*-like elements have been found in hagfish (Heierhorst *et al.*, 1992), salmonids (Goodier and Davidson, 1994; Radice *et al.*, 1994), zebra fish (Radice *et al.*, 1994; Izsvak *et al.*, 1995) and amphibians (Lam *et al.*, 1996b).

Most transposable elements from vertebrates are inactive because of extensive insertions/deletions and the presence of in-frame termination codons in their transposase genes (Ivics *et al.*, 1996, 1997; Lam *et al.*, 1996a,b; Karsi, 2001). The genetic conservation of transposase, the structure of the vertebrate *Tc1*-like elements and the remnants of these elements observed within active genes imply that they were once mobile (Heierhorst *et al.*, 1992; Henikoff, 1992; Goodier and Davidson, 1994; Radice *et al.*, 1994; Lam *et al.*, 1996a,b). Lam *et al.* (1996b) detected the emergence of 11 new *Tzf* loci from 25 offspring, suggesting movement of the elements in amphibians. If the existence of these active transposases is confirmed, they would be extremely useful as genetic tools for mutation analysis by sequence tagging, transgenesis and other gene manipulations.

In contrast to *P* elements, *Tc1*-like transposases seem to have fewer requirements for various cellular factors for transposition (Loukeris *et al.*, 1995; Plasterk, 1996; Gueiros-Filho and Beverley, 1997) and therefore should function in a wider variety of hosts (Karsi, 2001). The first *Tc1*-like element in vertebrates, *TcIp1*, was discovered from channel catfish (Henikoff, 1992), despite a very limited number of sequenced genes from channel catfish (Liu

et al., 1997; Karsi *et al.*, 1998), indicating that channel catfish may be a rich source of these elements. The sequence of *TcIp1* is the most distinctive *Tc1*-like element identified in teleosts, suggesting a unique position in the molecular evolution of the *Tc1* transposons in vertebrates (Liu *et al.*, 1999c).

Channel catfish have multiple families of *Tc1*-like transposons, *Tip1* and *Tip2* (Liu *et al.*, 1999c). *Tip1* and *Tip2* are similar in structural organization to other *Tc1*-like elements isolated from teleosts, but sequence analysis indicated that each one belongs to a different subfamily within the *Tc1/mariner* superfamily. *Tip1* resembles *Tss* and *Tdr* elements isolated from salmonids and zebra fish, while *Tip2* is similar to *Tc1* elements from invertebrates, such as fruit flies and nematodes. *Tip1* includes 1568 bp and harbours a short IR 30 bp long, while *Tip2* is 1011 bp and contains a long IR of 176 bp (Karsi, 2001). Both elements have terminal repeats sharing some similarity to the *Tss* and *Tdr* elements, and have extensive mutations in their transposase gene, but are different in that they show sequence divergence within the IRs at regions important for transposase binding.

Tip1 and *Tip2* are different from *TcIp1*; therefore *Tip1*, *Tip2* and *TcIp1* may represent three subfamilies of the *Tc1/mariner* superfamily. *TcIp1* (Henikoff, 1992; Liu *et al.*, 1999c) and *Tip2* are not simply deleted forms of *Tip1*, and multiple sequence alignments of *TcIp1* (Henikoff, 1992), *Tip1* and *Tip2* did not indicate any significant sequence similarities among them (Karsi, 2001). *TcIp1* (Henikoff, 1992) seems to be the most divergent member of the *Tc1*-like elements identified from various fish species (Radice *et al.*, 1994; Izsvak *et al.*, 1995), whereas *Tip1* exhibits high levels of similarity with most other *Tc1*-like elements isolated from teleosts. High similarities were observed between *Tip1* and *Tc1*-like elements from salmon (*Tss1*), carp (*Tcc*), goldfish (*Tca*), zebra fish (*Tdr1*) and northern pike (*Tel*) (Ivics *et al.*, 1996; Liu *et al.*, 1999c; Karsi, 2001). High similarities between *Tip2* and other *Tc1*-like elements from teleost species were primarily due to the terminal IR and codon usage being similar among teleosts and different from those of invertebrates such as *C. elegans* and *Drosophila* (Karsi, 2001). The amino acid blocks of *Tip2*, with the exception of the *Tes* element isolated from hagfish, were similar to *Tc1* elements from nematodes or fruit flies. Curiously, the amino acid identity of *Tip2* is similar to that of invertebrate transposons, followed by *Tc1*-like elements from amphibians such as *txz* and *Txr*.

Channel catfish *Tip2* is an ancient member of the *Tc1* superfamily, more homologous to *Tc1*-like elements from invertebrates than to other elements, and *Tip1* is an element common in teleosts highly homologous to *Tc1*-like elements from zebra fish and salmonids (Liu *et al.*, 1999c). *TcIp1* is more divergent from any *Tc1*-like elements thus far identified from invertebrates or from teleosts (Karsi, 2001). Karsi (2001) concludes that three *Tc1*-like elements – *TcIp1*, *Tip1* and *Tip2* – may belong to three divergent families in the *Tc1/mariner* superfamily.

In addition to the *Tip1* and *Tip2* families of transposons, at least four more fragments were amplified using the same PCR primer amplifying *Tip1* and *Tip2*; thus multiple families of *Tc1*-like elements coexist in channel catfish (Liu *et al.*, 1999c). *Tip1* elements have approximately 150 copies and *Tip2* 4000 copies per haploid genome. They are linked with active genes, indicating their previous mobility and potential importance in evolution and gene expression. The coding of *Tip2* potentially contains amino acid blocks similar to those in a variety of cellular and viral genes, such as that for the human immunodeficiency virus (HIV) envelope glycoprotein. Therefore, their previous activity may have had both positive and negative effects: these elements may have been used by viruses against the host organism and transposable elements may have had a large role in shaping the genomic structure and organization of vertebrates (Karsi, 2001).

The first vertebrate *Tc1*-like element, *TcIp1* from channel catfish, was not initially recognized as a transposable element because it was localized in the fifth intron of the IgM gene (Wilson *et al.*, 1990). Sequences similar to *Tip1* are associated with globin genes in Atlantic salmon (McMorrow *et al.*, 1996), the Atlantic salmon ependymin gene (Mueller-Schmid *et al.*, 1992) and the rainbow trout methycholanthrene-responsive gene (Berndtson and Chen, 1994). The *Tip1*-like sequences in lake trout were associated with SnAluI-8 repetitive elements (Reed and Phillips, 1995), suggesting a high abundance of *Tip1*-like sequences in lake trout (Karsi, 2001). *Tip2*-like sequences from channel catfish (Karsi, 2001) were associated with Atlantic salmon globin genes (McMorrow *et al.*, 1996) and the glutathione *S*-transferase gene in plaice (Leaver *et al.*, 1997).

Alignments of deduced amino acid sequences among homologous *Tc1*-like elements from various organisms suggest the existence of four highly conserved amino acid blocks (Liu *et al.*, 1999c; Karsi, 2001) with all four domains located in the C-terminal half of the transposases. Three of four conserved domains contain the previously identified DDE box (Doak *et al.*, 1994); the other conserved domain possesses the glycine-rich box (Ivics *et al.*, 1997), and the bipartite nuclear localization signal (Vos and Plasterk, 1994; Ivics *et al.*, 1996) is also conserved, but not as highly as the other four blocks. The DDE box is the catalytic centre of the transposase activity (Vos and Plasterk, 1994; Craig, 1995). A GAA → AAA mutation in *Tip2* changes the glutamic acid (E) to lysine (K) in the DDE box, indicating that the E residue in the DDE box might not be required for transposase activity in *Tip1* transposase in channel catfish as it is in other *Tc1* transposases. In contrast, *Tip2* was inactivated at this functional DDE box prior to its inactivation in other regions. The occurrence of the latter phenomenon should have been less likely because functional domains are often protected from mutation by selective pressures (Karsi, 2001). A similar mutation exists in the *SALT* element of Atlantic salmon (Goodier and Davidson, 1994). These divergences are consistent with a study that found the conserved DDE box to be non-functional (Lohe *et al.*, 1997). The DDE and glycine-rich boxes appear to be well conserved but completely conserved residues in the four domains are difficult to find (Karsi, 2001). Correlated sequences may make complete conservation unnecessary, however, as several highly conserved sub-blocks exist within the four domains for *Tc1*-like elements (Karsi, 2001). Three regions of leucine/isoleucine/valine–tryptophan (LW) appear well conserved, in addition to the two D regions in the original DDE box. The conserved triple-LW boxes, along with the completely conserved double-D boxes, suggest their important functionality in a wide range of transposases in the *Tc1/mariner* superfamily. Additionally, a histidine residue (H) in the third block and an isoleucine residue (I) in the fourth block are completely conserved among all *Tc1*-like elements (Karsi, 2001).

Additional *Tc1*-related genomic sequences also exist (Karsi, 2001), but these genomic sequences cannot be easily classified as transposon sequences because of their low identity to *Tc1* elements. Blocks of transposon-related sequences may have served as genetic materials, building blocks and pools of diversity for both hosts and transposons. Reassortment of functional domains and horizontal transmission between species have been proposed as mechanisms for the formation and spread of new types of transposable elements.

Although not statistically significant, the sequences of *A. thaliana* chromosome 4 within the BAC clone F28A23 (accession AL021961) and of human chromosome 19 cosmid clone F18547 (accession AC003682) have noticeable DNA sequence identity to *Tip2* from channel catfish (Karsi, 2001), and amino acid blocks of many proteins, such as black-beetle virus coat protein (accession P04329, 30% identity and 51% similarity in a block of 71 amino acids), contain homologous amino acid blocks to *Tip2*. High levels of amino acid similarities between *Tc1*-like transposase and *Pax* paired-domain family of transcriptional regulators also exist (Ivics *et al.*, 1996). An amino acid identity of 44% and similarity of 52% exist between the conceptual *Tip2* transposase and a 34-amino acid block of the HIV1 envelope glycoprotein

(Karsi, 2001), and 45–52% similarity still exists when this block of amino acids is extended to 55 amino acids from both sides. This block of sequences also encodes the highly conserved amino acid blocks of transposase, although a different reading frame is used. The similarity of HIV envelope protein to *Tip2* transposase was higher than that to several transposases from the fungus *Fusarium oxysporium* (accession number S75106), the biting fly *Haematobia* (accession number U13806) and the biting fly *Stomoxys* (accession number U13824). HIV-homologous sequences have also been observed in IS-encoded transposases and bacterial recombinases (Zuerner, 1994); however, in these cases, the sequences are homologous to the HIV reverse transcriptase, which is functionally related to the IS-encoded transposases.

The fact that *Tip2* exhibits homologous sequences to the HIV envelope glycoprotein provides additional evidence of horizontal transmission of transposase sequences (Karsi, 2001). Alternatively, many of these sequences could have evolved by chance and gained similarities independent of the *Tc1* elements, but the possibility that they could have been horizontally transmitted among hosts via parasites, viruses, bacteria and the transposons cannot be excluded (Houck *et al.*, 1991; Kidwell, 1992). Horizontal DNA transfers require at least an ecological relationship between the species concerned, but in reality vast arrays of organisms are interconnected ecologically. Horizontal transfer of a *Tc1*-like transposon from the *Cydia pomonella* host into its baculovirus, the granulovirus, has been confirmed (Jehle *et al.*, 1995, 1998), so this is a plausible explanation for some of the various similarities of nucleotide sequences and amino acid sequences among various transposons and genes.

Ribosomes

Ribosomes are responsible for protein synthesis, and the structure of ribosomal proteins and their interactions with RNAs have been thoroughly studied (Wool, 1979, 1986;

Wool *et al.*, 1995, 1996; Draper and Reynaldo, 1999; Patterson, 2001; Karsi *et al.*, 2002a; Patterson *et al.*, 2003). Ribosomal protein genes are highly expressed in cells because they are obviously critical for protein synthesis; therefore, ribosomal protein genes are abundant in cDNA libraries constructed for EST analysis in various tissues and fish, such as channel catfish and Japanese flounder (Adams *et al.*, 1991; Ju *et al.*, 2000; Cao *et al.*, 2001; Karsi, 2001). Due to this high representation, it is relatively easy to compile a complete set of ribosomal protein cDNA sequences using EST analysis (Karsi, 2001).

Ribosomes are responsible for protein synthesis in all cells and thus link transcriptomes with proteomes (Karsi, 2001). Ribosomal proteins have been studied to understand the post-transcriptional regulation of gene expression (Perry and Meyuhas, 1990; Aloni *et al.*, 1992; Meyuhas *et al.*, 1996; Meyuhas, 2000).

Mammalian ribosomes are derived from a total of 79 proteins and four RNAs (Wool, 1995; Warner and Nierras, 1998). The 60S ribosome is composed of three rRNAs and 47 ribosomal proteins, whereas the 40S contains the 18S rRNA and 32 ribosomal proteins (Wool, 1979; Wool *et al.*, 1995). The large number of ribosomal proteins and the fact that all ribosomal proteins studied to date have large pseudogene families (Wool *et al.*, 1996) have complicated the full understanding and deciphering of their gene structures and genomic organizations (Karsi, 2001).

The basic structural and functional features of ribosomes are evolutionarily conserved. Initial research examined the small subunit rRNA (Sogin, 1991), but rRNA-based phylogenies were difficult to determine because of drastic differences in GC content among taxa (Loomis and Smith, 1990; Galtier and Gouy, 1995). Translation of ribosomal protein genes into amino acid sequences overcame the problem of GC-content differences, and large numbers of ribosomal proteins provide a large set of homologous sequences for analysis; thus they have been used in more recent phylogenetic studies (Liao and Dennis, 1994;

Veuthey and Bittar, 1998; Yang, D. *et al.*, 1999).

In the case of fish, ribosomal genes have been thoroughly studied in the channel catfish. All 79 channel catfish 40S ribosomal protein genes were obtained from EST analysis of brain, head kidney and skin tissues (Ju *et al.*, 2000; Cao *et al.*, 2001; Karsi *et al.*, 2002b), and so were all 47 60S ribosomal protein genes (Patterson *et al.*, 2003). The nomenclature of the rat (Wool *et al.*, 1996) was followed because of the numerous synonyms existing in the GenBank databases and in the literature (Karsi, 2001). Thirty-four large and 25 small ribosome subunits were sequenced for channel catfish. Each ribosome contained 50 distinct proteins, which must be made, theoretically, at exactly the same rate (Karsi, 2001). Although ribosomal proteins are proportionally required for the assembly of ribosomes, large differences were observed in the abundance of ESTs for the ribosomal proteins (Karsi, 2001). Perhaps some are used in a greater number of ribosomes than others. The primary control of ribosomal protein synthesis focuses on translation of the mRNA, not on its synthesis; thus the level of translational regulation is dramatic and potentially complex.

The most abundant ribosomal protein gene products were L41 (18 clones), L24 (ten clones), S27 (nine clones), L35 (eight clones), L21 and L22 (both seven clones), L5b and L32 (both six clones), ribosomal protein large P2, L35a, L37a and L38 (all five clones), and L7a, L11, L39, S9 and S24 (all four clones) (Karsi *et al.*, 2002b). Three clones were obtained for L10a, L12, L15, L18, L30, L31, L36, S2, S3, S8, S15, S16, S18, S20, S21 and S22. Two clones were observed for S4, S10, S11, S12, S13, S19 and S23. Only one clone was found for the remaining 16 ribosomal genes. The expression level of ribosomal proteins varied 18-fold. Several ribosomal protein genes, such as L41, S24 and S27, must have strong promoters, and translational control must account for more than ten to 20 times the adjustment in RNA abundance for the end-products to be the same level of ribosomal proteins (Karsi *et al.*, 2002b). Despite high levels of evolutionary conservation of the 40S ribosomal protein genes, the channel catfish ribosomal protein genes exhibited several unique features, including high levels of alternative polyadenylation, differential splicing and multifunctional genes encoding for two ribosomal protein mRNAs (Karsi, 2001).

All 32 40S ribosomal proteins of channel catfish initiate at the first AUG, which is similar to the rat, where all but L5 initiate at the first AUG; however, UAA is the most frequently used termination codon in the catfish ribosomal protein genes (73.53%) (Karsi et al., 2002b), which is different from the most frequently used termination codon, UGA, in other vertebrates (Cavener and Ray, 1991). The 3-UTRs were highly AT-rich, and the catfish genome is AT-rich. Of the 34 mRNAs (including two different mRNAs for S26 and S27), 33 have the typical AAUAAA polyadenylation signals (Proudfoot, 1991), and only S11 has the AAUUAA poly-(A)+ signal (Karsi et al., 2002b). The poly-(A)+ tract begins eight to 21 nucleotides from the poly-(A)+ signal, with the exception being S8, which has the typical AAUAAA poly-(A)+ signal 49 bases upstream of the poly-A sites. However, a non-typical AUUAAA was found 16 bases from the poly-(A)+ sites, and possibly this second poly-(A)+ signal was used by channel catfish.

S19 had two typical poly-(A)+ signals of AAUAAA in the 3-UTR located 126 bases apart, and both poly-(A)+ signals were used, generating S19–1 and S19–2 mRNAs (Karsi, 2001). In contrast, S21 also produces two mRNAs, but has only one poly-(A)+ signal, AATAAA, located 15 bases and six bases upstream from the poly-(A)+ tails of S21–1 and S21–2, respectively. The only difference between S21–1 and S21–2 mRNAs was the presence of two bases (TG) after a stretch of As within S21–1, and the differential mRNAs may have been a result of alternative polyadenylation or by non-specific addition of bases into poly-(A)+ tails (Karsi, 2001).

In eukaryotes, a conserved AAUAAA and a variable GU-rich element located ten to 40 nucleotides from the 3′ end (Wahle and Keller, 1996; Colgan and Manley, 1997; Beaudoing et al., 2000), coupled with endonucleolytic cleavage, followed by poly-(A)+ synthesis, result in effective polyadenylation and mature mRNAs (Karsi, 2001). A significant fraction of mRNAs contains multiple poly-(A)+ (Gautheret et al., 1998); however, alternative polyadenylation may occur more frequently in channel catfish than in other eukaryotes (Karsi et al., 2002b). Patterson et al. (2003) also found that ribosomal protein L31 has three types of mRNA from differential polyadenylation, and alternative polyadenylation may play a role in the post-transcriptional regulation of the ribosomal protein gene expression by affecting the stabilities of the ribosomal protein mRNAs (Shatkin and Manley, 2000; Karsi, 2001; Macdonald, 2001).

Differentially spliced forms of ribosomes may be another post-translational mechanism (Karsi et al., 2002b). For example, S3–1 is the longest version of the S3 cDNA in channel catfish, while S3–2 and S3–3 are differentially spliced forms (Karsi, 2001). S3–2 has a deletion of 48 bp while S3–3 has a deletion of 85 bp; typical AG and GT splicing junction sequences exist in the S3–2 but are absent in S3–3. S3–1 and S3–2 possess proper open reading frames, but S3–3 is out of frame due to a deletion of 85 bp (non-integral of triplet codon) (Karsi, 2001). S3–1, the major spliced form (85%), encodes a protein of 245 amino acids, while S3–2 encodes a protein of 229 amino acids. The shorter version of the S3 may or may not be biologically active, but the production of the aberrant transcripts may reduce the overall expression of the S3 protein and serve as one mechanism of post-transcriptional regulation of its expression.

Channel catfish ribosomal protein genes have zinc-finger domains (Karsi et al., 2002b) similar to zinc-finger structures from rat (Chan et al., 1993). The conservation of these structural features suggests that zinc may have a role in the binding to rRNA (Wool et al., 1996).

The majority of the 40S channel catfish ribosomal proteins were highly similar to their mammalian counterparts for deduced amino acid sequences, and the number of amino acids was highly conserved evolutionarily (Karsi *et al.*, 2002b). Channel catfish 40S ribosomal proteins contain basic and acidic amino acid clusters, also consistent with findings in the rat ribosomal proteins (Wool *et al.*, 1996). Channel catfish S27, S27a and S30 have basic amino acid clusters, such as KKKKK and KKRKKK, whereas S9 has an acidic amino acid cluster, DDEEED (Karsi *et al.*, 2002b). Sa contains KEE repeats, and S3a has a polar SSSS repeat. Again, similar to the rat, the channel catfish ribosomal proteins S27a and S30 are also expressed as carboxyl extensions of ubiquitin-like proteins. Of the 32 40S ribosomal protein cDNAs, 21 had open reading frames with identical numbers of amino acids to those in the rat, two had one extra amino acid, three had one fewer amino acid and four had two extra amino acids.

The overall similarity of the channel catfish 40S ribosomal proteins to those of the mammals was 94.3% (Karsi *et al.*, 2002b). The most conserved 40S ribosomal proteins were S18, S14 and S23, and the most divergent ones were S19, S21 and S25.

All of the channel catfish 40S ribosomal proteins had only one type of mRNA, except S26 and S27, with each of these having two (Karsi, 2001). S26–1 and S26–2 were highly divergent at the nucleotide level, sharing only 72.3% identity, but encoded highly similar proteins, having 94.8% identity, and S27–1 and S27–2 were even more divergent at nucleotide level with only 55.3% identity, but their proteins differed only by one amino acid. These data are again indicative of strong selection pressure to retain specific amino acid sequences, despite base substitutions at the nucleotide level.

The different cDNAs for both S26 and S27 may be transcripts of two different genes, since SNPs within coding regions are generally about 1–2% in channel catfish (Z. Liu, unpublished results); however, two functional genes are rare for ribosomal protein genes in mammalian species (Karsi, 2001). Most mammalian ribosomal genes have multiple copies – presumably retroposon pseudogenes – but prior to this example with channel catfish, only human S4

encoded separate functional genes (Fisher *et al.*, 1990; Wool *et al.*, 1996). Patterson *et al.* (2003) also found that two different genes encode channel catfish L5; therefore at least three ribosomal proteins might be encoded by two loci in channel catfish.

The 40S ribosomal protein genes were highly expressed, accounting for 5.33–11.42% of the cDNA clones in head kidney, brain and skin (Karsi, 2001). This high level of expression would be expected for proteins involved in all cellular translation.

Despite the specific stoichiometric ratio of ribosomal proteins required by the ribosome structure (Wool, 1979), differential gene expression was observed within a single tissue/organ or among the tissues for the various ribosomal proteins in channel catfish (Karsi *et al.*, 2002b). Strong tissue-specific expression was observed for some of the highly expressed ribosomal proteins, such as S2, which was highly expressed in the head kidney and skin, but had relatively low expression in the brain. Sa was strongly expressed in the head kidney and S9, S20, S24 and S30 in the skin, while S27 was more highly expressed in the brain than in the head kidney or skin. The highly variable mRNA levels of various ribosomal protein genes within and among tissues strongly suggest significant regulation at the post-transcriptional levels to meet the equimolar requirement of the ribosomal proteins in the ribosomes (Karsi, 2001).

Proteomics

The entire protein counterpart of the genome, the proteome, can be examined for its changes in expression in response to specific treatments, challenges and environmental conditions (Dunham *et al.*, 2001). The collective methodology is defined as proteomics. Protein expression can be efficiently examined utilizing two-dimensional protein gel electrophoresis, followed by analysis with powerful computer software packages.

Proteins, the genetic end-product, are of ultimate functional importance, which can more accurately reflect real functional and physiological differences because of the occasional deviation between the measured expression of RNAs and proteins. Protein expression signatures can be obtained and displayed for thousands of proteins under specific sets of environmental conditions. The disadvantages of proteomics are that molecular studies after the initial identification of proteins and protein differences are difficult, and sequencing of proteins is extremely difficult, requiring purification of proteins. Lastly, protein sequences can be decoded into only degenerated sequences, which often specify too many oligonucleotides to be highly useful for further research on the corresponding genes (Dunham *et al.*, 2001).

16

Gene-transfer Technology

Recombinant DNA technology and genetic engineering allowed a new biotechnology that became feasible for application in aquaculture species during the 1980s. The first successful gene transfer – genetic engineering – for fish was reported in China in 1985. Since that time, gene transfer in aquaculture species has been accomplished in many countries. Most of the research has focused on GH-gene transfer.

Genetic engineering (gene transfer) techniques have been developed that may complement traditional breeding programmes for the improvement of quantitative and qualitative traits. Individual genes from one species are isolated, linked to promoters (regulatory DNA sequences or on/off switches), cloned and multiplied, primarily in plasmids, but bacteriophages and cosmids may be used for specific cases. These genes are then transferred into genomes of other species by viral vectors, microinjection, electroporation, sperm-mediated transport or gene-gun bombardment. Organisms containing foreign genes, homologous genes or DNA sequences inserted artificially are termed transgenic. Five objectives must be met to have a successful gene transfer. The appropriate gene needs to be isolated and cloned. The foreign gene must be transferred to the fish and be integrated in the host's genome. The transgene must be expressed, as transfer of the gene does not guarantee that it will express and function. A positive biological effect must result from expression of the foreign DNA and with no adverse biological or commercial effects. The gene must be inherited in subsequent generations.

Transgenic fish have been developed that have improved growth, colour, disease resistance, survival in cold and body composition, and that can produce pharmaceutical proteins. Transgenes elicit pleiotropic effects, some positive and a few negative, but most of the negative effects appear to lower fitness traits which is positive for biological containment. Transgenic fish appear to pose little environmental risk, but this research is not fully conclusive. Transgenic zebra fish with altered coloration have been commercialized and GH-transgenic salmon, carp and tilapia are near commercialization. To expedite commercialization and minimize environmental risk, transgenic sterilization needs to be developed. When transgenic fish research was initiated, a large percentage of the work was actually conducted on commercial aquaculture species, but now an increasing amount of research is conducted with model species. A potential positive impact of transgenic fish appears likely in many arenas.

Norman Maclean and S. Talawar of Southampton University, UK, were the first researchers to inject cloned genes into fish (rainbow trout) eggs (Maclean and Talawar, 1984). This was followed by Zuoyan Zhu at the Institute of Hydrobiology in China reporting production of a transgenic fish (Zhu *et al.*, 1985). Twenty-six years later, transgenic fish application sits not only on the cutting edge, but on the regulatory edge of making its first as well as major impact. The first application has actually been in the ornamental fish industry rather than in aquaculture. GloFish®, a transgenic of the zebra fish (*Danio rerio*) containing fluorescent protein genes GFP, YFP and RFP (green, yellow and red), is now commercialized. These fish were actually an output of experiments to develop transgenic fish for environmental monitoring, but were an obvious choice to create a marketing niche in the ornamental fish trade.

Transgenic fish can provide insights into the mechanisms of development, gene regulation, actions of oncogenes and the intricate interactions within the immune system (Chen and Powers, 1990). The transfer of exogenous DNA in fish represents a powerful strategy for studying the regulation of gene expression *in vivo* (Volckaert *et al.*, 1994).

Fish and many aquatic organisms produce large quantities of eggs, which are usually fertilized and incubated externally (Dunham, 1990a). The embryos are relatively easy to obtain, manipulate and incubate, and hatch rapidly in warm-water species of fish and shellfish. Because fish undergo external fertilization, the potential transgenic embryos do not require complex manipulations, such as *in vivo* culturing of embryos and transferring of embryos into foster-mothers, manipulations essential in mammalian systems (Powers *et al.*, 1992). These features make the fish a good organism for the application of gene-transfer technology.

The foundation for gene-transfer research was actually laid as early as 1910, when embryologists experimented with injecting cellular material into frog eggs (Gurdon and Melton, 1981). By the early 1970s, it was apparent that gene-transfer technology could provide great insight into the function of DNA sequences (Gurdon and Melton, 1981). Technological advances in the isolation and proliferation of eukaryotic genes, coupled with the development of microinjection procedures for amphibian eggs, resulted in the rapid expansion of gene-transfer research. Continued research improved microinjection techniques and the techniques used to detect the expression of transferred genes (Joyce, 1989). By the late 1970s, the focus of gene transfer shifted to mammalian tissue-culture cells and, soon after, mammalian embryos.

The first widely publicized work was the transfer of mRNA and DNA into mouse eggs (Palmiter and Brinster, 1986). Gordon *et al.* (1980) were among the first to microinject a series of recombinant molecules into the pronuclei of mouse embryos at the one-cell stage of development. This pioneering, landmark research in mice provided the impetus for the initiation of genetic-engineering research with fish, which followed in 4–5 years.

Originally, fusion genes – constructs produced by splicing DNA sequences from different sources – with origins from mammals, birds, insects, bacteria and viruses were transferred to fish embryos by microinjection. Prior to the early 1990s, very few fish genes had been isolated. With the advent of functional genomics and new molecular genetics techniques, such as EST analysis, and the generation of huge gene databases, thousands of fish genes have now been isolated, and a long list of fish genes is growing for the study of gene expression and potential transfer. These genes may have application for the enhancement of growth, reproduction, disease resistance, carcass yield, cold tolerance and other economic traits.

Gene-transfer Technique in Fish

Gene-transfer research with fish began in the mid-1980s, utilizing microinjection (Zhu *et al.*, 1985; Chourrout *et al.*, 1986b;

Dunham *et al.*, 1987). Transgenic individuals of several farmed fish species, including goldfish (Zhu *et al.*, 1985), rainbow trout (Chourrout *et al.*, 1986b), channel catfish (Dunham *et al.*, 1987) and Nile tilapia (Brem *et al.*, 1988), were produced. Transgenic research was also initiated in non-commercial species, such as the loach, *Misgurnus anguillicaudatus* (Maclean *et al.*, 1987a), and medaka, *Oryzias latipes* (Ozato *et al.*, 1986).

Zhu *et al.* (1985) published the first report of transgenes microinjected into the fertilized eggs of goldfish. In almost all fish gene transfer, the foreign gene was microinjected (Fig. 16.1) into the cytoplasm (Hayat, 1989). Zhu *et al.* (1985) injected a linear DNA fragment from the recombinant plasmid pBPMG-b into the germinal disc of the egg. Chourrout *et al.* (1986b) and Dunham *et al.* (1987) successfully transferred human growth hormone gene (hGHg) constructs into rainbow trout and channel catfish, respectively.

Transgenic fish containing bacterial genes, β-galactosidase (McEvoy *et al.*, 1988), neomycin resistance (goldfish) (Yoon *et al.*, 1990), hygromycin resistance (Stuart *et al.*, 1988) and chloramphenicol transacetylase (tilapia) (Indiq and Moav, 1988), were produced. The primary purpose of these studies was to develop gene-transfer technology and develop systems for the rapid and easy study of gene transfer, as these genes obviously have little commercial importance.

Ozato *et al.* (1986) had a slightly different approach, and injected the oocytes of the medaka, which had been removed from the ovaries 9 h before ovulation. The chicken δ-crystallin gene was injected and found in four of the eight medaka embryos examined (Ozato *et al.*, 1986); however, when they injected into the cytoplasm of medaka eggs, they had no success.

Once the foreign DNA is microinjected, it appears to replicate rapidly in the cytoplasm of the developing embryo, and then begins to disappear as development proceeds (Houdebine and Chourrout, 1991). This phenomenon has probably led to overestimation of the number of transgenic individuals in several experiments where embryos or fry were evaluated, because positive individuals probably often possessed non-integrated concatemers in the cytoplasm rather than integrated sequences in the

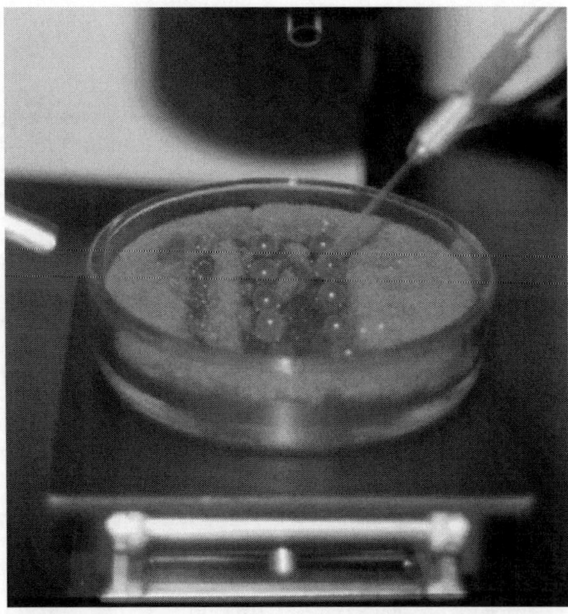

Fig. 16.1. Microinjection of DNA into salmon eggs. (Photograph by Robert Devlin.)

genome. An average of about 5% of the surviving injected embryos integrated the foreign DNA (although this figure varies widely) at the two-cell stage or beyond (integration at the one-cell stage has never been observed), resulting in mosaic transgenic individuals when microinjection was used.

Regardless of survival or integration rate, microinjection is a tedious and slow procedure (Powers *et al.*, 1992) and can result in high egg mortality (Dunham *et al.*, 1987). After the initial development of microinjection, new techniques, such as electroporation, retroviral integration, liposomal reverse-phase evaporation, sperm-mediated transfer and high-velocity microprojectile bombardment, followed (Chen and Powers, 1990), which can more efficiently produce large quantities of transgenic individuals in a shorter time period. Mass transfer systems are especially beneficial for recombinant DNA research in fish and shellfish because of the high fecundity and high embryo mortality of some of these aquatic species (Powers *et al.*, 1992).

Electroporation involves placing the eggs in a buffer solution containing DNA and applying short electrical pulses to create transient openings of the cell membrane, allowing the transfer of genetic material from the solution into the cell (Figs 16.2 and 16.3). However, the exact behaviour of the cell membrane under the influence of electroporation is not known. The efficiency of the electroporation is affected by a variety of factors including voltage, number of pulses and frequency of pulses.

The first successful gene transfer utilizing electroporation produced integration rates and survival similar to those for microinjection (Inoue *et al.*, 1990). Powers *et al.* (1992) then demonstrated that electroporation can be more efficient than microinjection, although even electroporation has limitations, including the number of embryos that can be electroporated at a time. Higher rates of integration – sometimes as much as 30–100% – were obtained using electroporation rather than microinjection of DNA (Powers *et al.*, 1992). When compared with microinjection, electroporation is an excellent method for the transfer of DNA into a large number of fish embryos in a short time period (Powers *et al.*, 1992).

In the case of channel catfish, Walker (1993) found that hatching rates were higher

Fig. 16.2. Equipment for electroporating eggs of channel catfish, *Ictalurus punctatus*, for gene transfer.

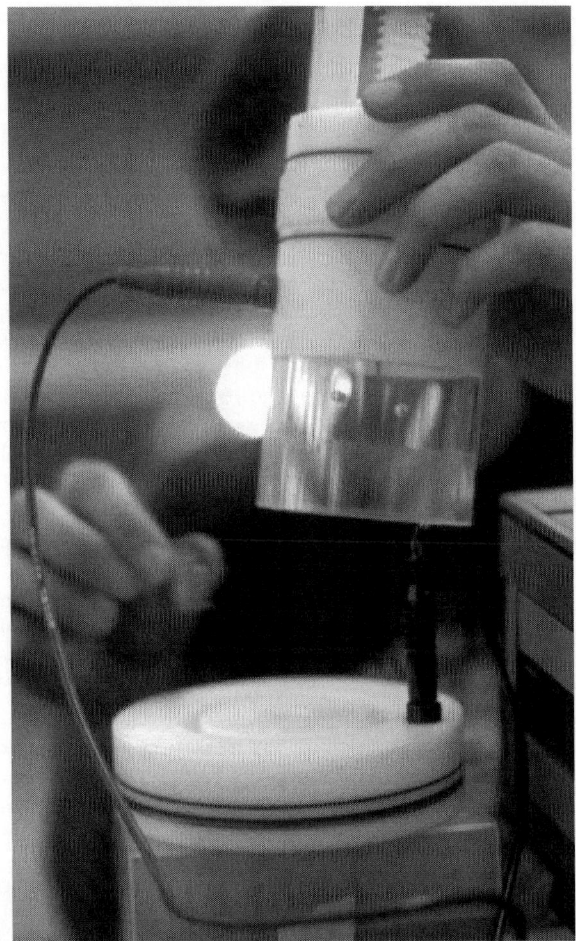

Fig. 16.3. Electroporation of fish eggs. Eggs (15–50) are placed in a DNA solution inside the cuvette and then placed between two electrodes to receive rapid pulses of high voltage.

for electroporated embryos than for those that were microinjected. Post-fertilization electroporation treatments had higher hatching rates than electroporation of sperm and then eggs prior to fertilization. Electroporation of sperm followed by fertilization of non-manipulated eggs also resulted in high hatch rates. Embryos electroporated with GH cDNA (5.2 kb) had higher hatching rates and greater survival to 15 g than those manipulated with the larger immunoglobulin genes (20 kb). GH genes (5.2 kb) could be more easily transferred with electroporation than with microinjection. However, when using the larger immunoglobulin-gene constructs (20 kb), no integration was obtained for either microinjection or electroporation. Apparently, these larger DNA fragments can have an adverse effect on hatch and subsequent survival. This is consistent with results obtained for the induction of gynogenesis, where low radiation levels for sperm inactivation resulted in the introduction of large supernumerary chromosome fragments and lower hatching rates than in embryos with smaller supernumerary chromosomes. Large DNA fragments reduce viability.

The efficiency of gene transfer is determined by several factors, including hatching percentage, gene integration frequency, the number of eggs that can be manipulated in a given amount of time and the quantity of effort required to manipulate the embryos. Powers *et al.* (1992) indicated that electroporation resulted in higher integration rates than microinjection; 15 to 100 catfish eggs can be electroporated every 4 s simply by depressing a button, while the tedious microinjection procedure is more dependent upon the skill of the operator, with a maximum speed of about one egg microinjected per minute. Electroporation is one of the most promising and powerful techniques for the mass generation of P_1 transgenic fish.

Promoters

The recombinant gene (transgene or fusion gene) needs to be fused to a promoter sequence which regulates or allows the expression of the recombinant DNA. If the gene were introduced without a promoter, it is highly unlikely that it would be integrated near an endogenous (already in the host genome) promoter that would allow expression. In reality, one of the keys to transgenic technology is the use of artificial regulatory sequences. The use of artificial promoters may allow circumvention of natural regulatory mechanisms that may inhibit and regulate expression. The efficient production of transgenic fish with enhanced performance relies on the development of expression vectors that both have high integration rates and can maintain reliable and stable transgene expression in the transgenic progeny.

The promoters most commonly evaluated in early studies of transgenic fish were of viral and mammalian origin. The viral promoters pose no known or logical biological or food-safety risk. However, less research is conducted today with viral promoters because of poor public perception of the word virus, which would probably make the future commercialization of transgenic fish extremely difficult.

Numerous promoters have been evaluated in transgenic fish and they vary greatly in their ability to allow expression of foreign DNA. Cytomegalovirus (CMV), Rous sarcoma virus long terminal repeat (RSV-LTR), β-actin and chicken δ-crystallin are constitutive promoters that allow expression of foreign DNA in the host species. Mouse metallothionein (MT) and rainbow trout MT are inducible promoters that have allowed expression in transgenic fish. However, they are leaky and appear to allow expression even when not induced. Deletions, modifications and rearrangements can drastically strengthen or weaken the ability of these regulatory sequences to produce mRNA. Transgenic fish commonly produce the foreign gene product in tissues and cells where expression is not normally found, since the artificial regulatory sequences apparently circumvent natural regulatory mechanisms and biological feedback mechanisms.

A variety of promoters have been shown to be active in fish cells, and initial investigations used promoters derived from non-piscine vertebrates and their viruses. Reporter-gene activity was detected in fish or fish tissue-culture cells when the inserted sequence was driven by the promoters RSV-LTR, simian virus type 40 (SV40), CMV-tk, CMV-IE and mouse mammary tumour virus (MMTV), polyoma viral promoters, human and mouse MT, and human heat-shock protein 70 (hsp70) promoters (Hackett, 1993).

Piscine genes and their promoters were the next sequences utilized in fish transgenesis and in cell transfection studies. Promoters from flounder antifreeze, carp β-actin and salmonid MT-B and histone H3 have been found to be active (Liu *et al.*, 1990; Gong *et al.*, 1991; Chan and Devlin, 1993). In general, it appears that many eukaryotic promoters are able to function in fish cells, although, if derived from non-homologous sources, the level of expression may be somewhat reduced.

Expression of uninterrupted coding regions from prokaryotic and eukaryotic sources has been successful in fish cells (Du *et al.*, 1992). However, expression in fish cells of gene constructs containing mammalian introns may be inefficient due to difficulties in RNA processing to yield functional mRNA (Bearzotti *et al.*, 1992; Bétancourt *et al.*, 1993). At present, it is not possible to generalize about the activities of various gene promoters and gene constructs in different fish species because of lack of data.

GFP has been utilized to study promoters and expression. Yoshizaki *et al.* (2000) cloned and characterized rainbow trout vasa-like gene-regulatory regions, whose transcripts are restricted to PGCs. Transgenic rainbow trout were produced containing the vasa promoter fused with the GFP gene. Green fluorescence was first observed in the mid-blastula stage, but no cell-specific expression was detected at this time. At the eyed stage, about 30% of the transgenic embryos expressed GFP in the PGCs, and this increased to 70% at hatching. GFP-expressing cells were located on the genital ridge. Kinoshita and Tanaka (2002) obtained similar results in medaka, utilizing medaka vasa promoter and GFP. GFP expression was detected at the ventrolateral region of the intestine at the blood-circulation stage. By hatching, the GFP-expressing cells had moved to the gonadal region. Utilization of the GFP promoter has sometimes been criticized for its inconsistent or weak activity. However, Gibbs and Schmale (2000) developed a GFP cassette with a modified β-actin promoter, boundary elements and insulators yielding stronger and more consistent GFP expression.

Several tissue-specific promoters have been developed from zebra fish (Gong *et al.*, 2002), including epidermis-specific keratin 8, fast-muscle-specific myosin light polypeptide 2 and pancreatic exocrine-cell-specific elastase B. Two-colour transgenic zebra fish have been developed by introducing epidermis-specific keratin 8 promoter–GFP and muscle-specific myosin light polypeptide–RFP (Gong *et al.*, 2002).

For some applications, inducible promoters may be desirable to allow induction of transgene expression at specific developmental life stages. The inducible hsp70 gene, which encodes an enzyme that plays an essential role in protein metabolism and has been isolated and characterized from *Oreochromis mossambicus*, dramatically increased its rate of mRNA transcription when fish were exposed to a transient heat shock (Molina *et al.*, 2000). The entire isolated regulatory region was able to mediate heat-shock-inducible expression of the reporter gene, with no preference for a particular tissue, in microinjected zebra fish embryos.

Integration

Some researchers feel that low integration rates and mosaicism hinder research on transgenic fish and the development of potentially valuable commercial lines of transgenic fish. The basis for this thinking is that most of the DNA constructs introduced via microinjection or other transfer procedures are lost during the first 10 days after delivery into host embryos. A million copies of the gene are introduced into each embryo and only about 0.0001% of the constructs take up permanent residence in the fish genome. Of course, if the opposite were true and genes were easily integrated, the addition of too many multiple copies could be problematic, so some middle ground is needed. However, if gene transfer were highly efficient, the number of copies introduced per embryo could be reduced, greatly reducing the need to grow and purify huge quantities of the constructs. Screening for transgenic individuals is extremely time-consuming and expensive, so efficient transformation would be a great benefit. Integration rates between 2 and 30% have been reported (Chen and Powers, 1990; Hacket, 1993), although some batches of embryo can have integration rates of 100%.

Delayed integration also causes mosaicism (Stuart *et al.*, 1988, 1990; Culp *et al.*, 1991; Hayat *et al.*, 1991; Gross *et al.*, 1992; Hackett, 1993): not all tissues contain the transgene and not all cells within the transgenic tissues harbour the transgene. This phenomenon has been reported in all transgenic fish studies, including those involving common carp (Hayat, 1989), medaka (Ozato *et al.*, 1986) and zebra fish (Stuart *et al.*, 1988). All tissues examined in transgenic medaka had foreign DNA, but only half of the cells examined contained foreign DNA.

The mouse metallothionein–human growth hormone fusion gene (MThGHg) was transferred to the channel catfish. The integration rate of the foreign DNA was 20% for fish analysed at 3 weeks of age (Dunham *et al.*, 1987). When the fish were resampled at 3 months of age, only 4% of the fish contained the MThGHg. One of three 3-month-old transgenic channel catfish survived to sexual maturation. This female was mated to a non-transgenic male. None of the resulting 65 progeny analysed inherited the MThGHg. Although this female possessed MThGHg in fin tissue, she was mosaic and did not have MThGHg in her germ line.

The percentage of individuals detected as transgenic was lower when individual fish were analysed as fingerlings compared with fry, which is consistent with other studies (Dunham, 1990a). Individuals possessing these foreign DNA constructs may be sub-viable and experience differential mortality compared with non-transgenic siblings. Two more likely explanations exist. DNA may persist for several cell divisions or possibly for a few weeks without integration or degradation, resulting in false positives being detected when the fish are analysed as fry. Alternatively, mosaicism may occur in embryos microinjected with foreign DNA. Since whole fry are homogenized and assayed when analysing fish at this stage, most mosaics would be detected. When tissues are biopsied from older, larger fish, many mosaics may not be detected unless all tissues are assayed.

The entire MThGHg fusion gene was incorporated in two of the three transgenic channel catfish. The third individual had a fraction of the fusion gene deleted during chromosomal insertion. All three individuals were smaller than non-transgenic siblings. No conclusion on growth rate can be drawn because of the low number of transgenic individuals and because no expression data were obtained from these three individuals.

Mosaicism is a common phenomenon in transgenic fish produced by microinjection of recombinant DNA. One of 20 transgenic zebra fish possessed foreign DNA in its germ line and transmitted it to its offspring (Stuart *et al.*, 1988). Although 75% of transgenic common carp transmitted their

foreign construct to their progeny, 50% of these parental transgenic carp were probably mosaics (Zhang *et al.*, 1990).

The microinjected DNA can persist and replicate for several cell divisions in fish prior to degradation. This has been demonstrated in rainbow trout and Atlantic salmon (Rokkones *et al.*, 1985, 1989), loach, *M. anguillicaudatus* (Maclean *et al.*, 1987a), zebra fish (Stuart *et al.*, 1988) and tilapia (Phillips, 1989). Stuart *et al.* (1988) reported that the microinjected DNA was actually amplified in early embryos prior to degradation and non-integrated DNA was detectable in low copy number after 3 weeks. Copy numbers can range from one to several thousand at a single locus, and, in contrast to the head-to-tail organization observed in the mouse system, in some cases the DNA can also be found organized in all possible concatameric forms (Tewari *et al.*, 1992), suggesting random end-to-end ligation of the injected DNA prior to integration.

Microinjected DNA can persist in protozoa for 2 years as extrachromosomal DNA (Garvey and Santi, 1986). Theoretically, the DNA forms circular concatamers, although it has not been proved that the foreign DNA is unable to persist and replicate in linear form. Zhu *et al.* (1985) indicated that injected DNA was replicating in circular form in goldfish embryos. Evidently, integration can occur after one or more cell divisions, resulting in mosaics.

Apparently, it is common for recombinant genes microinjected into fish embryos not to be integrated until some time after the first cell division, resulting in a large percentage of mosaics. Even individuals transmitting their foreign DNA to progeny in normal Mendelian ratios could lack the recombinant gene in other tissues. Probably due to the cytoplasmic nature of DNA injection, virtually all founder transgenic fish are mosaic and the integrated DNA is found only in a subset of developmental cell lineages. Mosaicism has been demonstrated in somatic tissues based on molecular tests, and can also be inferred for the germ line based on the observed frequencies of transgene transmission to F_1 progeny being less than at Mendelian ratios. For salmonids, the frequency of transgene transmission from founder animals averages about 15%, suggesting that integration of the foreign DNA occurs on average at the two- to four-cell stage of development (Devlin, 1997b). Transmission of transgenes to F_2 or later progeny occurs at Mendelian frequencies (Shears *et al.*, 1991), indicating that the DNA is stably integrated into the host genome and passes normally through the germ line.

Low integration rate coupled with mosaicism causes inefficiency, results in an extra generation of research to generate a product, and increases the effort needed for producing transgenic fish. If parental transgenic (P_1) fish are mosaic, only a fraction of transgenic P_1 fish transmit the transgene to F_1, and even then usually at lower than expected rates. Then transgenic F_1 fish usually transmit the transgene to the F_2 generation in expected Mendelian ratios (Stuart *et al.*, 1988; Shears *et al.*, 1991; Chen *et al.*, 1993, 1995; Gibbs *et al.*, 1994; Moav *et al.*, 1995). Achieving early and high integration, therefore, might enhance efficiency of production of transgenic fish by reducing the screening required of the original fish and possibly by eliminating mosaicism in P_1. Additionally, inefficient integration results in few lines and genetic backgrounds to choose from when attempting to select for lines with optimal gene expression. Higher integration rates would potentially increase the number of lines generated, with a greater number of genotypes to evaluate and select from, as well as reducing potential problems with inbreeding and founder effects from initiating transgenic populations from a limited number of founders. Of course, the inbreeding and lack of genetic variation could be corrected by crossbreeding, but this complicates evaluation of performance because of potential variation in combining abilities in the crossbreeds.

Mosaicism is not a critical problem if sufficient fish-culture facilities are available. Spawning of large numbers of pairs or mass spawning of potentially transgenic fish almost guarantees the generation of F_1 transgenics that harbour the transgenes in every cell. Detailed research and commercialization can readily follow from this point. Alternatively, a considerable amount of research has been directed towards developing constructs that can enhance integration rates and minimize the position effects on gene expression. Transposon sequences, retroviral sequences and border elements have been studied to overcome integration inefficiencies and position effects on expression.

Retroviral vectors containing the envelope protein of vesicular stomatitis virus have been developed (Burns *et al.*, 1994) and used to produce transgenic fish (Lin *et al.*, 1994; Yee *et al.*, 1994; Lu *et al.*, 1996). Integration rates may be increased because of active infection. Unfortunately, these vectors are prone to unstable expression or even complete silencing of transgene expression. Ivics *et al.* (1993) used a retroviral integrase protein to improve gene integration in zebra fish; however, the integrase activity was short-lived in fish cells, which, if this held true for whole fish embryos, would limit its usefulness. However, Sarmasik *et al.* (2001a,b) successfully utilized retroviral constructs to produce transgenic crayfish and topminnows, *Poeciliposis lucida*.

Standard techniques for inserting foreign genes have been difficult to apply to shrimp, crustaceans and live-bearing fish, because embryos are released from their mothers at a relatively advanced stage. Thus, newly fertilized eggs are unavailable for microinjection or electroporation. Sarmasik *et al.* (2001a,b) developed pantropic retroviral vectors derived from the hepatitis B virus and the vesicular stomatitis virus, a pathogen similar to foot-and-mouth disease which infects mammals, insects and possibly plants. The vector sticks to most cell membranes of any species. Transgenic crayfish and topminnows were produced by injecting immature gonads of the crayfish and topminnow with a solution of the vector about 1 month before the normal age of first reproduction. Matured injected individuals were mated with normal individuals and produced 50% transgenic offspring. Integration, expression and transmission of the pantropic retroviral reporter transgene were observed for at least three generations. This is a very good gene-transfer technique for live-bearers, but, again, the introduction of viral sequences – in this case, retrovirus sequences – into food fish may not be accepted by the public.

The use of transposases to enhance integration rates may be a more viable option than retroviral vectors for oviparous aquatic organisms, but does not solve the problem of live-bearers. Several sequences similar to transposable elements have been characterized from fish (Henikoff, 1992; Radice *et al.*, 1994; Izsvak *et al.*, 1995; Liu *et al.*, 1999c). However, no functioning transposable elements have been found in any fish. The *Tc1/mariner* superfamily of transposons has a wide phylogenetic distribution, may not require specific host factors and therefore might possibly be used to develop gene constructs to enhance gene transfer (Karsi, 2001). Unfortunately, all *Tc1/mariner* elements found in vertebrates have been defective (Henikoff, 1992; Heierhorst *et al.*, 1992; Goodier and Davidson, 1994; Radice *et al.*, 1994; Izsvak *et al.*, 1995; Ivics *et al.*, 1996; Lam *et al.*, 1996a,b; Liu *et al.*, 1999c), and thus have no potential without modification to enhance gene-transfer efficiency.

Two options exist for utilizing transposable elements: the introduction of transposons into heterologous systems, or, since all of these vertebrate elements discovered to date have been inactive, the reconstitution of active transposons based on sequences (Karsi, 2001). Both of these approaches have had success. The *mariner* element from *Drosophila mauritiana* was transposed in the parasitic flagellate protozoan, *Leishmania* (Gueiros-Filho and Beverley, 1997), and reconstitution of the transposability of a synthetic transposon, *Sleeping Beauty* (Ivics *et al.*, 1997), was accomplished in carp EPC cells, mouse LMTK cells and human HeLa cells (Ivics *et al.*, 1997; Izsvak *et al.*, 1997; Luo *et al.*, 1998). *Sleeping Beauty* enhanced integration two- to 20-fold when tested in these cell types (Ivics *et al.*, 1997). Ivics *et al.* (1996, 1997) developed the artificial transposon system, *Sleeping Beauty*, by reconstructing open reading frames using a comparative phylogenetic approach. The basis of the approach is the ability of the transposase to enhance integration of DNA placed between the inverted repeats. Karsi (2001) also had success transforming cells with *Sleeping Beauty*, as co-transfection of the *Sleeping Beauty* expression plasmid pCMVSB with pTSVNeo increased the number of resistant colonies by more than threefold for NIH3T3 cells as compared with the control without the transposase.

Sleeping Beauty was reconstructed based on 1.6 kb *Tc1*-like elements. The utility of this construct will partially be based on its ability to insert constructs of variable sizes. Karsi *et al.* (2001) found that insert size affected the transposition efficiency of *Sleeping Beauty*. Increasing insert size decreases the transformation efficiency of *Sleeping Beauty* constructs, and transformation efficiency was inversely correlated with the insert sizes. Insertion of stuffing fragments of 1.4–3.4 kb resulted in increasingly reduced transformation efficiency, and no enhancement in transformation occurred with the insertion of a 6.9 kb stuffing fragment. *Sleeping Beauty* transposase enhanced integration for 5.6 kb constructs, but was unable to transpose 9.1 kb transposons. The *Sleeping Beauty* transposon is effective for transferring genes of small sizes into cell lines, but may not be useful for transferring genes of large sizes. The correlation between transposing ability in cell lines and whole embryos needs to be determined to confirm the practicality of these results. Although there is an upper limit in DNA size for which *Sleeping Beauty* provides increased integration, most eukaryotic cDNAs are smaller than 5 kb; thus this transposon system will be potentially useful for most genetic-engineering development in aquatic organisms (Karsi *et al.*, 2001).

Transmission of Transgenes

Initial P_1 transgenic fish produced via any transfer technique are all mosaics, which do not possess the foreign DNA in every cell or tissue. However, many of these fish still transmit the stably integrated DNA to their progeny, but at less than the expected Mendelian ratios because of the mosaicism. During the late 1980s, the first data were generated demonstrating that transgenes could be inherited and transmitted to future generations. The resulting F_1 transgenics have the foreign gene in all of their somatic cells and the germ cells, and usually transmit the foreign DNA to their progeny in expected Mendelian ratios (Stuart *et al.*, 1990; Shears *et al.*, 1991), except in some families of transgenic common carp (Chen *et al.*, 1993), which either transmitted the introduced gene in less than expected ratios or experienced differential mortality. Perhaps the mechanism that allows the insertion of foreign DNA may act in reverse if the foreign DNA has been integrated in an unstable location of the genome.

Stuart *et al.* (1988) showed the replication, integration and transmission of the hygromycin resistance gene to F_1 and F_2 zebra fish. Guyomard *et al.* (1989a,b) demonstrated persistence, integration and germ-line transmission but no expression of rat GH gene or hGHg microinjected into fertilized rainbow trout eggs. Zhang *et al.* (1990) have shown gene transfer, expression and inheritance of rainbow trout GH (rtGH) cDNA in the common carp. Chen *et al.* (1990) demonstrated the expression and inheritance of GH gene in carp and loach. Culp *et al.* (1991) demonstrated a high level of germ-line transmission in zebra fish inserted with RSV–galactosidase.

Transgene Expression of Growth Hormone and Reporter Genes

The initial emphasis for transgenic fish research was the transfer of GH genes and reporter genes. At first, the most commonly transferred gene was the hGHg (Zhu *et al.*, 1985; Chourrout *et al.*, 1986b; Dunham *et al.*, 1987; Maclean *et al.*, 1987b; Brem *et al.*, 1988) because of the exciting results for enhancement of mouse growth with this gene and because fish genes including GH were not available.

Zhu *et al.* (1985, 1986) introduced hGHg into the germinal disc of goldfish, carp and loach, and reported a transformation rate of 75% and up to a 4.6-fold increase in body weight when GH gene was injected, compared with uninjected control fish. However, integration and expression data were not generated. Chourrout *et al.* (1986b) showed that hGH cDNA injected with the cytoplasm of fertilized trout eggs was integrated into the genome of 30-day-old embryos at a rate of about 33%; however, expression and enhanced growth were not obtained. Ozato *et al.* (1986) and Inoue *et al.* (1989) reported the introduction and expression of chicken δ-crystallin gene in 7-day-old medaka embryos, and the transformation rate was 16%. The data that they presented did not exclude transient expression of the δ-crystallin gene. Maclean *et al.* (1987a,b) and Penman *et al.* (1990) introduced hGHg into loach and rainbow trout, and rat GH (Rahman and Maclean, 1992) into tilapia. Dunham *et al.* (1987) reported the microinjection of mouse MThGHg into channel catfish and the persistence of the gene in head-to-tail tandem array in 3-week-old fish. Brem *et al.* (1988) injected MThGH vector into tilapia and obtained a low rate of integration of the DNA.

After the initial focus on the transfer of the hGHg, various GH genes were isolated and became available for transgenic fish research. The bovine (Schneider *et al.*, 1989), rat (Maclean *et al.*, 1987a; Penman *et al.*, 1987; Guyomard *et al.*, 1988; Rokkones *et al.*, 1989), rainbow trout (Zhang *et al.*, 1990) and coho salmon (Hayat *et al.*, 1991) GH genes have also been transferred.

Rokkones *et al.* (1989) demonstrated the expression of hGHg in 1-year-old Atlantic salmon. Rokkones *et al.* (1989) also found that MT allowed expression of the hGHg in Atlantic salmon and rainbow trout embryos. Human GH RNA was found (1–4 ng/egg) in salmonid embryos. Higher levels of RNA were found in embryos microinjected with circular DNA than with linear DNA, suggesting that non-integrated concatamers were transcribing more efficiently than integrated or non-integrated linear DNA or, more probably, the concatamers were much more abundant than the integrated sequences. Human GH was found in one of seven and three of seven groups of embryos injected with circular and linear DNA, respectively, perhaps suggesting higher integration rates when linearized DNA was introduced and the expression being associated with integrated sequences. Contradictorily, hGH was not detected in the blood of 1-year-old transgenic salmonid fingerlings.

Crucian and silver carp possessing the MThGHg construct also expressed hGH (Chen *et al.*, 1989). Fifty per cent of the transgenic carp evaluated had hGH. Zhu *et al.* (1988) also reported expression of hGH in progeny of transgenic carp. Mouse MT promoter appears to be effective in fish, but again utilization of mouse DNA could lead to negative advertisement campaigns by competitors towards a public that has little knowledge concerning genetics and genetic manipulation.

A promoter from SV40 has also been fused to hGHg, and transferred to rainbow trout (Chourrout *et al.*, 1986b). The SV40–hygromycin fusion gene was not transcribed or translated in transgenic zebra fish (Stuart *et al.*, 1988). SV40 did drive the expression of the chloramphenicol transacetylase gene in transgenic Nile tilapia (Indiq and Moav, 1988).

Ozato *et al.* (1986) observed the expression in five of six medaka embryos possessing the C-18 plasmid chicken δ-crystallin gene. The quantity of δ-crystallin ranged from 20 to 400 pg/embryo. Expression was highest in the brain, muscle and gills and was also observed in the spinal cord, retina and lens, which is a broader range of tissue expression than found in the chicken for this gene, again illustrating how the fusion genes overcome normal gene regulation.

Zhang *et al.* (1990) fused the RSV-LTR – RSV is a chicken virus – to rtGH cDNA, and transferred this fusion gene to common carp. All nine transgenic common carp evaluated produced rtGH. The level of expression ranged from 8 to 89 pg/g protein in the erythrocytes (Fig. 16.4), and was not correlated with the copy number integrated per individual. Expression of the rtGH gene

was obtained, despite the presence of plasmid vector sequences on each end of the recombinant GH genes (Zhang *et al.*, 1990). Prokaryotic vector sequences inhibited expression in transgenic mice (Townes *et al.*, 1985; Palmiter and Brinster, 1986).

The rtGH was found in erythrocytes but not the serum of these transgenic carp (Zhang *et al.*, 1990). Normally, GH is produced in the pituitary and secreted into the serum. The transferred cDNA lacked a signal peptide sequence that codes for a portion of the protein necessary for secretion, which explains the absence of rtGH in serum. Gene expression was observed in a cell – the erythrocyte – where expression is usually absent, because of the artificial regulatory mechanism, the RSV-LTR promoter. The rtGH cannot reach target tissues via the serum when the signal peptide sequence is absent; however, the rtGH was found in several tissues. Since these transgenic common carp grew faster than controls, this natural mechanism of GH distribution must not be necessary if target cells are producing their own elevated levels of GH.

Similar results were obtained for transgenic GH tilapia. Transgenic *Oreochromis hornorum urolepis* were produced containing one copy per cell of the tilapia GH (tiGH) cDNA under the regulatory sequences derived from the human CMV (Martinez

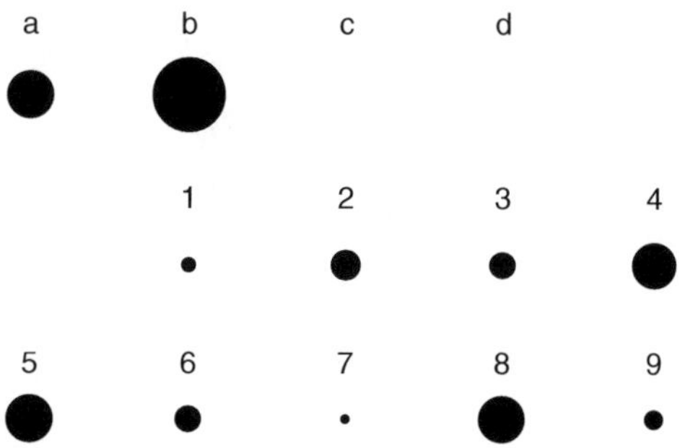

Fig. 16.4. Variable expression of rainbow trout growth hormone (rtGH) in growth-hormone-transgenic common carp, *Cyprinus carpio*. Control wells a and b are positive controls; c and d are negative controls. Numbered wells are individual transgenic common carp with different levels of rtGH expression based on radioimmunoassay. (Adapted from Zhang *et al.*, 1990.)

et al., 1999). The transgene was transmitted to F_1–F_4 generations in a Mendelian fashion, and there was low-level expression of tiGH in brain, heart, gonad, liver and muscle cells, where GH expression is not normally found.

RSV promoter also allowed the transcription of the neomycin resistance gene in goldfish (Yoon *et al.*, 1990) and the production of bovine GH in northern pike, *Esox lucius* (Schneider *et al.*, 1989). Gross *et al.* (1992) showed the integration and expression of salmon and bovine GH cDNA in northern pike. The RSV-LTR promoter appears to be another strong, useful promoter for fish transgenic research, except again it has negative commercial connotations.

Very high levels of transgene expression have been found in wild transgenic salmon (Mori and Devlin, 1999). Mori and Devlin (1999) examined the expression of the sockeye salmon MTB–sockeye GH1 in transgenic coho salmon, utilizing RT-PCR. These fish have dramatically elevated growth and serum GH levels 40-fold greater than normal. GH was found in all tissues examined – liver, kidney, skin, intestine, stomach, muscle, spleen and pyloric caeca. Surprisingly, because of the sensitivity of this technique, GH expression was also detected in the intestine of the control salmon. GH expression was greater in younger transgenic coho salmon (20–21 g) compared with older transgenic salmon (400–500 g). Although transgenic fry grew much faster than controls, transgene expression was not detectable. Effects on the pituitary were dramatically different in the transgenics compared with the controls. When similar-size transgenics and controls (older) were compared, GH mRNA levels were higher in controls, pituitary glands were larger in controls and relative pituitary size decreased with increasing body weight in transgenic coho salmon, but not in non-transgenic coho salmon. The site of GH production was the same in transgenic and control pituitary glands.

Most studies have not found a correlation between copy number and expression. One exception is found for transgenic Nile tilapia. Three lines of transgenic tilapia were generated with a construct containing a β-galactosidase (*lacZ*) reporter gene spliced to a 4.7 kb, 5-regulatory region of a carp β-actin gene (Rahman *et al.*, 2000). The three lines contained different copy numbers of transgenes, and the levels of *lacZ* expression were related to transgene copy number. Mosaic patterns of somatic *lacZ* expression were observed in these three lines of tilapia, which differed between lines but were consistent within a line. The expression of the reporter gene in homozygous transgenic tilapia was approximately twice that of hemizygous transgenics, and analysis of reporter-gene expression on a tissue-to-tissue basis demonstrated that *lacZ* expression of the reporter gene in stably transformed F_1 and F_2 tilapia was variable in different organs and tissues and was also sometimes variable in different cells of the same tissue.

Transgenic *O. hornorum urolepis* containing one copy per cell of the tiGH cDNA under regulatory sequences derived from the human CMV also appeared to exhibit a gene-dosage effect at 4 months of age (Martinez *et al.*, 1999). Heterozygote GH transgenics appeared to grow faster than homozygous transgenics, and both grew faster than controls in an experiment with limited replication.

Performance of Transgenic Fish

Growth

The greatest amount of work has focused on transfer of GH genes. Positive biological effects have been obtained by transferring transgenes to fish in some, but not all, cases. Initially, this research focused on the transfer of foreign GH-gene constructs into fish. Due to the lack of available piscine gene sequences, transgenic fish research in the mid-1980s employed existing mammalian GH-gene constructs, and growth enhancement was reported for some fish species examined (Zhu *et al.*, 1986; Enikolopov *et al.*, 1989; Gross *et al.*, 1992; Lu *et al.*, 1992; Zhu, 1992; Wu *et al.*, 1994).

Mammalian gene constructs of mouse metallothionein/rat growth hormone (mMT/rGH) failed to affect the growth of salmonids (Guyomard *et al.*, 1989a,b; Penman *et al.*, 1991), despite the fact that salmonids are very responsive to growth stimulation by exogenously administered mammalian GH protein (McLean and Donaldson, 1993). Gene constructs containing fish GH sequences driven by non-piscine promoters elicited growth enhancement in transgenic carp, catfish, zebra fish and tilapia (Zhang *et al.*, 1990; Dunham *et al.*, 1992a; Chen *et al.*, 1993; Zhao *et al.*, 1993; Martinez *et al.*, 1996). Growth-stimulatory effects observed with the above constructs have ranged from no effect to twofold increases in weight relative to controls, and provided the first convincing data demonstrating that growth enhancement in fish can be achieved by transgenesis.

However, subsequent experiments demonstrated that growth can be enhanced through transgenesis by 10% up to an incredible 30-fold in some conditions (Figs 16.5 to 16.7). Like other breeding programmes, sometimes no enhancement is obtained. The results are basically consistent. Several species, including loach, common carp, crucian carp, Atlantic salmon, channel catfish, tilapia, medaka and northern pike, containing human, bovine or salmonid GH genes are reported to grow 10–80% faster than non-transgenic fish in aquaculture conditions, if the proper promoters are utilized. Chen *et al.* (1990) have shown integration of hGHg in loach and significant increases in the length and weight of these fish. Du *et al.* (1992) used an all-fish GH-gene construct to make transgenic Atlantic salmon. They report a two- to sixfold increase of the transgenic fish growth rate.

Similar results have been obtained for transgenic *Oreochromis niloticus* possessing one copy of an eel (ocean pout)

Fig 16.5. Growth-hormone-transgenic common carp, *Cyprinus carpio*, and control (bottom fish), illustrating a 50% growth increase. (Photograph by Rex Dunham.)

Fig. 16.6. Growth-hormone-transgenic Nile tilapia, illustrating a two- to fourfold body-weight enhancement compared with non-transgenic controls. (Photograph by Norman Maclean.)

promoter–chinook salmon GH fusion: they grew 2.5- to fourfold faster and converted feed 20% better than their non-transgenic siblings (Rahman *et al.*, 1998, 2001; Rahman and Maclean, 1999). At 7 months, the mean body weight of transgenic tilapia was 653 g compared with 260 g for non-transgenic siblings. These were heterozygous lines of GH-transgenic Nile tilapia, and the accelerated growth was obtained in the F_1 and F_2 generations. However, F_1 fish transgenic for a construct consisting of a sockeye salmon MT promoter spliced to a sockeye salmon GH gene exhibited no growth enhancement (Rahman *et al.*, 1998), although salmon transgenic for this construct show greatly enhanced growth. The growth-enhanced transgenic lines of Nile tilapia were strongly positive for the salmon GH in their serum, whereas the non-growth-enhanced lines were negative. Attempts to induce expression from the MT promoter by exposing fish to increased levels of zinc failed. Preliminary results indicated that homozygous transgenic Nile tilapia produced from the ocean pout antifreeze–chinook salmon GH construct have growth similar to that of the hemizygous transgenics.

Insertion of other GH constructs into tilapia has also yielded positive results, but not as dramatic as those with the salmon GH constructs. Two possible explanations for the difference in results are that the type of construct and the type of tilapia studied were different. Introduction of a CMV–tiGH construct into a hybrid *O. hornorum* resulted in a 60–80% growth acceleration (Martinez *et al.*, 1996; Estrada *et al.*, 1999), depending on the culture conditions. Different patterns and levels of ectopic expression of tiGH and tilapia insulin-like growth factor (IGF) were detected in organs of four lines of transgenic tilapia by RNA or protein analysis (Hernandez *et al.*, 1997). The two lines with lower ectopic tiGH mRNA levels were the only ones that exhibited growth acceleration, suggesting that the expression of ectopic tiGH promoted growth only at low expression levels. Higher ectopic tiGH levels resulted in a low condition factor. Overexpression of tiGH had no positive or negative effects, similar to the result observed in GH-transgenic pigs but opposite to what was observed in GH-transgenic salmon exhibiting hyperlevels of GH.

One of the most thorough studies of GH-gene transfer is that of the transfer of the rtGH cDNA driven by the RSV-LTR promoter into channel catfish and common

Fig. 16.7. Growth-hormone-transgenic salmon and controls illustrating a ten- to 30-fold increase in growth rate. (Photograph by Robert Devlin, adapted from Devlin *et al.*, 1994b.)

carp. Transgenic individuals of some families of carp and catfish grow 20–60% faster than their non-transgenic full siblings, but in some families no differences exist. Differences in genetic background, epistasis, copy number of the foreign gene, insertion site and level of expression are logical explanations for these results, which also illustrate the fact that a combination of traditional breeding programmes, such as selection, along with gene transfer will probably be necessary to develop the best genotypes for aquaculture.

Zhang *et al.* (1990) have shown the gene transfer, expression and inheritance of rtGH cDNA in the common carp. Progeny that inherited the transgene from transgenic common carp possessing the RSV–rtGH cDNA grew 20–40% faster than full siblings that did not inherit the gene (Zhang *et al.*, 1990). Thirty to fifty per cent of the transgenic progeny were larger than the largest non-transgenic sibling. The coefficient of variation for body weight was similar for transgenic and non-transgenic siblings, indicating that the population distribution was the same for body weight. The percentage of individuals that were deformed was not different between transgenic and non-transgenic progeny.

Transgenic common carp and silver crucian carp possessing MThGHg grew 11 and 78% faster than non-transgenic controls, respectively (Chen *et al.*, 1989). The different response of the two species can be explained by the variable expression of individuals, differing insertion sites and therefore differing regulation and expression, or small sample sizes.

The transgenic common and silver crucian carp had a coefficient of variation twice that of non-transgenic controls (Chen *et al.*, 1989). This is in contrast to the results of Zhang *et al.* (1990), and may be due to the less consistent expression of the transgenic carp in the experiments of Chen *et al.* (1989). All transgenic carp evaluated that possessed RSV–rtGH cDNA expressed recombinant GH, while only 50% of the transgenic carp that possessed MThGHg expressed recombinant GH. Variable expression should lead to variable growth. Loach containing MThGHg (Maclean *et al.*, 1987a) and northern pike containing RSV–bovine GH gene (Schneider *et al.*, 1989) also exhibited more variable growth than controls. Combining selection with genetic engineering procedures might reduce this variation.

Transgenic salmon illustrate the most dramatic results obtained in fish genetic engineering (Devlin *et al.*, 1994b). The GH-gene constructs utilized were comprised entirely of piscine gene sequences using either an ocean pout antifreeze promoter (opAFP) driving a chinook salmon GH cDNA, or a sockeye salmon MT promoter driving the full-length sockeye GH1 gene. When introduced into salmonids, these gene constructs elevated circulating GH levels by as much as 40-fold (Devlin *et al.*, 1994b; Devlin, 1997b), resulted in up to five- to 30-fold increase in weight after 1 year of growth (Du *et al.*, 1992; Devlin *et al.*, 1994b, 1995a,b), and allowed precocious development of physiological capabilities necessary for marine survival (smoltification). The largest of these P_1

transgenics were mated and produced offspring with extraordinary growth.

The extraordinary accelerated growth was obtained in a number of salmonid species. The opAFP–chinook salmon GH construct accelerated growth in coho salmon by ten to 30 times (Devlin *et al.*, 1995a). Parr–smolt transformation occurred 6 months early in the transgenic fish compared with the control fish. Results varied among species and families and might be related to different gene constructs, coding regions, chromosome positions and copy numbers. Insertion of opAFP–chinook salmon GH1 cDNA increased growth 3.2-fold in rainbow trout, whereas the sockeye MTB–sockeye GH1 accelerated growth tenfold (Devlin *et al.*, 1995a; Devlin, 1997b). The opAFP–chinook salmon GH1 cDNA construct improved growth tenfold in cutthroat trout, *Oncorhynchus clarki*, and 6.2-fold in chinook salmon (Devlin *et al.*, 1995a). Growth at 5 months was better than growth at 15 months; again illustrating that, on a relative basis, genetic improvement for growth is usually more impressive at younger ages than at older ages. Condition factor, *K*, was lower for transgenic fish because length changed more rapidly than weight. Some families that had 30-fold increased growth exhibited acromegaly in the jaw, skull and opercular area (Fig. 16.8) and, by 15 months, the growth of these fish slowed and they died. As was seen with transgenic common carp and channel catfish, the effect of GH-gene insertion was variable among families, and multiple insertion sites and multiple copies of the gene were observed. Transgenic rainbow trout experienced early maturation at 2 years of age, but in the same season as the controls.

Sockeye MTB–sockeye GH1 cDNA introduced into coho salmon increased growth 11- to 37-fold and increased GH expression by 40-fold in cold temperatures, when GH expression is normally low (Devlin *et al.*, 2001). Results with Atlantic salmon are not quite as impressive as with coho salmon. Transgenic Atlantic salmon containing the opAFP–chinook salmon GH1 cDNA construct had a three- to sixfold accelerated growth rate compared with non-transgenic

Fig. 16.8. Growth-hormone-transgenic salmon exhibiting both hyperlevels of growth and acromegaly. (Photograph by Robert Devlin, adapted from Devlin *et al.*, 1995b.)

salmon (Du *et al.*, 1992; Cook *et al.*, 2000a). Insertion of sockeye MTB–sockeye GH1 cDNA (Devlin, 1997b) produced a similar result, fivefold growth enhancement.

Prior to first feeding, the transgenic progeny were found to be 21.2% heavier and 11.9% longer than their non-transgenic siblings, suggesting that the expression of GH in early development can influence the rate or efficiency of conversion of yolk energy reserves (Devlin *et al.*, 1995a,b). Magnification effects can explain some of the growth differences between transgenic and control salmon; however, specific growth rates of the transgenic coho salmon were approximately 2.7-fold higher than those of older non-transgenic animals of similar size and 1.7-fold higher than those of their non-transgenic siblings (Devlin *et al.*, 2000), indicating that the transgenic salmon are growing at a faster rate at numerous sizes and life stages. GH levels were increased dramatically (19.3- to 32.1-fold) relative to size-control salmon, but IGF-I levels were only modestly affected, being slightly enhanced in one experiment and slightly reduced in another. Insulin levels in transgenic animals did not differ from same-size controls, but were higher than in non-transgenic siblings, and thyroxine levels in transgenic salmon were intermediate between levels found in size and age controls.

GH transgenic fish had always grown faster controls, but not reached abnormaly large sizes. However, GH transgenic loach have been developed in Korea that reach sizes 30× larger than the normal adult size.

Dramatic Growth of Transgenic Fish: Explanations and Limitations

Growth enhancement varies greatly among different transgenic fish systems (Devlin, 1997b; Dunham and Devlin, 1998), and salmonids in particular have shown the greatest response to stimulation to date. Several potential explanations exist that indicate it may be difficult to duplicate these results in other fish species.

It is possible that completely homologous gene constructs, derived only from the same species or from piscine sources, such as those used successfully in salmonids, are expressed in fish more efficiently than are gene constructs derived from other vertebrates. While this probably plays a role in efficient expression, it is also very likely that the differences in growth response observed in different transgenic systems are due to the vastly different physiologies and life-history characteristics that exist among the fish species examined (Dunham and Devlin, 1998).

The biology of salmon and their unique physiological adaptations probably play an important role in the dramatic growth enhancement observed in transgenic individuals. Growth in salmonids is normally relatively slow throughout the year, and is extremely low when water temperatures are low and food resources in nature are scarce. This low growth rate appears to be controlled at least in part by the level of circulating GH, and can be stimulated dramatically with exogenous GH protein and sufficient food. The dramatic growth stimulation observed in transgenic salmonids may arise partially from the seasonal deregulation of GH expression (Devlin *et al.*, 1994b, 1995a,b) to allow high growth rates during winter months when control salmon have very slow growth rates. This accelerated winter growth may also give the transgenic fish a large advantage that can later be magnified (Moav and Wohlfarth, 1974a,b). Additionally, salmonids are anadromous, and accelerated growth in transgenics allows them to reach a size at which they smolt earlier than controls. Growth in the smolt stage is naturally increased, providing transgenic individuals with a further advantage to increase the relative growth difference from controls (Dunham and Devlin, 1998).

Non-salmonid fish species often display more rapid growth, and consequently may be much more difficult to stimulate by expression of GH in transgenic organisms (Devlin, 1997a,b). These high growth rates occur naturally in some species such as tilapia, whereas others have been enhanced through genetic selection and through many years of domestication. Strains or species that have been selected to near-maximal growth rates presumably have had many of their metabolic and physiological processes optimized, and might be expected to be more difficult to stimulate by a single factor such as GH.

Domestication is also important in transgenic growth responses, and Devlin *et al.* (2001) observed that salmonid GH-gene constructs that have a dramatic effect on growth in wild rainbow trout strains (with naturally low growth rates) have little or no effect in strains where growth rate has been enhanced by selection over many years. P_1 and F_1 rainbow trout derived from a slow-growing wild strain and containing salmon metallothionein growth hormone (OnMTGH1) grew 17-fold faster than controls. Transgenic males and females eventually reached 8.2 and 14.2 kg, compared with 220 and 171g for non-transgenic males and females, respectively. These wild transgenic rainbow trout grew no faster than a fast-growing non-transgenic, domestic rainbow trout. This is indicative of the tremendous progress that domestication and selection have had on some aquaculture species and implies a possible ten- to 20-fold improvement of growth through selection.

Introduction of the transgene into the domesticated P_1 increased growth by only 4.4% (Devlin *et al.*, 2001); however, replication was extremely low, and fast-growing families could have been missed. These data indicate that the effects of selection and transgenesis are similar, but not additive in any way. Supportive of the conclusions reached by Devlin *et al.* (2001), rainbow trout treated with exogenous GH protein exhibited similar effects (Maclean *et al.*, 1987b) to those in the wild and domestic transgenic rainbow trout. Body weight of untreated wild-strain rainbow trout increased at a modest specific growth rate of 0.68%/day and untreated domestic rainbow trout grew at 2.18%/day; however, specific growth rates of hormone-treated wild-strain trout were enhanced 2.7-fold, similar to the untreated domestic fish, and domestic rainbow trout had an increase of only 9%. However, when we look at absolute rather than specific growth rate, the treated domestic fish were the largest group and were 25% larger than the untreated domestic fish, which is very close to and analogous to the transgenic GH results for domestic channel catfish and common carp. Consistent with these observations, the dramatically growth-responsive salmon previously observed (Du *et al.*, 1992; Devlin *et al.*, 1994b, 1995a,b) were also derived from wild strains. Apparently, slow-growing wild strains can benefit much more from GH insertion than fish that already have growth enhancement from selective breeding.

Similarly, dramatic growth stimulation in the mammalian system using GH transgenes has been observed in mice, but not in selected mice and domestic livestock that have had many centuries of genetic selection (Palmiter *et al.*, 1982; Pursel *et al.*, 1989). In these domesticated and selected lines, the capacity for further growth improvement by GH may now be restricted by limitations in other physiological pathways, and other methods, including traditional selective breeding methods, may yield the greatest gains. For aquacultural species, which have a much shorter history of domestication and selection, future genetic improvement will probably be accomplished by utilizing a combination of both approaches simultaneously (Dunham and Devlin, 1998).

In contrast and in comparison, GH transgenic catfish derived from domesticated and selectively bred strains exhibit only a moderate growth enhancement (41%). However, if we extrapolate from a series of experiments starting with slow-growing wild strains of channel catfish and then improve their growth through domestication (Dunham, 1996), followed by further improvement from selective breeding (Padi, 1995) and then further increases from inter-specific hybridization (Jeppsen, 1995) or gene transfer, the overall growth enhancement is approximately tenfold, comparable to that observed with transgenic wild salmon. Thus it appears that growth of wild fish can apparently be improved in one or two generations with the insertion of GH genes to the extent that would take many generations of selective breeding to achieve.

However, additional data on transgenic rainbow trout (Devlin *et al.*, 2001) refute this hypothesis of the effect of wild and domestic genetic backgrounds on response to GH transgene insertion. When OnMTGH1 was transferred to another wild rainbow trout strain, F77, growth was enhanced sevenfold, almost fourfold greater growth than that observed in a non-transgenic domestic rainbow trout. In this case, the wild transgenic rainbow trout is actually superior to the domestic selected strain, indicating that genetic engineering can have a greater effect than, rather than an equivalent effect to, domestication and selection. Perhaps strain effects in general, epistasis and genetic background are more significant with regard to affecting transgene response, rather than the domestic or wild nature of the fish. When F77 was crossbred with the domestic strain, growth of the crossbreed was intermediate to the parent strains, a typical result (Dunham and Devlin, 1998). However, the transgenic wild × domestic crossbreed was by far the largest genotype, 18 times larger than the non-transgenic wild parent, 13 times larger than the non-transgenic wild × domestic crossbreed, nine times larger than the non-transgenic domestic parent and more than 2.5 times larger than the wild F77 transgenic parent (Devlin *et al.*, 2001). The combined effects of transgenesis and crossbreeding had a much greater growth enhancement than crossbreeding or transgenesis alone. Additionally, a transgenic with 50% of its heritage from domestic sources was much larger than a wild transgenic, so great response from some domestic genotypes is possible.

Cold tolerance

Most efforts in transgenic fish have been devoted to growth enhancement, although there are reports of improvement in cold and disease resistance (Fletcher and Davies, 1991; Shears *et al.*, 1991; Anderson *et al.*, 1996; Liang *et al.*, 1996). Early research also involved the transfer of the antifreeze protein gene of the winter flounder (Fletcher *et al.*, 1988). The primary purpose of this research was to produce salmon that could be farmed under Arctic conditions, but expression levels obtained have been inadequate for increasing the cold tolerance of salmon. However, preliminary results with goldfish show some promise for increasing survival within the normal cold temperature range. Initial thrusts at improving cold tolerance via gene transfer have been minimally successful.

Disease resistance

Transgenic fish with enhanced disease resistance would increase profitability, production, efficiency and the welfare of the cultured fish. Genetic gain is also possible through traditional selective breeding, but it appears that the rate of genetic improvement and the consistency of genetic improvement may be greater with the transgenic approach for disease resistance. Selective breeding may also have the drawback that the disease organism may well respond to selective forces too, negating some of the selection response in the fish.

Momentum is being gained in transgenic enhancement of disease resistance. Expression of viral coat protein genes (Anderson *et al.*, 1996) or antisense expression of viral early genes may improve virus resistance, although bacterial diseases are often greater threats to the major aquaculture species, and bacterial disease resistance may be easier to genetically engineer than resistance for diseases caused by other classes of pathogen. Expression of viral coat protein genes or antisense expression of viral early genes has improved viral resistance in rainbow trout. Shrimp have been genetically engineered with antisense Taura syndrome virus-coat protein gene resulting in a doubling of the resistance to this disease.

One potential mechanism for improving disease resistance is the production of transgenic aquatic organisms containing lytic peptide genes. A great deal of information is available concerning antibacterial peptides; the number of structural families into which they fall is very large, it seems likely that they occur ubiquitously (Bevins and Zasloff, 1990; Lehrer *et al.*, 1993; Boman, 1996; Hoffmann *et al.*, 1996; Hancock, 1997) and organisms containing these genes should exhibit enhanced disease resistance. The earliest discovered and most thoroughly studied antibacterial peptides are the cecropins, small cationic peptides found originally in the moth *Hyalophora cecropia* (Steiner *et al.*, 1981). Antimicrobial peptides other than cecropin have been identified in many invertebrates and vertebrates.

Cecropins are initially translated into precursors of 62–64 amino acid residues, and are then processed intracellularly into mature peptides of 35–37 amino acid residues (Boman *et al.*, 1991; Boman, 1995). The unique structural features of cecropins and other antimicrobial peptides allows them to readily incorporate into cellular membranes of bacteria, fungi and parasites, resulting in the formation of pores on the membrane and leading to the inevitable death of pro- and eukaryotic pathogens (Bechinger, 1997). Cecropin analogues can be designed and synthesized that are as effective as, or even more potent than, the native compounds against animal and plant bacterial pathogens (Kadono-Okuda *et al.*, 1995; Merrifield *et al.*, 1995; Vunnam *et al.*, 1995) and protozoa (Rodriguez *et al.*, 1995). Cecropin genes and their analogues have been used to produce transgenic plants, such as potato and tobacco, with increased resistance to infection by bacterial or fungal pathogens (Hassan *et al.*, 1993; Jia *et al.*, 1993; Huang *et al.*, 1997).

Several studies have demonstrated the *in vitro* effectiveness of cecropins. Cecropins are of potential benefit in aquaculture because they show a broad spectrum of activity against Gram-negative bacteria (Kelly *et al.*, 1990; Kjuul *et al.*, 1999), which include most of the major bacterial pathogens of ictalurid catfish (Thune, 1993); they are non-toxic to eukaryotic cells (Jaynes *et al.* 1989; Kjuul *et al.*, 1999); they are found in mammals, as described for the pig (Lee, J.Y. *et al.*, 1989), as well as insects; and there is an extensive literature on the physicochemical properties and mode of action of cecropins which will facilitate the redesign of cecropin derivatives, if necessary. Passively administered cecropin derivatives can confer some protection against infection with *Edwardsiella ictaluri* (Kelly *et al.*, 1993), a major pathogen of cultured catfish (Thune, 1993). Chiou *et al.* (2002) examined the *in vitro* effectiveness of native cecropin B and a synthetic analogue, CF17, for killing several fish viral pathogens: IHN virus, VHS virus, snakehead rhabdovirus (SHRV) and IPN virus. When these peptides and viruses were co-incubated, the viral titres yielded in fish cells were reduced from several- to 104-fold. Direct disruption of the viral envelope and disintegration of the viral capsids may be the explanation for the inhibition of viral replication by the peptides. These antimicrobial peptides were more effective on enveloped viruses than on non-enveloped viruses, and may act at different stages during viral infection to inhibit viral replication (Chiou *et al.*, 2002). Three mechanisms – direct inactivation of viral particles by perturbing the lipid bilayers of the viral envelopes (Daher *et al.*, 1986), prevention of viral penetration into the host cell by inhibiting viral–cellular membrane fusion (Srinivas *et al.*, 1990; Baghian *et al.*, 1997) and inhibition of viral replication in infected cells by suppressing viral gene expression (Wachinger *et al.*, 1998) – have been proposed to explain the antiviral action. The same antimicrobial peptides act on different viruses through different mechanisms (Daher *et al.*, 1986; Srinivas *et al.*, 1990; Aboudy *et al.*, 1994; Marcos *et al.*, 1995; Baghian *et al.*, 1997; Wachinger *et al.*, 1998).

The *in vitro* studies indicate that transgenesis utilizing cecropin constructs should improve disease resistance in aquatic organisms. Bacterial disease resistance may be improved up to three- to four-fold through gene transfer. Insertion of the lytic peptide cecropin-B construct enhanced resistance to bacterial diseases two- to fourfold in channel catfish (Dunham *et al.*, 2002c). There was no pleiotropic effect on growth. P_1 transgenic catfish containing the cecropin-B construct were spawned and the transgene was transmitted to the F_1 generation. Transgenic and non-transgenic full siblings containing the cecropin-B construct were challenged in tanks with *E. ictaluri*. Both genotypes experienced mortality, but survival of the transgenic individuals was twice that of the controls. Transgenic channel catfish containing the preprocecropin-B construct and their full-sibling controls experienced a natural epizootic of columnaris, *Flavobacterium columnare*. No cecropin-transgenic fish were among the mortalities, and only control fish died. Both transgenic and control individuals were among the survivors. In this case, the transgene appears to have

imparted complete resistance. Catfish transgenic for cecropin constructs show enhanced resistance to both deliberate and natural challenge with pathogenic bacteria.

Similar results were obtained for cecropin-transgenic medaka (Sarmasik *et al.*, 2002). F_2 transgenic medaka from different families and controls were challenged with *Pseudomonas fluorescens* and *Vibrio anguillarum*, killing about 40% of the control fish by both pathogens, but only 0–10% of the F_2 transgenic fish were killed by *P. fluorescens* and about 10–30% by *V. anguillarum*. When challenged with *P. fluorescens*, zero mortality was found in one transgenic fish family carrying preprocecropin B and two families with porcine cecropin P1, and 0–10% cumulative mortality was found for five transgenic families with procecropin B and two families with cecropin B. When challenged with *V. anguillarum*, the cumulative mortality was 40% for non-transgenic control medaka, 20% in one transgenic family carrying preprocecropin B, between 20 and 30% in three transgenic families with procecropin B and 10% in one family with porcine cecropin P1. RT-PCR analysis confirmed that transgenic fish from most of the F_2 families expressed cecropin transgenes except those in three F_2 families.

Family variation was observed, and family variation can be extreme for transgenic fish potentially because of differences in genetic background, variable insertions sites, copy number, epistasis and other factors. This necessitates coupling of selection with gene transfer to obtain maximum genetic gain from the gene transfer.

Other antimicrobial peptides have shown promise for the protection of fish against pathogens and might be utilized with transgenic approaches. A cecropin–melittin hybrid peptide (CEME) or pleurocidin amide, a C-terminally amidated form of the natural flounder peptide, was delivered continuously using miniosmotic pumps placed in the peritoneal cavity, followed 12 days after pump implantation with intraperitoneal injections, into juvenile coho salmon infected with *V. anguillarum*, the causative agent of vibriosis (Jia *et al.*, 2000). Juvenile coho salmon that received 200 μg of CEME/day had lower mortality (13%) than the control groups (50–58%), and the fish that received pleurocidin amide 250 μg/day had lower mortality (5%) than control groups (67–75%).

Grass carp, *Ctenopharyngodon idellus*, have been transfected with carp β-actin–human lactoferrin gene, resulting in P_1 individuals that were more resistant to *Aeromonas*, exhibited enhanced phagocytosis and more viral resistance than controls. Japanese flounder keratin promoter linked to both the hen egg white (HEW) lyoszyme gene and GFP gene, and transferred to zebra fish, resulted in F_2 transgenic zebra fish with lytic activity of protein extracts from the liver 1.75 times higher than in the wild-type zebra fish. In a challenge experiment, 65% of the F_2 transgenic fish survived an infection of *F. columnare* and 60% survived an infection of *Edwardsiella tarda*, whereas 100% of the control fish were killed by both pathogens.

To date, two types of transgenic process have been successful for improving disease resistance: blocking viruses with antisense and overexpressing antibacterial compounds from distant taxa. This is the extent of reported research using transgenesis to directly improve disease resistance. However, transfer of other genes can indirectly affect disease resistance through pleiotropy both in a positive and negative way. If the pleiotropic effects are known, they could be intentionally manipulated for genetic gain in a manner analogous to indirect selection.

Body composition

Transgenic alteration of the nutritional characteristics of fish could be beneficial for consumers, and it is now possible to directly alter body composition via transgenesis. Zebra fish transfected with β-actin–salmon desaturase genes had enhanced levels of the *n*-3 fatty acids, docosahexaenoic acid (DHA) and eicosapentaenoic acid (EPA), in their flesh.

Transgenic Production of Pharmaceuticals

Another use of transgenic organisms is as biological factories to produce highly valuable pharmaceutical proteins. With the advent of diseases such as acquired immune deficiency syndrome (AIDS) and hepatitis, it became even more important to develop alternatives for extracting compounds such as blood-clotting factors from human blood. Transgenic goats, cows and other livestock have been developed that produce valuable biomedical products. Usually, the transgenically produced protein is extracted from the milk. Such technology is especially important in the modern world since human extracted products have the potential to be contaminated with HIV, hepatitis viruses and other human pathogens. These products can also be quite expensive. Transgenically produced biomedical compounds should be safe from human pathogens, should eventually be less expensive and more widely available.

Fish have potential advantages as bioreactors compared with mammals. These advantages include a short generation interval, low cost of maintenance of the animals, easy maintenance, large numbers of individuals, high-density culture and mammalian viruses and prions are not found in fish. Several examples are now available demonstrating the potential of fish as bioreactors for medical products, as well as for compounds that can be used in fish spawning.

CMV–human coagulation factor VII was produced in transgenic zebra fish, African walking catfish and Nile tilapia eggs. Clotting activity was detected, indicating proper post-translational modifications. Proteins could be collected in eggs, serum or possibly different proteins in different tissues for other types of gene.

Transgenic Nile tilapia secreted human insulin in Brockmann bodies. Islet tissue was used for xenotransplantation and successfully transferred to diabetic nude mice, reversing the effects of diabetes.

Single-chain goldfish LH gene was injected into rainbow trout eggs. At 4 days of age goldfish LH was isolated from the eggs and the recombinant LH injected into goldfish. Testosterone levels were elevated in male goldfish after the injections, proving biological activity.

Reverse Genetics/Gene Knockout/ Gene-Knockdown Technology

Theoretically, inactivation of a gene can be accomplished by knockout of the gene, by replacing the original gene with a mutated copy of the gene, homologous recombination, or by disruption of gene expression using the antisense approaches, ribozyme technology or interference RNA (RNAi). Homologous recombination can also be used to knockout in genes. The knockout approach is the ultimate method for gene inactivation because it will eliminate the transcript completely. Disadvantages of the antisense and ribozyme approaches are potential problems with position effects after integration and transgene inactivation in F_1 or later generations due to methylation of the transgenes.

Genes can also be 'knocked out' or mutated to study gene expression using chemical or irradiation mutagenesis. This has been a useful programme in zebra fish and many mutant varieties have been studied. However, chemical and irradiation mutation is extremely tedious and slow compared with insertional mutagenesis, and even when distinctive mutants are observed, it may be difficult to ascertain which gene has been mutated.

Gene knockout has not yet been accomplished directly with aquatic embryos. Gene knockout efficiency would probably be low since the selection methodology has not been developed to screen for the proper genotypes. A second approach using pluripotent embryonic stem (ES) cells or PGCs has been employed for the production of knockout transgenics in mice (Evans and Kaufman, 1981; Gossler et al., 1986; Matsui et al., 1992; Nagy et al., 1993; Labosky et al., 1994). Because ES cells and PGCs are pluripotent and contribute to the germ-cell lineage when transplanted into host embryos, these cells, if transformed, can serve as a vector for the introduction of foreign DNA into the germ line of the organism. In vivo pluripotency has been demonstrated by transmission of ES-cell genotypes to chimeric offspring in rabbit and pig (Wheeler, 1994; Schoonjans et al., 1996; Shim et al., 1997) and in fish (Hong et al., 2000). Additionally, pluripotency of embryo-derived cells has been demonstrated by bovine conceptus development and the birth of live lambs after nuclear transfer (Campbell et al., 1996; Stice et al., 1996).

The use of cultured cells as a vector for the production of transgenic aquatic organisms is advantageous because it eliminates the problem of mosaicism, since in vitro selection follows transfer of foreign DNA into the cultured cells to isolate cell clones that have stably integrated and properly expressed the transgene. Additionally, homologous recombination can be used to inactivate or replace endogenous genes by targeted insertion (Thompson et al., 1988; Capecchi, 1989), which has resulted in the production of knockout mice for the examination of specific gene function without changing any other genomic or physiological condition (Ernfors et al., 1994; Olson et al., 1996; McPherron et al., 1997).

Homologous recombination activity has been observed in zebra fish embryos (Hagmann et al., 1998), indicating the potential for knockout and knock-in technology in fish. Chen et al. (2002) developed homologous recombination vectors and positive–negative selection procedures for fish cells. The positive–negative selection procedures were functional, but the gene targeting was not achieved.

Despite the advantages of using PGCs and ES cells for transgenic research, this technology has not been fully developed for use with aquatic organisms. The primary disadvantage of this technology is that the founder knockout stock would be derived from an extremely small genetic base. This could be corrected by crossbreeding, but it would take two generations to develop an outbred population that was homozygous for the knockout construct.

The first step for developing this technology is the isolation of ES cells from aquatic organisms, and some initial work has been conducted with zebra fish (Collodi *et al.*, 1992a,b). Pluripotent embryonic cells exhibiting *in vitro* characteristics of ES cells have been cultured (Sun *et al.*, 1995a,b,c), and they contributed to tissues derived from all three germ layers when introduced into developing host embryos (Sun *et al.*, 1995a,b,c; Ghosh *et al.*, 1997; Speksnijder *et al.*, 1997). These cultures were obtained from blastulae. These zebra fish embryonic cells exhibited a similar morphology to ES cells, and were induced to differentiate into multiple cell types in culture, including melanocytes and muscle and nerve cells. Since this earlier work, ES-like cell lines have been developed from medaka (Wakamatsu *et al.*, 1994; Hong and Schartl, 1996; Hong *et al.*, 1996; Chen *et al.*, 2002) and seabream (Bejar *et al.*, 1997, 1999). The zebra fish cells stained positive for ALP activity, an indicator of pluripotency.

The medaka cell line, MES1, has shown pluripotency, retains the aneudiploid genotype and forms viable chimeras when injected into blastulae that later contribute to all three germ layers (Hong *et al.*, 1996, 1998a,b, 2000; Chen *et al.*, 2002). Genetic background has a key role in establishing these ES cell lines (Chen *et al.*, 2002).

ES-like cells have contributed to the germ-line cells of a host embryo. Both of these cell types, PGCs and spermatogonia A (SSCs), have been transplanted from a donor species to a related host species (Takeuchi *et al.*, 2003; Okutsu *et al.*, 2006a, 2007, Saito *et al.*, 2008) with the recipient species producing sperm and eggs (originating from testicular PGCs or SSCs) of the target species (Okutsu *et al.*, 2006b). This procedure can be successful utilizing fresh, cultured or cryopreserved donor cells, opening up the possibility of many potential applications.

Alternative methods also have the potential to allow xenogenesis and autogenesis via PGC transplantation. Saito *et al.* (2008) transplanted a single PGC from pearl danio, *Danio albolineatus*, into the blastula of a zebra fish whose native PGC production had been knocked out by an antisense morpholino oligonucleotide against dead end. The donated PGC formed a single testis that produced pearl danio sperm. Xenogenic pearl danio males were sex-reversed to femaleness and mated with untreated males to produce normal, fertile pearl danio offspring. Similarly, the zebra fish host was able to develop goldfish, *Carassius auratus*, and loach, *M. anguillicaudatus*, testis that produced donor sperm from the injection of a single donor PGC. Normally, a few dozen PGCs are needed to form gonads containing germ cells (Saito *et al.*, 2006). This study showed that one PGC and perhaps a single SSC are capable of producing a single testis.

Shikina *et al.* (2008) developed a technique to isolate and culture type A spermatogonia from the testes of immature rainbow trout. The cultured spermatogonia were maintained for more than a month, and were able to colonize recipient gonads and proliferate upon transplantation into embryos near hatching. This could be a vital link to making cell-mediated gene transfer possible in fish, as knockout mice have been produced from spermatogonial cell lines (Kanatsu-Shinohara *et al.*, 2006). If these cultured cells can be transformed and continue to be cultured, transgenic embryos could be generated through xenogenesis. The advantage of this system is that the initial transgenics would not be mosaic in contrast to transgenics produced traditionally by DNA microinjection, thus greatly speeding transgenic research.

In vitro selection can also be conducted on PGCs to establish cell lines containing genomes with specific genes knocked out. The strategy of knockout is to produce transgenic fish that are homozygous for null genes, which have negative effects on commercially important traits. One difficulty is identifying and marking PGCs. ALP activity can be used as a marker for germ cells, as can vasa, a DEAE box protein family and a homologue of *Drosophila* vasa, which has been cloned from mouse and *Xenopus* (Fujiwara *et al.*, 1994; Komiya *et al.*, 1994). Vasa mRNA is visible in embryos at the four-cell stage and, by mid-blastula, four distinct small groups of PGCs can be visualized by *in situ* hybridization using a vasa probe. At the early somite stage the PGCs should aggregate on the ventral side of the embryo near the interface with the developing yolk-sac.

An alternative method for identifying PGCs is to examine the cells for their ability to produce well-differentiated embryoid bodies similar to pluripotent ES cells (Matsui *et al.*, 1990; Labosky *et al.*, 1994). PGCs are homologously transfected with the gene-targeting vectors and selected using drugs such as G418 (neomycin is used in the vector). Prior to reintroduction into host embryos by microinjection, transformed cells need to be karyotyped to ensure that they still contain the full chromosome complement.

Theoretically, the transformation of the embryos can be accomplished by combining androgenesis and microinjection (Yoon *et al.*, 1990; Moav *et al.*, 1992a,b; Dunham, 1996) or by traditional nuclear transfer. In the case of the androgenetic approach, eggs are irradiated with gamma rays or UV rays to enucleate or destroy the maternal DNA. The irradiated eggs are activated with UV-irradiated sperm so that there is no paternal contribution. Transformed PGCs are microinjected into the enucleated eggs for the production of transgenic fish.

Knockout technology has been utilized to develop transgenic mice with greatly altered growth and body composition. Disruption of the myostatin gene, GDF-8, resulted in GDF-8-null mice with a 262% increase in muscle, a 25% increase in growth rate and a 27% decrease in fat percentage compared with controls. Homozygous mutants of the knockout mice were viable and fertile.

Additional techniques for gene silencing or knockout exist. Not all of them can be used for transgenesis. Some of the techniques do not have an effect on the germ line, but can be used to study gene expression, perhaps leading to future gene manipulations.

The insertion of a transgene can sometimes have a paradoxical effect and actually silence rather than enhance gene expression, apparently due to the effects of the mRNA that is generated. Another technique for post-transcriptional gene silencing is the utilization of single-stranded RNA antisense constructs; however, Fire *et al.* (1998) found that double-stranded RNA was more effective at gene silencing – RNA interference – than single-stranded RNA, and some earlier success of single-stranded RNA was actually due to contamination with double-stranded RNA. For RNA interference to work, an organism apparently has to have specific genes present. RNA-dependent RNA polymerases may need to be present, and this phenomenon may be associated with the evolution of defence mechanisms against RNA viruses. RNA interference can be accomplished in vertebrates. Both double-stranded RNA and antisense RNA were effective in disrupting the expression of GFP in transgenic zebra fish (Gong *et al.*, 2002).

Utilization of antisense oligos is another alternative for gene silencing. Antisense oligos need to possess the following traits: they must achieve efficacy in the cell at reasonable concentrations, they should inhibit the target sequences without attacking other sequences, they need to be stable extracellularly and within the cell, and they must be deliverable to the cellular compartments containing the target RNA (Summerton and Weller, 1997). Antisense oligos need to be as structurally similar to DNA as possible without coming under attack by nucleases.

The original antisense oligos were made from natural genetic material, with artificial crosslinking moieties added to irreversibly bind the antisense oligos to their target RNA, but these first-generation oligos were usually degraded naturally (Summerton *et al.*, 1997). Methylphosphonate-linked DNA oligos were then designed which resisted enzymatic degradation, but they had poor efficacy and poor aqueous solubility. Phosphorothionate-linked DNAs (S-DNAs) were developed next, resulting in greater efficacy and solubility; however, they had the disadvantages of a narrow range of efficacious concentrations and various interactions with non-target proteins. Morpholine-modified oligonucleotides (morpholinos) have overcome most of these problems. They contain 6-membered morpholine backbone moieties, joined by non-ionic phosphorodiamidate intersubunit linkages. Morpholinos are designed for their excellent RNA-binding ability and their specificity. The design and construction of morpholinos are quite complex. This is a useful tool for studying gene expression, but production of a morpholino-transgenic fish is probably extremely difficult to accomplish.

Targeted gene knockdown or knockout has been accomplished in zebra fish (Nasevicius and Ekker, 2000). Antisense morpholinos blocked the ubiquitous expression of the GFP transgene in transgenic zebra fish. This approach also generated phenocopies of the mutations of the genes no tail, chordin, one-eyed-pinhead, nacre and sparse, and knocked out or reduced expression of the uroporphyrinogen decarboxylase, sonic hedgehog and tiggy-winkle hedgehog.

Casanova-mutant zebra fish embryos lack endoderm and develop cardia bifidia, and Dickmeis *et al.* (2001) were able to duplicate this condition by knocking out the HMG box-containing gene, 10J3, with morpholine oligonucleotides.

Manipulating and transplanting nuclei to accomplish gene knockout is difficult and has yet to produce a transgenic aquatic organism. Manipulating blastomeres to either add or knock out genes may be a future possibility. Takeuchi *et al.* (2002) were able to produce germ-line chimeras of rainbow trout. Orange mutant embryos were used for donor cells to wild-type embryo donors, and cells were transplanted utilizing microinjection. Mid-blastula and early blastula were optimum developmental times for donor and host embryos. Survival of manipulated embryos was 12%, and 50% of these fish exhibited mutant coloration. When mated with wild-type individuals, 32% of the chimeras transmitted the orange colour to progeny at a rate of 0.414%; therefore the orange mutant must be a dominant mutant.

An increasing amount of research focuses on RNA interference (RNAi) to accomplish gene knockdown and knockout. Theoretically, short RNAs can block the function of large mRNA. Various forms of short RNA exist or can be developed, including siRNA (silencing RNA), miRNA (micro RNAs associated with binding of the 3′-ends of mRNA) and shRNA (short hairpin RNA) for systems to degrade mRNA. These are all processed with the dicer/RISC (RNA-induced silencing complex) system to degrade the mRNA. Animal cells possess a defence mechanism that recognizes double-stranded mRNA, attacks and degrades it. This system likely exists because double-stranded RNA can be associated with viral infection. The antisense strand binds with the sense strand forming double-stranded mRNA, which is the mechanism by which the older traditional antisense RNA system should have functioned. Theoretically, this would block translation via two mechanisms. The double-stranded RNA cannot be translated and it is also destroyed by the dicer/RISC system. However, most antisense experiments have failed.

Approaches using the smaller RNAs have more potential. Dicer is an RNase III-like enzyme that recognizes double-stranded RNA and cleaves it into double-stranded pieces comprising 21–23 nucleotides. Denaturation of the double-stranded RNA strands allows them to bind with complementary RNA to form base-pair sequences that are recognized by RISC.

These systems have shortcomings also. The problem with gene knockdown is that as little as 1–5% normal activity or expression can still allow normal function of the gene, protein and thus normal physiology (Larson *et al.*, 2004). Another problem is that if RNAi is effective, it can accidently knock down multiple, non-targeted genes, affecting additional traits that should not be knocked down (Jackson *et al.*, 2003; Behlke, 2006).

A recent development that may lead to more effective gene knockout is utilization of zinc-finger protein (ZFP) constructs.

ZFPs are transcription factors having a finger-like domain and unique, highly conserved consensus amino acid sequences. Breakage of genomic DNA at key locations can increase the ratio of homologous recombination. Therefore, a key to homologous recombination is to successfully introduce a cleavage site for the locus of interest. Utilization of zinc-finger nuclease, ZFN, an endonuclease, provides an efficient genomic DNA cleavage tool (Lloyd *et al.*, 2005).

ZFN can recognize DNA because its DNA-binding domain has a surface structure that can complement the double helix of DNA. The surface structure of ZFN can enter the major groove of the DNA double helix, using the α helix to have contact with basic groups of the DNA. ZFN can be artificially modified to have a specific DNA-binding domain with non-specific endonuclease activity to exert the shearing action, to introduce the break at the binding site to enhance the homologous recombination. If a gene of interest is supplied in conjunction with the shearing of the DNA, the chromosome will attempt to repair itself by splicing in the template provided that matches the end sequences regardless of what is in the centre of the gene template.

The DNA-binding domain of ZFN contains a minimum of three to four tandem Cys_2His_2 zinc fingers. Each zinc finger can identify and bind one specific triplet base group. Several zinc fingers connected together in series can form a zinc finger group which can identify a specific sequence of basic groups and have great specificity (Pabo *et al.*, 2001; Hurt *et al.*, 2003; Segal *et al.*, 2003; Kumar *et al.*, 2006).

The amino acids can be changed to obtain new DNA-binding specificities (Dreier *et al.* 2001), thus ZFN has both good DNA identification specificity and good plasticity. The non-specific cleavage domain from the type II restriction endonuclease *Fok*I is usually used as the cleavage domain (Kim *et al.*, 1996). The cleavage domain must dimerize to cleave DNA (Bitinaite *et al.*, 1998). Two ZFNs are used to bind opposite strands of DNA with their C-termini a defined distance apart. Recently, Cathomen and Joung (2008) modified the structure of *Fok*I which allows enzyme to cut activity only when two heterotype ZFNs work together. This decreases the ratio of shear dislocation (Szczepek *et al.*, 2007).

ZFNs can be used to excise dominant mutations in heterozygous individuals by producing double-strand breaks in the DNA. Multiple pairs of ZFNs can be utilized to remove entire large sequences in the genome (Lee, M.S. *et al.*, 1989).

The *gol* pigment gene and the *ntl* gene in zebra fish were modified with ZFN, creating mutations that were transmitted to further generations. Similar success was achieved for the *kdr* gene in zebra fish encoding the vascular endothelial growth factor-2 receptor. Homologous recombination knock-in was also accomplished in zebra fish using ZFN technology.

Potential weaknesses are similar to other gene knockdown technologies, such as cleavage of DNA at non-target locations. This may compromise the individual or result in random rather than targeted gene integration.

Pleiotropic Effects of Transferred Genes

When a gene is inserted with the objective of improving a specific trait, that gene may affect more than one trait. One gene affecting more than one trait is pleiotropy. Since pleiotropic effects could be positive or negative, it is important to evaluate several commercially important traits in transgenic fish in addition to the trait of intended alteration. Pleiotropic effects and correlated responses should be measured in any genetic improvement programme. The true breeding value of the fish and its value to the private sector are not completely known without measuring its performance for a variety of traits. The improvement of one trait may be offset by a decrease in performance of another trait.

Additionally, in the case of transgenic fish, it is important to measure numerous traits and potential pleiotropic effects because of environmental concerns, risks and speculations (Hallerman and Kapuscinski,

1995). The transfer of GH genes has pleiotropic effects: body composition, body shape, feed-conversion efficiency, disease resistance, reproduction, tolerance of low oxygen, carcass yield, swimming ability and predator avoidance can be altered.

The increased growth rate of transgenic individuals could be a result of increased food consumption, feed-conversion efficiency or both. Fast-growing transgenic common carp and channel catfish containing rtGH gene had lower feed-conversion efficiency than controls (Chatakondi, 1995; Dunham and Liu, 2002). Salmonids injected with GH also had improved feed conversion. Various transgenic common carp families had increased, decreased or no change in food consumption. Transgenic Nile tilapia had a 20% improvement in feed-conversion efficiency, and were better utilizers of protein and energy compared with controls (Rahman *et al.*, 2001). Transgenic tilapia expressing the tiGH cDNA under the control of human CMV regulatory sequences exhibited about 3.6 times less food consumption than non-transgenic controls, and feed-conversion efficiency was 290% better for the transgenic tilapia (Martinez *et al.*, 2000). Efficiency of growth, synthesis retention, anabolic stimulation and average protein synthesis were higher in transgenic than in control tilapia. Martinez *et al.* (2000) observed differences in hepatic glucose and in the level of enzymatic activities in target organs in the transgenic and control tilapia.

In GH-transgenic coho salmon, *Oncorhynchus kisutch*, the surface area of the intestine was 2.2 times that of control salmon and the growth rate was about twice that of controls. It seems likely that the enhanced intestinal surface area is a compensatory feature that is manifested in fast-growing salmonids (Stevens and Devlin, 2000b). The relative intestinal length was the same in transgenic and control salmon, but the surface area was greater for transgenics as a result of an increased number of folds. These differences could be related to the level of food consumption or GH may have a direct effect on intestinal growth. This increased intestinal surface area was found in both Atlantic and coho salmon. This change in intestinal surface area could also be associated with the increased feed-conversion efficiency of transgenic salmon. A digestibility trial suggested that transgenic tilapia were more efficient utilizers of protein, dry matter and energy (Rahman *et al.*, 2001).

The insertion of the rtGH gene alters the survival of common carp. The number of F_2 progeny inheriting this transgene was much less than expected. Differential mortality – a true pleiotropic effect – or loss of the recombinant gene during meiosis are likely explanations. These individuals were identified as transgenic or non-transgenic after reaching fingerling size. From that point on, survival of the remaining transgenic individuals was higher than that of controls when subjected to a series of stressors and pathogens, such as low oxygen, anchor worms, *Lernia*, *Aeromonas* and dropsy. GH-transgenic common carp had higher lysozyme activity in the serum compared with age-matched non-transgenic control fish. The serum bactericidal activity in the transgenics was 20% higher than in the controls. Values for leucocrit and phagocytic percentage of macrophages in head kidney were higher in transgenics than controls, but the phagocytic indices and relative spleen weights in the transgenics and the controls were not different. GH-transgene expression apparently not only stimulated growth, but also the non-specific immune functions of common carp.

Conversely, GH-transgenic salmon were more sensitive to *Vibrio* than controls, so GH-gene transfer does not always confer increased disease resistance. Survival among GH salmon families was sometimes improved, sometimes decreased and sometimes unchanged relative to controls. These differences in salmon may be caused by altered GH expression affecting a myriad of disease-related genes. GH appears to have pleiotropic effects and causes a cascade of events in a large number of biochemical pathways. Haem oxygenase, acyl-CoA binding protein, NADH dehydrogenase, mannose-binding lectin-associated serine protease, haemopexin-like protein, leucocyte-derived chemotaxin 2 (LECT2) and many other genes had enhanced expression in hepatic tissue of immature transgenic salmon, while complement C3-1, lectin, rabin, alcohol dehydrogenase, *Tc1*-like transposase and pentraxin genes had decreased expression compared with non-transgenic controls. Gene-expression pattern changed when transgenic salmon approached maturation, with haemopexin-like protein, haem oxygenase, inter α-trypsin inhibitor, LECT2, GTP cyclohydrolase I feedback regulatory protein (GFRP) and bikunin having enhanced expression and lectin, apolipoprotein and pentraxin exhibiting depressed expression. Lectin was found to be highly suppressed in all F_2 and immature F_3 salmon. Serum lysozyme activity of the innate immunity system was decreased in both generations of GH-transgenic fish. GH-transgenic amago salmon had altered hepatic gene expression relating to iron metabolism and innate immunity.

GH transgenesis can have significant metabolic costs. Cardiac function was enhanced by GH transgenesis, but universal up-regulation of post-smolt (adult) cardiorespiratory physiology did not occur in GH-transgenic salmon. Differences in arterial oxygen transport such as cardiac output and blood oxygen-carrying capacity are important for aerobic capacity; however, diffusion-limited processes may be bottlenecks that would need to be enhanced to achieve substantial improvements in metabolic and swimming performance. These diffusion-related limiting factors associated with gill function and morphology may explain differences in results from one study to another.

GH-gene transfer alters respiration and metabolism in many ways, which could affect the ability to overcome diseases either in an enhanced or detrimental manner. The results from common carp indicate GH transfer could be used as an indirect method to transgenically enhance disease resistance. The salmon situation may be different because of their different life history, the fact that they are a cold- rather than a warm-water fish and their GH enhancement is much more dramatic. The extent of pleiotropic effects is likely a product of the magnitude of the change in the primary target trait and the associated expression strength of the transgene.

In one example, both domestic and wild GH-transgenic rainbow trout had reduced survival, and all domestic transgenics died before sexual maturation (Devlin *et al.*, 2001). From an ecological standpoint this is beneficial, but, from an aquaculture or genetic improvement standpoint, this prevents utilization of the technology and F_1 and production populations cannot be produced.

Pleiotropy of GH gene for oxygen tolerance characteristics varies from one species to another. GH tilapia have a 58% higher metabolism than controls, compensate for oxygen consumption and have the same maximum swim speed as non-transgenics. GH-transgenic tilapia tolerate hypoxia equally as well as controls despite higher demand for oxygen.

GH-transgenic salmon have an increased need for dissolved oxygen; however, after 4 days of starvation, GH individuals had the same oxygen uptake as controls. After feeding, GH transgenics had 40–70% increased oxygen demand even when controls consumed equivalent amounts of feed. Adult transgenics had higher oxygen demand, poorer swimming ability and longer recovery time compared with ocean ranched salmon.

When subjected to low dissolved oxygen (0.4 ppm), mean absolute survival was the same for transgenic and control common carp. However, when mean survival time was calculated for all fish dead or alive, the transgenic individuals had longer mean survival time than the non-transgenic full siblings (Chatakondi, 1995; Dunham *et al.*, 2002b). The absolute mean survival of transgenic common carp subjected to low oxygen (0.4 mg/1) was higher than that of control carp in two of the eight families of common carp tested. The mean survival time in minutes for the transgenic genotype was greater in four of the eight families assessed. Transgenic common carp in some families had higher and longer survival than control common carp when subjected to low oxygen. The pleiotropic effect of pRSV–rtGH1 cDNA on survival under low oxygen in common carp has important implications for intensive fish culture.

GH can interact with a variety of other hormones, including thyroid hormone (Brent *et al.*, 1991). The consequence of additive effects of small changes in levels of numerous rate-limiting enzymes can be the stimulation of oxygen consumption (Brent *et al.*, 1991); thus tolerance of low dissolved oxygen should be a trait studied in GH-transgenic fish. One expectation is that fish with higher growth rates consume more oxygen and therefore would be more sensitive to low oxygen. In fact, channel catfish selected for increased growth rate exhibit greater sensitivity to low dissolved oxygen than controls (Rezk, 1993; Padi, 1995). Ventilation rate could be a possible explanation for the slightly better tolerance of low oxygen exhibited by the transgenic common carp. Transgenic channel catfish with the same rtGH construct as the common carp have a lower ventilation rate when subjected to low dissolved oxygen, compared with controls. Also, GH has a critical role in osmoregulation (Tang *et al.*, 2001), and this could be related to response under conditions of oxygen stress. There was a wide variation in the tolerance of low oxygen in various families of common carp tested. The most tolerant transgenic common carp would be produced by utilizing a combination of family and mass selection with gene transfer after the introduction of the foreign gene.

In the case of mammals, GH-gene insertion has a dramatic effect on body composition. Ebert *et al.* (1988) produced transgenic pigs containing a Moloney murine leukaemia virus–rat somatotrophin fusion gene. Elevated levels of somatotrophin resulted in increased growth of skeletal tissue, with a drastic reduction in fat deposition, without improvement in growth. Transgenic pigs grew faster, were more efficient feed converters and were leaner than littermate controls (Pursel *et al.*, 1989). Lower lipid levels were observed in transgenic mice (Pomp *et al.*, 1992; Knapp *et al.*, 1994), pigs (Ebert *et al.*, 1988; Pursel *et al.*, 1990; Wieghart

et al., 1990) and lambs (Nancarrow *et al.*, 1991). Machlin (1972) reported that exogenous porcine GH increased the muscle growth and decreased the fat deposition in pigs. Transgenic pigs containing bovine GH gene had a reduction of 41% in back-fat thickness (Wieghart *et al.*, 1990) and a 20% reduction in body fat (Pursel *et al.*, 1990).

Nancarrow *et al.* (1991) reported a 20–50% reduction of fat in transgenic lambs containing sheep MT–sheep GH. Reduction of fat in transgenic animals is attributed to hyperglycaemia and glucosuria (Rexroad *et al.*, 1991) or the direct action of GH (Etherton *et al.*, 1986; Ebert *et al.*, 1988). In mammals, GH is known to stimulate fat mobilization or activate the lipase to catalyse triglyceride hydrolysis. Transgenic sheep containing the callipyge gene directed the nutrients towards muscle-fibre growth and not towards fat deposition. Transgenic mammals with recombinant GH genes also show elevated levels of protein.

GH-transgenic fish also exhibit body-composition changes, but they are not as dramatic as those observed in mammals. However, the potential for similar dramatic body-composition changes exists for transgenic fish, as the treatment of salmonids with sustained-release recombinant porcine somatotrophin decreased carcass fat by 42–50% (McLean *et al.*, 1994). The body composition of transgenic common carp containing rtGH gene was different from that of controls. The transgenic common carp had more protein, less fat and less moisture than non-transgenic full siblings (about a 10% change). Body composition also changes in transgenic salmon, as the data of McLean *et al.* (1994) would lead one to predict. Moisture content in GH-transgenic Atlantic salmon was higher, relative to protein and ash, than in normal controls (Cook *et al.*, 2000a). GH promotes the synthesis of protein over fat, with the elevated levels of GH increasing this ratio in transgenic fish.

Dunham *et al.* (2002c) examined body-composition changes in GH-transgenic common carp over two generations, F_1 and F_2. The carcass composition of transgenic muscle had a lower percentage of lipids and higher protein in both generations. Moisture was lower in F_1 transgenic muscle, but unchanged in F_2 transgenic individuals. Sheridan *et al.* (1987) reported that endogenous somatotrophin stimulated the release of free fatty acids from coho salmon liver *in vivo*. Smith *et al.* (1988) suggested that slower-growing genotypes of fish direct their dietary energy into fat deposition and less to protein synthesis. This is not always true, as fast-growing catfish hybrids have increased fat deposition (Yant *et al.*, 1975). The faster-growing transgenic common carp may direct a greater proportion of their energy towards protein synthesis. Transgenic *O. hornorum urolepis* containing one copy per cell of tiGH cDNA under the regulatory sequences derived from the human CMV had lower levels of cholesterol, free alanine and aspartic acid in the muscle compared with controls (Martinez *et al.*, 1999).

The GH transgene not only improved the target trait but also improved correlated traits, which contribute to the economic merit of the carcass composition (Smith *et al.*, 1988). The body-composition data provide the expected change in nutrient requirement resulting from transgene manipulations (Steele and Pursel, 1990). The expression of the transgene had a significant effect on the proximate composition of transgenic fish. A 7.5% increase in the protein and a 13% decrease in the fat of a transgenic fish muscle result in the superior quality of transgenic common carp muscle compared with control common carp muscle and would be considered a healthier meat for human consumption. An increase in GH level of fish resulted in improved growth rate and an alteration of carcass composition. The reduced fat in the fish muscle could improve the shelf-life of the carcass, reduce the tendency to absorb off-flavours and improve the fish's ability to tolerate low dissolved oxygen (Chatakondi, 1995) under intensive farming systems.

Transgenic channel catfish with the same rtGH cDNA also had more protein, less fat and less moisture in their edible muscle than non-transgenic full siblings (about a 10% change). GH promotes the synthesis of protein over fat, and the elevated levels of GH in transgenic fish thus increase the protein/lipid ratio.

The increased level of protein in transgenic common carp and channel catfish muscle also results in increased levels of amino acids. However, amino acid ratios and fatty acid ratios were virtually identical in control and transgenic common carp and channel catfish, although some amino acids increase in proportion slightly more than do others.

Six of the 18 amino acids analysed in F_1 transgenic common carp muscle were higher than in the control genotype; however, amino acid ratios were minimally changed. Most of the amino acids were higher in transgenic common carp muscle: the amino acids aspartic acid, cystine, glutamic acid, histidine, lysine and threonine were higher than in the control carp muscle (Chatakondi *et al.*, 1995). Overall, the essential amino acid ratios were similar in transgenic and control genotypes; however, the differences observed were parallel to the essential amino acid content of the fish muscle (Chatakondi *et al.*, 1995). The computed essential amino acid ratios in transgenic carp muscle reflected increased histidine and lysine levels compared with control fish. These ratios are the basis for developing fish diets (Gatlin, 1987). The provision of the minimum dietary protein level with the proper essential amino acid pattern in the protein should satisfy the amino acid requirements of transgenic genotypes. Higher histidine and lysine ratios in the diet are recommended for the maximum growth and health of transgenic common carp in intensive culture systems based on essential amino acid ratios.

The fatty acid profiles of transgenic and control common carp were not different. No differences in *n*-3, *n*-6, total saturated fatty acids, unsaturated fatty acids and polyunsaturated fatty acids were observed in transgenic and control genotypes (Chatakondi *et al.*, 1995). The fatty acid contents of the transgenic and the control genotypes were comparable and the absence of major highly unsaturated fatty acids in both genotypes suggested that the grain-based diet dictates the fatty acid profile of the fish.

GH transgenesis also affects muscle characteristics and activity. GH-transgenic channel catfish had increased numbers of mitochondria in the cell, increased numbers of glycogen globules and increased numbers of muscle fibres, but a reduced number of fat globules. Muscle-fibre size was unchanged. Perhaps due to these changes in amino acid levels and ratios, changes in fat and ultrastructure of the muscle, the flavour and texture of transgenic catfish flesh were slightly better than those of non-transgenic controls.

Heterozygous GH-transgenic coho salmon had higher numbers of small-diameter fibres in somite muscles (Hill *et al.*, 2000). Both the dorsal and lateral regions of the somitic muscle were affected, suggesting that transgenic salmon grow by greater rates of hyperplasia relative to slower-growing non-transgenic fish. Higher levels of activity were found for phosphofructokinase and cytochrome oxidase in the white muscle of transgenic fish, indicating a higher glycolytic and aerobic requirement in the muscle of transgenic fish. The GH-gene insertion affected expression of several other genes, and many of the additional mRNAs in the transgenic fish were specifying myosin light chain 2, consistent with a high level of expression in the early stages of muscle-fibre construction.

Zhu (1992) reported an increase in muscle thickness and body width in transgenic common carp containing the hGHg. This observation was not quantified and was observed in only a few individuals. The effect of rtGH1 cDNA on body shape, dress-out yield and body composition were assessed in the F_1 and F_2 generations of transgenic common carp (Chatakondi *et al.*, 1994, 1995; Dunham *et al.*, 2002d). All measurements were compared with non-transgenic full-sibling common carp in their respective families and were communally evaluated in earthen ponds. The body weight and body length were highly correlated in both genotypes in all of the families. The correlation between head morphometric measurements and length or weight for F_1 and F_2 generations was negative (Chatakondi, 1995), indicating that the

fish's head does not grow proportionately to its length or weight. Various head, body and caudal traits grew disproportionately faster than total body length and this effect was greater in transgenic fish in both generations compared with control common carp. The transgenic individuals have relatively larger heads and deeper and wider bodies and caudal areas compared with controls. Quantitatively, the transgenic phenotype was stockier or more truncate, although this was not visually obvious. In both generations, as the difference in growth between transgenic and control families increased, the relative body-size differences increased and plateaued and then a small decline in relative difference was observed (Fig. 16.9). Similar changes are seen in GH-transgenic Nile tilapia, as the head/total length ratio, viscera somatic index and hepatosomatic index increased in transgenic fish relative to controls (Rahman *et al.*, 2001).

The condition factor K, which indicates the robustness of the fish, was proportionately higher in transgenic common carp in most of the families (Chatakondi, 1995). However, families 1 and 7 of the F_1 generation and 69 and 70 of the F_2 generation had a lower condition factor than their controls, despite a higher weight increase. The altered body shape of transgenic fish resulted in improved dress-out percentage in the F_2 generation, and a similar result was obtained for transgenic channel catfish containing the same GH construct. Dress-out percentage was higher in all the transgenic families, with the exception of family 18 of the F_1 generation.

Transgenic wild-strain rainbow trout had a slender body shape similar to that of wild controls, but their final size at sexual maturity was much larger than non-transgenic wild rainbow trout (Devlin *et al.*, 2001); thus no pleiotropic effect on body shape was seen for these fish, in contrast to the stockier, more truncate body shape of GH-transgenic domestic common carp compared with non-transgenics. However, the domestic transgenic rainbow trout derived from a deep-bodied strain, despite their minimal growth enhancement, had an even deeper body depth than the controls, caused

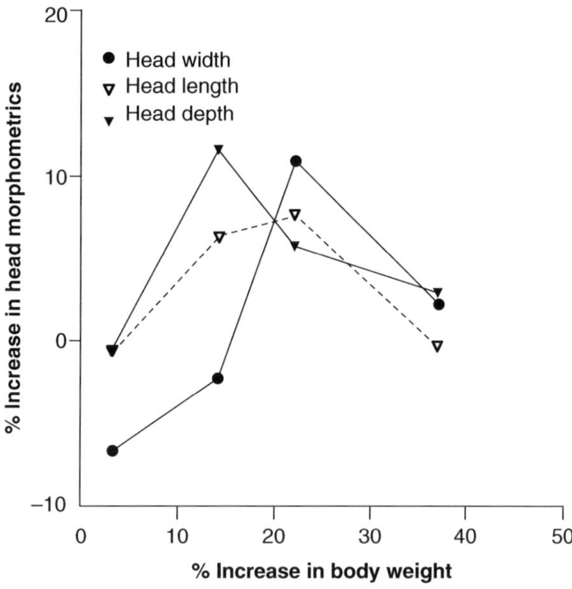

Fig. 16.9. Relationship of change in relative head size with body-weight improvement in growth-hormone-transgenic common carp, *Cyprinus carpio*. (Adapted from Chatakondi, 1995.)

by increased muscle or tremendous visceral fat deposits or both.

Change in body shape as a result of GH-gene transfer is common in transgenic fish. The P_1 generation of transgenic Pacific salmon, containing chinook salmon GH gene, had an impressive growth rate, with a slightly lower condition factor (Devlin *et al.*, 1995a). However, the excessive levels of GH resulted in morphological abnormalities in head, fin, jaw and operculum as a result of excessive cartilage and bone growth of the fastest-growing transgenic fish. Insertion of a plasmid OnMTGH1 gene construct into coho salmon altered centroid size, and the dorsal caudal peduncle and abdominal regions were also distinctly enhanced in transgenic fish when compared with controls (Ostenfeld *et al.*, 1998). Morphological changes of both whole body and syncranium were prominent.

GH transgenesis also affects gill morphology. Transgenic Atlantic salmon (Stevens and Sutterlin, 1999) and Pacific salmon (Stevens and Devlin, 2000a) had different gill morphology from that of controls, but the difference was expressed in different ways in the two species. Pacific transgenic salmon had gill filaments similar to those of controls in length but had smaller lamellar spacing. Atlantic transgenics had longer gill filaments than controls but with similar lamellar spacing to controls. This illustrates that the pleiotropic effects from GH transgenesis can be dissimilar for even closely related species.

Transgenic salmon represent one of the most dramatic results obtained in fish genetic engineering (Devlin *et al.*, 1994b). P_1 and F_1 salmon containing extra copies of salmon GH genes driven by recombinant promoters can grow two to 30 times faster than normal. The largest of these P_1 transgenics have been mated and produce offspring with extraordinary growth. Unfortunately, these fish are sub-viable and virtually all die. The endocrine stimulation has been elevated to pathological levels in these GH-transgenic salmon, and excessive, deleterious deposition of cartilage was observed (Devlin *et al.*, 1995a,b), analogous to the mammalian acromegaly syndrome. This effect can be sufficiently severe for impaired feeding and respiration to result in reduced growth and poor viability. Consequently, salmon that ultimately display the greatest growth enhancement as adults are those that have been only moderately stimulated (Devlin *et al.*, 1995a,b). Similar problems have been observed in transgenic pigs that expressed hyperlevels of recombinant GH. Progeny from transgenic parents with more moderate accelerated growth do not exhibit reduced survival or increased skeletal anomalies.

Fast-growing wild GH-transgenic rainbow trout and GH-transgenic domestic rainbow trout with minimal growth enhancement both exhibited cranial deformities (Devlin *et al.*, 2001), and Devlin *et al.* (2001) suggested this is evidence that transgenesis affects growth pathways outside the range supported by the homeostatic processes that maintain normal morphology and viability. The fact that growth was not enhanced but certain tissue types responded seems odd. Devlin *et al.* (2001) do not report the status of the mosaicism in these fish, and uneven expression of GH throughout the body of mosaic individuals could cause different rates of growth in various tissues, resulting in deformity and mortality. However, this hypothesis is not supported by the data of Maclean *et al.* (1987b), as domestic rainbow trout receiving exogenous GH showed modest increases in growth (9%), but also had cranial abnormalities and silver body coloration, whereas controls did not have these characteristics.

The deformities could be a species-specific phenomenon. Despite much more significant growth acceleration compared with the slow-growing rtGH transgenics, P_1, F_1, F_2, F_3 and F_4 GH-transgenic common carp and channel catfish do not exhibit deformities. Additionally, no abnormalities were apparent in rapidly growing GH-transgenic Nile tilapia, although minor changes to skull shape were observed in some fish (Rahman *et al.*, 1998).

The range of phenotypic variation as a consequence of integrating foreign DNA into the fish genome needs to be evaluated to better assess how to utilize these fish, as well as for understanding the potential environmental impact of genetically modified fish. Dunham *et al.* (1992a) evaluated the range in body weight of individual transgenic channel catfish containing the salmonid GH gene that had been microinjected at various developing egg stages. The phenotypic variations for body weights for transgenic members of families were generally less than in their non-transgenic full siblings, resulting in uniform growth rate. This may be surprising since these P_1 individuals are mosaic, which, conversely, could have induced phenotypic variation.

The increased production of GH in transgenic common carp (rtGH and common carp GH) could result in uniform levels of growth rate or mask the effects of other loci and genes affecting growth-rate uniformity. The range of body size of F_1 and F_2 generations of transgenic common carp containing rtGH cDNA was evaluated in earthen ponds and aquaria. Zhang *et al.* (1990) observed that the smallest and largest transgenic individuals were larger than the corresponding smallest and largest full-sibling control common carp in the F_1 generation. The coefficient of variation for body weights was smaller for transgenic fish than for non-transgenic fish in the families in which the mean body weight of the transgenic common carp was greater than that of the control common carp. The growth response was variable in transgenic and non-transgenic common carp when analysed by generation, by sex, by age and by environment.

However, transgenic individuals had the highest maximum body weights in only half of the families (Dunham *et al.*, 2002a). This relationship changed, and as the fish aged, maximum body weight was usually obtained by a transgenic full sibling. F_1 transgenic common carp, all heterozygous for the transgene, had more uniform growth than control siblings. However, F_2 non-transgenic individuals were less variable for body weight than transgenic full siblings in ponds. A GE interaction for growth variability was observed in the F_2 generation, as non-transgenic common carp were more variable for body weight in aquaria compared with transgenic full siblings.

The growth data may be used to formulate ecosystem risk analyses. The response of the transgenic common carp to the insertion of rtGH gene appears to be variable. The overall population variation was large for both the transgenic and the non-transgenic common carp; however, the variation was variable and may be a result of different sites of foreign gene insertion, copy number of the foreign gene and level of expression.

Evaluation of several families and large numbers of fish is important in transgenic research to overcome the great natural variation in fish, and provides further insight into the potential of transgenic fish in natural systems. The results of Dunham *et al.* (2002a) were more variable than those found by Zhang *et al.* (1990). The largest F_1 transgenic common carp weighed more than the largest control in two of four families and in three of eight for the F_2 generation. However, minimum body weight was highest for the transgenic genotype in all four F_1 families but only five of eight F_2 families. Maximum length was greater in four of four F_1 transgenic families and four of eight F_2 transgenic families. Minimum length was longest for three of four F_1 transgenic families but only three of eight F_2 families. These relationships changed with age and size and in F_2 families almost all largest minimum and maximum weights were found in the transgenic genotype. Sex also affected body-weight range. In the F_2 both transgenic males and females tended to have the maximum size but the smallest males were also transgenic. This was not the case for transgenic females. Insertion of pRSV–rtGH1 cDNA increased the absolute maximum size after two generations. The longest fish in the F_1 generation was a transgenic individual but the heaviest was a control. However, the longest and heaviest fish in the F_2 was a transgenic individual.

Reproductive traits have not been greatly affected by GH transgenesis. Fecundity is not affected by insertion of rtGH in common carp. Precocious sexual development was not observed in transgenic common carp. However, GH-transgenic male tilapia had reduced sperm production. Female GH-transgenic Nile tilapia had a lower GSI than non-transgenic siblings in both mixed and separate culture conditions (Rahman *et al.*, 2001). The transgenic males' GSI was higher in mixed culture and lower in separate culture than that of their non-transgenic siblings.

Colour changes in the GH-transgenic coho salmon (Devlin *et al.*, 1995b; Devlin, 1997b). Individuals containing opAFP or OnMT salmon GH constructs have lighter skin pigmentation and this is a reliable marker to identify transgenic salmon prior to first feeding (Devlin *et al.*, 1995b). Control fish possessed the normal brown coloration typical of coho salmon alevins, whereas the GH transgenics had a distinct green coloration and exhibited signs of cranial deformities and opercular overgrowth.

The most important pleiotropic effect, which is one of the major explanations for the growth differences in transgenic and control salmon, is the accelerated smoltification of the transgenics. The transgenics smolt up to 2 years early and display enhanced silver coloration and osmoregulatory ability (Devlin, 1997b).

Potential Role of Mitochondrial DNA in Gene Transfer

Optimizing accelerated growth rates of transgenic fish may depend on utilizing the proper combination of the specific transgene, mtDNA-encoded subunits and subunits encoded by nuclear DNA (nDNA). Mitochondrial DNA contains several important genes that in some manner can interact with the nuclear genome. Cytochrome c oxidase, encoded by a mitochondrial gene, couples the reduction of oxygen to water with the production of ATP as the final step in aerobic respiration. Since cytochrome c oxidase is a limiting step at the end of the respiratory chain, rates of respiration and chemical energy conversion are proportional to cytochrome c oxidase activity (Kadenbach *et al.*, 1987, 1988). The three subunits at the core of cytochrome c oxidase are encoded by mtDNA, and up to ten additional subunits are encoded by nDNA.

Subunits encoded by nDNA may modulate catalytic activity, be tissue specific and be specific to certain developmental stages. Subunits VIa, VIIb, VIc and VIII are not found in yeast, and may have a catalytic role in higher eukaryotes through their interaction with hormones, growth factors or neurotransmitters. Because of interactions between the mtDNA genes and the nuclear genes, maximum performance from transgenics could require transfer of both foreign nDNA and its conspecific mtDNA. Interspecific transfer of mtDNA would allow *in vivo* study of these interactions and confirm their importance.

Liepens and Hennen (1977) studied cytochrome oxidase deficiency in nucleo-cytoplasmic hybrids of two species of the genus *Rana* and found that, for regular mitochondrial metabolism to take place, at least a haploid set of nDNA from the same species as the mtDNA had to be present. Abramova *et al.* (1979) successfully transferred mitochondria isolated from loach embryos or frog heart into oocytes or fertilized eggs of the loach, newt, toad and frog, and this did not affect normal development. In another experiment, mitochondria isolated from a variety of sources have been successfully microinjected into oocytes or fertilized eggs of the loach, *Misgurnus fossilis* (Abramova *et al.*, 1979, 1980, 1983, 1989), and these are the only successful transfer of whole mitochondria in fish. Abramova *et al.* (1989) demonstrated the utility of induced heteroplasmy in the study of mitochondrial biogenesis, regulation and function by transferring mitochondria isolated from two lines of mouse fibroblast cell cultures, one resistant and one sensitive to chloramphenicol, into fertilized loach eggs. When incubated in the presence of chloramphenicol, only those embryos containing the transferred mitochondria from the resistant line survived, indicating that the injected mitochondria were retained and functioned in the developing embryos. Using isotopically labelled exogenous and endogenous mitochondria, Abramova *et al.* (1983) also demonstrated that the number of mitochondria or their mass is a regulated value

that may be due to an increase in the degradation of mitochondrial proteins compared with their synthesis. There was no selection against the exogenous mitochondria. The long-term stability and function of these mtDNA transgenics are not known as the experiments were concluded when the loach embryos were at the larval stage.

Induced heteroplasmy may also lead to elucidation of the mechanism causing the phenomenon of paternal predominance observed in catfish hybrids (Dunham *et al.*, 1982a), which may be caused by pleiotropic enhancement of paternal nDNA by maternally inherited mtDNA. Growth rates of reciprocal catfish hybrids appear to be maternally influenced and may be linked to interactions between mtDNA and nDNA. The ability to induce heteroplasmy would allow production of sufficient numbers of fish to enable comparison of their production characteristics with those of their monoplasmic half siblings and might also be used to enhance the production of androgenic fish by replacing destroyed mtDNA. Finally, the study of nDNA–mtDNA interactions may lead to an explanation of the phenomenon of F_2 breakdown observed in F_2 and Fn interspecific hybrids.

Whitehead (1994) attempted to transfer whole mitochondria and purified mtDNA between species of ictalurid catfish, using the electroporation parameters of Powers *et al.* (1992) to transfer salmonid GH genes into channel catfish. The mtDNA molecule is approximately three times the size of GH genes, but Ozato *et al.* (1986) successfully transferred an nDNA segment almost as large as mtDNA via electroporation. Whole mitochondria are much larger, usually 1–2 µm long by 0.3–0.7 µm wide, and transfer of intact mitochondria into fish eggs by electroporation has not been previously accomplished. Electroporation has been used to transfer protein macromolecules into mouse oocytes (Zhao *et al.*, 1989).

The transfer of mtDNA among ictalurids was apparently unsuccessful (Whitehead, 1994); however, if the exogenous mtDNA represented a minor fraction of the total mtDNA, the transferred mtDNA may have gone undetected because of analytical limitations. The fate of the exogenous mtDNA molecule once in the host cell is uncertain. The naked DNA may be degraded in the cytoplasm; it may linearize, enter the nucleus and integrate within a chromosome; it may replicate autonomously or as a concatemer; or it may enter a mitochondrion and replicate normally. Counter to earlier hypothesis, the exogenous mtDNA could encounter transcriptional or translational problems resulting in the production of faulty proteins due to differences in genetic codes, thus reducing the viability of the transgenic fish. The transfer of whole mitochondria might overcome these problems.

Even if the exogenous mtDNA were not eliminated by selection, it could still be eliminated by chance, especially if the copy number is low as compared with endogenous counterparts. The testing of electroporated embryos or fry may demonstrate successful short-term transfer not seen in older fish. The success of future electroporation attempts would probably be greatly enhanced by artificially selecting for the transferred material, such as exploiting differential sensitivity to antibiotics. Another possibility is that low numbers of transferred mtDNA or mitochondria could have resulted in mosaic catfish, which is the universal phenomenon in transgenic fish (Dunham *et al.*, 1992a), again making detection of mitochondrial transgenics difficult.

17
Combining Genetic Enhancement Programmes

Although much progress has been made in the genetic improvement of cultured fish via traditional genetic approaches, the potential remains for additional improvement through biotechnology. Maximum progress will probably be made by combining the tools of selective breeding and molecular genetics; utilizing more than one of these programmes together simultaneously can maximize genetic gain. The best genotypes for application in aquaculture in the future will be developed by using a combination of traditional selective breeding and the new biotechnologies (Fig. 17.1). Initial experiments indicate great promise for this approach. There is at least one example each illustrating that mass selection and crossbreeding, genetic engineering and selection, genetic engineering and crossbreeding, and sex reversal and polyploidy work more effectively in combination than alone.

Sex Reversal and Triploidy

The system of sex reversal and breeding implemented in Scotland for the production of monosex female populations of rainbow trout is being evaluated in combination with triploid induction to produce monosex triploid female populations. Lincoln and Scott (1984) first made sterile, triploid, all-female (XX) rainbow trout. It is hoped that these fish will have both superior growth rate and flesh quality resulting from both the sex reversal and the triploidy (Bye and Lincoln, 1986), and this has now been realized (Sheehan *et al.*, 1999). Selection, sex reversal and polyploidy are now routinely used to maximize the overall performance and value of salmonids (Dunham, 2004).

Genetic Engineering and Crossbreeding

In Israel, Hinits and Moav (1999) were able to improve common carp growth by using genetic engineering with crossbreeding, more than by using crossbreeding alone. Similarly, when salmon MT promoter–salmon GH1 cDNA (OnMTGH1) was transferred to another wild rainbow trout strain, F77, growth was enhanced sevenfold, which was almost fourfold more than that of a domestic rainbow trout (Devlin *et al.*, 2001). In this case, the wild transgenic is actually superior to the domestic selected strain, indicating that genetic engineering can have a greater effect than, rather than an equivalent effect to, domestication and selection. Perhaps strain effects in general, epistasis and genetic background are more

Fig. 17.1. Utilization of multiple genetic improvement programmes; hand-stripping eggs from striped bass, *Morone saxatilis*, to increase performance via interspecific hybridization with white bass, *Morone chrysops*, and sterilization from triploid induction.

significant in regard to affecting transgene response rather than the domestic or wild nature of the fish. When F77 was crossbred with the domestic strain, growth of the crossbreed was intermediate to that of the parent strains, a typical result (Dunham and Devlin, 1998). However, the transgenic wild × domestic crossbreed was by far the largest genotype, 18 times larger than the non-transgenic wild parent, 13 times larger than the non-transgenic wild × domestic crossbreed, nine times larger than the non-transgenic domestic parent and more than 2.5 times larger than the wild F77 transgenic (Devlin *et al.*, 2001). The combined effects of transgenesis and crossbreeding had a much greater growth enhancement than crossbreeding or transgenesis alone.

Genetic Engineering, Selection, Crossbreeding, Strains and Hybrids

Dunham and Liu (2002) were able to improve growth of channel catfish more by combining selection and genetic engineering than by selection alone. Combining selection and crossbreeding improves growth more than using one of these programmes singly for both common carp and channel catfish (Gupta and Acosta, 2001; R.A. Dunham, unpublished results). The best strains of blue catfish and channel catfish produce faster-growing interspecific hybrids.

Channel catfish transgenic for rtGH exhibited a moderate growth enhancement, 41%, and were derived from domestic, selectively bred catfish. If we extrapolate from a series of experiments starting with slow-growing wild strains of channel catfish and then improve their growth through domestication (Dunham, 1996), followed by further improvement from selective breeding (Padi, 1995) and then further increases from interspecific hybridization (Jeppsen, 1995) or gene transfer, the overall growth enhancement is tenfold, comparable to that observed with wild transgenic salmon (Devlin *et al.*, 1994b). This illustrates the potential and value of combining various genetic enhancement programmes.

Selection, Crossbreeding and Sex Reversal

Combining selective breeding and biotechnology has the potential to address weaknesses within genetic biotechnology programmes. For instance, reciprocal recurrent selection has the potential to identify and propagate YY males and XX females that consistently produce the desired 100% male XY fry. Mair and Abucay (2001) found an increase in the overall proportions of males in the progeny testing of YY genotypes derived from crosses of selected (on the basis of the 100% male sex ratios in initial progeny tests) YY males, indicating some form of response to this selection and thus a genetic basis to the occurrence of these aberrant females (Dunham *et al.*, 2001). Similarly, the higher than average proportions of males in repeat matings of selected (produced >96% male sex ratios in initial progeny tests) YY males provided further support for this hypothesis of a genetic basis for the unexpected existence of females in putative all-male progenies.

Then Mair and Abucay (2001) utilized selection to improve the growth performance and sex ratio (per cent male) of genetically male Nile tilapia (GMT). YY males and YY females from different families were progeny-tested by mating them to randomly chosen normal XX females and sex-reversed XX male genotypes, respectively. YY males and females that produced 100% male progeny were selected to produce the next generation of YY brood stock, and those that did not were culled. The percentage of males sired by YY males increased from 92.3 ± 7.5 in the base population to 97.4 ± 7.7 and 98.3 ± 3.6 after one and two generations of selection, respectively. The percentage of males produced by YY females was consistently above 99.4% in all generations.

Males and females were divergently selected for growth, utilizing a within-family selection in a rotational mating scheme with five original strains of Nile tilapia. Selection for sex ratio for the growth-selected females crossed to YY males was conducted in the base, first and third generations. Mixed-sex progeny from high-growth selected lines were 37–102% larger than progeny from low-selected lines when tested in a range of environments after three generations of divergent selection. Heritability estimates ranged from 0.08 to 0.4. Unfortunately this is not proof of growth improvement, as bidirectional selection responses can be asymmetric. Theoretically, the difference between the two lines could be entirely from decreases in body weight in the low line. In fact, previously one generation of bidirectional selection in Nile tilapia resulted in no increased body weight in the high line and a significant drop in body weight for the low line (Teichert-Coddington and Smitherman, 1988). GMT produced by crossing high-line females to YY males showed superior growth performance compared with the best crossbred GMT and the original intrastrain GMT, illustrating the value of combining selection, crossbreeding, sex reversal and breeding, but not substantiating selection response.

Other tests also indicate the value of combining crossbreeding with GMT technology, as crossbred GMT produced growth rates and yields 20–40% greater than the original GMT, indicating heterosis for growth above and beyond the improvement from all-maleness (Mair *et al.*, 1997). The sex ratio was more variable in some crossbred GMT, which is not surprising since sex ratio varies in strains of both normal and GMT Nile tilapia.

Gynogenesis, Selection and Hybridization

Another example of combining a biotechnological approach with traditional breeding was also utilized to increase the percentage of males in tilapia hybrids. A meiogynogenetic line of blue tilapia was established and gynogenetically propagated for five generations in Israel (Dunham *et al.*, 2001).

Viability increased successively over these five generations, perhaps indicating a gradual elimination of lethal genes. Mitogynogenetic blue tilapia were produced (Shirak *et al.*, 1998), using third-generation meiogynogenetic females from this stock, and three generations of gynogenetic Nile tilapia were also produced. Males from the gynogenetic blue tilapia line were hybridized with gynogenetic Nile tilapia females, resulting in the consistent production of 100% male hybrids (Hulata *et al.*, 1983, 1993).

Wiegertjes *et al.* (1994) demonstrated the potential of combining gynogenesis and selection to enhance immune responses. Crossbred common carp from a single mating were immunized with the hapten-carrier complex DNP-KLH, and then divergent selection for antibody response was conducted at 12 and 21 days post-immunization, followed by the propagation of homozygous mitogens of early/high or late/low responders (Wiegertjes *et al.*, 1994). Upon immunization with DNP-KLH, the antibody response was higher in the early/high-responder homozygous off-spring. The homozygosity of the high-responder offspring apparently caused a lower and slower antibody response compared with the base population – inbreeding depression. However, the differences between the high- and low-responder offspring do indicate a genetic basis for antibody response. The realized heritability for antibody production was 0.37 ± 0.36, indicating the potential to combine gynogenesis with selection. However, some caution is needed concerning expectations for the ultimate outcome of such a programme. Often genetic improvement after inbreeding allows the inbreds to approach or reach the performance of the base population, which is the suggested scenario from these common carp results. In reality, no genetic gain is accomplished. However, if the cumulative selection response eventually surpasses that of the original base population, true genetic gain or enhancement is accomplished.

Bongers *et al.* (1997) addressed selection theory in the light of the changed genetic variation and population structure of 100% homozygous clonal populations. Their analysis indicates that the additive genetic variation V_A in homozygous populations doubles compared with that in the base population, and they examined the possibility of using gynogenesis and androgenesis for estimating V_A. Additive genetic relationships and the distribution of V_A between and within homozygous gynogenetic and androgenetic families were derived. Derivations indicated that within-gynogenetic-family variance equals V_A + environmental variation V_E and between-family variance equals V_A. This theory was used to analyse experimental data on gonad development and fertility in homozygous gynogenetic common carp. Five gynogenetic families were produced from full-sib, outbred female parents, and they demonstrated large between- and within-family variance for GSI and egg-quality parameters. Heritabilities for GSI at 13 months and the percentage of normal larvae obtained after fertilization of egg samples at 19 months were 0.71 and 0.72, respectively, using Gibbs sampling. These heritabilities indicate significant levels of V_A and that homozygous individuals can be successfully selected from these common carp families to produce early- or late-maturing homozygous gynogenetic inbred strains with increased egg quality. Again, it will be important to determine if the amount of improvement will surpass the level of performance demonstrated by the base population. These heritabilities may be overestimated, as between-family variance should have both V_E and dominance genetic (V_D) components of variation. These are important because within a single family there should be no V_A, as these individuals are 100% homozygous and within-family selection should not be successful.

The difference between families should be due to $V_E + V_A + V_D$ + variation due to epistasis (V_I). Between-family selection should be successful, but is limited in potential by how many families are generated to choose from. A selection plateau should be reached in a single generation unless additional families are evaluated or selection intensity is increased, because, again, there is no genetic variation within the family. The selection plateau could be overcome by crossing the selected families, all theoretically possessing the best genes for the trait of interest.

Inada *et al.* (1997) suggest that useful quantitative characters can be fixed via two generations of gynogenesis. The range of the variation for agglutinating antibody titre following an intraperitoneal injection of formalin-killed *Vibrio anguillarum* was greater in gynogenetic (first-generation meiogens) ayu, *Plecoglossus altivelis*, than that of normal diploids. Gynogenetic and normal diploid fish resistant to the disease were selected for two generations from challenges and perpetuated via gynogenesis. When challenged, survival and number of individuals showing high titre were higher in the gynogenetic group than for normal diploid or non-selected control groups in both generations. Gynogenesis combined with selection was more effective than selection alone for developing fish resistant to vibroilosy.

Similarly, heritability for resistance to VHS virus in rainbow trout was 0.63 ± 0.26 and selected lines had 0–10% mortality, while mortality of the controls was 70–90% (Dorson *et al.*, 1995). The meiotic gynogenetic progeny of select-line females also demonstrated high resistance (mortality less than 10%). The virus replicated poorly in fins from resistant fish compared with fins of susceptible rainbow trout.

18

Genotype–Environment Interactions

Aquaculture genetics research is of reduced utility unless it results in an impact on the farm or the industry. For application of genetics research intended to improve aquatic organisms for aquaculture, the GE (genotype–environment) interaction is critical, as the best genotype for one environment is not necessarily the best genotype for another environment. Thus, fish and shellfish genetically improved in the research environment are not necessarily the best for the commercial environment and these relationships have to be verified before the application of research stocks in the farm environment. In general, the GE interactions increase for aquacultured aquatic organisms with increasing genetic distance and increasing environmental differences (Fig. 18.1), especially for species such as carp or tilapia, which can be cultured simply and low on the food chain or with complete artificial feeds. The most valuable genetics research is that which is conducted in environments that most closely simulate commercial environments. GE interactions are most likely to occur when there is a large change in the environment or when comparing fish with large genetic distances (more distantly related).

Two types of GE interaction exist. One occurs when the rank of two or more genotypes changes when compared in two or more environments (Fig. 18.2). In the second case the rank does not change, but the magnitude of the difference between the two genotypes changes significantly when compared across environments, and this change in magnitude of the difference would have economic implications (Fig. 18.3).

Traditional Breeding

GE interactions are prevalent in aquaculture, and interactions are more important for crossbreeds and hybrids of fishes than for strains or select lines. Strains and select lines of channel catfish performed similarly in aquaria, cages and ponds (Smitherman and Dunham, 1985). Channel catfish selected for increased body weight at one stocking density in ponds also grew faster than their control populations at other stocking densities (Brummett, 1986). Similar results have been obtained for common carp. Relative growth rates of several strains of common carp remained constant in three widely different environments (Suzuki *et al.*, 1976). Gunnes and Gjedrem (1978, 1981) found little evidence for GE interactions among strains of both Atlantic salmon and rainbow trout. On the other hand, distinct interactions among strains of rainbow trout were observed for different stocking

Fig. 18.1.   Genetic evaluation of channel catfish, *Ictalurus punctatus*, in cages. The photograph depicts two major environments, cages and earthen ponds, for genetic evaluations of fish and shellfish for which genotype–environment interactions could occur. (Photograph by R.O. Smitherman.)

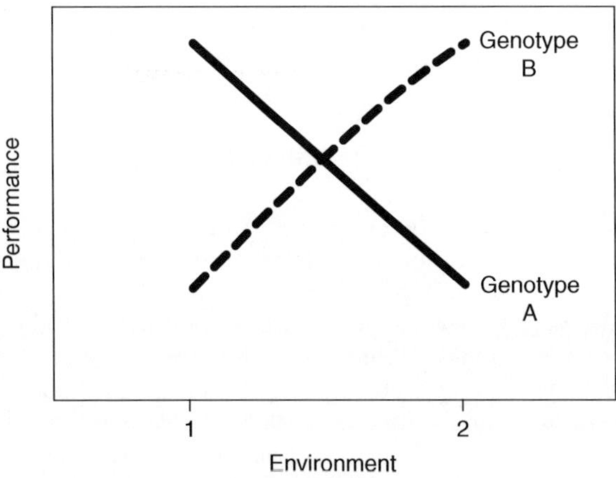

Fig. 18.2.   Genotype–environment interaction occurs when the rank of genotypes changes.

densities (Gall, 1969), different ponds (Ayles, 1975) and different culture units, cages and ponds (Klupp *et al.*, 1978). However, genetic rank did not change in different ponds (Ayles, 1975), and the same strain grew the fastest in both cages and ponds (Klupp *et al.*, 1978).

GE interactions are large and significant when comparing the growth of different species, intraspecific crossbreeds, interspecific hybrids or polyploids of catfish. The best genotype for ponds, the channel catfish female × blue catfish male, has mediocre growth in aquaria, tanks and cages (Fig. 18.4). The behaviours nervousness and aggressiveness are the factors causing GE interactions for the channel × blue hybrid and triploid channel catfish, respectively. Culture unit size is an alternative explanation. Additionally, GE interactions can be related to low oxygen levels when comparing channel × blue hybrids and their parents.

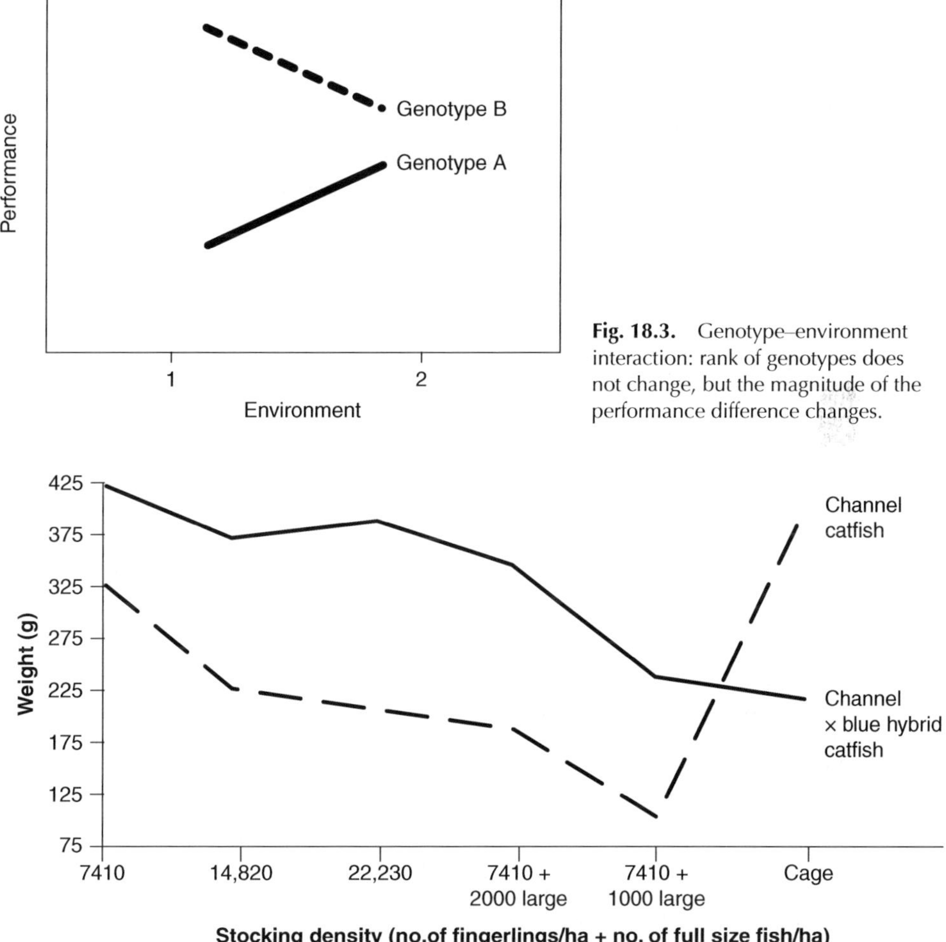

Fig. 18.3. Genotype–environment interaction: rank of genotypes does not change, but the magnitude of the performance difference changes.

Fig. 18.4. Genotype–environment interaction for channel catfish (*Ictalurus punctatus*) and channel catfish × blue catfish (*Ictalurus furcatus*) hybrids grown in cages and ponds. (Adapted from Dunham *et al.*, 1990.)

Steffens (1974) also found GE interactions for crossbred common carp and Klupp *et al.* (1978) found interactions for crossbred rainbow trout grown in cages and ponds.

GE interactions affect the h^2 values obtained for growth in European oysters. GE interactions can affect other traits. Environments with different oxygen levels can result in GE interactions when comparing channel catfish and channel × blue hybrid catfish for angling vulnerability. Chum salmon exhibit GE interactions for incubation rates.

GE interactions are more likely to occur for cultured fish that can be grown low on the food chain with natural foods or with artificial feeds. This type of interaction has been observed in both strains of common carp (Moav *et al.*, 1975; Wohlfarth *et al.*, 1983) and strains of Nile tilapia (Khater, 1985) when the fish were grown in ponds fertilized with manure or in ponds receiving pelleted fish feed.

Polyploidy and Transgenics

Altered ploidy level changes behaviour, and this appears to cause GE interactions

(Wolters *et al.*, 1991). In tanks, triploid channel catfish were not as active as diploids (Wolters *et al.*, 1982b). However, this lack of activity and aggressiveness did not affect their food consumption in the tank environment where the water was clear, sinking pellets were utilized and the feed was presented in a manner that required minimal effort to find the food. Triploid channel catfish grown in ponds were difficult to train to feed and did not exhibit aggressive feeding behaviour in that environment, requiring more effort to find the feed because of plankton turbidity and the use of floating feed (Wolters *et al.*, 1991). These differences in behaviour led to GE interactions for growth rate in triploid channel catfish, as triploids grew faster in tanks but slower in earthen ponds after 18 months (454 g) (Wolters *et al.*, 1981a, 1991).

Ploidy level also affects behaviour and performance in grass carp (Cassani and Caton, 1986b). When in competition with diploids in a variety of environmental conditions varying in stocking density and quantities of food, triploids grew more slowly than diploids and had lower condition factors. If triploids were grown to satiation in separate ponds with duck weed, their growth, condition factor, feed conversion, rate of food consumption and ability to control aquatic vegetation were similar to those of diploids. Since in competition the diploids grew faster, this size difference could be used as a tool to sort and eliminate diploids from the triploid population prior to screening by Coulter Counter Channelyzer for commercial sales and commercial-scale verification of ploidy levels. In contrast to diploid channel catfish and grass carp, which were more aggressive and more competitive for food, diploid and triploid brook trout had similar food intake and growth when food was limiting (O'Keefe and Benfey, 1999).

GE interactions occur when comparing triploid and diploid Chinese catfish with regard to a combination of temperature and feed type (Qin *et al.*, 1998). GE interactions were observed in respect of feeding level and carcass yield in triploid African catfish (Henken *et al.*, 1987). However, GE interactions did not occur for growth, feed conversion or GSI in relation to these different feeding levels. GE interactions were observed for accumulation of fat in the muscle for triploid and diploid ayu in relation to fat levels in the diet (Watanabe *et al.*, 1988).

Feeding frequency and type of feed can cause GE interactions. Triploid blue tilapia were produced and compared with diploids when fed organic manure (Byamungu *et al.*, 2001). The triploids were 80% female and the diploids had a 1:1 sex ratio. As expected, diploids grew faster than triploids in both tanks and ponds since males grow much faster than females in blue tilapia and the triploids were primarily female. When female triploids were compared with female diploids in tanks, there was no difference in growth when the fish were fed 7 days per week, but triploids grew more than 30% faster when the fish were fed 5 days per week.

GE interactions are common in oyster culture. Usually, the local genotype has the best performance, but is not the best oyster when transferred from its geographical region. Additionally, triploid oysters cannot express their increased growth potential in oligotrophic environments. The advantage of the triploid oyster is minimized or lost in culture situations where food is limiting (Dunham, 1986). This is an important point, which may be relevant to other genetic improvement programmes of shellfish and finfish. However, some studies are contradictory. Triploid Pacific oysters grew faster than diploids in environments with reduced levels of suspended particulate matter, and were apparently more efficient filter feeders than diploids; but in more productive environments growth differences were not observed (Davis, 1988a, 1989).

Depth of culture can result in GE interactions. Downing (1988b,c) made all possible diploid and triploid crosses between *Crassostrea gigas* and *Crassostrea rivularis*, and grew them on long lines or suspended off a dock in California. No differences in growth rate were observed except that crosses with at least two chromosome sets from *C. rivularis* grew faster in bottom culture.

Temperature may lead to GE interactions. Triploids were larger than diploids for Pacific oysters at two locations in Washington (Davis, 1989). However, the relative advantage of the triploid was greater in the warmer environment.

Fluctuating severe versus constant salinity did not cause major GE interactions when comparing triploid, tetraploid and diploid Pacific oysters (Brooks and Langdon, 1998). Five tetraploid families had the best growth, followed by three diploid families and then the triploids. However, triploid eastern oysters grew faster than triploid Pacific oysters at low salinities, the same at medium salinities and more slowly at high salinities (Calvo *et al.*, 1999a). Triploid eastern oysters had higher production than triploid Pacific oysters in Chesapeake Bay, but the reverse was true along the Atlantic coast of Virginia (Calvo *et al.*, 1998). The Pacific oyster triploid was more susceptible to the mud worm, *Polydora* spp., than triploid eastern oysters at low and medium salinities, but there was no difference at high salinities. Similarly, at low salinities survival of triploid eastern oysters was higher than that of triploid Pacific oysters, but at medium and high salinities there was no difference in survival.

GE interactions occur for the growth of transgenic channel catfish (Dunham *et al.*, 1995) containing salmonid GH genes, and the transgenics grew 33% faster than normal channel catfish in aquaculture conditions with supplemental feeding. However, there was no significant difference in growth performance between transgenic and non-transgenic channel catfish in ponds without supplemental feeding, indicating equal foraging abilities and the inability of transgenic catfish to exhibit their growth potential with limited feed (Chitminat, 1996). The foraging ability of transgenic and control catfish is similar under these conditions of competition and natural food sources, and, as is the case for most genetic improvement programmes, genetically engineered fish need adequate food to express their potential.

GE interactions are now recognized as a key component in determining environmental risk of transgenic aquatic organisms (Sundström *et al.*, 2007). GE interactions and transgene by genetic background interactions occur in salmonids and could affect the ability of a transgenic organism to establish itself in the natural environment. Thus, Sundström *et al.* (2007) suggest that environmental risk assessment of transgenic aquatic organisms be conducted in a variety or range of potential natural environmental conditions.

GE interactions should be considered in all genetics and breeding programmes. Manipulated fish should be evaluated in the commercial environment as well as the laboratory before being recommended for wide use.

19
Commercial Application of Fish Biotechnology

Genetic enhancement programmes and improved germplasms are being applied in both developed and developing countries and are having an impact. Select lines of several aquatic species are utilized in many countries and crossbreeding, particularly of common carp, has been applied in many Asian, European and Middle Eastern countries. Interspecific hybridization, polyploidy, monosex application and transgenic fish are being utilized in various countries worldwide. Breeding companies and aquatic biotechnology companies have been established. Great progress has been made since the early 1970s, but much more can be done.

Selection

Selection for improvement of aquatic organisms dates back 2000 years although the early practitioners did not fully understand the genetic principles behind their selection programmes. In Europe, deep-bodied common carp are utilized which have an appearance that is quite distinctive compared with wild common carp. Lines of common carp selected for increased body weight are utilized around the world, particularly in Asia and Europe. The Norwegian salmon breeding programme and the Auburn University channel catfish breeding programme in the USA both now have a 40-year history of selecting for improved growth rate in Atlantic salmon and channel catfish, respectively. Both the rainbow trout and shrimp industries have conducted intensive efforts in selection to improve a variety of traits. The oyster industry has placed large efforts in improving disease resistance. The GIFT Nile tilapia breeding programme and other Nile tilapia breeding programmes have produced select lines of tilapia widely used in the Philippines, across Asia and globally. The ornamental fish industry has benefitted greatly from selection for anomalies in colour, scalation, finnage and body conformation, particularly the goldfish and koi industries. Selection is being conducted in several new aquaculture species. Selection has had a major impact on global aquaculture production, and this progress is probably under-recognized and underappreciated.

Intraspecific Crossbreeding

Intraspecific crossbreeding is not as widely utilized in commercial aquaculture as selection. However, it has had major impact as well. Crossbred common carp were at one time the basis for carp aquaculture in Europe, Israel, China and Vietnam. Cross-breeding has been utilized to a minor extent in the American catfish industry with excellent results. Maintenance of multiple strains and lines requires dedication and execution lacking on many commercial farms.

Interspecific Hybridization

Interspecific hybridization rarely results in an aquatic organism that is improved and suitable for aquaculture or fisheries application. However, there are some unique and valuable exceptions. Hybridization is the basis of the walking catfish industry in Thailand and extending into Vietnam. The best example of genetic improvement in an aquacultured organism is the hybrid between channel catfish females and blue catfish males. This hybrid exhibits heterosis for multiple traits, at least eight or more. The use of this hybrid is increasing gradually in the US catfish industry. In the year 2000 approximately 1–2 million fry were produced. By 2004 that number increased to 2–5 million. In 2009 and 2010, 65 million and 85 million hybrid catfish fry were produced, respectively, and the 2011 projection is more than 100 million hybrid fry. The hybrid between the striped bass and the white bass is also an example of a beneficial interspecific hybridization impacting not only the Morone bass aquaculture industry, but also resulting in a fish heavily used for recreational fishing. In some cases, an F_2 hybrid can be a beneficial composite fish.

The Israeli tilapia industry was based heavily on the F_2 hybrid between Nile tilapia and blue tilapia until a severe snowstorm hit Israel several years ago. The red coloration of Mossambique tilapia has been introgressed into other tilapia species that are more suitable for aquaculture, and these red tilapia hybrids are widely used in aquaculture. The Philippines has used hybridization and introgression to produce tilapia with greater cold tolerance or better salinity tolerance.

Polyploidy

Sex-reversal and polyploidy technologies are impacting aquaculture production. This is a distinct advantage for the genetic improvement of aquatic species since terrestrial animals are less plastic genetically and cannot currently be improved via polyploidy. These technologies are on the threshold of advancing global food security significantly, and have been adopted and applied in both developed and developing countries.

Several examples of the successful application of triploidy exist in aquaculture. Triploid salmon, trout, grass carp and Pacific oysters are commercially produced.

Triploid grass carp are widely utilized in the USA to control aquatic vegetation (Cassani and Caton, 1986b; Fig. 19.1). Many states have banned the use of fertile diploid grass carp since it is an exotic species which, if established, might have a detrimental impact on native species and on local aquatic plant ecology. Many states

Fig. 19.1. Application of triploidy to sterilize and utilize an exotic species, grass carp, *Ctenonpharyngodon idella*, in the USA.

have legalized triploid grass carp because of its sterility and public pressure to utilize these beneficial fish. The triploid genotype results in functional sterility, which allows application of this exotic species for aquatic weed control where diploid grass carp are still illegal (Wattendorf, 1986; Wattendorf and Anderson, 1986). States that allow the use of triploid grass carp require that only triploids be introduced. Screening of most or all potential triploids for ploidy level is required, which makes it desirable for an induction procedure that yields a high percentage of, if not 100%, triploids.

In Europe, particularly the UK, mono-sex female rainbow trout that are triploids are cultured. The European market requires a larger trout than the American market, and the triploid fish continue to grow and surpass the diploids in size after onset of sexual-maturation effects slows the growth of the diploids. Additionally, as sexual maturation emerges, flesh quality decreases in diploids, whereas triploid salmonids have superior flesh quality at this time. There is also a preference for stocking triploid rather than diploid brown trout in European rivers and lakes for fishery enhancement.

Commercialization of triploid Pacific oysters began on the west coast of the USA in 1985 (Nell, 2002) and they are now widely utilized, accounting for 30% of oyster production. Similar to the example with salmon, the triploid oysters grow faster and have superior flesh quality. The triploidy again counteracts the sexual-maturation effects of decreased growth rate and flesh quality. Triploid oysters are now used in Brittany (Legraien and Crosaz, 1999), and the demand for triploid Sydney rock oysters is greater than supply (Nell *et al.*, 1999). The only oyster with wide-spread commercial application is the Pacific oyster (Nell, 2002).

Geographical variation in culturists' attitudes affects the acceptance and utilization of triploid oysters. Triploid Pacific oysters, *Crassostrea gigas*, are commonly cultured on the west coast of the USA and also in France and Australia; however, triploid induction is seldom utilized for eastern oyster, *Crassostrea virginica*, culture on the

east coast of the USA, although this has started to change.

One of the keys to the widespread utilization of triploid Pacific oysters is hatchery production of seed (Nell, 2002). There is almost total dependence on hatcheries in the Pacific Northwest for triploid oyster seed. Use of tetraploid males to produce triploid spat is now common (Nell, 2002). Triploids are especially preferred over diploids in the summer because of the decreased marketability of diploids in spawning condition. Triploid oyster utilization was virtually non-existent in France until sperm from tetraploids was made available in 1999/2000, resulting in 10–20% utilization of triploid spat. Triploid oysters are now a viable option in the UK (Mallia *et al.*, 2006). China has produced triploid Pacific oysters since 1997. Worldwide, the application of triploid oysters is minimal except where hatcheries have been established to provide triploid spat.

Sex Reversal and Breeding

Major examples of sex reversal and breeding include the commercial aquaculture of salmonids, Nile tilapia and silver barb farming. Monosex female culture is now being applied in salmonid and silver barb culture, and monosex male culture for Nile tilapia.

Sex-reversed XX male rainbow trout are mated with normal XX female rainbow trout to produce 100% XX monosex females. This technology is commonly used in the UK. The advantages of this system are the faster growth and higher flesh quality of female salmonids. This technology is being adopted by many American rainbow trout farmers. The sex-reversal technology has been adopted, along with triploid technology, in Europe to take advantage of the growth and carcass-quality characteristics of salmonids produced by both of these programmes. Multiple generations of sex reversal and progeny testing of *Oreochromis niloticus* have resulted in both YY males and YY females. When mated, they produce 100% all-male YY progeny. The advantage of this system is the increased growth of the males and the lack of reproduction of the monosex populations. The progeny of the YY males ('genetically male tilapia' (GMT) distinguishes them from sex-reversed male tilapia) have been tested in commercial field trials in the Philippines. Results from on-station trials indicate that GMT application increased yields by up to 58% compared with mixed-sex tilapia of the same strain (Mair *et al.*, 1995), which were consistently greater than those for sex-reversed male tilapia. In addition to the negligible recruitment of females in GMT populations, GMT have more uniform harvest-size distribution, higher survival and better food-conversion ratios. Economic analysis based on the results of these field trials indicated a major impact on profitability for Philippine tilapia farms from the use of monosex YY populations. Another advantage of monosex technology is the elimination of the need for sex hormones to produce the monosex populations. The reliability of this system is also superior to hybrid technology, where cross-contamination of brood stock of different species living near each other is problematic, resulting in the production of less than 100% male populations. The GMT technique is environmentally friendly, species/strain purity is maintained and the fish produced for culture are normal genetic males.

Although the development process takes several years and is labour-intensive, once developed the production of monosex males can be maintained through occasional feminization of YY genotypes (Dunham *et al.*, 2001). Assuming that brood-stock purity can be maintained, GMT production can be applied in existing hatchery systems without any special facilities or labour requirements. Disregarding the initial development costs, additional costs for the utilization of GMT technology at the hatchery level should be minimal, while the potential economic advantage to growers is substantial.

The outputs of the research on the YY male technology, GMT and GMT-producing brood stock have been widely disseminated in the Philippines since 1995, in Thailand since 1997 and to a lesser extent in a number of other areas, including Vietnam, China, Fiji, Europe and the USA. In the Philippines and Thailand, the strategy for dissemination was to produce and distribute brood stock from breeding centres to accredited hatcheries. As of 2000, 32 accredited hatcheries existed in the Philippines, producing an estimated 40 million GMT, representing 5–10% of fingerling production (Dunham *et al.*, 2001). In Thailand, conservative estimates indicate that GMT now accounts for more than 15% of fry produced. The XY system was proved effective in maintaining quality control over the technology, but limits the scale of the dissemination. This dissemination strategy is being carried out in a financially viable way, which is considered essential to the long-term sustainability of the technology and its impact.

Dissemination of GMT has reached a scale where the technology has a significant impact upon worldwide tilapia production. This impact will grow as availability of further improved GMT increases, along with the increased likelihood of restrictions on the use of hormones in aquaculture (combined with consumer resistance).

Wide-scale application of XX all-female (neomale) silver barb has occurred in Thailand. Hatchery trials on station and pilot-scale commercial production in private hatcheries in Thailand demonstrated that neomale brood stock performed satisfactorily. Monosex female fingerlings from neomale brood stock are now produced on a significant scale in commercial hatcheries in Thailand, and research is ongoing in other countries in the region.

Genetic Engineering

Genetic engineering will probably be one of the best tools to improve disease resistance and tolerance of low oxygen in fish in the future. Faster-growing, genetically engineered fish will soon be available for commercial use. However, environmental concerns, public fear and government regulation may slow the commercialization of genetically engineered fish. Transgenic

Fig. 19.2. GloFish®, transgenic zebra fish containing fluorescent pigment genes, the first commercial application of transgenic fish. (Photograph from www.glofish.com.)

fish must be shown to be safe for the environment and for consumption. If this is demonstrated, the general public, leaders and legislators must be educated on the safety and benefits of genetically engineered food, and scientists, industry and policy makers need to work together to ensure realistic and reasonable government regulations.

In the past, salmon have been produced in Europe, New Zealand and North America that contain additional salmon GH genes, which enhance levels of GH and rate of growth (Reichhardt, 2000). Efforts in Europe and New Zealand have ceased as approval and consumption of transgenic salmon will probably be long in coming in Europe owing to the strong anti-transgenic sentiment there and because of company-related decisions in New Zealand. Otter Ferry Salmon in Scotland initiated a growth trial with transgenic Atlantic salmon in 1996 in a closed system; the fish were grown for 18 months and then destroyed. The Scottish Salmon Association distanced itself from the experiments, fearing consumer protests in the market. Later, when it was learned in the House of Commons that the government had approved the privately funded experiment with transgenic salmon, there was great concern and opposition to this government support in the UK. The

trials were successful as the transgenic salmon grew at four times the rate of controls; however, nine salmon-growing countries agreed to a ban on genetically modified fish subsequent to this trial (Seafood Datasearch, 1999).

In contrast, in North America the marketing of transgenic salmon may be close, following submission of an application by A/F Protein, AquaBounty Farms, Waltham, Massachusetts, to the US Food and Drug Administration (FDA) to gain approval to sell GH-transgenic salmon which contain genetically modified GH genes (Niiler, 1999). FDA approval for consumption of these fish has yet to occur, but is expected during 2011.

It had been rumoured sometime close to 2000 that transgenic carp have been commercialized in China and transgenic Nile tilapia in Cuba, but officials in these countries indicated this has never occurred. Approval of transgenic fish for aquaculture is apparently pending in both of these countries.

The only transgenic fish approved for commercialization has been the GloFish® (Fig. 19.2). Zebra fish containing fluorescent pigment genes from jellyfish have been commercialized and are sold at various locations in Asia and the USA in the ornamental fish industry.

20
Environmental Risk of Aquatic Organisms from Genetic Biotechnology

Theoretical Risks

Commercialization of the GloFish®, a transgenic zebra fish, for the aquarium trade has taken place, but transgenic aquatic organisms have not yet been approved for food fish aquaculture. Commercialization of transgenic aquatic organisms on a large scale may have a variety of ecological implications (Hallerman and Kapuscinski, 1992a,b, 1993). Eventual escape of transgenic aquatic organisms from confinement will occur from a commercial facility, and the range of receiving ecosystems is broader.

Much concern exists concerning the potential food safety, ethics of utilization and ecological impacts of transgenic fish (Dunham *et al.*, 2001). In addition to benefits, aquatic genetically modified organisms (GMOs) may also pose environmental and food-safety hazards (FAO, 2001). Potential ecological hazards include adverse interactions with a range of species with which a GMO interacts in the accessible ecosystem, and genetic hazards to conspecific natural populations. Ecological hazards include the possibility of increased predation or competition, colonization by GMOs in ecosystems outside the native range of the species and, possibly, alteration of population or community dynamics due to the activities of GMOs (Dunham *et al.*, 2001; FAO, 2001).

Fertile GMOs could interbreed with natural populations, and any genetic or evolutionary impacts, positive, negative or neutral, would depend on the fitness of the new genotypes in the wild. Risk would exist when fitness relative to the wild type is high, and also when maladaptive traits and genes might be introduced into native populations, although, logically, these would be selected against.

In the case of mammals, the transfer of recombinant genes can result in insertional mutagenesis. These gene disruptions are usually recessive so they are evident only in homozygous individuals, can remain in the population through carriers and result in genetic load for the population. Again, logically, selection should minimize the effects of these insertional mutants and they would likely be eliminated in time.

Integration of transgenes can also cause activation of genes at or in the vicinity of the transgene because of the insertion of promoters and enhancers, gene fusion and the introduction of elements that stabilize mRNA. This can effect expression of genes large distances from the insertion site in cell lines because of alteration of methylation. This phenomenon has not yet been observed in transgenic animals and fish.

Insertion of a transgene can also result in adverse pleiotropic effects. Purification of transgenes can be imperfect, leading to the potential of insertion of bacterial or plasmid sequences from the gene preparation. This could lead to insertional mutagenesis, and also has the potential to recombine to form novel infectious viruses. If constructs are utilized containing marker genes, there is the remote possibility of some of them aiding in the development of antibiotic-resistant pathogens.

Use of viral vectors could result in new types of virus that could be horizontally transmitted to other species. Use of transposons could pose similar risks regarding pathogens and also gene disruption.

Some of the concerns listed above have been observed in transgenic animals, and others have yet to be observed. Many have a low probability of occurring. The majority of these problems can be eliminated by using safe construct design and avoiding or inactivating some elements. Additionally, many of these problems can be eliminated by conducting selection upon the transgenic population that is generated as natural selection would correct some of the adverse genotypes, and it is hoped the either one of these or both would be done prior to commercialization.

The ecological, genetic and evolutionary impacts of GMOs in the range of relevant aquatic and marine systems, and the ecological and genetic risk pathways and end-points posed by commercial-scale application of aquaculture biotechnology, need to be thoroughly studied (Bartley and Hallerman, 1995). Interdisciplinary approaches to environmental risk assessment and monitoring will be needed to totally understand the effects of transgenic aquatic organisms in the environment (Kapuscinski and Hallerman, 1995). Risk will need to be evaluated on a temporal basis. Rapidly occurring phenomena, such as large-scale escapement, need to be compared with slowly occurring phenomena, such as the adaptive evolution of transgenic aquatic organisms in the environment. Spatial connections among aquatic ecosystems will also need consideration.

Concerns about hazards posed by aquatic GMOs have been inferred on the basis of ecological principles (Kapuscinski and Hallerman, 1990b, 1991; Hallerman and Kapuscinski, 1992a,b, 1993); however, the totality of the impact and how it might occur are more complex than what is presented in these papers based on principles. Existing experimental evidence (Dunham, 1996; Farrell *et al.*, 1997; Devlin *et al.*, 1999; Guillen *et al.*, 1999) indicates that most transgenics pose little ecological risk and a subset of transgenics may pose a potential risk, although these data were derived from models in highly artificial situations. Questions concerning these issues are beginning to be answered. Transgenic salmon may be approved for human consumption in the USA in 2011, but approval for the growth of these fish in the USA will not be granted at the same time. A non-food fish, transgenic zebra fish containing fluorescent pigment genes, has been commercialized in both Asia and the USA. However, in locations such as Europe and Japan, conservative approaches to the development of transgenic fish will prevail politically for many more years. Because of these concerns, transgenic fish will probably be utilized commercially to a greater extent in developing countries than in developed countries in the short term (Bartley and Hallerman, 1995).

Transgenic fish, assuming they are derived from domestic strains, may not have any more genetic impact on natural populations than domestic conspecifics. However, genetic modifications that would allow expansion of a species' range – essentially the development of an exotic species – would probably have the greatest ecological impact. For instance, the development of a cold-resistant tilapia or a cold-resistant salmon with antifreeze protein would allow these fish to expand their geographical range. As an exotic species, they would interact with local biota and have the potential for ecological impact, as alteration in species composition is considered detrimental.

The majority of introductions of exotic fish are unsuccessful (Courtenay and Stauffer, 1984). Successful introductions are more likely to occur in temperate rather than tropical habitats, in environmentally stressed fish communities, in simple rather than diverse fish communities and in environmentally stressed habitats. About 10%

of attempted introductions are successful and, of these, 10–20% result in species introductions that are considered to have adverse ecological effects. However, when they occur, the adverse impacts can be severe. One example is the introduction of Nile perch in Lake Victoria resulting in the extinction of several cichlids. The introduction of predatory species of transgenics may have the potential for larger effects than the introduction of prey species.

The US Department of Agriculture (USDA) has developed performance standards for conducting research on transgenic fish and assessing their risk (ABRAC, 1995; Hallerman *et al.*, 1998). Changes in a genetically modified fish's metabolism, tolerance of physical factors, behaviour, resource/substrate use, population regulatory factors, reproduction, morphology and life history could lead to ecological impacts. This document concludes that escapement of aquatic GMOs into environments containing threatened or endangered species, introgression with conspecifics with a likely lowering of fitness, low-resiliency environments (lack of species diversity or perturbed environment), genetically modified predators and genetic modifications with a large likelihood of allowing the transgenic fish to alter the ecosystem structure would pose the greatest environmental risks.

Because of these concerns about transgenic aquatic organisms, research on food safety and potential environmental impact, including the measurement of fitness traits, such as predator avoidance, foraging ability, swimming ability and reproduction, is needed to allow educated decisions on the risk of utilizing specific transgenic fish (Dunham *et al.*, 2001). These data will be necessary for the application of transgenic fish in North America and Europe.

The impact of domestic aquacultured organisms, interspecific hybrids, polyploids and genetically engineered fish on the genetic variation of conspecifics, population numbers and performance of conspecifics, and the ecosystem in general, is currently being questioned, debated and researched. Data are building up concerning the interactions between domestic and wild populations

and the fitness of genetically enhanced aquatic organisms such as to allow policy development and management application to be based solely on scientific fact and principle.

Certain triploid aquatic organisms may also pose risk under particular circumstances. Triploid male grass carp and salmonids undergo sexual maturation and may seek matings, which would result in loss of the resulting broods, posing demographic risk to the small natural populations they may encounter (Dunham *et al.*, 2001). Triploid Pacific oysters, *Crassostrea gigas*, can exhibit reversion to diploidy (Hallerman, 1996), although it is not yet known whether this restores fertility. The effectiveness of triploidy as a means of limiting risks associated with introductions of non-native species, genotypes or transgenics needs further evaluation.

Assuming that environmental risks associated with transgenic fish exist, benefit–risk analysis will be required (Bartley and Hallerman, 1995). Ecologically, the primary concerns regarding the utilization of transgenic fish are loss of genetic diversity, loss of biodiversity and changes in the relative abundance of species upon release or escape followed by establishment of transgenic fish in the natural environment. Conversely, the utilization of transgenic fish in aquaculture could actually enhance genetic diversity and biodiversity by increasing food production and production efficiency, thus relieving pressure on commercial harvests of natural populations and decreasing pressure on land and water use for agriculture and aquaculture.

In the past, society decided to increase population size, expand agriculture, exploit natural populations, dam rivers to generate electricity, feed people and increase quality of life at the expense of biodiversity. Although we need to be mindful and cautious concerning the effect of aquaculture gene pools on native conspecific gene pools, and we need to determine the relative benefit and risks of genetically enhanced fish to society and to the environment, the primary dangers to genetic diversity and biodiversity that need to be addressed most urgently with more effort are the effects of overexploitation, pollution, habitat alteration and stocking of exotic species on genetic diversity. If properly implemented, aquaculture and genetic improvement have the potential to increase production and fish availability and to decrease pressure on wild stocks, thus preserving natural genetic diversity. Captive stocks, if properly managed, could also be utilized to preserve genetic diversity. Cost–benefit analysis will be necessary to assess whether or not transgenic fish application is warranted.

The extent of phenotypic change from the introduction of a fusion gene could be analogous to the development of an exotic species, a select line or a domestic strain, depending on the magnitude of the phenotypic change. Transgenic fish could express phenotypes that are analogous to the formation of a new exotic species. Research on gene transfer that has a high probability of such a result should be avoided, since it is well documented that exotic species can cause major changes in ecological and population balance and lead to the elimination of native species in the invaded environment. Approximately 11% of introductions of exotic species actually become established and of these about 10% have negative ecological impacts (Welcomme, 1998).

The genetic impact of genetically improved aquaculture fish could have neutral, positive or negative effects on wild populations, in the short term or long term. Transgenic fish could escape and the transgene become part of the gene pool. This could add genetic diversity to the population, lower or raise fitness, or have no phenotypic or ecological effect. If there is a lowering of fitness, it should be temporary as the transgene should be selected against. The mixing of the gene pools might enhance genetic resources, increase genetic variation or result in heterosis, all potentially positive results. The wild fish may outcompete and eliminate the domestic fish or the domestic fish may have no long-term impact on the performance of the population – neutral or non-effects. Negative impacts could result from outbreeding depression, which would theoretically be temporary, or from the elimination of wild genotypes through competition. Although the concern that the accidental or intentional release of domestic and genetically improved fish will have a damaging effect on native gene pools is legitimate and requires careful scrutiny, the available data indicate that the potential damage of domestic fish to native conspecific gene pools is quite small (see Chapter 13). First, it is difficult to change the genetic make-up of an established native gene pool with even the intentional stocking of wild conspecifics from another watershed. Second, wild fish almost always outcompete their domestic genotypes in natural settings.

Most data indicate that wild fish are more competitive than domestic fish (Dunham, 1996), resulting in the elimination of the domestic fish and their potential positive or negative impacts. However, recent evidence from salmonid research indicates that there are situations where domestic fish can have a genetic impact on wild populations. When repeated large-scale escapes of domestic fish occur, genetic impact can occur just from the swamping effect of sheer force of numbers. Transgenic fish could make an impact in this scenario of large-scale escape, but again the consequences should not vary much from that of fish genetically altered by other means.

Most types of transgenic fish, including those that are GH-gene transformants, are more analogous to a selected line or domestic strain. The change in phenotype is similar to what would be observed in or what would be the goal of strain selection, selection, intraspecific crossbreeding, interspecific hybridization, sex reversal or gynogenesis. If a fourfold increase in growth is possible through traditional breeding (and such gains are possible; Dunham, 1996), ecological impacts would be the same regardless of the mechanism of phenotypic alteration – traditional or biotechnological.

Transgenic fish should be analogous to select lines and domestic strains in a second manner. Transgenic fish will probably be generated from select lines and domestic strains since these fish are generally more suited for aquaculture and already have increased performance in the aquaculture environment (Dunham, 1996).

Obviously the primary benefits of transgenic fish would be increased aquaculture production and profitability. However, other potential benefits exist in addition to those mentioned earlier.

Escaped transgenic fish could add genetic diversity to populations. This would be artificially induced genetic diversity, which some sectors of society would value and to which others would be opposed. This artificial genetic diversity could actually increase fitness in some endangered populations or species and make such genetic units more viable. For example, natural and human-induced factors (if we consider humans an unnatural aspect of global ecology) have apparently reduced genetic variation in the cheetah to the point where reduced reproductive performance threatens its existence. Gene transfer would be an option to restore reproductive performance and save this species.

Evidence exists that when humans began exploiting fish populations much more efficiently and intensively in the last 200 years, certain traits, such as size, were genetically selected against (Ricker, 1975, 1981). It is likely that we have permanently eliminated important growth alleles and perhaps alleles for other traits from some fish species. Gene transfer is an option to restore phenotypes that have been artificially eliminated. However, Swain *et al.* (2007) studied wild cod that were under fishing pressure for 30 years, and found that in some generations this selective pressure resulted in positive growth response and others negative. Thus, the long-term impact of humans on genetic variation for growth in natural populations is not clear.

The escaped transgenic fish could replace the natural population. Depending on the existence or absence of this genotype, genetic diversity would be lost. The long-term survival of that species or population at the location could be enhanced, decreased or unchanged. Environmental risk data to date, however, indicate that the above scenario, replacement of the natural population, is unlikely (Devlin *et al.*, 1995a,b; Dunham, 1995; Dunham *et al.*, 1995; Chitmanat, 1996).

Transgenic fish could become established and hybridize with other species, spreading the transgene to other species. This scenario is unlikely since reproductive isolating mechanisms usually restrict permanent gene flow between species of fish (Argue and Dunham, 1999).

All available data indicate that transgenic fish are less fit than non-transgenic fish and would probably have little, if any, environmental impact. Additionally, domesticated transgenic fish would be expected to have less environmental risk than wild transgenic fish, based on the discussion above. However, the greatest environmental risk that a transgenic fish would have is when the gene insert would allow the transgenic genotype to expand its geographical range, essentially becoming equivalent to an exotic species. About 1% of such releases of exotics result in adverse environmental consequences.

Altering temperature or salinity tolerance would be analogous to the development of an exotic species since this would allow the expansion of a species outside its natural range. This type of transgenic research and application should be avoided. Antifreeze-protein genes from winter flounder have been introduced into Atlantic salmon in an attempt to increase their cold tolerance (Shears *et al.*, 1991). If this research were successful, a real possibility of environmental impact exists. Similarly, if tilapia were made more cold tolerant, a strong possibility of detrimental environmental impact exists. Sterilization could reduce risk, but genetic means of sterilization, such as triploidy, decrease performance (Dunham, 1996). Additionally, fertile brood stock is necessary, so risk is minimized but not eliminated. Transgenic sterilization, to be discussed later, is potentially a much better option than triploidy.

Currently, it is common practice in Asia to introduce exotic species to address shortcomings in the aquaculture performance of native species. Again, the introduction of exotic species has the greatest potential to adversely affect biodiversity and, consequently, genetic diversity (Welcomme, 1988). The utilization of transgenic fish derived from the indigenous aquaculture species is more likely to be an environmentally safe means of addressing the perceived aquaculture shortcomings of native species and is less likely to decrease biodiversity and genetic diversity compared with the continued practice of exotic-species introduction.

Interspecific hybrids pose similar environmental risks to those of transgenic aquatic organisms. In some cases, usually where parent species have limited geographical distributions, but very rarely, hybrids introgress permanently with parent species in a natural setting. It appears to be a greater problem in maintaining aquaculture populations pure for hybridization programmes.

Environmental Risk Data on Transgenic Fish

Efforts should be organized to evaluate the potential environmental risk of transgenic fish. Reproductive performance, foraging ability, swimming ability and predator avoidance (Fig. 20.1) are the key factors determining fitness of transgenic fish and should be a standard measurement prior to commercial application. Most ecological data on transgenic fish gathered to date indicate a low probability of environmental impact. Extremely fast-growing salmon and loach have low fitness and die (Devlin *et al.*, 1994b, 1995a,b).

Several models have been developed that estimate and indicate genetic risk of transgenic fish. Muir and Howard (1999) evaluated a model and described the Trojan gene effect: the extinction of a population due to mating preferences for large transgenic males with reduced fitness. Therefore, reduced fitness as well as increased fitness has potential adverse ecological effects. This modelling was based on experimental results of medaka in aquaria.

Hedrick (2001) developed a deterministic model indicating that, if a transgene has a male-mating advantage and a general viability disadvantage, analogous to the Trojan gene effect of Muir and Howard (1999), then in 66.7% of the possible combinations for possible mating and viability parameters for its invasion in a natural population, the transgene increases in frequency and for 50% of the combinations – the possible combinations of the possible mating and viability parameters – the transgene goes to fixation. The increase in the frequency of the transgene reduces the viability of the

Fig. 20.1. Dragonfly predation on young fish. Most fish in the natural environment die of predation or starvation prior to reaching sexual maturation. Therefore, predator avoidance is a key trait for evaluation fitness to determine the environmental or genetic risk of genetically improved fish.

natural population, increasing the probability of extinction of the natural population.

In another modelling exercise, Muir and Howard (2001) again conclude that a transgene is able to spread to a wild population even if the gene markedly reduces a component of fitness, based on data from a laboratory population of medaka harbouring a regulatory sequence from salmon fused to the coding sequence for human GH. The juvenile survival of transgenics was reduced in the laboratory but the growth rate increased, resulting in changes in the development rate and size-dependent female fecundity. The important factors in the model were the probabilities of the various genotypes mating, the number of eggs produced by each female genotype, the probability that the eggs would be fertilized by the sperm of each male genotype (male fertility), the probability that an embryo would be a specific genotype given its parental genotypes, the probability that the fry would survive and parental survival. Muir and Howard's (2001) interpretation was that transgenes would increase in populations despite high juvenile viability costs if transgenes also had sufficiently high positive effects on other fitness traits. Sensitivity analyses indicated that transgene effects on age at sexual maturity should have the greatest impact on transgene allele frequency. Juvenile viability had the second greatest impact. A defect in the simulation was the fact that the effect of predation in the wild could not be included in the model, biasing viability estimates (Muir and Howard, 2001).

Although these modelling experiments based on laboratory data on small model species illustrate the potential risk of transgenic fish, some weaknesses in the analysis exist. The environment was artificial, the mating preference does not exist for many fish, including catfish, the data were not put into the model to account for GE interactions, which are likely, predation is absent, as Muir and Howard (2001) indicate, and the overall performance of the fish is not accounted for.

Body size does not necessarily result in mating advantages. Rakitin *et al.* (2001) utilized allozymes and minisatellites to determine that male size, condition factor and total or relative body-weight loss over the season were not correlated with the estimated proportion of larvae sired by each Atlantic cod male during the spawning season. Similar results were observed in salmon (Doyle, 2003). However, Atlantic cod male reproductive success was affected by female size, with males larger (>25% total length) than females siring a smaller proportion of larvae (Rakitin *et al.*, 2001). In this case, large size was reproductively disadvantageous.

The Trojan gene effect models also do not account for another possibility. In the case of GH-transgenic common carp in China, non-transgenic males became sexually mature at 6 months of age and only 1.1 kg while transgenic males were still not mature at 7 months of age and 3.1 kg. Size selection would not matter in this case as the largest males were yet to mature sexually. This also illustrates that models for one species may not be relevant to another species at all.

Although there may be cases where size increases reproductive fitness of both sexes, cultured transgenic fish may not be allowed to grow large enough to have a mating advantage as escapees (Doyle, 2003), and data on transgenics of larger aquaculture species have not shown any reproductive advantage for the transgenics. Fast-growing transgenic tilapia have reduced sperm production. Transgenic channel catfish and common carp have a similar reproduction and rate of sexual maturity compared with controls (Dunham *et al.*, 1992a; Chen *et al.*, 1993; Chatakondi, 1995). The spawning success of transgenic channel catfish and controls appeared similar. When the two genotypes were given a choice in a mixed pond, the mating was random and the spawning ability of transgenic and control channel catfish was equal (Dunham *et al.*, 1995). In a second experiment with GH-transgenic common carp (Wu *et al.*, 2003), there was no difference in fertility and hatchability of transgenic and non-transgenic individuals. GSIs of the transgenic individuals were smaller than those of controls.

GE interactions are important and occur for the growth of transgenic channel catfish (Dunham *et al.*, 1995). Transgenic channel catfish containing salmonid GH genes grew 33% faster than normal channel catfish in aquaculture conditions with supplemental feeding. However, there was no significant difference in growth performance between transgenic and non-transgenic channel catfish in ponds without supplemental feeding, indicating equal foraging abilities and the inability of transgenic catfish to exhibit their growth potential with limited feed (Chitmanat, 1996). The foraging ability of transgenic and control catfish is similar under these conditions of competition and natural food sources and growth is no different between transgenic and control catfish in these more natural conditions. When grown under natural conditions where food is limiting, the transgenic channel catfish has a slightly lower survival than controls and grows at the same rate as non-transgenic controls. As is the case in most genetic improvement programmes, genetically altered fish need adequate food to express their potential.

GE interactions are now recognized as a key component in determining environmental risk of transgenic aquatic organisms (Sundström *et al.*, 2007). GE interactions and transgene by genetic background interactions occur in salmonids and could affect the ability of a transgenic organism to establish itself in the natural environment. GH coho salmon did not affect the growth of non-transgenic coho salmon when food abundance was high (Devlin *et al.*, 2004). However, when food abundance was low, dominant individuals – usually transgenic – would emerge, suppress the growth of the other fish, exhibit cannibalism and drive the population to extinction. In another study (Sundström *et al.*, 2007), GH coho salmon grew more than three times faster than controls in the hatchery environment, but only 20% faster in a simulated stream environment. Additionally, the transgenic coho salmon were much more predatory than controls in the hatchery environment and, again, this was much reduced in the simulated stream environment. Thus, Sundström *et al.* (2007) suggest that environmental risk assessment of transgenic aquatic

organisms be conducted in a variety or range of potential natural environmental conditions.

The faster-growing transgenic fish could have impaired swimming, leading to predator vulnerability, problems in capturing prey, reduced mating ability for some species and reduction in competitiveness for any trait requiring speed. Selection for swimming ability may be one of the primary mechanisms limiting the genetic increase in size of fish and preventing fish from evolving to larger and larger sizes.

Silversides, *Menidia menidia*, from Nova Scotia ate more food, had more efficient feed conversion and grew faster than a population from South Carolina (Billerbeck *et al.*, 2001). However, the maximum prolonged and short-term swimming speeds of Nova Scotia strain were lower than those of the South Carolina strain, and the swimming speeds of fast-growing phenotypes/genotypes were lower than those of slow-growing phenotypes/genotypes within each strain. Slow swimming speed has a fitness cost: vulnerability to predation. The Nova Scotia strain was more vulnerable to predation than the South Carolina strain, and predation increased with growth rate and feeding rate both within and between strains (Billerbeck *et al.*, 2000). Maximizing energy intake and growth rate engenders fitness costs in the form of increased vulnerability to predation (Doyle, 2003).

In fact, initial experiments all indicate that the predator avoidance of transgenic fish is inferior compared with controls, as would be predicted by the experiment with the silversides. Predator avoidance was slightly better for non-transgenic catfish fry and fingerlings when exposed to largemouth bass, *Micropterus salmoides*, and green sunfish, *Lepomis cyanellus*, than for transgenic channel catfish (Dunham, 1995; Dunham *et al.*, 1995, 1999). Data on salmon also indicate that they probably have reduced fitness for non-aquaculture, the natural environment. GH-transgenic salmon have an increased need for dissolved oxygen (Stevens *et al.*, 1998; Cook *et al.*, 2000b,c), a reduced swimming ability (Farrell *et al.*, 1997; Stevens *et al.*, 1998) and a lack of fear of natural predators (Abrahams and Sutterlin, 1999). GH Atlantic salmon fed in the presence of predators, whereas controls totally avoided the obviously dangerous area. However, Vandersteen Tymchuk *et al.* (2005) found no difference in mortality to predators associated with feeding behaviour in GH coho salmon at both the fry and parr stage. Again, perhaps species differences, gene construct or gene insertion site effects occur, which would require risk assessment on a case-by-case basis.

On an absolute speed basis, transgenic coho salmon swam no faster at their critical swimming speed than smaller non-transgenic controls, and much more slowly than older non-transgenic controls of the same size (Farrell *et al.*, 1997). Again, as was found with the silversides, a marked trade-off was observed between growth rate and swimming performance. Farrell *et al.* (1997) hypothesized that the decreased swimming ability may be a result of some physiological change due to the hyperlevels of GH excretion. However, Ostenfeld *et al.* (1998) offer an alternative explanation. Coho salmon containing the *Oncorhynchus* MT–GH1 plasmid (pOnMTGH1) had an altered body contour and centroid size and enhanced caudal peduncle and abdominal regions compared with controls. The most prominent alterations were the change in the syncranium and that the head of the transgenics was less elliptical. The opercular series were shifted, with an enlargement of the branchiostegal and augmentation of both the opercular and the cleitrum regions. The overall body shape is less fusiform for the transgenic coho salmon. Therefore, the decrease in swimming ability may be a result of loss of hydrodynamics and increased drag coefficients caused by the altered body shape. This change in body shape might also alter leverage or efficiency of the muscle movements for swimming. The inferior swimming ability of the transgenic salmon should cause them to have inferior predator avoidance, inferior ability to capture food and inferior ability to migrate to reach the sea or return to reproduce.

Transgenic fish could be more competitive in seeking feed. Devlin *et al.* (1999) examined the ability of F_1 coho salmon

(250 g) containing a sockeye MT-B promoter fused to the type 1 GH gene-coding region to compete for food through higher feeding motivation. The consumption of contested food pellets was determined by matching pairs of one sibling control or by size-matching pairs of one control (1 year older non-transgenic coho salmon) and one GH-transgenic coho salmon, and then determining which fish captured the first three pellets presented one at a time at each feeding trial. The transgenic coho salmon consumed 2.5 times more contested pellets than the sibling controls and the transgenic fish consumed 2.9 times more pellets than the non-transgenic size controls, indicating the high feeding motivation of the transgenic fish throughout the feeding trials. The shortcomings are that this is a highly artificial environment and a food type that will not be encountered under natural conditions.

F_2 transgenic Atlantic salmon contained a salmon GH gene that was continuously expressed in the liver, enhancing growth 2.62- to 2.85-fold over the size range 8–55 g and improving feed-conversion efficiency by 10% (Cook *et al.*, 2000a). These transgenic fish had higher metabolic rates, but they consumed 42% less total oxygen between hatching and smolt size and, when starved, the rate of oxygen consumption declined more rapidly in the transgenic Atlantic salmon. The starved transgenic Atlantic salmon also lost protein, dry matter, lipid and energy more quickly than controls. The persistence of transgenic Atlantic salmon in maintaining a higher metabolic rate, combined with their lower initial endogenous energy reserves, suggests that the likelihood of growth-enhanced transgenic salmon achieving maximum growth or even surviving outside intensive culture conditions may be lower than that of non-transgenic salmon (Cook *et al.*, 2000a,b,c). These hypotheses are consistent with the data of Dunham *et al.* (1999) with GH-transgenic catfish, which did not come near to the dramatic phenotypic alterations observed in GH-transgenic salmon. These transgenic catfish grew at the same rate as controls under natural conditions, perhaps due to lack of food, higher metabolism coupled with lack of food or inferior swimming ability to allow capture of prey. These fish also exhibited higher mortality under the natural conditions – again, possibly being related to the lack of food and inability to capture food, coupled with higher metabolism and more rapid loss of nutrient stores, as well as possible differential mortality due to predation by aquatic insects and the potential slower swimming of the transgenics.

All transgenic fish evaluated to date have fitness traits that are either the same or weaker compared with controls (Dunham and Devlin, 1998). The increased vulnerability to predators, lack of increased growth when foraging and unchanged spawning percentage of these transgenic fish examples indicate that some transgenic fish may not compete well under natural conditions or cause major ecological or environmental damage. Although transgenic fish may be released to nature by accident, ecological effects should be unlikely because of these examples of reduced fitness. However, implementation of physical and biological (sterilization) containment methods may reduce further potential interaction between transgenic and wild fish populations (Devlin and Donaldson, 1992).

Common Goals of Aquaculture and Genetic Conservation

The preservation of genetic diversity is a common goal for both aquaculture breeders and managers of natural populations. Wild populations and their genes represent a living gene bank that is needed for future resources for genetic improvement. Therefore, transgenic fish research as well as aquaculture genetics research should be conducted in a manner that minimizes genetic impact on natural populations for sound ecological reasons, as well as protecting future resources for exploitation, except in situations where genetic impact on the natural population is desirable.

Production of transgenic fish and aquatic invertebrates is an extremely promising approach to enhancing global food security and efficiency by developing high-performance aquatic organisms. Transgenic fish may actually provide better protection of natural genetic resources by relieving pressure on natural exploitation and decreasing the need for the destruction of habitat for increased food production. Early evidence indicates that high-performance transgenic fish may actually have low fitness, decreasing the likelihood of their establishment in the wild and of associated potential impacts.

Transgenic fish and aquatic invertebrate research is now conducted in many countries (Dunham, 1999). Transgenic fish development is inevitable. The organized development of these programmes would help ensure that environmental risk and fitness traits, as well as food-safety issues, are addressed. The establishment of collaborative networks to develop protocols for and to conduct sound and safe research on transgenic aquatic organisms would help ensure that the benefits rather than the detriments are the product of aquaculture gene-transfer research throughout the world.

Genetic Sterilization

Genetic enhancement of farmed fish has advanced to the point that it is now having an impact on aquaculture worldwide; however, potential maximum improvement in overall performance is not close to being achieved (Dunham *et al.*, 2001). Examples exist indicating that greater genetic gain can be obtained in one or more traits by simultaneously utilizing more than one genetic strategy. Overall performance can probably be maximized by combining the advantages of selection, intraspecific crossbreeding, interspecific hybridization, polyploidy and genetic engineering. There are environmental concerns regarding the application of domestic, hybrid and transgenic aquatic organisms.

Additionally, the aquaculture industry's ability to capitalize on market potential is hampered by community and scientific concern over escapees from aquaculture facilities and their impact on aquatic ecosystems (Naylor *et al.*, 2000). Debate over the costs/benefits of these issues has divided communities and resulted in the development of strict environmental policies and legislation that often preclude growers from producing the aquaculture product of choice (Thresher *et al.*, 2009). Utilization of exotic species is especially unacceptable for many government and stakeholder groups because of the perception and sometimes reality of the high risk of escapees establishing feral populations and having a negative impact on local ecosystems.

Aquacultured organisms, such as Pacific oysters in Australia and Atlantic salmon in British Columbia, and recreational/commercial species, such as Nile perch, have established destructive feral populations, creating environmental problems (Thresher *et al.*, 2009). Additionally, stocked domestic and wild conspecifics have the potential to alter the allele frequencies of established native populations, limiting management options for natural-resources agencies. Concern about these potential environmental and genetic effects has led to restrictions on industry development at some locations (Thresher *et al.*, 2009). Concern over these issues is likely to grow as demand for genetically improved stock escalates to fulfil production requirements.

One option is physical containment, but in reality protection of native stocks from escapees of aquaculture production facilities by the use of physical containment cannot be guaranteed in most cases. The ultimate safeguard would be a mechanism that prevented breeding in the wild of domestic, exotic, highly selected or transgenic stocks. Such a mechanism would prevent cultured aquatic organisms from establishing feral populations, preventing genetic pollution of local strains from domestic conspecifics and protecting the intellectual property invested in highly selected lines or genetically enhanced populations.

Combining variations of chromosome manipulation, monosex and transgenic techniques may produce sterile individuals of only one sex. For example, manipulating GnRH-gene expression may cause sterility, which could be accomplished, theoretically, through antisense RNA constructs, ribozyme approaches or gene knockout. Sterilization with polyploidy, hybridization, transgenesis or combinations of these is the ultimate method to diminish these concerns and environmental risk.

However, in many cases, hybridization reduces, but does not eliminate, reproduction (Dunham and Argue, 2000), so it has limitations, including the fact that fertile parental stocks must be maintained, which could escape. A better sterilization procedure is the induction of triploidy. However, the disadvantages are that triploidy also requires the existence of completely fertile brood stock, and triploidy can have adverse effects on performance (Wolters *et al.*, 1991; Lilyestrom *et al.*, 1999), at least partially negating the genetic gain from the primary enhancement programme. Additionally, triploidy is not feasible or commercially feasible for some species, batch-to-batch variation does not guarantee that all individuals produced are triploid (Reichhardt, 2000) and triploid oysters and some triploid fish have the ability to produce small numbers of diploid progeny. Therefore, triploidy is not 100% effective for some species and triploidy would reduce the rate at which feral populations become established, but would not prevent them.

Monosex approaches could be successful for some, but not all, species. However, they would only be effective for application of exotic species, not conspecifics. Additionally, in the case of applying this technology for exotic species, it is effective only if the exotic species is not already present and, in the event of escape of the monosex/monogenetic organisms of both sexes, short-term impacts are possible for two generations until both sexes die out. The transgenic approach is the ultimate approach.

The Commonwealth Scientific Industrial Research Organisation (CSIRO), Australia, has initiated research on a transgenic sterilization approach, utilizing the insertion of gene constructs that reversibly interrupt development (Thresher *et al.* 2009), resulting in functional sterility (Thresher *et al.* 2009). This technique by itself or in combination with others is the ideal solution for inducing sterility and protecting the environment. Theoretically, aquatic organisms would only be able to complete their life cycle under culture conditions, and escapees from captivity would be unable to breed or produce viable offspring. Progeny from fish that escape die without the intervention of humans, effectively meaning that the escaped parents are sterile. In the hatchery, a simple repressor compound is applied at a particular embryonic stage, allowing the aquatic organism to live and eventually breed. If successful, this technique has profound implications for preventing gene introgression from domestic translocations and escapes into wild gene pools, for importing exotics for culture without establishment or long-term effects on biodiversity and for controlling nuisance species via the sterile screwfly approach (Klassen, 2003).

This technology is related to a procedure to sterilize genetically modified plants: the terminator gene or technology protection system (TPS) developed by the Delta and Pine Land Company (D&PL) (Oliver *et al.*, 1998). The method stops the seeds of certain plants from germinating, and utilizes a transiently active promoter operably linked to a toxic gene, but separated from the toxic gene by a blocking sequence that prevents the lethal gene's expression. A second gene encodes a recombinase, which, upon expression, excises the blocker sequence, and a third gene encodes a tetracycline-controllable repressor of the recombinase.

Plants transgenic for all three genes grow normally and are fertile. Seeds sold to farmers are treated with tetracycline, which activates expression of the recombinase, which then excises the blocker sequence. These seeds germinate, but once the blocker sequence is removed expression of the toxin gene occurs, causing these plants to produce seeds that are sterile (Thresher *et al.*, 2009). Theoretically this technology works, but its existence has not been verified. This technology may be difficult to duplicate exactly in aquatic organisms, as few recombinases have been identified that will function in animals, and those that have been identified, Cre and Flp recombinase, function in only a limited number of species. This technology

has met great opposition because it forces the farmer to repeatedly return to the vendor for seeds, and the controversy has been so great that the technology has been temporarily abandoned for plants (Service, 1998; Niiler, 1999; Thresher *et al.*, 2009). However, because of the tremendous potential mobility of and the concern for transgenic aquatic organisms and their perceived potential impact, such technology for aquatic organisms should be welcomed, in contrast to the situation with less mobile plants. Other strategies, such as trait-specific genetic use restriction technology (T-Gurt), have also been suggested to provide intellectual-property protection, but these technologies do not prevent gene transfer to wild stocks (Masood, 1999).

In the CSIRO approach, genes have been identified that are crucial for and activated only during embryonic development and/or gametogenesis. DNA constructs have been made containing a blocker to development or gametogenesis and a genetic switch to control its function. Thresher *et al.* (2009) have developed promoters, blockers and repressors for zebra fish, *Brachydanio rerio*, and Pacific oyster. Usually native promoters and blocking sequences were utilized for each species. Sterile feral gene constructs developed from these components prevent production of functional gametes or cause mortality in offspring produced by escapees mating outside a controlled hatchery environment (Thresher *et al.*, 2009). Components of the sterile feral construct are fused so that a species-specific promoter is coupled to a repressible element, which in turn drives expression of a blocker gene. In captivity, the blocker can be inactivated by triggering the promoter to allow production of the repressor protein with materials such as zinc or an antibiotic, via diet or in soluble form, so that fertilization can occur or embryos can complete development. In the wild, where the compounds to activate the repressor are not available, the blocker remains active and disrupts the function of the critical gene, either preventing fertility or causing lethality in embryos. The promoter that drives expression of the blocker gene has a narrow spatial and temporal window of activity. Thus, addition of a specific repressor molecule to the food or water in the hatchery is required only for a brief period to repress transcription and subsequent knockout in the resulting offspring. Outside this temporal window, even in the absence of the repressor molecule, the promoter is inactive and the blocker gene is not transcribed. This permits hatchery-reared offspring to survive and remain free once placed into the farm environment for grow-out. However, the promoter is expressed in the absence of the repressor molecule in any offspring that are produced outside the hatchery conditions and they die. The active promoter transcribes the blocker sequence, which leads to disruption of critical gene function and eventual mortality. The blocker gene functions as a dominant allele and thus escapees cannot produce viable offspring even if they interbreed with wild-type fish.

The same approach as above can be used to block gametogenesis. Without delivery of a repressor activator, gamete production does not occur and escapees cannot breed. With the delivery of the repressor via feed, injection or implant, gamete production is allowed, but the offspring of escapees are sterile as no repressor is available in the natural environment. Again, the blocker must act as a dominant allele so that the offspring of transgenic aquatic organisms mated with wild types are sterile.

One option for the blocker mechanism for embryogenesis is disruption of normal cellular activity to cause embryonic mortality via production of a cytotoxic protein. However, production of aquatic animals possessing a toxic gene, although biologically safe, would probably not be accepted by consumers.

An approach to achieving embryonic gene disruption, or knockout, is the use of mRNA specifically targeted to interference of gene function. Mechanisms used to achieve mRNA knockout, including the expression of ribozymes, antisense mRNA and double-stranded mRNA, have succeeded in several organisms (Izant and Weintraub, 1984; Xie *et al.*, 1997; Fire *et al.*, 1998; Waterhouse *et al.*, 1998; Bosher and Labouesse, 2000;

Yin-Xiong *et al.*, 2000). Because of the high specificity of mRNA targeting, the targets in aquaculture species are unlikely to have any close homologies in humans, which eliminates this as an issue should it ever become of concern (Thresher *et al.*, 2009). Currently, there are efforts to transgenically sterilize tilapia by preventing expression of GnRH and LH via knockout, antisense and ribozymes, and there is some preliminary evidence of reduced fertility in these fish (Maclean *et al.*, 2002a,b).

Repression of the knockout function is achieved in zebra fish by coupling a developmental stage-specific promoter to components derived from the commercially available Tet-Off controllable expression system, marketed by CLONTECH (http://www.clontech. com/products/detail.asp?product_id= 157247&tabno=2; Thresher *et al.*, 2009). The Tet-Off system is also functional in oyster primary cell cultures, and has also proved effective at repressing the action of a transgene system in *Drosophila* (Thomas *et al.*, 2000). Tetratracycline or doxycycline is used as the repressor molecule, and both can be easily administered (Thresher *et al.*, 2001) either in hatchery water to allow embryonic development or in food or by injection to allow gametogenesis in brood stock. Tetracycline is a routinely used antibiotic in fish and shellfish cultures (Stoffreggen *et al.*, 1996) and therefore should be acceptable from a food-safety perspective.

Thresher *et al.* (2009) describe the sterile feral construct as follows: the sterile feral construct accomplishes repression of the blocker gene by tetracycline (Tet) or doxycycline (Dox) acting upon the regulatory protein (transactivator protein, tTA), a fusion of TetR and VP16 as derived from the pTet-Off regulatory plasmid. A zebra fish promoter drives expression of the tTA, and tTA regulates the Tet-responsive human cytomegalovirus promoter (PhCMV*-1), which controls expression of the blocker gene. PhCMV*-1 contains the Tet-responsive element (TRE), which consists of seven copies of the tet operator sequence (tetO) located just upstream of the minimal cytomegalovirus promoter (PminCMV). PhCMV*-1 does not function in the absence of the binding of tTA to the tetO. The tetracycline-sensitive element is described by Gossen and Bujard (1992) for Tet-Off and Tet-On is described by Gossen *et al.* (1995) and Kistner *et al.* (1996). In the Tet-Off system, the addition of Tet or Dox, a Tet derivative, prevents the binding of tTA to the TRE. Expression of the blocker gene is controlled by TRE, and is repressed until Tet is removed from the incubation water. In the absence of Tet or Dox, the blocker gene is transcribed as long as the zebra fish promoter continues to express tTA, thus killing the embryo (Fig. 20.2).

The *lac* operon system may provide another transgenic mechanism for sterilization. The essential parts of the *Escherichia coli* repressor system were altered to function at high efficiency in mice, and then transferred into mice to control the production of tyrosinase (Cronin *et al.*, 2001). When fed a standard laboratory diet, the transgenic mice were albino, as the repressor protein of the transgene blocks the operation of the mouse tyrosinase gene. When a lactose analogue is added to the diet, the repressor protein changes its shape and disconnects from the DNA, and the tyrosinase gene begins functioning, turning the mice brown. When the lactose feeding ceases, the mice revert to albinism when they run out of tyrosinase. The modified *lac* operon might be used to control many other types of vertebrate gene, such as those that are

normally lethal early in embryonic development (Cronin *et al.*, 2001; Doyle, 2003).

An alternative, although even the initial steps of building such construct have yet to be initiated, is to produce a GABA enhancement system to disrupt GnRH production and also induce sterility. The background and rationale follow.

GABA (Mananos *et al.*, 1999) has shown great promise as a potential sterilizing agent during embryogenesis. GnRH is the main regulator of gonadal development, and disruption of GnRH production, which can theoretically be accomplished with GABA, could result in enhanced growth, enhanced nutrient utilization, increased carcass yield and improved flesh quality.

GnRH is produced in the neurons of the brain and then secreted into the pituitary gland. This causes the secretion of LH and FSH into the circulatory system, resulting in steroidogenesis, gametogenesis and growth from the target gonads. Control of GnRH production could be of great utility as it is the key hormone for reproduction (Amano *et al.*, 1997; Zohar and Mylonas, 2001) and therefore could also affect other traits that interact with hormone production, sexual maturation, gonadal development and reproduction, such as growth and nutrient utilization.

Production of GnRH is at maximum levels in the pituitary during gamete maturation (Amano *et al.*, 1992; Holland *et al.*,

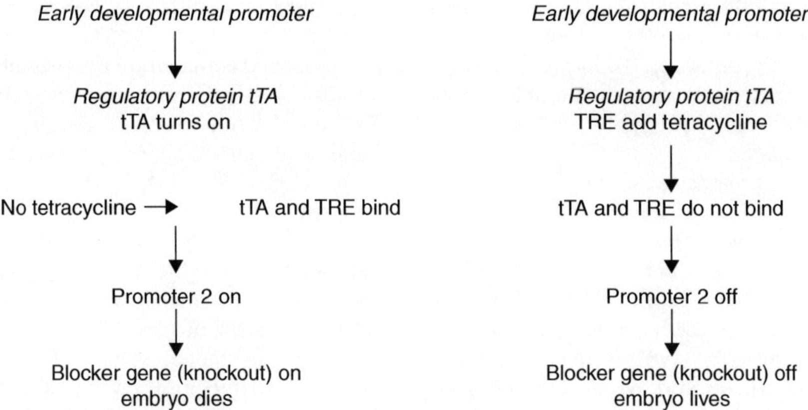

Fig. 20.2.   Sterile feral mechanism. tTA, transactivator protein; TRE, tetracycline-responsive element.

1998, 2001). GnRH production is stimulated when salmonids become sexually mature (Okuzawa *et al.*, 1990; Amano *et al.*, 1992; Lewis *et al.*, 1992), and administration of GnRH can induce oocyte maturation and spawning of fish under artificial conditions (Zohar and Mylonas, 2001).

Perciform fish have three molecular forms of GnRH – chicken GnRH II, salmon GnRH and seabream GnRH (Gothilf *et al.*, 1995, 1997) – and these multiple forms have two distinct GnRH neuronal systems, resulting in specific expression in different areas of the brain (Montero and Dufour, 1996; Amano *et al.*, 1997): forebrain for the fish forms of GnRH and midbrain for the chicken form. Within the forebrain, the seabream GnRH is expressed in the preoptic area of the anterior hypothalamus (Gothilf *et al.*, 1995, 1997; Okuzawa *et al.*, 1997; Parhar and Sakuma, 1997; Parhar *et al.*, 1998; White and Fernald, 1998a,b; Munoz-Cueto *et al.*, 2000), and the GnRH neurons located here innervate the fish pituitary gland (Kah *et al.*, 1993), which is correlated with germ-cell differentiation (Chiba *et al.*, 1999). Apparently, the salmon GnRH serves another function, as it is expressed in the terminal nerve of the olfactory bulb, and the GnRH nerves here affect sexual behaviour, gonad function (Demski, 1993) and salmon migration (Parhar *et al.*, 1994; Kudo *et al.*, 1996), which is reproductively influenced.

The forebrain and midbrain GnRH neurons have different embryonic-stem origins (Schwanzel-Fukuda and Pfaff, 1989; Wray *et al.*, 1989; Northcutt and Muske, 1994; Quanbeck *et al.*, 1997; Amano *et al.*, 1998; Daikoku and Koide, 1998; Parhar *et al.*, 1998; White and Fernald, 1998a,b; Kim *et al.*, 1999). In mammalian and avian embryos GnRH neurons migrate out of the nasal region into the forebrain, establish their final location and develop projections to the pituitary (Murakami *et al.*, 1991, 1998; Norgren and Lehman, 1991; Wray *et al.*, 1994; Takada *et al.*, 1995; Daikoku and Koide, 1998; Mulrenin *et al.*, 1999). Initial studies indicate that the development of these systems in fish is similar to that of mammals and birds. Cells of the preoptic area that express seabream GnRH in the

African cichlid and cells from the terminal nerve of the olfactory bulb that express salmon GnRH both originate from the nasal placode; however, there are temporal differences in expression, with salmon GnRH expression initiated earlier than seabream GnRH (White and Fernald, 1998a,b). In mammals and birds, the migration process of GnRH neurons during development takes a few days and is mediated by nerve cells, chemical products and enhancing and inhibiting factors, including GABA, a naturally occurring endogenous compound (Schwanzel-Fukuda *et al.*, 1992a,b; Norgren and Brackenbury, 1993; Wray *et al.*, 1994; Takada *et al.*, 1995; Mulrenin *et al.*, 1999; Gao *et al.*, 2000; Kramer and Wray, 2000).

GABA, $C_4H_9NO_2$, molecular weight 103.12, is an amino acid that functions as a neurotransmitter (Budavari *et al.*, 1996) and is a decarboxylation product of glutamate (Zubay, 1983). GABA is a major factor controlling chemotaxic and chemokinetic processes in vertebrate brain development, mediated through GABA-A, GABA-B and GABA-C receptor subtypes on cells of the central nervous system (Behar *et al.*, 1996). In mouse, rat and human, GABA-ergic neurons are associated with, are spatially and temporally located near and migrate in a parallel fashion to GnRH neurons (Tobet *et al.*, 1996; Wray *et al.*, 1996). Axonal projections from GABA neurons migrate across the nasal placode and terminate at the cribriform plate, appearing at the nasal side of the cribriform plate at exactly the same time that the GnRH neurons pause at the nasal/ forebrain junction. However, at this point the GnRH neurons migrate into the forebrain while the GABA neurons remain behind.

GABA has an inhibitory effect on GnRH neuronal migration. *In vitro* administration of GABA agonists to mouse embryonic nasal explants decreased GnRH gene expression (Fueshko *et al.*, 1998a) and inhibited normal GnRH neuronal migration to the forebrain (Fueshko *et al.*, 1998b) by activating GABA-A receptors. GABA directly acts on GABA-A-type receptors to provide a migratory stop signal during mammalian GnRH neuronal development; the migrational pause is required for proper organization of the

GnRH neurons in the forebrain, as *in vivo* administration of GABA-A agonists to early mouse embryos caused decreased migration of GnRH neurons out of the nasal placode, and antagonism of GABA-A receptors caused disorganized distribution of GnRH neurons within the forebrain (Bless *et al.*, 2000).

Adult goldfish (Kah *et al.*, 1992), rainbow trout (Mananos *et al.*, 1999) and Atlantic croaker (Khan and Thomas, 1999) respond to *in vivo* administration of GABA with increased plasma gonadotrophin during early gonadal recrudescence; GABA had no effect on gonadotrophin release from dispersed pituitary cells *in vitro*, but pituitary GnRH nerve terminals exhibited an increase in gonadotrophin release, indicating that GABA acts directly on GnRH nerve terminals to potentiate GnRH secretion and subsequent gonadotrophin release. This is further confirmed by the fact that GABA stimulates the release of seabream GnRH from red seabream hypothalamic explants (Senthilkumaran *et al.*, 2001). The stimulatory mechanism of GABA on GnRH secretion in fish is mediated by GABA-A-type receptors (Khan and Thomas, 1999; Trudeau *et al.*, 2000; Senthilkumaran *et al.*, 2001), just as is observed in mammals.

GABA inhibits activated GnRH neurons in immature mammals (Lamberts *et al.*, 1983; Donoso, 1988; Mitsushima *et al.*, 1994) and in immature rainbow trout, which do not show an increase in gonadotrophin release unless GABA is co-administered with gonadal steroids (Mananos *et al.*, 1999). Data to date indicate that interactions of the GABA and GnRH neuronal systems in fish (Medina *et al.*, 1994; Ekstroem and Ohlin, 1995; Anglade *et al.*, 1998; Doldan *et al.*, 1999) are similar to those in mammals and birds.

In summary, the key factors are that GABA inhibits GnRH neuron migration during embryogenesis in higher vertebrates, and GnRH is critical for gonadal development. This is further substantiated by GnRH mutations in mice (Mason *et al.*, 1986) and humans (Schwanzel-Fukuda *et al.*, 1989), both resulting in hypogonadism and infertility. Introduction of a complete GnRH gene restores fertility in the mutant mice, and administration of synthetic GnRH restores fertility in the mutant humans. Similarly, more recent data indicate that reproductive dysfunction of captive fish can be overcome and early sexual maturation can be induced in fish by the administration of GnRH and GnRH analogues (Hassin *et al.*, 2000; Fornies *et al.*, 2001; Mylonas and Zohar, 2001; Zohar and Mylonas, 2001).

The sterilization strategy would require the insertion of two genes. Glutamate is the substrate for the synthesis of GABA. A glutamate construct would need to be transferred to increase levels of glutamate in the developing embryo. Glutamate decarboxylase action would need to be enhanced as well, to convert the glutamate to GABA. An artificial glutamate decarboxylase gene would also need to be introduced. The increased expression of both should result in elevated levels of GABA, which would disrupt GnRH neuron migration, and subsequently the production of GnRH. Fertility of brood fish would be restored by the artificial application of GnRH. The gene action in this case would differ from the sterile feral gene action. Brood stock would need to be homozygous for each transgene. Although both constructs would act as dominant genes, matings of heterozygous individuals would result in some individuals being homozygous recessive for one or both transgenes and these individuals would be fertile.

Preliminary results where the GnRH antisense approach was utilized to transgenically sterilize fish have been very promising. Antisense is most effective when rare messages are targeted and the antisense construct is driven by a strong promoter. A carp β-actin–tilapia salmon-type GnRH antisense construct was injected into Nile tilapia (N. Maclean, University of Southampton, 2004, personal communication). Transgenic females were crossed with wild-type males. A reduction in fertility of about half that of non-transgenic control females was observed. Fertility was much more greatly reduced in transgenic males crossed to control females. In some cases, 0% fertility was obtained, with an average of about an 80% reduction in fertility. Limited data on transgenic females crossed with transgenic males indicated near-zero fertility.

A tilapia β-actin–tilapia seabream GnRH antisense construct was injected into Nile tilapia. In this case, no reduction in the fertility of heterozygous transgenic males and females was observed.

Limited data on transgenic females crossed with transgenic males indicated no reduction in fertility. Reciprocal crosses between seabream and salmon GnRH antisense transgenics gave hatch rates that appeared to be dictated by the salmon GnRH antisense parent. Apparently, salmon-type GnRH has a more critical role in fertility than seabream-type GnRH. This type of result has been confirmed in transgenic rainbow trout. Transgenic rainbow trout containing salmon-type antisense GnRH from Atlantic salmon, driven by either the GnRH or histone-3 promoter, had reduced levels of GnRH and appeared to be sterile (Uzbekova *et al.*, 2000a,b). Preliminary data indicated that spermiation of transgenic males was obtained after only prolonged treatment with salmon pituitary extract, whereas control males spermiated naturally. Data are still needed for the females.

Potential problems exist for knocking down GnRH in fish for transgenic sterilization. GnRH is related to increased levels of GH, and is involved in the regulation of GH. Knockdown of GnRH expression in zebra fish embryos caused a defective mid–hind-brain boundary, underdeveloped eyes and, in some cases, defective hearts (Sherwood and Wu, 2005). This may not be a surprising result in light of GnRH neuron migration in the early embryonic brain as discussed above. Because of these problems Wong and Van Eenennaam (2008) suggest that the LH β subunit may be a better target for knock-out to control reproduction as its main role is more specific and is responsible for gonadal development and maturation. This knockout approach was successful in mice, and was reversible (Ma *et al.*, 2004).

Wong and Van Eenennaam (2008) also discuss the hammerhead ribozyme (*trans*-acting antisense RNA molecules that contain short regions with enzymatic activities that bind selectively to and cleave specific target RNAs; Takagi *et al.*, 2001, as cited by Wong and Van Eenenaam, 2008) approach to gene knockdown and describe it as follows. RNA secondary structures endow ribozymes with the ability to act as biological catalysts. The antisense sequences embedded in the ribozyme allow the ribozyme to target the complementary RNA and to enzymatically cleave the target message. The hammerhead name is derived from shape of the secondary structure of ribozymes, which cleave their targets at NHH sites where N is any base and H is any base except G. The sequence most likely to be cleaved by a ribozyme cleavage is GUC, and a shortcoming of the hammerhead ribozyme approach is identifying the best target site in an mRNA for cleavage. Ribozymes have been successfully used to create and mimic known mutations in zebra fish (Xie *et al.*, 1997).

Wong and Van Eenennaam (2008) also describe the approach to use recombinases such as Cre and Flp to excise transgenes from gonadal tissue while leaving the transgene in the somatic cells. These recombinases excise, insert and invert DNA sequences at specific recognition sites. The transgene of interest is removed by crossing individuals with a recognition-site-flanked transgene to individuals with the recombinase gene protein. Gonad-specific promoters are used to allow excision in the gonads while leaving the transgene in the somatic cells.

Cre recombinase catalyses site-specific recombination between 34 bp loxP-recognition sites; the concept has proved highly effective in excising transgenes in zebra fish by injecting Cre mRNA or plasmids producing Cre mRNA. Hu *et al.* (2006) report a successful crossing of transgenic lines to excise a transgene in zebra fish, but provided little detail.

The outcome of the Cre recombinase approach is that if transgenic production fish escaped they would not be able to spread the transgene to the wild population. However, the problem with the Cre recombinase approach is that two fertile transgenic brood stocks are required. If brood stock escape, they can readily transmit the transgenes to wild populations. This technology has the same shortcoming of triploidy: fertile transgenic brood stock must be present that can escape, breed and introduce the transgene to wild populations. In fact, Wong and Van Eenennaam (2008) argue that escaped recombinase-expressing transgenic fish could be a greater threat to wild fish populations than other transgenic fish because long-term recombinases can create chromosomal rearrangements and reduce the fertility of recombinase-expressing lines because of 'pseudo-recognition' sites in the genome of eukaryotes (Thyagarajan *et al.*, 2000).

If successful, transgenic sterilization has many applications and would allow domestic, transgenic, hybrid or exotic aquatic organisms to be cultured in any watershed without any potential for affecting native gene pools or local biodiversity. Even conspecific wild strains could be moved from one watershed to another for various purposes without the risk of genetic impact. However, short-term ecological damage could occur from the escape of any of these fish until the time of their death. On the other hand, this is the only technology that would assure no genetic impacts. This is the best-case scenario and will provide the best benefit–cost ratio. Another strategy would be to utilize multiple sterile feral constructs for redundant sterility, virtually ensuring no escapement (Thresher *et al.*, 2009). This technology also allows the protection of proprietary germplasm.

Another transgenic sterilization application is the disruption of reproduction of pest species by the intentional release of sterile individuals into the environment, the 'screwfly approach'. The sterile individuals, males, mate with fertile females, resulting in infertile egg clutches and thus reducing the number of individuals in a population or totally eliminating the population. The release of large numbers of sterile or otherwise unsuitable mates has been used for many years to control insect pests by disrupting reproduction, resulting in infertile eggs (Doyle, 2003). S.A. Davis *et al.* (2000) analysed classical deterministic genetic models to determine the result of flooding a target population with individuals carrying a transgene that can be switched on at will by a chemical spray in the watershed. The chemical would induce expression of a transgene that causes death or sterility in fish carrying the transgene. The population growth of the target population is reduced, and the triggering chemical would be environmentally harmless (Doyle, 2003). The model indicated that the population level of the target species can be reduced to any desired level but, if the inducer is applied too often, selection for transgene-free individuals will nullify the effect.

21

Food Safety of Transgenic Aquatic Organisms

In addition to environmental concerns, the expression of transgenes in foods has brought about concerns and debate regarding food-safety issues. Food safety and education are also critical issues, particularly in regard to consumption of transgenic aquatic organisms. The general public has little understanding of biology and the vagaries of how their food is grown and where it comes from, so public education on the positive aspects of transgenic food and its risks is lacking and is needed (FAO, 2001). Food-safety issues posed by transgenic fish are discussed by Berkowitz and Kryspin-Sorensen (1994). Concerns have been voiced over the possible risks of consumption of transgenes and their resulting protein, the potential production of toxins by aquatic transgenic organisms, changes in the nutritional composition of foods, the activation of viral sequences and the allergenicity of transgenic products. These risks have been analysed and, while the majority of genetic modifications to foodstuffs will be safe, the greatest potential risk and harm is allergenicity.

The most common accepted measure of food safety for transgenic or genetically manipulated organisms is the substantial equivalence principle. This concept was first suggested by the Organisation for Economic Co-operation and Development in 1993 (OECD, 1993). The World Health Organization (WHO) then recommended in 1995 and 1996 that the substantial equivalence principle should be the basic guidance for determining food safety of all GMOs (WHO, 1994). If the body composition of a transgenic organism is essentially the same as its non-transgenic counterpart, then there would be no greater food-safety risk posed by the transgenic organism.

FAO/WHO (2003a,b,c,d,e) has a series of recommendations concerning the food safety of transgenic fish and animals. No unnecessary DNA sequences, marker genes and sound construct design should be used to minimize potential unexpected adverse pleiotropic effects leading to health concerns. Substantial equivalence should be determined, including food-intake assessment and, if appropriate, full risk characterization. When appropriate, post-market surveillance should be conducted. There is a need for worldwide accessible databases on the natural variation in key compositional constituents in animal products to allow substantial equivalence evaluation.

In Chapter 20 on environmental risk, the potential adverse effects of reporter sequences, bacterial contamination, transposons, virions and viral vectors as well as gene disruption and adverse pleiotropic effects were discussed. FAO/WHO (2003f) recommends that careful construct design be used and also selection be utilized in the development of transgenic lines to minimize or avoid these problems with regard to how they might affect health of the animal or the consumer. For example, *Prn-p* genes can increase an organism's susceptibility to prions, and these genes should never be inserted into food animals or fish.

On the other hand, transgenesis might actually be used to improve food safety. For instance, if *Prn-p* genes were knocked out in cattle and sheep, it might increase resistance to mad cow disease.

Kok and Kuiper (2003) suggest that comparative safety assessment (CSA) be used for safety assessment of a genetically modified (GM) food, rather than substantial equivalence. First there is a thorough comparison of the transgenic food item and its non-transgenic counterpart to reveal any differences that could cause a food-safety risk. Both compositional analysis and phenotypic characters are defined. In the second step, nutritional and toxicological evaluation is conducted upon any differences found between the transgenic and conventional product.

The first step of CSA includes detailed molecular assessment of the transgenic organism/food item. The information gathered includes the transformation process, sequence of the transgene both before and after insertion, copy number and insertion site(s) of insertion, sequence analysis of the flanking regions, stability of the integration, safety of newly expressed proteins, assessment of allergenicity, determination of pleiotropic effects and their consequences, the role of the new GM animal food in the diet, and any influence of processing or spoilage on the GM food product (FAO/WHO, 2003f). In the case of expression of any novel proteins, sequence analysis would be conducted to see if there are any homologies with known toxins, and if it is an unknown protein, a full toxicological analysis should be conducted. It was also suggested that GM animals or fish that are not intended for food should also be evaluated because of the possibility of them accidentally entering the market.

There is no known parameter that allows prediction of allergenicity of a protein or food item. FAO/WHO (2001) and FAO/WHO (2003d,e) suggest serum testing of patients known to be allergic to the source organism and its related species from which the transgene DNA was derived, measurement of pepsin resistance, determination of the prevalence of the trait and assessment using animal models.

A transgenic soybean has been developed expressing a gene from Brazil nut to increase its protein content and this transgenic soybean was allergenic to some humans (Nordlee *et al.*, 1996). Labelling laws concerning transgenic foods are also currently being debated and some companies argue that they would be unfairly discriminated against for being compartmentalized as transgenic; however, the potential allergic reactions to transgenic proteins from the donor organisms is one of the strongest arguments for the enactment of some type of labelling (Hallerman, 1992, 1993, 2001b; FAO, 2001).

FAO/WHO (2003f) recommends food-intake assessment. Traditional risk assessment utilizes exposure assessment to elucidate hazards. Food-intake assessment is more complicated and examines complex foods rather than individual chemical compounds, and how their entry into the food supply can affect diets and also overall consumption patterns of the transgenic food source. This assessment determines the amount of food or food ingredient an individual or population group may consume, thus anticipated dosage relates to potential toxicity and adverse effects. Processing, preparation, cooking or lack of cooking could have effects on food safety. If tracking is implemented, post-market consumption data can be determined. Consumption patterns among different groups are important since subsets such as pregnant or lactating women or specific patient groups may have different risk potential.

Another food-safety consideration is the expression of bioactive proteins and their possibility of continued bioactivity after ingestion. Antibacterial proteins, for example, might alter the balance of intestinal flora in the human gut or result in the development of resistant strains of human pathogens.

In general, one might expect fewer risks when the transgene is derived from another food animal as toxins from animals are rare, but are more common in plant-derived foods. A frequently mentioned concern is that of inadvertent activation of potential fish toxins such as tetrodotoxin in puffer fish. However, there are no known such toxins produced by the genomes of fish: the tetrodotoxin in puffer fish is actually produced by symbiotic microorganisms in the fish. Puffers hatched and grown in aquaria do not develop the symbiotic relationship or produce the toxin. Another concern is the activation or addition of anti-nutrient genes such as thiaminase from the viscera of certain fish species. A key aspect of the potential risk or safety of transgenic fish as food is that proteins are usually broken down into amino acids in the gut and most food items are cooked or pasteurized, denaturing the proteins.

Basis of some fears of transgenic food and the need for public education

People do not understand where their food comes from or how it is grown and do not necessarily have a logical perspective, partly because of the influence of the popular media. For instance, people who regularly consume the enzyme chymosin from transgenic bacteria in their Parmesan cheese (the natural source is from calf stomach) or who would use genetically engineered insulin for diabetes are sometimes opposed to the consumption of transgenic aquatic organisms (Dunham *et al.*, 2001). Hoban and Kendall (1993) surveyed North Carolinians in the USA concerning their attitudes towards transgenic foods. Having hypothesized the importance of the perception of biotechnological foods put forward by the media and various watch groups and the importance of the lack of knowledge for influencing opinion, Hoban and Kendall (1993) included two biological questions to allow a proper interpretation of the answers. The first was 'Have you ever eaten a hybrid fruit or vegetable?' and the second was 'Do you think it is ethical to eat a hybrid fruit or vegetable?' Approximately 60% of the respondents answered no to both questions. Of course, almost all Americans consume hybrid fruit and vegetables on a regular basis, as most fruits and vegetables are hybrids. There is a great need for public education on biotechnology's advantages, disadvantages, benefits and risks for the benefit of society and to restore faith in science, industry, government and environmental organizations.

International Guidelines

Internationally, the European Union (EU) and the Codex Alimentarius Commission (CAC) have taken the lead roles in voicing concerns about the consumption of transgenic foods and the need for labelling and regulation (Dunham *et al.*, 2001; FAO, 2001). CAC is an intergovernmental body established by the FAO and WHO, and has a membership of most countries of the world. CAC has developed Codex Standards, which address all food-safety considerations, descriptions of essential food hygiene and quality characteristics, labelling, methods of analysis and sampling, and systems for inspection and certification (Bartley, 1999). Codex standards, guidelines and recommendations are not binding on member countries, but are a point of reference for international law (General Assembly Resolution 39/248; Agreement on the Application of Sanitary and Phytosanitary Measures; Agreement on Technical Barriers to Trade) (D. Bartley, FAO, 2003, personal communication).

At the 23rd session in July 1999, the CAC established an Ad Hoc Intergovernmental Task Force on Foods Derived from Biotechnology to develop standards, guidelines or recommendations for foods derived from biotechnology or traits introduced into food by biotechnology. This was to be accomplished on the basis of scientific evidence and risk analysis and having regard to other legitimate factors relevant to the health of consumers and the promotion of fair trade practices.

In the USA, the FDA regulates transgenic foods under the purview that the introduction of an exogenous gene is analogous to the introduction of a drug (Bartley, 1999; Hallerman, 2000, 2001a). The consumption of transgenic plants has been approved, but no approval has been granted for the marketing of transgenic aquatic organisms. As stated in Chapter 19, an application from AquaBounty Farms to market GH-transgenic salmon in the USA is under consideration by the FDA, and a decision may be reached in 2011.

Labelling

The CAC also preliminarily adopted an amendment to the General Standard for the Labelling of Prepackaged Foods (D. Bartley, FAO, 2003, personal communication). This amendment addresses the need to label foods developed through biotechnology that are substantially different from usual foods. In the EU, the 1997 EU Novel Foods and Novel Food Ingredients Regulation 258/97 dictates mandatory premarket approval and labelling for all foods without a history of consumption in the EU or for food obtained from GMOs (Dunham *et al.*, 2001). For foods that are substantially equivalent, only a simple notification is required to speed approval if the GMO or product is equivalent to the non-GMO counterpart, and then no extra legislation or oversight is needed. These include cases when there is no evidence of any specific health hazards. The Proposed Draft Recommendations for the Labelling of Foods Obtained through Biotechnology from the CAC states:

> When a food produced by biotechnology is not substantially equivalent to any existing food in the food supply and no conventional comparator exists, the labelling shall indicate clearly the nature of the product, its nutritional composition, its intended use and any other essential characteristic necessary to provide a clear description of the product.

Labelling of GMOs or products from GMOs has become controversial, and major trade conflicts were and are waged between Europe and the USA over labelling of genetically modified crops, such as maize and soybeans (Bartley, 1999; Dunham *et al.*, 2001). This is extremely important to the USA as 50–70% of its maize and soybean crop is now transgenic. Some countries think labelling is impractical and ambiguous at best; however, most of Europe thinks it is necessary for informed consumer decisions and proper public relations. Recently, the USA and Europe agreed to the importation of transgenic crops into Europe if they are labelled.

Labels can be a potentially positive marketing tool, as in the examples of 'dolphin-friendly tuna' and 'organically grown food'. However, the labels could also be used for negative advertising as well. It will be critical to have potentially allergenic transgenic products properly labelled. Additionally, issues such as the extent of the information on the label, verification of the authenticity of labels and enforcement will need to be resolved. The International Federation of Organic Agriculture Movements (IFOAM) has produced standards required for their certification (Dunham *et al.*, 2001; www.ifoam.org): vaccines are allowed, but genetically engineered vaccines are not; feeds may not contain GMOs or their products; triploids and genetically engineered species or breeds are also not allowed for organic certification. For many transgenic products, these restrictions do not make biological sense and may be overly restrictive.

Testing of growth-hormone-transgenic fish

Cuba and China have evaluated the food safety of GH-transgenic tilapia and common carp, respectively. Cuba fed GH tilapia to monkeys long-term and to people short-term, and measured various physiological and blood parameters. They found no effects from eating GH-transgenic tilapia.

In the case of the People's Republic of China, GH-transgenic common carp were fed to mice for 6 weeks. Growth performance, biochemical analysis of blood, histochemical assay of 12 organs and reproductive ability of the mice were unaffected (Zhang, F. *et al.*, 2000).

22

A Case Example: Safety of Consumption of Transgenic Salmon Potentially Containing Elevated Levels of Growth Hormone and Insulin-like Growth Factor

Transgenic Atlantic salmon containing the ocean pout antifreeze promoter–salmon GH gene construct have been developed by AquaBounty Farms, and these fish have a four- to sixfold accelerated growth rate compared with non-transgenic salmon (Cook *et al.*, 2000a). This is an impressive phenotypic change, but these fish are far from abnormal regarding growth enhancement. The growth rate of channel catfish, *Ictalurus punctatus*, has been improved up to tenfold over many years from the cumulative effect of many genetic enhancement programmes, mostly traditional, such as domestication, strain selection, mass selection, intraspecific crossbreeding and interspecific hybridization, as well as from gene transfer (Dunham and Devlin, 1998; Dunham and Liu, 2002). Additionally, many species of wild fish have growth rates far in excess of transgenic farmed salmon.

A discussion of the safety of consuming transgenic salmon potentially containing elevated levels of GH and IGF follows. Safety is defined as the absence (minimal probability) of known harm (Klassen *et al.*, 1996). Berkowitz and Kryspin-Sorensen (1994) and Dunham (1999) have discussed and analysed the food-safety issues for transgenic fish. Allergenicity is probably the most important and, in most cases, the only real potential health risk, but only in unusual cases where genes from groundnuts, maize, shellfish or other foods to which people are allergic might be inserted into a fish and the fish produces the appropriate antigen. In the case of the transgenic salmon containing a recombinant gene and producing a protein of totally salmonid origin, no foreign protein is produced. Therefore, there is no change in allergenicity, so that is not an issue for these fish.

Because of a lack of understanding (Hoban and Kendall, 1993), there is public concern regarding food safety due to potentially elevated levels of GH and IGF in transgenic salmon. Addressing this safety issue will be the focus of this chapter.

GH/Salmonid GH

Human GH (hGH) (somatotrophin) is a small protein containing 191 amino acids in mammals. Somatotrophin is a naturally occurring peptide hormone normally produced in the pituitary gland of humans and animals, including fish. The sequence and number of the amino acids vary among species and, because there is species variation in the structure, there is species specificity for its biological activity (Bauman, 1992).

GH causes the growth of all tissues capable of growing by promoting increased cell size and an increased number of cells. GH increases the rate of protein synthesis, the mobilization of fatty acids from adipose tissue and the use of fatty acids for energy, and decreases the rate of glucose utilization, with the overall effect of increasing growth, enhancing body protein, utilizing fat stores and conserving carbohydrates (Guyton, 1981).

Salmonid GH is quite different from hGH or mammalian GH (Nicoll *et al.*, 1987; Agellon *et al.*, 1988). The salmonid GH gene spans a region of approximately 4 kb, nearly twice that of mammalian GH genes. The salmonid GH gene is comprised of six exons, in contrast to the five exons in mammals (Agellon *et al.*, 1988). The additional intron in the fish gene interrupts translated regions that are analogous to the last exon of the mammalian counterpart. Additionally, the alleged internally repeating sequence in mammalian GH is not present in the predicted polypeptide sequence of fish GH. The direct repeats that flank exons I, III and V of the mammalian GH genes are absent in the fish GH gene.

The salmon GH polypeptide (sGH) has 210 amino acid residues, including the 22 amino acid residues of the signal peptide (Sekine *et al.*, 1989). The number of amino acids in the actual hormone, 188, is slightly smaller than the 191 for humans. Salmon GH I has two disulfide bonds, Cys 49–Cys 161 and Cys 178–Cys 186, as observed in mammalian GHs (Sugimoto *et al.*, 1991; Vestling *et al.*, 1991). This is analogous to the big-loop–little-loop pattern found in hGH. Recombinant sGH I, as well as natural sGH I, is an α-helix-rich protein, as indicated by its circular dichroism spectrum. The homology between sGH and hGH is very low – approximately 30% (Watanabe *et al.*, 1992).

IGF/Salmonid IGF

IGFs, including IGF-I and IGF-II, are single-chain polypeptides having structural homology to proinsulin. IGF-I and IGF-II are essential for fetal and postnatal growth in mammals (Jones and Clemmons, 1995), and IGF-I mediates many of the growth-promoting effects of GH. IGF-I is synthesized in the liver under the regulation of GH, circulates in the blood and acts on distant target tissues in an endocrine fashion, but is also produced in a wide variety of cell types and acts locally in a paracrine and autocrine fashion (Duan, 1998). IGF-II is expressed predominantly during the fetal stage in multiple tissues, and IGF-I is expressed in a wide variety of tissues during fetal and postnatal stages. Shortly after birth, the liver becomes the predominant site for endocrine IGF-I production, under the regulation of GH (Duan, 1998).

The expression patterns of IGF in teleost fish seem to be similar to those in mammals (Duan, 1998). The biological actions of IGF are diverse and IGF may stimulate cell growth, stimulate the expression of differentiated functions and inhibit apoptosis (Jones and Clemmons, 1995). Biological actions of IGF are mediated by the IGF-I receptors, which are expressed in a broad array of cell types. The IGF-I receptor has a heterotetrameric structure with a tyrosine kinase domain in the cytoplasmic portion of the β subunit (Czech, 1989). In mammals, a second transmembrane IGF receptor, the IGF-II/mannose 6-phosphate receptor, also exists and preferentially binds to IGF-II over IGF-I, but this causes the internalization and degradation of IGF-II (Oka *et al.*, 1985).

Most of the IGF present in circulation and throughout the extracellular fluids is bound to members of a family of high-affinity IGF-binding proteins (IGFBPs), which are critical for transport and bioactivity (Duan, 1998). These proteins can act as carrier proteins in the bloodstream and control the efflux of IGF from the vascular space. The IGF/IGFBP complexes prolong the half-lives of IGF and buffer the acute hypoglycaemic effects of IGF (Duan, 1998). More importantly, because IGFs bind to IGFBPs with higher affinities than to the IGF receptors, IGFBPs may provide a means of localizing IGF on target cells and can alter their biological activity by regulating their interaction with IGF receptors (Jones and Clemmons, 1995).

Cao *et al.* (1989) deduced the amino acid sequence of coho salmon preproIGF-I, which contains 176 amino acids, including a 44-amino-acid signal peptide, a 70-amino-acid mature IGF-I and a 62-amino-acid E peptide. Nucleotide sequences of IGF-I cDNAs have been determined in coho salmon, Atlantic salmon, chinook salmon, rainbow trout, carp, catfish, seabream and shark (Cao *et al.*, 1989; Duguay *et al.*, 1992, 1995, 1996; Shamblott and Chen, 1992; Wallis and Devlin, 1993; McRory and Sherwood, 1994; Liang *et al.*, 1996). The amino acid sequence of the coding regions is well conserved among these fishes, as well as among a variety of organisms including humans. The predicted amino acid sequence of salmon IGF-I (sIGF-I) is 80% identical to that of human IGF-I (hGF-I) (Duan, 1998). Fish IGF-II is also very similar to mammalian IGF-II (Shamblott and Chen, 1992; Duguay *et al.*, 1995, 1996), and the amino acid sequence of trout IGF-II is 80% identical to hIGF-II.

The IGF receptors would also be expected to be highly conserved. Seabream, *Sparus auratus*, IGF-I receptors are very similar to those of mammals, and the amino acid sequence identity between seabream and human IGF-I receptors is greater than 70% (Duan, 1998). Fish and mammalian IGFBPs are also similar (Duan, 1998).

GH Levels in Non-transgenic Salmon and Fish

GH levels fluctuate widely and naturally, depending on the size and age of the fish, the environmental conditions and the species. There may be large peaks for GH expression stimulated by environmental change, stress or reproduction. Karsi *et al.* (1998) found that, during ovulation of channel catfish, GH expression increases by 393%. Therefore, assuming similar natural changes in salmon, the GH levels in a farm-raised transgenic salmon may in fact be similar to those in wild-caught female salmon on spawning runs, which are consumed by native Americans and others. The spawning cycle and spawning result in five- to tenfold increase in GH production in rainbow trout (Le Bail *et al.*, 1991; Sumpter *et al.*, 1991b),

which is in excess of that observed by Karsi *et al.* (1998) for channel catfish.

Plasma GH levels in juvenile Atlantic salmon can range from 0.7 to 11 ng/ml (Björnsson *et al.*, 2000). GH is responsive to photoperiod in salmonids and increases in March as photoperiod increases (Björnsson *et al.*, 1989, 1995; McCormick *et al.*, 1995), which is consistent with the GH shifts seen due to temperature in catfish by Karsi *et al.* (1998). Plasma GH levels ranged from 1 to 20 ng/ml in cultured juvenile Atlantic salmon and were affected by photoperiod, salinity and onset of smoltification (Björnsson *et al.*, 1998). Similar to what is discussed below for wild chinook salmon, free-living, migratory Atlantic salmon smolts have very high levels of plasma GH (McCormick and Björnsson, 1994), and it is well documented that GH increases dramatically when salmon smolt (Björnsson *et al.*, 1989, 1995; McCormick *et al.*, 1995).

Temperature, sexual-maturation effects, size and salinity can all cause changes in plasma GH levels, and these factors cause two- to fivefold levels of change in channel catfish (Tang *et al.*, 2001). In general, GH expression was highest in fingerling fish, decreased by half in food-size fish and increased to equivalent levels found in fingerlings prior to the onset of spawning for older fish. Increase in salinity stimulated increases in GH in catfish and other species. GH levels were minimal for large catfish in winter. Stress can cause a five- to tenfold increase in GH levels in rainbow trout (Pickering *et al.*, 1991).

Salinity also induces GH changes in salmon. GH ranged from 1 to 3 ng/ml in Atlantic salmon parr, increased slightly when the fish were stressed (McCormick *et al.*, 1998) and ranged from 1 to 14 ng/ml in response to salinity and temperature changes (Handeland *et al.*, 2000). GH is also elevated during exercise (Sweeting and McKeown, 1987), so GH levels can be naturally altered and manipulated by providing the right environmental conditions.

Food consumption also affects GH levels, and fasting rainbow trout demonstrate high levels of GH (Farbridge and Leatherland, 1992). In both gilthead seabream (*S. auratus*) and rainbow trout, rapidly growing fish have low levels of GH and high levels of IGF (Storebakken *et al.*, 1991; Perez-Sanchez *et al.*, 1995). Nutritional status has a large impact on the GH–IGF-I axis in fish. In salmonids, prolonged starvation causes cessation of growth, but significant elevation of plasma GH concentrations (Sumpter *et al.*, 1991a; Duan and Plisetskaya, 1993). This phenomenon has been documented in many vertebrate species, including humans, sheep, dogs and chickens (Thissen *et al.*, 1994).

IGF Levels in Non-transgenic Salmon and Fish

IGF is expressed primarily in the liver (Duan *et al.*, 1993). In the case of coho salmon, expression of IGF in the muscle, as indicated by the presence of mRNA, is 12% of that found in liver (Duan *et al.*, 1993). Similar to what was observed with catfish and GH, increasing plasma IGF levels are associated with gametogenesis in rainbow trout (Le Gac and Loir, 1993). Salmon that are in the midst of spawning migrations are traditionally consumed as food, and would be expected to have elevated IGF levels.

Chinook salmon parr have plasma IGF-I values that fluctuate from 25 to 60 ng/ml and then soar to 70–110 ng/ml when they smolt (Beckman *et al.*, 2000). These changes in IGF-I associated with smoltification in wild fish were much more dramatic than previously measured for fish reared in hatcheries, laboratories and aquaculture environments (Beckman and Dickhoff, 1998; Beckman *et al.*, 1998, 1999; Silverstein *et al.*, 1998). Apparently, salmon in the wild may express higher IGF values than salmon in captivity, actually making transgenic salmon more comparable with wild fish regarding IGF than with other domestic salmon. Chinook salmon demonstrate another peak of IGF in the autumn, which is associated either with autumn smoltification or possibly with redistribution movements (Beckman *et al.*, 2000).

Myers *et al.* (1998) found that faster-growing species of salmon and larger individuals within a species had higher IGF levels. This implies that IGF levels produced by transgenic salmon are not outside the natural ranges expected for fish and that consumption of fast-growing and large fish, regardless of species or genotype, probably results in the consumption of an organism with high IGF levels. IGF-I ranged from 100 to 130 ng/ml in Atlantic salmon parr and increased slightly when the fish were stressed (McCormick *et al.*, 1998). Aquacultured barramundi, *Lates calcarifer*, had low levels of plasma IGF-I – 20–30 ng/ml (Nankervis *et al.*, 2000).

As in mammals, food deprivation causes reduction of circulating levels of IGF-I in coho salmon (Moriyama *et al.*, 1994). In a closely related species, rainbow trout, 4 weeks of starvation caused a significant decrease in the circulating levels of IGF-like peptide(s) (Niu *et al.*, 1993). Perez-Sanchez *et al.* (1995) showed a positive correlation between dietary protein content and plasma IGF-I concentrations in the gilthead seabream. Increasing feeding ration size resulted in an increase in plasma IGF-I concentrations in these fish. Both protein and energy intakes are important in the regulation of circulating IGF-I concentrations in fish (Duan, 1998).

IGF and GH can both fluctuate greatly from one year to the next or from one environmental condition to the next. For transgenic salmon, domestic salmon and wild salmon of the same genotypes, both GH and IGF levels varied from one year to the next by as much as five- to sevenfold, demonstrating the sensitivity of these two hormones to environmental change (Devlin *et al.*, 2000).

GH Levels in Transgenic Salmon and Fish

The increased production of GH can be as much as 19- to 40-fold greater for transgenic coho salmon compared with non-transgenic salmon (Devlin *et al.*, 1994b, 2000). Similar results were obtained with different constructs – *Oncorhynchus* MT–GH1 (OnMTGH1) and ocean pout antifreeze promoter–GH (opAFPGHc) (Devlin *et al.*, 1994b; Devlin, 1997b). OpAFPGHc accelerated growth similarly and dramatically in both coho salmon (Devlin *et al.*, 2000) and Atlantic salmon (Du *et al.*, 1992); however, this construct dramatically increased GH in coho salmon, but not in Atlantic salmon. Transgenic coho salmon containing the GH construct opAFPGHc (Du *et al.*, 1992) had a 19.3- to 32.1-fold (65–410 ng/ml serum) increase in GH (Devlin *et al.*, 2000) compared with controls (0–15 ng/ml) for fish weighing about 250 g. However, GH expression decreases in larger transgenic salmon (400–500 g) as they approach harvest size (Mori and Devlin, 1999). This would be expected because, as fish grow larger, their relative rate of growth slows, since rate of growth is dependent on size. Therefore, it is likely that the plasma GH levels of transgenic and non-transgenic fish become more similar when measured at equal weights (which would correspond to different ages). The lower-end values of GH for transgenic salmon and fish are in the normal range found in the cyclical fluctuations for humans, as discussed below.

Estimates of GH production in transgenic carps containing exogenous GH genes can be quite variable – 200–500 ng/ml (Zhang *et al.*, 1990), 1–4 ng/ml (Chen *et al.*, 1992a,b) – and extremely high in small 0.2 g carp – 26–50 ng/fish. Analogous to the transgenic situation, ovine GH implants resulted in GH levels of 19 ng/ml in rainbow trout and increased their growth rate (Foster *et al.*, 1991).

IGF Levels in Transgenic Salmon and Fish

Transgenic coho salmon containing the GH construct opAFPGHc (Du *et al.*, 1992) had only slight changes in IGF-I levels, both increasing and decreasing (Devlin *et al.*, 2000). Transgenic coho salmon containing the GH construct opAFPGHc (Du *et al.*, 1992) had plasma IGF-I levels of 75–280 ng/ml (Devlin *et al.*, 2000) compared with 35–400 ng/ml in controls for fish weighing about 250 g. Analogous to transgenic fish, when Silverstein *et al.* (2000) injected channel catfish with bovine GH, the plasma IGF-I levels increased from 4–8 ng/ml to 8–12 ng/ml and differences were attributed to the injection, temperature and strain of fish. IGF levels in transgenic and non-transgenic fish do not appear to be very different.

GH and IGF Levels in Humans

Serum GH levels in human children fluctuate throughout the day and average about 5 ng/ml (Bright *et al.*, 1983). Peaks as high as 20–70 ng/ml can be reached during a 24 h period, and can be spontaneous or associated with sleep and exercise (Nindl *et al.*, 2001). Total daily production of GH in humans is estimated to be 100,000–1,000,000 ng, based on 24 h secretion (Hartman *et al.*, 1991, 1992; Pralong *et al.*, 1991; Weissberger *et al.*, 1991; Cuneo *et al.*, 1995; Friend *et al.*, 1997; de la Motte *et al.*, 2001) and total plasma volume (Guyton, 1981).

In humans the mean blood concentration of IGF-I is 200 ng/ml in adults and 100 ng/ml in infants (Juskevich and Guyer, 1990; Blum *et al.*, 1993); in children it is reported to be 20–60 ng/ml (Frost *et al.*, 1996). The plasma concentration of IGF-I and II has also been reported as 315±27 ng/ml (Costigan *et al.*, 1988) in humans. The normal range is 50–800 ng/ml. Total daily production of IGF-I in adults is 10–13 million ng (Guier *et al.*, 1989). To consume this much IGF-I from a meal of transgenic salmon would be nearly impossible, even if IGF were not broken down when the meat was cooked. Obviously, with this amount of production, the metabolism and excretion of IGF-I are also enormous. Gastrointestinal (GI) (saliva, gastric juice, intestinal secretions, pancreatic juice and bile) secretion of IGF-I by adult humans is 357,400 ng/day (Vander *et al.*, 1990; Chaurasia *et al.*, 1994).

GH and IGF Levels in Mammals

Serum GH levels can fluctuate between 1.8 and 5.7 ng/ml in small mammals (Lauterio *et al.*, 1988). In large mammals, serum IGF-I can range from 46 to 158 ng/ml and serum IGF-II from 128 to 228 ng/ml. IGF-I in heifers can vary between 5 and 30 ng/ml, with a surge 2 h after feeding. In cows values can vary between 5 and 20 ng/ml and in sheep between 2.5 and 4 ng/ml, with increased IGF in response to cold (Trenkle, 1978).

Rats can have bursts of GH production resulting in GH blood concentrations greater than 200 ng/ml associated with compensatory gain after fasting (Mosier *et al.*, 1985). Somatostatin withdrawal can result in dogs expressing plasma GH levels as high as 15.3 times greater than normal (Cowan *et al.*, 1984).

Bioavailability of sGH and sIGF in the Upper Gastrointestinal Tract

Foods such as milk, meat and eggs naturally contain trace amounts of GH and IGF from the species of origin. Proteins, including GH and IGF, are denatured and hydrolysed to amino acids and short, inactive peptide fragments when eaten – hydrolysed in the stomach and both hydrolysed and enzymatically degraded in the small intestine (Hammond *et al.*, 1990; Bauman, 1992; Pontiroli, 1998). This is true for all large protein hormones in all species. However, if, for any reason, some were to escape digestion, only a miniscule amount has been shown to be absorbed intact across the mucosa (Ziv and Bendayan, 2000).

Insulin is another example of a large protein hormone that is not biologically available or active when taken orally and, like GH, must be injected into the bloodstream to be active (Bauman, 1992). Bovine GH (bGH) is active when injected into rats, and a variety of GHs are active when injected into fish. However, when ingested, GH has no availability or activity. Even at neutral pH, GH is proteolytically degraded by organ preparations *in vitro* (Wroblewski *et al.*, 1991).

Additionally, the fact the sGH and sIGF would be destroyed by cooking makes the probability that these compounds would be biologically available even less likely. Even in the case of consumption of raw salmon, the GH and IGF would be destroyed and denatured in the gut. The lack of bioavailability of sGH and sIGF precludes bioactivity in the human and therefore would not cause any biological effects.

When the fish is cooked, the polypeptides (IGF-I and IGF-II) are denatured by heat. Cooking temperatures in excess of 90°C for several minutes denature even the most stable of proteins, bacterial proteins (Morita, 1980). Not all cooked fish is subject to such high temperatures and some protein possibly escapes thermal denaturation. Both IGF-I and IGF-II (Bell *et al.*, 1995a,b), relatively more thermostable proteins, can survive Holder pasteurization (62.5°C for 30 min) (Ford *et al.*, 1977; Eyres *et al.*, 1978; Donovan *et al.*, 1991). Thus, the two growth factors may be present (Chaurasia *et al.*, 1994) in partially cooked salmon. Further, any IGF-I and IGF-II escaping the effect of proteolysis could retain biological activity. Evidence that milk-borne IGFs, like epidermal-derived growth factor (EDGF), are probably biologically active while in the neonate GI tract for as long as 30 min after ingestion (Britton *et al.*, 1988; Philipps *et al.*, 1995, 2000) supports this idea; however, physiological doses (8.6 ng) of IGF-I administered orally to adult rats were rapidly metabolized (proteolysis) in the stomach (half-life, $t_{1/2}$ = 2.5–8 min), duodenum and ileum ($t_{1/2}$ = 2 min) (Xian *et al.*, 1995), and pharmacological doses – 2–12.5 µg/g three times daily (Steeb *et al.*, 1998) or 2 mg/kg (Fholenhag *et al.*, 1997) – of IGF-I administered orally to suckling or weaned rat pups, respectively, were not mitogenic in the gut; nor did any dose cause enzyme induction in the gut. The differences in these experiments are explained by the fact that gastric proteolysis is 50-fold greater in weanling rats when compared with suckling rats (Britton and Koldovsky, 1988). This latter point is important, as it would be unreasonable to expect neonatal humans to consume the proposed product, especially in the raw form.

This principle that GH is digested in the gut and is not bioavailable is not a phenomenon unique to humans. Further substantiation of this principle is the fact that GH is also digested and not bioavailable in other organisms. Rats fed doses of bGH did not have detectable bGH in their serum or any growth response when fed doses up to

40,000 µg/kg body weight (Seaman *et al.*, 1988). Further, huge doses of bGH – 40,000 µg/kg/day (Seaman *et al.*, 1988) or 50,000 µg/kg/day (Hammond *et al.*, 1990) for 90 days – administered orally to rats have been shown to have no effect on body-weight gain (a sensitive assay). Additionally, very large doses (200–2000 µg/kg/day for 14 days) of bovine IGF (bIGF-I) administered orally to rats were not bioavailable or active (Hammond *et al.*, 1990). In another study, three treatment groups were given 60, 600 or 6000 µg of bGH zinc/kg body weight daily (FDA/CVM, 2001). Animals were observed daily and a number of parameters were examined, including body weight and food consumption, haematology, blood chemistry, urinalysis, post-mortem examination, organ weights and histopathology. No adverse effects were observed in test animals. Body weights, food consumption, clinical parameters, organ weights and gross and microscopic pathology were unaffected at all dosage levels. Therefore, the highest dosage tested, 6000 µg/kg body weight, was considered as having no observable effect and no bioavailability. IGF-I, as previously mentioned, is digested in the GI tract like other dietary proteins and none is absorbed (NIH, 1990; Houle *et al.*, 1995; Philipps *et al.*, 1995). Because it is not orally active, experiments to determine the effects of IGF in rats are always conducted by injecting the rats (Svanberg *et al.*, 1998).

Additionally, there is no reason to suppose that sIGF is different from other proteins with respect to digestion and inactivation (Matthews and Laster, 1965; Rindi, 1966; Fisher, 1967; Goldberg *et al.*, 1968; Gardner *et al.*, 1970a,b; Nixon and Mawer, 1970a,b; Peters, 1970; Adamson *et al.*, 1988; Britton and Koldovsky, 1988; Britton *et al.*, 1988; Nikolaevskaia, 1989; Hammond *et al.*, 1990; Draghia-Akli *et al.*, 1999; Frenhani and Burini, 1999). The three-dimensional structure of GH and IGF is necessary for it to have biological activity (Bauman, 1992). As a result, consumption of foods containing small amounts of peptide or growth-promoting compounds is safe because the compound is destroyed in the digestive system. Therefore, the consumption of foods

containing GH produced endogenously through normal endocrinological function, endogenously by alteration of genomic DNA to induce the production of GH or IGF characteristic of another food animal or present in the animal from exogenous injection or elevated levels within an animal is safe because the bioactivity and physical structure of the GH or IGF are both destroyed by enzymatic function in the digestive tract, making them unavailable biologically.

Additional evidence that IGF-I is not bioavailable when orally delivered was obtained from experimentation with rats (FDA/CVM, 2001). IGF-I was administered by daily oral gavage to groups of 20 male and 20 female rats for 16 consecutive days. Oral-gavage treatment groups were given 20, 200 or 2000 µg IGF-I/kg body weight daily. The positive control groups were administered either IGF-I (50 or 200 µg/day) or alanyl porcine somatotrophin (APS) (4000 µg/day) via implanted osmotic pumps for 14 consecutive days. Animals were observed daily and a number of parameters were examined, including body weight, food consumption, haematology, blood clinical chemistry, blood IGF-I levels, urinalysis, post-mortem examination, organ weights, histological examination of the GI tract, tibia length and tibial epiphyseal width.

No adverse effects were observed in this study. Body weights and food consumption were not affected by oral administration of IGF-I in male and female rats. There were no dose-related changes in organ weights, blood IGF-I levels or other measured clinical parameters in male or female gavage-dosed rats. The GI tract of gavage-treated rats was normal as indicated by microscopic examination. Rats administered IGF-I systemically (pump), especially at the higher dose, exhibited increased body-weight gain, changes in clinical parameters and various organ weights, elevated blood IGF-I and increased tibial epiphyseal width in contrast to there being no effect in the rats receiving bGH via the gut. The highest oral-gavage dosage of 2000 µg/kg body weight was considered a no-observed-effect level, since no significant effect or toxicity was observed in rats at this level of exposure.

Binding proteins modulate plasma IGF-I bioavailability in humans (Mauras *et al.*, 1999). IGF-I/II-binding proteins are not present in gut exocrine secretions in humans (Chaurasia *et al.*, 1994). This is one reason why IGF is not absorbed across the mucosa in significant quantities. Assuming some amount of protein possibly escapes digestion due to the presence of achlorhydria, hypermotility or pancreatic insufficiency, systemic exposure is limited by bioavailability and first-pass effects. Two factors, mucosal (also liver) first-pass metabolism and the ability to cross the mucosa, by either diffusion or active transport, determine bioavailability and hence systemic exposure. In this respect, absorption across the intestinal epithelium of peptides and proteins still able to retain their biological activity has been discovered (Webb, 1986; Ziv *et al.*, 1987; Bendayan *et al.*, 1990, 1994; Milstein *et al.*, 1998; Ziv and Bendayan, 2000); however, only minute amounts of active protein are absorbed in this manner (Ziv and Bendayan, 2000). Hence, dose and bioavailability as a result of intestinal absorption would be essentially non-existent.

Even hGH is not biologically available when ingested. Human GH is an essential therapeutic drug for the treatment of human individuals with GH deficiency (Iyoda *et al.*, 1999). Since GH of all types, including sGH and hGH, cannot be effectively delivered via the gut, the delivery system for therapy is the extremely inconvenient technique of injection, and alternative forms of delivery have focused on nasal absorption for potentially replacing the daily subcutaneous injections (Laursen *et al.*, 1994; Moller *et al.*, 1994). However, this alternative route of delivery has still not replaced injections (Iyoda *et al.*, 1999; Muller *et al.*, 1999). Additionally, hGH can now be utilized for correcting muscle weakness, muscle protein kinetics in cancer patients, reconditioning respiratory muscle after lung surgery and various other medical applications (Berman *et al.*, 1999; Felbinger *et al.*, 1999; Janssen *et al.*, 1999). Once again, even though GH therapy is now 50 years old, the method of administration in these new medical applications is still injections, since GH and hGH are not bioavailable orally.

Thus, it can safely be assumed, without investigating other parameters, that any sGH and sIGFs absorbed from the adult gut would not result in a concentration high enough to produce any systemic physiological effect. Any measurable physiological effect obtained from such a brief exposure (single meal) to either of the growth factors seems highly implausible.

In summary, therapeutic proteins and peptides do not survive in the gut of humans, requiring administration via parenteral routes. The fact that sGH and sIGF would be destroyed by cooking, or, in the case of consumption of raw salmon, destroyed and denatured in the gut, makes them not biologically available, and they would not cause any biological effects. Therefore, there are no risks and no detriment to health from the consumption of transgenic salmon. Lack of bioavailability provides one level of safety.

Dosage from Consumption

If we were to make the unrealistic assumption that sGH and sIGF were totally bioavailable and active when orally ingested, we could calculate the maximum dose from consumption of the transgenic product. The amount of raw fish in one serving would be approximately 60–120 g (restaurant standard). If one assumes there is no effect of autolysis (Fricker, 1969), then IGF concentration in the raw fish is identical to that in the live fish. Autolytic enzymes may reduce the expected dose of both GH and IGF. Therefore, assuming a worst-case (and unrealistic) scenario that sGH and sIGF are totally available biologically, the dose delivered to the bloodstream from a single meal could be calculated as follows.

Blood volume in salmon is approximately 50 ml/kg (Olson, 1992). The erythrocytes account for 30–35% of that volume, so plasma volume is approximately 35 ml/kg. We shall assume that the majority of GH and IGF is located in the plasma, and the plasma remains in the flesh upon processing. Plasma volume in the red and white muscle is 16 and 6 ml/kg, respectively. Blood volumes are difficult to measure and, if individual tissues are measured and summed, blood-volume estimates can double. The total plasma volume in a 1 kg rainbow trout has been estimated to be 45, 3 and 0.5 ml for skin, white muscle and red muscle, respectively (Olson, 1992). The plasma volume for a 70-kg man is 3 litres (Guyton, 1981).

As discussed earlier, a transgenic salmon may have a maximum value of 400 ng GH/ml and 300 ng IGF/ml in its plasma. If a man were to consume a huge portion of flesh (more than the restaurant standard), nearly 600 g, skin (a rare practice) and all the muscle from a 1 kg fish, he would ingest, assuming no degradation or denaturation of the protein, a total of 19,400 ng GH and 14,550 ng IGF, or 1400 ng GH and 1050 ng IGF if he ate only the muscle.

Total GH present in the body at any one time ranges from 15,000 to 210,000 ng. Therefore, if the protein was not digested and 100% was absorbed and bioavailable, the man would be consuming anywhere from 9 to 130% of his current GH level in the case where he had this huge portion and ate all of the skin. If he did not eat the skin, he would be consuming between 0.6 and 9% of his current level of plasma GH. Based on the total daily GH production of humans, consumption of the large salmon meal with skin would be equivalent to 1.9–19% of the daily production of GH, and consumption of the meat would equal only 0.1–1.0% of the total daily production.

Similarly, as the daily production of IGF is 10 million ng, if he consumed the skin and muscle of this huge portion, he would consume an amount equivalent to only 0.2% of his daily production of IGF. If he did not eat the skin, he would consume an amount equivalent to 0.01% of his daily IGF production. If we were to be more conservative and assume that the standard restaurant-size portion is consumed, all of these dose estimates are reduced another five- to tenfold.

Even if sGH and sIGF were bioavailable and bioactive, their consumption via transgenic salmon flesh would represent an extremely small percentage of daily human GH and IGF production even in the case where the salmon are producing maximum amounts of hormone and the human is consuming maximum amounts of flesh. The contribution of ingested GH and IGF is physiologically insignificant when compared with the total daily production of these hormones, and also with what is usually present at any one time in a human, and the physiological significance of the ingested GH and IGF would be questionable even if they were biologically available, which they are not.

Bioactivity of Salmon/Non-primate GH in Humans

GH contains 191 amino acids in mammals, and the sequence of the 191 amino acids varies among species. Because there is species variation in the structure, there is species specificity for its biological activity. For instance, bGH and porcine GH are recognized by the human immune system as foreign proteins, and the human body destroys bovine and porcine somatotrophin if it is given systemically (Bauman, 1992). Orally administered GH is not active in mammals (Albert *et al.*, 1993; Stoll *et al.*, 2000). GH must be injected to be biologically active (Juskevich and Guyer, 1990; Bauman, 1992). It has been known for more than four decades that only primate GH is effective in primates, and GH from other organisms is not active in primates and humans (Lauterio *et al.*, 1988; Behncken *et al.*, 1997). Clinical studies in the 1950s uniformly demonstrated that non-primate GH was biologically inactive in humans (Bauman and Vernon, 1993).

In the 1960s, it was determined that certain forms of human dwarfism were a result of inadequate production of GH. Human GH was not abundantly available at that time and extensive research was conducted in the hope that injection or ingestion of non-human GH would be biologically active and correct the dwarfism (Bauman, 1992). However, only primate GH was active in humans, and GH from all non-primate organisms tested failed to elicit any biologically activity and response in humans (Wallis, 1975;

Kostyo and Reagan, 1976; Juskevich and Guyer, 1990).

Again, there is convincing medical evidence that non-human GH is not bioactive in humans (Raben, 1959). Therefore, sGH would not have an effect on humans even in the implausible event that sGH was bioavailable after being exposed to the digestive system.

Additional supporting evidence that fish GH is not active in mammals comes from studies with fish prolactin. Prolactins and GH are derived from a common ancestral protein and are quite similar. Tilapia prolactin has no potency on mouse mammary gland, and fish prolactins, closely related to fish GH, satisfy structural requirements for activity in fish only (Doneen, 1976), again demonstrating that fish hormones are not bioactive in or effective in humans.

In summary, the physiological response to any fraction of salmonid hormone surviving the gut would be minimal in humans, since human receptors for these hormones exhibit a high degree of species specificity with respect to ligand, unlike the receptors of lower vertebrates. Human GH has bioactivity in lower vertebrates but the reverse is not true. This provides a second level of food safety.

Bioactivity of Fish IGF

The IGF system appears highly conserved between teleost fish and mammals. There are only 14 amino acid differences out of 70 between sIGF-I and hIGF-I. The C-terminal E domain of salmon proIGF-I, which is presumed to be proteolytically cleaved during biosynthesis, also shows striking amino acid sequence homology with its mammalian counterpart, except for an internal 27-residue segment that is unique to salmon proIGF-I (Duan, 1998).

The biological potency of IGF-I is remarkably conserved. However, the data on fish IGF bioactivity on mammalian cells are contradictory. In general, the evidence indicates that fish and salmon IGF are bioactive on mammalian and human cells,

but are much less potent than hIGF. However, Upton *et al.* (1996) showed that hIGF-I and sIGF-I are equally potent in binding to the hIGF-I receptors and in stimulating protein synthesis in mammalian cells. Human and fish IGF-I can be equally potent in mammalian and fish bioassay systems (Duan, 1998).

Rat, kangaroo, chicken, salmon and barramundi IGF-I proteins differ from hIGF-I by 3, 6, 8, 14 and 16 amino acids, respectively (Upton *et al.*, 1996, 1998). Of these organisms, the fish IGFs are the most distant from those of humans. IGF-I proteins exhibit similar biological activities and type-I IGF receptor-binding affinities, regardless of whether mammalian, avian or piscine cell lines are tested (Upton *et al.*, 1998); however, there was a trend suggesting that the fish IGFs were most effective in studies using a homologous system.

In contradiction to the study previously discussed, sIGF-I was not as potent as hIGF-I in bioassays in mammalian cells, but was as effective as hIGF-I in piscine cells (Upton *et al.*, 1998). When tested in rat myoblasts, fish IGF, including sIGF, required two- to threefold more protein to obtain one-half of the effect. Fish (barramundi) IGF-I could compete equally well with hIGF-I for binding to rat IGF-I receptors, but the fish IGF-I potency was sixfold less (Upton *et al.*, 1998). Fish IGF-I was less effective than hIGF-II for binding to type II IGF receptors of rat. Although not potent, fish IGF-I was more competitive for binding to IGF-II receptors than human and mammalian IGF-I.

In vitro assessment of recombinant hagfish IGF in cultured cells indicates that hagfish IGF shares functional properties with mammalian IGFs (Upton *et al.*, 1997). Thus, hagfish IGF stimulates protein synthesis in rat myoblasts, but 20- and fivefold more peptide, respectively, is required to achieve the same half-maximal responses as with hIGF-I or hIGF-II. Hagfish IGF also competes for binding to the type I IGF receptor present both on rat myoblasts and on salmon embryo fibroblasts, though with somewhat lower affinity than either hIGF-I or hIGF-II. In general, fish IGF appears to be less potent in mammalian systems than hIGF.

Explanation for Primate Specificity for GH Bioactivity

The hGH receptor (hGHR) only recognizes primate GH (Hammond *et al.*, 1990; Juskevich and Guyer, 1990; Baumann *et al.*, 1994; Clackson and Wells, 1995; Souza *et al.*, 1995; Wells, 1995, 1996; Behncken *et al.*, 1997; Clackson *et al.*, 1998; Pearce *et al.*, 1999). And unlike IGF, GH, because of variation in its amino acid sequences (Machlin, 1976; Holladay and Puett, 1977; Li *et al.*, 1986; Nicoll *et al.*, 1987; Hammond *et al.*, 1990), has failed to retain its bioactivity across species (Souza *et al.*, 1995; Behncken *et al.*, 1997). Non-primate GHs bind and activate non-primate GHRs but have very limited activity for primate GHRs (Carr and Friesen, 1976; Lesniak *et al.*, 1977). The ligand selectivity of the primate GHRs was first recognized in 1956 (Knobil *et al.*, 1956), and more recently has been explored at the structural level, using mutagenic studies with recombinant GHs and receptors. The presence of complementary residues – arginine 43 (Arg-43) on the primate receptor and aspartate 171 (Asp-171) on the ligand (primate GH) – is required for binding and subsequent activity. There is a second specific binding domain. hGH binding to hGHRs is species- and ligand-specific, and is facilitated by the binding of GH to its two receptors and their interactions, and probably requires a conformational change in the receptor (Behncken and Waters, 1999). Both sites 1 and 2 must bind to activate the receptor, thus fragments of GH are not active.

The GHR1 is a member of the haematopoietic cytokine receptor family, and shares common structural and functional features with receptors for prolactin, erythropoietin, granulocyte and granulocyte–macrophage colony-stimulating factors, many interleukins, thrombopoietin, ciliary neurotrophic factor, oncostatin M and leptin (Takahashi *et al.*, 1996; Wells, 1996). The interaction of GH with its receptor is well characterized, as extensive structure/function studies have been conducted on both GH and the GHR (Wells, 1996). The crystal structure of the hormone–receptor complex is known (De Vos *et al.*, 1992). Binding of hGH to its receptor is required for bioactivity and the regulation of normal human growth and development. The 2.8 Å crystal structure of the complex between the hormone and the extracellular domain of its receptor (hGH-binding protein, hGHbp) consists of one molecule of GH per two molecules of receptor (De Vos *et al.*, 1992).

The hormone is a four-helix bundle with an unusual topology. The binding protein has two distinct domains. Both hGHbp domains contribute residues that participate in hGH binding. In the complex, both receptors donate essentially the same residues to interact with the hGH, despite the fact that the two binding sites on hGH have no structural similarity.

In addition to the hormone–receptor interfaces, a substantial contact surface exists between the carboxyl-terminal domains of the receptors. The relative extents of the contact areas support a sequential mechanism for dimerization, which is possibly crucial for signal transduction. Both crystal structure and solution studies support the concept that two identical receptor subunits bind the helix-bundle hormone through similar loop determinants on the receptor-sandwich structures. The hormone is caught by receptor 1 through binding to determinants located in a 900 Å^2 patch encompassing helices 1 and 4 and the unstructured loop between helices 1 and 2. Eight key residues are responsible for 85% of the binding energy, with electrostatic interactions governing the approach of hormone to the receptor binding site (Cunningham and Wells, 1993; Clackson and Wells, 1995). Additionally, electrostatic interactions are important specificity determinants since five of the seven residues that are modified to allow prolactin to bind to the GHR with high affinity involve charged residues (Cunningham and Wells, 1991). Of residues within the five major loops involved in hGH binding (Takahashi *et al.*, 1996), the interaction between Arg-43 of the human receptor and Asp-171 of the hGH is remarkable because in non-primate receptors this position is replaced by leucine and histidine replaces aspartate in non-primate GH (the same position in porcine and bovine GH is 170)

(Souza *et al.*, 1995; Behncken *et al.*, 1997). There is unfavourable charge repulsion/ steric hindrance between GH His-170 and receptor Arg-43, rather than a favourable salt bridge between this arginine and primate GH Asp-171, and this is probably an important element in the inability of non-primate hormones to bind to the human receptor.

Behncken *et al.* (1997) found that the single interaction between Arg-43 primate GHR and non-primate complementary GH residue His-170/171 explains most of the primate GH species specificity and this is in agreement with the crystal structure of GH. In the case of non-primate hormones, the steric hindrance resulting from the incompatibility of histidine at GH residue 170 and arginine at receptor residue 43 presumably increases because of repulsive interactions between these basic residues.

Even when the appropriate mutation is induced for these key receptors in the laboratory, the non-primate GH has only 5% of the affinity for the hGHR and 5% of the potency of hGH (Behncken *et al.*, 1997). Such a mutation is highly improbable given that this has not already occurred naturally.

Although the evidence presented here supports a central role for His-170 in determining binding specificity, other species-specific determinants may reside in the unstructured loop between helices 1 and 2, based on the homologue scanning mutagenesis study of Cunningham *et al.* (2001). They found that, while substituting porcine growth hormone (pGH) residues 164–191 into hGH eliminated binding to the hGHR binding domain, substitution of pGH residues 54–74 resulted in a 17-fold decrease in affinity. These data can be interpreted as possibly indicating that other species determinants exist.

These determinants were probably found by Peterson and Brooks (2000). Their data indicate that the Asp-171 along with DeltaPhe-44 or Delta-32–46 residues give hGH its specificity and the acceptance of hGHRs to only primate GH. The cooperative interaction of these two distant motifs determines the species specificity of GH activity in humans (Peterson and Brooks, 2000).

Potential Toxic Effects of GH/IGF and Food Safety

The potential toxic effects of megadoses of IGF and GH have been studied in rats. In these studies daily oral administration of bIGF-I for 16 days at doses of 20, 200 and 2000 µg/kg/day, or even higher doses up to 6000 µg/kg/day for up to 4 weeks, did not produce any adverse clinical signs or any GI pathology indicative of a preneoplastic effect. When rats were fed a dose of bGH equivalent to a human ingesting 2.3 million times more GH than what would be found in five glasses of milk, there was no activity or effect (Sechen, 1989; Juskevich and Guyer, 1990). Adverse effects from high concentrations of GH in foods, particularly milk, are not reported in the large body of scientific literature examined (PubMed, EMBASE, Current Contents, Biosis and Medline databases). WHO has also examined and reviewed potential health risks of GH (bGH) and IGF (Ungemach and Weber, 1998). It also concluded that megadoses of these two compounds posed no health risks and had no or minimal bioavailability and bioactivity. Elevated levels of IGF in the plasma are associated with growth-factor tumours (Ungemach and Weber, 1998), and therefore there is concern regarding the cancer risk of IGF. However, WHO concluded that the exposure to IGF from biotechnological applications was miniscule compared with endogenous natural production and exposure, and posed no health/cancer risk. The data presented earlier in this chapter demonstrated that the IGF levels in transgenic salmon were essentially the same as in normal salmon, the concentration of IGF in transgenic salmon flesh is minuscule compared with that in the human system and the sIGF is not bioavailable; therefore, it is illogical for there to be any cancer risk or increase in cancer risk from consuming this product.

The food safety of GH and specifically of bGH or bovine somatotrophin has been carefully analysed. A variety of organizations, including the American Medical Association, American Academy of Pediatrics, American Cancer Society, Council of Agricultural Science and Technology, Food and Nutrition Science Alliance, FAO and WHO, all concluded that GH posed no health or safety concern for consumers.

No adverse effects from the ingestion of GH or IGF, including megadoses of these hormones, have ever been found.

Studies on GH-Transgenic Fish Food Safety

Human food-safety data have been collected in an example or model relevant to the case of the GH-transgenic salmon. Specific experimental evidence that teleost GH is not active in primates was obtained by Guillen *et al.* (1999). Juvenile monkeys, *Macaca fascicularis* (macaques), were injected with recombinant tilapia GH at a dose of 1000 ng/kg/day for 30 days, equivalent to administering 70,000 ng/day to a 70-kg human. Blood parameters examined included haemoglobin, serum total proteins, blood glucose, packed-cell volume, total leucocytes and total erythrocytes. Body weight, rectal temperature, heart rate and respiratory rate were recorded daily. Head-to-tail length, interscapular cutaneous pleat, left-flank cutaneous pleat, cranial circumference and cranial diameter were measured. At the end of the experiment, the animals were sacrificed, autopsies conducted, and the organs and tissues macroscopically examined and weighed. Histopathological analysis was conducted for all organs and tissues. Tilapia GH did not affect animal behaviour pattern or food intake. Body weight, temperature, heart rate and respiratory rate were unaffected by tilapia GH administration to macaques. The blood profiles and somatic growth of tilapia-GH-treated macaques and controls were no different. Autopsies revealed that all organs, tissues and cavities were normal, and no changes relative to controls were detected for common targets of GH, such as tongue, palate plate concavity, liver, muscle, heart, kidneys and others. Subcutaneous and abdominal fat, mesenteric fat and peritoneal fat were unchanged, normal and of the usual colour. No histopathological or morphological changes were observed.

Additionally, doses of recombinant tilapia GH did not affect sulfate uptake in rabbit cartilage at doses up to 20 µg/ml. However, tilapia GH did stimulate sulfate uptake in tilapia cartilage with doses as low as 1 µg/ml. Bovine insulin and hGH stimulated sulfate uptake in rabbit cartilage. The results reported here and those of others suggest that mammalian (rabbit) GHRs have no or a very low affinity for teleost GH (Guillen *et al.*, 1999).

Twenty-two humans were fed tilapia (transgenic hybrid *Oreochromis hornorum*) that contained and expressed tilapia GH transgene (Guillen *et al.*, 1999). These tilapia grew twice as fast as non-transgenic controls. The humans were fed transgenic or control tilapia for five consecutive days, twice daily. Haemoglobin, total serum proteins, glucose, creatinine, cholesterol, leucocytes and erythrocytes were measured. No clinical or biochemical parameters and no blood profiles of humans evaluated before and after onset of experimentation were affected by consuming transgenic tilapia.

The fact that tilapia (teleost) GH did not promote modifications of blood glucose values and total protein and creatinine, as well as having no effect on growth, target tissues, lipolysis and protein synthesis in the muscle and no contra-insulin effects, is indicative and confirms that fish GH is not bioactive in primates. GH should stimulate erythropoiesis and lymphopoiesis and increase spleen and kidney weight (Gluckman *et al.*, 1991) and is associated with stimulating fluid retention, growth and changes in blood volume and blood characteristics (Ho and Kelly, 1991), but none of these phenomena were observed.

When clinical and pathology data were evaluated, there was no evidence for any activity or harm to primates that could be attributed to the additional fish GH. Tilapia (fish) GH had no effect when administered to the non-human primates. There were no effects on the humans who participated in consumption studies of transgenic GH tilapia. At the end of the experiment, comprehensive clinical, pathology, necropsy and histopathological examinations were conducted. No effect of the teleost GH was detected in any parameters expected to change, including body weight and adipose tissue.

Conclusions on Human Food Safety

To summarize with respect to food safety: levels of GH and IGF expressed by transgenic GH salmon are not always outside the range or much greater than the upper limit of GH and IGF secretion for other fish, food animals or humans; sGH and sIGF are not bioavailable when ingested orally (cooked or raw; adequate cooking would denature the proteins); even if they were totally bioavailable, the dose from one meal would only be a small fraction of total daily human production of GH and IGF; GH is not active orally in higher species; sGH is not bioactive in humans; the primate GHR binds only primate GH and it requires both binding sites to be occupied; initial studies indicate that fish GH has no biological effect on primates and short-term ingestion of GH-transgenic fish has no biological effect on humans. The lack of oral activity of GH and IGF-I and the non-toxic nature of the residues of these compounds, even at exaggerated doses, demonstrates that salmonid GH and IGF-I present no human-safety concern when consumed orally. Therefore, there is no need to establish a safe concentration of total residue or a residue method in transgenic salmon meat. Thus, viewed from a number of aspects, any increased concentrations of GH or IGF in edible salmon skeletal muscle or skin is not hazardous to human health.

Government Regulation of Transgenic Fish and Biotechnology Products

The fisheries-science community (Hallerman and Kapuscinski, 1990a,b, 1992a,b; Kapuscinski and Hallerman, 1990b) recognized the potential hazards posed by research, development work and commercialization of aquatic GMOs, published position papers and testified at government hearings (Hallerman, 1991). The American Fisheries Society (AFS) adopted a position statement (Kapuscinski and Hallerman, 1990a) that recommends a cautious approach to research and development of aquatic GMOs. Voluntary research guidelines have been adopted by the US federal government that have funding ramifications if not followed. Several fisheries and aquaculture professional communities have since adopted position statements (Hallerman and Kapuscinski, 1995).

Historically, worldwide, policies for the research and marketing of transgenic food organisms range from non-existent to rather strict, as in the EU (Bartley, 1999; Kapuscinski *et al.*, 1999; Dunham *et al.*, 2001; FAO, 2001). Government regulation of transgenic aquacultured species based on sound scientific data is needed; however, those data are lacking.

International instruments, some legally binding and others voluntary, cover a broad range of issues associated with GMOs in aquaculture: the introduction (transboundary movements) and release into the environment, international trade, human health, food safety, labelling, intellectual-property rights (IPR) and ethics (Bartley, 1999; Dunham *et al.*, 2001; FAO, 2001). Much of what is presented below comes from the plant sector, but aquaculture may expect similar processes, opportunities and problems in the further development of aquatic GMOs; it will be prudent to follow developments in the crop sector. The importation of a transgenic aquatic organism and its environmental release are addressed by European Community Directives, United Nations (UN) Recommendations on the Transport of Dangerous Goods (1995), the Convention on Biological Diversity (CBD), the FAO Code of Conduct for Responsible Fisheries (CCRF), the International Council for the Exploration of the Sea (ICES) and the FDA in the USA. Common denominators of the legislation or guidelines are licensing for field trials and release of GMOs, notification that a GMO is being exported/imported and released, and an environmental impact assessment. Members of the EU agreed to uniform licensing procedures for testing and sales of GMOs, directing that, if a GMO was licensed in one country, field testing, release and sale should also be allowed in the other countries. However, European countries have not adhered to this policy, some operating strictly under home rule and others overturning approvals granted earlier by previous administrations. Politics and protectionism appear to have major roles in the decision-making process.

In 1992, the CBD (UNCED, 1994), which includes 175 countries, requested the establishment of 'means to regulate, manage or control the risks associated with the use and release of LMOs [living modified organisms] ... which are likely to have adverse environmental impacts' (Article 8g). CBD also requests legislative, policy or administrative measures to support biotechnological research, especially in those countries that provide genetic resources (Article 19). Additionally, Article 19 (3) directs participating countries to consider the establishment of internationally binding protocols for the safe transfer, handling and utilization of LMOs that have a potentially adverse effect on the conservation and sustainable use of biological diversity.

The Convention seeks to promote biosafety regimes to allow the conservation of biodiversity as well as the sustainable use of biotechnology (Bartley, 1999; Dunham et al., 2001; FAO, 2001). Article 14 mandates each signatory to require environmental impact assessments of proposed projects that are likely to have adverse effects on biological diversity, with the goal of avoiding or minimizing such effects. A Working Group on Biosafety developed a biosafety protocol, and each country was encouraged to develop central units capable of collecting and analysing data and to require industry to share their technical knowledge with governments. A worldwide databank was suggested that would collect information on GMOs and manage research on their environmental effects. Some, but not all, of the participating countries agreed that industry has a responsibility to inform national governments when it wishes to produce or export GM foods. The sixth meeting of the Working Group on Biosafety was held in February 1999 to finalize the International Protocol on Biosafety on transport between countries of living GMOs. The meeting was not successful, with a few nations blocking final approval (Pratt, 1999). Delegates were divided along issues such as: whether or not

to have the precautionary principle drive biosafety decisions; liability in the case of negative effects on human health or biodiversity; possible social and economic impacts on rural cultures; whether or not to regulate movement across borders of products derived from genetically engineered crops; and whether to segregate and label genetically engineered crops and, possibly, their derived products. The negotiations, development and ratification of these protocols are ongoing (www.biodiv.org).

Article 9.3 of the FAO CCRF addresses the 'Use of aquatic genetic resources for the purposes of aquaculture including culture-based fisheries' (Bartley, 1999; Kapuscinski et al., 1999; Dunham et al., 2001; FAO, 2001). The key components of this article recommend the conservation of genetic diversity and ecosystem integrity, minimization of the risks from non-native species and genetically altered stocks, creation and implementation of relevant codes of practice and procedures, and adoption of appropriate practices in the genetic improvement and selection of brood stock and their progeny. Article 9.2.3 instructs: 'States should consult with their neighbouring States, as appropriate, before introducing non-indigenous species into transboundary aquatic ecosystems.' The Technical Guidelines on Aquaculture Development indicate: 'Consultation on the introduction of genetically modified organisms should also be pursued.' The definition of non-indigenous in this case is not conventional and includes domesticated, selected breeding, chromosome-manipulated, hybridized, sex-reversed and transgenic organisms.

Aquatic GMOs will also eventually come under the purview of the FAO Commission on Genetic Resources for Food and Agriculture (CGRFA) (Bartley, 1999; Kapuscinski et al., 1999; Dunham et al., 2001; FAO, 2001). This Commission is the only permanent UN intergovernmental agency working on the conservation and utilization of genetic resources and technologies for food and agriculture. The Commission established working groups for plant genetic resources and farm-animal genetic resources, but not for aquatic resources.

To date, CGRFA has focused primarily on treaties regarding plants (FAO, 2011).

The ICES has guidelines for GMOs, and recognizes them as a form of non-native species, which again is scientifically questionable (Bartley, 1999; Kapuscinski *et al.*, 1999; Dunham *et al.*, 2001; FAO, 2001). Like the CCRF of FAO, the ICES protocols are voluntary; however, they have been adopted by numerous regional fishery bodies, by the International Network for Genetics in Aquaculture (International Center for Living Aquatic Resources (ICLARM) secretariat (now the World Fish Centre)) and by national governments, including the Philippines.

With regard to food safety, the CAC established the Codex Principles for the Risk Analysis of Foods Derived from Modern Biotechnology in 2003. This organization provides a framework for conducting risk analysis of the safety and nutritional aspects of foods derived from biotechnology.

The Cartagena Protocol on Biosafety, which came into force in September 2003, is a legally binding international instrument that regulates the transboundary movement of LMOs motivated by the desire to protect the environment. The primary document utilized is the advance informed agreement (AIA). The importing country must give consent prior to the shipment, receipt and introduction of an LMO into their boundaries and environment. Another important aspect of this protocol is that it provides a mechanism to provide relevant information to the importing country, so it can make educated decisions regarding the importation of the LMO.

Currently, the EU policy regarding agricultural biotechnology and transgenic organisms is quite variable from one Member State, or country, to another (Henard *et al.*, 2008). Virtually all Member States abide by EU Directive 2001/18 concerning regulations on traceability and labelling. Many EU members (Belgium, Czech Republic, Germany, Hungary, Portugal, Romania and Slovakia) have initiated national coexistence frameworks for biotech and non-biotech crops or (France, Spain and UK) are currently preparing coexistence rules.

EC Directive 94/15/EC of 15 April 1994 (Council Directive 90/220/EEC) establishes Europe's regulatory framework for biotechnology and requires notification that a GMO is to be deliberately released into the environment (Bartley, 1999; Dunham *et al.*, 2001; FAO, 2001). The Directive includes requirements for impact assessment, control and risk assessment. Requirements are different for higher plants and other organisms, and the definition of release includes placement on the market. These rules were significantly strengthened by Council Directive 2001/18/EC, which has been adopted by each EU Member State (Henard *et al.*, 2008). This Directive includes procedural guidelines for experimental field trials, commercialization into the market, environmental risk assessment, mandatory post-market (environmental) monitoring, mandatory dissemination of public information, mandatory labelling and traceability of the biotechnology product at all stages of the market, and the establishment of a molecular registry. These renewable authorizations are granted for a maximum of 10 years.

The Directives, which are enacted by approval of the national legislatures, are accompanied by a series of EU Regulations which do not need to be authorized by national legislatures, are immediately enforceable law and govern the approval and use of genetically engineered products.

US Performance Standards

The US government approach to oversight of biotechnology was established in the Coordinated Framework for the Regulation of Biotechnology (OSTP, 1985, 1986), under which the USDA was given jurisdiction over research and development activities with multicellular agricultural organisms (Bartley, 1999; Kapuscinski *et al.*, 1999; Dunham *et al.*, 2001; FAO, 2001). USDA procedures and actions must comply with the National Environmental Policy Act (NEPA). The USDA's Office of Agricultural Biotechnology (OAB), which no longer exists, determined that a strong environmental review procedure should be developed. The OAB organized a Working Group on Aquaculture Biotechnology and

Fig. 23.1. Outdoor confinement facilities at Auburn University. Designed by Auburn University, a team of US scientists and aquaculture engineers and environmental organizations, and then approved by the Office of Agricultural Biotechnology, US Department of Agriculture. (Photograph by Rex Dunham.)

Environmental Safety, which, with public and scientific-community input, produced the Performance Standards for Safely Conducting Research with Genetically Modified Fish and Shellfish (ABRAC, 1995).

The Performance Standards contain decision-support flowcharts to achieve biosafety while allowing flexibility in how to achieve safety (Fig. 23.1) (ABRAC, 1995). The Performance Standards ask questions concerning fish and shellfish that are to be altered by gene transfer, chromosome-set manipulation or interspecific hybridization. If a specific risk(s) is identified, the user is led to risk management; if not, the experiment exits the Performance Standards. If no summary finding is reached, the user proceeds to the section to evaluate ecosystem effects.

Potential impacts assessed are introgression, non-reproductive interference, reproductive interference and effects on ecosystem structure and process. If no risk pathways are identified, the conclusion is that there is reason to believe the GMO is safe, and the Performance Standards are exited. If a specific risk is identified or if a judgement cannot be reached, the user proceeds to the risk management and the Performance Standards offer technical and procedural guidance. This procedure was adopted by USDA in 1996 and exists in interactive electronic versions (Hallerman *et al.*, 1998), both on disk and the Internet (www.nbiap.vt.edu). The Performance Standards have limitations as they are voluntary and scope is limited to small-scale research and development with fish and shellfish. The Performance Standards do not address large-scale commercial use of GMOs, nor do they address aquatic organisms beyond fish and shellfish. Although voluntary, adverse consequences would probably result if they were not followed. This system of oversight and voluntary control, which in reality existed prior to the USDA Performance Standards, and which is still in existence today as individual institutional biosafety committees at universities and in research organizations, has successfully prevented any release of transgenic aquatic organisms in the USA. The commercial use of transgenic aquatic organisms is controlled by the FDA.

In the USA, under the authority of the Federal Food, Drugs, and Cosmetics Act, the FDA has jurisdiction for approval of both commercial production and marketing of transgenic aquatic organisms. Transgenic fish and shellfish expressing an introduced GH gene are considered the same as new animal drugs (Matheson, 1999), with transgenesis defined as a means for delivering GH to the fish or shellfish. Under the authority of NEPA, the FDA has the mandate to consider the environmental effects of production of transgenic aquatic organisms, and it appears that, in cooperation with the US Fish and Wildlife Service, they will exercise this mandate.

In addition to federal control, some individual states, such as North Carolina and Minnesota, have enacted their own regulation of aquatic GMOs. California has enacted some of the most stringent laws restricting GMOs (Van Eenennaam and Olin, 2006).

US policy on research and development work with aquatic GMOs has influenced that in other countries, and national policies have been adopted in Canada (DFO, 1998), Norway, China, Denmark and other countries (Hallerman and Kapuscinski, 1995; Dunham, 1999). Other countries, such as Chile, remain without policy and regulation on aquaculture biotechnology. Further, considering the growing scale of international trade and the connectivity of aquatic and marine environments, national policies have inherent limitations.

International Performance Standards

The Performance Standards were the impetus for another larger international effort by the Edmonds Institute to approach the problems of risk assessment and risk management at the commercialization scale (Bartley, 1999; Kapuscinski *et al.*, 1999; Dunham *et al.*, 2001; FAO, 2001). A manual was developed to assess ecological and human health effects for all taxa of GMOs (Klinger, 1998). The Scientists' Working Group on Biosafety (1998) of this effort identified four major goals of biosafety: determination in advance of hazards to human health and natural systems if any specific GMO is released into the environment; anticipation of when a specific GMO or any of its products(s) will be harmful if it becomes human food; discernment of actual benefit(s) of a GMO as designed; and ascertaining that hazards will not occur when GMOs are transported, intentionally or unintentionally, among different ecosystems and nations. These four objectives were developed into a set of decision trees designed to allow the user to identify, assess and manage specific risks for specific GMOs and applications. This manual builds upon the US Performance Standards and is more extensive, as flowcharts cover environmental and human-food safety as well as additional aquatic taxa, including algae,

vascular aquatic plants and aquatic micro-organisms (www.edmonds-institute.org).

Canada

In Canada legislation has proposed to transfer responsibility for food safety from the Health Department to the Agriculture Department (Bartley, 1999; Kapuscinski *et al.*, 1999; Dunham *et al.*, 2001; FAO, 2001). This was perceived by some as being a direct conflict of interest, as there is great mistrust, some justified and some unjustified, of government and specifically agriculture and aquaculture sectors. Natural-resources agencies, on the other hand, can be overly restrictive in the other direction, may have the opposite conflicts of interest and even be anti-aquaculture. The goals of aquaculture and natural-resource management/conservation should be the same, and both can benefit the goals of the other. Some balance based on scientific fact is needed.

In Canada, government regulation also appears to be working. AquaBounty Farms indicate that its hatchery is in full compliance with Canadian guideline regulations (DFO, 1998). Licensees of AquaBounty are required to produce their fish in closed aquaculture systems to effectively confine the GMOs, and those growing transgenic salmon in net pens are required to produce triploids.

The Royal Society of Canada recommended a more conservative approach concerning application of transgenic fish. It concluded that the consequences of the escape of transgenic fish would have uncertain effects on wild populations, and that the effectiveness of sterilizing transgenic fish was uncertain. The report recommended a moratorium on the rearing of GM fish in marine pens and suggested that transgenic fish be raised only in land-locked facilities.

United Kingdom

Regulations concerning the application of transgenic animals in Europe are strict. In general, public opinion is negative concerning GM foods. The Royal Society (UK) has studied the issue of GM animals, including fish, and has issued a report and policy statement (Royal Society (UK), 2001).

In summary, the Royal Society (UK) concluded that the debate concerning GMOs should not just revolve around moral and social issues, but should stress sound scientific evidence. It recognized the importance of GM animals as tools and models for medical research on human disease and the critical need for GM animals for producing medical compounds and preventing the extraction of such substances from human tissues that could be harbouring viruses. The Society believed that much of the technology for agriculture application was still at an early stage, more research was needed and the benefits of the transgenic approaches, cloning and other biotechnologies needed to be compared with those of traditional selective breeding to ascertain benefit. It recommended the sharing of knowledge, currently restricted under patenting and licensing agreements, and assisting developing countries with GMO technology, particularly in the area of disease control. The importance of developing the transgenic transformation of insects that normally carry human disease so that they are incapable of transmitting disease and replacing natural populations with the modified insects was stressed.

The Royal Society (UK) cited the report on biotechnology by the Royal Society of Canada (discussed above) which concluded that the consequences of the escape of transgenic fish would have uncertain effects on wild populations and that the effectiveness of sterilizing transgenic fish was uncertain. The Royal Society of London endorsed the recommendation of a moratorium on the rearing of GM fish in marine pens and suggested that transgenic fish be raised only in land-locked facilities.

All GM animals and fish developed in the UK, regardless of their application, must be evaluated by a comprehensive framework of committees and legislation that regulate and provide advice on GM animals. This policy is administered by the newly formed Agriculture and Environment Biotechnology Commission (AEBC), which is independent of the UK government. The Cabinet Biotechnology Committee has the responsibility of coordinating government policy and legislation, which is quite complex. This policy is confusing to the point that the Society recommends the development of a simple handbook, so that scientists can understand the procedure for obtaining permission to conduct research.

Although the Society recognized the great potential benefits of genetic engineering and new biotechnologies, it expressed concerns over the hazards of such organisms, the cost–benefit analyses and the effect of such genetic modifications on animal welfare. However, it felt that the same animal-welfare issues probably existed regardless of whether the approach was traditional or a new biotechnology. The Society recommended open, frank, public debate concerning GMO and transgenic issues.

Laws concerning genetic engineering are extensive in the UK (Royal Society (UK), 2001). There are nine Acts dating back as far as 1968 that affect work in or employment of transgenic applications and eight European Directives, not all ratified by the UK. Upwards of four UK agencies regulate these policies. All scientific research with animals in the UK must be licensed under the Animals (Scientific Procedures) Act 1986, analogous to the animal-welfare regulations in the USA. GM animals and fish must be demonstrated not to be likely to suffer pain and distress under this Act. Every laboratory conducting gene-transfer research in the UK must be registered with the Health and Safety Executive (HSE) under the Genetically Modified Organisms (Contained Use) Regulations 2000. Approximately nine committees review the application, and HSE inspectors ensure compliance with regulations. The research must not cause harm to humans. Environmental risk assessment of the research must be conducted.

Application of GMOs is regulated under the Environmental Protection Act 1990. Under Part VI of this Act, it is illegal to purposely or accidentally release GMOs into the environment without the consent of the Secretary of State. If GM animals are to be released outside a contained facility, an application of consent must describe the GMO and assess the risks to human health and safety and to the environment under the Genetically Modified Organisms (Deliberate Release) Regulations 1992, 1995, 1997, which are administered by the Department of the Environment, Transport and the Regions and reviewed by the nine committees. There have been no applications to release or market GM animals in the UK.

Application and approval for commercial use of GM animals or fish would also have to be made to the Food Standards Agency to ascertain the safety of transgenic animals for consumption, and, of course, no applications have been made. Lastly, insertion of transgenes should not cause adverse effects on animal welfare, and this is regulated by a variety of Acts dating back to 1912.

The UK stance on GM foods may be changing. The UK is leading efforts that would lead to wider use of GM products in Europe.

Nomenclature

One problem that hinders responsible development of legislation and government regulation is the lack of a common, understandable nomenclature for GMOs. This is further hampered by jargon and the development of new terms, and a new language called 'genetics' (Bartley, 1999; Kapuscinski *et al.*, 1999; Dunham *et al.*, 2001; FAO, 2001). Aquaculture geneticists at the Fourth Meeting of the International Association of Geneticists in Aquaculture refused to draft a technical definition, and the CBD has yet to develop a definition of LMO (D. Bartley, FAO, 2003, personal communication). Generally, international legal organizations and the industry restrict the definition of GMOs to transgenics, whereas some voluntary guidelines adopt a wider definition that includes additional genetic modifications, such as hybridization, chromosome manipulations, sex reversal and selective breeding.

The ICES defines a GMO as:

An organism in which the genetic material has been altered anthropogenically by

means of genes and cell technologies. Such technologies include isolation, characterization, and modification of genes and their introduction into living cells of viruses or DNA as well as techniques in the production involving cells of new combination of genetic material by infusion of two or more cells. (Bartley, 1999; Kapuchinski *et al.*, 1999; Dunham *et al.*, 2001; FAO, 2001)

USDA Performance Standards on conducting research on GMOs apply to deliberate gene changes, including changes in genes, transposable elements, non-coding DNA (including regulatory sequences), synthetic DNA sequences and mitochondrial DNA; deliberate chromosome manipulations, including manipulation of chromosome numbers and chromosome fragments; and deliberate interspecific hybridization (except for non-applicable species), referring to human-induced hybridization between taxonomically distinct species (Bartley, 1999; Kapuscinski *et al.*, 1999; Dunham *et al.*, 2001; FAO, 2001). The USDA defines non-applicable organisms as intraspecific selectively bred species and widespread and well-known interspecific hybrids that do not have adverse ecological effects.

The CBD (GMOs have become LMOs in the language of the CBD) definition of a 'living modified organism' is any living organism that possesses a novel combination of genetic material obtained through the use of modern biotechnology (Bartley, 1999; Kapuscinski *et al.*, 1999; Dunham *et al.*, 2001; FAO, 2001). 'Living organism' means any biological entity capable of transferring or replicating genetic material, including sterile organisms, viruses and viroids. 'Modern biotechnology' means the application of: (i) *in vitro* nucleic acid techniques, including recombinant DNA and direct injection of nucleic acid into cells or organelles; and (ii) fusion of cells beyond the taxonomic family that overcome natural physiological reproductive or recombination barriers and that are not techniques used in traditional breeding and selection.

The EU defines a GMO as 'an organism in which the genetic material has been altered in a way that does not occur naturally by mating and/or natural recombination'

(Bartley, 1999). 'Genetically modified micro-organisms are organisms in which genetic material has been purposely altered through genetic engineering in a way that does not occur naturally.'

It is obvious that the definition of GMO varies considerably, and the scope can vary from one user to another (Bartley, 1999; Kapuscinski *et al.*, 1999; Dunham *et al.*, 2001; FAO, 2001).

International Trade

The primary body governing international trade, the World Trade Organization (WTO), organized through the General Agreement on Tariffs and Trade (GATT), was created in 1995 following the Uruguay Round of trade negotiations and currently has 134 member countries (Bartley, 1999; Kapuscinski *et al.*, 1999; Dunham *et al.*, 2001; FAO, 2001). Its responsibilities include administration of international trade, resolving international trade disputes, promoting trade liberalization, which includes non-discrimination, and establishing conditions for stable, predictable and transparent trade. The Center for International Environmental Law (CIEL) has concluded that, with the 'interlocking relationships between trade and other issues, including environmental protection, WTO activities now have more extensive ramifications' (http://www.igc.apc.org/ciel/shmptur.html). WTO's objectives are to minimize and remove trade barriers, to promote international commerce and to establish guidelines for intellectual-property protection, patenting and labelling for aquatic transgenic organisms.

Not surprisingly, global cooperation on issues of biotechnology is not unified. Countries that are party to the CBD and involved in the WTO are divided on key issues, such as transport of transgenic organisms between countries, precautionary principles driving bisosafety decisions, liability in the case of negative effects on human health or biodiversity, possible social and economic impacts on rural cultures, regulation of transgenic products across borders, food safety, and protection of transgenic trade goods.

Some conflict exists within the WTO, and key issues concerning biotechnology deadlocked talks between trade ministers at the WTO meeting in Seattle (Bartley, 1999; Kapuscinski *et al.*, 1999; Dunham *et al.*, 2001; FAO, 2001). The USA wanted to create a WTO working group on GM goods with the goal of establishing rules that would protect trade in these goods, but Europe refused, citing that the safety of such products had not been proved (Kaiser and Burgess, 1999). Eventually a WTO group was formed to study international trade in GM foods (Pearlstein, 1999). The EU members of the WTO refused to accept transgenic crops from the USA. Later this was partially resolved, with the EU accepting imports if labelled as transgenic.

Countries involved in the WTO took the first steps in resolving some of their differences in 2001. International legislation, guidelines and codes of conduct have been and continue to be established to address these issues. The EU regulations have significant impact on international trade (Henard *et al.*, 2008). The WTO Dispute Settlement Body determined that the EU had violated Article 8 of the Sanitary and Phytosanitary Measures Agreement by instituting a *de facto* moratorium on the approval of biotech products in 2006. Ongoing conflict and negotiations continue between the European Commission and the USA regarding normalization of trade in GM products because of inconsistencies in European policies. Inconsistent worldwide policies affect availability and prices and can eliminate markets.

Intellectual-Property Rights

IPR concerning GM aquatic organisms are controversial (Bartley, 1999; Kapuscinski *et al.*, 1999; Dunham *et al.*, 2001; FAO 2001). Some feel that it is not ethical to patent life-forms and that such systems restrict access or make access too expensive for farmers in developing countries. The other opinion is that the research investment will not be made unless protection of IPR is maintained and that this protection and commercialization ultimately lead to greater access to genetically enhanced stocks in developing countries. There is also some debate about ownership when biological or genetic resources are removed from one country to another and what the cost of access should be for countries where the genetic resource originated.

The WTO initiated the first global system to establish guidelines for IPR for biological diversity, specifically referring to plants, to protect the inventors of products and to promote innovation (Bartley, 1999; Kapuscinski *et al.*, 1999; Dunham *et al.*, 2001; FAO, 2001). The Agreement on Trade-related Aspects of Intellectual Property Rights (TRIPS) requires members to form intellectual monopoly rights on, *inter alia*, certain food and living organisms. TRIPS Article 27(3)(b) allows the patenting of life-forms and mandates that systems for IPR be developed by 2000 in developing countries and by 2005 in the least developed countries. Article 27 of TRIPS indicates that patentable life-forms must meet the criteria of novelty, inventiveness (non-obvious) and industrial applicability. TRIPS has loopholes as TRIPS countries have discretion on whether or not to protect plants or animals with patents or a *sui generis* system and whether or not to recognize such patents.

DG XII of the EC has similar guidelines for IPR protection, except the EC states that 'plant and animal varieties and essentially biological processes for the production of plants or animals, including crossing or selection, are not patentable' nor are they biotechnological processes (Bartley, 1999; Kapuscinski *et al.*, 1999; Dunham *et al.*, 2001; FAO, 2001). Plant variety protection and the registration of pure breeds partly fill this void of protection in the USA. Genetic sterilization will eventually provide the ultimate IPR protection. Biotechnological inventions are more strongly protected in the EU because common rules for patent law are established in Directive 98/44/EC (http://europa.eu.int/).

Patenting and intellectual-property protection from various organizations and countries are complicated and in conflict (Bartley, 1999; Kapuscinski *et al.*, 1999; Dunham

et al., 2001). Some countries do not have patent laws, do not recognize international patent law, are difficult to work with or do not have enforcement. Countries like Vietnam, where there are no animal and plant IPR, are losing access to some genetic resources. The WTO and the USA allow patenting of living organisms, whereas the EC does not (http://www.uspto.gov/web/offices/pac/doc/general/what.htm), so trade is complicated and divisive. The FAO is also becoming involved in this arena as the FAO CGRFA is establishing and negotiating a Code of Conduct on Biotechnology, which has a component on IPR.

A conflict exists between those who want open access to the world's indigenous genetic resources and those who want to restrict ownership so that benefits can return to the developing countries where useful genes commonly originate (Charles, 2001a,b; Doyle, 2003). Multinational agrobiotechnology firms and some international agricultural development agencies desire the open access, while national governments, environmentalists and native-rights groups want restrictions. The CBD inspired the enactment in more than 50 nations of laws restricting the export of plant seeds and other local genetic materials (including, in some instances, human genes) to restrict exploitation of developing countries by multinationals; however, one negative outcome is that non-profit research and gene banking, which greatly benefit the development of agriculture in poor countries, are stopping (Charles, 2001a; Doyle, 2003). The legal walls that Third World governments are building around seed banks force farmers and researchers to reduce their options and restrict their access to diversity, and it is as irresponsible as the exploitation of these genetics resources (Charles, 2001a,b). For example, a strain of drought-resistant maize was developed in Kenya at the International Centre for the Improvement of Maize and Wheat and will be distributed free to farmers in southern Africa (Charles, 2001a). This could not have happened without free access to seed from local maize landraces in Latin America, for which access is now impossible. Genetic resources now flow mainly from 'North' to 'South' rather than vice versa. Recently, for every single seed sample that developing nations sent to international gene banks, those gene banks sent about 60 samples back (Charles, 2001a; Doyle, 2003). Farmers in poor nations now depend on seeds held by gene banks located in or funded by developed nations and, if poor nations create a world in which they have to bargain for access to the genetic resources in these banks, they might lose (Charles, 2001a; Doyle, 2003). Obviously, this has implications for the future of the development, sharing and utilization of genetic resources from aquatic organisms.

Government Approval of Transgenic Fish/Animals/Plants

The governments of China, Cuba and the USA are considering the approval of GH-transgenic fish for human consumption. All of these applications are pending. The USA has approved certain transgenic animals for rendering (FDA, 2006).

With regard to transgenic crops (plants), the following countries grow them: Australia, Canada, Czech Republic, Germany, Portugal, Slovakia, Spain and the USA. However, the following countries (states, provinces) ban transgenic crops: Austria, Bulgaria, France, Greece, Hungary, Ireland, Italy, Japan, Luxemburg, New Zealand, Poland, South Australia and Tasmania.

Mendocino, Trinity and Marin counties, in California, banned the production of GMOs and clones. These bans are not based on science, and the wording of the laws contains blatant scientific errors. In fact, if the letter of the law was followed all grapes in Napa Valley would need to be destroyed as indeed grape varieties are clones.

Although Europe does not grow many transgenic crops, it is a large consumer of transgenic soybeans. Biotechnology research in Europe is declining because of political pressure.

24
Strategies for Genetic Conservation, Gene Banking and Maintaining Genetic Quality

Gene banking or genetic conservation may be accomplished by protecting native populations (Fig. 24.1), by artificially propagating a variety of genotypes, lines and strains and/or through the use of cryopreservation. Cryopreservation of sperm from several fish species has been accomplished and is fairly routine; however, the technology for the cryopreservation of eggs and embryos of aquatic species does not yet exist. An alternative would be to cryopreserve cells from the blastula stage of the embryo. After thawing, these cells can be placed into a donor embryo of the same species. If the donor is triploidized, the only viable cells in the gonads will be from the cryopreserved blastulas and, reproductively, the entire genome of an individual has been preserved and banked, including the mitochondria and the cytoplasm. Of course, in the case of cryopreserved sperm, even if the individual is regenerated via androgenesis, the cytoplasm and the mtDNA from the original fish are lost, so the exact total genome and its potential interactions with the cytoplasm have been lost.

Currently, long-term storage of fish eggs or embryos is not possible, severely limiting the effectiveness of cryopreservation for gene banking. However, androgenesis is a potential mechanism to recover diploid genotypes from cryopreserved sperm, which is, of course, a viable technology. There are major drawbacks to this strategy since androgens have low viability, all individuals would be 100% homozygous and the mtDNA portion of the genome would be different from the original cryopreserved individuals unless donor eggs came from exactly the same family or genotype.

Cloning of individuals from somatic cells is another option for regenerating individuals. The problem is that, in the case of cloning, although it would theoretically be superior to the androgenesis approach because androgens are 100% homozygous and clones would preserve the genetic variation from every locus of the individual, clones, like the androgens, have the cytoplasm and mtDNA of the recipient cell and not their own. Cloning has become a reality.

Another potential method of conserving a species may be the isolation of SSCs, if they can be adequately frozen, and transferring these cells into other species, which have been made experimentally, sterilized and depleted of cells in their testis. The SSCs then colonize in the host male and differentiate into sperm cells, which may then be used for fertilization. Stem cells have been cryopreserved and revived and they behave similarly to those of non-cryopreserved controls. The xenogenesis strategy discussed in Chapter 10 utilizes stem-cell technology. When the SSCs are transferred

Fig. 24.1. Giant Mekong catfish, *Pangasius gigas*, in Thailand. An example of a potential species for genetic conservation.

to a triploid female embryo host, ovaries and ova from the donor will develop. Thus, lines, strains, species can be regenerated exactly with progeny produced that have the donor nDNA, mtDNA and cytoplasm, a huge advantage compared with cryopreservation of sperm. In fact, if we have enough foresight to create testes cryobanks, if a species or genetic line becomes extinct, it can be resurrected from extinction. Similarly, ovarian stem cells could be transplanted to regenerate ovarian tissues, and theoretically this would not have the disadvantage of loss of original mtDNA genotype and cytoplasm.

Another potential technique to gene-bank individuals or transfer genes might be the transfer of genes into cultured blastula cells and then reinserting transgenic or normal blastula cells into embryos via nuclear transplantation. Alternatively, diploid blastulas, transgenic or normal, could be inserted into triploids and, if they became part of the germ line, a fertile individual would be regenerated. Chen *et al.* (1986) were able to transplant 59th-generation cultured blastula cells into crucian carp. Nuclear-transplant gastrulae were obtained at a rate of 7.9%. Cells from these gastrulae were then serially transplanted into additional enucleated crucian carp cells and one adult fish was obtained that lived for 3 years. Unfortunately, this fish was aneuploid, illustrating some of the problems of nuclear transplantation with cultured cells. The gonads of this fish did not develop.

The same researchers also transferred cultured kidney cells in a similar manner and produced a fertile female; however, they do not provide sufficient information to confirm that the injected nuclei directed development. Gasaryan *et al.* (1979) irradiated and then transferred blastulae of loach, *Misgurnus fossilis*, donors, and the resulting nuclei were 1N, 2N, 3N and 4N. Nuclear transplantation is somewhat problematic.

Tufto (2001) addresses the possibility of re-establishing an extinct population using animals bred in captivity, considering that those animals may have strayed from the local fitness optimum. The supplementation of wild stocks by release of domesticated, gene-banked animals could damage the fitness of the remnant wild population. The deterministic model he developed had complex results, dependent on selection intensity, immigration rate, recombination rate, the presence or absence of competitors, the basal intrinsic rate of increase and the carrying capacity (Doyle, 2003). When immigration and selection were both small, the population was reduced below carrying capacity only if the immigrating maladapted animals were far from the optimum, indicating that it is important to prevent as much domestication in the gene bank as possible. When selection is strong but density dependence is weak, as it might be in the early stages of reintroduction, it may be feasible to use maladapted individuals to initiate a population (Doyle, 2003). If the ongoing supplementation (immigration) continues at a low, constant rate, the new population will adapt sufficiently and quickly reach a stable population equilibrium (Tufto, 2001).

25
Ethics

This chapter is a section from FAO/WHO (2003f) that is reprinted with the permission of FAO/WHO. Matthias Kaiser was the lead author of this section of FAO/WHO (2003f).

The production of GM animals raises a variety of legitimate ethical concerns. The need to address these concerns openly and without prejudice is further amplified by sections of a critical public that question some of the achievements of modern biotechnology. Responsible decision making and policy needs to integrate ethics among the salient factors in the preparatory stages of risk analysis.

Ethics is rooted both in the world religions and in secular philosophies, while a sense of morality and moral value is common for everybody. Ethics comprises both a positive dimension relating to our conceptions of the good life/society and a negative dimension relating to our judgements of what is morally wrong. For instance, owing to religion-based food practices relating to eating pork, the utilization of genetic material from pigs could present problems to some cultures and religions.

Environmental Ethics and Animal Welfare

In large parts of mainstream Western ethics, the objects of our moral concerns have been human beings. With the advance of our knowledge, many people have come to realize that there is ample reason to extend the realm of moral concern to animals and perhaps even to ecosystems. This has led to discussions on the moral status of animals, as well as to increased attention being given to animal-welfare issues, also related to transgenic animals. For instance, one set of early experiments with growth-enhanced salmon showed some individuals with cranial deformities (Devlin *et al.*, 1995a). As a general, but not unexceptional, rule, the intended use of GM animals for food production bespeaks the interest of the producer to ensure or improve the health of animals and good animal welfare. Therefore, aspects of animal welfare of transgenic animals need to be evaluated on a case-by-case basis by competent bodies.

Uncertainty

Ethically responsible decision making demands, *inter alia*, both the utilization of the best available knowledge and an awareness of the relevant uncertainties involved. While it is widely acknowledged that good risk assessment demands a measure of uncertainty, the common instruments to make these uncertainties visible are still limited. However, research on this topic has made significant progress during the last decade, and valuable and useful instruments to represent the relevant uncertainties are now available (Walker *et al.*, 2003). This is ethically significant because, in some cases, the uncertainties relate to our state of knowledge (thus often indicating the need for more research) but, in other cases, the uncertainties relate to inherent characteristics of the system under study, such as chaos or complexity, or multiple states of equilibrium without linear and deterministic state change. In the latter cases of inherently limited predictability, we have to adopt a responsible scheme for the management of these uncertainties. This point is important both for the use of precaution in risk management and for the provision and presentation of scientific findings as the basis for such management. Precaution does not assume an unrealistic notion of zero risk. Sometimes the best precautionary action is carefully controlled, monitored and stepwise development. There is a need to address explicitly the uncertainties involved in our assessments, and to adopt schemes for their responsible management.

Transparency and Public Deliberation

The recognition of consumer autonomy and the right to free and informed market choices is an aspect of responsible management. Another aspect relates to the worry in sections of the public that the new genetic technologies may not be used for the 'right' or ethically justified ends. The distribution of risks and benefits may be morally problematic, and the scope of benefits may be wanting. At present, even the lack of data on such distributions is a concern.

Similar concerns relate to the technological divide and the unbalanced distribution of benefits and risks between developed and developing countries. Often the problem becomes even more acute through the existence of IPR and patenting that places an advantage on the strongholds of scientific and technological expertise. Equity and fairness issues are thus obviously important.

All these considerations point towards the positive dimension of ethics, i.e. a discussion of purpose, benefits and risks. There is a societal need to address these issues upfront, and in the early stages of development (Sagar *et al.*, 2000; Kapuscinski *et al.*, 2003). Proactive assessment is indicated, and the need for scientific data to inform such assessments should be described. Risk managers and decision makers should shoulder this task in collaboration with stakeholders.

The Role of Ethical Principles in Assessments

In relation to human health and medicine there is already a tradition of carrying out practical ethical assessments. Four principles have been established as fundamental in the biomedical field: respect for autonomy, beneficence, non-maleficence and justice (Beauchamp and Childress, 2001). These principles seem to be widely accepted, represent important ethical theories and cover most of the problems appearing in the biomedical field.

Within the field of GM animals a similar framework needs to be introduced if ethics is indeed to become an integral part of regulation and guided policy advice. Extensions of the principle approach in biomedicine to other technological and environmental issues have been carried out, for example, in the ethical matrix approach (Mepham, 1996; Kaiser and Forsberg, 2000; Schroeder and Palmer, 2003). The basic idea in this framework is to combine the use of a variety of principles with the interest-related perspectives of the various stakeholders and other potentially affected organisms and their environment. The rationale of these frameworks and approaches is to make ethical assessment more transparent and more methodical, and thus amenable to quality assurance.

Table 25.1.   Simplified ethical matrix (for illustrative purposes only – italicized items in the matrix also relate directly to the scientific description of safety and benefit assessment of genetically modified animals).

Ethical matrix for GM fish	Welfare as eliminating negative utilities	Welfare as promoting positive utilities	Dignity/autonomy	Justice/fairness
Small producers	Dependence on nature and corporations	Adequate income and work security	Freedom to adopt or not to adopt	Fair treatment in trade
Consumers	*Safe food*	*Nutritional quality*	Respect for consumer choice (labelling)	General affordability of food product
Treated fish	*Proper animal welfare*	*Improved disease resistance*	Behavioural freedom	Respect for natural capacities (telos)
Biota	*Pollution and strain on natural resources*	*Increasing sustainability*	Maintenance of biodiversity	No additional strain on regional natural resources

A Schematic Ethical Assessment

Assuming that we want to assess the ethical aspects of a certain genetic modification of a fish species for food production in a region, following the ethical matrix approach we would first address the issue of who the relevant stakeholders are. We also need to agree upon potentially affected organisms and their components of the environment, for example fish and other biota. A proper set of ethical principles then needs to be established, for instance justice/fairness, dignity/autonomy and welfare considerations as comprising both the elimination of negative welfare and the increase of positive welfare. Once a common understanding of these principles is ensured, it is important to specify the principles for each interest perspective.

Through the presentation of an example ethical matrix (see Table 25.1) it becomes clear that some of the items (italicized entries in Table 25.1) relate directly to the scientific description in the safety and benefit assessments of GM animals. Thus there is an overlap between the ethical assessment and the risk assessment and management. A description of specific consequences of the new technology as well as uncertainties in our knowledge is now possible within the matrix. This enables a broader evaluation of the issues.

26
Constraints and Limitations of Genetic Biotechnology

During the International Conference on Aquaculture in the Third Millennium (2000), a genetics working group convened to define the issues, constraints and limitations facing aquaculture genetics in the near future. Working-group members included Rex A. Dunham (chair), Kshitish Majumdar (co-chair), Zhanjiang (John) Liu, Eric Hallerman, Gideon Hulata, Trgve Gjedrem, Peter Rothlisberg, David Penman, Nuanmanee Pongthana, Ambekar Eknath, Modadugu Gupta, Graham Mair, P.V.G.K. Reddy, Janos Bakos, S. Ayyappan, Devin Bartley and Gabriele Hoerstgen-Schwark. The following is adapted from Dunham *et al.* (2001) and this was the output from this working group regarding these issues. The progress over the past 10 years since this conference in addressing these issues is also discussed.

Several constraints and limitations will need to be overcome for aquaculture genetics to have its maximum impact and benefit in the coming years. These include environmental issues, such as biodiversity, genetic conservation and the environmental risk of genetically altered aquatic organisms; research issues, such as funding, training of scientists and impact assessment; economic issues, such as proprietary rights, dissemination, food safety and consumer perceptions; and political issues, such as government regulation and global cooperation.

Research Issues

Manpower – the lack of trained traditional and molecular aquaculture geneticists – is a potential constraint that needs to be addressed. Impact assessment is another void that needs more intense consideration. This is needed to ensure that the research and germplasm developed through research are appropriate for the commercial sector and are applied properly and disseminated properly to achieve maximum impact. Research on impact assessment should be influential in the design of the most effective breeding programmes, extension development and dissemination strategy.

In the past 10 years, training of graduate students in molecular genetics has been intense and this void is being adequately addressed. Partially due to the popularity, perception and power of genomic technologies and lack of funding, student interest in the area of traditional genetics and breeding has lagged far behind. Now that genomics is maturing and a tremendous amount of data has been generated, practical application of genomics is needed for real impact in aquaculture. Scientists trained in both genomics and traditional quantitative genetics are needed to bridge the gap between these two related disciplines and to accelerate practical utilizations of genomics knowledge.

In the USA, large-scale retirement of senior scientists in the agricultural and aquacultural sciences as well as fisheries science and fisheries genetics will occur in the near future. There is a great need to train young scientists to fill this coming void.

General Recommendations

The genetic improvement of cultured fish and shellfish should be given higher priority by government, non-government and commercial organizations. This will increase productivity and turnover rate, result in better utilization of resources and reduce production costs. Multiple-trait selection programmes need to be further developed. Efficient breeding plans should be developed where selection is combined with other genetic technologies. Better genetic controls need to be developed for monitoring the progress of breeding programmes. More education and training programmes are needed for the further development of aquaculture geneticists, especially in developing countries. The establishment of national and international genetic controls, including homozygous and heterozygous clonal populations for some key species, would help in comparing genetic results and germplasm from different research institutions, increase global cooperation and enhance research efficiency. Domestication of some wild-cultured organisms, such as shrimp, is needed.

Ten years later, the situation is basically the same except the training of scientists in developing countries has gained momentum and is making great progress. However, Africa is lagging behind in this regard, and there is a great need on that continent for additional expertise in aquaculture and fish genetics. Good progress has been made in domesticating shrimp and other species.

Development Issues

There is a greater need for intervention and collaboration in developing hatchery management and breeding programmes for low-value/low-input species in developing countries. Good brood-stock management needs to be promoted to counter the negative genetic impacts of inbreeding. Species and traits relevant to low-input systems need to be prioritized for genetic enhancement programmes that better address the food security issue.

Networking (e.g. the International Network on Genetics in Aquaculture (INGA)) as a way to circumvent the lack of resources (human and infrastructure) in developing countries should be strengthened to coordinate sharing of information, expert opinion, education and research, and to assist in obtaining funding. Efforts should be made to design and promote equitable dissemination strategies to ensure that genetic enhancements have positive impacts on aquaculture and food security and enhance livelihoods. Research should be carried out to assess the impact of research, development and dissemination of genetically improved stocks, including IPR issues. Species and traits need to be prioritized for genetic enhancement programmes that better address global food security.

During the past 10 years, there has been progress in genetic improvement of low-input fishes in Asia, but not much progress in other developing countries. Not much progress has been made on IPR issues. The developmental issues still need to be more strongly addressed.

Biodiversity Issues

Aquatic biodiversity needs to be characterized and protected. The population genetics of many key species requires closer examination. Interactions of wild and domesticated species needs much closer study, including modelling. There should be an intensification of live, frozen and molecular gene-banking efforts. More research is needed in the area of effective sterilization techniques for domesticated and transgenic aquatic organisms. There is a need for greater controls of transboundary movements of aquatic germplasm.

Research on transgenic aquatic organisms should continue because of their potential benefits (especially in developing countries); however, much greater understanding of potential environmental impacts is necessary. Linkages should be formed among civil society, organizations, scientists, industry and governments to address genetic issues and to support the development of practical regulations and sound policy. Dissemination of transgenic aquatic organisms for aquaculture should only be carried out within the framework of adequate regulations and policy.

These are still needs. With the exception of the GloFish®, transgenic zebra fish used in the aquaria trade, transgenic fish are yet to be approved or disseminated. Progress is gradually being made with respect to protecting genetic biodiversity, but again much more could be done. Molecular gene banking, of course, has been a major area of emphasis and tremendous progress has been made in this area. Germplasm preservation and whole-organism gene banking is lagging behind what should be done to protect the future of genetic biodiversity of aquatic organisms.

Political Issues

Worldwide, policies for research and marketing of transgenic food organisms range from non-existent to stringent, as in the EU. Government regulation of transgenic aquacultured species, based on sound scientific data, is lacking and much needed. Not surprisingly, global cooperation on issues of biotechnology is not unified. Countries party to the CBD and involved in the WTO are divided on key issues, such as transport of transgenic organisms between countries, precautionary principles driving biosafety decisions, liability in the case of negative effects on human health or biodiversity, possible social and economic impacts on rural cultures, regulation of transgenic products across borders, food safety and protection of transgenic trade goods. International legislation, guidelines and codes of conduct have been, or are being, established to help address these areas of concern. These statements are still true today.

Economic Issues

Among the key issues are economic ones, particularly those revolving around proprietary-rights issues. There are many aspects to this, including those related to biodiversity and molecular genetics, as organisms found in almost any country have genes that are potentially valuable to another organism in a different country. Ownership in cases of international germplasm transfer is an issue. A genetics research and breeding programme requires financial support. Appropriate, equitable dissemination and ownership of germplasm developed with tax monies or donor funding, with the goal of having a positive impact on the impoverished in developing countries, is a complex and controversial topic. This is an increasingly difficult problem as, with the initiation of private breeding companies and biotechnology companies, alternative options to government dissemination, with both impact and income generation for research opportunities, exist. The most cost-efficient dissemination strategies with the highest impact have not been completely defined and evaluated. These issues are still with us in 2011.

Glossary

***ab initio* gene discovery** A method for identifying likely genes in a stretch of genomic sequence that does not depend on prior information such as similarity to a gene from another species or identity to another transcript. Most *ab initio* approaches use a hidden Markov model to search for sequence motifs that are commonly found in genes, such as long open reading frames, intron–exon boundary signatures and conserved upstream regulatory motifs.

acentric chromosome A chromosome with no centromere.

acentric fragment Generated by breakage of a chromosome, lacks a centromere and is lost at cell division.

acridine A chemical mutagen that intercalates between the bases of a DNA molecule, causing single-base insertions or deletions.

acrocentric chromosome A chromosome with the centromere near one end.

adaptation Any characteristic of an organism that improves its chance of survival and reproduction in its environment; the evolutionary process by which a species undergoes progressive modification favouring its survival and reproduction in a given environment.

addition rule The principle that the probability that any one of a set of mutually exclusive events is realized equals the sum of the probabilities of the separate events.

additive genetic variance Genetic or hereditary variance dependent on additive gene effects, i.e. a gene has a given plus or minus effect regardless of which other member of the pair or allelic series may be present.

adenine (A) A nitrogenous purine base found in DNA and RNA.

adjacent segregation Type of segregation from a heterozygous reciprocal translocation in which a structurally normal chromosome segregates with a translocated chromosome. In adjacent-1 segregation, homologous centromeres go to opposite poles of the first-division spindle; in adjacent-2 segregation, homologous centromeres go to the same pole of the first-division spindle.

alignment The process of lining up two or more DNA or protein sequences so as to maximize the number of identical nucleotides or residues while minimizing the number of mismatches and gaps.

allele frequency The relative proportion of all alleles of a gene that are of a designated type.

allele/allelomorph One of two or more alternative forms of a gene occupying a given locus on a chromosome; member of a pair (or series) of different hereditary factors that may occupy a given locus on a specific chromosome and that segregate in the formation of gametes.

allopolyploid A polyploid formed by hybridization between two different species.

allozyme Any of the alternative electrophoretic forms of a protein coded by different alleles of a single gene.

alpha satellite Highly repetitive DNA sequences associated with mammalian centromeres.

alternate segregation Segregation from a heterozygous reciprocal translocation in which both parts of the reciprocal translocation separate from both non-translocated chromosomes in the first meiotic division.

alternative splicing The combination of different sets of exons to produce two or more primary mature messenger RNAs from the same primary transcript. Commonly observed in higher eukaryotes, with the result that a single gene can generate multiple protein isoforms.

amber codon The nucleotide triplet UAG, one of three codons that cause termination of protein synthesis.

amber mutation Any change in DNA that creates an amber codon at a site previously occupied by a codon representing an amino acid in a protein.

amber suppressors Mutant genes that code for transfer RNAs (tRNAs) whose anticodons have been altered so that they can respond to UAG codons as well as or instead of their previous codons.

amino acid Any one of a class of organic molecules that have an amino group and a carboxyl group; 20 different amino acids are the usual components of proteins.

aminoacylated transfer RNA (tRNA) A tRNA covalently attached to its amino acid; charged tRNA; tRNA carrying an amino acid; the covalent linkage is between the NHQ group of the amino acid and either the 3′- or the 2′-OH group of the terminal base of the tRNA.

aminoacyl-tRNA synthetases Enzymes responsible for covalently linking amino acids to the 2′- or 3′-OH position of transfer RNA (tRNA).

amino terminus The end of a polypeptide chain at which the amino acid bears a free amino group ($-NH_2$).

amplification of genes The production of additional copies of a chromosomal sequence, found as intrachromosomal or extrachromosomal DNA.

amplified fragment length polymorphism (AFLP) markers Genomic DNA is digested with two restriction enzymes, *Eco*RI and *Mse*I, suitable adaptors are ligated to the fragments and the ligated DNA fragments are selectively amplified with different primer combinations; then the products are resolved by gel electrophoresis. AFLPs are highly polymorphic, and the molecular bases of AFLP polymorphism are base substitutions at the restriction sites, insertion or deletion between the two restriction sites, base substitution at the preselection and selection bases, and chromosomal rearrangements. Bands are inherited in a dominant fashion.

anaphase The stage of mitosis or meiosis in which chromosomes move to opposite ends of the spindle. In anaphase I of meiosis, homologous centromeres separate; in anaphase II, sister centromeres separate.

ancestor Animal of a previous generation that has passed on genes through a line of descent.

androgen An individual whose nuclear DNA is all paternally inherited. Also a term for hormones that stimulate activity of accessory sex organs and sexual characteristics in males. Testosterone is one of these hormones. They are often termed male sex hormones.

aneuploid A cell or organism in which the chromosome number is not an exact multiple of the haploid number; more generally, aneuploidy refers to a condition in which particular genes or chromosomal regions are present in extra or fewer copies compared with the wild type; the constitution differs from the usual diploid constitution by loss or duplication of chromosomes or chromosomal segments.

annotation Linking a gene's sequence to genetic data about function, expression, mutant phenotypes of the protein associated with that locus, as well as to comparative data from homologous proteins in other species. In the context of genome sequencing, annotation refers to the identification of putative genes using a combination of *ab initio* methods, homology searches, and physical evidence.

antibody A blood protein produced in response to a specific antigen and capable of binding with the antigen.

anticoding strand A template of duplex DNA used to direct the synthesis of RNA that is complementary to it.

anticodon The three bases in a transfer RNA (tRNA) molecule that are complementary to the three-base codon in mRNA.

antigen A substance able to stimulate the production of antibodies.

antiparallel The chemical orientation of the two strands of a double-stranded nucleic acid molecule; the 5′-to-3′ orientations of the two strands run in opposite directions.

antitermination proteins Allow RNA polymerase to transcribe through certain terminator sites.

AP endonuclease An endonuclease (nuclease enzyme) that cleaves a DNA strand at any site at which the deoxyribose lacks a base; make incisions in DNA on the 5′ side of either apurinic or apyrimidinic sites.

apoinducer Protein that binds to DNA to switch on transcription by RNA polymerase.

apoptosis Genetically programmed cell death, especially in embryonic development.

aporepressor Protein converted into a transcriptional repressor by binding with a particular molecule.

atavism Reappearance of an ancestral trait or character after a skip of one or more generations; also referred to as reversion; usually results from recessive genes being present in the homozygous condition in an individual after having been masked in ancestors by their dominant alleles.

attached X chromosome Chromosome in which two X chromosomes are joined to a common centromere; also called a compound X chromosome.

attenuation/attenuator Regulatory base sequence near the beginning of an mRNA molecule at which transcription can be terminated; when an attenuator is present, it precedes the coding sequences.

autonomous determination Cellular differentiation determined intrinsically and not dependent on external signals or interactions with other cells.

autopolyploid organism Organism whose cells contain more than two basic sets of homologous chromosomes; intraspecific polyploid.

autoregulation Regulation of gene expression by the product of the gene itself.

autosomes Chromosome pairs which are alike in both sexes; all chromosomes other than the sex chromosomes; a diploid cell has two copies of each autosome.

backcross Cross of an F_1 heterozygote with a partner that has the same genotype as one of its parents.

bacterial artificial chromosome (BAC) Plasmid vector with regions derived from the F plasmid, which contains a large fragment of cloned DNA.

balancer chromosomes Chromosomes that have been engineered to carry multiple inversions that suppress crossing over, and so can be used to maintain recessive mutations in genetic stock. Balancers usually also carry a recessive lethal and a dominant visible genetic marker.

bands of polytene chromosomes Visible as condensed regions that contain the majority of DNA; bands of normal mitotic chromosomes are relatively much larger and are generated in the form of regions that retain a stain on certain chemical treatments.

Barr body A darkly staining body found in the interphase nucleus of certain cells of female mammals; consists of the condensed, inactivated X chromosome.

base Single-ring (pyrimidine) or double-ring (purine) component of a nucleic acid.

base analogue Chemical so similar to one of the normal bases that it can be incorporated into DNA.

base pair Pair of nitrogenous bases, most commonly one purine and one pyrimidine, held together by hydrogen bonds in a double-stranded region of a nucleic acid molecule; commonly abbreviated bp, the term is often used interchangeably with the term nucleotide pair. The normal base pairs in DNA are A–T and G–C.

base-substitution mutation Incorporation of an incorrect base into a DNA duplex.

biochemical pathway Diagram showing the order in which intermediate molecules are produced in the synthesis or degradation of a metabolite in a cell.

bivalent A pair of homologous chromosomes, each consisting of two chromatids, associated in meiosis I; this structure contains all four chromatids (two representing each homologous chromosome) at the start of meiosis.

blastoderm Structure formed in the early development of an insect larva; the syncytial blastoderm is formed from repeated cleavage of the zygote nucleus without cytoplasmic division; the cellular blastoderm is formed by migration of the nuclei to the surface and their inclusion in separate cell membrane.

blastula A hollow sphere of cells formed early in animal development.

block in a biochemical pathway Stoppage in a reaction sequence due to a defective or missing enzyme.

blocked reading frame Cannot be translated into protein because it is interrupted by termination codons.

blunt ends Ends of a DNA molecule in which all terminal bases are paired; the term usually refers to termini formed by a restriction enzyme that does not produce single-stranded ends.

blunt-end ligation A reaction that joins two DNA duplex molecules directly at their ends.

breakage and reunion The mode of genetic recombination in which two DNA duplex molecules are broken at corresponding points and then rejoined crosswise (involving formation of a length of heteroduplex DNA around the site of joining).

breed Group of animals having a common origin and identifying characters that distinguish them as belonging to a breeding group.

breeding value Genetic worth of an animal's genotype for a specific trait.

broad-sense heritability The ratio of genotypic variance to total phenotypic variance.

C banding A technique for generating stained regions around centromeres.

C value The total amount of DNA in a haploid.

cAMP–CRP complex A regulatory complex in prokaryotes consisting of cyclic AMP (cAMP) and the CRP protein; the complex is needed for transcription of certain operons. cAMP–CRP activates transcription by binding to specific sites on DNA, where it interacts directly with RNA polymerase. This process is mediated by the CRP protein. (CRP stands for 'catabolite regulation protein'. This protein is also sometimes called CAP for 'catabolite activator protein'.)

candidate gene A gene proposed to be involved in the genetic determination of a trait because of the role of the gene product in the cell or organism.

cap A complex structure at the 5′ termini of most eukaryotic mRNA molecules, having a 5′–5′ linkage instead of the usual 3′–5′ linkage; the structure at the 5′ end of eukaryotic mRNA, introduced after transcription by linking the terminal phosphate of 5′-guanosine triphosphate (GTP) to the terminal base of the mRNA. The added G (and sometimes some other bases) is methylated, giving a structure of the form 7MeG5′ppp5′Np.

carboxyl terminus The end of a polypeptide chain at which the amino acid has a free carboxyl group (–COOH).

CARG box Part of a conserved sequence located upstream of the start-points of eukaryotic transcription units; it is recognized by a large group of transcription factors.

carrier A heterozygote for a recessive allele.

cDNA clone A duplex DNA sequence representing an RNA, carried in a cloning vector.

cell cycle The growth cycle of a cell; in eukaryotes, it is subdivided into Gl (gap 1), S (DNA synthesis), G2 (gap 2) and M (mitosis).

cell fate The pathway of differentiation that a cell normally undergoes.

cell lineage The ancestor–descendant relationships of a group of cells in development.

cellular oncogene (proto-oncogene) A gene coding for a cellular growth factor whose abnormal expression predisposes to malignancy.

centimorgan (cM) A unit of distance in the genetic map equal to 1% recombination; also called a map unit; the standard unit of genetic distance – an expression of the percentage of progeny in which a recombination event has occurred between two markers.

central dogma The concept that genetic information is transferred from the nucleotide sequence in DNA to the nucleotide sequence in an RNA transcript to the amino acid sequence of a polypeptide chain (now proved to have exceptions – see reverse transcriptase).

centrioles Small hollow cylinders consisting of microtubules which become located near the poles during mitosis. They reside within the centrosomes.

centromere The region of the chromosome that is associated with spindle fibres and that participates in normal chromosome movement during mitosis and meiosis; a constricted region of a chromosome that includes the site of attachment to the mitotic or meiotic spindle (*see also* kinetochore).

centrosomes The regions from which microtubules are organized at the poles of a mitotic cell. In animal cells, each centrosome contains a pair of centrioles surrounded by a dense amorphous region to which the microtubules attach.

chain elongation The process of addition of successive amino acids to the growing end of a polypeptide chain.

chain initiation The process by which polypeptide synthesis is begun.

chain termination The process of ending polypeptide synthesis and releasing the polypeptide from the ribosome; a chain-termination mutation creates a new stop codon, resulting in premature termination of synthesis of the polypeptide chain.

chain termination sequencing The most commonly used method for sequencing clones of DNA up to one kilobase in length, based on a method first devised by Fred Sanger, in which molecules of all possible lengths are produced by random termination of DNA polymerization when a dideoxynucleotide (ddNTRP) is incorporated.

charged tRNA A transfer RNA molecule to which an amino acid is linked; acylated tRNA.

chiasma (*pl.* chiasmata) The cytological manifestation of crossing over; the cross-shaped exchange configuration between non-sister chromatids of homologous chromosomes that is visible in prophase I of meiosis; a site at which two homologous chromosomes appear to have exchanged material during meiosis.

chimeric gene A gene produced by recombination, or by genetic engineering, that is a mosaic of DNA sequences from two or more different genes.

chromatid Either of the linear subunits produced by chromosome replication; the copies of a chromosome produced by replication. The name is usually used to describe them in the period before they separate at the subsequent cell division.

chromatid interference In meiosis, the effect that crossing over between one pair of non-sister chromatids may have on the probability that a second crossing over in the same chromosome will involve the same or different chromatids; chromatid interference does not generally occur.

chromatin The aggregate of DNA and histone proteins that makes up a eukaryotic chromosome; complex of DNA and protein in the nucleus of the interphase cell. Individual chromosomes cannot be readily distinguished in it except during mitosis or meiosois. It was originally recognized by its reaction with stains specific for DNA.

chromocentre An aggregate of heterochromatin from different chromosomes.

chromomere A tightly coiled, bead-like region of a chromosome most readily seen during the pachytene substage of meiosis; the beads are in register in a polytene chromosome, resulting in the banded appearance of the chromosome; densely staining granules visible in chromosomes under certain conditions, especially early in meiosis, when a chromosome may appear to consist of a series of chromomeres.

chromosome complement The set of chromosomes in a cell or organism.

chromosome map A diagram showing the locations and relative spacing of genes along a chromosome.

chromosome painting Use of differentially labelled, chromosome-specific DNA strands for hybridization with chromosomes to label each chromosome or area of a chromosome with a different colour.

chromosome theory of heredity The theory that chromosomes are the cellular objects that contain the genes.

chromosome walking The sequential isolation of clones carrying overlapping sequences of DNA, allowing large regions of the chromosome to be spanned. Walking is often performed in order to reach a particular locus of interest.

chromosomes Darkly staining bodies in cell nuclei which carry the hereditary material. They occur in pairs in somatic cells, with the number of pairs being characteristic of the species. In eukaryotes, a DNA molecule that contains genes in linear order to which numerous proteins are bound and that has a telomere at each end and a centromere; in prokaryotes, the DNA is associated with fewer proteins, lacks telomeres and a centromere, and is often circular; in viruses, the chromosome is DNA or RNA, single-stranded or double-stranded, linear or circular, and often free of bound proteins; discrete unit of the genome carrying many genes. Each chromosome consists of a very long molecule of duplex DNA and an approximately equal mass of proteins. It is visible as a morphological entity only during cell division.

***cis*-acting locus** Affects the activity only of DNA sequences on its own molecule of DNA; this property usually implies that the locus does not code for protein.

***cis*-acting protein** Has the exceptional property of acting only on the molecule of DNA from which it was expressed.

***cis* configuration** Two sites on the same molecule of DNA. The arrangement of linked genes in a double heterozygote in which both mutations are present in the same chromosome – for example, b1b2/++; also called coupling.

***cis*-dominant** Of or pertaining to a mutation that affects the expression of only those genes on the same DNA molecule.

***cis/trans* test** Assays the effect of relative configuration on expression of two mutations. In a double heterozygote, two mutations in the same gene show mutant phenotype in *trans* configuration, wild type in *cis* configuration.

cistron A DNA sequence specifying a single genetic function as defined by a complementation test; a nucleotide sequence coding for a single polypeptide. The genetic unit defined by the *cis/trans* test; equivalent to gene in comprising a unit of DNA representing a protein.

cleavage division Mitosis in the early embryo.

clone Individual genetically identical to another; a collection of organisms derived from a single parent and, except for new mutations, genetically identical to that parent. In genetic engineering, also a segment of recombinant DNA and its vector after insertion in a vector, commonly a plasmid; the linking of a specific gene or DNA fragment to a replicable DNA molecule, such as a plasmid or phage DNA.

cloned DNA sequence A DNA fragment inserted into a vector and transfected into a host organism.

cloned gene A DNA sequence incorporated into a vector molecule capable of replication in the same or a different organism.

cloning The process of producing cloned genes.

cloning vector A plasmid or phage that is used to carry inserted foreign DNA for the purposes of producing more material or a protein product.

closed reading frame Contains termination codons that prevent its translation into protein.

coding region The part of a DNA sequence that codes for the amino acids in a protein.

coding sequence A region of a DNA strand with the same sequence as is found in the coding region of a messenger RNA, except that T is present in DNA instead of U.

co-dominance The expression of both alleles in a heterozygote.

co-dominant alleles Both contribute to the phenotype; neither is dominant over the other.

codon A sequence of three adjacent nucleotides (triplet) in an mRNA molecule, specifying or representing either an amino acid or a stop signal in protein synthesis.

coefficient of variation The standard deviation of a trait in all individuals of a population expressed as a percentage of the mean and multiplied by 100.

cohesive end A single-stranded region at the end of a double-stranded DNA molecule that can adhere to a complementary single-stranded sequence at the other end or in another molecule.

collateral relatives Individuals not related directly, e.g. aunts, uncles, cousins.

collinearity The linear correspondence between the order of amino acids in a polypeptide chain and the corresponding sequence of nucleotides in the DNA molecule.

colony A visible cluster of cells formed on a solid growth medium by repeated division of a single parental cell and its daughter cells.

colony hybridization assay A technique for identifying colonies that contain a particular cloned gene; many colonies are transferred to a filter, lysed and exposed to radioactive DNA or RNA complementary to the DNA sequence of interest, after which colonies that contain a sequence complementary to the probe are located by autoradiography.

combinatorial control Strategy of gene regulation in which a relatively small number of time- and tissue-specific positive and negative regulatory elements are used in various combinations to control the expression of a much larger number of genes.

combining or nicking ability The ability of two strains, lines or breeds to produce heterosis when crossed.

complementary DNA (cDNA) A DNA molecule made by copying RNA with reverse transcriptase; a single-stranded DNA complementary to an RNA, synthesized from it by reverse transcription *in vitro*.

complementation The phenomenon in which two recessive mutations with similar phenotypes result in a wild-type phenotype when both are heterozygous in the same genotype; complementation means that the mutations are in different genes; complementation refers to the ability of independent (non-allelic) genes to provide diffusible products that produce a wild phenotype when two mutants are tested in *trans* configuration in a heterozygote.

complementation group A group of mutations that fail to complement one another; a series of mutations unable to complement when tested in pairwise combinations in *trans*; defines a genetic unit (the cistron) that might better be called a non-complementation group.

complementation test A genetic test to determine whether two mutations are alleles (are present in the same functional gene).

complex trait A multifactorial trait influenced by multiple genetic and environmental factors, each of relatively small effect, and their interactions.

concatamer DNA consisting of a series of unit sequences repeated in tandem.

concatenated circles DNA interlocked like rings on a chain.

concerted evolution The ability of two related genes to evolve together as though constituting a single locus.

conditional mutation A mutation that results in a mutant phenotype under certain (restrictive) environmental conditions but results in a wild-type phenotype under other (permissive) conditions.

conformation Externally visible or measurable variations in shape or body proportions of animals (*see also* type).

consensus sequence A generalized base sequence derived from closely related sequences found in many locations in a genome or in many organisms; each position in the consensus sequence consists of the base found in the majority of sequences at that position; an idealized sequence in which each position represents the base most often found when many actual sequences are compared.

conservative recombination Breakage and reunion of pre-existing strands of DNA without any synthesis of new stretches of DNA.

conservative transposition The movement of large elements, originally classified as transposons, but now considered to be episomes. The mechanism of movement resembles that of phage lambda.

conserved sequence A base or amino acid sequence that changes very slowly in the course of evolution.

constant antibody region The part of the heavy and light chains of an antibody molecule that has the same amino acid sequence among all antibodies derived from the same heavy-chain and light-chain genes.

constitutive genes Are expressed as a function of the interaction of RNA polymerase with the promoter, without additional regulation; sometimes also called housekeeping genes in the context of describing functions expressed in all cells at a low level.

constitutive heterochromatin The inert state of permanently non-expressed sequences, usually satellite DNA.

constitutive mutant A mutant in which synthesis of a particular mRNA molecule (and the protein that it encodes) takes place at a constant rate independent of the presence or absence of any inducer or repressor molecule; constitutive mutations cause genes that are usually regulated to be expressed without regulation.

contig A set of cloned DNA sequence fragments overlapping at the fragment ends in such a way as to provide unbroken coverage of a contiguous region of the genome; a contig contains no gaps.

continuous trait A trait in which the possible phenotypes have a continuous range from one extreme to the other, rather than falling into discrete classes.

coordinate gene Any of a group of genes that establish the basic anterior–posterior and dorsal–ventral axes of the early embryo.

coordinate regulation Control of synthesis of several proteins by a single regulatory element; in prokaryotes, the proteins are usually translated from a single mRNA molecule.

copy number variation (CNV) Polymorphism in the number of copies of a stretch of DNA, including deletions and duplications of whole genes.

core particle The aggregate of histones and DNA in a nucleosome, without the linking DNA.

co-repressor A small molecule that binds with an aporepressor to create a functional repressor molecule; small molecule that triggers repression of transcription by binding to a regulator protein.

correlated response Change of the mean in one trait in a population accompanying selection for another trait.

correlation Association between characteristics of individuals. The correlation coefficient is a statistical measure of degree of association and varies from −1.0 to +1.0 (*see also* regression).

correlation coefficient A measure of association between pairs of numbers, equalling the covariance divided by the product of the standard deviations.

cosmids Plasmids into which phage lambda cos sites have been inserted; as a result, the plasmid DNA can be packaged *in vitro* in the phage coat.

co-transduction Transduction of two or more linked genetic markers by one transducing particle.

coupling *cis* configuration.

covariance Variation that is common between two traits. It may result from joint hereditary or environmental influences. A measure of association between pairs of numbers that is defined as the average product of the deviations from the respective means.

CpG islands Stretches of vertebrate DNA typically between 1 and 2 kb long that contain ten times higher frequency of the doublet nucleotide CG than occurs elsewhere in the genome. CpG islands are commonly found in 5′ end of genes.

Cre-lox recombination system A combination of site specific recombinase (Cre) and its recognition site (lox) from the bacteriophage P1 that has been engineered into yeast and eukaryotic genomes to facilitate targeted recombination.

crossbreed/crossbred An animal produced by crossing two or more pure breeds, strains or lines. Crossbred individuals are from intraspecific matings.

crossbreeding Mating systems in which hereditary material from two or more pure breeds, strains or lines is combined.

crossing over A process of exchange between non-sister chromatids of a pair of homologous chromosomes that results in the recombination of linked genes; describes the reciprocal exchange of material between chromosomes that occurs during meiosis and is responsible for genetic recombination.

crossover fixation A possible consequence of unequal crossing over that allows a mutation in one member of a tandem cluster to spread through the whole cluster (or to be eliminated).

crossover interference The phenomenon resulting when a single crossover event during meiosis tends to diminish or restrict the probability that an adjacent crossover event will occur within the same region. Complete positive interference indicates that at most there will be a single crossover event per chromosome arm.

cruciform The structure that can potentially be produced at inverted repeats of DNA if the repeated sequence pairs with its complement on the same strand (instead of with its regular partner in the other strand of the duplex).

cryptic satellite A satellite DNA sequence not identified as such by a separate peak on a density gradient; that is, it remains present in main-band DNA.

cryptic splice site A potential splice site not normally used in RNA processing unless a normal splice site is blocked or mutated.

cyclin One of a group of proteins that participates in controlling the cell cycle. Different types of cyclins interact with the p34 kinase subunit and regulate the G1/S and G2/M transitions. The proteins are called cyclins because their abundance rises and falls rhythmically in the cell cycle.

cytogenetics The study of the physical appearance of chromosomes.

cytokinesis The final process involved in separation and movement apart of daughter cells at the end of mitosis.

cytological band An area of the chromosome that stains differently from areas around it.

cytological map The banding patterns observed through a microscope on stained chromosome spreads; a type of chromosome map whereby genes are located on the basis of cytological findings obtained with the aid of chromosome mutations; also a diagrammatic representation of a chromosome.

cytoplasm The material between the plasma membrane and the nucleus.

cytoplasmic inheritance A property of genes located in mitochondria or chloroplasts (or possibly other extranuclear organelles).

cytoplasmic protein synthesis The translation of mRNAs representing nuclear genes; it occurs via ribosomes attached to the cytoskeleton.

cytosine (C) A nitrogenous pyrimidine base found in DNA and RNA.

D loop A region within mitochondrial DNA in which a short stretch of RNA is paired with one strand of DNA, displacing the original partner DNA strand in this region. The same term is used also to describe the displacement of a region of one strand of duplex DNA by a single-stranded invader in the reaction catalysed by RecA protein.

dam Female parent, the mother of an animal.

data normalization The process of removing systematic biases from microarray data that otherwise contributes to misinterpretation of apparent differences in transcript abundance.

daughter strand A newly synthesized single DNA or chromosome strand.

degeneracy In the genetic code refers to the lack of an effect of many changes in the third base of the codon on the amino acid that is represented.

deleterious genes Genes which in either the homozygous or heterozygous state has undesirable effects on an individual's viability or usefulness.

deletion Loss of a segment of the genetic material from a chromosome; also called deficiency. Generated by removal of a sequence of DNA, the regions on either side being joined together.

denaturation (DNA or RNA) Conversion from the double-stranded to the single-stranded state; separation of the strands is most often accomplished by heating.

denaturation (protein) Conversion from the physiological conformation to some other (inactive) conformation.

deoxyribose The five-carbon sugar present in DNA.

depurination Removal of purine bases from DNA.

derepressed state A gene that is turned on. It is synonymous with induced when describing the normal state of a gene; it has the same meaning as constitutive in describing the effect of mutation.

diakinesis The substage of meiotic prophase I, immediately preceding metaphase I, in which the bivalents attain maximum shortening and condensation.

dicentric chromosome A chromosome with two centromeres; the product of fusing two chromosome fragments, each of which has a centromere. It is unstable and may be broken when the two centromeres are pulled to opposite poles in mitosis.

dideoxy sequencing method Procedure for DNA sequencing in which a template strand is replicated from a particular primer sequence and terminated by the incorporation of a nucleotide that contains dideoxyribose instead of deoxyribose; the resulting fragments are separated by size via electrophoresis.

dideoxyribose A deoxyribose sugar that lacks the 3' hydroxyl group; when incorporated into a polynucleotide chain, it blocks further chain elongation.

dihybrid Heterozygous at each of two loci; progeny of a cross between true-breeding, or homozygous, strains that differ genetically at two loci.

diploid Cells with two members of each pair of chromosomes. This is termed the 2N condition and is characteristic of the body cells of most fish species. A cell or organism with two complete sets of homologous chromosomes; a set of chromosomes containing two copies of each autosome and two sex chromosomes.

diplotene The substage of meiotic prophase I, immediately following pachytene and preceding diakinesis, in which pairs of sister chromatids that make up a bivalent (tetrad) begin to separate from each other and chiasmata become visible.

direct repeat Copies of an identical or very similar DNA or RNA base sequence in the same molecule and in the same orientation; identical (or related) sequences present in two or more copies in the same orientation in the same molecule of DNA; they are not necessarily adjacent.

discontinuous replication The synthesis of DNA in short (Okazaki) fragments that are later joined (ligated) into a continuous strand.

disjunction The movement of members of a chromosome pair to opposite poles during cell division. At mitosis and the second meiotic division, disjunction applies to sister chromatids; at first meiotic division, it applies to sister chromatid pairs.

distribution In quantitative genetics, the mathematical relationship that gives the proportion of members in a population that has each possible phenotype.

divergence The per cent difference in nucleotide sequence between two related DNA sequences or in amino acid sequences between two proteins.

divergent transcription The initiation of transcription at two promoters on opposite DNA strands facing in the opposite direction, so that transcription proceeds away in both directions from a central region.

DNA (deoxyribonucleic acid) The basic hereditary material of all living matter. It is composed of basic units or nucleotides, each of which contains an organic base, a sugar and a phosphate. It is a chemically complex substance with gigantic molecules in a spiral, double-helix configuration capable of virtually infinite numbers of structural variations; the macromolecule, usually composed of two polarized and anti-parallel polynucleotide chains in a double helix, that is the carrier of the genetic information in all cells and many viruses.

DNA-binding motif A short stretch of DNA usually between 8 and 12 nucleotides, that is thought to be recognized by its DNA-binding protein. These motifs can be represented by a profile of the frequency of occurrence of each of the four nucleotides at each position in the sequence. They typically identify sequences important for the regulation of gene expression.

DNA chip A small plate of silicon, glass or other material containing an array of oligonucleotides to which DNA samples can be hybridized.

DNA library A collection of hundreds of thousands of clones, each of which contains a different piece of genomic or cDNA. If the clone is oriented in such a way that the clone can be transcribed and translated, the library is called an expression library.

DNA ligase An enzyme that catalyses the formation of a covalent bond between adjacent 5′-P and 3′-OH termini in a broken polynucleotide strand of double-stranded DNA.

DNA looping A mechanism by which enhancers that are distant from the immediate proximity of a promoter can still regulate transcription; the enhancer and promoter, both bound with suitable protein factors, come into indirect physical contact by means of the looping out of the DNA between them. The physical interaction stimulates transcription.

DNA methylase An enzyme that adds methyl groups ($-CH_3$) to certain bases, particularly cytosine.

DNA polymerase Any enzyme that catalyses the synthesis of DNA from deoxynucleoside 5′-triphosphates, using a template strand; enzyme that synthesizes daughter strands of DNA (under direction from a DNA template). May be involved in repair or replication.

DNA repair Any of several different processes for restoration of the correct base sequence of a DNA molecule into which incorrect bases have been incorporated or whose bases have been chemically modified.

DNA replicase A DNA-synthesizing enzyme required specifically for replication.

DNA replication The semi-conservative copying of a DNA molecule.

DNase An enzyme that attacks bonds in DNA.

DNA typing Electrophoretic identification of individual persons by the use of DNA probes for highly polymorphic regions of repetitious DNA in the genome, such that the genome of virtually every person exhibits a unique pattern of bands; sometimes called DNA fingerprinting.

DNA uracyl glycosylase An enzyme that removes uracil bases when they occur in double-stranded DNA.

dominance variance That portion of the hereditary or genetic variance which is due to dominance. That portion of the hereditary or genetic variance over and above that which can be accounted for by additive effects and which is due to dominance. Alternatively, can be defined as due to dominance deviations from a description based upon assumed additive effects.

dominant gene Allelic form of gene that has an observable effect when present in only one member of a chromosome pair. An allele whose presence in a heterozygous genotype results in a phenotype characteristic of the allele and the phenotype of the alternative allele is not observed.

dosage compensation A mechanism regulating X-linked genes such that their activities are equal in males and females; in mammals, random inactivation of one X chromosome in females results in equal amounts of the products of X-linked genes in males and females.

double-stranded DNA A DNA molecule consisting of two antiparallel strands that are complementary in nucleotide sequence.

down promoter mutations Decrease the frequency of initiation of transcription.

downstream Identifies sequences proceeding further in the direction of expression – for example, the coding region is downstream of the initiation codon.

duplex DNA A double-stranded DNA molecule.

duplication A chromosome aberration in which a chromosome segment is present more than once in the haploid genome; if the two segments are adjacent, the duplication is a tandem duplication.

ectopic expression The expression of a gene in a tissue in which it is not usually expressed – for example, expression in atypical tissues in a transgenic animal.

editing function The activity of a DNA polymerase that removes incorrectly incorporated nucleotides; also called the proofreading function.

electrophoresis A technique used to separate molecules on the basis of their different rates of movement in response to an applied electric field, typically through a gel.

elongation factors (EF in prokaryotes, eEF in eukaryotes) Proteins that associate with ribosomes cyclically, during addition of each amino acid to the polypeptide chain.

embryonic stem cells Cells in the blastocyst that give rise to the body of the embryo.

3′ end The end of a DNA or RNA strand that terminates in a sugar and so has a free hydroxyl group on the number 3 carbon.

5′ end The end of a DNA or RNA strand that terminates in a free phosphate group not connected to a sugar further along.

end labelling The addition of a radioactively labelled group to one end (5′ or 3′) of a DNA strand.

endocytic vesicles Membranous particles that transport proteins through endocytosis; also known as clathrin-coated vesicles.

endocytosis Process by which proteins at the surface of the cell are internalized, being transported into the cell within membranous vesicles.

endonuclease An enzyme that breaks internal phosphodiester bonds in a single- or double-stranded nucleic acid molecule; usually specific for either DNA or RNA; cleaves bonds within a nucleic acid chain; may be specific for RNA or for single-stranded or double-stranded DNA.

endoreduplication Doubling of the chromosome complement because of chromosome replication and centromere division without nuclear or cytoplasmic division.

end-product inhibition The ability of a product of a metabolic pathway to inhibit the activity of an enzyme that catalyses an early step in the pathway.

enhancer A base sequence in eukaryotes and eukaryotic viruses that increases the rate of transcription of nearby genes; the defining characteristics are that it need not be adjacent to the transcribed gene; a *cis*-acting sequence that increases the utilization of (some) eukaryotic promoters, and can function in either orientation and in any location (upstream or downstream) relative to the promoter.

enhancer trap A transposable element that has been modified so that when it inserts into the genome adjacent to a gene, the enhancers that drive expression of that gene also drive expression of a reporter gene carried on the enhancer trap transposable element.

environment All the external factors within which an animal's genotype acts to determine its phenotypic traits.

environmental variance The variance, in absolute terms, of any character in a population which is due to environmental influences. The part of the phenotypic variance that is attributable to differences in environment.

enzyme A protein that catalyses a specific biochemical reaction and is not itself altered in the process.

epigenetic changes Influence the phenotype without altering the genotype. They consist of changes in the properties of a cell that are inherited but do not represent a change in genetic information.

epistasis Genetic effects due to interactions among two or more pairs (or series) of non-allelic genes (genes at different loci). A term referring to an interaction between non-allelic genes in their effects on a trait. Generally, epistasis means any type of interaction in which the genotype at one locus affects the phenotypic expression of the genotype at another locus. In a more restricted sense, it refers to a situation in which the genotype at one locus determines the phenotype in such a way as to mask the genotype present at a second locus; expression of one gene wipes out the phenotypic effects of another gene at another locus.

epistatic variance That residual portion of the hereditary variance due to non-allelic gene interactions not accounted for by additive or dominance effects.

equational division Term applied to the second meiotic division because the haploid chromosome complement is retained throughout.

essential gene Gene whose deletion is lethal to the organism.

euchromatin A region of a chromosome that is relatively uncondensed during interphase and has normal staining properties and undergoes the normal cycle of mitotic condensation; relatively uncoiled in the interphase nucleus (compared with condensed chromosomes), it apparently contains most of the genes.

euploid A cell or an organism having a chromosome number that is an exact multiple of the haploid number.

evolution Cumulative change in the genetic characteristics of a species through time.

excisionase An enzyme that is needed for prophage excision; works together with an integrase.

excision repair Type of DNA repair in which segments of a DNA strand that are chemically damaged are removed enzymatically and then resynthesized, using the other strand as a template; the action of removing a single-stranded sequence of DNA containing damaged or mispaired bases and replacing it in the duplex by synthesizing a sequence complementary to the remaining strand.

exon The sequences in a gene that are retained in the messenger RNA after the introns are removed from the primary transcript; any segment of an interrupted gene that is represented in the mature RNA product.

exon shuffle The theory that new genes can evolve by the assembly of separate exons from pre-existing genes, each coding for a discrete functional domain in the new protein.

exonuclease An enzyme that removes a terminal nucleotide in a polynucleotide chain by cleavage of the terminal phosphodiester bond; nucleotides are removed successively, one by one; usually specific for either DNA or RNA and for either single-stranded or double-stranded nucleic acids. A 5′-to-3′ exonuclease cleaves successive nucleotides from the 5′ end of the molecule; a 3′-to-5′ exonuclease cleaves successive nucleotides from the 3′ end; cleaves nucleotides, one at a time from the end of a polynucleotide chain; may be specific for either the 5′ or 3′ end of DNA or RNA.

expressed sequence tag (EST) A partial or complete complementary DNA (cDNA) sequence that is a short, single-pass cDNA sequence; both the upstream and the downstream segments of cDNAs reverse-transcribed from mRNAs, and generated from randomly selected cDNA library clones.

expression library A library of cDNA clones in a vector that allows the gene products to be expressed (transcribed and translated) in a controlled manner.

expression vector A cloning vector designed so that a coding sequence inserted at a particular site will be transcribed and translated into protein.

expressivity The degree to which a trait is observed in affected individuals.

extranuclear genes Reside outside the nucleus in organelles such as mitochondria and chloroplasts.

F Inbreeding coefficient (values 0–1), 0 indicates no inbreeding.

F_1 The hybrid offspring, crossbred offspring or first filial generation from a given mating; the first generation of descent from a given mating.

F_2 Offspring of $F_1 \times F_1$ matings or second filial generation; the second generation of descent from a given mating, produced by intercrossing or self-fertilizing F_1 organisms.

F_3 Offspring of $F_2 \times F_2$ matings or third filial generation.

F_n Extension of foregoing.

family In animal breeding sometimes used to denote a line of descent (similar to family names in people) but more often to represent a group of animals having a genetic relationship. In fish genetics, commonly refers to full siblings from a single spawning between an individual male and female.

30 nm fibre The level of compaction of eukaryotic chromatin resulting from coiling of the extended, nucleosome-bound DNA fibre.

fingerprint (DNA) A pattern of polymorphic restriction fragments that differ between individual genomes.

fingerprint (protein) The pattern of fragments (usually resolved on a two-dimensional electrophoretic gel) generated by cleavage with an enzyme such as trypsin.

first meiotic division The meiotic division that reduces the chromosome number; sometimes called the reduction division.

first-division segregation Separation of a pair of alleles into different nuclei in the first meiotic division; happens when there is no crossing over between the gene and the centromere in a particular cell.

fitness A measure of the average ability of organisms with a given genotype to survive and reproduce.

fixed allele An allele whose allele frequency equals 1.0.

fold-back DNA Consists of inverted repeats that have renatured by intrastrand reassociation of denatured DNA.

folding domain A short region of a polypeptide chain within which interactions between amino acids result in a three-dimensional conformation that is attained relatively independently of the folding of the rest of the molecule.

footprinting A technique for identifying the site on DNA bound by some protein by virtue of the protection of bonds in this region against attack by nucleases.

forward genetics Genetic analysis that starts with a phenotype and moves toward isolation of the gene that caused the phenotype.

forward mutation A change from a wild-type allele to a mutant allele. Inactivates a wild-type gene.

founder effect The presence in a population of many individuals all with the same chromosome (or region of a chromosome) derived from a single ancestor.

fragile-X chromosome A type of X chromosome containing a site towards the end of the long arm that tends to break in cultured cells that are starved for DNA precursors; causes fragile-X syndrome.

frameshift mutation A mutational event caused by the insertion or deletion of one or more nucleotide pairs in a gene, resulting in a shift in the reading frame of all codons following the mutational site; arises by deletions or insertions that are not a multiple of 3 bp; changes the frame in which triplets are translated into protein.

frequency of co-transduction The proportion of transductants carrying a selected genetic marker that also carry a non-selected genetic marker.

frequency of recombination The proportion of gametes carrying combinations of alleles that are not present in either parental chromosome.

full sibs or siblings Individuals with the same sire and dam, full brothers or full sisters or a full brother–sister pair.

functional genomics A branch of genomics that studies the relationship of genome expression and genome functions.

fusion gene Gene resulting from the joining of transcriptional control sequences from one gene to structural sequences from a second gene.

G1 The period of the eukaryotic cell cycle between the last mitosis and the start of DNA replication.

G2 The period of the eukaryotic cell cycle between the end of DNA replication and the start of the next mitosis.

G banding A technique that generates a striated pattern in metaphase chromosomes that distinguishes the members of a haploid set.

gain-of-function mutation Mutation in which a gene is overexpressed or inappropriately expressed.

gamete A mature reproductive or germ cell. In animals, the male gamete is the sperm or spermatozoan and the female gamete is the egg or ovum. Gametes carry the reduced, 1N, or haploid number of chromosomes.

gap In DNA, the absence of one or more nucleotides in one strand of the duplex.

gene (cistron) The classical term for the basic unit of heredity. Functionally, the gene is equivalent to the cistron. The hereditary unit defined experimentally by the complementation test. At the molecular level, a region of DNA containing genetic information, usually transcribed into an RNA molecule, which is processed and either functions directly or is translated into a polypeptide chain; a gene can mutate to various forms, called alleles; segment of DNA involved in producing a polypeptide chain; it includes regions preceding and following the coding region (leader and trailer), as well as intervening sequences (introns) between individual coding segments (exons).

gene amplification A process in which certain genes undergo differential replication either within the chromosome or extrachromosomally (temporarily or permanently), increasing the number of copies of the gene.

gene cluster A group of adjacent genes that are identical or related.

gene conversion The phenomenon in which the products of a meiotic division in an Aa heterozygous genotype are in some ratio other than the expected 1A:1a – for example, 3A:1a, 1A: 3a, 5A:3a or 3A:5a; the alteration of one strand of a heteroduplex DNA to make it complementary with the other strand at any position(s) where there were mispaired bases.

gene dosage Number of gene copies; gives the number of copies of a particular gene in the genome.

gene expression The multistep process by which a gene is regulated and its product synthesized, thus making a contribution to the phenotype.

gene family Set of genes whose exons are related; the members were derived by duplication and variation from some ancestral gene.

gene frequency The proportion in a population of the loci of a given allelic series occupied by a particular gene or the frequency of a gene at a locus for a population.

gene knock-in Replacement of the endogenous gene with a different functional piece of DNA such that the inserted gene is expressed instead of the original gene.

gene knockout A mutation that targets a specific gene, produced by using homologous recombination, to replace an exon of the target gene with a piece of foreign DNA. Insertional mutations can also cause gene knockouts.

gene pool The totality of genetic information in a population of organisms.

gene product A term used for the polypeptide chain translated from an mRNA molecule transcribed from a gene; if the RNA is not translated (for example, ribosomal RNA), the RNA molecule is the gene product.

gene regulation Process by which gene expression is controlled in response to external or internal signals.

gene targeting Disruption or mutation of a designated gene by homologous recombination.

gene transfer Usually refers to the transfer of a specific gene or piece of DNA from one species or organism to another, involving recombinant DNA technology.

general transcription factor A protein molecule needed to bind with a promoter before transcription can proceed; transcription factors are necessary, but not sufficient, for transcription, and they are often shared among many different promoters.

generation interval Average age of parents when their offspring are born that become the next generation's brood stock.

genetic architecture Specification of the genetic and environmental factors that contribute to a complex trait and their interactions.

genetic code The correspondence between triplets in DNA (or RNA) and amino acids in protein; the set of 64 triplets of bases (codons) that correspond to the 20 amino acids in proteins and the signals for initiation and termination of polypeptide synthesis.

genetic correlation Association among traits of individuals due to genetic influences.

genetic drift Changes in gene or allele frequency in a population due to chance variations in proportions of gametes that are formed carrying specific genes or that succeed in accomplishing fertilization. Also called random genetic drift and is caused by small reproducing population numbers.

genetic engineering Genetic manipulation programme utilizing recombinant DNA technology and gene transfer. The linking of two DNA molecules by *in vitro* manipulations for the purpose of generating a novel organism with desired characteristics.

genetic heterogeneity The observation that the phenotype can have multiple different genetic causes. If this occurs within a single locus, the effect is allelic heterogeneity.

genetic map A description of the relative order of genetic markers in linkage groups in which the distance between markers is expressed in units of recombination.

genetic marker Any pair of alleles or DNA sequence whose inheritance can be traced through a mating or through a pedigree.

genetic variance Phenotypic variance attributable to genetic factors.

genetics The science concerned with determining the mode of inheritance or the transmission of biological properties from generation to generation in plants, animals and lower organisms. The study of biological heredity.

genome The total complement of genes contained in a cell or virus; commonly used to refer to all genes present in one complete haploid set of chromosomes in eukaryotes.

genome-wide association study (GWAS) A study to scan the entire genome for single nucleotide polymorphisms (SNPs) and copy number variations (CNVs) that are associated with a trait. Typically, 500,000 different genetic variants are measured or observed.

genomics The study and development of genetic and physical maps, large-scale DNA sequencing, gene discovery and computer-based systems for managing and analysing genomic data; the study of genes and their functions, study of the genome, molecular characterization of all genes in a species, comprehensive study of the genetic information of a cell or organism, study of the structure and function of a large number of genes simultaneously; study of the structure, content and evolution of genomes and the science of the study of the genome.

genotype The complete genetic make-up of an individual. Can also be partial make-up or specific to one locus; refers to make-up of both alleles possessed by an individual. Complete genetic make-up is also referred to as the genome.

genotype-by-sex interaction Genetic determination that differs between the sexes to result in different phenotypes for the same genotype depending on the sex of the individual.

genotype–environment association The condition in which genotypes and environments are not in random combinations.

genotype–environment (GE) interaction When the rank or value of a genotype changes relative to other genotypes when the environment changes.

genotype frequency The proportion of members of a population that are of a particular prescribed genotype.

genotypic variance The part of the phenotypic variance that is attributable to differences in genotype.

germ cell A cell that gives rise to reproductive cells.

GT–AG rule The presence of these constant dinucleotides at the first two and last two positions of introns of nuclear genes.

guanine (G) A nitrogenous purine base found in DNA and RNA.

guide RNA The RNA template present in telomerase.

gynogen An individual whose DNA is all maternally inherited.

hairpin A double-helical region formed by base pairing between adjacent (inverted) complementary sequences in a single strand of RNA or DNA.

half sib One of a pair of animals having one common parent; half-brother or half-sister.

haploid Cells with one member of each chromosome pair. This is termed the 1N condition and is often referred to as the reduced chromosome number. The reproductive cells or gametes have a haploid number of chromosomes. A cell or organism of a species containing the set of chromosomes normally found in gametes. Set of chromosomes containing one copy of each autosome and one sex chromosome; the haploid number N is characteristic of gametes of diploid organisms.

haplotype The particular combination of alleles in a defined region of some chromosome, in effect the genotype in miniature. Originally used to describe combinations of major histocompatibility complex (MHC) alleles, it may now be used to describe particular combinations of restriction fragment length polymorphisms (RFLPs) and mitochondrial DNA (mtDNA) genotypes; a set of polymorphisms found on a single chromosome.

Hardy–Weinberg law A law which states that in a population mating at random, with no migration, selection or mutation, the proportion of different types of zygotes produced for any allelic pair or series is directly proportional to the square of their respective gametic frequencies.

hemizygote A diploid individual that has lost its copy of a particular gene (for example, because a chromosome has been lost) and which therefore has only a single copy. Also, a transgenic organism with a novel introduced gene on one chromosome only.

hereditary A condition controlled or influenced to some degree by gene action. This is in contrast to characters that are entirely controlled by environmental variables.

hereditary variance The variance, in absolute terms, for any trait in a population that is due to genetic influences, additive, dominance and epistatic effects (*see also* heritability).

heritability (h^2) That portion or fraction of the total variance for any trait in a population which is due to additive genetic effects (narrow-sense definition).

heterochromatin Chromatin that remains condensed and heavily stained during interphase; commonly present adjacent to the centromere and in the telomeres of chromosomes. Some chromosomes are composed primarily of heterochromatin. Describes regions of the genome that are permanently in a highly condensed condition and are frequently not expressed genetically. May be constitutive or facultative subunit of chromatin.

heteroduplex (hybrid) All or part of a double-stranded nucleic acid molecule in which the two strands have different hereditary origins; produced either as an intermediate in recombination or by the *in vitro* annealing of single-stranded complementary molecules; DNA is generated by base pairing between complementary single strands derived from the different parental duplex molecules; it occurs during genetic recombination.

heteroduplex DNA A double-stranded DNA molecule containing a polymorphism, formed by renaturation of the PCR products from two different alleles.

heterogametic Having sex chromosomes or genes that are not alike. Heterogametic sex has the diploid chromosome constitution 2A + XY.

heterogeneous nuclear RNA (hnRNA) Transcript of nuclear genes made by RNA polymerase II; it has a wide size distribution and low stability, and is the precursor of messenger RNA.

heterokaryon A cell containing two (or more) nuclei in a common cytoplasm, generated by fusing somatic cells.

heteromultimeric proteins Non-identical subunits (coded by different genes).

heterosis Positive or negative differences in performance of progeny from the average of the parental types in crossbred or hybrid matings (*see also* hybrid vigour). The superiority of hybrids over either parent in respect of one or more traits.

heterozygote (*adj.* heterozygous) An individual in which a given locus in a chromosome pair carries unlike members or a pair or series of alleles; carrying dissimilar alleles of one or more genes; not homozygous; an individual with different alleles at some particular locus.

heterozygote superiority Gene pairs (or series) in which heterozygous individuals are superior to any homozygote of the pair or series. Technically called overdominance. The condition in which a heterozygous genotype has greater fitness than either of the homozygotes.

hexaploid A cell or organism with six complete sets of chromosomes.

highly repetitive DNA The first component to reassociate; is equated with satellite DNA.

histones Conserved DNA-binding proteins of eukaryotes that form the nucleosome.

homoeobox The conserved sequence that is part of the coding region of *Drosophila melanogaster* homoeotic genes; it is also found in some vertebrate genes, and expressed in early embryonic development.

homoeotic genes Defined by mutations that convert one body part into another; for example, an insect leg may replace an antenna.

homogametic Having sex chromosomes or genes that are alike. Homogametic sex has the diploid chromosome constitution 2A + XX.

homologs Highly conserved loci coming from a common ancestor.

homologues Chromosomes carrying the same genetic loci; a diploid cell has two copies of each homologue, one derived from each parent.

homomultimeric protein Consists of identical subunits.

homozygote (*adj.* homozygous) An individual in which both members of a chromosome pair carry the same gene at a specific locus. Homozygotes are therefore genetically pure for a given pair or series of hereditary factors. An individual with the same allele at corresponding loci on the homologous chromosomes.

horizontal gene transfer Transfer of a gene from the genome of one individual to that of another by means other than sexual or vertical transmission.

hot spot A site at which the frequency of mutation (or recombination) is very much increased.

hox genes Clusters of mammalian genes containing homoeoboxes; the individual members are related to the genes of the complex loci AN7-C and BX-C in *Drosophila melanogaster.*

housekeeping (constitutive) genes Those expressed in all cells because they provide basic functions needed for the sustenance of all cell types.

hybrid Technically refers to the offspring of parents that are each genetically pure (homozygous) for one or more pairs of hereditary factors, but with the two parents being homozygous for different members of allelic pairs or series. In practice the term has been extended to include offspring of species crosses, to progeny of crosses of inbred lines and in some cases to breed crosses. An organism produced by the mating of genetically unlike parents. Since intra- and interspecific crosses of fish can be routinely made, we strongly suggest that this term only be used for interspecific crosses or hybrids to reduce confusion.

hybrid (DNA) A duplex nucleic acid molecule produced of strands derived from different sources.

hybrid-arrested translation A technique that identifies the complementary DNA (cDNA) corresponding to an mRNA by relying on the ability to base-pair with the RNA *in vitro* to inhibit translation.

hybrid vigour Increased vigour or productivity often observed in hybrid, crossbred or cross-line individuals as compared with that of the average of the parental types (*see also* heterosis). The evolutionary definition is different and refers to the case where F_1 hybrids or crossbreeds contribute greater proportions of genes to the next generation than parental types.

ideogram A diagrammatic representation of the G-banding pattern of a chromosome.

imprinting A process of DNA modification in gametogenesis that affects gene expression in the zygote; a probable mechanism is the methylation of certain bases in the DNA; describes a change in a gene that occurs during passage through the sperm or egg with the result that the paternal and maternal alleles have different properties in the very early embryo.

inbreeding A system of mating in which mates are more closely related than average individuals of the population to which they belong. Mating between relatives.

inbreeding coefficient A measure of the genetic effects of inbreeding in terms of the proportionate reduction in heterozygosity in an inbred organism compared with the heterozygosity expected with random mating.

inbreeding depression Decreased performance due to inbreeding.

incomplete penetrance Condition in which a mutant phenotype is not expressed in all organisms with the mutant genotype.

indel An insertion–deletion polymorphism. Indels range in size from one or a few bases to several kilobases. Large indels often involve transposable elements.

independent assortment Behaviour at meiosis of genes located on different chromosome pairs. Random distribution of unlinked genes into gametes, as with genes in different (non-homologous) chromosomes or genes that are so far apart on a single chromosome that the recombination frequency between them is 1/2.

indirect selection Selection for one trait by selecting a second trait.

individual selection Selection based on each organism's own phenotype.

induced mutation A mutation formed under the influence of a chemical mutagen or radiation.

inducer A small molecule that inactivates a repressor, usually by binding to it and thereby altering the ability of the repressor to bind to an operator; small molecule that triggers gene transcription by binding to a regulator protein.

inducible transcription Transcription of a gene, or a group of genes, only in the presence of an inducer molecule.

induction Activation of an inducible gene; prophage induction is the derepression of a prophage that initiates a lytic cycle of phage development.

initiation The beginning of protein synthesis.

initiation factors (IF in prokaryotes, eIF in eukaryotes) Proteins that associate with the small subunit of the ribosome specifically at the stage of initiation of protein synthesis.

inosine (I) One of a number of unusual bases found in transfer RNA.

insertional mutagenesis The process of creating mutations by controlled insertion of a transposable element in the vicinity of a gene of interest.

insertions Identified by the presence of an additional stretch of base pairs in DNA.

***in situ* hybridization** Performed by denaturing the DNA of cells squashed on a microscope slide so that a reaction is possible with an added single-stranded RNA or DNA; the added preparation is radioactively or immunologically labelled and its hybridization is followed by autoradiography or specific staining. Detection of a specific mRNA in a tissue sample by hybridization of a section of a whole mount of the tissue to a DNA or RNA probe that is complementary to the mRNA, and which is labelled with a fluorescent or radioactive group or with a small compound that can be recognized by an antibody.

integration A DNA sequence's insertion into a host genome as a region covalently linked on either side to the host sequences.

interallelic complementation The change in the properties of a heteromultimeric protein brought about by the interaction of subunits coded by two different mutant alleles; the mixed protein may be more or less active than the protein consisting of subunits only of one or the other type.

interbands The relatively dispersed regions of polytene chromosomes that lie between the bands.

intercalation Insertion of a planar molecule between the stacked bases in duplex DNA.

intercistronic region The distance between the termination codon of one gene and the initiation codon of the next gene.

interference The tendency for crossing over to inhibit the formation of another crossover nearby.

intermediate component(s) Of a reassociation reaction are those reacting between the fast (satellite DNA) and slow (non-repetitive DNA) components; contain moderately repetitive DNA.

interphase The period between mitotic cell divisions, extending from the end of telophase of one division to the beginning of prophase of the next division; divided into G1, S and G2.

interval mapping A method for quantitative trait locus (QTL) mapping that uses the genotypes of two adjacent genetic markers to estimate the likely genotype at each point in the interval between the markers.

intervening sequence (intron) A non-coding DNA sequence in a gene that is transcribed but is then excised from the primary transcript in forming a mature mRNA molecule; found primarily in eukaryotic cells.

introgression Introduction of a small portion of one genome into the genetic background of another genome, by repeated backcrossing with selection for the region of interest.

intron A non-coding segment of DNA that is transcribed and represented in the heterogeneous nuclear RNA, but removed from within the transcript by splicing together the sequences (exons) on either side of it.

inversion A chromosomal change in which a segment has been rotated by 180° relative to the regions on either side and reinserted. A structural aberration in a chromosome in which the order of several genes is reversed from the normal order. A pericentric inversion includes the centromere within the inverted region, and a paracentric inversion does not include the centromere.

inversion loop Loop structure formed by the synapsis of homologous genes in a pair of chromosomes, one of which contains an inversion.

inverted repeat Either of a pair of base sequences present in the same molecule that are identical or nearly identical but are oriented in opposite directions; often found at the ends of transposable elements.

inverted terminal repeats The short related or identical sequences present in reverse orientation at the ends of some transposons.

isogenic Homozygous for the entire portion of the genome under consideration.

isozymes Multiple molecular forms of individual enzymes. These multiple forms can be alleles of one another at a single locus – allozymes – or can be products of different loci where there are multiple copies of genes making the same enzyme or enzyme subunits.

karyotype The chromosome complement of a cell or organism; often represented by an arrangement of metaphase chromosomes according to their lengths and the positions of their centromeres.

kilobase pair (kbp) Unit of length of a duplex DNA molecule; equal to 1000 base pairs.

kinetochore The cellular structure, formed in association with the centromere, to which the spindle fibres become attached in cell division. The kinetochore is the structural feature of the chromosome to which microtubules of the mitotic spindle attach (*see also* centromere).

lariat structure Structure of an intron immediately after excision in which the 5′ end loops back and forms a 5′–3′ linkage with another nucleotide.

leader The non-translated sequence at the 5′ end of mRNA that precedes the initiation codon.

leader polypeptide A short polypeptide encoded in the leader sequence of some operons coding for enzymes in amino acid biosynthesis; translation of the leader polypeptide participates in regulation of the operon through attenuation.

leader sequence A short N-terminal sequence of a protein responsible for passage into or through a membrane.

leading strand The DNA strand whose complement is synthesized as a continuous unit; synthesized continuously in the 5′–3′ direction.

leaky mutations Allow some residual level of gene expression.

left splicing junction The boundary between the right end of an exon and the left end of an intron.

leptotene The initial substage of meiotic prophase I during which the chromosomes become visible in the light microscope as unpaired thread-like structures.

lethal gene A gene that results in the death of an individual at some stage of life. Lethal genes may be dominant and exert their effect in heterozygotes. Such genes are comparatively rare and difficult to study since they are rapidly eliminated from a population unless their effects occur late in life after affected individuals have produced offspring. Most lethal genes are recessive and exert their effects only when homozygous.

lethal locus Any gene in which a lethal mutation can be obtained (usually by deletion of the gene).

library A set of cloned fragments together representing the entire genome.

ligand The molecule that binds to a specific receptor.

ligation The formation of a phosphodiester bond to link two adjacent bases separated by a nick in one strand of a double helix of DNA. (The term can also be applied to blunt-end ligation and to joining of RNA.)

line A population or genetic group developed by directed breeding usually involving some type of inbreeding, line-breeding or selection.

line-breeding A form of inbreeding in which an effort is made to maintain high relationships in subsequent generations with a favoured ancestor.

linkage Refers to gene pairs (or series), members of which are on the same chromosome, that tend to remain together at meiosis more frequently than would be expected if they segregated independently. The tendency of genes located in the same chromosome to be associated in inheritance more frequently than expected from their independent assortment in meiosis; the tendency of genes to be inherited together as a result of their location on the same chromosome; measured by per cent recombination between loci.

linkage disequilibrium A situation in which some combinations of genetic markers occur more or less frequently in the population than would be expected from their distance apart. It implies that a group of markers has been inherited coordinately. It can result from reduced recombination in the region or from a founder effect, in which there has been insufficient time to reach equilibrium since one of the markers was introduced into the population; non-random associations between two sites.

linkage group The set of genes present together in a chromosome. A linkage group includes all loci that can be connected (directly or indirectly) by linkage relationships; equivalent to a chromosome.

linkage map (genetic map) A diagram of the order of genes in a chromosome in which the distance between adjacent genes is proportional to the rate of recombination between them; also called a genetic map.

linker DNA In genetic engineering, synthetic DNA fragments that contain restriction-enzyme cleavage sites used to join two DNA molecules (*see also* nucleosome).

local population A group of organisms of the same species occupying an area within which most individual members find their mates; synonymous terms are deme and Mendelian population.

locus The site/position on a chromosome where a specific gene pair or allelic series is located; the position on a chromosome at which the gene for a particular trait resides; a locus may be occupied by any one of the alleles for the gene.

LOD score A measure of genetic linkage, defined as the $\log_{10}$ ratio of the probability that the data would have arisen if the loci were linked to the probability that the data could have arisen from unlinked loci. The conventional threshold for declaring linkage is a LOD score of 3.0, i.e. a 1000:1 ratio (which must be compared with the 50:1 probability that any random pair of loci will be unlinked).

loss-of-function mutation A mutation that eliminates gene function; also called a null mutation.

lost allele An allele no longer present in a population; its frequency is 0.

LTR An abbreviation for long terminal repeat, a sequence directly repeated at both ends of a retroviral DNA.

luxury genes Genes coding for specialized functions synthesized (usually) in large amounts in particular cell types.

major groove In B-form DNA, the larger of two continuous indentations running round the outside of the double helix.

major histocompatibility locus (MHC) A large chromosomal region containing a giant cluster of genes that code for transplantation antigens and other proteins found on the surfaces of lymphocytes.

map-based cloning A strategy of gene cloning based on the position of a gene in the genetic map; also called positional cloning.

map distance Measured as centimorgans (cM), which is per cent recombination (sometimes subject to adjustments).

mapping function The mathematical relationship between the genetic map distance across an interval and the observed percentage of recombination in the interval. The mathematical function converts recombination frequencies into genomic map distances by accounting for the incidence of double crossovers between markers. Two commonly used functions are Haldane and Kosambi. Kosambi also adjusts the data for interference.

map unit A unit of distance in a linkage map that corresponds to a recombination frequency of 1%. Technically, the map distance across an interval in map units equals one-half the average number of crossovers in the interval, expressed as a percentage. Map units are sometimes called centimorgans (cM).

masked mRNA Messenger RNA (mRNA) that cannot be translated until specific regulatory substances are available; present in eukaryotic cells, particularly eggs; storage mRNA.

marker (genetic) Any allele or DNA sequence of interest in an experiment.

marker-assisted selection (MAS) Selection for a trait based on DNA markers that are associated with the desirable phenotypes of the trait rather than selection based solely on the phenotype.

maternal effect The impact made by the size, age and condition of the female upon the quality of the eggs and upon the growth and viability characteristics of the embryo after fertilization; the environmental influence that is contributed by the mother to the phenotypes of her offspring.

maternal-effect gene A gene that influences early development through its expression in the mother and the presence of the gene product in the oocyte.

maternal inheritance Extranuclear inheritance of a trait through cytoplasmic factors or organelles contributed by the female gamete; also describes the preferential survival in the progeny of genetic markers provided by one parent.

mean Average of all measurements of a given character in a population.

megabase pair Unit of length of a duplex nucleic acid molecule; equal to 1 million base pairs.

meiocyte A germ cell that undergoes meiosis to yield gametes in animals or spores in plants.

meiosis Cell division during germ cell formation in which chromosome number is reduced, with each daughter cell receiving only one member of each chromosome pair. The process of nuclear division in gametogenesis or sporogenesis in which one replication of the chromosomes is followed by two successive divisions of the nucleus to produce four haploid nuclei. Meiosis occurs by two successive divisions (meiosis I and II) that reduce the starting number of 4N chromosomes to 1N in each of four product cells. Products may mature to germ cells (sperm or eggs).

meiotic gynogen Gynogen produced by blocking extrusion of the second polar body.

Mendelian genetics The mechanism of inheritance in which the statistical relationships between the distribution of traits in successive generations result from: (i) particulate hereditary determinants (genes); (ii) random union of gametes; and (iii) segregation of unchanged hereditary determinants in the reproductive cells.

messenger RNA (mRNA) An RNA molecule transcribed from a DNA sequence and translated into the amino acid sequence of a polypeptide. In eukaryotes, the primary transcript, heterogeneous nuclear RNA, undergoes elaborate processing to become the mRNA.

metabolic pathway A set of chemical reactions that take place in a definite order to convert a particular starting molecule into one or more specific products.

metabolite Any small molecule that serves as a substrate, an intermediate or a product of a metabolic pathway.

metacentric chromosome A chromosome with its centromere near to the middle so that the arms are equal or almost equal in length.

metaphase In mitosis, meiosis I or meiosis II, the stage of nuclear division in which the centromeres of the condensed chromosomes are arranged in a plane between the two poles of the spindle.

metaphase plate Imaginary plane, equidistant from the spindle poles in a metaphase cell, on which the centromeres of the chromosomes are aligned by the spindle fibres.

microarray A collection of gene probes that have been spotted on to a glass slide or synthesized as oligonucleotides on a silicon chip or collection of beads.

microsatellite A stretch of repetitive DNA made of a variable number of several to one hundred or more tandem repeats of a small number of nucleotides, most commonly di- or trinucleotides.

migration In a genetic sense, the introduction of genes into a population from a source outside the population. Movement of organisms among subpopulations; also, the movement of molecules in electrophoresis.

minor groove In B-form DNA, the smaller of two continuous indentations running around the outside of the double helix.

mismatch repair Removal of one nucleotide from a pair that cannot properly hydrogen-bond, followed by replacement with a nucleotide that can hydrogen-bond.

missense mutation An alteration in a coding sequence of DNA that results in an amino acid replacement in the polypeptide.

mitochondrial DNA (mtDNA) Circular DNA found in the mitochondria that is almost always only inherited maternally.

mitosis Cell division in which each chromosome duplicates itself and the daughter cells each have the same number of chromosomes as the parent cell. The process of nuclear division in which the replicated chromosomes divide and the daughter nuclei have the same chromosome number and genetic composition as the parent nucleus.

mitotic gynogen Gynogen produced by blocking first-cell division.

mode Class with the highest frequency when measurements of a given trait in a population are tabulated.

modified bases All those except the usual four from which DNA (T, C, A, G) or RNA (U, C, A, G) are synthesized; they result from post-synthetic changes in the nucleic acid.

molecular genetics The science having to do with genetic variation at the molecular level.

monocistronic mRNA Codes for one protein.

monohybrid A genotype that is heterozygous for one pair of alleles; the offspring of a cross between genotypes that are homozygous for different alleles of a gene.

monoploid The basic chromosome set that is reduplicated to form the genomes of the species in a polyploid series; the smallest haploid chromosome number in a polyploid series.

monosomic Condition of an otherwise diploid organism in which one member of a pair of chromosomes is missing.

mosaic An organism composed of two or more genetically different types of cell.

motif A short conserved sequence of nucleotides or amino acids, often suggesting conservation of function.

multifactorial trait A trait determined by the combined action of many factors, typically some genetic and some environmental.

multimeric proteins Consist of more than one subunit.

multiple alleles The presence in a population of more than two alleles of a gene. A series of more than two genes which can occupy a particular locus on a chromosome.

multiple cloning site The site in a plasmid into which foreign DNA is inserted at one of a number of unique restriction enzyme sites, also termed a polylinker.

multiple gene heredity Hereditary situations in which more than one gene pair (or series) influences a specific character of an animal or plant.

multiplex PCR Simultaneous amplification of multiple different fragments of DNA by using several pairs of gene-specific primers in the polymerase chain reaction.

mutagen An agent that is capable of increasing the rate of mutation.

mutant Any heritable biological entity that differs from the wild type, such as a mutant DNA molecule, mutant allele, mutant gene, mutant chromosome, mutant cell, mutant organism or mutant heritable phenotype; also, a cell or organism in which a mutant allele is expressed.

mutation A sudden, heritable change in genetic material. Chemically, a mutation is due to a change in DNA at a particular point in a chromosome. A heritable alteration in a gene or chromosome; also, the process by which such an alteration happens.

mutation rate The probability of a new mutation in a particular gene, either per gamete or per generation.

muton The smallest genetic unit capable of change or mutation. A single base in a nucleotide.

narrow-sense heritability The ratio of the additive genetic variance to the total phenotypic variance.

natural selection The process of evolutionary adaptation in which the genotypes genetically best suited to survive and reproduce in a particular environment give rise to a disproportionate share of the offspring and so gradually increase the overall ability of the population to survive and reproduce in that environment.

negative complementation When interallelic complementation allows a mutant subunit to suppress the activity of a wild-type subunit in a multimeric protein.

negative regulation Regulation of gene expression in which mRNA is not synthesized until a repressor is removed from the DNA of the gene. Negative regulators function by switching off transcription or translation.

negative supercoiling The twisting of a duplex of DNA in space in the opposite direction to the turns of the strands in the double helix.

neutral substitutions In a protein, those changes of amino acids that do not affect activity.

neutral theory The null hypothesis explaining the distribution of molecular variation in natural populations in the absence of natural selection. Factors affecting rates of neutral evolution include mutation pressure, migration rate, population size, breeding structure and recombination rate.

next-generation sequencing Pryrosequencing, reversible terminator technology, ligation sequencing which are more rapid and economical than traditional dideoxy-based methods of DNA sequencing.

nick A single-strand break in a DNA molecule. A nick in duplex DNA is the absence of a phosphodiester bond between two adjacent nucleotides on one strand.

nicking A situation in which offspring are superior to either parent or in which unexpectedly favourable results are obtained from crosses of two breeds or strains (*see also* specific combining ability).

nick translation The ability of *Escherichia coli* DNA polymerase I to use a nick as a starting point from which one strand of a duplex DNA can be degraded and replaced by resynthesis of new material; used to introduce radioactively labelled nucleotides into DNA *in vitro*.

non-autonomous controlling elements Defective transposons that can transpose only when assisted by an autonomous controlling element of the same type.

non-disjunction Failure of chromatids (duplicate chromosomes) to separate (disjoin) and move to opposite poles of the division spindle during mitosis or meiosis; the result is loss or gain of a chromosome.

non-replicative transposition When transposons move non-replicatively; thus do not generate more copies of themselves in the host chromosome. The transposon excises itself from the chromosome, moves and then re-integrates (through the action of transposase).

nonsense codon Any one of three triplets (UAG, UAA, UGA) that cause termination of protein synthesis. (UAG is known as amber, UAA as ochre.)

nonsense mutation A mutation that changes a codon specifying an amino acid into a stop codon, resulting in premature polypeptide chain termination; also called a chain-termination mutation; any change in DNA that causes a (termination) codon to replace a codon representing an amino acid.

nonsense suppressor A gene coding for a mutant transfer RNA (tRNA) able to respond to one or more of the termination codons.

non-transcribed spacer The region between transcription units in a tandem gene cluster.

normal distribution A symmetrical bell-shaped distribution characterized by the mean and the variance; in a normal distribution, approximately 68% of the observations are within 1 standard deviation from the mean, and approximately 95% are within 2 standard deviations.

Northern blotting A technique for transferring RNA from an agarose gel to a nitrocellulose filter, on which it can be hybridized to a complementary DNA.

nuclear DNA (nDNA) DNA found in the nucleus of the cell.

nuclease An enzyme that breaks phosphodiester bonds in nucleic acid molecules.

nucleic acid A polymer composed of repeating units of phosphate-linked five-carbon sugars to which nitrogenous bases are attached; DNA and RNA.

nucleic acid hybridization The formation of duplex nucleic acid from complementary single strands.

nucleolar organizer The region of a chromosome carrying genes coding for ribosomal RNA (rRNA).

nucleolus (*pl.* nucleoli) Nuclear organelle in which ribosomal RNA (rRNA) is made and ribosomes are partially synthesized; usually associated with the nucleolar organizer region (NOR). A nucleus may contain several nucleoli. A nucleolus is a discrete region of the nucleus created by the transcription of rRNA.

nucleoside A purine or pyrimidine base covalently linked to a sugar.

nucleosome The basic repeating subunit of chromatin, consisting of a core particle composed of two molecules each of four different histones around which a length of DNA containing about 145 nucleotide pairs is wound, joined to an adjacent core particle by about 55 nucleotide pairs of linker DNA associated with a fifth type of histone, styled H1. The nucleosome is the basic structural subunit of chromatin, consisting of approximately 147 bp of DNA and an octamer of histone proteins.

nucleotide A nucleoside phosphate.

nucleotide analogue A molecule that is structurally similar to a normal nucleotide and that is incorporated into DNA.

nucleotide diversity The average proportion of nucleotide differences between all pairs of sequences in a sample.

null mutation Completely eliminates the function of a gene, usually because it has been physically deleted.

ochre codon The triplet UAA, one of three codons that cause termination of protein synthesis.

ochre mutation Any change in DNA that creates a UAA codon at a site previously occupied by another codon.

ochre suppressor A gene coding for a mutant transfer RNA (tRNA) able to respond to the UAA codon to allow continuation of protein synthesis; ochre suppressors also suppress amber codons.

oestrogen A generic term for substances with biological effects characteristic of oestrogenic hormones. Often called female sex hormones. They are involved in many reproductive functions.

Okazaki fragment Any of the short strands of DNA produced during discontinuous replication of the lagging strand of DNA; also called a precursor fragment.

oligonucleotide primer A short, single-stranded nucleic acid synthesized for use in DNA sequencing or as a primer in the polymerase chain reaction.

oncogene A gene that can initiate tumour formation, especially when mutated or when its expression pattern is disturbed.

open reading frame (ORF) In the coding strand of DNA or in mRNA, a region containing a series of at least 20 codons uninterrupted by stop codons and therefore capable of coding for a polypeptide chain.

operator A regulatory region in DNA of prokaryotes that interacts with a specific repressor protein in controlling the transcription of adjacent structural genes. The operator is the site on DNA at which a repressor protein binds to prevent transcription from initiating at the adjacent promoter.

operon A collection of adjacent structural genes in prokaryotic organisms regulated by an operator and a repressor. *See* open reading frame. An operon is a unit of bacterial gene expression and regulation, including structural genes and control elements in DNA recognized by regulator gene product(s).

orphans Genes of unknown function.

orphons Isolated individual genes found in isolated locations, but related to members of a gene cluster.

orthologues (true homologues) Two genes in separated species that are derived from a common ancestor without gene duplication.

outbreeding A system of mating in which mates are less related than average individuals of the population being intermated.

outcrossing Mating unrelated animals within the same pure breed. Often, 'unrelated' is interpreted to mean no common ancestors in the first four to six generations of their pedigrees.

overdominance A genetic situation in which individuals heterozygous for a gene pair (or series) are superior in some manner to any homozygote of the pair or series. Negative overdominance occurs when the F_1 is inferior to both parents.

***P* transposable element** A *Drosophila* transposable element used for the induction of mutations, germ-line transformation, and other types of genetic engineering.

P1 artificial chromosome A plasmid vector containing regions of the bacteriophage P1 and a large inserted DNA fragment.

P_1 generation The parents used in a cross or the original parents in a series of generations; also called the P generation if there is no chance of confusion with the grandparents or more remote ancestors.

pachytene The middle substage of meiotic prophase I, in which the homologous chromosomes are closely synapsed.

palindrome A sequence of DNA that is the same when one strand is read left to right or the other is read right to left; consists of adjacent inverted repeats.

paracentric inversion An inversion that does not include the centromere.

paralogues Similar genes that arose from duplication of an ancestral gene in one or both lineages prior to an evolutionary split.

parental combination Alleles present in an offspring chromosome in the same combination as that found in one of the parental chromosomes.

parent strand The strand that served as the template in a newly formed duplex in DNA replication.

pedigree A record of the animals from which a given individual is descended. The definition is often extended to include animals that are collaterally related to an individual. In animal breeding the term 'pedigree information' includes identification of ancestors and collateral relatives and information on their performance or progeny records. A diagram representing the familial relationships among relatives.

penetrance The proportion of organisms having a particular genotype that actually express the corresponding phenotype. If the phenotype is always expressed, penetrance is complete; otherwise, it is incomplete.

permissive condition An environmental condition in which the phenotype of a conditional mutation is not expressed; contrasts with the non-permissive or restrictive condition.

phenocopy An environmentally induced mutation that mimics a known genetic mutation.

phenotype The external appearance, performance or some other observable or measurable characteristic of an individual. The observable properties of a cell or an organism that result from the genotype, the environment and the interaction of the genotype and the environment.

phenotypic variance Total variance including that due to both environmental and hereditary influences.

phosphodiester bond In nucleic acids, the covalent bond formed between the 5'-phosphate group (5-P') of one nucleotide and the 3'-hydroxyl group (3'-OH) of the next nucleotide in line; these bonds form the backbone of a nucleic acid molecule.

photoreactivation The enzymatic splitting of pyrimidine dimers produced in DNA by ultraviolet light; requires visible light and the photoreactivation enzyme.

phylogenetic footprinting The process of aligning the sequences of a stretch of DNA from multiple divergent species, typically for the purpose of detecting evolutionarily conserved elements that may encode genes or other important DNA sequences.

phylogenetic shadowing The process of aligning the sequences of a stretch of DNA from several closely related species, typically for the purpose of detecting unusually highly conserved DNA elements that may encode regulatory elements.

phylogenetic tree A diagram showing the genealogical relationships among a set of genes or species.

physical map A diagram showing the relative positions of physical landmarks in a DNA molecule; common landmarks include the positions of restriction sites and particular DNA sequences. A physical map is an assembly of contiguous stretches of chromosomal DNA, contigs.

plasmid An extrachromosomal genetic element commonly found in prokaryotes that replicates independently of the host chromosome; it may exist in one or many copies per cell and may segregate in cell division to daughter cells in either a controlled or a random fashion. Some plasmids, such as the F factor, may become integrated into the host chromosome.

pleiotropy Genetic situations in which one gene affects more than one qualitative or quantitative character or trait of an individual.

ploidy The number of copies of the chromosome set present in a cell; a haploid has one copy, a diploid has two copies, etc.

point mutations In DNA, changes involving single base pairs.

polarity The 5'-to-3' orientation of a strand of nucleic acid.

poly-A tail The sequence of adenines added to the 3' end of many eukaryotic mRNA molecules in processing.

polycistronic mRNA An mRNA molecule from which two or more polypeptides are translated; found primarily in prokaryotes. Polycistronic mRNA includes coding regions representing more than one gene.

polylinker A short DNA sequence that is present in a vector and that contains a number of unique restriction sites suitable for gene cloning.

polymerase chain reaction (PCR) Repeated cycles of DNA denaturation, renaturation with primer oligonucleotide sequences and replication, resulting in exponential growth in the number of copies of the DNA sequence located between the primers. PCR describes a technique in which cycles of denaturation, annealing with primer and extension with DNA polymerase are used to amplify the number of copies of a target DNA sequence by >10^6 times.

polymorphic gene A gene for which there is more than one relatively common allele in a population.

polymorphism The case where a locus has two or more alleles. The presence in a population of two or more relatively common forms of a gene, chromosome or genetically determined trait.

polynucleotide chain A polymer of covalently linked nucleotides.

polypeptide/polypeptide chain A polymer of amino acids linked together by peptide bonds.

polyploid A general term applied to cells with three or more times the haploid number of chromosomes. The condition of a cell or organism with more than two complete sets of chromosomes.

polysome A complex of two or more ribosomes associated with an mRNA molecule and actively engaged in polypeptide synthesis (translation); a polyribosome.

polysomy The condition of a diploid cell or organism that has three or more copies of a particular chromosome or of many, but not all, chromosomes in the set.

polytene chromosome A giant chromosome consisting of many identical strands laterally apposed and in register, exhibiting a characteristic pattern of transverse banding.

population A group of organisms of the same species inhabiting a single locality and forming a single unit.

population genetics A field of enquiry in which genetics as related to a group or population is considered in contrast to the genetics of individuals. Application of Mendel's laws and other principles of genetics to entire populations of organisms.

population substructure Organization of a population into smaller breeding groups between which migration is restricted. Also called population subdivision.

position effect A phenomenon often observed in transgenic animals and plants, in which the site of insertion has a large effect on the level of expression of the transgene.

positional cloning A strategy of gene cloning based on the position of a gene in the genetic map; also called map-based cloning.

positional information Developmental signals transmitted to a cell by virtue of its position in the embryo.

positive regulation Mechanism of gene regulation in which an element must be bound to DNA in an active form to allow transcription. Positive regulation contrasts with negative regulation, in which a regulatory element must be removed from DNA.

post-replication repair DNA repair that takes place in non-replicating DNA or after the replication fork is some distance beyond a damaged region.

precursor fragment *See* Okazaki fragment.

primary transcript An RNA copy of a gene; in eukaryotes, the transcript must be processed to form a translatable mRNA molecule.

primer A short RNA or single-stranded DNA segment that functions as a growing point in polymerization. A primer is a short sequence (often of RNA) that is paired with one strand of DNA and provides a free 3′-OH end at which a DNA polymerase starts synthesis of a deoxyribonucleotide chain.

primosome The enzyme complex that forms the RNA primer for DNA replication in eukaryotic cells.

probe A radioactive DNA and RNA molecule used in DNA–RNA and DNA–DNA hybridization assays.

processed pseudogene An inactive gene copy that lacks introns, contrasted with the interrupted structure of the active gene. Such genes presumably originate by reverse transcription of mRNA and insertion of a duplex copy into the genome.

product molecule The end result of a biochemical reaction or a metabolic pathway.

progeny Young or offspring of given individuals.

progeny test Estimate of the genetic value or make-up of an individual through measuring or observing the performance, appearance or other characteristics of a group of progeny.

programmed cell death Cell death that happens as part of the normal developmental process (*see also* apoptosis).

promoter Regulatory DNA that starts and stops expression of a gene. A DNA sequence at which RNA polymerase binds and initiates transcription.

prophase The initial stage of mitosis or meiosis, beginning after interphase and terminating with the alignment of the chromosomes at metaphase; often absent or abbreviated between meiosis I and meiosis II.

pseudo-autosomal region A small region of the X and Y chromosome containing homologous genes in mammals.

pseudogenes Inactive but stable components of the genome derived by mutation from an ancestral active gene.

puff An expansion of a band of a polytene chromosome associated with the synthesis of RNA at some locus in the band.

pure-breed/pure-bred An animal both of whose parents are duly registered in the herd, flock or studbook of a given breed.

purine An organic base found in nucleic acids; the predominant purines are adenine and guanine.

pyrimidine An organic base found in nucleic acids; the predominant pyrimidines are cytosine, uracil (in RNA only) and thymine (in DNA only).

pyrimidine dimer Two adjacent pyrimidine bases, typically a pair of thymines, in the same polynucleotide strand, between which chemical bonds have formed; the most common lesion formed in DNA by exposure to ultraviolet light.

qualitative inheritance Heredity relating to traits for which populations can be divided into discrete classes and whose phenotypic expression is controlled by environment and by the action of one pair or a few pairs of alleles of genes.

quantitative inheritance Heredity relating to traits for which populations exhibit a continuous array of variability and whose phenotypic expression is affected by environment and by the action of several pairs (or series) of genes. Effects of individual genes can seldom be detected.

quantitative trait A trait – typically measured on a continuous scale, such as height or weight – that results from the combined action of several or many genes in conjunction with environmental factors.

quantitative trait locus (QTL) A locus segregating for alleles that have different, measurable effects on the expression of a quantitative trait.

quaternary structure Multimeric constitution of a protein.

radiation hybrid mapping A method of assembly of genetic maps of vertebrates or plants in which fragments of the genome of one species are propagated in hybrid cell lines with another species. Co-segregation of sequences in multiple lines indicates that the two sequences are physically linked.

random amplified polymorphic DNA (RAPD) markers Polymorphic DNA sequences separated by gel electrophoresis after PCR, using one or a pair of short (8–12 bp) random oligonucleotide primers. Polymorphisms are a result of base changes in the primer binding sites or sequence length changes caused by insertions, deletions or rearrangements. When two binding sites are close enough (3000 bp or less), an RAPD band is produced on the gel. Each RAPD primer usually amplifies several bands, and RAPD markers are expressed and scored as dominant alleles.

random genetic drift Fluctuation in allele frequency from generation to generation resulting from restricted population size.

random mating A breeding situation in which any male or any female has an equal probability of mating with any other individual of opposite sex in the population regardless of similarity or dissimilarity of appearance, measurable characteristics or parentage. System of mating in which mating pairs are formed independently of genotype and phenotype.

reading frame The phase in which successive triplets of nucleotides in mRNA form codons; depending on the reading frame, a particular nucleotide in an mRNA could be in the first, second or third position of a codon. The reading frame actually used is defined by the AUG codon that is selected for chain initiation. The reading frame is one of three possible ways of reading a nucleotide sequence as a series of triplets.

reassociation The pairing of complementary single strands to form a double helix of DNA, following strand separation by melting.

receptor A transmembrane protein, located in the plasma membrane, that binds a ligand in a domain on the extracellular side and, as a result, has a change in activity of the cytoplasmic domain. (The same term is sometimes used also for the steroid receptors, which are transcription factors that are activated by binding ligands that are steroids or other small molecules.)

recessive allele Genes that have no observable effect unless present in both members of a chromosome pair; an allele, or the corresponding phenotypic trait, expressed only in homozygotes; obscured in the phenotype of a heterozygote by the dominant allele.

recessive lethal A recessive allele that is lethal when the cell is homozygous for it.

reciprocal cross A cross in which the sexes of the parents are the reverse of those in another cross.

reciprocal recombination The production of new genotypes with the reverse arrangements of alleles according to maternal and paternal origin.

reciprocal translocation Interchange of parts between non-homologous chromosomes. Reciprocal translocation exchanges part of one chromosome with part of another chromosome.

recombinant A chromosome that results from crossing over and that carries a combination of alleles differing from that of either chromosome participating in the crossover; the cell or organism that contains a recombinant chromosome.

recombinant DNA A DNA molecule composed of one or more segments from other DNA molecules.

recombinant progeny Have a different genotype from that of either parent.

recombination Occurrence in offspring of genetic combinations not found in parents. Exchange of parts between DNA molecules or chromosomes; recombination in eukaryotes usually entails a reciprocal exchange of parts, but in prokaryotes it is often non-reciprocal.

recombination nodules (nodes) Dense areas present on the synaptonemal complex; could be involved in crossing over.

recombination repair A mode of filling a gap in one strand of duplex DNA by retrieving a homologous single strand from another duplex.

recruitment The process in which a transcriptional activator protein interacts with one or more components of the transcription complex and attracts it to the promoter.

reductional division Term applied to the first meiotic division because the chromosome number (counted as the number of centromeres) is reduced from diploid to haploid.

redundancy The feature of the genetic code in which an amino acid corresponds to more than one codon; also called degeneracy.

redundant gene A gene whose function can be supplied by another gene or genes if mutated.

regression Amount of change in one trait associated with a unit change in another trait in a population (*see also* correlation).

regulatory gene Codes for an RNA or protein product whose function is to control the expression of other genes.

related A term indicating that two individuals have one or more common ancestors or that one is a descendant of the other. In ordinary usage, animals are usually considered to be related only if they have common ancestry in the first four to six generations of their pedigrees.

relationship The degree to which individuals are more highly related than the average of individuals for the population to which they belong.

relative fitness The fitness of a genotype expressed as a proportion of the fitness of another genotype.

repeatability The tendency for an individual to repeat its performance, e.g. a dairy cow in successive lactations, a ewe in weaning weights of successive lambs, linear measurements or gains of any animal in successive periods, etc. In statistical terms, it is the proportion of total variance in a population that is due to similarity of performance of individuals when all are measured or evaluated more than once.

replacement sites In a gene, those sites at which mutations alter the amino acid that is coded.

replication fork In a replicating DNA molecule, the region in which nucleotides are added to growing molecular strands.

replication origin The base sequence at which DNA synthesis begins.

replication slippage The process in which the number of copies of a small tandem repeat can increase or decrease during replication.

replicative transposition A mechanism of transposition in molecular biology in which the transposable element is duplicated during the reaction, so that the transposing entity is a copy of the original element. Replicative transposition is characteristic to retrotransposons and occurs from time to time in class II transposons.

replicon A DNA molecule that has a replication origin.

reporter gene A coding unit whose product is easily assayed (such as chloramphenicol transacetylase); it may be connected to any promoter of interest so that expression of the gene can be used to assay promoter function.

repressible transcription A regulatory process in which a gene is temporarily rendered unable to be transcribed.

repressor A protein that binds specifically to a regulatory sequence adjacent to a gene and blocks transcription of the gene.

repulsion *See trans* configuration.

restriction endonuclease A nuclease that recognizes a short nucleotide sequence (restriction site) in a DNA molecule and cleaves the molecule at that site; also called a restriction enzyme.

restriction enzyme *See* restriction endonuclease.

restriction fragment A segment of duplex DNA produced by cleavage of a larger DNA molecule by a restriction enzyme.

restriction fragment length polymorphism (RFLP) Genetic variation in a population associated with the size of restriction fragments that contain sequences homologous to a particular probe DNA; the polymorphism results from the positions of restriction sites flanking the probe, and each variant is essentially a different allele. RFLP refers to inherited differences in sites for restriction enzymes (for example, caused by base changes in the target site) that result in differences in the lengths of the fragments produced by cleavage with the relevant restriction enzyme. RFLPs are utilized for genetic mapping to link the genome directly to a conventional genetic marker.

restriction map A diagram of a DNA molecule showing the positions of cleavage by one or more restriction endonucleases. A restriction map is a linear array of sites on DNA cleaved by various restriction enzymes.

restriction site The base sequence at which a particular restriction endonuclease makes a cut.

restrictive condition A growth condition in which the phenotype of a conditional mutation is expressed.

retroposon A transposon that mobilizes via an RNA form; the DNA element is transcribed into RNA and then reverse-transcribed into DNA, which is inserted at a new site in the genome.

retrovirus One of a class of RNA animal viruses that cause the synthesis of DNA complementary to their RNA genomes on infection. An RNA virus that propagates via conversion into duplex DNA. The duplex DNA synthesized from the RNA genome of a retrovirus may be integrated into the host genome.

reverse genetics Procedure in which mutations are deliberately produced in cloned genes and introduced back into cells or the germ line of an organism.

reverse mutation A mutation that undoes the effect of a preceding mutation.

reverse transcriptase An enzyme that makes complementary DNA from a single-stranded RNA template.

reverse transcriptase PCR (RT-PCR) Amplification, using an RNA template, of a duplex DNA molecule originally produced by reverse trascriptase.

reverse transcription Synthesis of DNA on a template of RNA; accomplished by reverse transcriptase enzyme.

reverse translation A technique for isolating genes (or mRNAs) by their ability to hybridize with a short oligonucleotide sequence prepared by predicting the nucleic acid sequence from the known protein sequence.

reversion Restoration of a mutant phenotype to the wild-type phenotype by a second mutation.

ribose The five-carbon sugar in RNA.

ribosomal RNA (rRNA) RNA molecules that are components of the ribosomal subunits; in eukaryotes, there are four rRNA molecules – 5S, 5.8S, 18S and 28S; in prokaryotes, there are three – 5S, 16S and 23S.

ribosome The cellular organelle on which the codons of mRNA are translated into amino acids in protein synthesis. Ribosomes consist of two subunits, each composed of RNA and proteins. In prokaryotes, the subunits are 30S and 50S particles; in eukaryotes, they are 40S and 60S particles.

ribosome-binding site The base sequence in an mRNA molecule to which a ribosome can bind to initiate protein synthesis; also called the Shine–Dalgarno sequence.

ribosomal translocation Movement of the ribosome along a molecule of messenger RNA in translation.

ribosome tRNA-binding sites The transfer RNA (tRNA)-binding sites on the ribosome to which tRNA molecules are bound. The aminoacyl site receives the incoming charged tRNA, the peptidyl site holds the tRNA with the nascent polypeptide chain, and the exit site holds the outgoing uncharged tRNA.

ribozyme An RNA molecule able to catalyse one or more biochemical reactions.

ring chromosome A chromosome whose ends are joined; one that lacks telomeres.

RNA (ribonucleic acid) A nucleic acid in which the sugar constituent is ribose; typically, RNA is single-stranded and contains the four bases adenine, cytosine, guanine and uracil.

RNA polymerase An enzyme that makes RNA by copying the base sequence of a DNA strand.

RNA processing The conversion of a primary transcript into an mRNA, rRNA or tRNA molecule; includes splicing, cleavage, modification of termini and (in tRNA) modification of internal bases.

RNA splicing Excision of introns and joining of exons.

Robertsonian translocation A chromosomal aberration in which the long arms of two acrocentric chromosomes become joined to a common centromere.

rolling-circle replication A mode of replication in which a circular parent molecule produces a linear branch of newly formed DNA.

S phase The restricted part of the eukaryotic cell cycle during which synthesis of DNA occurs.

S1 nuclease An enzyme that specifically degrades unpaired (single-stranded) sequences of DNA.

satellite DNA Eukaryotic DNA that forms a minor band at a different density from that of most of the cellular DNA in equilibrium density-gradient centrifugation; consists of short sequences repeated many times in the genome (highly repetitive DNA) or of mitochondrial or chloroplast DNA. Satellite DNA consists of many tandem repeats (identical or related) of a short basic repeating unit.

saturation density The density to which cultured eukaryotic cells grow *in vitro* before division is inhibited by cell–cell contacts.

scaffold A protein-containing material in chromosomes, believed to be responsible in part for the compaction of chromatin; the scaffold of a chromosome is a proteinaceous structure in the shape of a sister chromatid pair, generated when chromosomes are depleted of histones. Also a set-up of contigs constituting a whole-genome sequence.

second-division segregation Segregation of a pair of alleles into different nuclei in the second meiotic division, the result of crossing over between the gene and the centromere of the pair of homologous chromosomes.

second meiotic division The meiotic division in which the centromeres split and the chromosome number is not reduced; also called the equational division.

segment Any of a series of repeating morphological units in a body plan.

segmentation gene Any of a group of genes that determines the spatial pattern of segments and para-segments in *Drosophila* development.

segment-polarity gene Any of a group of genes that determines the spatial pattern of development within the segments of *Drosophila* larvae.

segregation Separation of members of a pair of hereditary factors at meiosis in germ cell formation. Separation of the members of a pair of alleles into different gametes in meiosis.

selection Any external influence in a population, either naturally or artificially imposed, that enhances opportunities of individuals of some genotypes to contribute genetic material to subsequent generations and thereby to change gene frequencies. Selection, imposed by a breeder, in which organisms of only certain phenotypes are allowed to breed.

selection coefficient The amount by which relative fitness is reduced or increased.

selection index A system of weighing values for several traits to arrive at a single score or numerical expression for use in determining which of a given group of animals to select for breeding use and which to cull.

selection limit or plateau The condition in which a population no longer responds to artificial selection for a trait.

selection pressure The degree or intensity of selection for or against a trait in the selection process.

selectively neutral mutation A mutation that has no (or negligible) effects on fitness.

selfish DNA DNA sequences that do not contribute to the fitness of an organism but are maintained in the genome through their ability to replicate and, in some cases, transpose.

semi-conservative replication The usual mode of DNA replication, in which each strand of a duplex molecule serves as a template for the synthesis of a new complementary strand and the daughter molecules are composed of one old (parental) and one newly synthesized strand.

semi-lethal gene Gene with detrimental effects on the viability of individuals carrying them but which may not cause death in favourable environments.

sequence-tagged site (STS) A DNA sequence, present once per haploid genome, that can be amplified by the use of suitable oligonucleotide primers in the polymerase chain reaction in order to identify clones that contain the sequence.

serial analysis of gene expression (SAGE) A method of profiling gene expression based on the sequencing of very large numbers of unique tags corresponding to each gene in the genome.

sex chromosomes Chromosomes that segregate as if they were members of the same pair but which are morphologically different in the two sexes. They, or factors carried in them, are partially or wholly responsible for sex determination.

sex-determining chromosome Chromosomes that may not be morphologically distinct but have a major influence on determining sex.

sex linkage Inheritance dependent upon hereditary factors located in the sex chromosomes. Sex linkage is a pattern of inheritance shown by genes carried on a sex chromosome (usually the X).

sex ratio The ratio of males to females at a specific life stage such as at birth.

sex reversal The process of making phenotypic sex the opposite of genetic sex through the administration of sex hormones at the appropriate stages of development.

shot gun sequencing Determination of the sequence of a long stretch of DNA by randomly breaking it into a redundant set of small clones that are sequenced *en masse* so that each fragment is represented between five and ten times. The contig is then assembled by computer alignment of the overlapping sequences.

sib/sibling A brother or sister, each having the same parents.

sibship A group of brothers and sisters.

signal sequence/leader sequence The region of a protein (usually N-terminal) responsible for co-translational or post-translational insertion into membranes of the endoplasmic reticulum.

signal transduction The process by which a receptor interacts with a ligand at the surface of the cell and then transmits a signal to trigger a pathway within the cell.

silent mutation A mutation that has no phenotypic effect.

simple-sequence DNA Satellite DNA.

simple tandem-repeat polymorphism (STRP) A DNA polymorphism in a population in which the alleles differ in the number of copies of a short, tandemly repeated nucleotide sequence.

SINES A class of retroposons found as short interspersed repeats in mammalian genomes; derived from transcripts of RNA polymerase III.

single nucleotide polymorphism (SNP) Caused by base variation among individuals at any site of the genome; a site in the DNA occupied by a different nucleotide pair among individuals in a population.

single-stranded DNA A DNA molecule that consists of a single polynucleotide chain.

sire Male parent, the father of an animal.

sister chromatids Chromatids produced by replication of a single chromosome. Sister chromatids are the copies of a chromosome produced by its replication.

small nuclear ribonucleoprotein particles (snRNP) Small nuclear particles that contain short RNA molecules and several proteins. They are involved in intron excision and splicing and in other aspects of RNA processing; any of several classes of small ribonucleoprotein particles involved in RNA splicing.

somatic cell Any cell of a multicellular organism other than the gametes and the germ cells from which gametes develop. Somatic cells are all the cells of an organism except those of the germ line.

somatic mutation A mutation arising in a somatic cell. Somatic mutation is a mutation occurring in a somatic cell, and therefore affecting only its descendants; it is not inherited.

Southern blot A nucleic acid hybridization method in which, after electrophoretic separation, denatured DNA is transferred from an agarose gel to a nitrocellulose filter and then exposed to radioactive DNA or RNA (or DNA or RNA labelled non-radioactively) under conditions of renaturation; the radioactive regions locate the homologous DNA fragments on the filter.

species A group of animals or plants possessing in common one or more distinctive characteristics and which are fully fertile when intermated. They are kept genetically distinct through various forms of reproductive isolation from other species.

specific combining ability Ability of two breeds, lines or strains to produce specific effects (favourable or unfavourable) in progeny when crossed.

spindle A structure composed of fibrous proteins on which chromosomes align during metaphase and move during anaphase.

splice acceptor The 5′ end of an exon.

splice donor The 3′ end of an exon.

spliceosome An RNA–protein particle in the nucleus through the activity of which introns are removed from RNA transcripts.

splicing The removal of introns and joining of exons in RNA; thus introns are spliced out, while exons are spliced together. The ligation of separate DNA molecules by ligase activity.

splicing junctions The sequences immediately surrounding the exon–intron boundaries.

spontaneous mutation A mutation that happens in the absence of any known mutagenic agent. Spontaneous mutations are those that occur in the absence of any added reagent to increase the mutation rate.

staggered cuts In duplex DNA, these are made when two strands are cleaved at different points near each other.

standard deviation The square root of the variance for a trait measured in all individuals of a population (*see also* variance).

start codon An mRNA codon, usually AUG, at which polypeptide synthesis begins.

start point (start site) The position on DNA corresponding to the first base incorporated into RNA.

sticky end A single-stranded end of a DNA fragment produced by certain restriction enzymes capable of reannealing with a complementary sequence in another such strand. Sticky ends are complementary single strands of DNA that protrude from opposite ends of a duplex or from ends of different duplex molecules; can be generated by staggered cuts in duplex DNA.

stop codon One of three mRNA codons – UAG, UAA and UGA – at which polypeptide synthesis stops.

strain A breeding unit within a species with the same origin and history that possess at least one unique trait different from other strains.

structural gene Codes for any RNA or protein product other than a regulator.

submetacentric chromosome A chromosome whose centromere divides it into arms of unequal length.

subpopulation Any of the breeding groups within a larger population between which migration is restricted.

substrate molecule A substance acted on by an enzyme.

supernumerary chromosome fragments Small chromosome fragments generated by irradiation introduced into androgens or gynogens by the irradiated donor gamete.

super-repressed Uninducible.

suppression The occurrence of changes that eliminate the effects of a mutation without reversing the original change in DNA.

suppressor (extragenic) Usually a gene coding for a mutant transfer RNA (tRNA) that reads the mutated codon either in the sense of the original codon or to give an acceptable substitute for the original meaning.

suppressor (intragenic) A compensating mutation that restores the original reading frame after a frameshift.

synapsis The pairing of homologous chromosomes or chromosome regions in the zygotene substage of the first meiotic prophase. Synapsis describes the association of the two pairs of sister chromatids representing homologous chromosomes that occurs at the start of meiosis; the resulting structure is called a bivalent.

synaptonemal complex The morphological structure of synapsed chromosomes.

syntenic genetic loci Lie on the same chromosome.

synteny Conservation of gene order on a chromosome across species (two or more).

tandem duplication A pair of identical or closely related DNA sequences that are adjacent and in the same orientation.

tandem repeats Multiple copies of the same sequence lying in series.

TATA-binding protein (TBP) A protein that binds to the TATA motif in the promoter region of a gene.

TATA box The base sequence 5'-TATA-3' in the DNA of a promoter. The TATA box is a conserved AT-rich septamer found about 25 bp before the start-point of each eukaryotic RNA polymerase II transcription unit; may be involved in positioning the enzyme for correct initiation.

telomerase An enzyme that adds specific nucleotides to the tips of the chromosomes to form the telomeres. Telomerase is the ribonucleoprotein enzyme that creates repeating units of one strand at the telomere, by adding individual bases.

telomere The natural end of a chromosome; the DNA sequence consists of a simple repeating unit with a protruding single-stranded end that may fold into a hairpin; the tip of a chromosome, containing a DNA sequence required for stability of the chromosome end.

telophase The final stage of mitotic or meiotic nuclear division.

temperature-sensitive mutation A conditional mutation that causes a phenotypic change only at certain temperatures.

template A strand of nucleic acid whose base sequence is copied in a polymerization reaction to produce either a complementary DNA or an RNA strand.

termination codon One of three triplet sequences, UAG (amber), UAA (ochre) or UGA, that cause termination of protein synthesis; they are also called nonsense codons.

terminator A sequence of DNA, represented at the end of the transcript, that causes RNA polymerase to terminate transcription.

tertiary structure The organization in space of a polypeptide chain.

test cross Involves crossing an unknown genotype to a recessive homozygote so that the phenotypes of the progeny correspond directly to the chromosomes carried by the parent of unknown genotype. Alternatively, a heterozygote tester is used, which is less accurate for determining the genotype of the unknown parent.

testis-determining factor (TDF) Genetic element on the mammalian Y chromosome that determines maleness.

tetrad The four chromatids that make up a pair of homologous chromosomes in meiotic prophase I and metaphase I; also, the four haploid products of a single meiosis.

tetraploid Cells or individuals with four members of each set of chromosomes. A cell or organism with four complete sets of chromosomes; in an autotetraploid, the chromosome sets are homologous; in an allotetraploid, the chromosome sets consist of a complete diploid complement from each of two distinct ancestral species.

three-point cross Cross in which three genes are segregating; used to obtain unambiguous evidence of gene order.

threshold trait A trait with a continuously distributed liability or risk; organisms with a liability greater than a critical value (the threshold) exhibit the phenotype of interest, such as a disorder.

thymine (T) A nitrogenous pyrimidine base found in DNA.

topoisomerase An enzyme that introduces or eliminates either underwinding or overwinding of double-stranded DNA. It acts by introducing a single-strand break, changing the relative positions of the strands and sealing the break.

total variance Summation of all sources of genetic and environmental variation.

trailer A non-translated sequence at the 3′ end of an mRNA following the termination codon.

trait Any aspect of the appearance, behaviour, development, biochemistry or other feature of an organism.

***trans* configuration** The arrangement in linked inheritance in which a genotype heterozygous for two mutant sites has received one of the mutant sites from each parent – that is, a,+/+az. Also, *trans* configuration of two sites refers to their presence on two different molecules of DNA (chromosomes).

transcribed spacer The part of a ribosomal RNA (rRNA) transcription unit that is transcribed but discarded during maturation; that is, it does not give rise to part of rRNA.

transcript An RNA strand that is produced from, and is complementary in base sequence to, a DNA template strand.

transcription The process by which the information contained in a template strand of DNA is copied into a single-stranded RNA molecule of complementary base sequence. Transcription is the synthesis of RNA on a DNA template.

transcription complex An aggregate of RNA polymerase (consisting of its own subunits) with other polypeptide subunits that makes transcription possible.

transcription unit The distance between sites of initiation and termination by RNA polymerase; may include more than one gene.

transcriptional activator protein Positive control element that stimulates transcription by binding with particular sites in DNA.

transcriptome The complete set of transcripts expressed in a particular cell or tissue under defined conditions.

transfer RNA (tRNA) A small RNA molecule that translates a codon into an amino acid in protein synthesis; it has a three-base sequence, called the anticodon, complementary to a specific codon in mRNA, and a site to which a specific amino acid is bound.

transformation Change in the genotype of a cell or organism resulting from exposure of the cell or organism to DNA isolated from a different genotype; also, the conversion of an animal cell, whose growth is limited in culture, into a tumour-like cell whose pattern of growth is different from that of a normal cell.

transgenic Individual possessing a specific piece of foreign or exogenous DNA. The source can be a different species, same species or artificial. An animal or plant in which novel DNA has been incorporated into the germ line, created by introducing new DNA sequences into the germ line via addition to the sperm, unfertilized egg, zygote or early embryo.

transient expression Activation of gene expression for a limited period of time.

transition mutation A mutation resulting from the substitution of one purine for another purine or that of one pyrimidine for another pyrimidine.

translation The process by which the amino acid sequence of a polypeptide is synthesized on a ribosome according to the nucleotide sequence of an mRNA molecule. Translation is the synthesis of protein on the mRNA template.

translocation (chromosome) A rearrangement in which part of a chromosome is detached by breakage and then becomes attached to some other chromosome.

translocation (gene) The appearance of a new copy of a gene sequence at a location in the genome elsewhere from the original copy.

translocation (protein) Its movement across a membrane.

translocation (ribosome) Its movement one codon along mRNA after the addition of each amino acid to the polypeptide chain.

translocation interchange Of parts between non-homologous chromosomes; also, the movement of mRNA with respect to a ribosome during protein synthesis.

transmembrane receptor A receptor protein containing amino acid sequences that span the cell membrane.

transmitting ability The average genetic superiority or inferiority that is transmitted by a parent to its offspring.

transposable element A DNA sequence capable of moving (transposing) from one location to another in a genome.

transposase Protein necessary for transposition. Transposase is the enzyme involved in the insertion of transposon at a new site.

transposition The movement of a transposable element. Transposition refers to the movement of a transposon to a new site in the genome. *See also* non-replicative transposition, replicative transposition and conservative transposition.

transposition immunity The ability of certain transposons to prevent others of the same type from transposing to the same DNA molecule.

transposon A transposable element that contains bacterial genes – for example, for antibiotic resistance; also used loosely as a synonym for transposable element. A transposon is a DNA sequence able to insert itself at a new location in the genome (without any sequence relationship with the target locus).

transposon tagging Insertion of a transposable element that contains a genetic marker into a gene of interest.

transvection The ability of a locus to influence the activity of an allele on the other homologue only when two chromosomes are synapsed.

transversion A mutation in which a purine is replaced by a pyrimidine or vice versa.

transversion mutation A mutation resulting from the substitution of a purine for a pyrimidine or that of a pyrimidine for a purine.

trinucleotide repeat A tandemly repeated sequence of three nucleotides; genetic instability in trinucleotide repeats is the cause of a number of human hereditary diseases.

triplet code A code in which each codon consists of three bases.

triploid Cells or individuals with three members of each set of chromosomes. A cell or organism with three complete sets of chromosomes.

trisomic A diploid organism with an extra copy of one of the chromosomes.

trivalent Structure formed by three homologous chromosomes in meiosis I in a triploid or trisomic chromosome when each homologue is paired along part of its length with first one and then the other of the homologues.

true-breeding A strain, breed or variety of organism that yields progeny like itself; homozygous.

truncation point The value of the phenotype that determines which organisms will be retained for breeding and which will be culled in artificial selection.

tumour-suppressor gene A gene whose absence predisposes to malignancy; also called an anti-oncogene.

type A word used in animal husbandry relative to the appearance of animals but having several connotations. It is sometimes used more or less synonymously with the word conformation. It is also used to indicate distinctive kinds of animals, e.g. beef versus dairy, large versus small, fine wool versus coarse wool, etc. (*see also* conformation).

unequal crossing over Crossing over between non-allelic copies of duplicated or other repetitive sequences – for example, in a tandem duplication, between the upstream copy in one chromosome and the downstream copy in the homologous chromosome. Unequal crossing over describes a recombination event in which the two recombining sites lie at non-identical locations in the two parental DNA molecules.

univalent Structure formed in meiosis I in a monoploid or a monosomic when a chromosome has no pairing partner.

upstream Identifies sequences proceeding in the opposite direction from expression; for example, the bacterial promoter is upstream from the transcription unit, the initiation codon is upstream of the coding region.

uracil (U) A nitrogenous pyrimidine base found in RNA.

URF An open (unidentified) reading frame, presumed to code for protein, but for which no product has been found.

variable expressivity Differences in the severity of expression of a particular genotype.

variance Average squared deviations from the mean for a trait measured in all individuals of a population. A measure of the spread of a statistical distribution; the mean of the squares of the deviations from the mean.

variation Differences among individuals in measurable or observable traits. Variation may be continuous (quantitative) or discontinuous (qualitative) in nature.

vector A DNA molecule, capable of replication, into which a gene or DNA segment is inserted by recombinant DNA techniques; a cloning vehicle.

Watson–Crick base pairing Base pairing in DNA or RNA in which A pairs with T (or U in RNA) and G pairs with C.

wild type The most common phenotype or genotype in a natural population; also, a phenotype or genotype arbitrarily designated as a standard for comparison.

wobble The acceptable pairing of several possible bases in an anticodon with the base present in the third position of a codon.

wobble hypothesis Accounts for the ability of a transfer RNA (tRNA) to recognize more than one codon by unusual (non-GC, AT) pairing with the third base of a codon.

X chromosome A chromosome that plays a role in sex determination and that is present in two copies in the homogametic sex and in one copy in the heterogametic sex.

xenogenesis A xenogenic organism is comprised of elements typically foreign to its species. Xenogenesis is a method of reproduction in which successive generations differ from each other.

X-linked gene A gene located in the X chromosome; X-linked inheritance is usually evident from the production of non-identical classes of progeny from reciprocal crosses.

Y chromosome The sex chromosome present only in the heterogametic sex; in mammals, the male-determining sex chromosome.

yeast artificial chromosome (YAC) In yeast, a cloning vector that can accept very large fragments of DNA; a chromosome introduced into yeast derived from such a vector and containing DNA from another organism.

zinc-finger protein Has a repeated motif of amino acids with characteristic spacing of cysteines that may be involved in binding zinc; is characteristic of some proteins that bind to DNA and/or RNA.

zygote The cell formed at fertilization by the union of the sperm and ovum. The product of the fusion of a female gamete and a male gamete in sexual reproduction; a fertilized egg.

zygotene The substage of meiotic prophase I in which homologous chromosomes synapse.

zygotic gene Any of a group of genes that control early development through their expression in the zygote.

References and Further Reading

Abdelhamid, A.A. (1988) Genetic homogeneity of seven populations of *Tilapia nilotica* in Africa, Central America and Southeast Asia. MSc thesis, Auburn University, Auburn, Alabama.

Abella, T.A., Palada, M.S. and Newkirk, G.F. (1990) Within family selection for growth rate with rotation mating in *Oreochromis niloticus*. In: Hirano, R. and Hanyu, I. (eds) *Proceedings of The Second Asian Fisheries Forum, Tokyo, Japan, 17–22 April 1989*. Asian Fisheries Society, Manila, pp. 515–518.

Aboudy, Y., Mendelson, E., Shalit, I., Bessalle, R. and Fridkin, M. (1994) Activity of two synthetic amphiphilic peptides and magainin-2 against herpes simplex virus types 1 and 2. *International Journal of Peptide and Protein Research* 43, 573–582.

ABRAC (1995) Performance Standards for Safely Conducting Research with Genetically Modified Fish and Shellfish. Agricultural Biotechnology Research Advisory Committee – US Department of Agriculture; available at www.nbiap.vt.edu/perfstands.

Abrahams, M.V. and Sutterlin, A. (1999) The foraging and anti-predator behaviour of growth-enhanced transgenic Atlantic salmon. *Animal Behaviour* 58, 933–942.

Abramova, N.B., Burakova, T.A., Korzh, V.P. and Neifakh, A.A. (1979) Injection of mitochondria into oocytes and fertilized eggs (English abstract). *Ontogenez* 10, 401–405.

Abramova, N.B., Korzh, V.P. and Neifakh, A.A. (1980) Regulation of oxygen uptake and cytochrome oxidase activity after artificial increase in number of mitochondria in fish embryos. *Biokhimiya* 45, 855–861.

Abramova, N.B., Korzh, V.P. and Neifakh, A.A. (1983) Regulation of number and function of mitochondria after artificial increase in their mass in fish embryos. *Biokhimiya* 48, 1096–1102.

Abramova, N.B., Bueverova, E.I. and Neifakh, A.A. (1989) Activity of mammalian mitochondria injected into fish embryos (English abstract). *Ontogenez* 20, 320–323.

Abucay, J.S., Mair, G.C., Skibinski, D.O.F. and Beardmore, J.A. (1999) Environmental sex determination: the effect of temperature and salinity on sex ratio in *Oreochromis niloticus* L. *Aquaculture* 173, 219–234.

Adams, M.D., Kelley, J.M., Gocayne, J.D., Dubnick, M., Polymeropoulos, M.H., Xiao, H., Merril, C.R., Wu, A., Olde, B., Moreno, R.F., Kerlavage, A.R., McCombie, W.R. and Venter, J.C. (1991) Complementary DNA sequencing: expressed sequence tags and human genome project. *Science* 252, 1651–1656.

Adamson, I., Esangbedo, A., Okolo, A.A. and Omene, J.A. (1988) Pepsin and its multiple forms in early life. *Biology of the Neonate* 53, 267–273.

Agellon, L.B., Davies, S.L., Chen, T.T. and Powers, D.A. (1988) Structure of a fish (rainbow trout) growth hormone gene and its evolutionary implications. *Proceedings of the National Academy of Sciences USA* 85, 5136–5140.

Agnese, J.F., Adepo-Gourne, B., Abban, E.K. and Fermon, Y. (1997) Genetic differentiation among natural populations of Nile tilapia *Oreochromis niloticus* (Teleostei, Cichlidae). *Heredity* 79, 88–96.

Agresti, J.J., Seki, S., Cnaani, A., Poompnang, S., Hallerman, E.M., Umiel, N., Hulata, G., Gall, G.A.E. and May, B. (2000) Breeding new strains of tilapia: development of an artificial center of origin and linkage map based on AFLP and microsatellite loci. *Aquaculture* 185, 43–56.

Aida, T. (1921) On the inheritance of color in a freshwater fish, *Aplocheilus latipes*, Temmick and Schlegel. *Genetics* 6, 554–573.

Akhtar, N. (1984) Anesthesia, abdominal surgery and efficacy of ovariectomy and subsequent androgen treatments in inducement of sex reversal in *Tilapia nilotica*. PhD dissertation, Auburn University, Auburn, Alabama.

Akhter, T., Islam, M.S., Uematsu, K., Islam, M.S. and Hussain, M.G. (2004) Studies of morphometric and meristic characters and early growth of different strains and crossbred of silver barb, *Barbodes gonionotus* Bleeker. *Journal of Aquaculture in the Tropics* 19, 81–92.

Al-Ahmad, T.A. (1983) Relative effects of feed consumption and feed efficiency on growth of catfish from different genetic backgrounds. PhD dissertation, Auburn University, Auburn, Alabama.

Albert, R., Marbach, P., Bauer, W., Briner, U., Fricker, G., Bruns, C. and Pless, J. (1993) SDZ CO 611: a highly potent glycated analog of somatostatin with improved oral activity. *Life Sciences* 53, 517–525.

Alcivar-Warren, A., Xu, Z., Meehan, D., Fan, Y. and Song, L. (2002) Shrimp genomics: development of a genetic map to identify QTLs responsible for economically important traits in *Litopenaeus vannamei*. In: Shimizu, N., Aoki, T., Hirono, I. and Takashima, F. (eds) *Aquatic Genomics: Steps Toward a Great Future*. Springer-Verlag, New York, New York.

Al-Daham, N.K. (1970) The use of chemosterilants, sex hormones, radiation and hybridization for controlling reproduction in *Tilapia* species. PhD dissertation, Auburn University, Auburn, Alabama.

Aldridge, F.J., Marston, R.Q. and Shireman, J.V. (1990) Induced triploids and tetraploids in bighead carp, *Hypophthalmichthys nobilis*, verified by multi-embryo cytofluorometric analysis. *Aquaculture* 87, 121–131.

Aliah, R.S., Yamaoka, K., Inada, Y. and Taniguchi, N. (1990) Effects of triploidy on tissue structure of some organs in ayu. *Bulletin of the Japanese Society of Scientific Fisheries* 56, 569–575.

Aliyeva, E., Metz, A.M. and Browning, K.S. (1996) Sequences of two expressed sequence tags (EST) from rice encoding different cap-binding proteins. *Gene* 180, 221–223.

Allen, S.K. Jr (1987) Reproductive sterility in triploid shellfish and fish. PhD dissertation, University of Washington, Seattle, Washington.

Allen, S.K. Jr (1988a) Cytology of gametogenesis in triploid Pacific oyster, *Crassostrea gigas*. *Journal of Shellfish Research* 7, 107.

Allen, S.K. Jr (1988b) Induced triploidy improves the physiology, not the genetics, of cultured shellfish. *Journal of Shellfish Research* 7, 146.

Allen, S.K. Jr (1988c) Triploid oysters ensure year-round supply. *Washington Sea Grant Oceanus* 31, 58–63.

Allen, S.K. Jr (2000) Research and development on suminoegaki, *Crassostrea ariakensis*, for aquaculture in Virginia, and other activities with non-natives. *Journal of Shellfish Research* 19, 612.

Allen, S.K. Jr and Downing, S.L. (1985) A review of the rationale and potential of triploid shellfish in the Pacific Northwest. *Journal of Shellfish Research* 5, 49.

Allen, S.K. Jr and Downing, S.L. (1986) Performance of triploid Pacific oysters, *Crassostrea gigas* (Thunberg). I. Survival, growth, glycogen content, and sexual maturation in yearlings. *Journal of Experimental Marine Biology and Ecology* 102, 197–208.

Allen, S.K. Jr and Downing, S.L. (1990) Performance of triploid Pacific oysters, *Crassostrea gigas*: gametogenesis. *Canadian Journal of Fisheries and Aquatic Sciences* 47, 1213–1222.

Allen, S.K. Jr and Downing, S.L. (1991) Consumers and 'experts' alike prefer the taste of sterile triploid over gravid diploid Pacific oysters (*Crassostrea gigas*, Thunberg, 1793). *Journal of Shellfish Research* 10, 19–22.

Allen, S.K. Jr and Stanley, J.G. (1981a) Mosaic polyploid Atlantic salmon (*Salmo salar*) induced by cytochalasin B: the early life history of fish. *Recent Studies* 178, 509–510.

Allen, S.K. Jr and Stanley, J.G. (1981b) Polyploidy and gynogenesis in the culture of fish and shellfish. *International Council for the Exploration of the Sea, Cooperative Research Report Series B* 28, 1–18.

Allen, S.K. Jr and Wattendorf, R.J. (1987) Triploid grass carp: status and management implications. *Fisheries* 12, 20–24.

Allen, S.K. Jr, Hidu, H. and Stanley, J.G. (1986a) Abnormal gametogenesis and sex ratio in triploid soft-shell clams *Mya arenaria*. *Biological Bulletin Marine Biological Laboratory, Woods Hole* 170, 198–210.

Allen, S.K. Jr, Thiery, R.G. and Hagstrom, N.T. (1986b) Cytological evaluation of the likelihood that triploid grass carp will reproduce. *Transactions of the American Fisheries Society* 115, 841–848.

Allendorf, F.W. and Danzmann, R.G. (1997) Secondary tetrasomic segregation of MDH-B and preferential pairing of homeologues in rainbow trout. *Genetics* 145, 1083–1092.

Allendorf, F.W. and Phelps, S.R. (1980) Loss of genetic variation in a hatchery stock of cutthroat trout. *Transactions of the American Fisheries Society* 109, 537–543.

Allendorf, F.W. and Phelps, S.R. (1981) Use of allelic frequencies to describe population structure. *Canadian Journal of Fisheries and Aquatic Sciences* 38, 1507.

Allendorf, F.W. and Seeb, L.W. (2000) Concordance of genetic divergence among sockeye salmon populations at allozyme, nuclear DNA, and mitochondrial DNA markers. *Evolution* 54, 640–651.

Allendorf, F.W. and Thorgaard, G.H. (1984) Tetraploidy and the evolution of salmonid fishes. In: Turner, B.J. (ed.) *Evolutionary Genetics of Fishes (Monographs in Evolutionary Biology)*. Plenum Press, New York, pp. 1–53.

Allendorf, F.W. and Utter, F.M. (1979) Population genetics. In: Hoar, W.S., Randall, D.J. and Britt, J.R. (eds) *Fish Physiology*. Vol. 8. *Bioenergetics and Growth*. Academic Press, New York, pp. 407–454.

Allendorf, F.W., Ryman, N., Stenneck, A. and Stahl, G. (1976) Genetic variation in Scandinavian brown trout (*Salmo trutta* L.): evidence of distinct sympatric populations. *Hereditas* 83, 73–82.

Allendorf, F.W., Stahl, G. and Ryman, N. (1984) Silencing of duplicate genes: a null polymorphism for lactate dehydrogenase in rainbow trout (*Salmo trutta*). *Molecular Biology and Evolution* 1, 238–248.

Allendorf, F.W., Gellman, W.A. and Thorgaard, G.H. (1994) Sex-linkage of two enzyme loci in *Oncorhyncus mykiss* (rainbow trout). *Heredity* 72, 498–507.

Alm, G. (1959) *Connection between Maturity, Size and Age in Fishes*. Report 33. Institute of Freshwater Research, Drottningholm, Sweden.

Aloni, R., Peleg, D. and Meyuhas, O. (1992) Selective translational control and nonspecific posttranscriptional regulation of ribosomal protein gene expression during development and regeneration of rat liver. *Molecular and Cellular Biology* 12, 2203–2212.

Alonso, M., Tabata, Y.A., Rigolino, M.G. and Tsukamoto, R.Y. (2000) Effect of induced triploidy on fin regeneration of juvenile rainbow trout, *Oncorhynchus mykiss*. *Journal of Experimental Zoology* 287, 493–502.

Alonso-Blanco, C., Peeters, A.J.M., Koornneef, M., Lister, C., Dean, C., Van Der Bosch, N., Pot, J. and Kupper, M.T.R. (1998) Development of an AFLP based linkage map of Ler, Col and Cvi *Arabidopsis thaliana* ecotypes and construction of a Ler/Cvi recombinant inbred line population. *Plant Journal* 14, 259–271.

Amano, M., Aida, K., Okumoto, N. and Hasegawa, Y. (1992) Changes in salmon GnRH and chicken GnRH-II contents in the brain and pituitary, and GTH contents in the pituitary in female masu salmon, *Oncorhynchus masou*, from hatching through ovulation. *Zoological Science* 9, 375–386.

Amano, M., Urano, A. and Aida, K. (1997) Distribution and function of gonadotropin-releasing hormone (GnRH) in the teleost brain. *Zoological Science* 14, 1–11.

Amano, M., Oka, Y., Kitamura, S., Ikuta, K. and Aida, K. (1998) Ontogenic development of salmon GnRH and chicken GnRH-II systems in the brain of masu salmon (*Oncorhynchus masou*). *Cell and Tissue Research* 293, 427–434.

Amos, W., Wilmer, J.W., Fullard, K., Burg, B., Croxall, J.P., Bloch, D. and Coulson, T. (2001) The influence of parental relatedness on reproductive success. *Proceedings of the Royal Society of London, Series B: Biological Sciences* 268, 2021–2027.

Anders, E. (1990) *Experimental Polyploidization in Rainbow Trout*. Mariculture Commission ICES Council Meeting 1990 (Collected Papers). International Council for the Exploration of the Sea, Copenhagen.

Anderson, C.E. and Smitherman, R.O. (1978) Production of normal male and androgen sex-reversed *Tilapia aurea* and *T. nilotica* fed a commercial catfish diet in ponds. In: Smitherman, R.O., Shelton, W.L. and Grover, J.H. (eds) *Culture of Exotic Fishes Symposium Proceedings*. Fish Culture Section, American Fisheries Society, Auburn University, Auburn, Alabama, p. 34.

Anderson, E.D., Mourich, D.V. and Leong, J.C. (1996) Genetic immunization of rainbow trout (*Oncorhynchus mykiss*) against infectious hematopoietic necrosis virus. *Molecular Marine Biology and Biotechnology* 5, 114–122.

Anderson, J.W. and Rouse, D.B. (1998) Performance of triploid eastern oysters (*Crassostrea virginica*) in Mobile Bay, Alabama. *Journal of Shellfish Research* 17, 317.

Anderson, R.B. (1962) A comparison of returns from fall and spring-stocked hatchery-reared lake trout in Maine. *Transactions of the American Fisheries Society* 91, 425–427.

Anderson, T.A., Bonnet, L.R., Conlon, M.A. and Owens, P.C. (1993) Immunoreactive and receptor-active insulin-like growth factor-I and IGF-binding protein in plasma from the freshwater fish *Macquaria ambigua* (golden perch). *Journal of Endocrinology* 136, 191–198.

Anglade, I., Douard, V., Le Jossic-Corcos, C., Mananos, E.L., Mazurais, D., Michel, D. and Kah, O. (1998) The GABAergic system: a possible component of estrogenic feedback on gonadotropin secretion in rainbow trout (*Oncorhynchus mykiss*). In: Prunet, P., Bagliniere, J.L., Breton, B., Kah, O., Pakdel, F. and Saglio, P. (eds) *Facteurs de l'Environnement et Biologie des Poissons, Rennes, France, 23–25 September 1997*. Colloque Institut Federatif de Recherche, Biologie et Ecologie des Poissons, No. 350–351, pp. 647–654.

Anglade, I., Mazurais, D., Douard, V., le Jossic-Corcos, C., Mañanos, E.L., Michel, D. and Kah, O. (1999) Distribution of glutamic acid decarboxylase mRNA in the forebrain of the rainbow trout as studied by *in situ* hybridization. *Journal of Comparative Neurology* 410, 277–289.

Ankorion, Y., Moav, R. and Wohlfarth, G.W. (1992) Bidirectional mass selection for body shape in common carp. *Genetics, Selection, Evolution* 24, 43–52.

Aoki, T., Tucker, C. and Hirono, I. (2002) Expressed sequence tag analyses of the Japanese flounder, *Paralichthys olivaceus*. In: Shimizu, N., Aoki, T., Hirono, I. and Takashima, F. (eds) *Aquatic Genomics: Steps Toward a Great Future*. Springer-Verlag, New York, pp. 102–114.

Aparicio, S., Chapman, J., Stupka, E., Putnam, N., Chia, J.M., Dehal, P., Christoffels, A., Rash, S., Hoon, S., Smit, A., Gelpke, M.D., Roach, J., Oh, T., Ho, I.Y., Wong, M., Detter, C., Verhoef, F., Predki, P., Tay, A., Lucas, S., Richardson, P., Smith, S.F., Clark, M.S., Edwards, Y.J., Doggett, N., Zharkikh, A., Tavtigian, S.V., Pruss, D., Barnstead, M., Evans, C., Baden, H., Powell, J., Glusman, G., Rowen, L., Hood, L., Tan, Y.H., Elgar, G., Hawkins, T., Venkatesh, B., Rokhsar, D. and Brenner, S. (2002) Whole-genome shotgun assembly and analysis of the genome of *Fugu rubripes*. *Science* 297, 1301–1310.

AquaBounty Farms (1996) News from the farm. *The AQUAdvantage* 1, 1–4.

Arai, K. (1988) Viability of allotriploids in salmonids. *Nippon Suisan Gakkaishi/Bulletin of the Japanese Society of Scientific Fisheries* 54, 1695–1701.

Arai, K. (2001) Genetic improvement of aquaculture finfish species by chromosome manipulation techniques in Japan. *Aquaculture* 197, 205–228.

Arai, K., Taniura, K. and Zhang, Q. (1999) Production of second generation progeny of hexaploid loach. *Tokyo Fisheries Science* 65, 186–192.

Araki, H., Arden, W. R., Olsen, E., Cooper, B. and Blouin, M. S. (2007) Reproductive success of captive-bred steelhead trout in the wild: evaluation of three hatchery programs in the Hood River. *Conservation Biology* 21, 181–190.

Araneda, C., Neira, R., and Iturra, P. (2005) Identification of a dominant SCAR marker associated with colour traits in coho salmon (*Oncorhyncus kisutch*). *Aquaculture* 247, 67–73.

Argue, B.J. (1996) Performance of channel catfish, *Ictalurus punctatus*, blue catfish, *I. furcatus*, and their F_1, F_2, F_3 and backcross hybrids. PhD dissertation, Auburn University, Auburn, Alabama.

Argue, B. and Dunham, R.A. (1999) Hybrid fertility, introgression and backcrossing in fish. *Reviews in Fisheries Science* 7, 137–195.

Armbruster, P., Hutchinson, R.A. and Linvell, T. (2000) Equivalent inbreeding depression under laboratory and field conditions in a tree-hole-breeding mosquito. *Proceedings of the Royal Society of London, Series B: Biological Sciences* 267, 1939–1945.

Arnason, E. and Rand, D.M. (1992) Heteroplasmy of short tandem repeats in mitochondrial DNA of Atlantic cod, *Gadus morhua*. *Genetics* 132, 211–220.

Artini, P.G., de Micheroux, A.A. and D'Ambrogio, G. (1996) Growth hormone co-treatment with gonadotropins in ovulation induction. *Journal of Endocrinological Investigation* 19, 763–779.

Ashby, K.R. (1957) The effect of steroid hormones on the brown trout *Salmo trutta* L. during the period of gonadal differentiation. *Journal of Embryology and Experimental Morphology* 5, 249–255.

Aspinwall, N. (1974) Genetic analysis of North American populations of the pink salmon, *Oncorhynchus gorbuscha*, possible evidence for the neutral mutation–random drift hypothesis. *Evolution* 28, 295.

Astaurov, B.L. (1940) *Artificial Parthenogenesis in the Silkworm* (Bombyx mori L.). Izdatelstvo Akademii Nauk, Moscow. Cited by Lewis, W.H. (1980) *Polyploidy, Biological Relevance*. Plenum Press, New York, New York.

Atkins, M., O'Keefe, R.A. and Benfey, T.J. (2000) The effect of exercise on the growth and hematology of diploid and triploid brook trout (*Salvelinus fontinalis*). *Bulletin of the Aquaculture Association of Canada* 4, 19–21.

Austin, D.F. and Lee, M. (1996) Genetic resolution and verification of quantitative trait loci for flowering and plant height with recombinant inbred lines of maize. *Genome* 39, 957–968.

Austin, D.F., Lee, M., Veldboom, L. and Hallauer, A.R. (2000) Genetic mapping in maize with hybrid progeny across testers and generations: grain yield and grain moisture. *Crop Science* 40, 30–39.

Avers, C.J. (1980) *Genetics*. D. Van Nostrand, New York, New York.

Avise, J.C. and Van Den Avyle, M.J. (1984) Genetic analysis of reproduction of hybrid white bass striped bass in the Savannah River. *Transactions of the American Fisheries Society* 113, 563–570.

Avise, J.C., Arnold, J., Ball, R.M., Bermingham, E., Lamb, T., Neigel, J.E., Reeb, C.A. and Saunders, N.C. (1987) Intraspecific phylogeography: the bridge between population genetics and systematics. *Annual Reviews of Ecological Systems* 18, 489.

Avtalion, R.R. and Hammerman, I.S. (1978) Sex determination in *Sarotherodon* (*Tilapia*). I. Introduction to a theory of autosomal influence. *Israeli Journal of Aquaculture/Bamidgeh* 30, 110–115.

Ayala, F.J. (2000) Neutralism and selectionism: the molecular clock. *Genetics* 261, 27–33.

Ayles, G.B. (1975) Influence of the genotype and the environment on growth and survival of rainbow trout (*Salmo gairdneri*) in central Canadian aquaculture lakes. *Aquaculture* 6, 181–188.

Ayles, G.B. and Baker, R.F. (1983) Genetic differences in growth and survival between strains and hybrids of rainbow trout (*Salmo gairdneri*) stocked in aquaculture lakes in the Canadian prairies. *Aquaculture* 33, 269–280.

Baghian, A., Jaynes, J., Enright, F. and Kousoulas, K.G. (1997) An amphipathic alpha-helical synthetic peptide analogue of melittin inhibits herpes simplex virus-1 (HSV-1)-induced cell fusion and virus spread. *Peptides* 18, 177–183.

Bakos, J. (1979) Crossbreeding Hungarian races of common carp to develop more productive hybrids. In: Pillay, T.V.R. and Dill, W.A. (eds) *Advances in Aquaculture*. Fishing News Books Ltd, Farnham, UK, pp. 633–635.

Bakos, J. and Gorda, S. (1995) Genetic improvement of common carp strains using intraspecific hybridization. *Aquaculture* 129, 183–186.

Balloux, F., Brünner, H., Lugon-Moulin, N., Hausser, J. and Goudet, J. (2000) Microsatellites can be misleading: an empirical and simulation study. *Evolution* 54, 1414–1422.

Ballvora, A., Hesselbach, J., Niewohner, J., Leister, D., Salamini, F. and Gebhardt, C. (1995) Marker enrichment and high-resolution map of the segment of potato chromosome VII harboring the nematode resistance gene Gro l. *Molecular and General Genetics* 249, 82–90.

Balon, E.K. (1995) Origin and domestication of the wild carp, *Cyprinus carpio*: from Roman gourmets to the swimming flowers. *Aquaculture* 129, 3–48.

Bandin, I., Dopazo, C.P., Van Muiswinkel, W.B. and Wiegertjes, G.F. (1997) Quantitation of antibody-secreting cells in high and low antibody responder inbred carp (*Cyprinus carpio* L.) strains. *Fish and Shellfish Immunology* 7, 487–501.

Bao, B., Peatman, E., Peng, X., Baoprasertkul, P., Wang, G. and Liu, Z. (2006) Characterization of 23 CC chemokine genes and analysis of their expression in channel catfish (*Ictalurus punctatus*). *Developmental and Compaative Immunology* 30, 783–796.

Baoprasertkul, P., He, C., Peatman, E., Zhang, S., Li, P. and Liu, Z. (2005) Constitutive expression of three novel catfish CXC chemokines; homeostatic chemokines in teleost fish. *Molecular Immunology* 42, 1355–1366.

Baras, E., Prignon, C., Gohoungo, G. and Mélard, C. (2000) Phenotypic sex differentiation of blue tilapia under constant and fluctuating thermal regimes and its adaptive and evolutionary implications. *Journal of Fish Biology* 57, 210–223.

Baroiller, J.-F. and Jalabert, B. (1989) Contribution of research in reproductive physiology to the culture of tilapias. *Aquatic Living Resources* 2, 105–116.

Baroiller, J.-F., Chourrout, D., Fostier, A. and Jalabert, B. (1995a) Temperature and sex chromosomes govern sex-ratios of the mouthbrooding cichlid fish *Oreochromis niloticus*. *Journal of Experimental Zoology* 271, 1–8.

Baroiller, J.-F., Clota, F. and Geraz, E. (1995b) Temperature sex determination in two tilapia, *Oreochromis niloticus* and the red tilapia (red Florida strain): effect of high or low temperature. In: *Proceedings of the Fifth International Symposium on the Reproductive Physiology of Fish, University of Texas at Austin, 2–8 July 1995*. University of Texas Press, Austin, Texas, pp. 158–160.

Bart, A. (1988) Effects of anesthesia, surgical procedure, temperature and season on survival of channel and blue catfish after surgical removal of testes. MSc thesis, Auburn University, Auburn, Alabama.

Bart, A.N. (1994) Effects of sperm concentration, egg number, fertilization method, nutrition and cryo-preservation of sperm on fertilization efficiency with channel catfish eggs and blue catfish sperm. PhD dissertation, Auburn University, Auburn, Alabama.

Bart, A.N. and Dunham, R.A. (1990) Factors affecting survival of channel catfish after surgical removal of testes. *Progressive Fish-Culturist* 52, 241–246.

Bartley, D.M. (1999) International instruments for the responsible use of genetically modified organisms. *FAO Aquaculture Newsletter* 23, 17–22.

Bartley, D.M. and Hallerman, E.M. (1995) Global perspective on the utilization of genetically modified organisms in aquaculture and fisheries. *Aquaculture* 137, 1–7.

Basavaraju, Y., Devaraj, K.V. and Ayyar, S.P. (1995) Comparative growth of reciprocal carp hybrids between *Catla catla* and *Labeo fimbriatus*. *Aquaculture* 129, 187–191.

Basavaraju, Y., Mair, G.C., Mohan Kumar, H.M., Pradeep Kumar, S., Keshavappa, G.Y. and Penman, D.J. (2002) An evaluation of triploidy as a potential solution to the problem of precocious sexual maturation in common carp, *Cyprinus carpio*, in Karnataka, India. *Aquaculture* 204, 407–418.

Bauman, D.E. (1992) Bovine somatotropin: review of an emerging animal technology. *Journal of Dairy Science* 75, 3432–3451.

Bauman, D.E. (1995) *IGF-I Fact Sheet*. Department of Animal Science, Cornell University, New York, New York.

Bauman, D.E. (1999) Bovine somatotropin and lactation: from basic science to commercial application. *Domestic Animal Endocrinology* 17, 101–116.

Bauman, D.E. and Vernon, R.G. (1993) Effects of exogenous bovine somatotropin on lactation. *Annual Reviews in Nutrition* 13, 437–461.

Baumann, G., Lowman, H.B., Mercado, M. and Wells, J.A. (1994) The stoichiometry of growth hormone-binding protein complexes in human plasma: comparison with cell surface receptors. *Journal of Clinical Endocrinology and Metabolism* 78, 1113–1118.

Bayless, J.D. (1972) *Artificial Propagation and Hybridization of Striped Bass, Morone saxatilis (Walbaum)*. South Carolina Wildlife and Marine Resources Department, Columbia, South Carolina.

Beacham, T.D., Withler, R.E. and Gould, A.P. (1985) Biochemical genetic stock identification of pink salmon (*Oncorhynchus gorbuscha*) in southern British Columbia and Puget Sound. *Canadian Journal of Fisheries and Aquatic Sciences* 42, 1474–1483.

Beardmore, J.A., Mair, G.C. and Lewis, R.J. (2001) Monosex male production in finfish as exemplified by tilapia: applications, problems, and prospects. *Aquaculture* 197, 283–301.

Bearzotti, M., Perrot, E., Michard-Vanhée, C., Jolivet, G., Attal, J., Théron, M.C., Puissant, C., Dreano, M., Kopchick, J.J., Powell, R., Gannon, F., Houdebine, L.M. and Chourrout, D. (1992) Gene expression following transfection of fish cells. *Journal of Biotechnology* 26, 315–325.

Beauchamp, T.L. and Childress, J.F. (2001) *Principles of Biomedical Ethics*, 5th edn. Oxford University Press, Oxford, UK.

Beaudoing, E., Freier, S., Wyatt, J.R., Claverie, J.M. and Gautheret, D. (2000) Patterns of variant polyadenylation signal usage in human genes. *Genome Research* 10, 1001–1010.

Beaver, J.A., Sneed, K.E. and Dupree, H.K. (1966) The difference in growth of male and female channel catfish in hatchery ponds. *Progressive Fish-Culturist* 28, 47–50.

Bechinger, B. (1997) Structure and functions of channel-forming peptides: magainins, cecropins, melitin and alamethicin. *Journal of Membranes (United States)* 156, 197–211.

Beck, M.L., Biggers, C.J. and Barker, C.J. (1984) Chromosomal and electrophoretic analyses of hybrids between grass carp and bighead carp (Pisces: Cyprinidae). *Copeia* 1984, 337–342.

Becker, J., Vos, P., Kuiper, M., Salamini, F. and Heun, M. (1995) Combined mapping of AFLP and RFLP markers in barley. *Molecular and General Genetics* 249, 65–73.

Beckman, B.R. and Dickhoff, W.W. (1998) Plasticity of smolting in spring chinook salmon: relation to growth and insulin-like growth factor-I. *Journal of Fish Biology* 53, 808–826.

Beckman, B.R., Larsen, D.A., Moriyama, S., Lee-Pawlak, B. and Dickhoff, W.W. (1998) Insulin-like growth factor-I and environmental modulation of growth during smoltification of spring chinook salmon (*Oncorhynchus tshawystscha*). *General and Comparative Endocrinology* 109, 325–335.

Beckman, B.R., Dickhoff, W.W., Zaugg, W.S., Sharpe, C., Hirtzel, S., Schrock, R., Larsen, D.A., Ewing, R.D., Palmisano, A., Schreck, C.B. and Mahnken, C.V.W. (1999) Growth, smoltification, and smolt-to-adult return of spring chinook salmon from hatcheries on the Deschutes River, Oregon. *Transactions of the American Fisheries Society* 128, 1125–1150.

Beckman, B.R., Larsen, D.A., Sharpe, C., Lee-Pawlak, B., Schreck, C.B. and Dickhoff, W.W. (2000) Physiological status of naturally reared juvenile spring chinook salmon in the Yakima River: seasonal dynamics and changes associated with smolting. *Transactions of the American Fisheries Society* 129, 727–753.

Behar, T.N., Li, Y.X., Tran, H.T., Ma, W., Dunlap, V., Scott, C. and Barker, J.L. (1996) GABA stimulates chemotaxis and chemokinesis of embryonic cortical neurons via calcium dependent mechanisms. *Journal of Neuroscience* 16, 1808–1818.

Behlke, M.A. (2006) Progress towards *in vivo* use of siRNAs. *Molecular Therapy* 13, 644–670.

Behncken, S.N. and Waters, M.J. (1999) Molecular recognition events involved in the activation of the growth hormone receptor by growth hormone. *Journal of Molecular Recognition* 12, 355–362.

Behncken, S.N., Rowlinson, S.W., Rowland, J.E., Conway-Campbell, B.L., Monks, T.A. and Waters, M.J. (1997) Aspartate 171 is the major primate-specific determinant of human growth hormone: engineering porcine growth hormone to activate the human receptor. *Journal of Biological Chemistry* 272, 27077–27083.

Behrends, L.L. and Smitherman, R.O. (1984) Development of a cold-tolerant population of red tilapia through introgressive hybridization. *Journal of the World Mariculture Society* 14, 172–178.

Behrends, L.L., Kingsley, J.B. and Price, A.H. III (1988) Bidirectional-backcross selection for body weight in a red tilapia. In: Pullin, R.S.V., Bhukasawan, T., Tonguthai, K. and Maclean, J.L. (eds) *Proceedings of the Second International Symposium on Tilapia in Aquaculture, Bangkok, Thailand, 16–20 March 1987.* ICLARM Conference Proceedings No. 15. Department of Fisheries Thailand and ICLARM, Manila, pp. 124–134.

Bejar, J., Borrego, J.J. and Alvarez, M.C. (1997) A continuous cell line from the cultured marine fish gilt-head seabream (*Sparus aurata* L.). *Aquaculture* 150, 143–153.

Bejar, J., Hong, Y. and Alvarez, M.C. (1999) Towards obtaining ES cells in the marine fish species *Sparus aurata*; multipassage maintenance, characterization and transfection. *Genetic Analysis: Biomolecular Engineering* 15, 125–129.

Bell, E.T., Brown, J.B., Fotherby, K., Lorraine, J.A. and Robson, J.S. (1962) The effect of derivatives of dithio-carbamoylhydrazine on hormone excretion in postmenopausal women. *Journal of Endocrinology* 25, 221–231.

Bell, L.N., Hageman, M.J. and Bauer, J.M. (1995a) Impact of moisture on thermally induced denaturation and decomposition of lyophilized bovine somatotropin. *Biopolymers* 35, 201–209.

Bell, L.N., Hageman, M.J. and Muraoka, L.M. (1995b) Thermally induced denaturation of lyophilized bovine somatotropin and lysozyme as impacted by moisture and excipients. *Journal of Pharmaceutical Sciences* 84, 707–712.

Benchakan, M. (1979) Morphometric and meristic characteristics of blue, channel, white, and blue–channel hybrid catfishes. MSc thesis, Auburn University, Auburn, Alabama.

Bendayan, M., Ziv, E., Ben-Sasson, R., Bar-On, H. and Kidron, M. (1990) Morphocytochemical and biochemical evidence for insulin absorption by the rat ileal epithelium. *Diabetologia* 33, 197–204.

Bendayan, M., Ziv, E., Gingras, D., Ben-Sasson, R., Bar-On, H. and Kidron, M. (1994) Biochemical and morpho-cytochemical evidence for the intestinal absorption of insulin in control and diabetic rats: comparison between the effectiveness of duodenal and colon mucosa. *Diabetologia* 37, 119–126.

Benfey, T.J. (1989) *A Bibliography of Triploid Fish, 1943 to 1988.* Canadian Technical Report of Fisheries and Aquatic Sciences 1682. Department of Fisheries and Oceans, West Vancouver, British Columbia, Canada.

Benfey, T.J. (1999) The physiology and behavior of triploid fishes. *Reviews in Fisheries Science* 7, 39–67.

Benfey, T.J. and Biron, M. (2000) Acute stress response in triploid rainbow trout (*Oncorhynchus mykiss*) and brook trout (*Salvelinus fontinalis*). *Aquaculture* 184, 167–176.

Benfey, T.J. and Sutterlin, A.M. (1983) Production of triploid landlocked Atlantic salmon (*Salmo salar* L.) and the implications of their haemotology to oxygen utilization. In: *Salmonid Reproduction: An International Symposium.* Sea Grant Program, Washington University, Seattle, Washington, p. 10.

Benfey, T.J. and Sutterlin, A.M. (1984a) Triploidy induced by heat shock and hydrostatic pressure in land-locked Atlantic salmon (*Salmo salar* L.). *Aquaculture* 36, 359–367.

Benfey, T.J. and Sutterlin, A.M. (1984b) Oxygen utilization by triploid landlocked Atlantic salmon (*Salmo salar* L.). *Aquaculture* 42, 69–73.

Benfey, T.J. and Sutterlin, A.M. (1984c) Growth and gonadal development in triploid landlocked Atlantic salmon (*Salmo salar*). *Canadian Journal of Fisheries and Aquatic Sciences* 41, 1387–1392.

Benfey, T.J. and Sutterlin, A.M. (1984d) Binucleated red blood cells in the peripheral blood of an Atlantic salmon, *Salmo salar* L., alevin. *Journal of Fish Diseases* 7, 415–420.

Benfey, T.J., Sutterlin, A.M. and Thompson, R.J. (1984) Use of erythrocyte measurements to identify triploid salmonids. *Canadian Journal of Fisheries and Aquatic Sciences* 41, 980–984.

Benfey, T.J., Solar, I.I., De Jong, G. and Donaldson, E.M. (1986) Flow-cytometric confirmation of aneuploidy in sperm from triploid rainbow trout. *Transactions of the American Fisheries Society* 115, 838–840.

Benfey, T.J., Dye, H.E. and Donaldson, E.M. (1989) Estrogen-induced vitellogenin production by triploid coho salmon (*Oncorhynchus kisutch*), and its effect on plasma and pituitary gonadotropin. *General and Comparative Endocrinology* 75, 83–87.

Benfey, T.J., Dye, H.M. and Donaldson, E.M. (1990) Induced vitellogenesis in triploid coho salmon (*Oncorhynchus kisutch*). *Aquaculture* 85, 318.

Benfey, T.J., McCabe, L.E. and Pepin, P. (1997) Critical thermal maxima of diploid and triploid brook charr, *Salvelinus fontinalis*. *Environmental Biology and Fisheries* 49, 259–264.

Benoit, H.P. and Pepin, P. (1999) Interaction of rearing temperature and maternal influence on egg development rates and larval size at hatch in yellowtail flounder (*Pleuronectes ferrugineus*). *Canadian Journal of Fisheries and Aquatic Sciences* 56, 785–794.

Bentsen, H.B., Godrem, T. and Hao, N.V. (1996) *Breeding Plan for Silver Barb (Puntius gonionotus) in Vietnam.* INGA Report No. 3.

Bentsen, H.B., Eknath, A.E., Palada-de Vera, M.S., Danting, J.C., Bolivar, H.L., Reyes, R.A., Dionisio, E.E., Longalong, F.M., Circa, A.V., Tayamen, M.M. and Gjerde, B. (1998) Genetic improvement of farmed tilapias: growth performance in a complete diallel cross experiment with eight strains of *Oreochromis niloticus*. *Aquaculture* 160, 145–173.

Bentzen, P., Leggett, W.C. and Brown, G.C. (1988) Length and restriction site heteroplasmy in the mitochondrial DNA of American shad (*Alosa sapidissima*). *Genetics* 118, 509–518.

Berejikian, B.A., Tezak, E.P., Park, L., Latfood, E., Schroder, S.L. and Beall, E. (2001) Male competition and breeding success in captively reared and wild coho salmon (*Oncorhynchus kisutch*). *Canadian Journal of Fisheries and Aquatic Sciences* 58, 804–810.

Berkowitz, D.B. and Kryspin-Sorensen, I. (1994) Transgenic fish: safe to eat? A look at the safety considerations regarding food transgenics. *Bio/Technology* 12, 247–252.

Berman, R.S., Harrison, L.E., Pearlstone, D.B., Burt, M. and Brennan, M.F. (1999) Growth hormone, alone and in combination with insulin, increases whole body and skeletal muscle protein kinetics in cancer patients after surgery. *Annals of Surgery* 229, 1–10.

Bermingham, E., Lamb, T. and Avise, J.C. (1986) Size polymorphism and heteroplasmy in the mitochondrial DNA of lower vertebrates. *Journal of Heredity* 77, 249–252.

Bernatchez, L. and Duchesne, P. (2000) Individual-based genotype analysis in studies of parentage and population assignment: how many loci, how many alleles? *Canadian Journal of Fisheries and Aquatic Sciences* 57, 1–12.

Berndtson, A.K. and Chen, T.T. (1994) Two unique CYP1 genes are expressed in response to 3-methylcholanthrene treatment in rainbow trout. *Archives of Biochemistry and Biophysics* 310, 187–195.

Bétancourt, O.H., Attal, J., Théron, M.C., Puissant, C., Houdebine, L.M. and Bearzotti, M. (1993) Efficiency of introns from various origins in fish cells. *Molecular Marine Biology and Biotechnology* 2, 181–188.

Betina, M.I., Vassetzky, S.G. and Kondratieva, O.T. (1985) Influence of hydrostatic pressure on fertilized loach eggs. *Ontogenez* 16, 365–374.

Bettoli, P.W., Neill, W.H., Kelsch, S.W., Bettoli, P.W. and Kelsch, S.W. (1985) Temperature preference and heat resistance of grass carp, *Ctenopharyngodon idella* (Valenciennes), bighead carp, *Hypophthalmichthys nobilis* (Gray), and their F_1 hybrid. *Journal of Fish Biology* 27, 239–247.

Beukema, J.J. (1969) Angling experiments with carp (*Cyprinus carpio* L.) I. Differences between wild, domesticated and hybrid strains. *Netherlands Journal of Zoology* 19, 596.

Bevins, C.L. and Zasloff, M. (1990) Peptides from frog skin. *Annual Review of Biochemistry* 59, 395–414.

Bidwell, C.A., Chrisman, C.L. and Libey, G.S. (1985) Polyploidy induced by heat-shock in channel catfish. *Aquaculture* 51, 25–32.

Bieniarz, K., Koldras, M. and Mejza, T. (1997) Unisex and polyploid populations of cyprinid and silurid fish. *Archiwum rybactwa polskiego/Archives of Polish Fisheries, Olsztyn* 5, 31–36.

Billerbeck, J.M., Schultz, E.T. and Conover, D.O. (2000) Adaptive variation in energy acquisition and allocation among latitudinal populations of the Atlantic silverside. *Oecologia* 122, 210–219.

Billerbeck, J.M., Lankford, T.E. Jr and Conover, D.O. (2001) Evolution of intrinsic growth and energy acquisition rates. I. Trade-offs with swimming performance in *Menidia menidia*. *Evolution* 55, 1863–1872.

Bitinaite, J., Wah, D.A., Aggarwal, A.K. and Schildkraut, I. (1998) FokI dimerization is required for DNA cleavage. *Proceedings of the National Academy of Sciences USA* 95, 10570–10575.

Björnsson, B.T., Thorarensen, H., Hirano, T., Ogasawara, T. and Kristinsson, J.B. (1989) Photoperiod and temperature affect plasma growth hormone levels, growth, condition factor and hypoosmoregulatory ability of juvenile Atlantic salmon (*Salmo salar*) during parr–smolt transformation. *Aquaculture* 82, 77–91.

Björnsson, B.T., Stefansson, S.O. and Hansen, T. (1995) Photoperiod regulation of plasma growth hormone levels during parr–smolt transformation of Atlantic salmon: implications for hypoosmoregulatory ability and growth. *General and Comparative Endocrinology* 100, 73–82.

Björnsson, B.T., Stefansson, G.V., Berge, A.I. and Hansen, T. (1998) Circulating growth hormone levels in Atlantic salmon smolts following seawater transfer: effects of photoperiod regime, salinity, duration of exposure and season. *Aquaculture* 168, 121–137.

Björnsson, B.T., Hemre, G., Bjornevik, M. and Hansen, T. (2000) Photoperiod regulation of plasma growth hormone levels during induced smoltification of underyearling Atlantic salmon. *General and Comparative Endocrinology* 119, 17–25.

Blanc, J.M. and Chevassus, B. (1979) Interspecific hybridization of salmonid fish. I. Hatching and survival up to the 15th day after hatching in F_1 generation hybrids. *Aquaculture* 18, 21–34.

Blanc, J.M. and Poisson, H. (1988) Triploid hybridization between the rainbow trout and the Arctic char: incubation and stocking with young fish. *Cybium* 12, 229–238.

Blanc, J.M. and Vallee, F. (1999) Genetic variability of farming performances in some salmonid species and hybrids modified by triploidization. *Cybium* 23, 77–88.

Blanc, J.M., Chourrout, D. and Krieg, F. (1987) Evaluation of juvenile rainbow trout survival and growth in half-sib families from diploid and tetraploid sires. *Aquaculture* 65, 215–220.

Blanc, J.M., Vallee, F. and Dorson, M. (2000) Survival, growth and dressing traits of triploid hybrids between rainbow trout and three charr species. *Aquaculture Research* 31, 349–358.

Blatch, G.L., Lassle, M., Zetter, B.R. and Kundra, V. (1997) Isolation of a mouse cDNA encoding mSTI1, a stress-inducible protein containing the TPR motif. *Gene* 194, 277–282.

Bless, E.P., Westaway, W.A., Schwarting, G.A. and Tobet, S.A. (2000) Effects of γ-aminobutyric acid (A) receptor manipulation on migrating gonadotropin-releasing hormone neurons through the entire migratory route *in vivo* and *in vitro*. Endocrinology 141, 1254–1262.

Blom, J.H. and Dabrowski, K. (1996) Ascorbic acid metabolism in fish: is there a maternal effect on the progeny? *Aquaculture* 147, 215–224.

Blum, W.F., Albertson-Wikland, K., Rosberg, S. and Ranke, M.B. (1993) Serum levels of insulin-like growth factor I (IGF-I) and IGF binding protein 3 reflect spontaneous growth hormone secretion. *Journal of Clinical Endocrinology and Metabolism* 75, 1610–1616.

Boguski, M.S. and Schuler, G.D. (1995) Establishing a human transcript map. *Nature Genetics* 10, 369–371.

Bolivar, R.B., Bartolome, Z.P. and Newkirk, G.F. (1994) Response to within-family selection for growth in Nile tilapia (*Oreochromis niloticus*). In: Chou, L.M., Munro, A.D., Lam, T.J., Chen, T.W., Cheong, L.K.K., Ding, J.K., Hooi, K.K., Khoo, H.W., Phang, V.P.E., Shim, K.F. and Tan, C.H. (eds) *The Third Asian Fisheries Forum*. The Asian Fisheries Society, Manila.

Boman, H.G. (1995) Peptide antibiotics and their role in innate immunity. *Annual Review of Immunology* 13, 61–92.

Boman, H.G. (1996) Peptide antibiotics: holy or heretic grails of innate immunity. *Scandinavian Journal of Immunology* 43, 475–482.

Boman, H.G., Faye, I., Gudmundsson, G.H., Lee, J.Y. and Lindholm, D.A. (1991) Cell-free immunity in Cecropia. *European Journal of Biochemistry* 201, 23–31.

Bonar, S.A., Thomas, G.L. and Pauley, G.B. (1988) Evaluation of the separation of triploid and diploid grass carp, *Ctenopharyngodon idella* (Valenciennes), by external morphology. *Journal of Fish Biology* 33, 895–898.

Bondari, K. (1983) Response to bidirectional selection for body weight in channel catfish. *Aquaculture* 33, 73–81.

Bondari, K. and Dunham, R.A. (1987) Effects of inbreeding on economic traits of channel catfish. *Theoretical and Applied Genetics* 74, 1–9.

Bondari, K. and Joyce, J.A. (1980) Effects of size of male and female channel catfish on their spawn weight, hatchability score, hatching time, and swim up interval. In: *Summary of Papers, Research Workshop, Annual Conference*. Catfish Farmers of America, New Orleans, Louisiana.

Bondari, K., Dunham, R.A., Smitherman, R.O., Joyce, J.A. and Castillo, S. (1983) Response to bidirectional selection for body weight in blue tilapia. In: Fishelson, L. and Yaron, Z. (compilers) *Proceedings of the International Symposium on Tilapia in Aquaculture*. Tel Aviv University, Tel Aviv, Israel, pp. 300–310.

Bondari, K., Ware, G.O., Mullinix, B.G. Jr and Joyce, J.A. (1985) Influence of brood fish size on the breeding performance of channel catfish. *Progressive Fish-Culturist* 47, 21–26.

Bondioli, K.R., Westhusin, M.E. and Looney, C.R. (1990) Production of identical bovine offspring by nuclear transfer. *Theriogenology* 33, 165–174.

Boney, S.E. (1982) Monosexing grass carp by sex reversal and breeding. PhD dissertation, Auburn University, Auburn, Alabama.

Boney, S.W., Shelton, W.L., Yang, S.L. and Wilken, L.O. (1984) Sex reversal and breeding of grass carp. *Transactions of the American Fisheries Society* 113, 348–353.

Bongers, A.B.J., Veld, E.P.C., Abo, H.K., Bremmer, I.M., Eding, E.H., Komen, J. and Richter, C.J.J. (1994) Androgenesis in common carp (*Cyprinus carpio* L.), using UV-irradiation in a synthetic ovarian fluid and heat shocks. *Aquaculture* 122, 119–132.

Bongers, A.B.J., Bovenhuis, H., Van Stokkom, A.C., Wiegertjes, G.F., Zandieh-Doulabi, B., Komen, J. and Richter, C.J.J. (1997) Distribution of genetic variance in gynogenetic or androgenetic families. *Aquaculture* 153, 225–238.

Bonham, K. and Donaldson, L.R. (1972) Sex ratios and retardation of gonadal development in chronically gamma irradiated chinook salmon smolts. *Transactions of the American Fisheries Society* 101, 428–434.

Bosher, J.M. and Labouesse, M. (2000) RNA interference: genetic wand and genetic watchdog. *Nature Cell Biology* 2, 31–36.

Boshra, H., Li, J. and Sunyer, J.O. (2006) Recent advances on the complement system of teleost fish. *Fish and Shellfish Immunology* 20, 239–262.

Boudry, P., Collet, B., Cornette, F., Hervouet, V. and Bonhomme, F. (2002) High variance in reproductive success of the Pacific oyster (*Crassostrea gigas*, Thunberg) revealed by microsatellite-based parentage analysis of multifactorial crosses. *Aquaculture* 204, 283–296.

Bourget, D. (2000) Fluctuating asymmetry and fitness in *Drosophila melanogaster*. *Journal of Evolutionary Biology* 13, 515–521.

Brämick, U., Puckhaber, B., Langholz, H.-J. and Hörstgen-Schwark, G. (1995) Testing of triploid tilapia (*Oreochromis niloticus*) under tropical pond conditions. *Aquaculture* 137, 343–353.

Brawley, C. and Matunis, E. (2004) Regeneration of male germline stem cells by spermatogonial dedifferentiation *in vivo*. *Science* 304, 1331–1334.

Brem, G., Brenig, B., Horstgen-Schwark, G. and Winnacker, E.L. (1988) Gene transfer in tilapia (*Oreochromis niloticus*). *Aquaculture* 68, 209–219.

Brent, G.A., Moore, D.D. and Larsen, P.R. (1991) Thyroid hormone regulation of gene expression. *Annual Reviews of Physiology* 53, 17–35.

Breton, B. and Jalabert, B. (1973) Pituitary and plasma gonotrophins levels and spermatogenesis in the goldfish *Carassius auratus* after methallibure treatment. *Journal of Endocrinology* 59, 415–420.

Briggs, R. (1979) Genetics of cell type determination. *International Review of Cytology (Supplement)* 9, 107–125.

Briggs, R. and King, T.J. (1952) Transplantation of living nuclei from blastula cells into enucleated frog's eggs. *Proceedings of the National Academy of Sciences USA* 38, 455–463.

Briggs, R. and King, T.J. (1953) Factors affecting the transplantability of nuclei of frog embryonic cells. *Journal of Experimental Zoology* 122, 485–505.

Bright, G.M., Rogol, A.D., Johanson, A.J. and Blizzard, R.M. (1983) Short stature associated with normal growth hormone and decreased somatomedin-C concentrations: response to exogenous hormone. *Pediatrics* 71, 576–580.

Britton, J.R. and Koldovsky, O. (1988) Development of luminal protein digestion in suckling and weanling rats. *Biology of the Neonate* 53, 39–46.

Britton, J.R., George-Nascimento, C. and Koldovsky, O. (1988) Luminal hydrolysis of recombinant human epidermal growth factor in the rat gastrointestinal tract: segmental and developmental differences. *Life Sciences* 43, 1339–1347.

Bromage, N., Jones, J., Randall, C., Thrush, M., Davies, B., Springate, J., Duston, J. and Barker, G. (1992) Broodstock management, fecundity, egg quality and the timing of egg production in the rainbow trout (*Oncorhynchus mykiss*). *Aquaculture* 100, 141–166.

Brooks, C. and Langdon, C. (1998) The laboratory performance of Pacific oyster spat, *Crassostrea gigas*, exposed to varying salinities, predictive of performance in the field? *Journal of Shellfish Research* 17, 1279.

Brooks, M.J., Smitherman, R.O., Chappell, J.A. and Dunham, R.A. (1982) Sex–weight relations in blue, channel, and white catfishes: implications for brood stock selection. *Progressive Fish-Culturist* 44, 105–106.

Broussard, M.C. (1979) Evaluation of four strains of channel catfish, *Ictalurus punctatus*, and intraspecific hybrids under aquacultural conditions. Doctoral dissertation, Texas A&M University, College Station, Texas.

Brown, E.H. Jr, Eck, G.W., Foster, N.R., Horrall, R.M. and Coberly, C.E. (1981) Historical evidence for discrete stocks of lake trout (*Salvelinus namaycush*) in Lake Michigan. *Canadian Journal of Fisheries and Aquatic Sciences* 38, 1747–1758.

Brown, K.H. and Thorgaard, G.H. (2002) Mitochondrial and nuclear inheritance in an androgenetic line of rainbow trout, *Oncorhynchus mykiss*. *Aquaculture* 204, 323–335.

Brown, L.A. and Richards, R.H. (1979) Surgical gonadectomy of fish: a technique for veterinary surgeons. *Veterinary Record* 104, 215.

Brummett, R.E. (1986) Effects of genotype–environment interactions on growth, variability and survival of improved catfish. PhD dissertation, Auburn University, Auburn, Alabama.

Bryden, C.A. and Heath, D.D. (2000) Heritability of fluctuating asymmetry for multiple traits in chinook salmon (*Oncorhynchus tshawytscha*). *Canadian Journal of Fisheries and Aquatic Sciences* 57, 2186–2192.

Bryden, C.A., Heath, J.W. and Heath, D.D. (2004) Performance and heterosis in farmed and wild chinook salmon (*Oncorhynchus tshawytscha*) hybrid and purebred crosses. *Aquaculture* 235, 249–261.

Brzeski, V.J. and Doyle, R.W. (1995) A test of an on-farm selection procedure for tilapia growth in Indonesia. *Aquaculture* 137, 219–230.

Budavari, S., Smith, A., Heckelman, P.E., Kinneary, J.F. and O'Neil, M. (1996) *The Merck Index*, 12th edn. Merck, Whitehouse Station, New Jersey.

Buddle, C.R. (1984) Androgen-induced sex inversion of *Oreochromis* (Trewavas) hybrid fry stocked into cages standing in an earthen pond. *Aquaculture* 40, 233–239.

Buettner, H.J. (1962) Recoveries of tagged, hatchery-reared lake trout from Lake Superior. *Transactions of the American Fisheries Society* 90, 404–412.

Burch, E.P. (1986) Heritabilities for body weight, feed consumption and feed conversion and the correlations among these traits in channel catfish, *Ictalurus punctatus*. MSc thesis, Auburn University, Auburn, Alabama.

Burke, J., Wang, H., Hide, W. and Davison, D.B. (1998) Alternative gene form discovery and candidate gene selection from gene indexing projects. *Genome Research* 8, 276–290.

Burns, E.R., Anson, J.F., Hinson, W.G., Pipkin, J.L., Kleve, M.G. and Goetz, R.C. (1986) Effect of fixation with formalin on flow cytometric measurement of DNA in nucleated blood cells. *Aquaculture* 55, 149–155.

Burns, J.C., Matsubara, T., Lozinski, G., Yee, J.K., Friedmann, T., Washabaugh, C.H. and Tsonis, P.A. (1994) Pantropic retroviral vector-mediated gene transfer, integration, and expression in cultured newt limb cells. *Developmental Biology* 165, 285–289.

Buroker, N.E., Brown, J.R., Gilbert, T.A., O'Hara, P.J., Beckenbach, A.T., Thomas, W.K. and Smith, M.J. (1990) Length heteroplasmy of sturgeon mitochondrial DNA: an illegitimate elongation model. *Genetics* 124, 157–163.

Bury, D.D. (1989) Induction of polyploidy in percichthyid basses and ictalurid catfishes with hydrostatic pressure shocks. MSc thesis, Auburn University, Auburn, Alabama.

Busack, C.A. and Gall, G.A.E. (1983) An initial description of the quantitative genetics of growth and reproduction in the mosquitofish, *Gambusia affinis*. *Aquaculture* 32, 123–140.

Buth, D.G. (1979) Duplicate gene expression in tetraploid fishes of the tribe Moxostomatini (Cypriniformes, Catostomidae). *Comparative Biochemistry and Physiology, Part B: Biochemistry & Molecular Biology* 63B, 7–12.

Buth, D.G. (1982) Glucosephosphate isomerase expression in a tetraploid fish, *Moxostoma lachneri* (Cypriniformes, Catostomidae): evidence for 'retetraploidization'? *Genetica* 57, 171–175.

Buth, D.G. (1983) Duplicate isozyme loci in fishes: origins, distribution, phyletic consequences, and locus nomenclature. In: Rattazzi, M.C., Scandalios, J.G. and Whitt, G.S. (eds) *Isozymes: Current Topics in Biological and Medical Research*, Vol. 10. *Genetics and Evolution*. Alan R. Liss, New York, p. 381.

Buth, D.G., Burr, B.M. and Schenck, J.R. (1980) Electrophoretic evidence for relationships and differentiation among members of the percid subgenus *Microperca*. *Biochemical Systematics and Ecology* 8, 297–304.

Byamungu, N., Darras, V.M. and Kuhn, E.R. (2001) Growth of heat-shock induced triploids of blue tilapia, *Oreochromis aureus*, reared in tanks and in ponds in Eastern Congo: feeding regimes and compensatory growth response of triploid females. *Aquaculture* 198, 109–122.

Bye, V.J. and Lincoln, R.F. (1986) Commercial methods for the control of sexual maturation in rainbow trout (*Salmo gairdneri* R.). *Aquaculture* 57, 299–309.

Byon, J.Y., Ohira, T., Hirono, I. and Aoki, T. (2005) Use of a cDNA microarray to study immunity against viral hemorrhagic septicemia (VHS) in Japanese flounder (*Paralichthys olivaceus*) following DNA vaccination. *Fish and Shellfish Immunology* 18, 135–147.

Calhoun, W.E. III (1981) Progeny of sex ratios from mass spawnings of sex-reversed brood stock of *Tilapia nilotica*. MSc thesis, Auburn University, Auburn, Alabama.

Calhoun, W.E. and Shelton, W.L. (1983) Sex ratios of progeny from mass spawnings of sex-reversed brood-stock of *Tilapia nilotica*. *Aquaculture* 33, 365–371.

Calvo, G.W., Luckenbach, M.W. and Burreson, E.M. (1998) Non-native oyster studies in Virginia. *Journal of Shellfish Research* 17, 349.

Calvo, G.W., Luckenbach, M.W. and Burreson, E.M. (1999a) Evaluating the performance of non-native oyster species in Virginia. *Journal of Shellfish Research* 18, 303.

Calvo, G.W., Luckenbach, M.W., Allen, S.K. Jr and Burreson, E.M. (1999b) Comparative field study of *Crassostrea gigas* (Thunberg, 1793) and *Crassostrea virginica* (Gmelin, 1791) in relation to salinity in Virginia. *Journal of Shellfish Research* 18, 465–473.

Calvo, G.W., Luckenbach, M.W. and Burreson, E.M. (2000) High performance of *Crassostrea ariakensis* in Chesapeake Bay. *Journal of Shellfish Research* 19, 643.

Campbell, K.H.S., Loi, P. and Wilmut, I. (1994) Improved development to blastocyst of ovine nuclear transfer embryos reconstructed during the presumptive S-phase of enucleated activated oocytes. *Biology of Reproduction* 50, 1385–1393.

Campbell, K.H., McWhir, J., Ritchie, W.A. and Wilmut, I. (1996) Sheep cloned by nuclear transfer from a cultured cell line. *Nature* 380, 64–66.

Campos-Ramos, R., Harvey, S., Masabanda, J., Carrasco, L., Griffin, D., McAndrew, B., Bromage, N. and Penman, D. (2001) Identification of putative sex chromosomes in the blue tilapia, *Oreochromis aureus*, through synaptonemal complex and FISH analysis. *Genetica* 111, 143–153.

Cao, D., Kocabas, A., Ju, Z., Karsi, A., Li, P., Patterson, A. and Liu, Z.J. (2001) Transcriptome of channel catfish (*Ictalurus punctatus*): initial analysis of genes and expression profiles from the head kidney. *Animal Genetics* 32, 169–188.

Cao, Q.P., Duguay, S.J., Plisetskaya, E., Steiner, D.F. and Chan, S.J. (1989) Nucleotide sequence and growth hormone-regulated expression of salmon insulin-like growth factor I mRNA. *Molecular Endocrinology* 3, 2005–2010.

Capecchi, M.R. (1989) Altering the genome by homologous recombination. *Science* 244, 1288–1292.

Capili, J.B. (1995) Growth and sex determination in the Nile tilapia (*Oreochromis niloticus* L.). Doctoral dissertation, University of Wales, Swansea, UK.

Capili, J.C. and Skibinski, D.O.F. (1996) Mitochondrial DNA restriction endonuclease and isozyme analyses of three strains of *Oreochromis niloticus*. In: Pullin, R.S.V., Lazard, J., Legendre, M., Amon-Kottias, J.B. and Pauly, D. (eds) *Third International Symposium on Tilapia in Aquaculture*. ICLARM Conference Proceedings No. 41. ICLARM, Manila, pp. 266–272.

Carmichael, G.J., Williamson, J.H., Woodward, C.A.C. and Tomasso, J.R. (1988) Responses of northern, Florida, and hybrid largemouth bass to low temperature and low dissolved oxygen. *Progressive Fish-Culturist* 50, 225–231.

Carmichael, G.J., Schmidt, M.E. and Morizot, D.C. (1992) Electrophoretic identification of genetic markers in channel catfish and blue catfish by use of low-risk tissues. *Transactions of the American Fisheries Society* 121, 26–35.

Caroffino, D.C., Miller, L.M., Kapuscinski, A.R. and Ostazeski, J.J. (2008) Stocking success of local-origin fry and impact of hatchery ancestry: monitoring a new steelhead (*Oncorhynchus mykiss*) stocking program in a Minnesota tributary to Lake Superior. *Canadian Journal of Fisheries and Aquatic Sciences* 65, 309–318.

Carr, D. and Friesen, H.G. (1976) Growth hormone and insulin binding to human liver. *Journal of Clinical Endocrinology and Metabolism* 42, 484–493.

Carrasco, L.A.P., Penman, D.J. and Bromage, N. (1999) Evidence for the presence of sex chromosomes in the Nile tilapia (*Oreochromis niloticus*) from synaptonemal complex analysis of XX, XY and YY genotypes. *Aquaculture* 173, 207–218.

Cassani, J.R. and Caton, W.E. (1986a) Efficient production of triploid grass carp (*Ctenopharyngodon idella*) utilizing hydrostatic pressure. *Aquaculture* 55, 43–50.

Cassani, J.R. and Caton, W.E. (1986b) Growth comparisons of diploid and triploid grass carp under varying conditons. *Progressive Fish-Culturist* 48, 184–187.

Cassani, J.R., Caton, W.E. and Clark, B. (1984) Morphological comparisons of diploid and triploid hybrid grass carp, *Ctenopharyngodon idella*♀ × *Hypophthalmichthys nobilis*♂. *Journal of Fish Biology* 25, 269–278.

Cassani, J.R., Maloney, D.R., Allaire, H.P. and Kerby, J.H. (1990) Problems associated with tetraploid induction and survival in grass carp, *Ctenopharyngodon idella*. *Aquaculture* 88, 273–284.

Cathomen, T. and Joung, J.K. (2008) Zinc-finger nucleases: the next generation emerges. *Molecular Therapy* 16, 1200–1207.

Cavener, D.R. and Ray, S.C. (1991) Eukaryotic start and stop translation sites. *Nucleic Acids Research* 19, 3185–3192.

Chambers, R.C. and Leggett, W.C. (1996) Maternal influences on variation in egg sizes in temperate marine fishes. *American Zoologist* 36, 180–196.

Chambers, S.A.C. (1984) Sex reversal of Nile tilapia in the presence of natural food. MSc thesis, Auburn University, Auburn, Alabama.

Chamnankuruwet, A. (1996). Response to bidirectional mass selection for growth rate in *Clarias macrocephalus* Gunther. MSc thesis, Kasetsart University, Bangkok.

Chan, M.T. (1971) Experience with the determination of realized heritability in the tilapia (*Tilapia mossambica*). *Genetica* 7, 53–59.

Chan, W.K. and Devlin, R.H. (1993) Polymerase chain reaction amplification and functional characterization of sockeye salmon histone H3, metallothionein-B, and protamine promoters. *Molecular Marine Biology and Biotechnology* 2, 308–318.

Chan, Y.L., Suzuki, K., Olvera, J. and Wool, I.G. (1993) Zinc finger-like motifs in rat ribosomal proteins S27 and S29. *Nucleic Acids Research* 21, 649–655.

Chao, N.-H., Chen, S.-J. and Liao, I.-C. (1986) Triploidy induced by cold shock in cyprinid loach, *Misgurnus anguillicaudatus*. In: *Proceedings of the First Asian Fisheries Forum, Manila, Philippines, 26–31 May 1986*, pp. 101–104.

Chapman, D.C., Hubert, W.A. and Jackson, U.T. (1987) Phosphorus retention by grass carp (*Ctenopharyngodon idella*) fed sago pondweed (*Potamogeton pectinatus*). *Aquaculture* 65, 221–225.

Chapman, R.W., Stephens, J.C., Lansman, R.A. and Avise, J.C. (1982) Models of mitochondrial DNA transmission genetics and evolution in higher eukaryotes. *Genetical Research* 40, 41–57.

Chappell, J.A. (1979) An evaluation of twelve genetic groups of catfish for suitability in commercial production. PhD dissertation, Auburn University, Auburn, Alabama.

Charles, D. (2001a) Seeds of discontent. *Science* 294, 772–775.

Charles, D. (2001b) *Lords of the Harvest: Biotech, Big Money, and the Future of Food*. Perseus Publishing, Cambridge, Massachusetts.

Charoensit, P. (1995) Hybridization in commercial oyster. MSc thesis, Chulalongkorn University, Bangkok.

Charo-Karisa, H., Komen, H., Rezk, M.A., Ponzoni, R.W., van Arendonk, J.A.M. and Bovenhuis, H. (2006) Heritability estimates and response to selection for growth of Nile tilapia (*Oreochromis niloticus*) in low-input earthen ponds. *Aquaculture* 261, 479–486.

Chatakondi, N.G. (1995) Evaluation of transgenic common carp, *Cyprinus carpio*, containing rainbow trout growth hormone in ponds. PhD dissertation, Auburn University, Auburn, Alabama.

Chatakondi, N.G. and Yant, R.D. (2001) Application of compensatory growth to enhance production in channel catfish, *Ictalurus punctatus*. *Journal of the World Aquaculture Society* 32, 278–285.

Chatakondi, N., Ramboux, A.C., Nichols, A., Hayat, M., Duncan, P.L., Chen, T.T., Powers, D.A. and Dunham, R.A. (1994) The effect of rainbow trout growth hormone gene on the morphology, dressing percentage and condition factor in the common carp, *Cyprinus carpio*. *Proceedings V World Congress of Genetics and Applied Livestock Production* 17, 481–484.

Chatakondi, N., Lovell, R., Duncan, P., Hayat, M., Chen, T., Powers, D., Weete, T., Cummins, K. and Dunham, R.A. (1995) Body composition of transgenic common carp, *Cyprinus carpio*, containing rainbow trout growth hormone gene. *Aquaculture* 138, 99–109.

Chaurasia, O.P., Marcuard, S.P. and Seidel, E.R. (1994) Insulin-like growth factor I in human gastrointestinal exocrine secretions. *Regulatory Peptides* 50, 113–119.

Chen, F.Y. (1969) Preliminary studies on the sex-determining mechanism of *Tilapia mossambica* Peters and *T. hornorum* Trewavas. *International Association of Theoretical and Applied Limnology Proceedings* 17, 719–724.

Chen, H., Yi, Y., Chen, M. and Yang, X. (1986) Studies on the developmental potentiality of cultured cell nuclei of fish. *Acta Hydrobiologica Sinica/Shuisheng Shengwu Xuebao* 10, 1–7.

Chen, L., Zhao, Y., Zhou, Z., Tang, L., Chen, W., Du, N., Lai, W. and Wei, S.O. (1997a) Studies on the induction of polyploid in the Chinese mitten-handed crab, *Eriocheir sinensis*: 2. Induction of triploid embryos and tetraploid zoeas in *Eriocheir sinensis* by heat shock treatment. *Acta Zoologica Sinica* 43, 390–398.

Chen, L.Q., Zhao, Y.L., Wang, Y.F., Chen, W., Tang, L.Y., Du, N.S. and Lai, W. (1997b) Triploidy and tetraploidy induction in the Chinese mitten-handed crab (*Eriocheir sinensis*) by cytochalasin. *Asian Fisheries Science* 10, 131–137.

Chen, L., Zhao, Y., Wang, Y., Chen, W., Tang, L., Du, N. and Lai, W. (1997c) Studies on the polyploid induction in the Chinese mitten-handed crab, *Eriocheir sinensis*, 1. Induction of triploid and tetraploid embryos in *Eriocheir sinensis* by cytochalasin B. *Journal of Fisheries China* 21, 19–25.

Chen, M., Yang, X., Yu, X., Chen, H., Yi, Y. and Liu, H. (1997) Chromosome ploidy manipulation of allotetraploids and their fertility in Japanese phytophagous crucian carp (JPCC female)×red crucian carp (RCC male). *Acta Hydrobiologica Sinica/Shuisheng Shengwu Xuebao* 21, 197–206.

Chen, S., Lin, X.H., Xu, C.G. and Zhang, Q. (2000) Improvement of bacterial blight resistance of 'Minghui 63', an elite restorer line of hybrid rice, by molecular marker-assisted selection. *Crop Science* 40, 239–244.

Chen, S., Hong, Y. and Schartl, M. (2002) Development of a positive–negative selection procedure for gene targeting in fish cells. *Aquaculture* 214, 67–79.

Chen, T.T. and Powers, D.A. (1990) Transgenic fish. *Trends in Biotechnology* 8, 209–214.

Chen, T.T., Zhu, Z., Lin, C.M., Gonzalez-Villasenor, L.I., Dunham, R. and Powers, D.A. (1989) Fish *Genetic Engineering*: a novel approach in aquaculture. In: *Aquaculture '89 Proceedings*. National Shellfish Association, Los Angeles, California.

Chen, T.T., Lin, C.M., Zhu, Z., Gonzalez-Villasenor, L.I., Dunham, R.A. and Powers, D.A. (1990) Gene transfer, expression and inheritance of rainbow trout and human growth hormone genes in carp and loach. In: Church, R.B. (ed.) *Transgenic Models In Medicine and Agriculture*. UCLA Symposia on Molecular and Cellular Biology, Los Angeles, California, pp. 127–139.

Chen, T.T., Lin, C.M., Dunham, R.A. and Powers, D.A. (1992a) Integration, expression and inheritance of foreign fish growth hormone gene in transgenic fish. In: Hew, C.L. and Fletcher, G.L. (eds) *Transgenic Fish*. World Scientific Publishing Co., River Edge, New Jersey, pp. 164–175.

Chen, T.T., Lin, C.M., Gonzalez-Villasenor, L.I., Dunham, R., Powers, D.A. and Zhu, Z. (1992b) Fish *Genetic Engineering*: a novel approach in aquaculture. In: Rosenfield, A. and Mann, R. (eds) *Dispersal of Living Organisms into Aquatic Ecosystems*. University of Maryland Publications, College Park, Maryland, pp. 265–279.

Chen, T.T., Kight, K., Lin, C.M., Powers, D.A., Hayat, M., Chatakondi, N., Ramboux, A.C., Duncan, P.L. and Dunham, R.A. (1993) Expression and inheritance of RSVLTR-rtGH1 complementary DNA in the transgenic common carp, *Cyprinus carpio*. *Molecular Marine Biology and Biotechnology* 2, 88–95.

Chen, T.T., Lu, J.K., Shamblott, M.J., Chench, C.M., Lin, C.M., Burns, J.C., Reimschuessel, R., Chatakondi, N. and Dunham, R.A. (1995) Transgenic fish: ideal models for basic research and biotechnological applications. *Zoological Studies* 34, 215–234.

Chen, Y.-L., Tsai, H. and Jen, S.O. (1997) Effect of electroporation conditions on loach sperm for successful gene transfer and early development. *Fisheries Science* 63, 527–532.

Cheney, D.P., Elston, R. and MacDonald, B. (1998) Oyster summer mortality – an update on ongoing sea grant sponsored research. *Journal of Shellfish Research* 17, 1279.

Cheong, H.T., Takahashi, Y. and Kanagawa, H. (1993) Birth of mice after transplantation of early cell-cycle stage embryonic nuclei into enucleated oocytes. *Biology of Reproduction* 48, 958–963.

Cherfas, N.B., Kozinsky, O., Rothbard, S. and Hulata, G. (1990) Induced diploid gynogenesis and triploidy in ornamental (koi) carp, *Cyprinus carpio* L. 1. Experiments on the timing of temperature shock. *Israeli Journal of Aquaculture/Bamidgeh* 42, 3–9.

Cherfas, N.B., Gomelsky, B., Ben-Dom, N., Peretz, Y. and Hulata, G. (1994a) Assessment of triploid common carp (*Cyprinus carpio* L.) for culture. *Aquaculture* 127, 11–18.

Cherfas, N.B., Gomelsky, B.I., Emelyanova, O.V. and Recoubratsky, A.V. (1994b) Induced diploid gynogenesis and polyploidy in crucian carp, *Carassius auratus gibelio* (Bloch), common carp, × *Cyprinus carpio* L., hybrids. *Aquaculture and Fisheries Management* 25, 943–954.

Cherfas, N.B., Peretz, Y., Ben-Dom, N., Gomelsky, B. and Hulata, G. (1994c) Induced diploid gynogenesis and polyploidy in the ornamental (koi) carp, *Cyprinus carpio* L. 4. Comparative study on the effects of high- and low-temperature shocks. *Theoretical and Applied Genetics* 89, 193–197.

Cherfas, N.B., Peretz, Y., Ben-Dom, N., Gomelsky, B. and Hulata, G. (1994d) Induced diploid gynogenesis and polyploidy in the ornamental (koi) carp *Cyprinus carpio* L. 3. Optimization of heat-shock timing during the 2nd meiotic division and the 1st cleavage. *Theoretical and Applied Genetics* 89, 281–286.

Cherfas, N.B., Gomelsky, B., Ben-Dom, N., Joseph, D., Cohen, S., Israel, I., Kabessa, M., Zohar, G., Peretz, Y., Mires, D. and Hulata, G. (1996) Assessment of all-female common carp progenies for fish culture. *Israeli Journal of Aquaculture/Bamidgeh* 48, 149–157.

Chevassus, B. (1979) Hybridization in salmonids: results and perspectives. *Aquaculture* 17, 113–128.

Chevassus, B., Guyomard, R., Chourrout, D. and Quillet, E. (1983) Production of viable hybrids in salmonids by triploidization. *Genetics, Selection, Evolution* 15, 519–532.

Chiasson, M.A., Pelletier, C.S. and Benfey, T.J. (2009) Triploidy and full-sib family effects on survival and growth in juvenile Arctic charr (*Salvelinus alpinus*). *Aquaculture* 289, 244–252.

Chiba, H., Nakamura, M., Iwata, M., Sakuma, Y., Yamauchi, K. and Parhar, I.S. (1999) Development and differentiation of gonadotropin hormone-releasing hormone neuronal systems and testes in the Japanese eel (*Anguilla japonica*). *General and Comparative Endocrinology* 114, 449–459.

Childers, W.F. (1967) Hybridization of four species of sunfishes (Centrarchidae). *Illinois Natural History Survey Bulletin* 29, 159–214.

Chimits, P. (1957) La tilapia y su cultivo: segunda resena y bibliografia. *FAO Fisheries Bulletin* 10, 27.

Chiou, P.P., Lin, C.-M., Perez, L. and Chen, T.T. (2002) Effect of cecropin B and a synthetic analogue on propagation of fish viruses *in vitro*. *Marine Biotechnology* 4, 294–302.

Chitminat, C. (1996) Predator avoidance of transgenic channel catfish containing salmonid growth hormone genes. MSc thesis, Auburn University, Auburn, Alabama.

Cho, Y.G., Blair, M.W., Panaud, O. and McCouch, S.R. (1996) Cloning and mapping of variety-specific rice genomic DNA sequences: amplified fragment length polymorphisms (AFLP) from silver-stained poly-acrylamide gels. *Genome* 39, 373–378.

Choubert, G. and Blanc, J.-M. (1985) Flesh colour of diploid and triploid rainbow trout (*Salmo gairdneri* Rich.) fed canthaxanthin. *Aquaculture* 47, 299–304.

Choubert, G. and Blanc, J.-M. (1989) Dynamics of dietary canthaxanthin utilization in sexually maturing female rainbow trout (*Salmo gairdneri* Rich.) compared to triploids. *Aquaculture* 83, 359–366.

Choubert, G., Blanc, J.-M. and Vallee, F. (1997) Colour measurement, using the CIELCH colour space, of muscle of rainbow trout, *Oncorhynchus mykiss* (Walbaum), fed astaxanthin: effects of family, ploidy, sex, and location of reading. *Aquaculture Research* 28, 15–22.

Chouinard, M. and Boulanger, Y. (1988) The production of a second generation and back crosses between a triploid hybrid (*Salvelinus fontinalis* female × *Salmo gairdneri* male) and brook trout (*Salvelinus fontinalis*). In: *Proceedings of Aquaculture International Congress and Exposition, Vancouver Trade and Convention Centre, Vancouver, British Columbia, Canada, 6–9 September 1988*, p. 56.

Chourrout, D. (1980) Thermal induction of diploid gynogenesis and triploidy in the eggs of the rainbow trout (*Salmo gairdneri* Richardson). *Reproduction, Nutrition, Development* 20, 727–733.

Chourrout, D. (1982) Tetraploidy induced by heat shocks in rainbow trout (*Salmo gairdneri* R.). *Reproduction, Nutrition, Development* 22, 569–574.

Chourrout, D. (1984) Pressure induced retention of second polar body and suppression of first cleavage in rainbow trout: production of all-triploids, all-tetraploids and heterozygous and homozygous diploid gynogenetics. *Aquaculture* 36, 111–126.

Chourrout, D. (1986a) Techniques of chromosome manipulation in rainbow trout: a new evaluation with karyology. *Theoretical and Applied Genetics* 72, 627–632.

Chourrout, D. (1986b) Use of grayling sperm (*Thymallus thymallus*) as a marker for the production of gynogenetic rainbow trout (*Salmo gairdneri*). *Theoretical and Applied Genetics* 72, 633–636.

Chourrout, D. and Itskovich, J. (1983) Three manipulations permitted by artificial insemination in tilapia: induced diploid gynogenesis, production of all-triploid populations and intergeneric hybridization. In: Fishelson, L. and Yaron, Z. (compilers) *Proceedings of the International Symposium on Tilapia in Aquaculture*. Tel Aviv University, Tel Aviv, Israel, p. 246.

Chourrout, D. and Quillet, E. (1982) Induced gynogenesis in the rainbow trout: sex and survival of progenies. Production of all-triploid populations. *Theoretical and Applied Genetics* 63, 201–205.

Chourrout, D., Chevassus, B., Krieg, F., Happe, A., Burger, G. and Renard, P. (1986a) Production of second generation triploid and tetraploid rainbow trout by mating tetraploid males and diploid females: potential of tetraploid fish. *Theoretical and Applied Genetics* 72, 193–206.

Chourrout, D., Guyomard, R. and Houdebine, L. (1986b) High efficiency gene transfer in rainbow trout (*Salmo gairdneri* Rich.) by microinjection into egg cytoplasm. *Aquaculture* 51, 143–150.

Chrisman, C.L., Wolters, W.R. and Libey, G.S. (1983) Triploidy in channel catfish. *Journal of the World Mariculture Society* 14, 279–293.

Chulitskava, F.V. (1970) Desynchronization of cell divisions in the course of egg cleavage and attempt at experimental shift in its onset. *Journal of Embryology and Experimental Morphology* 23, 359–374.

Circa, A.V., Eknath, A.E. and Taduan, A.G. (1994) Genetic improvement of farmed tilapias: the growth performance of the GIFT srain of Nile tilapia in rice-fish environments. In: Proceedings of the Fifth International Symposium on Genetics in Aquaculture, Halifax, Canada.

Clackson, T. and Wells, J.A. (1995) A hot spot of binding energy in a hormone–receptor interface. *Science* 267, 383–386.

Clackson, T., Ultsch, M.H., Wells, J.A. and de Vos, A.M. (1998) Structural and functional analysis of the 1:1 growth hormone:receptor complex reveals the molecular basis for receptor affinity. *Journal of Molecular Biology* 277, 1111–1128.

Clemens, H.P. and Inslee, T. (1968) The production of unisexual broods by *Tilapia mossambica* sex-reversed with methyltestosterone. *Transactions of the American Fisheries Society* 97, 18–21.

Clifford, S.L., McGinnity, P. and Ferguson, A. (1997) Genetic changes in an Atlantic salmon population resulting from escaped juvenile farm salmon. *Journal of Fish Biology* 5, 118–127.

Clifford, S.L., McGinnity, P. and Ferguson, A. (1998) Genetic changes in Atlantic salmon (*Salmo salar*) populations of northwest Irish rivers resulting from escapes of adult farm salmon. *Canadian Journal of Fisheries and Aquatic Sciences* 55, 358–363.

Clippinger, D. and Osborne, J.A. (1984) Surgical sterilization of grass carp, a nice idea. *Aquatics* 6, 9.

Cnaani, A., Hallerman, E.M., Ron, M., Weller, J.I., Indelman, M., Kashi, Y., Gall, G.A.E., and Hulata, G. (2003) Detection of a chromosomal region with two quantitative trait loci, affecting cold tolerance and fish size, in an F_2 tilapia hybrid. *Aquaculture* 223, 117–128.

Cnaani, A., Zilberman, N., Tinman, S., Hulata, G. and Ron, M. (2004) Genome-scan analysis for quantitative trait loci in an F_2 tilapia hybrid. *Molecular Genetics and Genomics* 272, 162–172.

Cohen, S.A., Chang, C.Y., Boyer, H.W. and Helling, R.B. (1973) Construction of biologically functional bacterial plasmids *in vitro*. *Proceedings of the National Academy of Sciences USA* 70, 3240–3244.

Colgan, D.F. and Manley, J.L. (1997) Mechanisms and regulation of mRNA polyadenylation. *Genes and Development* 11, 2755–2766.

Collodi, P., Kamei, Y., Ernst, T., Miranda, C., Buhler, D. and Barnes, D. (1992a) Culture of cells from zebrafish embryo and adult tissues. *Cell Biology and Toxicology* 8, 43–60.

Collodi, P., Kamei, Y., Weber, D., Sharps, A. and Barnes, D. (1992b) Fish embryo cell cultures for derivation of stem cells and transgenic chimeras. *Molecular Marine Biology and Biotechnology* 1, 257–265.

Colombo, L., Barbaro, A., Francescon, A., Libertini, A., Bortolussi, M., Argenton, F., Dalla Valle, L., Vianell, S. and Belvedere, P. (1998) Towards an integration between chromosome set manipulation, intergeneric hybridization and gene transfer in marine fish culture. In: Bartley, D. and Basurco, B. (eds) *Genetics and Breeding of Mediterranean Aquaculture Species*. *Cahiers Options Méditerranéennes* 34, 77–122.

Concibido, V.C., Denny, R.L., Lange, D.A., Orf, J.H. and Young, N.D. (1996) RFLP mapping and marker-assisted selection of soybean cyst nematode resistance in PI 209332. *Crop Science* 36, 1643–1650.

Consuegra, S., Verspoor, E., Knox, D. and Garcia de Leaniz, C. (2005) Asymmetric gene flow and the evolutionary maintenance of genetic diversity in small, peripheral Atlantic salmon populations. *Conservation Genetics* 6, 823–842.

Cook, J.T., McNiven, M.A., Richardson, G.F. and Sutterlin, A.M. (2000a) Growth rate, body composition and feed digestibility/conversion of growth-enhanced transgenic Atlantic salmon (*Salmo salar*). *Aquaculture* 188, 15–32.

Cook, J.T., McNiven, M.A. and Sutterlin, A.M. (2000b) Metabolic rate of pre-smolt growth-enhanced transgenic Atlantic salmon (*Salmo salar*). *Aquaculture* 188, 33–45.

Cook, J.T., Sutterlin, A.M. and McNiven, M.A. (2000c) Effect of food deprivation on oxygen consumption and body composition of growth-enhanced transgenic Atlantic salmon (*Salmo salar*). *Aquaculture* 188, 47–63.

Cook, M.T., Hayball, P.J., Birdseye, L., Bagley, C., Nowak, B.F. and Hayball, J.D. (2003) Isolation and partial characterization of a pentraxin-like protein with complement-fixing activity from snapper (*Pagrus auratus*, Sparidae) serum. *Developmental and Comparative Immunology* 27, 579–588.

Cook, M.T., Hayball, P.J., Nowak, B.F. and Hayball, J.D. (2005) The opsonising activity of a pentraxin-like protein isolated from snapper (*Pagrus auratus*, Sparidae) serum. *Developmental and Comparative Immunology* 29, 703–712.

Cooper, K. and Guo, X. (1989) Polyploid Pacific oysters produced by inhibiting polar body I and II with cytochalasin B. *Journal of Shellfish Research* 8, 412.

Coote, T. and Bruford, M.W. (1996) Human microsatellites applicable for analysis of genetic variation in apes and Old World monkeys. *Journal of Heredity* 87, 406–410.

Cordon, A.J. and Nicola, S.I. (1970) Harvest of four strains of rainbow trout, *Salmo gairdneri*, from Beardsley reservoir, California. *California Fish and Game* 56, 271–287.

Costigan, D.C., Guyda, H.J. and Posner, B.I. (1988) Free insulin-like growth factor I (IGF-I) and IGF-II in human saliva. *Journal of Clinical Endocrinology and Metabolism* 66, 1014–1018.

Cotter, D., O'Donovan, V., O'Maoiléidigh, N., Rogan, G., Roche, N. and Wilkins, N.P. (2000a) An evaluation of the use of triploid Atlantic salmon (*Salmo salar* L.) in minimising the impact of escaped farmed salmon on wild populations. *Aquaculture* 186, 61–75.

Cotter, D., O'Donovan, V., Roche, N. and Wilkins, N.P. (2000b) Gonadotropin and sex steroid hormone profiles in ranched, diploid and triploid Atlantic salmon (*Salmo salar* L.). In: Norberg, B., Kjesbu, O.S., Taranger, G.L., Andersson, E. and Stefansson, S.O. (eds) *Reproductive Physiology of Fish 2000. 6th International Symposium on the Reproductive Physiology of Fish, Bergen, Norway, 4–9 July 1999*. Department of Fisheries and Marine Biology, University of Bergen, Bergen, Norway.

Courtenay, W.R. Jr and Stauffer, J.R. Jr (1984) *Distribution, Biology, and Management of Exotic Fishes*. Johns Hopkins University Press, Baltimore, Maryland.

Cowan, J.S., Gaul, P., Moor, B.C. and Kraicer, J. (1984) Secretory bursts of growth hormone secretion in the dog may be initiated by somatostatin withdrawal. *Canadian Journal of Physiology and Pharmacology* 62, 199–207.

Crabtree, C.B. and Buth, D.G. (1981) Gene duplication and diploidization in tetraploid catostomid fishes *Catostomus fumeiventris* and *C. santaanae*. *Copeia* 1981, 705–708.

Craig, N.L. (1995) Unity in transposition reactions. *Science* 270, 253–254.

Cregan, P.B., Mudge, J., Fickus, E.W., Danesh, D., Denny, R. and Young, N.D. (1999) Two simple sequence repeat markers to select for soybean cyst nematode resistance conditioned by the rhg1 locus. *Theoretical and Applied Genetics* 99, 811–818.

Crocos, P.J. and Preston, N.P. (1999) Genetic improvement of farmed prawns in Australia. *Global Aquaculture Advocate* 2(December), 62–63.

Cronin, C.A., Gluba, W. and Scrable, H. (2001) The lac operator–repressor system is functional in the mouse. *Genes and Development* 15, 1506–1517.

Cross, T.F. and King, J. (1983) Genetic effects of hatchery rearing in Atlantic salmon. *Aquaculture* 33, 33–40.

Crozier, W.W. (1993) Evidence of genetic interaction between escaped farmed salmon and wild Atlantic salmon (*Salmo salar* L.) in a Northern Irish river. *Aquaculture* 113, 19–29.

Crozier, W.W. (1997) Genetic heterozygosity and meristic character variance in a wild Atlantic salmon population and a hatchery strain derived from it. *Aquaculture International* 5, 407–414.

Crozier, W.W. and Moffett, I.J.J. (1989a) Application of an electrophoretically detectable genetic marker to ploidy testing in brown trout (*Salmo trutta* L.) triploidised by heat shock. *Aquaculture* 80, 231–239.

Crozier, W.W. and Moffett, I.J.J. (1989b) Amount and distribution of biochemical–genetic variation among wild populations and a hatchery stock of Atlantic salmon, *Salmo salar* L., from north-east Ireland. *Journal of Fish Biology* 35, 665–677.

Crozier, W.W. and Moffett, I.J.J. (1990) Inheritance of allozymes in Atlantic salmon (*Salmo salar* L.). *Aquaculture* 88, 253–262.

Crozier, W.W. and Moffett, I.J.J. (1995) Application of low frequency genetic marking at GPI-3* and MDH-B1,2* loci to assess supplementary stocking of Atlantic salmon, *Salmo salar* L., in a Northern Irish stream. *Fisheries Management and Ecology* 2, 27–36.

Culp, P., Nusslein-Volhard, C. and Hopkins, N. (1991) High-frequency germ-line transmission of plasmid DNA sequences injected into fertilized zebrafish eggs. *Proceedings of the National Academy of Sciences USA* 88, 7953–7957.

Cuneo, R.C., Hickman, P.E., Wallace, J.D., The, B.T., Ward, G., Veldhuis, J.D. and Waters, M.J. (1995) Altered endogenous growth hormone secretory kinetics and diurnal GH-binding protein profiles in adults with chronic liver disease. *Clinical Endocrinology* 43, 265–275.

Cunningham, B.C. and Wells, J.A. (1991) Rational design of receptor-specific variants of human growth hormone. *Proceedings of the National Academy of Sciences USA* 88, 3407–3411.

Cunningham, B.C. and Wells, J.A. (1993) Comparison of a structural and a functional epitope. *Journal of Molecular Biology* 234, 554–563.

Cunningham, B.C., Jhurani, P., Ng, P. and Wells, J.A. (1989) Receptor and antibody epitopes in human growth hormone identified by homolog-scanning mutagenesis. *Science* 243, 1330–1336.

Cunningham, B.C., Ultsch, M., de Vos, A.M., Mulkerrin, M.G., Clauser, K.R. and Wells, J.A. (1991) Dimerization of the extracellular domain of the human growth hormone receptor by a single hormone molecule. *Science* 254, 821–825.

Cunningham, E.P., Dooley, J.J., Splan, R.K. and Bradley, D.G. (2001) Microsatellite diversity, pedigree relatedness and the contributions of founder lineages to thoroughbred horses. *Animal Genetics* 32, 360–364.

Curtis, T.A., Sessions, F.W., Bury, D., Rezk, M. and Dunham, R.A. (1987) Induction of polyploidy in striped bass, white bass and their hybrids with hydrostatic pressure. *Proceedings Annual Conference Southeastern Association of Fish Wildlife Agencies* 41, 63–69.

Czech, M.P. (1989) Signal transmission by the insulin-like growth factors. *Cell* 59, 235–238.

Dadzie, S. (1972) Inhibition of pituitary gonadotropic activity in cichlid fishes of the genus *Tilapia* by a dithicarbamoylhydrazine derivative (I.C.I. 33, 828). *Ghana Journal of Science* 12, 41.

Daher, K.A., Selsted, M.E. and Lehrer, R.I. (1986) Direct inactivation of viruses by human granulocyte defensins. *Journal of Virology* 60, 1068–1074.

Daikoku, S. and Koide, I. (1998) Spatiotemporal appearance of developing LHRH neurons in the rat brain. *Journal of Comparative Neurology* 393, 34–47.

Danzmann, R.G. and Gharbi, K. (2007) Linkage mapping in aquaculture species. In: Liu, Z. (ed.) *Aquaculture Genome Technologies*. Blackwell Publishing, Ames, Iowa, pp. 139–167.

Danzmann, R.G., Jackson, T.R. and Ferguson, M.M. (1999) Epistasis in allelic expression at upper temperature tolerance QTL in rainbow trout. *Aquaculture* 173, 45–58.

Darvasi, A. and Soller, M. (1992) Selective genotyping for determination of linkage between a marker locus and a quantitative trait locus. *Theoretical and Applied Genetics* 85, 353–359.

Darvasi, A. and Soller, M. (1994) Selective DNA pooling for determination of linkage between a molecular marker and a quantitative trait locus. *Genetics* 138, 1365–1373.

Darwin, C. (1859) *On the Origin of Species by Means of Natural Selection, or the Preservation of Favoured Races in the Struggle for Life*. J. Murray, London.

Davis, G.P. and Hetzel, D.J.S. (2000) Integrating molecular genetic technology with traditional approaches for genetic improvement in aquaculture species. *Aquaculture Research* 31, 3–10.

Davis, J.P. (1988a) Energetics of sterile triploid oysters uncouple the reproductive and somatic effort of diploids. *Journal of Shellfish Research* 7, 114.

Davis, J.P. (1988b) Growth rate of sibling diploid and triploid oysters, *Crassostrea gigas*. *Journal of Shellfish Research* 7, 202.

Davis, J.P. (1988c) Physiology and energetics relating to weight loss and glycogen utilization during starvation in diploid and triploid Pacific oysters. *Journal of Shellfish Research* 7, 549.

Davis, J.P. (1989) Growth rate of sibling diploid and triploid oysters, *Crassostrea gigas*. *Journal of Shellfish Research* 8, 319.

Davis, K.B., Simco, B.A., Goudie, C.A., Parker, N.C., Cauldwell, W. and Snellgrove, R. (1990) Hormonal sex manipulation and evidence for female homogamety in channel catfish. *General and Comparative Endocrinology* 78, 218–223.

Davis, K.B., Morrison, J. and Galvez, J. (2000) Reproductive characteristics of adult channel catfish treated with trenbolone acetate during the phenocritical period of sex differentiation. *Aquaculture* 189, 351–360.

Davis, S.A., Catchpole, E.A. and Fulford, G.R. (2000) Periodic triggering of an inducible gene for control of a wild population. *Theoretical Population Biology* 58, 95–106.

Davisson, M.T., Wright, J.E. and Atherton, L.M. (1973) Cytogenetic analysis of pseudolinkage of LDH loci in the teleost genus *Salvelinus*. *Genetics* 73, 645–658.

D'Cotta, H., Fostier, A., Guiguen, Y., Govoroun, M. and Baroiller, J.F. (2001) Search for genes involved in the temperature-induced gonadal sex differentiation in the tilapia, *Oreochromis niloticus*. *Journal of Experimental Zoology* 290, 574–585.

de Gortari, M.J., Freking, B.A., Kappes, S.M., Leymaster, K.A., Crawford, A.M., Stone, R.T. and Beattie, C.W. (1997) Extensive genomic conservation of cattle microsatellite heterozygosity in sheep. *Animal Genetics* 28, 274–290.

Dehring, T.R., Brown, A.F., Daugherty, C.H. and Phelps, S.R. (1981) Survey of the genetic variation among eastern Lake Superior lake trout (*Salvelinus namaycush*). *Canadian Journal of Fisheries and Aquatic Sciences* 38, 1738–1746.

Deka, R., Shriver, M.D., Yu, L.M., Jin, L., Aston, C.E., Chakraborty, R. and Ferrell, R.E. (1994) Conservation of human chromosome 13 polymorphic microsatellite $(CA)_n$ repeats in chimpanzees. *Genomics* 22, 226–230.

de la Motte, S., Klinger, J., Kefer, G., King, T. and Harrison, F. (2001) Pharmacokinetics of human growth hormone administered subcutaneously with two different injection systems. *Arzneimittelforschung/ Drug Research* 51, 613–617.

del Valle, G., Taniguchi, N. and Tsujimura, A. (1996) Genetic differences in some haematological traits of communally reared clonal ayu, *Plecoglossus altivelis* Temminck & Schlegal, under stressed and non-stressed conditions. *Aquaculture Research* 27, 787–793.

Demski, L.S. (1993) Terminal nerve complex. *Acta Anatomica* 148, 81–95.

Deng, H.-W. (2002) Population admixture may appear to mask, change or reverse genetic effects of genes underlying complex traits. *Genetics* 159, 1319–1323.

Deng, H.-W., Chen, W.M. and Recker, R.R. (2001) Population admixture: detection by Hardy–Weinberg test and its quantitative effects on linkage-disequilibrium methods for localizing genes underlying complex traits. *Genetics* 157, 885–897.

Desvignes, J.F., Laroche, J., Durand, J.D. and Bouvet, Y. (2001) Genetic variability in reared stocks of common carp (*Cyprinus carpio* L.) based on allozymes and microsatellites. *Aquaculture* 194, 291–301.

Devlin, R.H. (1997a) Applications of molecular genetics in salmon aquaculture and fisheries biology. In: *Proceedings of the First Joint Korea–Canada Symposium in Aquatic Biosciences*. Institute of Fisheries Science, Pukyong National University, Pusan, South Korea, pp. 113–127.

Devlin, R.H. (1997b) Transgenic salmonids. In: Houdebine, L.M. (ed.) *Transgenic Animals: Generation and Use*. Harwood Academic Publishers, Amsterdam, pp. 105–117.

Devlin, R.H. and Donaldson, E.M. (1992) Containment of genetically altered fish with emphasis on salmonids. In: Hew, C.L. and Fletcher, G.L. (eds) *Transgenic Fish*. World Scientific, Singapore, pp. 229–265.

Devlin, R.H., McNeil, B.K., Solar, I.I. and Donaldson, E.M. (1994a) A rapid PCR-based test for Y-chromosomal DNA allows simple production of all-female strains of chinook salmon. *Aquaculture* 128, 211–220.

Devlin, R.H., Yesaki, T.Y., Blagi, C.A., Donaldson, E.M., Swanson, P. and Chen, W.K. (1994b) Extraordinary salmon growth. *Nature* 371, 209–210.

Devlin, R.H., Yesaki, T.Y., Donaldson, E.M., Du, S.-J. and Hew, C.L. (1995a) Production of germline transgenic Pacific salmonids with dramatically increased growth performance. *Canadian Journal of Fisheries and Aquatic Sciences* 52, 1376–1384.

Devlin, R.H., Yesaki, T.Y., Donaldson, E.M. and Hew, C.L. (1995b) Transmission and phenotypic effects of an antifreeze/GH gene construct in coho salmon (*Oncorhynchus kisutch*). *Aquaculture* 137, 161–169.

Devlin, R.H., Johnsson, J.I., Smailus, D.E., Biagi, C.A., Johnsson, E. and Bjornsson, B.T. (1999) Increased ability to compete for food by growth hormone transgenic coho salmon (*Oncorhynchus kisutch* Walbaum). *Aquaculture Research* 30, 1–4.

Devlin, R.H., Swanson, P., Clarke, W.C., Plisetskaya, E., Dickhoff, W., Moriyama, S., Yesaki, T.Y. and Hew, C.L. (2000) Seawater adaptability and hormone levels in growth-enhanced transgenic coho salmon, *Oncorhynchus kisutch*. *Aquaculture* 191, 367–385.

Devlin, R.H., Biagi, C.A., Yesaki, T.Y., Smailus, D.E. and Byatt, J.C. (2001) Growth of domesticated transgenic fish. *Nature* 409, 781–782.

Devlin, R.H., D'Andrade, M., Uh, M. and Biagi, C.A. (2004) Population effects of growth hormone transgenic coho salmon depend on food availability and genotype by environment interactions. *Proceedings of the National Academy of Sciences USA* 101, 9303–9308.

De Vos, A.M., Ultsch, M. and Kossiakoff, A.A. (1992) Human growth hormone and extracellular domain of its receptor: crystal structure of the complex. *Science* 255, 306–312.

DeWoody, J.A. and Avise, J.C. (2000) Microsatellite variation in marine, freshwater and anadromous fishes compared with other animals. *Journal of Fish Biology* 56, 461–473.

DFO (1998) *Draft Policy on Research with and Rearing of Transgenic Aquatic Organisms*. Aquaculture and Oceans Science Branch, Department of Fisheries and Oceans Canada, Ottawa.

DiBerardino, M.A. (1987) Genomic potential of differentiated cells analyzed by nuclear transplantation. *American Zoologist* 27, 623–644.

DiBerardino, M.A. (ed.) (1997) *Genomic Potential of Differential Cells*. Columbia University Press, New York, New York.

Dickmeis, T., Rastegar, S., Aanstad, P., Clark, M., Fischer, N., Korzh, V. and Straehle, U. (2001) Expression of the anti-dorsalizing morphogenetic protein gene in the zebrafish embryo. *Development Genes and Evolution* 211, 568–572.

Dickson, I.W. and Kramer, R.H. (1971) Factors influencing scope for activity and active and standard metabolism of rainbow trout (*Salmo gairdneri*). *Journal of Fisheries Research Board of Canada* 28, 587–596.

Dillon, J.C., Schill, D.J. and Teuscher, D.M. (2000) Relative return to creel of triploid and diploid rainbow trout stocked in eighteen Idaho streams. *North American Journal of Fisheries Management* 20, 1–9.

Dinesh, K.R., Lim, T.M., Chua, K.L., Chan, W.K. and Phang, V.P.E. (1993) RAPD analysis: an efficient method of DNA fingerprinting in fishes. *Zoological Science* 10, 849–854.

Disney, J.E. and Wright, J.E. Jr (1990) Extensive multivalent pairing occurs in male lake trout meiosis. *Genome* 33, 219–224.

Doak, T.G., Doerder, F.P., Jahn, C.L. and Herrick, G. (1994) A proposed superfamily of transposase genes: transposon-like elements in ciliated protozoa and a common 'D35E' motif. *Proceedings of the National Academy of Sciences USA* 91, 942–946.

Dobosz, S. and Goryczko, K. (1988) Effect of polyploidization on survival of sea, brook, and rainbow trout hybrids during incubation and early feeding period. *Acta Ichthyologica et Piscatoria* 18, 19–22.

Dobson, S.H. and Holmes, R.M. (1984) Compensatory growth in the rainbow trout, *Salmo gairdneri* Richardson. *Journal of Fish Biology* 25, 649–656.

Doldan, M.J., Prego, B., Holmqvist, B.I. and de Miguel, E. (1999) Distribution of GABA immunolabeling in the early zebrafish (*Danio rerio*) brain. *European Journal of Morphology* 37, 126–129.

Don, J. and Avtalion, R.R. (1986) The induction of triploidy in *Oreochromis aureus* by heat shock. *Theoretical and Applied Genetics* 72, 186–192.

Donaldson, E.M. (1973) Reproductive endocrinology of fishes. *American Zoologist* 13, 909–927.

Donaldson, E.M. and Hunter, G.A. (1982) Sex control in fish with particular reference to salmonids. *Canadian Journal of Fisheries and Aquatic Sciences* 39, 99–110.

Doneen, B.A. (1976) Biological activities of mammalian and teleostean prolactins and growth hormones on mouse mammary gland and teleost urinary bladder. *General and Comparative Endocrinology* 30, 34–42.

Donoso, A.O. (1988) Localized increase of GABA levels in brain areas of the rat and inhibition of the plasma LH rise following ovariectomy. *Neuroendocrinology* 48, 130–137.

Donovan, S.M., Hintz, R.L. and Rosenfeld, R.G. (1991) Insulin-like growth factors I and II and their binding proteins in human milk: effect of heat treatment on IGF and IGF binding protein stability. *Journal of Pediatric Gastroenterology and Nutrition* 13, 242–253.

Doronin, Y.K., Golichenkov, V.A., Graschk, M.A., Georgiev, G.P., Rubstov, P.M., Skryabin, K.G. and Baev, A.A. (1989) Integration and expression of human growth hormone gene in teleostei. *Genetica (Russia)* 25, 24–35.

Dorson, M. and Chevassus, B. (1985) Susceptibility of rainbow trout × coho salmon triploid hybrids to pancreatic necrosis and haemorrhagic septicaemia viruses. *Bulletin Français de la Pêche et de la Pisciculture* 296, 29–34.

Dorson, M., Chevassus, B. and Torhy, C. (1991) Comparative susceptibility of three species of char and rainbow trout × char triploid hybrids to several pathogenic salmonid viruses. *Diseases of Aquatic Organisms* 11, 217–224.

Dorson, M., Quillet, E., Hollebecq, M.G., Torhy, C. and Chevassus, B. (1995) Selection of rainbow trout resistant to viral haemorrhagic septicaemia virus and transmission of resistance by gynogenesis. *Veterinary Research* 26, 361–368.

Douglas, S.E., Gallant, J.W., Bullerwell, C.E., Wolff, C., Munholland, J. and Reith, M.E. (1999) Winter flounder expressed sequence tags: establishment of an EST database and identification of novel fish genes. *Marine Biotechnology* 1, 458–464.

Downing, S.L. (1988a) Optimal induction of triploidy in *Crassostrea gigas* depends on temperature. *Journal of Shellfish Research* 7, 115.

Downing, S.L. (1988b) Triploid and diploid hybrids between the oysters *Crassostrea gigas* and *C. rivularis*: production, detection and potential. *Journal of Shellfish Research* 7, 156.

Downing, S.L. (1988c) Comparing adult performance of diploid and triploid monospecific and interspecific *Crassostrea* hybrids. *Journal of Shellfish Research* 7, 549.

Downing, S.L. (1989a) Estimating polyploid percentages using oyster larvae: a valuable hatchery management and research tool. *Journal of Shellfish Research* 8, 320.

Downing, S.L. (1989b) Hybridization, triploidy and salinity effects on crosses with *Crassostrea gigas* and *Crassostrea virginica*. *Journal of Shellfish Research* 8, 447.

Downing, S.L., Allen, S.K. Jr and Beattie, J.H. (1985) Triploidy induced in the Pacific oyster by means of cytochalasin B. *Journal of Shellfish Research* 5, 51.

Doyle, R. (2003) Genetic computation limited. Available at www.genecomp.com

Doyle, R. (2008) Genetic computation limited. Available at www.genecomp.com

Doyle, R.W., Perez-Enriquez, R., Takagi, M. and Taniguchi, N. (2001) Selective recovery of founder genetic diversity in aquacultural broodstocks and captive endangered fish populations. *Genetica* 111, 291–304.

Draghia-Akli, R., Fiorotto, M.L., Hill, L.A., Malone, P.B., Deaver, D.R. and Schwartz, R.J. (1999) Myogenic expression of an injectable protease-resistant growth hormone-releasing hormone augments long-term growth in pigs. *Nature Biotechnology* 17, 1179–1183.

Draper, D.E. and Reynaldo, L.P. (1999) RNA binding strategies of ribosomal proteins. *Nucleic Acids Research* 27, 381–388.

Dreier, B., Beerli, R.R., Segal, D.J., Flippin, J.D. and Barbas, C.F. 3rd (2001) Development of zinc finger domains for recognition of the 5′-ANN-3′ family of DNA sequences and their use in the construction of artificial transcription factors. *Journal of Biological Chemistry* 276, 29466–29478.

Du, F.L., Zou, X.G., Xiang, X.R., Xu, S.F. and Du, M. (1995) Studies on recipient goat oocytes in serial nuclear transplantation. In: *Symposium of Sixty Years Anniversary of the Establishment of Chinese Society of Zoology*. Chinese Society of Zoology, pp. 251–262.

Du, S.J., Gong, Z., Fletcher, G.L., Shears, M.A., King, M.J., Idler, D.R. and Hew, C.L. (1992) Growth enhancement in transgenic Atlantic salmon by the use of an 'all fish' chimeric growth hormone gene construct. *Bio/Technology* 10, 176–181.

Duan, C. (1998) Nutritional and developmental regulation of insulin-like growth factors in fish. *Journal of Nutrition* 128 (2 Suppl.), 306S–314S.

Duan, C. and Plisetskaya, E.M. (1993) Nutritional regulation of insulin-like growth factor I mRNA expression in salmon tissues. *Journal of Endocrinology* 139, 243–252.

Duan, C., Duguay, S.J. and Plisetskaya, E.M. (1993) Insulin-like growth factor I (IGF-I) mRNA expression in coho salmon, *Oncorhynchus kisutch*: tissue distribution and effects of growth hormone/prolactin family proteins. *Fish Physiology and Biochemistry* 11, 371–379.

Duan, C., Plisetskaya, E.M. and Dickhoff, W.W. (1995) Expression of insulin-like growth factor I in normally and abnormally developing coho salmon (*Oncorhynchus kisutch*). *Endocrinology* 136, 446–452.

Dufy, C. and Diter, A. (1990) Polyploidy in the Manila clam *Ruditapes philippinarum*: chemical induction and larval performance of triploids. *Aquatic Living Resources* 3, 55–60.

Duguay, S.J., Park, L.K., Samadpour, M. and Dickhoff, W.W. (1992) Nucleotide sequence and tissue distribution of three insulin-like growth factor I prohormones in salmon. *Molecular Endocrinology* 6, 1202–1210.

Duguay, S.J., Chan, S.J., Mommsen, T.P. and Steiner, D.F. (1995) Divergence of insulin like growth factor I and II in the elasmobranch, *Squalus acanthias*. *FEBS Letters* 371, 69–72.

Duguay, S.J., Lai-Zhang, J., Steiner, D.F., Funkenstein, B. and Chan, S.J. (1996) Developmental and tissue regulated expression of insulin-like growth factor I and II mRNA in *Sparus aurata*. *Journal of Molecular Endocrinology* 16, 123–132.

Dunham, R.A. (1981) Response to selection and realized heritability for body weight in three strains of channel catfish grown in earthen ponds. Doctoral dissertation, Auburn University, Auburn, Alabama.

Dunham, R.A. (1986) Selection and crossbreeding responses for cultured fish. In: Dickerson, G.E. and Johnson, R.K. (eds) *Proceedings of the 3rd World Congress on Genetic Applications to Livestock Production*, Vol. 10. University of Nebraska, Lincoln, Nebraska, p. 391.

Dunham, R.A. (1990a) Production and use of monosex or sterile fishes in aquaculture. *Reviews in Aquatic Sciences* 2, 1–17.

Dunham, R.A. (1990b) Genetic engineering in aquaculture. *AgBiotech News and Information* 2, 401–406.

Dunham, R.A. (1995) Predator avoidance, spawning and foraging ability of transgenic catfish. In: Levin, M., Grim, C. and Angle, J.S. (eds) *Proceedings of Biotechnology Risk Assessment Symposium, 22–24 June 1994*. USDA, EPA, College Park, Maryland, pp. 151–169.

Dunham, R.A. (1996) Contribution of genetically improved aquatic organisms to global food security. In: *International Conference on Sustainable Contribution of Fisheries to Food Security*. Government of Japan and FAO, Rome.

Dunham, R.A. (1999) Utilization of transgenic fish in developing countries: potential benefits and risks. *Journal of World Aquaculture Society* 30, 1–11.

Dunham, R.A. (2004) *Aquaculture and Fisheries Biotechnology: Genetic Approaches*. CAB International, Wallingford, UK.

Dunham, R.A. (2009) Transgenic fish resistant to infectious diseases, their risk and prevention of escape into the environment and future candidate genes for disease transgene manipulation. *Comparative Immunology, Microbiology and Infectious Diseases* 32, 139–181.

Dunham, R.A. (2011) *Aquaculture and Fisheries Biotechnology and Genetics*, 2nd edn. CAB International, Wallingford, UK.

Dunham, R.A. and Argue, B.J. (2000) Reproduction among channel catfish, blue catfish, and their F_1 and F_2 hybrids. *Transactions of the American Fisheries Society* 129, 222–231.

Dunham, R.A. and Brummett, R.E. (1999) Response of two generations of selection to increased body weight in channel catfish, *Ictalurus punctatus*, compared to hybridization with blue catfish, *I. furcatus*, males. *Journal of Applied Aquaculture* 9, 37–45.

Dunham, R.A. and Devlin, R. (1998) Comparison of traditional breeding and transgenesis in farmed fish with implications for growth enhancement and fitness. In: Murray, J.D., Anderson, G.B., Oberbauer, A.M. and McGloughlin, M.N. (eds) *Transgenic Animals in Agriculture*. CAB International, Wallingford, UK, pp. 209–229.

Dunham, R.A. and Liu, Z. (2002) Gene mapping, isolation and genetic improvement in catfish. In: Shimizu, N., Aoki, T., Hirono, I. and Takashima, F. (eds) *Aquatic Genomics: Steps Toward a Great Future*. Springer-Verlag, New York, pp. 45–60.

Dunham, R.A. and Smitherman, R.O. (1983a) Crossbreeding channel catfish for improvement of body weight in earthen ponds. *Growth* 47, 97–103.

Dunham, R.A. and Smitherman, R.O. (1983b) Response to selection and realized heritability for body weight in three strains of channel catfish, *Ictalurus punctatus*, grown in earthen ponds. *Aquaculture* 33, 88–96.

Dunham, R.A. and Smitherman, R.O. (1984) *Ancestry and Breeding Catfish in the United States*. Circular 273. Alabama Agricultural Experiment Station, Auburn University, Auburn, Alabama.

Dunham, R.A. and Smitherman, R.O. (1985) *Improved Growth Rate, Reproductive Performance and Disease Resistance of Crossbred and Selected Catfish from AU-M and AU-K Lines*. Circular 279. Alabama Agricultural Experiment Station, Auburn University, Auburn, Alabama.

Dunham, R.A. and Smitherman, R.O. (1987) *Genetics and Breeding of Catfish*. Regional Research Bulletin 325, Southern Cooperative Series. Alabama Agricultural Experiment Station, Auburn University, Auburn, Alabama.

Dunham, R.A., Philipp, D.P. and Whitt, G.S. (1980) Levels of duplicate gene expression in armoured catfishes. *Journal of Heredity* 71, 248–252.

Dunham, R.A., Smitherman, R.O., Brooks, M.J., Benchakan, M. and Chappell, J.A. (1982a) Paternal predominance in reciprocal channel–blue hybrid catfish. *Aquaculture* 29, 389–396.

Dunham, R.A., Smitherman, R.O., Chappell, J.A., Youngblood, P.N. and Bice, T.O. (1982b) Communal stocking and multiple rearing techniques for catfish genetics research. *Journal of the World Mariculture Society* 13, 261–267.

Dunham, R.A., Benchakan, M., Smitherman, R.O. and Chappell, J.A. (1983a) Correlations among morphometric traits of 11-month old blue, channel, white and hybrid catfishes and their relationship to dressing percentage at 18-month of age. *Journal of the World Mariculture Society* 14, 668–675.

Dunham, R.A., Smitherman, R.O., Horn, J.L. and Bice, T.O. (1983b) Reproductive performances of crossbred and pure strain brood stock. *Transactions of the American Fisheries Society* 112, 436–440.

Dunham, R.A., Smitherman, R.O. and Webber, C. (1983c) Relative tolerance of channel × blue hybrid and channel catfish to low oxygen concentrations. *Progressive Fish-Culturist* 45, 55–56.

Dunham, R.A., Joyce, J.A., Bondari, K. and Malvestuto, S.P. (1985) Evaluation of body conformation, composition and density as traits for indirect selection for dress-out percentage of channel catfish. *Progressive Fish-Culturist* 47, 169–175.

Dunham, R.A., Eash, J., Askins, J. and Townes, T.M. (1987) Transfer of the metallothionein–human growth hormone fusion gene into channel catfish. *Transactions of the American Fisheries Society* 116, 87–91.

Dunham, R.A., Norgren, K.G., Smitherman, R.O., Ober, R. and Rees, B. (1989) *Identification of Striped Bass Populations in Georgia*. Final Report Project F-42. Georgia Department of Natural Resources, Southeastern Cooperative Regional Project on Genetics and Breeding of Fishes, Atlanta, Georgia.

Dunham, R.A., Brummett, R.E., Ella, M.O. and Smitherman, R.O. (1990) Genotype–environment interactions for growth of blue, channel and hybrid catfish in ponds and cages at varying densities. *Aquaculture* 85, 143–151.

Dunham, R.A., Smitherman, R.O. and Bondari, K. (1991) Lack of inheritance of stumpbody and taillessness in channel catfish. *Progressive Fish-Culturist* 53, 101–105.

Dunham, R.A., Ramboux, A.C., Duncan, P.L., Hayat, M., Chen, T.T., Lin, C.M., Gonzalez-Villasenor, K.I. and Powers, D.A. (1992a) Transfer, expression and inheritance of salmonid growth hormone in channel catfish, *Ictalurus punctatus*, and effects on performance traits. *Molecular Marine Biology and Biotechnology* 1, 380–389.

Dunham, R.A., Turner, J. and Reeves, W.C. (1992b) Introgression of the Florida largemouth bass genome into native populations in Alabama public lakes. *North American Journal of Fisheries Management* 12, 494–498.

Dunham, R.A., Chen, T.T., Powers, D.A., Nichols, A., Argue, B. and Chiminat, C. (1995) Predator avoidance, spawning, and foraging ability of transgenic channel catfish with rainbow trout growth hormone gene. In: Levin, M., Grim, C. and Angle, J.S. (eds) *Proceedings of the Biotechnology Risk Assessment Symposium, 6–8 June 1995*. TechniGraphix, Reston, Virginia, pp. 127–139.

Dunham, R.A., Liu, Z., Wolters, W., Waldbeiser, G., Tiersch, T.R. and Winn, R. (1998) Summary of catfish mapping group. In: *Proceedings of the National Aquaculture Species Gene Mapping Workshop*. Publication No. NRAC 98–001. Northeastern Regional Aquaculture Center, Ithaca, New York, pp. 66–67.

Dunham, R.A., Chitminat, C., Nichols, A., Argue, B., Powers, D.A. and Chen, T.T. (1999) Predator avoidance of transgenic channel catfish containing salmonid growth hormone genes. *Marine Biotechnology* 1, 545–551.

Dunham, R.A., Majumdar, K., Hallerman, E., Bartley, D., Mair, G., Hulata, G., Liu, Z., Pongthana, N., Bakos, J., Penman, D., Gupta, M., Rothlisberg, P. and Hoerstgen-Schwark, G. (2001) Review of the status of aquaculture genetics. In: Subasinghe, R.P., Bueno, P., Phillips, M.J., Hough, C., McGladdery, S.E. and Arthur, J.R. (eds) *Technical Proceedings of the Conference on Aquaculture in the Third Millennium, Bangkok, Thailand, 20–25 February 2000*. NACA, Bangkok and FAO, Rome, pp. 129–157.

Dunham, R.A., Chatakondi, N., Kucuktas, H., Nichols, A. and Chen, T.T. (2002a) The effect of rainbow trout growth hormone gene on the phenotypic variations in the common carp, *Cyprinus carpio* (l.). *Marine Biotechnology* 4, 1–5.

Dunham, R.A., Chatakondi, N., Nichols, A., Chen, T.T., Powers, D.A. and Kucuktas, H. (2002b) Survival of F_2 transgenic common carp, *Cyprinus carpio* containing pRSVrtGH1 cDNA when subjected to low dissolved oxygen. *Marine Biotechnology* 4, 323–327.

Dunham, R.A., Warr, G., Nichols, A., Duncan, P.L., Argue, B., Middleton, D. and Liu, Z. (2002c) Enhanced bacterial disease resistance of transgenic channel catfish, *Ictalarus punctatus*, possessing cecropin genes. *Marine Biotechnology* 4, 338–344.

Dunham, R.A., Chatakondi, N., Nichols, A.J., Kucuktas, H., Chen, T.T., Powers, D.A., Weete, J.D., Cummins, K. and Lovell, R.T. (2002d) Effect of rainbow trout growth hormone complementary DNA on body shape, carcass yield, and carcass composition of F_1 and F_2 transgenic common carp (*Cyprinus carpio*). *Marine Biotechnology* 4, 604–611.

Dunham, R.A., Kucuktas, H., Liu, Z., Keefer, L., Spencer, M. and Robinson, S. (2002e) Biochemical genetics of brook trout in Georgia: management implications. *Southeastern Association of Fish and Wildlife Agencies* 55, 63–80.

Dunstan, G.A., Elliott, N.G., Appleyard, S.A., Holmes, B.H., Conod, N., Grubert, M.A. and Cozens, M.A. (2007) Culture of triploid greenlip abalone (*Haliotis laevigata* Donovan) to market size: commercial implications. *Aquaculture* 271, 130–141.

Durand, P., Wada, K.T. and Komaru, A. (1990) Triploidy induction by caffeine–heat shock treatments in the Japanese pearl oyster *Pinctada fucata martensii*. *Nippon Suisan Gakkaishi/Bulletin of the Japanese Society of Scientific Fisheries* 56, 1423–1425.

Dutta, O. (1979) Factors affecting gonadal differentiation in *Tilapia aurea*. Doctoral dissertation, Auburn University, Auburn, Alabama.

Eaton, D.L. and Klassen, C.D. (1996) Principles of toxicology. In: Klassen, C.D., Amdur, M.O. and Doull, J.D. (eds) *Toxicology*, 5th edn. McGraw Hill, New York, pp. 19–20.

Ebert, K.M., Low, E., Overstorm, F., Buonomo, T.M., Roberts, A., Lee, G., Mandel, G. and Goodman, R. (1988) A moloney MLV–rat somatotropin fusion gene produces biologically-active somatotropin in a transgenic pig. *Molecular Endocrinology* 2, 227–283.

Ehlinger, N.F. (1977) Selective breeding of trout for resistance to furunculosis. *New York Fish and Game Journal* 24, 25–36.

Einum, S. and Fleming, I.A. (1999) Maternal effects of egg size in brown trout (*Salmo trutta*): norms of reaction to environmental quality. *Proceedings of the Royal Society of London, Series B: Biological Sciences* 266, 2095–2100.

Einum, S. and Fleming, I.A. (2000a) Highly fecund mothers sacrifice offspring survival to maximize fitness. *Nature* 6786, 565–567.

Einum, S. and Fleming, I.A. (2000b) Selection against late emergence and small offspring in Atlantic salmon (*Salmo salar*). *Evolution* 2, 628–639.

Eipper, B.A. and Mains, R.E. (1980) Structure and function of pro-adrenocorticotropin/endorphin and related peptides. *Endocrine Reviews* 1, 1–27.

Ekstroem, P. and Ohlin, L.-M. (1995) Ontogeny of GABA-immunoreactive neurons in the central nervous system in a teleost, *Gasterosteus aculeatus*. *Journal of Chemical Neuroanatomy* 9, 271–288.

El-Gamal, A. (1987) Reproductive performance, sex ratios, gonadal development, cold tolerance, viability and growth of red and normally pigmented hybrids of *Tilapia aurea* and *T. nilotica*. Doctoral dissertation, Auburn University, Auburn, Alabama.

El-Ibiary, H.M., Andrews, J.W., Joyce, J.A., Page, J.W. and Deloach, H.L. (1976) Sources of variations in body size traits, dressout weight, and lipid content in channel catfish, *Ictalurus punctatus*. *Transactions of the American Fisheries Society* 105, 267–272.

El-Sherbini, A.E. (1990) Paternal predominance in channel catfish. PhD Dissertation. Auburn University, Auburn, Alabama.

Ellis, A.E. (1992) Fish immunology. In: Roitt, I.M. and Delves, P.J. (eds) *Encyclopedia of Immunology*, Vol. 3. Academic Press, San Diego, California, pp. 536–539.

Ellison, C.K. and Burton, R.S. (2008) Genotype-dependent variation of mitochondrial transcriptional profiles in interpopulation hybrids. *Proceedings of the National Academy of Sciences USA* 105, 15831–15836.

Elmerot, C., Arnason, U., Gojobori, T. and Janke, A. (2002) The mitochondrial genome of the pufferfish, *Fugu rubripes*, and ordinal teleostean relationships. *Gene* 295, 163–172.

Embody, G.C. and Hayford, C.D. (1925) The advantage of rearing brook trout fingerlings from selected breeders. *Transactions of the American Fisheries Society* 55, 135–138.

Engel, W., Schmidtke, J., Vogel, W. and Wolf, V. (1973) Genetic polymorphism of lactate dehydrogenase isoenzymes in the carp (*Cyprinius carpio*) apparently due to a 'null allele'. *Biochemical Genetics* 8, 281–289.

English, L.J., Nell, J.A., Maguire, G.B. and Ward, R.D. (2001) Allozyme variation in three generations of selection for whole weight in Sydney rock oysters (*Saccostrea glomerata*). *Aquaculture* 193, 213–225.

Enikolopov, G.N., Benyumov, A.O., Barmintsev, A., Zelenina, L.A., Sleptsova, L.A., Doronin, Y.K., Golichenkov, V.A., Grashchuk, M.A., Georgiev, G.P., Rubtsov, P.M., Skryabin, K.G. and Baev, A.A. (1989) Advanced growth of transgenic fish containing human somatotropin gene. *Doklady Akademii Nauk SSSR* 301, 724–727.

Ernfors, P., Lee, K., Kucera, J. and Jaenisch, J. (1994) Lack of neurotrophin 3 leads to deficiencies in the peripheral nervous system and loss of limb proprioceptive afferents. *Cell* 77, 503–514.

Ernst, D.H., Watanabe, W.O., Ellington, L.J., Wicklund, R.I. and Olla, B.L. (1991) Commercial-scale production of Florida red tilapia seed in low- and brackish-salinity tanks. *Journal of the World Aquaculture Society* 22, 36–44.

Estrada, M.P., Herrera, F., Cabezas, L., Martinez, R., Arenal, A., Tapanes, L., Vazquez, J. and de la Fuente, J. (1999) Culture of transgenic tilapia with accelerated growth under different intensive culture conditions. In: Nelson, J. and MacKinley, D. (eds) *Special Adaptations of Tropical Fish*. Department of Fisheries and Oceans, Vancouver, British Columbia, Canada, pp. 93–100.

Etherton, T.D. and Bauman, D.E. (1998) Biology of somatotropin in growth and lactation of domestic animals. *Physiological Reviews* 78, 745–761.

Etherton, T.D., Wiggins, J.P., Chung, C.S., Evock, C.M., Rebhun, J.F. and Walton, P.E. (1986) Stimulation of pig growth-performance by porcine growth-hormone and growth hormone-releasing factor. *Journal of Animal Science* 63, 1389–1399.

Eudeline, B. and Allen, S.K. (2000) Optimization of tetraploid induction in Pacific oysters, *Crassostrea gigas*, using first polar body as a natural indicator. *Aquaculture* 187, 73–84.

Eudeline, B., Allen, S.K. Jr and Guo, X. (2000) Delayed meiosis and polar body release in eggs of triploid Pacific oysters, *Crassostrea gigas*, in relation to tetraploid production. *Journal of Experimental Marine Biology and Ecology* 248, 151–161.

Evans, M. and Kaufman, M. (1981) Establishment in culture of pluripotential cells from mouse embryos. *Nature* 292, 154–156.

Ewart, K.V., Belanger, J.C., Williams, J., Karakach, T., Penny, S., Tsoi, S.C., Richards, R.C. and Douglas, S.E. (2005) Identification of genes differentially expressed in Atlantic salmon (*Salmo salar*) in response to infection by *Aeromonas salmonicida* using cDNA microarray technology. *Developmental and Comparative Immunology* 29, 333–347.

Eyres, R., Elliott, R.B., Howie, R.N. and Farmer, K. (1978) Low-temperature pasteurization of human milk. *New Zealand Medical Journal* 87, 134–135.

Fabbri, A., Jannini, E.A., Gnessi, L., Ulisse, S., Moretti, C. and Isidori, A. (1989) Neuroendocrine control of male reproductive function: the opioid system as a model of control at multiple sites. *Journal of Steroid Biochemistry* 32, 145–150.

Facchinetti, F., Genazzani, A.R., Pestarino, M., Vallarino, M., Pierantoni, R., Fasano, S., D'Antonio, M., Carnevali, O., Mosconi, G. and Polzonetti-Magni, A. (1992) Ovarian opioids and the reproductive cycle of the frog *Rana esculenta*. *Life Sciences* 50, 1389–1398.

Facchinetti, F., Genazzani, A.R., Vallarino, M., Pestarino, M., Polzonetti-Magni, A., Carnevali, O., Ciarcia, G., Fasano, S., D'Antonio, M. and Pierantoni, R. (1993) Opioids and testicular activity in the frog, *Rana esculenta*. *Journal of Endocrinology* 137, 49–57.

Fagerlund, V.H.M. and McBride, J.R. (1975) Growth increments and some flesh and gonad characteristics of juvenile coho salmon receiving diets supplemented with methyltestosterone. *Journal of Fish Biology* 7, 305–314.

Falconer, D.S. (1989) *Introduction to Quantitative Genetics*, 3rd edn. Longman Scientific and Technical, Harlow, UK.

Falconer, D.S. and Mackay, T.F.C. (1996) *Introduction to Quantitative Genetics*, 4th edn. Longman, Harlow, UK.

Fankhauser, G. (1945) The effects of changes in chromosome number on amphibian development. *Quarterly Review of Biology* 20, 20–78.

FAO (2001) *Genetically Modified Organisms, Consumers, Food Safety and the Environment*. Food and Agriculture Organization of the United Nations, Rome; available at http://www.fao.org/DOCREP/003/X9602E/X9602E00.HTM

FAO (2006) *State of World Aquaculture 2006*. FAO Fisheries Technical Paper No. 500. Food and Agriculture Organization of the United Nations, Rome.

FAO (2009) *The State of World Fisheries and Aquaculture 2008*. Food and Agriculture Organization of the United Nations, Rome.

FAO (2011) Meetings of the Commission on Genetic Resources for Food and Agriculture. Food and Agriculture Organization of the United Nations; available at http://www.fao.org/nr/cgrfa/cgrfa-meetings/cgrfa-comm/en/ (accessed April 2011).

FAO/WHO (2003a) Principles for the risk analysis of foods derived from modern biotechnology. Food and Agriculture Organization of the United Nations/World Health Organization; available at ftp://ftp.fao.org/es/esn/food/princ_gmfoods_en.pdf (accessed March 2004).

FAO/WHO (2003b) Guideline for the conduct of food safety assessment of foods derived from recombinant-DNA plants. Food and Agriculture Organization of the United Nations/World Health Organization; available at ftp://ftp.fao.org/es/esn/food/guide_plants_en.pdf (accessed March 2004).

FAO/WHO (2003c) Guideline for the conduct of food safety assessment of foods produced using recombinant-DNA microorganisms. Food and Agriculture Organization of the United Nations/World Health Organization; available at ftp://ftp.fao.org/es/esn/food/guide_mos_en.pdf (accessed March 2004).

FAO/WHO (2003d) Codex Alimentarius Commission. Report of the Twenty-Sixth Session, 30 June–7 July 2003. ftp://ftp.fao.org/codex/alinorm03/al03_41e.pdf (accessed March 2004).

FAO/WHO (2003e) Report of the Thirtieth Session of the Codex Committee on Food Labelling, 6–10 May 2002, Halifax, Canada. Food and Agriculture Organization of the United Nations/World Health Organization; available at ftp://ftp.fao.org/codex/alinorm03/Al03_22e.pdf (accessed March 2004).

FAO/WHO (2003f) FAO/WHO Expert Consultation on Safety Assessment of Foods Derived from Genetically Modified Animals including Fish. Food and Agriculture Organization of the United Nations/World Health Organization; available at http://www.fao.org/es/ESN/food/risk_biotech_animal_en.stm (accessed March 2004).

Farbridge, K.J. and Leatherland, J.F. (1992) Temporal changes in plasma thyroid hormone, growth hormone and free fatty acid concentrations, and hepatic 5-monodeiodinase activity, lipid and protein content during chronic fasting and refeeding in rainbow trout (*Oncorhynchus mykiss*). *Fish Physiology and Biochemistry* 10, 245–257.

Farbridge, K.J., Flett, P.A. and Leatherland, J.F. (1992) Temporal effects of restricted diet and compensatory increased dietary intake on thyroid function, plasma growth hormone levels and tissue lipid reserves of rainbow trout *Oncorhynchus mykiss*. *Aquaculture* 104, 157–174.

Faris, J.D., Laddomada, B. and Gill, B.S. (1998) Molecular mapping of segregation distortion loci in *Aegilops taushii*. *Genetics* 149, 319–327.

Farrell, A.P., Bennett, W. and Devlin, R.H. (1997) Growth-enhanced transgenic salmon can be inferior swimmers. *Canadian Journal of Zoology* 75, 335–337.

Fauconneau, B., Kaushik, S.J. and Blanc, J.M. (1989) Uptake and metabolization of dissolved compounds in rainbow trout (*Salmo gairdneri* R.) fry. *Comparative Biochemistry and Physiology, Part A: Comparative Physiology* 93A, 839–843.

Faustman, E.M. and Omenn, G.S. (1996) Risk assessment. In: Klassen, C.D., Amdur, M.O. and Doull, J.D. (eds) *Toxicology*. McGraw Hill, New York, p. 75.

FDA/CVM (2001) Freedom of information summary POSILAC® (sterile sometribove zinc suspension). NADA Number 140–872. US Food and Drug Administration/Center for Veterinary Medicine; available at http://www.fda.gov/downloads/AnimalVeterinary/Products/ApprovedAnimalDrugProducts/FOIADrugSummaries/ucm050022.pdf (accessed April 2011).

FDA (2006) Questions and answers about transgenic fish. US Food and Drug Administration; available at http://www.fda.gov/AnimalVeterinary/ResourcesforYou/ucm047112.htm (accessed April 2011).

Felbinger, T.W., Suchner, U., Goetz, A.E., Briegel, J. and Peter, K. (1999) Recombinant human growth hormone for reconditioning of respiratory muscle after lung volume reduction surgery. *Critical Care Medicine* 27, 1634–1638.

Felip, A., Zanuy, S., Carrillo, M. and Piferrer, F. (2000) The effects of induced triploidy on the reproductive physiology in the European sea bass, *Dicentrarchus labrax*. In: Norberg, B., Kjesbu, O.S., Taranger, G.L., Andersson, E. and Stefansson, S.O. (eds) *Reproductive Physiology of Fish 2000. 6th International Symposium on the Reproductive Physiology of Fish, Bergen, Norway, 4–9 July 1999*. Department of Fisheries and Marine Biology, University of Bergen, Bergen, Norway.

Felip, A., Piferrer, F., Carrillo, M. and Zanuy, S. (2001a) Comparison of the gonadal development and plasma levels of sex steroid hormones in diploid and triploid sea bass, *Dicentrarchus labrax* L. *Journal of Experimental Zoology* 290, 384–395.

Felip, A., Piferrer, F., Zanuy, S. and Carrillo, M. (2001b) Comparative growth performance of diploid and triploid European sea bass over the first four spawning seasons. *Journal of Fish Biology* 58, 76–88.

Felip, A., Zanuy, S., Carrillo, M. and Piferrer, F. (2001c) Induction of triploidy and gynogenesis in teleost fish with emphasis on marine species. *Genetica* 111, 175–195.

Felley, J.D. and Avise, J.C. (1980) Genetic and morphological variation of bluegill populations in Florida lakes. *Transactions of the American Fisheries Society* 109, 108–115.

Ferguson, A. and Mason, F.M. (1981) Allozyme evidence for reproductively isolated sympatric populations of brown trout *Salmo trutta* L. in Lough Melvin, Ireland. *Journal of Fish Biology* 18, 629–642.

Ferguson, A., Verspoor, E., Cross, T., Garcia Vazquez, Medale, F. and Garcia De Leaniz Caprile, C. (1998) An assessment of the genetic consequences of the deliberate and inadvertent introduction of non-active Atlantic salmon into natural populations. In: *Proceedings of Third European Marine Science and Technology Conference (MAST), Lisbon, 23–27 May 1998. Project Synopses, Vol. 5. Fisheries and Aquaculture (AIR: 1990–94) – Selected projects from the research programme for Agriculture and Agro-Industry including Fisheries.* pp. 162–165.

Ferguson, M.M. and Danzmann, R.G. (1999) Inter-strain differences in the association between mitochondrial DNA haplotype and growth in cultured Ontario rainbow trout (*Oncorhynchus mykiss*). *Aquaculture* 178, 245–252.

Ferris, S.D. and Whitt, G.S. (1977) Loss of duplicate gene expression after polyploidisation. *Nature* 265, 258–260.

Ferris, S.D. and Whitt, G.S. (1978a) Phylogeny of tetraploid catostomid fishes based on the loss of duplicate gene expression. *Systematic Zoology* 27, 189–206.

Ferris, S.D. and Whitt, G.S. (1978b) Genetic and molecular analysis of nonrandom dimer assembly of creatine kinase isozymes of fishes. *Biochemical Genetics* 16, 811–829.

Fholenhag, K., Arrhenius-Nyberg, V., Sjogren, I. and Malmlof, K. (1997) Effects of insulin-like growth factor I (IGF-I) on the small intestine: a comparison between oral and subcutaneous administration in the weaned rat. *Growth Factors* 14, 81–88.

Fields, R., Lowe, S.S., Kaminski, C., Whitt, G.S. and Philipp, D.P. (1987) Critical and chronic thermal maxima of northern and Florida largemouth bass and their reciprocal F_1 and F_2 hybrids. *Transactions of the American Fisheries Society* 116, 856–863.

Finnegan, D.J. (1989) Eukaryotic transposable elements and genome evolution. *Trends in Genetics* 5, 103–107.

Fire, A., Xu, S., Montogomery, M.K., Kostas, S.A., Driver, S.E. and Mello, C.C. (1998) Potent and specific genetic interference by double-stranded RNA in *Caenorhabditis elegans*. *Nature* 391, 806–811.

Fishback, A.G., Danzmann, R.G. and Ferguson, M.M. (2000) Microsatellite allelic heterogeneity among hatchery rainbow trout maturing in different seasons. *Journal of Fish Biology* 57, 1367–1380.

Fisher, E.M., Beer-Romero, P., Brown, L.G., Ridley, A., McNeil, J.A., Lawrence, J.B., Willard, H.F., Bieber, F.R. and Page, D.C. (1990) Homologous ribosomal protein genes on the human X and Y chromosomes: escape from X inactivation and possible implications for Turner syndrome. *Cell* 63, 1205–1218.

Fisher, R.B. (1967) Absorption of proteins. *British Medical Bulletin* 23, 241–246.

FitzSimmons, N.N., Moritz, C. and Moore, S.S. (1995) Conservation and dynamics of microsatellite loci over 300 million years of marine turtle evolution. *Molecular Biology and Evolution* 12, 432–440.

Fjalestad, K. T., Gjedrem, T., Carr, W. H. and Sweeney, J. N. (1997) *Final Report: The Shrimp Breeding Program. Selective Breeding of Penaeus vannamei.* Report No. 17/97. Akvaforsk Genetics Center AS, Sunndalsøra, Norway.

Flajshans, M. (1997) A model approach to distinguish diploid and triploid fish by means of computer-assisted image analysis. *Acta Veterinaria (Brno)* 66, 101–110.

Fleming, I.A., Hindar, K., Mjoelneroed, I.B., Jonsson, B., Balstad, T. and Lamberg, A. (2000) Lifetime success and interactions of farm salmon invading a native population. *Proceedings of the Royal Society of London, Series B: Biological Sciences* 267, 1517–1523.

Fletcher, G. and Davies, P.L. (1991) Transgenic fish for aquaculture. *Genetic Engineering* 13, 331–369.

Fletcher, G.L., Shears, M.A., King, M.J., Davies, P.L. and Hew, C.L. (1988) Evidence for antifreeze protein gene transfer in Atlantic salmon (*Salmo salar*). *Canadian Journal of Fisheries and Aquatic Sciences* 45, 352–357.

Fletcher, G.L., Davies, P.L. and Hew, C.L. (1992) Genetic engineering of freeze resistant Atlantic salmon. In: Hew, C.L. and Fletcher, G.L. (eds) *Transgenic Fish.* World Scientific Publishing, River Edge, New Jersey, pp. 190–208.

Flick, W.A. and Webster, D.A. (1962) Problems in sampling wild and domestic stocks of brook trout (*Salvelinus fontinalis*). *Transactions of the American Fisheries Society* 91, 140–144.

Flick, W.A. and Webster, D.A. (1964) Comparative first year survival and production in wild and domestic strains of brook trout, *Salvelinus fontinalis*. *Transactions of the American Fisheries Society* 93, 58–69.

Flick, W.A. and Webster, D.A. (1976) Production of wild, domestic and interstrain hybrids of brook trout (*Salvelinus fontinalis*) in natural ponds. *Journal of the Fisheries Research Board of Canada* 33, 1525–1539.

Flynn, M.B. (1973) The effect of methallibure and a constant 12 hours light:12 hours dark photoperiod on the gonadal maturation of pink salmon (*Oncorhynchus gorbuscha*). MSc thesis, University of British Columbia, Vancouver, British Columbia, Canada.

Foley, P.A. and Luckinbill, L.S. (2001) The effects of selection for larval behavior on adult life-history features in *Drosophila melanogaster*. *Evolution* 55, 2493–2502.

Folkertsma, R.T., Rouppe van der Voort, J.N., de Groot, K.E., van Zandvoort, P.M., Schots, A., Gommers, F.J., Helder, J. and Backer, J. (1996) Gene pool similarities of potato cyst nematode populations assessed by AFLP analysis. *Molecular Plant–Microbe Interactions* 9, 47–54.

Fontaine, P.-M. and Dodson, J.J. (1999) An analysis of the distribution of juvenile Atlantic salmon (*Salmo salar*) in nature as a function of relatedness using microsatellites. *Molecular Ecology* 8, 189–198.

Foo, C.L., Dinesh, K.R., Lim, T.M., Chan, W.K. and Phang, V.P. (1995) Inheritance of RAPD markers in the guppy fish, *Poecilia reticulata*. *Zoological Science* 12, 535–541.

Foolad, M.R., Chen, F.Q. and Lin, G.Y. (1998) RFLP mapping of QTLs conferring cold tolerance during seed germination in an interspecific cross of tomato. *Molecular Breeding* 4, 519–529.

Forbes, S.H. and Allendorf, F.W. (1989) Use of gametic disequilibrium to detect synteny in hybrid swarms of cutthroat trout. *Isozyme Bulletin* 22, 60.

Ford, J.E., Law, B.A., Marshall, V.M. and Reiter, B. (1977) Influence of the heat treatment of human milk on some of its protective constituents. *Journal of Pediatrics* 90, 29–35.

Fornies, M.A., Mañanos, E., Carrillo, M., Rocha, A., Laureau, S., Mylonas, C.C., Zohar, Y. and Zanuy, S. (2001) Spawning induction of individual European sea bass females (*Dicentrarchus labrax*) using different GnRHa-delivery systems. *Aquaculture* 202, 221–234.

Foster, A.R., Houlihan, D.F., Gray, C., Medale, F., Fauconneau, B., Kaushik, S.J. and Le Bail, P.Y. (1991) The effects of ovine growth hormone on protein turnover in rainbow trout. *General and Comparative Endocrinology* 82, 111–120.

Foster, R.F., Donaldson, L.R., Welander, A.D., Bonham, K. and Seymour, A.H. (1949) The effect on embryos and young of rainbow trout from exposing the parent fish to x-rays. *Growth* 13, 119–142.

Franco, G.R., Adams, M.D., Bento, S.M., Simpson, A.J.G., Venter, J.C. and Pena, S.D.J. (1995) Identification of new *Schistosoma mansoni* genes by the EST strategy using a directional cDNA library. *Gene* 152, 141–147.

Frankham, R., Gilligan, D.M., Morris, D. and Briscoe, D.A. (2001) Inbreeding and extinction: effects of purging. *Conservation Genetics* 2, 279–284.

Fransz, P., Armstrong, S., Alonso-Blanco, C., Fischer, T.C., Torres-Ruitz, R.A. and Jones, G. (1998) Cytogenetics for the model system *Arabidopsis thaliana*. *Plant Journal* 13, 867–876.

Fraser, J.M. (1972) Recovery of planted brook trout, splake and rainbow trout from selected Ontario lakes. *Journal of the Fisheries Research Board of Canada* 29, 129–142.

Fraser, J.M. (1974) An attempt to train hatchery-reared brook trout to avoid predation by the common loon. *Transactions of the American Fisheries Society* 103, 815–818.

Fraser, J.M. (1980) Survival, growth and food habits of brook trout and F_1 splake planted in Precambrian Shield lakes. *Transactions of the American Fisheries Society* 109, 491–501.

Fraser, J.M. (1981) Comparative survival and growth of planted wild hybrid and domestic strains of brook trout (*Salvelinus fontinalis*) in Ontario lakes. *Canadian Journal of Fisheries and Aquatic Sciences* 38, 1672–1684.

Fredholm, M. and Wintero, A.K. (1995) Variation of short tandem repeats within and between species belonging to the Canidae family. *Mammalian Genome* 6, 11–18.

Freeland, J.R. and Boag, P.T. (1999) The mitochondrial and nuclear genetic homogeneity of the phenotypically diverse Darwin's groundfinches. *Evolution* 53, 1553–1563.

Frenhani, P.B. and Burini, R.C. (1999) Mechanism of action and control in the digestion of proteins and peptides in humans. *Arquivos de Gastroenterologia* 36, 139–147.

Friar, E.A., Ladoux, T., Roalson, E.H. and Robichaux, R.H. (2000) *Argyroxiphium sandwicense* ssp. *sandwicense* (Asteraceae), and its implications for reintroduction. *Molecular Ecology* 9, 2027–2034.

Friars, G.W. (1993) *Breeding Atlantic Salmon: A Primer*. ASF, St Andrews, Nebraska.

Friars, G.W., McMillan, I., Quinton, V.M., O'Flynn, F.M., McGeachy, S.A. and Benfey, T.J. (2001) Family differences in relative growth of diploid and triploid Atlantic salmon (*Salmo salar* L.). *Aquaculture* 192, 23–29.

Frick, G.P., Tai, L.-R., Baumbach, W.R. and Goodman, H.M. (1998) hGH tissue distribution, turnover, and glycosylation of the long and short growth hormone receptor isoforms in rat tissues. *Endocrinology* 139, 2824–2830.

Fricker, A. (1969) Enzymatic transformations in frozen foods. *Zeitschrift für Ernahrungswissenschaft Supplement* 8, 63–72.

Friend, K., Iranmanesh, A., Login, I.S. and Veldhuis, J.D. (1997) Pyridostigmine treatment selectively amplifies the mass of GH secreted per burst without altering GH burst frequency, half-life, basal GH secretion or the orderliness of GH release. *European Journal of Endocrinology* 137, 377–386.

Frost, R.A., Nachman, S.A., Lang, C.H. and Gelato, M.C. (1996) Proteolysis of insulin-like growth factor-binding protein-3 in human immunodeficiency virus-positive children who fail to thrive. *Journal of Clinical Endocrinology and Metabolism* 81, 2957–2962.

Fueshko, S.M., Key, S. and Wray, S. (1998a) Luteinizing hormone releasing hormone (LHRH) neurons maintained in nasal explants decrease LHRH messenger ribonucleic acid levels after activation of GABA(A) receptors. *Endocrinology* 139, 2734–2740.

Fueshko, S.M., Key, S. and Wray, S. (1998b) GABA inhibits migration of luteinizing hormone releasing hormone neurons in embryonic olfactory explants. *Journal of Neuroscience* 18, 2560–2569.

Fu-hua, L., Ling-hua, Z., Jian-hai, X., Xu-dong, L. and Jin-zhao, Z. (1999) Triploidy induction with heat shocks to *Penaeus chinensis* and their effects on gonad development. *Chinese Journal of Oceanology and Limnology* 17, 57–61.

Fuji, K., Hasegawa, O., Honda, K., Kumasaka, K., Sakamoto, T. and Okamoto, N. (2007) Marker-assisted breeding of a lymphocystis disease-resistant Japanese flounder (*Paralichthys olivaceus*). *Aquaculture* 272, 291–295.

Fujiwara, M., Tanaka, M. and Okabe, M. (2005) Heterosis in the cockle *Fulvia mutica* in a crossbreeding experiment among inbred lines. *Kyoto Furitsu Kaiyo Senta Kenkyu Houkoku* (27), 25–30.

Fujiwara, Y., Komiya, T., Kawabata, H., Sato, M., Fujimoto, H., Furusawa, M. and Noce, T. (1994) Isolation of a DEAD-family protein gene that encodes a murine homolog of *Drosophila* vase and its specific expression in germ cell lineage. *Proceedings of the National Academy of Sciences USA* 91, 12258–12262.

Gaffney, P. (1999) Summary of the oyster mapping group. In: *Proceedings of the Aquaculture Species Genome Mapping Workshop, Dartmouth, Massachusetts, May 1997.* Publication NRAC 98–001/NRAES-124. Natural Resource, Agriculture, and Engineering Service, Ithaca, New York, pp. 68–71.

Gaffney, P. (2002) Genomic approaches to marker development and mapping in the eastern oyster, *Crassostrea virginica*. In: Shimizu, N., Aoki, T., Hirono, I. and Takashima, F. (eds) *Aquatic Genomics: Steps Toward a Great Future.* Springer-Verlag, New York, pp. 84–91.

Gajewski, M. and Voolstra, C. (2002) Comparative analysis of somatogenesis related genes of the hairy/Enhancer of split class in *Fugu* and zebrafish. *BMC Genomics* 5, 21.

Galbreath, P.F. and Samples, B.L. (2000) Optimization of thermal shock protocols for induction of triploidy in brook trout. *North American Journal of Aquaculture* 62, 249–259.

Galbreath, P.F. and Thorgaard, G.H. (1997) Saltwater performance of triploid Atlantic salmon *Salmo salar* L. × brown trout *Salmo trutta* L. hybrids. *Aquaculture Research* 28, 1–8.

Galbreath, P.F., Adams, K.J., Wheeler, P.A. and Thorgaard, G.H. (1997) Clonal Atlantic salmon × brown trout hybrids produced by gynogenesis. *Journal of Fish Biology* 50, 1025–1033.

Gall, G.A.E. (1969) Quantitative inheritance and environmental response of rainbow trout. In: Neuhaus, O.W. and Halver, J.E. (eds) *Fish Research.* Academic Press, New York, New York.

Gall, G.A.E. (1974) Influence of size of eggs and age of female on hatchability and growth in rainbow trout. *California Fish and Game* 60, 26–35.

Gall, G.A.E. (1986) Sexual maturation and growth rate. In: Dickerson, G.E. and Johnson, R.K. (eds) *Third World Congress on Genetics Applied to Livestock Production, 16–22 July 1986*, Vol. X. University of Nebraska, Lincoln, Nebraska, pp. 401–410.

Gall, G.A.E. and Gross, S.J. (1978) Genetic studies of growth in domesticated rainbow trout. *Aquaculture* 3, 225–234.

Galtier, N. and Gouy, M. (1995) Inferring phylogenies from DNA sequences of unequal base compositions. *Proceedings of the National Academy of Sciences USA* 92, 11317–11321.

Galvin, P., Taggart, J., Ferguson, A., O'Farrell, M. and Cross, T. (1996) Population genetics of Atlantic salmon (*Salmo salar*) in the River Shannon system in Ireland: an appraisal using single locus minisatellite (VNTR) probes. *Canadian Journal of Fisheries and Aquatic Sciences* 53, 1933–1942.

Gao, C., Noden, D.M. and Norgren, R.B. (2000) LHRH neuronal migration: heterotypic transplantation analysis of guidance cues. *Journal of Comparative Neurology* 42, 95–103.

Garant, D., Dodson, J.J. and Bernatchez, L. (2001) A genetic evaluation of mating system and determinants of individual reproductive success in Atlantic salmon (*Salmo salar* L.). *Journal of Heredity* 92, 137–145.

Garcia-Abiado, M.A.R., Dabrowski, K., Christensen, J.E., Czesny, S. and Bajer, P. (1999) Use of erythrocyte measurements to identify triploid saugeyes. *North American Journal of Aquaculture* 61, 319–325.

Gardner, J.D., Brown, M.S. and Laster, L. (1970a) The columnar epithelial cell of the small intestine: digestion and transport (first of three parts). *New England Journal of Medicine* 283, 1196–1202.

Gardner, J.D., Brown, M.S. and Laster, L. (1970b) The columnar epithelial cell of the small intestine: digestion and transport (third of three parts). *New England Journal of Medicine* 283, 1317–1324.

Garvey, E.P. and Santi, D.V. (1986) Stable amplified DNA in drug resistant *Leishmania* exists as extrachromosomal circles. *Science* 233, 535–540.

Gasaryan, K.G., Hung, N.M., Neyfakn, A.A. and Ivanenkov, V.V. (1979) Nuclear transplantation in teleost *Misgurnus fossilis* L. *Nature* 280, 585–587.

Gatlin, D.M. III (1987) Whole-body amino acid composition and comparative aspects of amino acid nutrition of the goldfish, golden shiner and fathead minnow. *Aquaculture* 60, 223–229.

Gautheret, D., Poirot, O., Lopez, F., Audic, S. and Claverie, J.M. (1998) Alternate polyadenylation in human mRNAs: a large-scale analysis by EST clustering. *Genome Research* 8, 524–530.

Gaylord, T.G. and Gatlin, D.M. III (2000) Assessment of compensatory growth in channel catfish *Ictalurus punctatus* R. and associated changes in body condition indices. *Journal of the World Aquaculture Society* 31, 326–336.

Gaylord, T.G. and Gatlin, D.M. III (2001a) Dietary protein and energy modifications to maximize compensatory growth of channel catfish (*Ictalurus punctatus*). *Aquaculture* 194, 337–348.

Gaylord, T.G. and Gatlin, D.M. III (2001b) Dietary protein and energy modifications to maximize compensatory growth of channel catfish *Ictalurus punctatus*. In: *Aquaculture 2001: Book of Abstracts*. Elsevier, Amsterdam, p. 248.

Gaylord, T.G., MacKenzie, D.S. and Gatlin, D.M. III (2001) Growth performance, body composition and plasma thyroid hormone status of channel catfish (*Ictalurus punctatus*) in response to short-term feed deprivation and refeeding. *Fish Physiology and Biochemistry* 24, 73–79.

Gendreau, S. and Grizel, H. (1990) Induced triploidy and tetraploidy in the European flat oyster, *Ostrea edulis* L. *Aquaculture* 90, 229–238.

Gerber, S., Mariette, S., Streiff, R., Bodénès, C. and Kremer, A. (2000) Comparison of microsatellites and amplified fragment length polymorphism markers for parentage analysis. *Molecular Ecology* 9, 1037–1048.

Gervai, J., Peter, S., Nagy, A., Horvath, L., Csanyi, V., Gervai, J. and Csanyi, V. (1980) Induced triploidy in carp, *Cyprinus carpio* L. *Journal of Fish Biology* 17, 667–671.

Gerwick, L., Corley-Smith, G. and Bayne, C.J. (2007) Gene transcript changes in individual rainbow trout livers following an inflammatory stimulus. *Fish and Shellfish Immunology* 22, 157–171.

Gestl, E.E., Kauffman, E.J., Moore, J.L. and Cheng, K.C. (1997) New conditions for generation of gynogenetic half-tetrad embryos in the zebrafish (*Danio rerio*). *Journal of Heredity* 88, 76–79.

Gharrett, A.J., Shirley, S.M. and Tromble, G.R. (1987) Genetic relationships among populations of Alaskan chinook salmon (*Oncorhynchus tshawytscha*). *Canadian Journal of Fisheries and Aquatic Sciences* 44, 765–774.

Ghosh, C., Liu, Y., Ma, C. and Collodi, P. (1997) Cell cultures derived from early zebrafish embryos differentiate *in vitro* into neurons and astrocytes. *Cytotechnology* 23, 221–230.

Gibbs, P.D. and Schmale, M.C. (2000) GFP as a genetic marker scorable throughout the life cycle of transgenic zebrafish. *Marine Biotechnology* 2, 107–125.

Gibbs, P.D.L., Gray, A. and Thorgaard, G. (1994) Inheritance of *P* element and reporter gene sequences in zebrafish. *Molecular Marine Biology and Biotechnology* 3, 317–326.

Gibson, G., and Muse, S.V. (2009) *A Primer of Genome Science*, 3rd edn. Sinauer Associates, Inc. Publishers, Sunderland, Massachusetts.

Gilligan, P., Brenner, S. and Venkatesh, B. (2002) *Fugu* and human sequence comparison identifies novel human genes and conserved non-coding sequences. *Gene* 294, 35–44.

Gitterle, T., Martinez, W., Marimon, F., Salazar, M. , Faillace, J., Suarez, A., Cock, J.H., Johnsen, H., and Rye, M. (2009) Commercial field performance as a measure of genetic improvement in the Pacific white shrimp *Penaeus (Litopenaeus) vannamei*. In: *Book of Abstracts of 10th International Symposium on Genetics in Aquaculture (ISGA X) 'Roles of Aquaculture Genetics in Addressing Global Food Crisis', 22–26 June 2009, Bangkok, Thailand*. Faculty of Fisheries, Kasetsart University, Bangkok, p. 90.

Gjedrem, T. (1979) Selection for growth rate and domestication in Atlantic salmon. *Zeitschrift für Tierzüchtung und Züchtungsbiologie* 96, 56–59.

Gjedrem, T. (1997) Selective breeding to improve aquaculture production. *World Aquaculture* 28, 33–45.

Gjedrem, T., Salte, R. and Gjøen, H.M. (1991) Genetic variation in susceptibility of Atlantic salmon to furunculosis. *Aquaculture* 97, 1–6.

Gjerde, B. (1984) Response to individual selection for age at sexual maturity in Atlantic salmon. *Aquaculture* 38, 229–240.

Gjerde, B. (1986) Growth and reproduction in fish and shellfish. *Aquaculture* 57, 37–55.

Gjerde, B. and Gjedrem, T. (1984) Estimates of phenotypic and genetic parameters for carcass traits in Atlantic salmon and rainbow trout. *Aquaculture* 36, 97–110.

Gjerde, B. and Korsvoll, A. (1999) Realised selection differentials for growth rate and early sexual maturity in Atlantic salmon. In: *Aquaculture Europe '99 – Towards Predictable Aquaculture. Abstracts of contributions presented at the International Conference 'Aquaculture Europe '99' held in Trondheim, Norway, 7–10 August.* Special Publication No. 27. European Aqauculture Society, Oostende, Belgium, pp. 73–74.

Gjerde, B., Evensen, O., Bentsen, H.B. and Storset, A. (2009) Genetic (co)variation of vaccine injuries and innate resistance to furunculosis (*Aeromonas salmonicida*) and infectious salmon anaemia (ISA) in Atlantic salmon (*Salmo salar*). *Aquaculture* 287, 52–58.

Glick, B.R. and Pasternak, J.J. (1998) *Molecular Biotechnology: Principles and Applications of Recombinant DNA.* ASM Press, Washington, DC.

Glover, K.A., Ottera, H., Olsen, R.E., Slinde, E., Taranger, G.L. and Skaala, O. (2009) A comparison of farmed, wild and hybrid Atlantic salmon (*Salmo salar* L.) reared under farming conditions. *Aquaculture* 286, 203–210.

Gluckman, D.P., Douglas, R.G., Ambler, G.R., Breier, H.B., Hodkinson, S.C., Koea, J.B. and Shaw, J.H.F. (1991) The endocrine role of insulin-like growth factor. I. *Acta Pediatrica Scandinavia (Supplement)* 372, 97–105.

Goetz, F.W., Donaldson, E.M., Hunter, G.A. and Dye, H.M. (1979) Effects of estradiol-17β and 17-methyltestosterone on gonadal differentiation in the coho salmon (*Oncorhynchus kisutch*). *Aquaculture* 17, 267–278.

Gold, J.R. and Richardson, L.R. (1990) Restriction site heteroplasmy in the mitochondrial DNA of the marine fish *Sciaenops ocellatus* (L.). *Animal Genetics* 21, 313–316.

Goldberg, D.M., McAllister, R.A. and Roy, A.D. (1968) Studies on human intestinal proteolysis. *Scandinavian Journal of Gastroenterology* 3, 193–201.

Gomelsky, B., Cherfas, N.B., Peretz, Y., Ben-Dom, N. and Hulata, G. (1994) Hormonal sex inversion in the common carp (*Cyprinus carpio* L.). *Aquaculture* 126, 265–270.

Gomelsky, B., Cherfas, N.B., Gissis, A. and Hulata, G. (1998) Induced diploid gynogenesis in white bass. *Progressive Fish-Culturist* 60, 288–292.

Gomelsky, B., Cherfas, N.B., Gissis, A. and Hulata, G. (1999) Hormonal sex inversion in striped bass and white bass × striped bass hybrids. *North American Journal of Aquaculture* 61, 199–205.

Gómez, A., Wellbrock, C. and Schartl, M. (2002) Constitutive activation and specific signaling of the X mrk receptor in *Xiphophorus* melanoma. *Marine Biotechnology* 4, 208–217.

Gong, Z. (1999) Zebrafish expressed sequence tags and their applications. *Methods in Cell Biology* 60, 213–233.

Gong, Z., Hew, C.L. and Vielkind, J.R. (1991) Functional analysis and temporal expression of promoter regions from fish antifreeze protein genes in transgenic Japanese medaka embryos. *Molecular Marine Biology and Biotechnology* 1, 64–72.

Gong, Z., Hu, Z., Gong, Z.Q., Kitching, R. and Hew, C.L. (1994) Bulk isolation and identification of fish genes by cDNA clone tagging. *Molecular Marine Biology and Biotechnology* 3, 243–251.

Gong, Z., Wan, H., Ju, B., He, J., Wang, X. and Yan, T. (2002) Generation of living color transgenic zebrafish. In: Shimizu, N., Aoki, T., Hirono, I. and Takashima, F. (eds) *Aquatic Genomics: Steps Toward a Great Future.* Springer-Verlag, New York, pp. 329–339.

Goodenough, U.I. and Levine, R.P. (1974) *Genetics.* Holt, Rinehart and Winston, New York, New York.

Goodier, J.L. (1981) Native lake trout (*Salvelinus namaycush*) stocks in the Canadian waters of Lake Superior prior to 1955. *Canadian Journal of Fisheries and Aquatic Sciences* 38, 1724–1737.

Goodier, J.L. and Davidson, W.S. (1994) *Tc1* transposon-like sequences are widely distributed in salmonids. *Journal of Molecular Biology* 241, 26–34.

Gordon, G.W., Scangos, G.A., Plotkin, D.J., Barbosa, J.A. and Ruddle, F.H. (1980) Genetic transformation of mouse embryos by injection of purified DNA. *Proceedings of the National Academy of Sciences USA* 77, 7380–7384.

Gorshkov, S., Gordin, H., Gorshkova, G. and Knibb, W. (1997) Reproductive constraints for family selection of the gilthead seabream (*Sparus aurata* L.). *Israeli Journal of Aquaculture/Bamidgeh* 49, 124–134.

Gorshkov, S., Gorshkova, G., Hadani, A., Gordin, H. and Knibb, W. (1998) Chromosome set manipulations and hybridization experiments in gilthead seabream (*Sparus aurata*). I. Induced gynogenesis and intergeneric hybridization using males of the red seabream (*Pagrus major*). *Israeli Journal of Aquaculture/Bamidgeh* 50, 99–110.

Gorshkova, G., Gorshkov, S., Hadani, A., Gordin, H. and Knibb, W. (1995) Chromosome set manipulations in marine fish. *Aquaculture* 137, 157–158.

Gossen, M. and Bujard, H. (1992) Tight control of gene expression in mammalian cells by tetracycline responsive promoters. *Proceedings of the National Academy of Sciences USA* 89, 5547–5551.

Gossen, M., Freundlieb, S., Bender, G., Muller, G., Hillen, W. and Bujard, H. (1995) Transcriptional activation by tetracycline in mammalian cells. *Science* 268, 1766–1769.

Gossler, A., Doetschman, T., Korn, R., Serfling, E. and Kemler, R. (1986) Transgenesis by means of blastocyst-derived embryonic stem cell lines. *Proceedings of the National Academy of Sciences USA* 83, 9065–9069.

Gothilf, Y., Elizur, A. and Zohar, Y. (1995) Three forms of gonadotropin-releasing hormone in gilthead seabream and striped bass: physiological and molecular studies. In: Goetz, F.W. and Thomas, P. (eds) *Proceedings of the Fifth International Symposium on the Reproductive Physiology of Fish, University of Texas at Austin, 2–8 July 1995.* University of Texas Press, Austin, Texas, pp. 52–54.

Gothilf, Y., Meiri, I., Elizur, A. and Zohar, Y. (1997) Preovulatory changes in the levels of three gonadotropin-releasing hormone-encoding messenger ribonucleic acids (mRNAs), gonadotropin b-subunit mRNAs, plasma gonadotropin, and steroids in the female gilthead seabream, *Sparus aurata. Biology of Reproduction* 57, 1145–1154.

Goudie, C.A. (1984) Tissue distribution and elimination of radiolabeled methyltestosterone administered in the diet to sexually undifferentiated and adult tilapia, *Tilapia (Oreochromis) aurea.* Doctoral dissertation, Auburn University, Auburn, Alabama.

Goudie, C.A., Redner, B.D., Simco, B.A. and Davis, K.B. (1983) Feminization of channel catfish by oral administration of steroid sex hormones. *Transactions of the American Fisheries Society* 112, 670–672.

Goudie, C.A., Khan, G. and Parker, N. (1985) Gynogenesis and sex manipulation with evidence for female homogameity in channel catfish (*Ictalurus punctatus*). In: *Annual Progress Report 1 October–30 September.* USFWS, Southeastern Fish Culture Laboratory, Marion, Alabama.

Goudie, C.A., Shelton, W.L. and Parker, N.C. (1986) Tissue distribution and elimination of radiolabelled methyltestosterone fed to sexually undifferentiated blue tilapia. *Aquaculture* 58, 215–226.

Goudie, C.A., Simco, B.A. and Davis, K.B. (1995) Failure of gynogenetically-derived male channel catfish to produce all-male offspring. In: Goetz, F.W. and Thomas, P. (eds) *Proceedings of the Fifth International Symposium on the Reproductive Physiology of Fish, University of Texas at Austin, 2–8 July 1995.* University of Texas Press, Austin, Texas, p. 118.

Graham, G.G. and Adrianzen, B. (1972) Related articles links late 'catch-up' growth after severe infantile malnutrition. *Johns Hopkins Medical Journal* 131, 204–211.

Graham, M.S., Fletcher, G.L. and Benfey, T.J. (1985) Effect of triploidy on blood oxygen content of Atlantic salmon. *Aquaculture* 50, 133–139.

Grant, W.S. (1985) Biochemical genetic stock structure of the southern African anchovy, *Engraulis capensis* Gilchrist. *Journal of Fish Biology* 27, 23–29.

Grattapaglia, D. and Sederoff, R. (1994) Genetic linkage maps of *Eucalyptus grandis* and *Eucalyptus urophylla* using a pseudo-testcross: mapping strategy and RAPD markers. *Genetics* 137, 1121–1137.

Greene, C.W. (1952) Results from stocking brook trout of wild and hatchery strains at Stillwater Pond. *Transactions of the American Fisheries Society* 81, 43–52.

Grimholt, U., Larsen, S., Nordmo, R., Midtlyng, P., Kjoeglum, S., Storset, A., Saebø, S. and Stet, R.J. (2003) MHC polymorphism and disease resistance in Atlantic salmon (*Salmo salar*); facing pathogens with single expressed major histocompatibility class I and class II loci. *Immunogenetics* 55, 210–219.

Groh, S., Khairallah, M.M., Gonzales-De-Leon, D., Willcox, M., Jiang, C., Hoisington, D.A. and Melchinger, A.E. (1998) Comparison of QTLs mapping in RILs and their test-cross progenies of tropical maize for insect resistance and agronomic traits. *Plant Breeding* 117, 193–202.

Gross, M.L., Schneider, J.F., Moav, N., Moav, B., Alvarez, C., Myster, S.H., Liu, Z., Hallerman, E.M., Hackett, P.B., Guise, K.S., Faras, A.J. and Kapuscinski, A.R. (1992) Molecular analysis and growth evaluation of northern pike (*Esox lucius*) microinjected with growth hormone genes. *Aquaculture* 103, 253–273.

Grupe, A., Germer, S., Usuka, J., Aud, D., Belknap, J.K. and Klein, R.F. (2001) *In silico* mapping of complex disease-related traits in mice. *Science* 292, 1915–1918.

Guan, G.J., Kobayashi, T. and Nagahama, Y. (2000) Sexually dimorphic expression of two types of DM (Doublesex/Mab-3)-domain genes in a teleost fish, the tilapia (*Oreochromis niloticus*). *Biochemical and Biophysical Research Communications* 272, 662–666.

Gueiros-Filho, F.J. and Beverley, S.M. (1997) Trans-kingdom transposition of the *Drosophila* element mariner within the protozan *Leishmania*. *Science* 276, 1716–1719.

Guerrero, R.D. (1974) The use of synthetic androgens for the production of monosex male *Tilapia aurea* (Steindachner). Doctoral dissertation, Auburn University, Auburn, Alabama.

Guerrero, R.D. III (1975) Use of androgens for the production of all male *Tilapia aurea* (Steindachner). *Transactions of the American Fisheries Society* 104, 342–348.

Guier, H.-P., Zapf, J., Schmid, C. and Froesch, E.R. (1989) Insulin-like growth factors I and II in healthy man. Estimations of half-lives and production rates. *Acta Endocrinologica* 121, 753–758.

Guillen, I., Berlanga, J., Valenzuela, C.M., Morales, A., Toledo, J., Estrada, M.P., Puentes, P., Hayes, O. and de la Fuente, J. (1999) Safety evaluation of transgenic tilapia with accelerated growth. *Marine Biotechnology* 1, 2–14.

Gunnes, K. and Gjedrem, T. (1978) Selection experiments with salmon IV. Growth of Atlantic salmon during two years in the sea. *Aquaculture* 15, 19–33.

Gunnes, K. and Gjedrem, T. (1981) A genetic analysis of body weight and length in rainbow trout reared in seawater for 18 months. *Aquaculture* 24, 161–174.

Guo, X. (1990) Growth and survival of intrastrain and interstrain rainbow trout (*Oncorhynchus mykiss*) triploids. *Journal of the World Aquaculture Society* 21, 250–256.

Guo, X. and Allen, S.K. Jr (1994a) Viable tetraploids in the Pacific oyster (*Crassostrea gigas* Thunberg) produced by inhibiting polar body 1 in eggs from triploids. *Molecular Marine Biology and Biotechnology* 3, 42–50.

Guo, X. and Allen, S.K. Jr (1994b) Reproductive potential and genetics of triploid Pacific oysters, *Crassostrea gigas* (Thunberg). *Biological Bulletin* 187, 309–318.

Guo, X. and Allen, S.K. Jr (1997) Sex and meiosis in autotetraploid Pacific oyster *Crassostrea gigas* (Thunberg). *Genome* 40, 397–405.

Guo, X. and Allen, S.K. Jr (1998) Tetraploid Shellfish. US Patent 5,824,841 (assigned to State University of New Jersey).

Guo, X., Hershberger, W.K., Chew, K.K., Downing, S.L. and Waterstrat, P. (1988) Cell fusion in the Pacific oyster, *Crassostrea gigas*: tetraploids produced by blastomere fusion. *Journal of Shellfish Research* 7, 549–550.

Guo, X., Cooper, K. and Hershberger, W. (1989) Aneuploid Pacific oyster larvae produced by treating with cytochalasin B during meiosis. 1. *Journal of Shellfish Research* 8, 448.

Guo, X., Hershberger, W.K. and Myers, J.M. (1990) Growth and survival of intrastrain and interstrain rainbow trout (*Oncorhynchus mykiss*) triploids. *Journal of the World Aquaculture Society* 21, 250–256.

Guo, X., DeBrosse, G.A. and Allen, S.K. Jr (1996) All-triploid Pacific oysters (*Crassostrea gigas* Thunberg) produced by mating tetraploids and diploids. *Aquaculture* 142, 149–161.

Guo, X., Ford, S., DeBrosse, G. and Smolowitz, R. (2000) Breeding for a superior eastern oyster for the northeastern region. *Journal of Shellfish Research* 19, 572.

Guo, X., Wang, Y. and Xu, Z. (2007) Genomic analyses using fluorescence *in situ* hybridization. In: Liu, Z. (ed.) *Aquaculture Genome Technologies*. Blackwell Publishing, Ames, Iowa, pp. 289–311.

Gupta, M.V. and Acosta, B.O. (eds) (2001) *Fish Genetics Research in Member Countries and Institutions of the International Network on Genetics in Aquaculture. Proceedings of the Fifth Steering Committee Meeting of INGA held on 3–5 March 1999, Kuala Lumpur, Malaysia.* ICLARM Conference Proceedings No. 64. ICLARM, Manila.

Gurdon, J.B. and Melton, D.A. (1981) Gene transfer in amphibian eggs and oocytes. *Annual Review of Genetics* 15, 189–218.

Guyomard, R., Chourrout, D. and Houdebine, L.M. (1988) Gene transfer by microinjection into fertilized eggs: efficient integration and germ line transmission (abstract). In: Gjedrem, T. (ed.) *Genetics in Aquaculture III. Proceedings of the Third International Symposium on Genetics in Aquaculture, Trondheim, Norway, 20–24 June 1988.* International Association for Genetics in Aquaculture.

Guyomard, R., Chourrout, D. and Houdebine, L. (1989a) Production of stable transgenic fish by cytoplasmic injection of purified genes. In: Beaudet, A., Mulligan, R. and Verma, I. (eds) *Gene Transfer and Gene Therapy*, Alan R. Liss, New York, New York, pp. 9–18.

Guyomard, R., Chourrout, D., Leroux, C., Houdebine, L.M. and Pourrain, F. (1989b) Integration and germ line transmission of foreign genes microinjected into fertilized trout eggs. *Biochimie* 71, 857–863.

Guyton, A.C. (1981) *Textbook of Medical Physiology*, 6th edn. W.B. Saunders, Philadelphia, Pennsylvania.

Hackett, P.B. (1993) The molecular biology of transgenic fish. In: Hochachka, P.W. and Mommsen, T.P. (eds) *Biochemistry and Molecular Biology of Fishes, Molecular Biology Frontiers*, Vol. 2. Elsevier, Amsterdam, pp. 207–240.

Hadley, N.H., Dillon, R.T. Jr and Manzi, J.J. (1991) Realized heritability of growth rate in the hard clam *Mercenaris mercenaria*. *Aquaculture* 93, 109–119.

Hagmann, M., Bruggmann, R., Xue, L., Georgiev, O., Schaffner, W., Rungger, D., Spaniol, P. and Gerster, T. (1998) Homologous recombination and DNA-end joining reactions in zygotes and early embryos of zebrafish (*Danio rerio*) and *Drosophila melanogaster*. *Journal of Biological Chemistry* 379, 673–681.

Haley, L.E., Newkirk, G.F., Waugh, D.W. and Doyle, R.W. (1975) A report on the quantitative genetics of growth and survivorship of the American oyster, *Crassostera virginica*, under laboratory conditions. In: Persoone, G. and Jaspers, E. (eds) *Proceedings of the 10th European Symposium on Marine Biology, Ostend, Belgium, 17–23 September 1975*, Vol. 1, *Research in Mariculture at Laboratory and Pilot Scale*. Universa Press, Wetteren, Belgium, pp. 221–228.

Hallerman, E.M. (1991) Review of current and proposed agricultural biotechnology regulatory authority and the Omnibus Biotechnology Act of 1990. In: *Hearing before the Subcommittee on Department Operations, Research and Foreign Agriculture of the Committee on Agriculture, House of Representatives, 101st Congress, Second Session, on H.R. 5312, 2 October 1990*. Serial No. 101–75, US Government Printing Office, Washington, DC, pp. 130–139.

Hallerman, E.M. (1992) Concerns associated with the introduction of exotic or genetically manipulated species. In: Kissil, G.W. and Saar, L. (eds) *Proceedings of the US–Israel Workshop on Mariculture and the Environment, Eilat, Israel, 8–10 June 1992*. National Centre for Mariculture, Eilat, Israel, pp. 100–123.

Hallerman, E.M. (1996) Genetically modified fish and shellfish. In: *Proceedings of MIT Sea Grant Exotic Species Workshop: Issues Related to Aquaculture and Biodiversity, Cambridge, Massachusetts, 8 February 1996*, pp. 18–22.

Hallerman, E.M. (2000) Administration announces sweeping assessment of federal agricultural and food biotech regulations. *ISB (Information Systems for Biotechnology) News Report* June, 15–17.

Hallerman, E.M. (2001a) Results of six-month review of Federal biotechnology policy released for comment. *ISB (Information Systems for Biotechnology) News Report* March.

Hallerman, E.M. (2001b) Societal issues posed by commercialization of aquatic GMOs. In: *Biotechnology–Aquaculture Interface: The Site of Maximum Impact Workshop, Shepperdstown, West Virginia, 5–7 March 2001*.

Hallerman, E.M. and Kapuscinski, A.R. (1990a) Transgenic fish and public policy: regulatory concerns. *Fisheries* 15, 12–20.

Hallerman, E.M. and Kapuscinski, A.R. (1990b) Transgenic fish and public policy: patenting of transgenic fish. *Fisheries* 15, 21–24.

Hallerman, E.M. and Kapuscinski, A.R. (1992a) Ecological and regulatory uncertainties associated with transgenic fish. In: Hew, C.L. and Fletcher, G.L. (eds) *Transgenic Fishes*. World Scientific, Singapore, pp. 209–228.

Hallerman, E.M. and Kapuscinski, A.R. (1992b) Ecological implications of using transgenic fishes in aquaculture. *ICES Marine Science Symposium* 194, 56–66.

Hallerman, E.M. and Kapuscinski, A.R. (1993) Potential impacts of transgenic and genetically manipulated fish on wild populations: addressing the uncertainties through field testing. In: Cloud, J.G. and Thorgaard, G.H. (eds) *Genetic Conservation of Salmonid Fishes*. Plenum Press, New York, pp. 93–112.

Hallerman, E.M. and Kapuscinski, A.R. (1995) Incorporating risk assessment and risk management into public policies on genetically modifed finfish and shellfish. *Aquaculture* 17, 9–17.

Hallerman, E.M., Dunham, R.A. and Smitherman, R.O. (1986) Selection or drift – isozyme allele frequency changes among channel catfish selected for rapid growth. *Transactions of the American Fisheries Society* 115, 60–68.

Hallerman, E.M., King, D. and Kapuscinski, A.R. (1998) A decision support software for safely conducting research with genetically modified fish and shellfish. *Aquaculture* 173, 309–318.

Halstrom, M.L. (1984) Genetic studies of a commercial strain of red tilapia. MSc thesis, Auburn University, Auburn, Alabama.

Hammerman, I.S. and Avtalion, R.R. (1979) Sex determination in *Sarotherodon (Tilapia)*. *Theoretical and Applied Genetics* 55, 177–187.

Hammond, B.G., Collier, R.J., Miller, M.A., McGrath, M., Hartzell, D.L., Kotts, C. and Vandaele, W. (1990) Food safety and pharmacokinetic studies which support a zero (0) meat and milk withdrawal time for use of sometribove in dairy cows. *Annales de Recherches Vétérinaires/Annals of Veterinary Research* 21 (Suppl. 1), 107S–120S.

Han, H.-S., Taniguchi, N. and Tsujimura, A. (1991) Production of clonal ayu by chromosome manipulation and confirmation by isozyme marker and tissue grafting. *Nippon Suisan Gakkaishi* 57, 825–832.

Hancock, R.E. (1997) Peptide antibiotics. *Lancet* 349, 418–422.

Hand, R.E. and Nell, J.A. (1999) Studies on triploid oysters in Australia XII. Gonad discoloration and meat condition of diploid and triploid Sydney rock oysters (*Saccostrea commercialis*) in five estuaries in New South Wales, Australia. *Aquaculture* 171, 181–194.

Hand, R.E., Nell, J.A. and Maguire, G.B. (1998a) Studies on triploid oysters in Australia. X. Growth and mortality of diploid and triploid Sydney rock oysters *Saccostrea commercialis* (Iredale and Roughley). *Journal of Shellfish Research* 17, 1115–1128.

Hand, R.E., Nell, J.A., Smith, I.R. and Maguire, G.B. (1998b) Studies on triploid oysters in Australia. XI. Survival of diploid and triploid Sydney rock oysters (*Saccostrea commercialis*) (Iredale and Roughley) through outbreaks of winter mortality caused by *Mikrocytos roughleyi* infestation. *Journal of Shellfish Research* 17, 1129–1136.

Hand, R.E., Nell, J.A., Reid, D.D., Smith, I.R. and Maguire, G.B. (1999) Studies on triploid oysters in Australia: effect of initial size on growth of diploid and triploid Sydney rock oysters, *Saccostrea commercialis* (Iredale and Roughley). *Aquaculture Research* 30, 35–42.

Handeland, S.O., Berge, A., Bjornsson, B.T. and Lie, S.O. (2000) Seawater adaptation by out-of-season Atlantic salmon (*Salmo salar* L.) smolts at different temperatures. *Aquaculture* 181, 377–396.

Handler, A.M., Gomez, S.P. and O'Brochta, D.A. (1993) A functional analysis of the *P* element gene transfer vector in insect. *Archives of Insect Biochemistry and Physiology* 22, 373–384.

Hansen, M.M., Ruzzante, E.D., Nielsen, E.E. and Mensberg, D.K. (2000) Microsatellite and mitochondrial DNA polymorphism reveals life-history dependent interbreeding between hatchery and wild brown trout (*Salmo trutta* L.) *Molecular Ecology* 5, 583–594.

Hansen, M.M., Ruzzante, D.E., Nielsen, E.E. and Mensberg, K.L.D. (2001) Brown trout (*Salmo trutta*) stocking impact assessment using microsatellite DNA markers. *Ecological Applications* 11, 148–160.

Hanson, L.H. and Manion, P.J. (1978) *Chemosterilization of the Sea Lamprey (*Petromyzon marinus*)*. Technical Report No. 29. Great Lakes Fisheries Commission, Ann Arbor, Michigan.

Hanson, T.R., Smitherman, R.O., Shelton, W.L. and Dunham, R.A. (1983) Growth comparisons of monosex tilapia produced by separation of sexes, hybridization and sex reversal. In: Fishelson, L. and Yaron, Z. (compilers) *Proceedings of the International Symposium on Tilapia in Aquaculture*. Tel Aviv University, Tel Aviv, Israel, pp. 564–573.

Happe, A., Quillet, E. and Chevassus, B. (1988) Early life history of triploid rainbow trout (*Salmo gairdneri* Richardson). *Aquaculture* 71, 107–118.

Hardy, R.W. (1999) Collaborative opportunities between fish nutrition and other disciplines in aquaculture: an overview. *Aquaculture* 177, 217–230.

Harris, H. and Hopkinson, D.A. (1976) *Handbook of Enzyme Electrophoresis in Human Genetics*. North Holland, Amsterdam.

Hartman, M.L., Faria, A.C., Vance, M.L., Johnson, M.L., Thorner, M.O. and Veldhuis, J.D. (1991) Temporal structure of *in vivo* growth hormone secretory events in humans. *American Journal of Physiology* 260 (1 Pt 1), E101–E110.

Hartman, M.L., Veldhuis, J.D., Johnson, M.L., Leeberti, K.G., Samojlik, E. and Thorner, M.O. (1992) Augmented growth hormone (GH) secretory burst frequency and amplitude mediate enhanced GH secretion during a two-day fast in normal men. *Journal of Clinical Endocrinology and Metabolism* 74, 757–765.

Hartnell, G.F. (1995) Bovine somatotropin: production, management and United States experience. In: Ivan, M. (ed.) *Animal Science Research and Development: Moving Toward a New Century*. Center for Food and Animal Research, Agriculture and Agri-Food Canada, Ottawa, pp. 189–203.

Hartwell, L., Hood, L., Goldberg, M.L., Reynolds, A.E., Silver, L.M. and Veres, R.C. (2000) *Genetics: From Genes to Genomes*. McGraw Hill, Boston, Massachusetts.

Hassan, M., Sinden, S.L., Kobayashi, R.S., Nordeen, R.O. and Owens, L.D. (1993) Transformation of potato (*Solanum tuberosum*) with a gene for an antibacterial protein, cecropin. *Acta Horticulturae* 336, 127–131.

Hassin, S., Elizur, A. and Zohar, Y. (1995) Molecular cloning and sequence analysis of striped bass (*Morone saxatilis*) gonadotropins I and II subunits. *Journal of Molecular Endocrinology* 15, 23–35.

Hassin, S., Holland, M.C.H. and Zohar, Y. (2000) Early maturity in the male striped bass: FSH and LH gene expression and their regulation by GnRHa and testosterone. *Biology of Reproduction* 63, 1691–1697.

Hawkins, A.J.S., Magoulas, A., Heral, M., Bougrier, S., Naciri-Graven, Y., Day, A.J. and Kotoulas, G. (2000) Separate effects of triploidy, parentage and genomic diversity upon feeding behaviour, metabolic efficiency and net energy balance in the Pacific oyster *Crassostrea gigas*. *Genetical Research* 76, 273–284.

Hayat, M. (1989) Transfer, expression and inheritance of growth hormone genes in channel catfish (*Ictalurus punctatus*) and common carp (*Cyprinus carpio*). Doctoral dissertation, Auburn University, Auburn, Alabama.

Hayat, M., Joyce, C.P., Townes, T.M., Chen, T.T., Powers, D.A. and Dunham, R.A. (1991) Survival and integration rate of channel catfish and common carp embryos microinjected at various developmental stages. *Aquaculture* 99, 249–255.

He, M., Lin, Y., Shen, Q., Hu, J. and Jiang, W. (2000a) Production of tetraploid pearl oyster (*Pinctada martensii* Dunker) by inhibiting the first polar body in eggs from triploids. *Journal of Shellfish Research* 19, 147–151.

He, M., Shen, Q., Lin, Y., Hu, J. and Jiang, W. (2000b) Inducement of tetraploid *Pinctada martensii* in eggs from triploid. *Journal of Fisheries China/Shuichan Xuebao* 24, 22–27.

Heasman, K.G., Pitcher, G.C., McQuaid, C.D. and Hecht, T. (1998) Shellfish mariculture in the Benguela system: raft culture of *Mytilus galloprovincialis* and the effect of rope spacing on food extraction, growth rate, production, and condition of mussels. *Journal of Shellfish Research* 17, 33–39.

Heath, D.D., Fox, C.W. and Heath, J.W. (1999) Maternal effects on offspring size: variation through early development of chinook salmon. *Evolution* 53, 1605–1611.

Hedgecock, D. (2002) Genomic approaches to understanding heterosis and improving yield of Pacific oysters. In: Shimizu, N., Aoki, T., Hirono, I. and Takashima, F. (eds) *Aquatic Genomics: Steps Toward a Great Future*. Springer-Verlag, New York, pp. 73–83.

Hedgecock, D. and Davis, J.P. (2007) Heterosis for yield and crossbreeding of the Pacific oyster *Crassostrea gigas*. *Aquaculture* 272, S17–S29.

Hedgecock, D., McGoldrick, D.J. and Bayne, B.L. (1995) Hybrid vigor in Pacific oysters: an experimental approach using crosses among inbred lines. *Aquaculture* 137, 285–298.

Hedrick, P.W. (2001) Invasion of transgenes from salmon or other genetically modified organisms into natural populations. *Canadian Journal of Fisheries and Aquatic Sciences* 58, 841–844.

Hedrick, P.W., Hedgecock, D., Hamelberg, S. and Croci, S.J. (2000a) The impact of supplementation in winter-run chinook salmon on effective population size. *Journal of Heredity* 91, 112–116.

Hedrick, P.W., Rashbrook, V.K. and Hedgecock, D. (2000b) Effective population size of winter-run chinook salmon based on microsatellite analysis of returning spawners. *Canadian Journal of Fisheries and Aquatic Sciences* 57, 2368–2373.

Hedrik, P. (1992) Shooting the RAPDs. *Nature* 335, 679–680.

Heierhorst, J., Lederis, K. and Richter, D. (1992) Presence of a member of the *Tc1*-like transposon family from nematodes and *Drosophila* within the vasotocin gene of a primitive vertebrate, the Pacific hagfish *Eptatretus stouti*. *Proceedings of the National Academy of Sciences USA* 89, 6798–6802.

Hellberg, M.E., Balch, D.P. and Roy, K. (2001) Climate-driven range expansion and morphological evolution in a marine gastropod. *Science (Washington)* 292, 1707–1710.

Hemmer-Hansen, J., Nielsen, E.G., Groenkjaer, P. and Loeschcke, V. (2007) Evolutionary mechanisms shaping the genetic population structure of marine fishes; lessons from the European flounder (*Platichthys flesus* L.) *Molecular Ecology* 16, 3104–3118.

Henard, M.-C., Achilles, D. and Williams, B. (2008) *EU-27 Biotechnology Annual 2008 GAIN Report Number E48082*. USDA Foreign Agricultural Service, Washington, DC; available at http://www.fas.usda.gov/gainfiles/200810/146296234.pdf

Henikoff, S. (1992) Detection of *Caenorhabditis* transposon homologs in diverse organisms. *New Biology* 4, 382–388.

Henken, A.M., Brunink, A.M. and Richter, C.J.J. (1987) Differences in growth rate and feed utilization between diploid and triploid African catfish, *Clarias gariepinus* (Burchell 1822). *Aquaculture* 63, 233–242.

Herbinger, C.M., Doyle, R.W., Pitman, E.R., Paquet, D., Mesa, K.A., Morris, D.B., Wright, J.M. and Cook, D. (1995) DNA fingerprint-based analysis of paternal and maternal effects on offspring growth and survival in communally reared rainbow trout. *Aquaculture* 137, 245–256.

Hernandez, O., Guillen, I., Estrada, M.P., Cabrera, E., Pimentel, R., Pina, J.C., Abad, Z., Sanchez, V., Hidalgo, Y., Martinez, R., Lleonart, R. and De la Fuente, J. (1997) Characterization of transgenic tilapia lines with different ectopic expression of tilapia growth hormone. *Molecular Marine Biology and Biotechnology* 6, 364–375.

Hershberger, W.K. and Myers, J.M. (1990) Growth and survival of intrastrain and interstrain *Crassostrea gigas*: tetraploids produced by blastomere fusion. *Journal of Shellfish Research* 7, 549–550.

Hershberger, W.K., Bonham, K. and Donaldson, L.R. (1978) Chronic exposure of chinook salmon eggs and alevins to gamma irradiation: effects on their return to freshwater as adults. *Transactions of the American Fisheries Society* 107, 622–631.

Hershberger, W.K., Myers, J.M., Iwamoto, R.N., McAuley, W.C. and Saxton, A.M. (1990) Genetic changes in the growth of coho salmon (*Oncorhynchus kisutch*) in marine net-pens, produced by ten years of selection. *Aquaculture* 85, 187–197.

Hetzel, D.J.S., Crocos, P.J., Davis, G.P., Moore, S.S. and Preston, N.C. (2000) Response to selection and heritability for growth in the Kuruma prawn, *Penaeus japanicus*. *Aquaculture* 181, 215–223.

Hickling, C.F. (1967) On the biology of a herbivorous fish, the white amur, *Ctenopharyngodon idella* (Val.). *Proceedings of the Royal Society of Edinburgh* 70, 62–81.

Hidu, H., Mason, K.M., Shumway, S.E. and Allen, S.K. (1988) Induced triploidy in *Mercenaria mercenaria* L.: effects on performance in the juveniles. *Journal of Shellfish Research* 7, 202.

Hikima, J., Hirono, I. and Aoki, T. (2002) The lysozyme gene in fish. In: Shimizu, N., Aoki, T., Hirono, I. and Takashima, F. (eds) *Aquatic Genomics: Steps Toward a Great Future*. Springer-Verlag, New York, pp. 301–309.

Hill, J.A., Kiessling, A. and Devlin, R.H. (2000) Coho salmon (*Oncorhynchus kisutch*) transgenic for a growth hormone gene construct exhibit increased rates of muscle hyperplasia and detectable levels of differential gene expression. *Canadian Journal of Fisheries and Aquatic Sciences* 57, 939–950.

Hindar, K. (1999) Introductions at the level of genes and populations. In: Sandlund, O.T., Schei, P.J. and Viken, S. (eds) *Invasive Species and Biodiversity Management*. Kluwer, Dordrecht, The Netherlands, pp. 149–161.

Hindar, K., Ryman, N. and Utter, F. (1991) Genetic effects of cultured fish on natural fish populations. *Canadian Journal of Fisheries and Aquatic Sciences* 48, 945–957.

Hindar, K., Ferguson, A., Youngson, A. and Poole, R. (1998) Hybridisation between escaped farmed Atlantic salmon and brown trout: frequency, distribution, behavioural mechanisms and effects on fitness. In: *Proceedings of Third European Marine Science and Technology Conference (MAST), Lisbon, 23–27 May 1998. Project Synopses, Vol. 5. Fisheries and Aquaculture (AIR: 1990–94) – Selected projects from the research programme for Agriculture and Agro-Industry including Fisheries*. pp. 134–137.

Hines, R.S., Wohlfarth, G.W., Moav, R. and Hulata, G. (1974) Genetic differences in susceptibility to two diseases among strains of the common carp. *Aquaculture* 3, 187–197.

Hines, S.A., Philipp, D.P., Childers, W.F. and Whitt, G.S. (1983) Thermal kinetic differences between allelic isozymes of malate dehydrogenase (Mdh-B locus) of largemouth bass, *Micropterus salmoides*. *Biochemical Genetics* 21, 1143–1151.

Hinits, Y. and Moav, B. (1999) Growth performance studies in transgenic *Cyprinus carpio*. *Aquaculture* 173, 285–296.

Hirono, I. and Aoki, T. (1997) Expressed sequence tags of medaka (*Oryzias latipes*) liver mRNA. *Molecular Marine Biology and Biotechnology* 6, 345–350.

Hirono, I. and Aoki, T. (2002) Immune-related genes of the Japanese flounder, *Paralichthys olivaceus*. In: Shimizu, N., Aoki, T., Hirono, I. and Takashima, F. (eds) *Aquatic Genomics: Steps Toward a Great Future*. Springer-Verlag, New York, pp. 286–300.

Hirose, K. and Hibiya, T. (1968) Physiological studies on growth promoting effect of protein anabolic steroids on fish. II. Effects of 4-chlorotestosterone acetate on rainbow trout. *Bulletin of the Japanese Society of Scientific Fisheries* 34, 473.

Hishiki, T., Kowamoto, S., Morishita, S. and Okubo, K. (2000) Body map: a human and mouse gene expression database. *Nucleic Acids Research* 28, 136–138.

Hislop, J.R.G. (1988) The influence of maternal length and age on the size and weight of the eggs and the relative fecundity of the haddock, *Melanogrammus aeglefinus*, in British waters. *Journal of Fish Biology* 32, 923–930.

Ho, K.Y. and Kelly, J.J. (1991) Role of growth hormone in fluid homeostasis. *Hormone Research* 36 (Suppl. 1), 44–48.

Hoar, W.S., Wiebe, J. and Wai, E.H. (1967) Inhibition of the pituitary gonadotropic activity of fishes by a dithiocarbamoylhydrazine derivative (ICI 33, 828). *General and Comparative Endocrinology* 8, 101–109.

Hoban, T.J. and Kendall, P.A. (1993) *Consumer Attitudes about Food Biotechnology*. Department of Sociology and Anthropology, North Carolina State University, Raleigh, North Carolina.

Hoffmann, J.A., Reichart, J.M. and Hetru, C. (1996) Innate immunity in higher insects. *Current Opinion in Immunology* 8, 8–13.

Hoffmann, W.E. (1934) *Preliminary Notes on the Freshwater Fish Industry in South China, Especially Kwangtung Province.* Science Bulletin No. 5. Lignan University, Hong Kong.

Holladay, L.A. and Puett, D. (1977) Comparative conformational analysis of human choriomammotropin and somatotropin from several species. *International Journal of Peptide Protein Research* 10, 363–368.

Holland, M.C.H., Gothilf, Y., Meiri, I., King, J.A., Okuzawa, K., Elizur, A. and Zohar, Y. (1998) Levels of the native forms of GnRH in the pituitary of the gilthead seabream, *Sparus aurata*, at several characteristic stages of the gonadal cycle. *General and Comparative Endocrinology* 112, 394–405.

Holland, M.C.H., Hassin, S. and Zohar, Y. (2001) Seasonal fluctuations in pituitary levels of the three forms of gonadotropin-releasing hormone in striped bass, *Morone saxatilis* (Teleostei), during juvenile and pubertal development. *Journal of Endocrinology* 169, 527–538.

Hollebecq, M.G., Chourrout, D., Wohlfarth, G. and Billard, R. (1986) Diploid gynogenesis induced by heat shocks after activation with UV-irradiated sperm in common carp. *Aquaculture* 54, 69–76.

Hollebecq, M.G., Chambeyron, F. and Chourrout, D. (1988) Triploid common carp produced by heat shock. In: *Reproduction in Fish. Basic and Applied Aspects in Endocrinology and Genetics.* Colloque no. 44. Institut National de Recherches Agronomiques, Paris, pp. 207–212.

Hollister, A., Johnson, K.R. and Wright, J.E. Jr (1984) Linkage association in hybridized *Salvelinus* genomes: the duplicate loci encoding pepidase-D and glucosephosphate isomerase and the unduplicated sorbitol dehydrogenase locus. *Journal of Heredity* 75, 253–259.

Hollt, V., Przewlocki, R., Haarmann, I., Almeida, O.F., Kley, N., Millan, M.J. and Herz, A. (1986) Stress-induced alterations in the levels of messenger RNA coding for proopiomelanocortin and prolactin in rat pituitary. *Neuroendocrinology* 43, 277–282.

Hong, Y. and Schartl, M. (1996) Embryonic stem cells, cell transplantation and nuclear transplantation in fish. In: *Proceedings of the Conference of IIR Commission C2, Bordeaux Colloquium – Refrigeration and Aquaculture.* Institut International du Froid, Paris, pp. 475–489.

Hong, Y., Winkler, C. and Schartl, M. (1998a) Efficiency of cell culture derivation from blastula embryos and of chimera formation in the medaka (*Oryzias latipes*) depends on donor genotype and passage number. *Development Genes and Evolution* 208, 595–602.

Hong, Y., Winkler, C. and Schartl, M. (1998b) Production of medakafish chimeras from a stable embryonic stem cell line. *Proceedings of the National Academy of Sciences USA* 95, 3679–3684.

Hong, Y., Chen, S. and Schartl, M. (2000) Embryonic stem cells in fish: current status and perspectives. *Fish Physiology and Biochemistry* 22, 165–170.

Hong, Y.-K., Kim, Y.-T. and Kim, S.-K. (1996) Molecular characterization of seaweeds using RAPD and differential display. *Journal of the Korean Fisheries Society* 29, 770–778.

Hooe, M.L., Buck, D.H. and Wahl, D.H. (1994) Growth, survival, and recruitment of hybrid crappies stocked in small impoundments. *North American Journal of Fisheries Management* 14, 137–142.

Hopkins, K.D. (1977) Sex reversal of genotypic male *Sarotherodon aureus* (Cichlidae). MSc thesis, Auburn University, Auburn, Alabama.

Hopkins, K.D., Shelton, W.L. and Engle, C.R. (1979) Estrogen sex-reversal of *Tilapia aurea. Aquaculture* 18, 263–268.

Horton, G.M.J. and Holmes, W. (1978) Compensatory gain by beef cattle at grassland or on an alfalfa based diet. *Journal of Animal Science* 46, 297–303.

Houck, M.A., Clark, J.B., Peterson, K.R. and Kidwell, M.G. (1991) Possible horizontal transfer of *Drosophila* genes by the mite *Proctolaelaps regalis. Science* 253, 1125–1129.

Houdebine, L.M. and Chourrout, D. (1991) Transgenesis in fish. *Experientia* 47, 891–897.

Houle, V.M., Schroeder, E.A., Park, Y.K., Odle, J. and Donovan, S.M. (1995) Orally administered IGF-I stimulates intestinal development without exerting systemic effects (abstract). In: *Proceedings of the 77th Annual Meeting of the Endocrine Society, Washington, DC, 14–17 June 1995.* The Endocrine Society, Chevy Chase, Maryland, p. 518.

Howerton, R.D., Okimoto, D.K. and Grau, E.G. (1988) Changes in the growth rate of the tilapia, *Oreochromis mossambicus*, following treatment with the hormones, triodothyronine (T_3) and testosterone. In: Pullin, R.S.V., Bhukasawan, T., Tonguthai, K. and Maclean, J.L. (eds) *Proceedings of the Second International Symposium on Tilapia in Aquaculture, Bangkok, Thailand, 16–20 March 1987.* ICLARM Conference Proceedings No. 15. Department of Fisheries Thailand and ICLARM, Manila.

Hoyheim, B., Danzmaiin, R., Guyomard, R., Holm, L., Powell, R. and Taggart, J. (1998) Constructing a genetic map of salmonid fishes: Salmap. *Abstracts in Plant and Animal Genome* 6, 165.

Hu, W., Wang, Y. and Zhu, Z. (2006) A perspective on fish gonad manipulation for biotechnical applications. *Chinese Science Bulletin* 51, 1–6.

Huang, C.M. and Liao, I.C. (1990). Response to mass selection for growth rate in *Oreochromis niloticus*. *Aquaculture* 85, 199–205.

Huang, W.-B., Chiu, T.-S. and Shih, C.-T. (1999) Effects of material conditions on early life history traits of black porgy *Acanthopagrus schlegeli*. *Journal of Applied Ichthyology* 15, 87–92.

Huang, Y., Nordeen, R.O., Di, M., Owens, L.D. and McBeath, J.H. (1997) Expression of an engineered cecropin gene cassette in transgenic tobacco plants confers disease resistance to *Pseudomonas syringae* pv. *tabaci*. *Phytopathology* 87, 494–499.

Hubert, S., English, L.J., Landau, B.J., Guo, X. and Hedgecock, D. (2000) Microsatellite analysis of trisomic families in the Pacific oyster, *Crassostrea gigas* Thunberg. *Journal of Shellfish Research* 19, 615.

Hudson, T.J., Stein, L.D., Gerety, S.S., Ma, J., Castle, A.B., Silva, J., Slonim, D.K., Baptista, R., Kruglyak, L., Xu, S.H., Hu, X., Colbert, A.M., Rosenberg, C., Reeve-Daly, M.P., Rozen, S., Hui, L., Wu, X., Vestergaard, C., Wilson, K.M., Bae, J.S., Maitra, S., Ganiatsas, S., Evans, C.A., DeAngelis, M.M., Ingalls, K.A., Nahf, R.W., Horton, L.T. Jr, Anderson, M.O., Collymore, A.J., Ye, W., Kouyoumjian, V., Zemsteva, I.S., Tam, J., Devine, R., Courtney, D.F., Renaud, M.T., Nguyen, H., O'Connor, T.J., Fizames, C., Fauré, S., Gyapay, G., Dib, C., Morissette, J., Orlin, J.B., Birren, B.W., Goodman, N., Weissenbach, J., Hawkins, T.L., Foote, S., Page, D.C. and Lander, E.S. (1995) An STS-based map of the human genome. *Science* 270, 1945–1954.

Hughes, C.R. and Queller, D.C. (1993) Detection of highly polymorphic microsatellite loci in a species with little allozyme polymorphism. *Molecular Ecology* 2, 131–137.

Hukriede, N.A., Joly, L., Tsang, M., Miles, J., Tellis, P., Epstein, J.A., Barbazuk, W.B., Li, F.N., Paw, B., Postlethwait, J.H., Hudson, T.J., Zon, L.I., McPherson, J.D., Chevrette, M., Dawid, I.B., Johnson, S.L. and Ekker, M. (1999) Radiation hybrid mapping of the zebrafish genome. *Proceedings of the National Academy of Sciences USA* 96, 9745–9750.

Hukriede, N., Fisher, D., Epstein, J., Joly, L., Tellis, P., Zhou, Y., Barbazuk, B., Cox, K., Fenton-Noriega, L., Hersey, C., Miles, J., Sheng, X., Song, A., Waterman, R., Johnson, S.L., Dawid, I.B., Chevrette, M., Zon, L.I., McPherson, J. and Ekker, M. (2001) The LN54 radiation hybrid map of zebrafish expressed sequences. *Genome Research* 11, 2127–2132.

Hulata, G., Moav, R. and Wohlfarth, G. (1974) The relationship of gonad and egg size to weight and age in the European and Chinese races of the common carp, *Cyprinus carpio* L. *Journal of Fish Biology* 6, 745–758.

Hulata, G., Wohlfarth, G. and Rothbard, S. (1983) Progeny-testing selection of tilapia broodstocks producing all-male hybrid progenies – preliminary results. *Aquaculture* 33, 263–268.

Hulata, G., Wohlfarth, G.W. and Halevy, A. (1986) Mass selection for growth rate in Nile tilapia (*Oreochromis niloticus*). *Aquaculture* 57, 177–184.

Hulata, G., Wohlfarth, G.W., Karplus, I., Schroeder, G.L., Harpaz, S., Halevy, A. and Rothbard, S., Cohen, S., Israel, I. and Kavessa, M. (1993) Evaluation of *Oreochromis niloticus* × *O. aureus* hybrids, progeny of different geographical isolates, under varying management conditions. *Aquaculture* 115, 253–271.

Hunter, G.A. and Donaldson, E.M. (1983) Hormonal sex control and its application to fish culture. In: Hoar, W.S., Randall, D.J. and Donaldson, E.M. (eds) *Fish Physiology*. Academic Press, New York, p. 223.

Hunter, G.A., Donaldson, E.M., Goetz, F.W. and Edgell, P.R. (1982) Production of all-female and sterile coho salmon, and experimental evidence for male heterogamety. *Transactions of the American Fisheries Society* 111, 367–372.

Hunter, G.A., Donaldson, E.M., Stoss, J. and Baker, I. (1983) Production of monosex female groups of chinook salmon (*Oncorhynchus tshawytscha*) by the fertilization of normal ova with sperm from sex-reversed females. *Aquaculture* 33, 355–364.

Hunter, R. and Markert, C. (1957) Histochemical demonstration of enzymes separated by zone electrophoresis in starch gels. *Science* 125, 1294–1295.

Hurt, J.A., Thibodeau, S.A., Hirsh, A.S., Pabo, C.O. and Joung, J.K. (2003) Highly specific zinc finger proteins obtained by directed domain shuffling and cell-based selection. *Proceedings of the National Academy of Sciences USA* 100, 12271–12276.

Hussain, M.G. and Islam, M.S. (1999) Genetic improvement of carp species in Asia, Bangladesh component. In: *Meeting on Genetic Improvement of Carp Species in Asia, Kuala Lumpur, Malaysia*. ICLARM, Manila.

Hussain, M.G., McAndrew, B.J., Penman, D.J. and Sodsuk, P. (1994) Estimate gene–centromere recombination frequencies in gynogenetic diploids of *Oreochromis niloticus* (L.) using allozymes, skin colour and a putative sex-determination locus (SDL-2). In: Beaumont, A.R. (ed.) *Genetics and Evolution of Aquatic Organisms*. Chapman and Hall, London, pp. 502–508.

Hussain, M.G., Rao, G.P.S., Humayun, N.M., Randall, C.F., Penman, D.J., Kime, D., Bromage, N.R., Myers, J.M. and McAndrew, B.J. (1995) Comparative performance of growth, biochemical composition and endocrine profiles in diploid and triploid tilapia *Oreochromis niloticus* L. *Aquaculture* 138, 87–97.

Hutson, A.M. (2008) A QTL map for growth and morphometric traits using a channel catfish × blue catfish interspecific hybrid system. Doctoral dissertation. Auburn University, Auburn, Alabama.

Hynes, J.D., Brown, E.H. Jr, Helle, J.H., Ryman, N. and Webster, D.A. (1981) Guidelines for the culture of fish stocks for resource management. *Canadian Journal of Fisheries and Aquatic Sciences* 38, 1867–1876.

Illmensee, K. and Hoppe, P.C. (1981) Nuclear transplantation in *Mus musculus*: developmental potential of nuclei from preimplantation embryos. *Cell* 23, 9–18.

Inada, Y., Yoshimura, K. and Taniguchi, N. (1990) Study on the resistance to vibriosis in induced triploid ayu *Plecoglossus altivelis*. *Bulletin of the Japanese Society of Scientific Fisheries* 56, 1587–1591.

Inada, Y., Chikushi, Y., Tsujimura, A. and Taniguchi, N. (1997) Selection response of resistance to vibriosis in gynogenetic ayu *Plecoglossus altivelis*. *Bulletin of the Japanese Society of Scientific Fisheries* 63, 722–727.

Indiq, F.E. and Moav, B. (1988) A prokaryotic gene is expressed in fish cells and persists in tilapia embryos and adults following microinjection. In: *Reproduction in Fish: Basic and Applied Aspects of Endocrinology and Genetics*. INRA Press, Paris, pp. 221–225.

Ingram, V.M. (1963) *The Hemoglobins in Genetics and Evolution*. Columbia University Press, New York.

Inoue, K., Ozato, K., Kondoh, H., Iwamatsu, T., Wakamatsu, Y., Fujita, T. and Okada, T.S. (1989) Stage-dependent expression of the chicken delta-crystallin gene in transgenic fish embryos. *Cell Differentiation and Development* 27, 57–68.

Inoue, K., Yamashita, S., Hata, J.-I., Kabeno, S., Asada, S., Nagahisa, E. and Fujita, T. (1990) Electroporation as a new technique for producing transgenic fish. *Cell Differentiation and Development* 29, 123–128.

Inoue, S., Nam, B., Hirono, I. and Aoki, T. (1997) A survey of expressed sequence tags in Japanese flounder (*Paralichthys olivaceus*) liver and spleen. *Molecular Marine Biology and Biotechnology* 6, 376–380.

Ishikawa, T., Aoki, H., Isowa, K. and Komaru, A. (2009) Development of a new breeding method for the the Japanese pearl oyster, *Pinctada fucata*, using shell-closing strength – II: The estimation of heritability of shell-closing strength to establish a high survival strain. In: *Book of Abstracts of 10th International Symposium on Genetics in Aquaculture (ISGA X) 'Roles of Aquaculture Genetics in Addressing Global Food Crisis', 22–26 June 2009, Bangkok, Thailand*. Faculty of Fisheries, Kasetsart University, Bangkok, p. 131.

Isley, J.J., Noble, R.L., Koppelman, J.B. and Philipp, D.P. (1987) Spawning period and first-year growth of northern, Florida and intergrade stocks of largemouth bass. *Transactions of the American Fisheries Society* 116, 757–762.

Ivics, Z., Izsvak, Z. and Hackett, P.B. (1993) Enhanced incorporation of transgenic DNA into zebrafish chromosomes by a retroviral integration protein. *Molecular Marine Biology and Biotechnology* 2, 162–173.

Ivics, Z., Izsvak, Z., Minter, A. and Hackett, P.B. (1996) Identification of functional domains and evolution of *Tc1*-like transposable elements. *Proceedings of the National Academy of Sciences USA* 93, 5008–5013.

Ivics, Z., Hackett, P.B., Plasterk, R.H. and Izsvak, Z. (1997) Molecular reconstruction of *Sleeping Beauty*, a *Tc1*-like transposon from fish, and its transposition in human cells. *Cell* 91, 501–510.

Iyengar, A., Muller, F. and Maclean, N. (1996) Regulation and expression of transgenes in fish – a review. *Transgenic Research* 5, 147–166.

Iyoda, K., Moriwake, T., Seino, Y. and Niimi, H. (1999) The clinical usefulness of liquid human growth hormone (hGH) (Norditropin Simple Xx) in the treatment of GH deficiency. *Hormone Research* 51 (Suppl. 3), 113–115.

Izadyar, F., Colenbrander, B. and Bevers, M.M. (1996) *In vitro* maturation of bovine oocytes in the presence of growth hormone accelerates nuclear maturation and promotes subsequent embryonic development. *Molecular Reproduction and Development* 45, 372–377.

Izant, J.G. and Weintraub, H. (1984) Inhibition of thymidine kinase gene expression by anti-sense RNA: a molecular approach to genetic analysis. *Cell* 36, 1007–1015.

Izsvak, Z., Ivics, Z. and Hackett, P.B. (1995) Characterization of a *Tc1*-like transposable element in zebrafish (*Danio rerio*). *Molecular and General Genetics* 247, 312–322.

Izsvak, Z., Ivics, Z. and Hackett, P.B. (1997) Repetitive sequence elements in zebrafish and their use in gene mapping and transgenesis. *Biochemistry and Cell Biology* 75, 507–523.

Jackson, A.L., Bartz, S.R., Schelter, J., Kobayashi, S.V., Burchard, J., Mao, M., Li, B., Cavet, G. and Linsley, P.S. (2003) Expression profiling reveals off-target gene regulation by RNAi. *Nature Biotechnology* 21, 635–637.

Jackson, T.R., Ferguson, M.M., Danzmann, R.G., Fishback, A.G., Ihssen, P.E., O'Connell, M. and Crease, T.J. (1998) Identification of two QTL-influencing upper temperature tolerance in three rainbow trout (*Oncorhynchus mykiss*) half-sib families. *Heredity* 80, 143–151.

Jalabert, B., Billard, R. and Chevassus, B. (1975) Preliminary experiments on sex control in trout: production of sterile fishes and simultaneous self-fertilizable hermaphrodites. *Annales de Biologie Animale Biochimie Biophysique* 15, 19–28.

Janssen, Y.J., Doornbos, J. and Roelfsema, F. (1999) Changes in muscle volume, strength, and bioenergetics during recombinant human growth hormone (GH) therapy in adults with GH deficiency. *Journal of Clinical Endocrinology and Metabolism* 84, 279–284.

Jarimopas, P. (1990) Realized response of Thai red tilapia to 5 generations of size-specific selection for growth. In: Hirano, R. and Hanyu, I. (eds) *Proceedings of The Second Asian Fisheries Forum, Tokyo, Japan, 17–22 April 1989*. Asian Fisheries Society, Manila.

Jarimopas, P., Kumnan, A. and Wongchan, J. (1989) Mass selection of *Clarias macrocephalus* Gunther for growth (4 generations). In: Final Report, Fish Genetics Report (Phase II), submitted to the International Development Research Center, National Aquaculture Genetics Research Institute, Department of Fisheries, Thailand, pp. 177–191.

Jaynes, J.M., Xanthopoulos, K.G., Destefano-Beltran, L. and Dodds, J.H. (1987) Increasing bacterial disease resistance in plants utilizing antibacterial genes from insects. *BioEssays* 6, 263–270.

Jaynes, J.M., Julian, G.R., Jeffers, G.W., White, K.L. and Enright, F.M. (1989) *In vitro* cytocidal effect of lytic peptides on several transformed mammalian cell lines. *Peptide Research* 2, 157–160.

Jeffreys, A.J., Wilson, V. and Thien, S.L. (1985) Individual-specific 'fingerprints' of human DNA. *Nature* 316, 76–79.

Jehle, J.A., Fritsch, E., Nickel, A., Huber, J. and Backhaus, H. (1995) TCl4.7: a novel lepidopteran transposon found in *Cydia pomonella* granulosis virus. *Virology* 207, 369–379.

Jehle, J.A., Nickel, A., Vlak, J.M. and Backhaus, H. (1998) Horizontal escape of the novel *Tc1*-like lepidopteran transposon *Tcp3.2* into *Cydia pomonella* granulovirus. *Journal of Molecular Evolution* 46, 215–224.

Jensen, G.L. and Shelton, W.L. (1979) Effects of estrogens on *Tilapia aurea*: implications for production of monosex genetic male *Tilapia*. *Aquaculture* 16, 233–242.

Jensen, G.L. and Shelton, W.L. (1983) Gonadal differentiation in relation to sex control of grass carp, *Ctenopharyngodon idella* (Pisces: Cyprinidae). *Copeia* 1983, 749–755.

Jensen, G.L., Shelton, W.L., Yang, S.L. and Wilken, L.O. (1983) Sex reversal of gynogenetic grass carp by implantation of methyltestosterone. *Transactions of the American Fisheries Society* 112, 79–85.

Jensen, J., Dunham, R.A. and Flynn, J. (1983) *Producing Channel Catfish Fingerlings*. Circular ANR-327. Alabama Cooperative Extension Service, Auburn University, Auburn, Alabama.

Jeppsen, T.S. (1995) Comparison of performance of channel catfish, *Ictalurus punctatus*, blue catfish, *I. furcatus*, hybrids from Kansas select and Kansas random dams. MSc thesis, Auburn University, Auburn, Alabama.

Jia, S.R., Xie, Y., Tang, T., Feng, L.X., Cao, D.S., Zhao, Y.L., Yuan, J., Bai, Y.Y., Jiang, C.X. and Jaynes, J.M. (1993) Genetic engineering of Chinese potato cultivars by introducing antibacterial polypeptide gene. *Current Plant Science and Biotechnology in Agriculture* 15, 208–212.

Jia, X., Patrzykat, A., Devlin, R.H., Ackerman, P.A., Iwama, G.K. and Hancock, R.E.W. (2000) Antimicrobial peptides protect coho salmon from *Vibrio anguillarum* infections. *Applied and Environmental Microbiology* 66, 1928–1932.

Jiang, C., Edmeades, G.O., Armstead, I., Lafitte, H.R., Hayward, M.D. and Hoisington, D. (1999) Genetic analysis of adaptation differences between highland and lowland tropical maize using molecular markers. *Theoretical and Applied Genetics* 99, 1106–1119.

Jitpiromsri, N., (1990) Preliminary study on genetics of *Puntius gonionotus* (Bleeker). MS thesis, Kasetsart University, Bangkok.

Jo, J.Y., Smitherman, R.O. and Behrends, L.L. (1988) Effects of 17-methyltestosterone concentration in the diet on sex-reversal and growth of *Tilapia aurea*. In: Pullin, R.S.V., Bhukasawan, T., Tonguthai, K. and Maclean, J.L. (eds) *Proceedings of the Second International Symposium on Tilapia in Aquaculture, Bangkok, Thailand, 16–20 March 1987*. ICLARM Conference Proceedings No. 15. Department of Fisheries Thailand and ICLARM, Manila.

Jobling, M. and Koskela, J. (1996) Interindividual variations in feeding and growth in rainbow trout during restricted feeding and in a subsequent period of compensatory growth. *Journal of Fish Biology* 49, 658–667.

Jobling, M. and Koskela, J. (1997) Feeding and growth in rainbow trout during compensatory growth. In: Houlihan, D., Kiessling, A. and Boujard, T. (eds) *Voluntary Food Intake in Fish, First (Cost 827) Workshop, Aberdeen, UK, 3–6 April 1997*, p. 53.

Johnsen, B.O. and Ugedal, O. (1986) Feeding by hatchery-reared and wild brown trout, *Salmo trutta* L., in a Norwegian stream. *Aquaculture and Fisheries Management* 17, 281–287.

Johnson, K.R. and Wright, J.E. (1984) Linkage and pseudolinkage groups in salmonid fishes (trout, charr, salmon). In: *Genetic Maps 1984: A Compilation of Linkage and Restriction Maps of Genetically Studied Organisms*, Vol. 3. Cold Spring Harbor Laboratory Press, Cold Spring Harbor, New York, pp. 334–339.

Johnson, K.R., Wright, J.E. Jr and May, B. (1987) Linkage relationships reflecting ancestral tetraploidy in salmonid fishes. *Genetics* 116, 579–591.

Johnson, O.W., Rabinovich, P.R. and Utter, F.M. (1984) Comparison of the reliability of a Coulter counter with a flow cytometer in determining ploidy levels in Pacific salmon. *Aquaculture* 43, 99–103.

Johnson, O.W., Dickhoff, W.W. and Utter, F.M. (1986) Comparative growth and development of diploid and triploid coho salmon, *Oncorhynchus kisutch*. *Aquaculture* 57, 329–336.

Johnson, S.L., Midson, C., Ballinger, E.W. and Postlethwait, J.H. (1994) Identification of RAPD primers that reveal extensive polymorphisms between laboratory strains of zebrafish. *Genomics* 19, 152–156.

Johnson, S.L., Gates, M.A., Johnson, M., Talbot, W.S., Horne, S., Baik, K., Rude, S., Wong, J.R. and Postlethwait, J.H. (1996) Centromere-linkage analysis and consolidation of the zebrafish genetic map. *Genetics* 142, 1277–1288.

Johnsson, J.I., Hojesjo, J. and Fleming, I.A. (2001) Behavioural and heart rate responses to predation risk in wild and domesticated Atlantic salmon. *Canadian Journal of Fisheries and Aquatic Sciences* 58, 788–794.

Johnston, I.A. (2000) Genetic and environmental mechanisms regulating muscle growth in fish. In: *Abstracts of the International Symposium 'A Step Toward the Great Future of Aquatic Genomics', 10–12 November 2000*. Tokyo University of Fisheries, Tokyo, p. 56.

Johnston, I.A., Strugnell, G., McCracken, M.L. and Johnstone, R. (1999) Muscle growth and development in normal-sex-ratio and all-female diploid and triploid Atlantic salmon. *Journal of Experimental Biology* 202, 1991–2016.

Johnston, I.A., Hall, T.E. and Fernandez, D.A. (2002) Genes regulating the growth of myotomal muscle in teleost fish. In: Shimizu, N., Aoki, T., Hirono, I. and Takashima, F. (eds) *Aquatic Genomics: Steps Toward a Great Future*. Springer-Verlag, New York, pp. 153–166.

Johnstone, R., Simpson, T.H. and Youngson, A.F. (1978) Sex reversal in salmonid culture. *Aquaculture* 13, 115–134.

Johnstone, R., Simpson, T.H., Youngson, A.F. and Whitehead, C. (1979a) Sex reversal in salmonid culture. Part II. The progeny of sex-reversed rainbow trout. *Aquaculture* 18, 13–19.

Johnstone, R., Simpson, T.H. and Walker, A.F. (1979b) Sex reversal in salmonid culture. Part III. The production and performance of all-female populations of brook trout. *Aquaculture* 18, 241–252.

Johnstone, R., Macintosh, D.J. and Wright, R.S. (1983) Elimination of orally administered 17-methylestosterone by *Oreochromis mossambicus* (tilapia) and *Salmo gairdneri* (rainbow trout) juveniles. *Aquaculture* 35, 249–257.

Johnstone, R., Knott, R.M., MacDonald, A.G. and Walsingham, M.V. (1989) Triploidy induction in recently fertilized Atlantic salmon ova using anaesthetics. *Aquaculture* 78, 229–236.

Jonasson, J. (1993) Selection experiments in salmon ranching: I. Genetic and environmental sources of variation in survival and growth in freshwater. *Aquaculture* 109, 225–236.

Jones, I.J. and Clemmons, D.R. (1995) Insulin-like growth factors and their binding proteins: biological actions. *Endocrine Reviews* 16, 3–34.

Jones, M.W., McParland, T.L., Hutchings, J.A., Roy, R.G. and Danzmann, G. (2001) Low genetic variability in lake populations of brook trout (*Salvelinus fontinalis*): a consequence of exploitation? *Conservation Genetics* 2, 245–256.

Joyce, C.P. (1989) Effects of developmental stage, spawning period and gene construct on the survival of microinjected common carp and channel catfish embryos. MSc thesis, Auburn University, Auburn, Alabama.

Ju, Z. (2001) Development of expressed sequence tags (ESTS) and cDNA microarrays for analysis of differentially expressed genes in the channel catfish (*Ictalurus punctatus*) brain. Doctoral dissertation, Auburn University, Auburn, Alabama, USA.

Ju, Z., Karsi, A., Kocabas, A., Patterson, A., Li, P., Cao, D., Dunham, R. and Liu, Z. (2000) Transcriptome analysis of channel catfish (*Ictalurus punctatus*): genes and expression profile from the brain. *Gene* 261, 373–382.

Ju, Z., Dunham, R. and Liu, Z.J. (2002) Differential gene expression in the brain of channel catfish (*Ictalurus punctatus*) in response to cold acclimation. *Molecular Genetics and Genomics* 268, 87–95.

Juskevich, J.C. and Guyer, C.G. (1990) Bovine growth hormone: human food safety evaluation. *Science* 249, 875–884.

Kaattari, S.L. and Piganelli, J.D. (1996) The specific immune response: humoral defense: In: Iwama, G. and Nakanishi, T. (eds) *The Fish Immune System: Organism, Pathogen and Environment*. Academic Press, San Diego, California, pp. 207–254.

Kadenbach, B., Kuhn-Nentwig, L. and Buge, U. (1987) Evolution of a regulatory enzyme: cytochrome c oxidase (complex IV). *Current Topics in Bioenergetics* 15, 113–161.

Kadenbach, B., Reimann, A., Stroh, A. and Huther, F.J. (1988) Evolution of cytochrome c oxidase. *Progress in Clinical and Biological Research* 274, 653–668.

Kadono-Okuda, K., Taniai, K., Kato, Y., Kotani, E. and Yamakawa, M. (1995) Effects of synthetic *Bombyx mori* cecropin B on the growth of plant pathogenic bacteria. *Journal of Invertebrate Pathology USA* 65, 309–310.

Kafiani, C.A., Timofeeva, M.J., Neyfakh, A.A., Melnikova, N.L. and Rachkus, J.A. (1969) RNA synthesis is the early embryogenesis of a fish (*Misgurnus fossilis*). *Journal of Embryology and Experimental Morphology* 21, 295–308.

Kah, O., Trudeau, V.L., Sloley, B.D., Chang, J.P., Dubourg, P., Yu, K.-L. and Peter, R.E. (1992) Influence of GABA on gonadotropin release in the goldfish. *Neuroendocrinology* 55, 396–404.

Kah, O., Anglade, I., LeProtre, P., Dubourg, P. and de Monbrison, D. (1993) The reproductive brain in fish. *Fish Physiology and Biochemistry* 11, 85–98.

Kaiser, M. and Forsberg, E.-M. (2000) Assessing fisheries – using an ethical matrix in a participatory process. *Journal of Agricultural and Environmental Ethics* 14, 91–200.

Kaiser, R.G. and Burgess, J. (1999) A Seattle primer: how not to hold WTO talks: sticking points. *Washington Post*, Sunday 12 December, A40.

Kallman, K. and Borkoski, V. (1978) A sex linked gene controlling the onset of sexual maturity in female and male platyfish (*Xiphohorus maculatus*), fecundity in females and adult size in males. *Genetics* 89, 79–119.

Kamonrat, W. (1996) Spatial genetic structure of Thai silver barb *Puntius gonionotus* (Bleeker) populations in Thailand. Doctoral dissertation, Dalhousie University, Halifax, Nova Scotia, Canada.

Kanatsu-Shinohara, M., Ikawa, M., Takehashi, M., Ogonuki, N., Miki, H., Inoue, K., Kazui, Y., Lee, J., Toyokuni, S., Oshimura, M., Ogura, A. and Shinohara, T. (2006) Production of knockout mice by random or targeted mutagenesis in spermatogonial stem cells. *Proceedings of the National Academy of Sciences USA* 103, 8018–8023.

Kandeel, F.R. and Swerdloff, R.S. (1997) The interaction between beta-endorphin and gonadal steroids in regulation of luteinizing hormone (LH) secretion and sex steroid regulation of LH and proopiome-lanocortin peptide secretion by individual pituitary cells. *Endocrinology* 138, 649–656.

Kanis, E., Refstie, T. and Gjedrem, T. (1976) A genetic analysis of egg, alevin and fry mortality in salmon (*Salmo salar*), sea trout (*Salmo trutta*) and rainbow trout (*Salmo gairdneri*). *Aquaculture* 8, 259–268.

Kansaku, N., Shimada, K., Terada, O. and Saito, N. (1994) Prolactin, growth hormone, and luteinizing hormone-beta subunit gene expression in the cephalic and caudal lobes of the anterior pituitary gland during embryogenesis and different reproductive stages in the chicken. *General and Comparative Endocrinology* 96, 197–205.

Kapuscinski, A.R. and Hallerman, E.M. (1990a) AFS position statement: transgenic fishes. *Fisheries* 15 (1), 2–5.

Kapuscinski, A.R. and Hallerman, E.M. (1990b) Transgenic fish and public policy: anticipating environmental impacts of transgenic fish. *Fisheries* 15 (4), 2–11.

Kapuscinski, A.R. and Hallerman, E.M. (1991) Implications of introduction of transgenic fish into natural ecosystems. *Canadian Journal of Fisheries and Aquatic Sciences* 48 (Suppl. 1), 99–107.

Kapuscinski, A.R. and Hallerman, E.M. (1995) Benefits, environmental risks, social concerns, and policy implications of biotechnology in aquaculture. In: White, R. (ed.) *Aquaculture: Food and Renewable Resources from US Waters*. Office of Technology Assessment, US Congress, US Government Printing Office, Washington, DC.

Kapuscinski, A.R., Nega, T. and Hallerman, E.M. (1999) Adaptive biosafety assessment and management regimes for aquatic genetically modified organisms in the environment. In: Pullin, R.S.V., Bartley, D.M. and Kooiman, J. (eds) *Towards Policies for Conservation and Sustainable Use of Aquatic Genetic Resources*. ICLARM Conference Proceedings No. 59. ICLARM, Manila, pp. 225–251.

Kapuscinski, A.R., Goodman, R.M., Hann, S.D., Jacobs, L.R., Pullins, E.E., Johnson, C.S., Kinsey, J.D., Krall, R.L., La Viña, A.G.M., Mellon, M.G. and Ruttan, V.W. (2003) Making safety first a reality for biotechnology products. *Nature Biotechnology* 21, 599–601.

Karatzas, C.N., Guemene, D., Zadworny, D. and Kuhnlein, U. (1997) Changes in expression of the prolactin and growth hormone gene during different reproductive stages in the pituitary gland of turkeys. *Reproduction, Nutrition, Development* 37, 69–79.

Karsi, A. (2001) Development of genomic resources for genetic enhancement of channel catfish (*Ictalurus punctatus*). Doctoral dissertation, Auburn University, Auburn, Alabama.

Karsi, A., Li, P., Dunham, R.A. and Liu, Z.J. (1998) Transcriptional activities in the pituitaries of channel catfish (*Ictalurus punctatus*) before and after induced ovulation as revealed by expressed sequence tag analysis. *Journal of Molecular Endocrinology* 21, 121–129.

Karsi, A., Moav, B., Hackett, P. and Liu, Z.J. (2001) Effects of insert size on transposition efficiency of the *Sleeping Beauty* transposon in mouse cells. *Marine Biotechnology* 3, 241–245.

Karsi, A., Cao, D., Li, P., Patterson, A., Kocabas, A., Feng, J., Ju, Z., Mickett, K. and Liu, Z.J. (2002a) Transcriptome analysis of channel catfish (*Ictalurus punctatus*): initial analysis of gene expression and microsatellite-containing cDNAs in the skin. *Gene* 285, 157–168.

Karsi, A., Patterson, A., Feng, J. and Liu, Z.J. (2002b) Translational machinery of channel catfish: I. A transcriptomic approach to the analysis of 32 40S ribosomal protein genes and their expression. *Gene* 291, 177–186.

Kashi, Y., Hallerman, E. and Soller, M. (1990) Marker-assisted selection of candidate bulls for progeny testing programmes. *Animal Production* 51, 63–74.

Katz, Y., Abraham, M. and Eckstein, B. (1976) Effects of adrenosterone on gonadal and body growth in *Tilapia nilotica* (Telostei, Cichlidae). *General and Comparative Endocrinology* 29, 414–418.

Kavumpurath, S. and Pandian, T.J. (1990) Induction of triploidy in the zebrafish, *Brachydanio rerio* (Hamilton). *Aquaculture and Fisheries Management* 21, 299–306.

Kawamura, K., Ueda, T., Aoki, K. and Hosoya, K. (1999) Spermatozoa in triploids of the rosy bitterling *Rhodeus ocellatus ocellatus*. *Journal of Fish Biology* 55, 420–432.

Kawamura, T. and Nishioka, M.J. (1963a) Reciprocal diploid nucleocytoplasmic hybrids between two species of Japanese pond frogs and their offspring. *Journal of Science Hiroshima University Series, B Division 1 (Zoology)* 21, 65–84.

Kawamura, T. and Nishioka, M.J. (1963b) Nucleo-cytoplasmic hybrid frog between two species of Japanese brown frog and their offspring. *Journal of Science Hiroshima University Series, B Division 1 (Zoology)* 21, 107–134.

Keele, J.W., Schackelford, S.D., Kappes, S.M., Koohmaraie, M. and Stone, R.T. (1999) A region on bovine chromosome 15 influences beef longissimus tenderness in steers. *Journal of Animal Science* 77, 1364–1371.

Keim, P., Kalif, A., Schupp, J., Hill, K., Travis, S.E., Richmond, K., Adair, D.M., Hugh-Jones, M., Kuske, C.R. and Jackson, P. (1997a) Molecular evolution and diversity in *Bacillus anthracis* as detected by amplified fragment length polymorphism markers. *Journal of Bacteriology* 179, 818–824.

Keim, P., Schupp, J.M., Travis, S.E., Clayton, K., Zhu, T., Shi, L., Ferreira, A. and Webb, D.M. (1997b) A high-density soybean genetic map based on AFLP markers. *Crop Science* 37, 537–543.

Keller, L.F., Jeffery, K.J., Arcese, P., Beaumont, M.A., Hochachka, W.M., Smith, J.N.M. and Bruford, M.W. (2001) Immigration and the ephemerality of a natural population bottleneck: evidence from molecular markers. *Proceedings of the Royal Society of London, Series B: Biological Sciences* 268, 1387–1394.

Keller, M., Karutz, C., Schmid, J.E., Stamp, P., Winzeler, M., Keller, B. and Messmer, M.M. (1999) Quantitative trait loci for lodging resistance in a segregating wheat spelt population. *Theoretical and Applied Genetics* 98, 1171–1182.

Keller, M., Kollmann, J. and Edwards, P.J. (2000) Genetic introgression from distant provenances reduces fitness in local weed populations. *Journal of Applied Ecology* 37, 647–659.

Kelly, D., Wolters, W.R. and Jaynes, J.M. (1990) Effect of lytic peptides on selected fish bacterial pathogens. *Journal of Fish Diseases* 13, 317–321.

Kelly, D.G., Wolters, W.R., Jaynes, J.M. and Newton, J.C. (1993) Enhanced disease resistance to enteric septicemia in channel catfish, *Ictalurus punctatus*, administered lytic peptide. *Journal of Applied Aquaculture* 3, 25–34.

Kempinger, J.J. and Churchill, L.S. (1970) *Contribution of Native and Stocked Walleye Fingerlings to the Anglers' Catch, Escanaba Lake, Wisconsin*. Project No. Wis. F-083-R-05/WK.PL.21/Job 08. Wisconsin Conservation Department, Madison, Wisconsin.

Kerby, J.H., Everson, J.M., Harrell, R.M., Geiger, J.G., Starling, C.C. and Revels, H. (2002) Performance comparisons between diploid and triploid sunshine bass in fresh water ponds. *Aquaculture* 211, 91–108.

Khan, F.A., Margolis, R.L., Loev, S.L., Sharp, A.H., Li, S.H. and Ross, C.A. (1996) cDNA cloning and characterization of an atrophin-1 (DRPLA disease gene)-related protein. *Neurobiological Disease* 3, 121–128.

Khan, H.A., Gupta, S.D., Reddy, P.V.G.K., Tantia, M.S. and Kowtal, G.V. (1990) Production of sterile intergeneric hybrids and their utility in aquaculture and stocking. In: Keshavanath, P. and Radhakrishnan, K.V. (eds) *Carp Seed Production Technology*. Special Publication No. 2. Asian Fisheries Society, Indian Branch, Mangalore, India, pp. 41–48.

Khan, I. and Thomas, P. (1999) GABA exerts stimulatory and inhibitory influences on gonadotropin II secretion in the Atlantic croaker (*Micropogonias undulatus*). *Neuroendocrinology* 69, 261–268.

Khan, J., Simon, R., Bittner, M., Chen, Y., Leighton, S.B., Pohida, T., Smith, P.D., Jiang, Y., Gooden, G.C., Trent, J.M. and Meltzer, P.S. (1998) Gene expression profiling of alveolar rhabdomyosarcoma with cDNA microarrays. *Cancer Research* 58, 5009–5013.

Khan, J., Saal, L.H., Bittner, M.L., Chen, Y., Trent, J.M. and Meltzer, P.S. (1999) Expression profiling in cancer using cDNA microarrays. *Electrophoresis* 20, 223–229.

Khater, A.A. (1985) Identification and comparison of three *Tilapia nilotica* strains for selected aquacultural traits. Doctoral dissertation, Auburn University, Auburn, Alabama.

Khaw, H.L., Ponzoni, R.W. and Danting, M.J.C. (2008) Estimation of genetic change in the GIFT strain of Nile tilapia (*Oreochromis niloticus*) by comparing contemporary progeny produced by males born in 1991 or in 2003. *Aquaculture* 275, 64–69.

Khoo, S.K., Ozaki, A., Nakamura, F., Arakawa, T., Ishimoto, S., Nikolov, R., Sakamoto, T., Akatsu, T., Mochuzuki, M., Denda, I. and Okamoto, N. (2004) Identification of a novel chromosomal region associated with IHNV resistance in rainbow trout. *Fish Pathology* 39, 95–102.

Kidwell, M.G. (1992) Horizontal transfer. *Current Opinion in Genetics and Development* 2, 868–873.

Kim, D.S., Kim, I.-B. and Baik, Y.G. (1986) A report of triploid rainbow trout production in Korea. *Bulletin of the Korean Fisheries Society* 19, 575–580.

Kim, K.H., Patel, L., Tobet, S.A., King, J.C., Rubin, B.S. and Stopa, E.G. (1999) Gonadotropin releasing hormone immunoreactivity in the adult and fetal human olfactory system. *Brain Research* 826, 220–229.

Kim, M.K. and Lovell, R.T. (1995) Effect of restricted feeding regimens on compensatory weight gain and body tissue changes in channel catfish *Ictalurus punctatus* in ponds. *Aquaculture* 135, 285–293.

Kim, S. (2000) Expressed sequence tag analysis of reproductive hormones and muscle proteins and analysis of the α-actin promoter in channel catfish, *Ictalurus punctatus*. PhD dissertation, Auburn University, Auburn, Alabama.

Kim, S., Karsi, A., Dunham, R.A. and Liu, Z. (2000) The skeletal muscle alpha-actin gene of channel catfish (*Ictalurus punctatus*) and its association with piscine specific SINE elements. *Gene* 252, 173–181.

Kim, Y.G., Cha, J. and Chandrasegaran, S. (1996) Hybrid restriction enzymes: zinc finger fusions to Fok I cleavage domain. *Proceedings of the National Academy of Sciences USA* 93, 1156–1160.

Kime, D.E. and Dolben, I.P. (1985) Hormonal changes during induced ovulation of the carp, *Cyprinus carpio*. *General and Comparative Endocrinology* 58, 137–149.

Kime, D.E., Epler, P., Bieniarz, K., Motyka, K. and Mikolajczyk, T. (1989) Interactions between oocytes of different maturational stages in the carp *Cyprinus carpio*: effects on maturational and steroidogenic activity *in vitro*. *General and Comparative Endocrinology* 74, 45–49.

Kincaid, H.L. (1976a) Effects of inbreeding on rainbow trout populations. *Transactions of the American Fisheries Society* 105, 273–280.

Kincaid, H.L. (1976b) Inbreeding in rainbow trout (*Salmo gairdneri*). *Journal of the Fisheries Research Board of Canada* 33, 2420–2426.

Kincaid, H.L. (1981) *Trout Salmon Registry*. FWS/NFC-L/81–1. US Fish and Wildlife Service, Kearneysville, West Virginia.

Kincaid, H.L. (1983a) Inbreeding in fish populations used for aquaculture. *Aquaculture* 33, 215–227.

Kincaid, H.L. (1983b) Results from six generations of selection for accelerated growth rate in a rainbow trout population (abstract). In: *The Future of Aquaculture in North America*. Fish Culture Section of the American Fisheries Society, Bethesda, Maryland, pp. 26–27.

Kincaid, H.L., Bridges, W.R. and Von Limbach, B. (1977) Three generations of selection for growth rate in fall spawning rainbow trout. *Transactions of the American Fisheries Society* 106, 621–629.

Kinghorn, B. (1983) Genetic variation in food conversion efficiency and growth in rainbow trout. *Aquaculture* 32, 141–155.

Kinoshita, M. and Tanaka, M. (2002) Transgenic medaka as a model for fish biology and aquaculture. In: Shimizu, N., Aoki, T., Hirono, I. and Takashima, F. (eds) *Aquatic Genomics: Steps Toward a Great Future*. Springer-Verlag, New York, pp. 320–328.

Kirpichnikov, V.S. (1981) *Genetic Bases of Fish Selection*. Springer Verlag, Berlin.

Kirpichnikov, V.S., Ilyassov, Y.I., Shart, L.A., Vikhman, A.A., Ganchenko, M.V., Ostashevsky, L.A., Siminov, V.M., Tikhonov, G.F. and Tjurin, V.V. (1993) Selection of krasnodar common carp (*Cyprinus carpio* L.) for resistance to dropsy: principal results and prospects. *Aquaculture* 111, 7–20.

Kistner, A., Gossen, M., Zimmermann, F., Jerecic, J., Ullmer, C., Lubbert, H. and Bujard, H. (1996) Doxy-cycline-mediated quantitative and tissue-specific control of gene expression in transgenic mice. *Proceedings of the National Academy of Sciences USA* 93, 10933–10938.

Kjoeglum, S., Henryon, M., Aasmundstad, T. and Korsgaard, I. (2008) Selective breeding can increase resistance of Atlantic salmon to furunculosis, infectious salmon anaemia and infectious pancreatic necrosis. *Aquaculture Research* 39, 498–505.

Kjuul, A.K., Bullesbach, E.E., Espelid, S., Dunham, R.A., Jorgensen, T.O., Warr, G.W. and Styrvold, O.B. (1999) Effects of cecropin peptides on bacteria pathogenic to fish. *Journal of Fish Diseases* 22, 387–394.

Klassen, C.D., Amdur, M.O. and Doull, J.D. (1996) *Toxicology*. McGraw Hill, New York.

Klassen, W. (2003) Edward F. Knipling: titan and driving force in ecologically selective area-wide pest management. *Journal of the American Mosquito Control Association* 19, 94–103.

Klinger, T. (1998) Biosafety assessment of genetically engineered organisms in the environment. *Trends in Ecology and Evolution* 13, 5–6.

Klupp, R., Heil, G. and Pirchner, F. (1978) Effects of interaction between growth in rainbow trout. *Aquaculture* 32, 141–155.

Knapp, J.R., Chen, W.Y., Turner, N.D., Byers, F.M. and Kopchick, J.J. (1994) Growth-patterns and body-composition of transgenic mice expressing mutated bovine somatotropin genes. *Journal of Animal Science* 72, 2812–2819.

Knibb, W., Gorshkova, G. and Gorshkov, S. (1997a) Growth in strains of gilthead seabream (*Sparus aurata* L.). *Israeli Journal of Aquaculture/Bamidgeh* 49, 43–56.

Knibb, W., Gorshkova, G. and Gorshkov, S. (1997b) Selection for growth in the gilthead seabream (*Sparus aurata* L.). *Israeli Journal of Aquaculture/Bamidgeh* 49, 57–66.

Knibb, W., Gorshkova, G. and Gorshkov, S. (1998a) Genetic improvement of cultured marine finfish: case studies. In: De Silva, S. (ed.) *Tropical Mariculture*. Academic Press Ltd, New York, pp. 111–149.

Knibb, W., Gorshkova, G. and Gorshkov, S. (1998b) Selection and crossbreeding in Mediterranean cultured marine fish. In: Bartley, D.M. and Basurco, B. (eds) *Proceedings of the TECAM Seminar on Genetics and Breeding of Mediterranean Aquacultured Species, Zaragoza, Spain, 28–29 April 1997*, pp. 47–60.

Knight, M., Miller, A.N., Patterson, C.N., Rowe, C.G., Michaels, G., Carr, D., Richards, C. and Lewis, F.A. (1999) The identification of markers segregating with resistance to *Schistosoma mansoni* infection in the snail *Biomphalaria glabrata*. *Proceedings of the National Academy of Sciences USA* 96, 1510–1515.

Knobil, E., Wolf, R.C. and Greep, R.O. (1956) Some physiologic effects of primate pituitary growth-hormone preparations in hypophysectomized rhesus monkeys. *Journal of Clinical Endocrinology and Metabolism* 16, 916.

Kobayakawa, Y. and Kubota, H. (1981) Temporal patterns of cleavage and the onset of gastrulation in amphibian embryos developed from eggs with reduced cytoplasm. *Journal of Embryology and Experimental Morphology* 62, 83–94.

Kobayashi, S. and Mogami, M. (1958) Effects of X-irradiation upon rainbow trout (*Salmo irideus*). III. Ovary growth in the stages of fry and fingerlings. *Bulletin of the Faculty of Fisheries, Hokkaido University* 9, 89.

Kobayashi, T., Ide, A., Hiasa, T., Fushiki, S. and Ueno, K. (1994) Production of cloned amago salmon, *Oncorhynchus rhodurus*. *Fisheries Science, Tokyo* 60, 275–281.

Kobiyama, A., Nihei, Y. and Watabe, S. (2002) Temporal and spatial expression patterns of mRNAs encoding myosin heavy chain isoforms in association with those of related transcription factors during temperature acclimation of carp. In: Shimizu, N., Aoki, T., Hirono, I. and Takashima, F. (eds) *Aquatic Genomics: Steps Toward a Great Future*. Springer-Verlag, New York, pp. 167–184.

Kocabas, A. (2001) Development of genomic resources and cDNA microarray technology for functional genomics of catfish. Doctoral dissertation, Auburn University, Auburn, Alabama.

Kocabas, A., Kucuktas, H., Dunham, R.A. and Liu, Z.J. (2002a) Molecular characterization and differential expression of the myostatin gene in channel catfish (*Ictalurus punctatus*). *Biochimica et Biophysica Acta* 1575, 99–107.

Kocabas, A., Li, P., Cao, D., Karsi, A., He, C., Patterson, A., Ju, Z., Dunham, R. and Liu, Z.J. (2002b) Expression profile of the channel catfish spleen: analysis of genes involved in immune functions. *Marine Biotechnology* 4, 526–536.

Kocher, T.D., Lee, W.J., Sobolewska, H., Penman, D. and McAndrew, B. (1998) A genetic linkage map of a cichlid fish, the tilapia (*Oreochromis niloticus*). *Genetics* 148, 1225–1232.

Kocher, T.D., Albertson, R.C., Carleton, K.L. and Streelman, J.T. (2002) The genetic basis of biodiversity: genomic studies of cichlid fishes. In: Shimizu, N., Aoki, T., Hirono, I. and Takashima, F. (eds) *Aquatic Genomics: Steps Toward a Great Future*. Springer-Verlag, New York, pp. 35–44.

Kocour, M., Mauger, S., Rodina, M., Gela, D., Linhart, O. and Vandeputte, M. (2007a) Heritability estimates for processing and quality traits in common carp (*Cyprinus carpio* L.) using a molecular pedigree. *Aquaculture* 270, 43–50.

Kocour, M., Mauger, S., Rodina, M., Gela, D., Linhart, O., Flajshans, M. and Vandeputte, M. (2007b) Heritability estimates for growth and dress out traits in common carp (*Cyprinus carpio* L.) using a molecular pedigree. *Aquaculture* 272, S277–S278.

Koedprang, W. and Na-Nakorn, U. (2000) Preliminary study on performance of triploid Thai silver barb, *Puntius gonionotus*. *Aquaculture* 190, 211–221.

Kok, E.J. and Kuiper, H.A. (2003) Comparative safety assessment for biotech crops. *Trends in Biotechnology* 21, 439–444.

Koljonen, M.L. (1989) Electrophoretically detectable genetic variation in natural and hatchery stocks of Atlantic salmon in Finland. *Hereditas* 110, 23–26.

Komaru, A. and Wada, A.T. (1989) Gametogenesis and growth and growth of triploid scallops *Chlamys nobilis*. *Bulletin of the Japanese Society of Scientific Fisheries* 55, 447–452.

Komen, J., Eding, E.H., Bongers, A.B.J. and Richter, C.J.J. (1993) Gynogenesis in common carp (*Cyprinus carpio*). IV. Growth, phenotypic variation and gonad differentiation in normal and methyltestosterone-treated homozygous clones and F_1 hybrids. *Aquaculture* 111, 271–280.

Komiya, T., Itoh, K., Ikenishi, K. and Furusawa, M. (1994) Isolation and characterization of a novel gene of the DEAD box protein family which is specifically expressed in germ cells of *Xenopus laevis*. *Developmental Biology* 162, 354–363.

Konikoff, M. and Lewis, W.M. (1974) Variation in weight of cage-reared channel catfish. *Progressive Fish-Culturist* 36, 138–144.

Kono, T., Kwon, O.Y. and Nakahara, T. (1991) Development of enucleated mouse oocytes reconstituted with embryonic nuclei. *Journal of Reproduction and Fertility* 93, 165–172.

Korver, S., van der Steen, H.A.M., van Arendonk, J.A.M., Bakker, H., Brascamp, E.W. and Dommerholt, J. (1988) *Advances in Animal Breeding*. Pudoc, Wageningen, The Netherlands, 33 pp.

Kostyo, J.L. and Reagan, R.C. (1976) The biology of growth hormone. *Pharmacological Theriogenology B 2*, 591–604.

Kramer, P.R. and Wray, S. (2000) Novel gene expressed in nasal region influences outgrowth of olfactory axons and migration of luteinizing hormone-releasing hormone (LHRH) neurons. *Genes and Development* 14, 1824–1834.

Krasznai, Z. and Marian, T. (1985) Improving genetic capacity of European catfish. *Halászat* 31, 78 (in Hungarian).

Krasznai, Z. and Marian, T. (1986) Shock-induced triploidy and its effect on growth and gonad development of the European catfish, *Silurus glanis* L. *Journal of Fish Biology* 29, 519–527.

Krieg, F. and Guyomard, R. (1985) Population genetics of French brown trout (*Salmo trutta* L.): large geographical differentiation of wild populations and high similarity of domesticated stocks. *Genetics, Selection, Evolution* 17, 225–242.

Krisfalusi, M. (1999) Gonadal morphology of hormonally and chromosomally manipulated rainbow trout. *Dissertation Abstracts International Part B: Science and Engineering* 60, 2558.

Krisfalusi, M., Wheeler, P.A., Thorgaard, G.H. and Cloud, J.G. (2000) Gonadal morphology of female diploid gynogenetic and triploid rainbow trout. *Journal of Experimental Zoology* 286, 505–512.

Krueger, C.C. and May, B. (1987) Stock identification of naturalized brown trout in Lake Superior tributaries, differentiation based on allozyme data. *Transactions of the American Fisheries Society* 116, 785–794.

Krueger, C.C. and Spangler, G.R. (1981) Genetic identification of sea lamprey (*Petromyzon marinus*) populations from the Lake Superior basin. *Canadian Journal of Fisheries and Aquatic Sciences* 38, 1832–1837.

Krueger, C.C., Marsden, J.E., Kincaid, H.L. and May, B. (1989) Genetic differentiation among lake trout strains stocked into Lake Ontario. *Transactions of the American Fisheries Society* 118, 317–330.

Kubelik, A.R. and Szabo, L.J. (1995) High-GC primers are useful in RAPD analysis of fungi. *Current Genetics* 28, 384–389.

Kubota, Y., Shimada, A. and Shima, A. (1992) Detection of gamma-ray-induced DNA damages in malformed dominant lethal embryos of the Japanese medaka (*Oryzias latipes*) using AP-PCR fingerprinting. *Mutation Research* 283, 263–270.

Kucharczyk, D., Jankun, M. and Luczynski, M. (1997a) Ploidy level determination in genetically manipulated bream, *Abramis brama*, based on the number of active nucleoli per cell. *Journal of Applied Aquaculture* 7, 13–22.

Kucharczyk, D., Luczynski, M.J., Luczynski, M., Glogowski, J., Babiak, I. and Jankun, M. (1997b) Triploid and gynogenetic bream (*Abramis brama* L.) produced by heat shock. *Polish Archives of Hydrobiology* 44, 33–38.

Kucharczyk, D., Woznicki, P., Luczynski, M.J., Klinger, M. and Luczynski, M. (1999) Ploidy level determination in genetically manipulated northern pike based on the number of active nucleoli per cell. *North American Journal of Aquaculture* 61, 38–42.

Kucuktas, H., Wagner, B.K., Shopen, R., Gibson, M., Dunham, R.A. and Liu, Z. (2002) Genetic analysis of Ozark hellbenders utilizing RAPD markers. *Southeastern Association of Fish and Wildlife Agencies* 55, 126–137.

Kudo, H., Hyodo, S., Ueda, H., Hiroi, O., Aida, K., Urano, A. and Yamauchi, K. (1996) Cytophysiology of gonadotropin-releasing hormone neurons in chum salmon (*Oncorhynchus keta*) forebrain before and after upstream migration. *Cell and Tissue Research* 284, 261–267.

Kulzer, K.E., Noble, R.L. and Forshage, A.A. (1985) Genetic effects of Florida largemouth bass introductions into selected Texas reservoirs. *Proceedings Annual Conference Southeastern Association of Fish and Wildlife Agencies* 39, 56–64.

Kumar, S., Allen, G.C. and Thompson, W.F. (2006) Gene targeting in plants: fingers on the move. *Trends in Plant Science* 11, 159–161.

Kwok, S., Kellog, D.E., McKinney, N., Spasic, D., Goda, L., Levenson, C. and Sninsky, J.J. (1990) Effects of primer–template mismatches on the polymerase chain reaction: human immunodeficiency virus type 1 model studies. *Nucleic Acids Research* 18, 999–1005.

Labosky, P., Barlow, D. and Hogan, B. (1994) Mouse embryonic germ cell lines: transmission through the germline and differences in the methylation imprint of insulin-like growth factor 2 receptor gene compared with embryonic stem cell lines. *Development* 120, 3197–3204.

Lahav, M. and Lahav, E. (1990) The development of all-male tilapia hybrids in Nir David. *Israeli Journal of Aquaculture/Bamidgeh* 42, 58–61.

Lahogue, F., This, P. and Bouquet A. (1998) Identification of a codominant scar marker linker to the seedlessness character in grapevine. *Theoretical and Applied Genetics* 97, 950–959.

Laird, L.M., Ellis, A.E., Wilson, A.R. and Holiday, F.G.T. (1978) The development of the gonadal and immune systems in the Atlantic salmon (*Salmo salar* L.) and a consideration of the possibility of inducing autoimmune destruction of the testis. *Annales de Biologie Animale, Biochimie* 18, 1011–1106.

Laird, L.M., Wilson, A.R. and Holliday, E.G.T. (1981) Field trials of a method of induction of autoimmune gonad rejection in Atlantic salmon (*Salmo salar* L.). *Reproduction, Nutrition, Development* 20, 1781–1788.

Lakhaanantakun, A. (1992) The effects of triploidy on survival rate, growth rate and feed conversation ratio of walking catfish (*Clarias macrocephalus* Gunther). MSc thesis, Kasetsart University, Bangkok, 74 pp.

Lam, W.L., Lee, T.S. and Gilbert, W. (1996a) Active transposition in zebrafish. *Proceedings of the National Academy of Sciences USA* 93, 10870–10875.

Lam, W.L., Seo, P., Robison, K., Virk, S. and Gilbert, W. (1996b) Discovery of amphibian *Tc1*-like transposon families. *Journal of Molecular Biology* 257, 359–366.

Lamberts, R., Vijayan, E., Graf, M., Mansky, T. and Wuttke, W. (1983) Involvement of preoptic–anterior hypothalamic GABA neurons in the regulation of pituitary LH and prolactin release. *Experimental Brain Research* 52, 356–362.

Lande, R. and Thompson, R. (1990) Efficiency of marker-assisted selection in the improvement of quantitative traits. *Genetics* 124, 743–756.

Landry, C. and Bernatchez, L. (2001) Comparative analysis of population structure across environments and geographical scales at major histocompatability complex and microsatellite loci in Atlantic salmon (*Salmo salar*). *Molecular Ecology* 10, 2525–2539.

Landry, C., Garant, D., Duchesne, P. and Bernatchez, L. (2001) 'Good genes as heterozygosity': the major histocompatability complex and mate choice in Atlantic salmon. *Proceedings of the Royal Society of London, Series B: Biological Sciences* 268, 1279–1285.

Langston, A.L., Johnstone, R. and Ellis, A.E. (2001) The kinetics of the hypoferraemic response and changes in levels of alternative complement activity in diploid and triploid Atlantic salmon, following injection of lipopolysaccharide. *Fish and Shellfish Immunology* 11, 333–345.

Lankford, T.E. Jr, Billerbeck, J.M. and Conover, D.O. (2001) Evolution of intrinsic growth and energy acquisition rates. II. Trade-offs with vulnerability to predation in *Menidia menidia*. *Evolution* 55, 1873–1881.

Larsen, P.J. and Mau, S.E. (1994) Effect of acute stress on the expression of hypothalamic messenger ribonucleic acids encoding the endogenous opioid precursors preproenkephalin A and proopiomelanocortin. *Peptides* 15, 783–790.

Larson, J.D., Wadman, S.A., Chen, E., Kerley, L., Clark, K.J., Eide, M., Lippert, S., Nasevicius, A., Ekker, S.C., Hackett, P.B. and Essner, J.J. (2004) Expression of VE-cadherin in zebrafish embryos: a new tool to evaluate vascular development. *Developmental Dynamics* 231, 204–213.

Launey, S. and Hedgecock, D. (2001) High genetic load in the Pacific oyster *Crassostrea gigas*. *Genetics* 159, 255–265.

Laursen, T., Ovesen, P., Grandjean, B., Jensen, S., Jorgensen, J.O.L., Illum, P. and Christiansen, J.S. (1994) Nasal absorption of growth hormone in normal subjects: studies with four different formulations. *Annals of Pharmacotherapy* 28, 845–848.

Lauterio, T.J., Trivedi, B., Kapadia, M. and Daughaday, W.H. (1988) Reduced ^{125}I-hGH binding by serum of dwarf pigs but not by serum of dwarfed poodles. *Comparative Biochemistry and Physiology, Part A: Comparative Physiology* 91A, 15–19.

Leary, R.F., Allendorf, F.W. and Knudsen, K.L. (1983) Developmental stability and enzyme heterozygosity in rainbow trout. *Nature* 301, 71–73.

Leary, R.F., Knudsen, K.L., Allendorf, F.W. and Thorgaard, G.H. (1984) Developmental stability of gynogenetic diploid and triploid rainbow trout. *Genetics* 107, 60.

Leary, R.F., Allendorf, F.W. and Knudsen, K.L. (1985a) Developmental instability as an indicator of reduced genetic variation in hatchery trout. *Transactions of the American Fisheries Society* 114, 230–235.

Leary, R.F., Allendorf, F.W. and Knudsen, K.L. (1985b) Inheritance of meristic variation and the evolution of developmental stability in rainbow trout. *Evolution* 39, 308–314.

Leary, R.F., Allendorf, F.W., Knudsen, K.L. and Thorgaard, G.H. (1985c) Heterozygosity and developmental stability in gynogenetic diploid and triploid rainbow trout. *Heredity* 54, 219–225.

Leaver, M.J., Wright, J. and George, S.G. (1997) Structure and expression of a cluster of glutathione *S*-transferase genes from a marine fish, the plaice (*Pleuronectes platessa*). *Biochemical Journal* 321, 405–412.

Le Bail, P.Y., Sumpter, J.P., Carragher, J.F., Mourout, B., Niu, P.D. and Weil, C. (1991) Development and validation of a highly sensitive radioimmunoassay for chinook salmon (*Oncorhynchus tshayttscha*) growth hormone. *General and Comparative Endocrinology* 83, 75–85.

Leberg, P.L. and Firmin, B.D. (2008) Role of inbreeding depression and purging in captive breeding and restoration programmes. *Molecular Ecology* 17, 334–343.

Lee, B.Y., Penman, D.J. and Kocher, T.D. (2003) Identification of a sex-determining region in Nile tilapia (*Oreochromis niloticus*) using bulked segregant analysis. Animal Genetetics 34, 379–383.

Lee, B.Y., Hulata, G. and Kocher, T.D. (2004) Two unlinked loci controlling the sex of blue tilapia (*Oreochromis aureus*). *Heredity* 92, 543–549.

Lee, B.Y., Lee, W.J., Streelman, J.T., Carleton, K.L., Howe, A.E., Hulata, G., Slettan, A., Stern, J.E., Terai, Y. and Kocher, T.D. (2005) A second-generation genetic linkage map of tilapia (*Oreochromis* spp.). *Genetics* 170, 237–244.

Lee, E.-H., Chung, B.-G., Kim, J.-S., Ahn, C.-B. and Oh, K.-S. (1989a) Studies on the food components of triploid carp muscle. 1. The taste compounds of triploid carp muscle. *Bulletin of the Korean Fisheries Society* 22, 154–160.

Lee, E.-H., Chung, B.-G., Kim, J.-S., Ahn, C.-B., Joo, D.-S. and Oh, K.-S. (1989b) Studies on the food components of triploid carp muscle. 2. Lipid components of triploid carp muscle. *Bulletin of the Korean Fisheries Society* 22, 161–168.

Lee, G.M. and Wright, J.E. Jr (1981) Mitotic and meiotic analyses of brook trout, *Salvelinus fontinalis*. *Journal of Heredity* 72, 321–327.

Lee, J.Y., Boman, A., Sun, C.X., Anderson, M., Jornvall, H., Hutt, V. and Boman, H.G. (1989) Antibacterial peptides from pig intestine: isolation of a mammalian cecropin. *Proceedings of the National Academy of Sciences USA* 86, 9159–9162.

Lee, K.-Y., Huang, H., Ju, B., Yang, Z. and Shuo (2002) Cloned zebrafish by nuclear transfer from long-term-cultured cells. *Nature Biotechnology* 20, 795–799.

Lee, M.M. (1988) Abnormal gametogenesis in triploid American oysters *Crassostrea virginica*. *Journal of Shellfish Research* 7, 201–202.

Lee, M.S., Gippert, G.P., Soman, K.V., Case, D.A and Wright, P.E. (1989) Three-dimensional solution structure of a single zinc finger DNA-binding domain. *Science* 245, 635–637.

Lee, N., Lambremont, E.N. and Tiersch, T.R. (2000) Considerations for gamma irradiation of aquatic organisms. *North American Journal of Aquaculture* 62, 95–102.

Lee, W.J. and Kocher, T.D. (1996) Microsatellite DNA markers for genetic mapping in *Oreochromis niloticus*. *Journal of Fish Biology* 49, 169–171.

Leeds, T.D., Silverstein, J. T., Vallejo, R.L., Palti, Y., Rexroad, C.E. III, Evenhuis, J., Hadidi, S., Weber, G.M., Welch, T.J. and Wiens, G.D. (2009) Respone to selection for bacterial cold water disease in rainbow trout. In: *Book of Abstracts of 10th International Symposium on Genetics in Aquaculture (ISGA X) 'Roles of Aquaculture Genetics in Addressing Global Food Crisis', 22–26 June 2009, Bangkok, Thailand*. Faculty of Fisheries, Kasetsart University, Bangkok, p. 200.

Leeprasert, K. (1987) Genetic parameters of some quantitative traits in *Pangasius sutchi* Fowler. MSc thesis, Kasetsart University, Bangkok, 75 pp.

Le Gac, F. and Loir, M. (1993) Expression of insulin-like growth factor (IGF) I and action of IGF I and II in the trout testes. *Reproduction, Nutrition, Development* 33, 80–82.

Legraien, E.Y.G. and Crosaz, O. (eds) (1999) *Oyster Culture in Brittany: Farming's Techniques and Problems*. Ecole Nationale Veterinaire d'Alfort, Maisons Alfort, France, 179 pp.

LeGrande, W.H. (1981) Chromosomal evolution in North American catfishes (Siluriformes: Ictaluridae) with particular emphasis on the madtoms, *Noturus*. *Copeia* 1981, 33–52.

LeGrande, W., Dunham, R.A. and Smitherman, R.O. (1984) Comparative karyology of three species of North American catfishes (Siluriformes: Ictaluridae: Ictalurus) and four of their hybrid combinations. *Copeia* 1984, 873–878.

Lehrer, R.I., Lichtenstein, A.K. and Ganz, T. (1993) Defensins: antimicrobial and cytotoxic peptides of mammalian cells. *Annual Review of Immunology* 11, 105–128.

Lepage, O., Deverli, O., Petersson, E., Jaervi, T. and Winberg, S. (2000) Differential stress coping in wild and domesticated sea trout. *Brain, Behavior and Evolution* 56, 259–268.

Lerder, S.A., Chilocote, M.W. and Loch, J.J. (1984) Spawning characteristics of sympatric populations of steelhead trout, *Salmo gairdneri*, evidence for partial reproductive isolation. *Canadian Journal of Fisheries and Aquatic Sciences* 41, 1454–1462.

Lesniak, M.A., Gorden, P. and Roth, J. (1977) Reactivity of non-primate growth hormones and prolactins with human growth hormone receptors on cultured human lymphocytes. *Journal of Clinical Endocrinology and Metabolism* 44, 838–849.

Lester, L.J., Lawson, K.S., Abella, T.A. and Palada, M.S. (1989) Estimated heritability of sex ratio and sexual dimorphism in tilapia. *Aquaculture and Fisheries Management* 20, 369–380.

Lewis, K.A., Swanson, P. and Sower, S.A. (1992) Changes in brain gonadotropin-releasing hormone, pituitary and plasma gonadotropins, and plasma thyroxine during smoltification in chinook salmon (*Oncorhynchus tshawytscha*). *General and Comparative Endocrinology* 87, 461–470.

Lewontin, R.C. (1984) Detecting population differences in quantitative characters as opposed to gene frequencies. *American Naturalist* 123, 115–124.

Lewontin, R.C. (1985) Population genetics. *Annual Review of Genetics* 19, 81–102.

Li, C.H., Chung, D., Lahm, H.W. and Stein, S. (1986) The primary structure of monkey pituitary growth hormone. *Archives of Biochemistry and Biophysics* 245, 287–291.

Li, F., Zhou, L.-H., Xian, J.-H., Liu, X.-D., and Zhu, J.-Z. (1999) Triploidy induction with heat shocks to *Penaeus chinensis* and their effects on gonad development. *Chinese Journal of Oceanology and Limnology* 17, 57–61.

Li, J., Wang, G., Bai, Z. and Zheng, H. (2009) Research progress on genetic resource exploitation and utilization of freshwater pearl mussel. In: *Book of Abstracts of 10th International Symposium on Genetics in Aquaculture (ISGA X) 'Roles of Aquaculture Genetics in Addressing Global Food Crisis', 22–26 June 2009, Bangkok, Thailand*. Faculty of Fisheries, Kasetsart University, Bangkok, p. 37.

Li, S., Li, C., Dey, M., Gagalac, F. and Dunham, R. (2002) Cold tolerance of three strains of Nile tilapia, *Oreochromis niloticus*, in China. *Aquaculture* 213, 123–129.

Li, Y., Wilson, K.J., Byrne, K., Whan, V., Iglesis, D., Lehnert, S.A., Swan, J., Ballment, B., Fayazi, Z., Kenway, M., Benzie, J., Pongsomboon, S., Tassanakajon, A. and Moore, S.S. (2000) International collaboration on genetic maping of the black tiger shrimp, *Penaeus monodon*: progress update. In: *Plant and Animal Genome VIII, San Diego, California, 9–12 January 2000*, p. 8.

Liang, Y.H., Cheng, C.H.K. and Chan, K.M. (1996) Insulin-like growth factor Iea2 is the predominantly expressed form of IGF in common carp (*Cyprinus carpio*). *Molecular Marine Biology and Biotechnology* 5, 145–152.

Liao, D. and Dennis, P.P. (1994) Molecular phylogenies based on ribosomal protein L11, L1, L10, and L12 sequences. *Journal of Molecular Evolution* 38, 405–419.

Lie, Ø., Slettan, A., Lingaas, F., Olsaker, I., Hordvik, I. and Refstie, T. (1994) Haploid gynogenesis: a powerful strategy for linkage analysis in fish. *Animal Biotechnology* 5, 33–45.

Liepens, A. and Hennen, S. (1977) Cytochrome oxidase deficiency during development of amphibian nucleo-cytoplasmic hybrids. *Developmental Biology* 57, 284–292.

Lilyestrom, C.F., Wolters, W.R., Bury, D., Rezk, M. and Dunham, R.A. (1999) Growth, carcass traits, and oxygen tolerance of diploid and triploid catfish hybrids. *North American Journal of Aquaculture* 61, 293–303.

Lim, C., Leamaster, B. and Brock, J.A. (1993) Riboflavin requirement of fingerling red hybrid tilapia grown in seawater. *Journal of the World Aquaculture Society* 24, 451–458.

Lim, S. and Bailey, G. (1977) Gene duplication in salmonid fish: evidence for duplicated but catalytically equivalent A4 lactate dehydrogenases. *Biochemical Genetics* 15, 707–721.

Lin, B., Chen, S., Cao, Z., Lin, Y., Mo, D., Zhang, H., Gu, J., Dong, M., Liu, Z. and Xu, A. (2007) Acute phase response in zebrafish upon *Aeromonas salmonicida* and *Staphylococcus aureus* infection: striking similarities and obvious differences with mammals. *Molecular Immunology* 44, 295–301.

Lin, S., Long, W., Chen, J. and Hopkins, N. (1992) Germ line chimeras in zebrafish by cell transplants from genetically pigmented to albino embryos. *Proceedings of the National Academy of Sciences USA* 89, 4519–4523.

Lin, S., Gaiano, N., Culp, P., Burns, J.C., Friedmann, T., Yee, J.-K. and Hopkins, N. (1994) Integration and germ-line transmission of a pseudotyped retroviral vector in zebrafish. *Science* 265, 666–669.

Lincoln, R.F. (1981) Sexual maturation in triploid male plaice (*Pleuronectes platessa*) and plaice × flounder (*Platichthys flesus*) hybrids. *Journal of Fish Biology* 19, 415–426.

Lincoln, R.F. and Scott, A.P. (1984) Sexual maturation in triploid rainbow trout, *Salmo gairdneri* Richardson. *Journal of Fish Biology* 25, 385–392.

Linhart, O., Flajshans, M., Gela, D., Duda, P., Slechta, V. and Slechtova, V. (1998) Breeding programme of common carp in the Czech Republic. *Czech Journal of Animal Science* 43, 434. [XVIII-th Genetic Days, Ceske Budejovice].

Liu, C.Y. (1977) Aspects of reproduction and progeny testing in *Sarotherodon aureus* (Steindachner). MS thesis, Auburn University, Auburn, Alabama.

Liu, G., Zheng, W. and Chen, X. (2007) Molecular cloning of proteasome activator PA28-β subunit of large yellow croaker (*Pseudosciana crocea*) and its coordinated up-regulation with MHC class I α-chain and $β_2$-microglobulin in poly I:C-treated fish. *Moleular Immunology* 44, 1190–1197.

Liu, Q., Goudie, C.A., Simco, B.A., Davis, K.B. and Morizot, D.C. (1992) Gene-centromere mapping of six enzyme loci in gynogenetic channel catfish. *Journal of Heredity* 83, 245–248.

Liu, W., Heasman, M., Simpson, R., Dworjanyn, S. and Pirozzi, I. (2006) Growth and feeding in juvenile triploid and diploid blacklip abalone, *Haliotis rubra* (Leach, 1814), at two temperatures. *Aquaculture Nutrition* 12, 410–417.

Liu, W., Heasman, M. and Simpson, R. (2009) Growth and reproductive performance of triploid and diploid blacklip abalone, *Haliotis rubra* (Leach, 1814). *Aquaculture Research* 40 (Suppl. 2), 188–203.

Liu, Z. (2007) *Aquaculture Genome Technologies*. Blackwell Publishing, Ames, Iowa.

Liu, Z.J. and Dunham, R. (1998a) *Genetic Linkage and QTL Mapping of Ictalurid Catfish*. Circular Bulletin 321. Alabama Agricultural Experiment Station, Auburn University, Auburn, Alabamam 19 pp.

Liu, Z.J. and Dunham, R.A. (1998b) Progress report on the state of catfish gene mapping. In: Abstracts of the International Plant and Animal Genome VI, San Diego, California, 18–23 January 1998; available at http://www.intl-pag.org/6/abstracts/477.html

Liu, Z.J. and Feng, J. (2001) Gene mapping and marker-assisted selection in channel catfish (*Ictalurus punctatus*). *Acta Agriculturae Boreali-Occidentalis Sinica* 10, 110–114.

Liu, Z., Moav, B., Faras, A.J., Guise, K.S., Kapuscinski, A.R. and Hackett, P.B. (1990) Development of expression vectors for transgenic fish. *Biotechnology* 8, 1268–1272.

Liu, Z., Li, P., Argue, B. and Dunham, R.A. (1997) Gonadotropin a subunit glycoprotein from channel catfish (*Ictalurus punctatus*) and its expression during hormone-induced ovulation. *Molecular Marine Biology and Biotechnology* 6, 221–231.

Liu, Z., Li, P., Argue, B.J. and Dunham, R.A. (1998a) Inheritance of RAPD markers in channel catfish (*Ictalurus punctatus*), blue catfish (*I. furcatus*) and their F_1, F_2 and backcross hybrids. *Animal Genetics* 29, 58–62.

Liu, Z.J., Li, P. and Dunham, R.A. (1998b) Characterization of an A/T-rich family of sequences from the channel catfish (*Ictalurus punctatus*). *Molecular Marine Biology and Biotechnology* 7, 232–239.

Liu, Z.J., Nichols, A., Li, P. and Dunham, R. (1998c) Inheritance and usefulness of AFLP markers in channel catfish (*Ictalurus punctatus*), blue catfish (*I. furcatus*) and their F_1, F_2 and backcross hybrids. *Molecular and General Genetics* 258, 260–268.

Liu, Z.J., Tan, G., Kucuktas, H., Karsi, A., Li, P., Zheng, X., Kim, S. and Dunham, R.A. (1998d) Conservation of microsatellite loci between channel catfish (*Ictalurus punctatus*) and blue catfish (*I. furcatus*). In: Abstracts of the International Plant and Animal Genome VI, 18–23 January 1998; available at http://www.intl-pag.org/6/abstracts/471.html

Liu, Z., Karsi, A. and Dunham, R.A. (1999a) Development of polymorphic EST markers suitable for genetic linkage mapping of catfish. *Marine Biotechnology* 1, 437–447.

Liu, Z.J., Li, P., Argue, B.J. and Dunham, R.A. (1999b) Random amplified polymorphic DNA markers: usefulness for gene mapping and analysis of genetic variation of catfish. *Aquaculture* 174, 59–68.

Liu, Z.J., Li, P., Kucuktas, H. and Dunham, R.A. (1999c) Characterization of nonautonomous *Tcl*-like transposable elements of channel catfish (*Ictalurus punctatus*). *Fish Physiology and Biochemistry* 21, 65–72.

Liu, Z.J., Li, P., Kucuktas, H., Nichols, A., Tan, G., Zheng, X., Argue, B.J., Yant, R. and Dunham, R.A. (1999d) Development of AFLP markers for genetic linkage mapping analysis using channel catfish and blue catfish interspecific hybrids. *Transactions of the American Fisheries Society* 128, 317–327.

Liu, Z.J., Tan, G., Kucuktas, H., Li, P., Karsi, A., Yant, D.R. and Dunham, R.A. (1999e) High levels of conservation at microsatellite loci among ictalurid catfishes. *Journal of Heredity* 90, 307–312.

Liu, Z.J., Tan, G., Li, P. and Dunham, R.A. (1999f) Transcribed dinucleotide microsatellites and their associated genes from channel catfish, *Ictalurus punctatus*. *Biochemical and Biophysical Research Communications* 259, 190–194.

Liu, Z., Wang, Q., Zhang, Y. and Yang, A. (2000) Studies on excretion of triploid and diploid *Chlamys farreri*. *Journal of Fisheries Science China* 7, 10–13.

Liu, Z.J., Li, P., Kocabas, A., Ju, Z., Karsi, A., Cao, D. and Patterson, A. (2001) Microsatellite-containing genes from the channel catfish brain: evidence of trinucleotide repeat expansion in the coding region of nucleotide excision repair gene RAD23B. *Biochemical and Biophysical Research Communications* 289, 317–324.

Liu, Z.J., Karsi, A., Li, P., Cao, D. and Dunham, R.A. (2003) An AFLP-based genetic linkage map of channel catfish (*Ictalurus punctatus*) constructed by using an interspecific hybrid resource family. *Genetics* 165, 687–694.

Lloyd, A., Plaisier, C.L., Carroll, D. and Drews, G.N. (2005) Targeted mutagenesis using zinc-finger nucleases in *Arabidopsis*. *Proceedings of the National Academy of Sciences USA* 102, 2232–2237.

Lohe, A.R., De Aguiar, D. and Hartl, D.L. (1997) Mutations in the *mariner* transposase: the D, D(35)E consensus sequence is nonfunctional. *Proceedings of the National Academy of Sciences USA* 94, 1293–1297.

Loomis, W.F. and Smith, D.W. (1990) Molecular phylogeny of *Dictyostelium discoideum* by protein sequence comparison. *Proceedings of the National Academy of Sciences USA* 87, 9093–9097.

Lopez-Macias, J. (1980) Comparative efficacy of two androgens and two estrogens in functional sex reversal of *Tilapia aurea*. MSc thesis, Auburn University, Auburn, Alabama.

Lou, Y., Shen, H., Xu Q. and Xu, Y. (1995) Comparative studies on biochemical compositions in serum of Gaoyou hybrid crucian carp (*Carassius auratus auratus* × *Carassius auratus cuvieri*) and its parents. *Acta Hydrobiologica Sinica* 19, 157–163.

Lou, Y.D. and Purdom, C.E. (1984) Polyploidy induced hydrostatic pressure in rainbow trout, *Salmo gairdneri* Richardson. *Journal of Fish Biology* 25, 345–357.

Loukeris, T.G., Livadaras, I., Arca, B., Zabalou, S. and Savakis, C. (1995) Gene transfer into the medfly, *Ceratitis capitata*, with a *Drosophila hydei* transposable element. *Science* 270, 2002–2005.

Lovell, R.T. (1979) Factors affecting voluntary food consumption by channel catfish. *Proceedings Annual Conference Southeastern Association of Fish and Wildlife Agencies* 33, 563–571.

Lovell, T. (1996) Feed deprivation increases resistance of channel catfish to bacterial infection. *Aquaculture Magazine* 22, 65–67.

Lovshin, L.L. and Da Silva, A.B. (1976) Culture of monosex and hybrid tilapias. In: *FAO/CIFA Symposium on Aquaculture in Africa, Accra, Ghana, 30 September–6 October 1975*. CIFA/75/SR9. Food and Agriculture Organization of the United Nations, Rome.

Lovshin, L.L., DaSilva, A.B. and Fernandes, J.A. (1974) The intense culture of the all-male hybrid of *Tilapia hornorum* (male) × *T. nilotica* (female) in northeast Brazil. In: *FAO/CARPAS Symposium on Aquaculture in Latin America, Montevideao, Urugauy, 26 November–2 December 1974*. Food and Agriculture Organization of the United Nations, Rome.

Lu, J.K., Chen, T.T., Chrisman, C.L., Andrisani, O.M. and Dixon, J.E. (1992) Integration, expression, and germ-line transmission of foreign growth hormone genes in medaka (*Oryzias latipes*). *Molecular Marine Biology and Biotechnology* 1, 366–375.

Lu, J.K., Chen, T.T., Allen, S.K., Matsubara, T. and Burns, J.C. (1996) Production of transgenic dwarf surf-clams, *Mulinia lateralis*, with pantropic retroviral vectors. *Proceedings of the National Academy of Sciences USA* 93, 3482–3486.

Lucchesi, J.C. and Rawls, J.M. Jr (1973) Regulation of gene function: a comparison of enzyme activity levels in relation to gene dosage in diploids and triploids of *Drosophila melanogaster*. *Biochemical Genetics* 9, 41–51.

Lundin, L. (1993) Evolution of the vertebrate genome as reflected in paralogous chromosomal regions in man and the house mouse. *Genomics* 16, 1–19.

Luo, G., Ivics, Z. Izsvak, Z. and Bradley, A. (1998) Chromosomal transposition of a *Tcl/mariner*-like element in mouse embryonic stem cells. *Proceedings of the National Academy of Sciences USA* 95, 10769–10773.

Luquet, P., Oteme, Z. and Cisse, A. (1995) Evidence for compensatory growth and its utility in the culture of *Heterobranchus longifilis*. *Aquatic Living Resources* 8, 389–394.

Lutz, C.G. and Wolters, W.R. (1989) Estimation of heritabilities for growth, body size, and processing traits in red swamp crawfish, *Procambarus clarkii* (Girard). *Aquaculture* 78, 21–23.

Lyttle, T.W. (1991) Segregation distorters. *Annual Review of Genetics* 25, 511–557.

Ma, X., Dong, Y., Matzuk, M.M. and Kumar, T.R. (2004) Targeted disruption of luteinizing hormone β-subunit leads to hypogonadism, defects in gonadal steroidogenesis, and infertility. *Proceedings of the National Academy of Sciences USA* 101, 17294–17299.

Macaranas, J.M., Taniguchi, N., Pante, M.J.R., Capili, J.B. and Pullin, R.S.V. (1986) Electrophoretic evidence for extensive hybrid gene introgression into commercial *Oreochromis niloticus* (L.) stocks in the Philippines. *Aquaculture and Fisheries Management* 17, 249–258.

McBride, J.R., Fagerlund, V.H.M., Smith, M. and Tommnson, N. (1963) Resumption of feeding by and survival of adult sockeye salmon (*Oncorhynchus nerka*) following advanced gonad development. *Journal of Fisheries Research Board of Canada* 20, 95–100.

McCarter, N.H. (1988) Verification of the production of triploid grass carp (*Ctenopharyngodon idella*) with hydrostatic pressure. *New Zealand Journal of Marine and Freshwater Research* 22, 501–505.

McClelland, E.K. and Naish, K.A. (2007) What is the fitness outcome of crossing unrelated fish populations? A meta-analysis and an evaluation of future research directions. *Conservation Genetics* 8, 397–416.

McConnell, T.J., Godwin, U.B., Norton, S.F., Nairn, R.S., Kazianis, S. and Morizot, D.C. (1998) Identification and mapping of two divergent, unlinked major histocompatibility complex class II B genes in *Xiphophorus* fishes. *Genetics* 149, 1921–1934.

McCormick, S.D. and Björnsson, B. (1994) Physiological and hormonal differences among Atlantic salmon parr and smolts reared in the wild, and hatchery smolts. *Aquaculture* 121, 235–244.

McCormick, S.D., Björnsson, B.T., Sheridan, M., Eilertson, C., Carey, J.B. and O'Dea, M. (1995) Increased day length stimulates plasma growth hormone and gill Na super (+), K super (+)-ATPase in Atlantic salmon (*Salmo salar*). *Journal of Comparative Physiology, Part B: Biochemical, Systemic, and Environmental Physiology* 165, 245–254.

McCormick, S.D., Shrimpton, J.M., Carey, J.B., O'Dea, M.F., Sloan, K.E., Moriyama, S. and Björnsson, B.T. (1998) Repeated acute stress reduces growth rate of Atlantic salmon parr and alters plasma levels of growth hormone, insulin-like growth factor I and cortisol. *Aquaculture* 168, 221–235.

Macdonald, P. (2001) Diversity in translational regulation. *Current Opinion in Cell Biology* 13, 326–331.

Maceina, M.J., Murphy, B.R. and Isely, J.J. (1988) Factors regulating Florida largemouth bass stocking success and hybridization with northern largemouth bass in Aquilla Lake, Texas. *Transactions of the American Fisheries Society* 117, 221–231.

McEvoy, T., Stack, M., Kean, B., Barry, T., Sreenan, J. and Gannon, F. (1988) The expression of a foreign gene in salmon embryos. *Aquaculture* 68, 27–37.

McGinnity, P., Stone, C., Taggart, J.B., Cooke, D., Cotter, D., Hynes, R., McCamley, C., Cross, T. and Ferguson, A. (1997) Genetic impact of escaped farmed Atlantic salmon (*Salmo salar* L.) on native populations: use of DNA profiling to assess freshwater performance of wild, farmed and hybrid progeny in a natural river environment. *ICES Journal of Marine Science* 54, 998–1008.

McGinty, A.S. (1980) Survival, growth, and variation in growth of channel catfish fry and fingerlings. Doctoral dissertation, Auburn University, Auburn, Alabama.

McGrath, J. and Solter, D. (1984a) Inability of mouse blastomere nuclei transferred to enucleated zygotes to support development *in vitro*. *Science* 226, 1317–1319.

McGrath, J. and Solter, D. (1984b) Completion of mouse embryogenesis requires both the maternal and paternal genomes. *Cell* 37, 179–183.

Machlin, L.J. (1972) Effect of porcine growth hormone on growth and carcass composition of the pig. *Journal of Animal Science* 35, 794–800.

Machlin, L.J. (1976) Role of growth hormone in improving animal production. *Environmental Quality and Safety Supplement* 2, 43–55.

McKay, J.K., Bishop, J.G., Lin, J.Z., Richards, J.H., Sala, A. and Mitchell-Olds, T. (2001) Local adaptation across a climatic gradient despite small effective population size in the rare sapphire rockcress. *Proceedings of the Royal Society of London, Series B: Biological Sciences* 268, 1715–1721.

McKenzie, W.D. Jr, Crews, D., Kallman, K.D., Policansky, D. and Sohn, J.J. (1983) Age, weight and the genetics of sexual maturation in platyfish, *Xiphohorus maculatus. Copeia* 1983, 770.

Mackill, D.J., Zhang, Z., Redona, E.D. and Colowit, P.M. (1996) Level of polymorphism and genetic mapping of AFLP markers in rice. *Genome* 39, 969–977.

McLean, E. and Donaldson, E.M. (1993) The role of somatotropin in growth in poikilotherms. In: Schreibman, M.P., Scanes, C.G. and Pang, P.K.T. (eds) *The Endocrinology of Growth, Development and Metabolism in Vertebrates.* Academic Press, New York, pp. 43–71.

McLean, E., Donaldson, E.M., Mayer, I., Teskeredzic, Z., Pitt, C. and Souza, L.M. (1994) Evaluation of a sustained-release polymer-encapsulated form of recombinant porcine somatotropin upon long-term growth-performance of coho salmon, *Oncorhynchus kisutch. Aquaculture* 122, 359–368.

Maclean, J.A., Evans, D.O., Martin, N.V. and Desjardine, R.L. (1981) Survival, growth, spawning, distribution, and movements of introduced and native lake trout (*Salvelinus namaycush*) in two inland Ontario lakes. *Canadian Journal of Fisheries and Aquatic Sciences* 38, 1685–1700.

Maclean, N. and Talwar, S. (1984) Injection of cloned genes with rainbow trout eggs. *Journal of Embryology and Experimental Morphology* 82 (Suppl.) 187.

Maclean, N., Penman, D. and Talwar, S. (1987a) Introduction of novel genes into fish. *Biotechnology* 5, 257–261.

Maclean, N., Penman, D. and Talwar, S. (1987b) Introduction of novel genes into rainbow trout. In: *EIFAC/FAO Symposium on Selection, Hybridization and Genetic Engineering in Aquaculture,* Vol. II. Heenemann Verlagsgesellschaft, Berlin, pp. 325–334.

Maclean, N., Hwang, G.L. and Farahmand, H. (2002a) Exploiting transgenic tilapia and the tilapia genome. In: Shimizu, N., Aoki, T., Hirono, I. and Takashima, F. (eds) *Aquatic Genomics: Steps Toward a Great Future.* Springer-Verlag, New York, pp. 365–381.

Maclean, N., Rahman, M.A., Sohm, F., Hwang, G., Iyengar, A., Ayad, H., Smith, A. and Farahmand, H. (2002b) Transgenic tilapia and the tilapia genome. *Gene* 295, 265–277.

McMillen, I.C., Mercer, J.E. and Thorburn, G.D. (1988) Pro-opiomelanocortin mRNA levels fall in the fetal sheep pituitary before birth. *Journal of Molecular Endocrinology* 1, 141–145.

McMorrow, T., Wagner, A., Deryckere, F. and Gannon, F. (1996) Structural organisation and sequence analysis of the globin locus from Atlantic salmon. *DNA and Cell Biology* 15, 407–414.

McPherron, A.C. and Lee, S.J. (1996) In: LeRoith, D. and Bondy, C. (eds) *Growth Factors and Cytokines in Health and Disease.* JAI, Greenwich, Connecticut, pp. 357–393.

McPherron, A.C. and Lee, S.J. (1997) Double muscling in cattle due to mutations in the myostatin gene. *Proceedings of the National Academy of Sciences USA* 94, 12457–12461.

McPherron, A.C., Lawler, A.M. and Lee, S.-J. (1997) Regulation of skeletal muscle mass in mice by a new TGF-β superfamily member. *Nature* 387, 83–90.

McRory, J.E. and Sherwood, N.M. (1994) Catfish express two forms of insulin-like growth factor-I (IGF-I) in the brain: ubiquitous IGF-I and brain-specific IGF-I. *Journal of Biological Chemistry* 269, 18588–18592.

Magoulas, A. and Zouros, E. (1993) Restriction-site heteroplasmy in anchovy (*Engraulis encrasicolus*) indicates incidental biparental inheritance of mitochondrial DNA. *Molecular Biology and Evolution* 10, 319–325.

Maguire, G.B., Gardner, N.C., Nell, J.A., Kent, G.N. and Kent, A.S. (1994a) Studies on triploid oysters in Australia: II. Growth, condition index, glycogen content and gonad area of triploid and diploid Pacific oysters, *Crassostrea gigas* (Thunberg), in Tasmania. In: Nell, J.A. and Maguire, G.B. (eds) *Evaluation of Triploid Sydney Rock Oysters* (Saccostrea commercialis) *and Pacific Oysters* (Crassostrea gigas) *on Commercial Leases in New South Wales and Tasmania, Final Report to FRDC, September 1994.* NSW Fisheries, Port Stephens Research Centre, Taylors Beach, New South Wales, and University of Tasmania, Launceston, Tasmania, pp. 37–71.

Maguire, G.B., Boocock, B., Kent, G.N. and Gardner, N.C. (1994b) Studies on triploid oysters in Australia: IV. Sensory evaluation of triploid and diploid Pacific oysters, *Crassostrea gigas* (Thunberg), in Tasmania. In: Nell, J.A. and Maguire, G.B. (eds) *Evaluation of Triploid Sydney Rock Oysters (Saccostrea commercialis) and Pacific Oysters (Crassostrea gigas) on Commercial Leases in New South Wales and Tasmania, Final Report to FRDC, September 1994.* NSW Fisheries, Port Stephens Research Centre, Taylors Beach, New South Wales, and University of Tasmania, Launceston, Tasmania, pp. 178–193.

Mahapatra, K.D., Meher, P.K., Saha, J.N., Gjerde, B., Reddy, P.V.G.K., Jana, R.K., Sahoo, M. and Rye, M. (2000) Selection response of rohu, *Labeo rohita*, for two generations of selective breeding. In: *Abstracts of the Fifth Indian Fisheries Forum, Bhubaneswar, India, 17–20 January 2000.*

Mahmound, A.J., Thompson, F.N., Peck, D.D., Mininga, K.M., Leshin, L.S., Rund, L.A., Stuedemann, J.A. and Kiser, T.E. (1989) Difference in luteinizing hormone response to an opioid antagonist in beef heifers and cows. *Biology of Reproduction* 41, 431–437.

Mair, G.C. (1988) Studies on sex determining mechanisms in *Oreochromis* species. Doctoral thesis, University of Wales, Swansea, UK, 326 pp.

Mair, G.C. and Abucay, J.S. (2001) Selection approaches for improved genetically male tilapia (GMT registered) in *Oreochromis niloticus*. In: *Aquaculture 2001: Book of Abstracts.* Elsevier, Amsterdam.

Mair, G.C., Beardmore, J.A. and Skibinski, D.O.F. (1990) Experimental evidence for environmental sex determination in *Oreochromis* species. In: Hirano, R. and Hanyu, I. (eds) *Proceedings of The Second Asian Fisheries Forum, Tokyo, Japan, 17–22 April 1989.* Asian Fisheries Society, Manila, pp. 555–558.

Mair, G.C., Scott, A., Penman, D.J., Beardmore, J.A. and Skibinski, D.O.F. (1991a) Sex determination in the genus *Oreochromis* I: sex reversal, gynogenesis, and triploidy in *O. niloticus* L. *Theoretical and Applied Genetics* 82, 144–152.

Mair, G.C., Scott, A., Penman, D.J., Skibinski, D.O.F. and Beardmore, J.A. (1991b) Sex determination in the genus *Oreochromis* II: sex reversal, hybridisation, gynogenesis and triploidy in *O. aureus* Steindachner. *Theoretical and Applied Genetics* 82, 153–160.

Mair, G.C., Abucay, J.S., Beardmore, J.A. and Skibinski, D.O.F. (1995) Growth performance trials of genetically male tilapia (GMT) derived from 'YY' males in *Oreochromis niloticus* L.: on-station comparisons with mixed sex and sex reversed male populations. *Aquaculture* 137, 313–322.

Mair, G.C., Abucay, J.S., Skibinski, D.O.F., Abella, T.A. and Beardmore, J.A. (1997) Genetic manipulation of sex ratio for the large-scale production of all-male tilapia, *Oreochromis niloticus*. *Canadian Journal of Fisheries and Aquatic Sciences* 54, 396–404.

Mallia, J.V., Muthiah, P. and Thomas, P.C. (2006) Growth of triploid oyster, *Crassostrea madrasensis* (Preston). *Aquaculture Research* 37, 718–724.

Maluszynska, J. and Heslop-Harrison, J.S. (1991) Localization of randomly repeated DNA sequences in *Arabidopsis thaliana*. *Plant Journal* 1, 159–166.

Maluwa, A.O. and Gjerde, B. (2007) Response to selection for harvest body weight of *Oreochromis shiranus*. *Aquaculture* 273, 33–41.

Malven, P.V., Clemens, H.A. and Sawyer, C.H. (1971) Inhibition of ovarian compensatory hypertrophy in the rat by intrahypothalamic implantation of methallibure. *Endocrinology* 88, 511–513.

Mananos, E.L., Anglade, I., Chyb, J., Saligaut, C., Breton, B. and Kah, O. (1999) Involvement of aminobutyric acid in the control of GtH-1 and GtH-2 secretion in male and female rainbow trout. *Neuroendocrinology* 69, 269–280.

Manning, A.J. and Crim, L.W. (2000) The induction of triploidy in the yellowtail flounder, *Pleuronectes ferrugineus* (Storer). In: Norberg, B., Kjesbu, O.S., Taranger, G.L., Andersson, E. and Stefansson, S.O. (eds) *Reproductive Physiology of Fish 2000. 6th International Symposium on the Reproductive Physiology of Fish, Bergen, Norway, 4–9 July 1999.* Department of Fisheries and Marine Biology, University of Bergen, Bergen, Norway.

Marcos, J.F., Beachy, R.N., Houghten, R.A., Blondelle, S.E. and Perez-Paya, E. (1995) Inhibition of a plant virus infection by analogs of melittin. *Proceedings of the National Academy of Sciences USA* 92, 12466–12469.

Marshall, C.T., Yaragina, N.A., Lambert, Y. and Kjesbu, O.S. (1999) Total lipid energy as a proxy for total egg production by fish stocks. *Nature* 402, 288–290.

Martin, S.A., Blaney, S.C., Houlihan, D.F. and Secombes, C.J. (2006) Transcriptome response following administration of a live bacterial vaccine in Atlantic salmon (*Salmo salar*). *Molecular Immunology* 43, 1900–1911.

Martinez, R., Estrada, M.P., Hernandez, O., Cabrera, E., Garcia del Barco, D., Lleonart, R., Berlanga, J. and Pimentel, R. (1994) Towards growth manipulation in tilapia (*Oreochromis* sp.): generation of transgenic tilapia with chimeric constructs containing the tilapia growth hormone cDNA. In: *Programme, Abstracts and List of Participants of the Third International Marine Biotechnology Conference, Tromsø University, Tromsø, Norway, 7–12 August 1994*. International Advisory Committee of the International Marine Biotechnology Conference 1994, Tromsoe, Norway, p. 70.

Martinez, R., Estrada, M.P., Berlanga, J., Guillin, I., Hernandez, O., Cabrera, E., Pimentel, R., Morales, R., Herrera, F., Morales, A., Pina, J., Abad, Z., Sanchez, V., Melamed, P., Lleonart, R. and de la Fuente, J. (1996) Growth enhancement of transgenic tilapia by ectopic expression of tilapia growth hormone. *Molecular Marine Biology and Biotechnology* 5, 62–70.

Martinez, R., Arenal, A., Estrada, M.P., Herrera, F., Huerta, V., Vazquez, J., Sanchez, T. and de la Fuente, J. (1999) Mendelian transmission, transgene dosage and growth phenotype in transgenic tilapia (*Oreochromis hornorum*) showing ectopic expression of homologous growth hormone. *Aquaculture* 173, 271–283.

Martinez, R., Juncal, J., Zaldivar, C., Arenal, A., Guillen, I., Morera, V., Carrillo, O., Estrada, M., Morales, A. and Estrada, M.P. (2000) Growth efficiency in transgenic tilapia (*Oreochromis* sp.) carrying a single copy of an homologous cDNA growth hormone. *Biochemical and Biophysical Research Communications* 267, 466–472.

Martyniuk, C.J., Perry, G.M.L., Mogahadam, H.K., Ferguson, M.M., and Danzmann, R.G. (2003) The genetic architecture of correlations among growth-related traits and male age at maturation in rainbow trout. *Journal of Fish Biology* 63, 746–764.

Mason, A.J., Pitts, S.L., Nikolics, K., Szonyi, E., Wilcox, J.N., Seeburg, P.H. and Stewart, T.A. (1986) The hypogonadal mouse – reproductive functions restored by gene-therapy. *Science* 234, 1372–1378.

Mason, J.W., Brynildson, O.M. and Degurse, P.E. (1967) Comparative survival of wild and domestic strains of brook trout in streams. *Transactions of the American Fisheries Society* 96, 313–319.

Mason, K.M., Shumway, S.E., Allen, S.K. and Hidu, H. Jr (1988) Induced triploidy in the soft-shelled clam *Mya arenaria*: energetic implications. *Marine Biology* 98, 519–528.

Masood, E. (1999) Compromise sought on 'Terminator'. *Nature* 399, 721.

Matheson, J. (1999) Will transgenic fish be the first ag-biotech food producing animals? *FDA Veterinarian* 14 (3); available at www.fda.gov.cvm.fda/inflores/fdavet/1999/may.htm.

Matsuda, M. and Nagahama, Y. (2002) Positional cloning of the sex-determining region of medaka using a Y congenic strain. In: Shimizu, N., Aoki, T., Hirono, I. and Takashima, F. (eds) *Aquatic Genomics: Steps Toward a Great Future*. Springer-Verlag, New York, pp. 236–245.

Matsui, Y., Zsebo, K. and Hogan, B. (1990) Embryonic expression of a haematopoietic growth factor encoded by the S1 locus and the ligand for c-kit. *Nature* 347, 667–669.

Matsui, Y., Zsebo, K. and Hogan, B. (1992) Derivation of pluripotential embryonic stem cells from murine primordial germ cells in culture. *Cell* 70, 841–847.

Matsuyama, T., Fujiwara, A., Nakayasu, C., Kamaishi, T., Oseko, N., Hirono, I. and Aoki, T. (2007) Gene expression of leucocytes in vaccinated Japanese flounder (*Paralichthys olivaceus*) during the course of experimental infection with *Edwardsiella tarda*. *Fish and Shellfish Immunology* 22, 598–607.

Matthews, D.M. and Laster, L. (1965) Absorption of protein digestion products: a review. *Gut* 6, 411–426.

Matthews, M.A., Poole, W.R., Thompson, C.E., McKillen, J., Ferguson, A., Hindar, K. and Wheelan, K.F. (2000) Incidence of hybridization between Atlantic salmon, *Salmo salar* L., and brown trout, *Salmo trutta* L., in Ireland. *Fisheries Management and Ecology* 7, 337–347.

Matthews, S.G. and Challis J.R. (1995) Levels of pro-opiomelanocortin and prolactin mRNA in the fetal sheep pituitary following hypoxaemia and glucocorticoid treatment in late gestation. *Journal of Endocrinology* 147, 139–146.

Matthews, S.G., Han, X., Lu, F. and Challis, J.R.G. (1994) Developmental changes in the distribution of pro-opiomelanocortin and prolactin mRNA in the pituitary of the ovine fetus and lamb. *Journal of Molecular Endocrinology* 13, 175–185.

Mauras, N., Quarmby, V. and Bloedow, D.C. (1999) Pharmacokinetics of insulin-like growth factor I in hypopituitarism: correlation with binding proteins. *American Journal of Physiology* 277, E579–E584.

May, B. and Grewe, P.M. (1993) Fate of maternal mtDNA following ^{60}Co inactivation of maternal nuclear DNA in unfertilized salmonid eggs. *Genome* 36, 725–730.

May, B. and Krueger, C.C. (1990) Use of allozyme data for population analysis. In: Whitmore, D.H. (ed.) *Electrophoretic and Isoelectric Focusing Techniques in Fisheries Management*. CRC Press, Boca Raton, Florida, pp. 157–172.

May, B., Johnson, K.R. and Wright, J.E. Jr (1989) Sex linkage in salmonids: evidence from a hybridized genome of brook trout and Arctic charr. *Biochemical Genetics* 27, 291–302.

Mayer, I., McLean, E., Kieffer, T.J., Souza, L.M. and Donaldson, E.M. (1994) Antisomatostatin-induced growth acceleration in chinook salmon (*Oncorhynchus tshawytscha*). *Fish Physiology and Biochemistry* 13, 295–300.

Meagher, S.M., Penn, D.J. and Potts, W.K. (2000) Male–male competition magnifies inbreeding depression in wild house mice. *Proceedings of the National Academy of Sciences USA* 97, 3324–3329.

Medina, M., Reperant, J., Dufour, S., Ward, R., Le Belle, N. and Miceli, D. (1994) The distribution of GABA-immunoreactive neurons in the brain of the silver eel (*Anguilla anguilla* L.). *Anatomy and Embryology* 189, 25–39.

Mei, Q., Zou, X.G. and Du, M. (1993) Studies on the early development of nucleocytoplasmic hybrid embryos between mouse and rabbit by nuclear transplantation. *Acta Biologiae Experimentalis Sinica* 26, 389–397.

Meijer, A.H., Verbeek, F.J., Salas-Vidal, E., Corredor-Adámez, M., Bussman, J., van der Sar, A.M., Otto, G.W., Geisler, R. and Spaink, H.P. (2005) Transcriptome profiling of adult zebrafish at the late stage of chronic tuberculosis due to *Mycobacterium marinum* infection. *Molecular Immunology* 42, 1185–1203.

Mekhedov, S., de Ilarduya, O.M. and Ohlrogge, J. (2000) Toward a functional catalog of the plant genome: a survey of genes for lipid biosynthesis. *Plant Physiology* 122, 389–401.

Meksem, K., Leiser, D., Peleman, J., Zabeau, M., Salamini, F. and Gebhardt, C. (1995) A high-resolution map of the vicinity of the R1 locus on chromosome V of potato based on RFLP and AFLP markers. *Molecular and General Genetics* 249, 74–81.

Menotti-Raymond, M.A. and O'Brien, S.J. (1995) Evolutionary conservation of ten microsatellite loci in four species of Felidae. *Journal of Heredity* 86, 319–322.

Mepham, T.B. (ed.) (1996) Ethical analysis of food biotechnologies: an evaluative framework. In: *Food Ethics*. Routledge, London, pp. 101–119.

Merch, J., Shane, J.M. and Jafloten, F. (1975) The effects of methallibure on LH release in castrated male rats challenged with LHRH. *Endocrinology* 97, 493–495.

Merilä, J. and Crnokrak, P. (2001) Comparison of genetic differentiation at marker loci and quantitative traits. *Journal of Evolutionary Biology* 14, 892–903.

Meriwether, F.H. and Shelton, W.L. (1981) Observation on aquarium spawning of estrogen-treated and untreated tilapia. *Proceedings Annual Conference Southeastern Association of Fish and Wildlife Agencies* 34, 81–87.

Merrifield, E.L., Mitchell, S.A., Ubach, J., Boman, H.G., Andreu, D. and Merrifield, R.B. (1995) D-Enantiomers of 15-residue cecropin A-melitin hybrids. *International Journal of Peptide Protein Research* 46, 214–220.

Meuwissen, T. and Arendonk, J. (1992) Potential improvements in rate of genetic gain from marker-assisted selection in dairy cattle breeding schemes. *Journal of Dairy Science* 75, 1651–1659.

Meyuhas, O. (2000) Synthesis of the translational apparatus is regulated at the translational level. *European Journal of Biochemistry* 267, 6321–6330.

Meyuhas, O., Avni, D. and Shama, S. (1996) Translational control of ribosomal protein mRNAs in eukaryotes. In: Hershey, J.W.B., Mathews, M.B. and Sonenberg, N. (eds) *Translational Control*. Cold Spring Harbor Laboratory Press, Cold Spring Harbor, New York, pp. 363–388.

Mezzera, M. and Largiadèr, C.R. (2001) Evidence for selective angling of introduced trout and their hybrids in a stocked brown trout population. *Journal of Fish Biology* 59, 287–301.

Michelmore, R.W., Paran, I. and Kesseli, R.V. (1991) Identification of markers linked to disease resistance genes by bulked segregant analysis: a rapid method to detect markers in specific genomic regions by using segregating populations. *Proceedings of the National Academy of Sciences USA* 88, 9828–9832.

Migalovskii, L.P. (1971) Development of fish eggs and the early period of gametogenesis in the embryos and larvae of the Atlantic salmon under conditions of radioactive contamination of water. Tr. Polyarn Nauh-noissled. *AEC Translations* 36, 52.

Miklas, P.N., Afanador, L. and Kelly, J.D. (1996) Recombination-facilitated RAPD marker-assisted selection for disease resistance in common bean. *Crop Science* 36, 86–90.

Miller, J.C. and Tanksley, S.D. (1990) RFLP analysis of phylogenetic relationships and genetic variation in the genus *Lycopersicon*. *Theoretical and Applied Genetics* 80, 437–448.

Miller, K.M., Kaukinen, K.H., Beacham, T.D. and Withler, R.E. (2001) Geographic heterogeneity in natural selection on an MHC locus in sockeye salmon. *Genetica* 111, 237–257.

Miller, R.B. (1952) Survival of hatchery-reared cutthroat trout in an Alberta stream. *Transactions of the American Fisheries Society* 81, 35–42.

Milstein, S.J., Leipold, H., Sarubbi, D., Leone-Bay, A., Mlynek, G.M., Robinson, J.R., Kasimova, M. and Freire, E. (1998) Partially unfolded proteins efficiently penetrate cell membranes – implications for oral drug delivery. *Journal of Controlled Release* 53, 259–267.

Mires, D. (1969) Mixed culture of tilapia with carp and grey mullet in Ein Hamifratz fish ponds. *Israeli Journal of Aquaculture/Bamidgeh* 21, 25–32.

Mirza, J.A. and Shelton, W.L. (1988) Induction of gynogenesis and sex reversal in silver carp. *Aquaculture* 68, 1–14.

Mitsushima, D., Hei, D.L. and Terasawa, E. (1994) Gamma-aminobutyric acid is an inhibitory neurotransmitter restricting the release of luteinizing hormone-releasing hormone before the onset of puberty. *Proceedings of the National Academy of Sciences USA* 91, 395–399.

Moav, B., Liu, Z., Moav, N., Gross, M.L., Kapuscinski, A.R., Faras, A.J., Guise, K.S. and Hackett, P.B. (1992a) Expression of heterologous genes in transgenic fish. In: Hew, C.L. and Fletcher, G.L. (eds) *Transgenic Fish*. World Science Publishers, Singapore, pp. 120–141.

Moav, B., Liu, Z., Groll, Y. and Hackett, P.B. (1992b) Selection of promotors for gene transfer into fish. *Molecular Marine Biology and Biotechnology* 1, 338–345.

Moav, B., Hinits, Y., Groll, Y. and Rothbard, S. (1995) Inheritance of recombinant carp β-actin/GH cDNA gene in transgenic carp. *Aquaculture* 137, 179–189.

Moav, R. and Wohlfarth, G.W. (1973) Fish breeding in Israel. In: Moav, R. (ed.) *Agricultural Genetics Topics*. John Wiley & Sons, New York.

Moav, R. and Wohlfarth, G. (1974a) Carp breeding in Israel. In: Moav, R. (ed.) *Agricultural Genetics*. John Wiley & Sons, New York.

Moav, R. and Wohlfarth, G. (1974b) Magnification through competition of genetic differences in yield capacity in carp. *Heredity* 33, 181–202.

Moav, R. and Wohlfarth, G. (1976) Two-way selection for growth rate in the common carp, (*Cyprinus carpio* L.). *Genetics* 82, 83–101.

Moav, R., Wohlfarth, G. and Lahman, M. (1964) Genetic improvement of carp. VI. Growth rate of carp imported from Holland, relative to Israeli carp, and some crossbred progeny. *Israeli Journal of Aquaculture/Bamidgeh* 16, 142–149.

Moav, R., Hulata, G. and Wohlfarth, G. (1975) Genetic differences between the Chinese and European races of the common carp. I. Analysis of genotype–environment interactions for growth rate. *Heredity* 34, 323–340.

Moen, T., Agresti, J.J., Cnaani, A., Moses, H., Famula, T.R., Hulata, G., Gall, G.E. and May, B. (2004a) A genome scan of a four-way tilapia cross supports the existence of a quantitative trait locus for cold tolerance on linkage group 23. *Aquaculture Research* 35, 893–904.

Moen, T., Fjalestad, K. T., Munck, H., and Gomez-Raya, L. (2004b) A multistage testing strategy for detection of quantitative trait loci affecting disease resistance in Atlantic salmon. *Genetics* 167, 851–858.

Moffett, I.J.J. and Crozier, W.W. (1991) Genetic marker identification in Atlantic salmon (*Salmo salar* L.) by electrophoretic analysis of low risk tissues. *Journal of Fish Biology* 39, 605–611.

Molina, A., Biemar, F., Muller, F., Iyengar, A., Prunet, P., Maclean, N., Martial, J.A. and Muller, M. (2000) Cloning and expression analysis of an inducible HSP70 gene from tilapia fish. *FEBS Letters* 474, 5–10.

Moller, J., Lauersen, T., Mindeholm, L., Hoelgaard, A., Ovesen, P., Jørgensen, J.O. and Christiansen, J.S. (1994) Serum growth hormone (GH) profiles after nasally administered GH in normal subjects and GH deficient patients. *Clinical Endocrinology (Oxford)* 40, 511–513.

Monaco, P.J., Rasch, E.M. and Balsano, J.S. (1984) Apomictic reproduction in the Amazon molly, *Poecilia formosa* and its triploid hybrids. In: Turner, B.J. (ed.) *Evolutionary Genetics of Fishes*. Plenum Press, New York, pp. 311–327.

Montero, M. and Dufour, S. (1996) Gonadotropin-releasing hormones (GnRH) in fishes: evolutionary data on their structure, localization, regulation and function. *Zoological Studies* 35, 149–160.

Moore, S.S., Sargeant, L.L., King, T.J., Mattick, J.S., Georges, M. and Hetzel, D.J. (1991) The conservation of dinucleotide microsatellites among mammalian genomes allows the use of heterologous PCR primer pairs in closely related species. *Genomics* 10, 654–660.

Moran, C., Prendville, D.J., Quirke, J.F. and Roche, J.F. (1997) Effects of estradiol, zeranolor trenbolone acetate implants on puberty, reproduction and fertility in heifers. *Journal of Reproduction and Fertility* 89, 527–536.

Mori, T. and Devlin, R.H. (1999) Transgene and host GH gene expression in pituitary and nonpituitary tissues of normal and GH transgenic salmon. *Molecular and Cellular Endocrinology* 149, 129–139.

Morita, R.Y. (1980) Biological limits of temperature and pressure. *Origins of Life* 10, 215–222.

Moritz, C., Dowling, T.E. and Brown, W.M. (1987) Evolution of animal mitochrondrial DNA: relevance for population biology and systematics. *Annual Review of Ecology and Systematics* 18, 269–292.

Moriyama, S., Swanson, P., Nishi,M., Takahashi, A., Kawauchi, H., Dickhoff, W.W. and Plisetskaya, E.M. (1994) Development of a homologous radioimmunoassay for coho salmon insulin-like growth factor-I. *General and Comparative Endocrinology* 96, 149–161.

Morizot, D.C. (1990) Use of fish gene maps to predict ancestral chordate geneome structure. In: Ogita, Z. and Market, C.L. (eds) *Isozymes: Structure, Function, and Use in Biology and Medicine.* Liss-Wiley, New York, pp. 207–234.

Morizot, D.C. (1994) Reconstructing the gene map of the vertebrate ancestor. *Animal Biotechnology* 5, 113–122.

Morizot, D.C. and Siciliano, M.J. (1979) Polymorphisms, linkage and mapping of four enzyme loci in the fish genus *Xiphophorus* (Poeciliidae). *Genetics* 93, 947–960.

Morizot, D.C. and Siciliano, J.J. (1984) Gene mapping in fishes and other vertebrates. In: Turner, B.J. (ed.) *Evolutionary Genetics of Fishes.* Plenum Press, New York, pp. 173–233.

Morizot, D.C., Wright, D.A. and Siciliano, M.J. (1977) Three linked enzyme loci in fishes: implications in the evolution of vertebrate chromosomes. *Genetics* 86, 645–656.

Morizot, D.C., Schultz, R.J. and Wells, R.S. (1989) Assignment of six enzyme loci to multipoint linkage groups in fishes of the genus *Poeciliopsis* (Poeciliidae): designation of linkage groups III–V. *Biochemical Genetics* 28, 83–95.

Morizot, D.C., Slaugenhaupt, S.A., Kallman, K.D. and Chakravarti, A. (1991) Genetic linkage map of fishes of the genus *Xiphophorus* (Teleostei: Poeciliidae). *Genetics* 127, 399–410.

Morizot, D.C., Harless, J., Nairn, R.S., Kallman, K.D. and Walter, R.B. (1993) Linkage maps of non-salmonid fishes. In: O'Brien, S.J. (ed.) *Genetic Maps: Locus Maps of Complex Genomes,* 5th edn. Cold Spring Harbor Laboratory Press, Cold Spring Harbor, New York.

Morizot, D., Schmidt, M. and Carmichael, G. (1994) Joint segregation of allozymes in catfish genetic crosses: designation of *Ictalurus punctatus* linkage group I. *Transactions of the American Fisheries Society* 123, 22–27.

Morizot, D.C., McEntire, B.B., Coletta, L.D., Kazianis, S., Schartl, M. and Nairn, R.S. (1998) Mapping of tyrosine kinase gene family members in a *Xiphophorus* melanoma model. *Molecular Carcinogenesis* 22, 150–157.

Morrison, R.N., Cooper, G.A., Koop, B.F., Rise, M.L., Bridle, A.R., Adams, M.B., Nowak, B.F. (2006) Transcriptome profiling the gills of amoebic gill disease (AGD)-affected Atlantic salmon (*Salmo salar* L.): a role for tumor suppressor p53 in AGD-pathogenesis? *Physiological Genomics* 26, 15–34.

Mosier, H.D. Jr, Jansons, R.A. and Dearden, L.C. (1985) Increased secretion of growth hormone in rats undergoing catch-up growth after fasting. *Growth* 49, 346–353.

Moyle, P.B. (1969) Comparative behaviour of young brook trout of domestic and wild origin. *Progressive Fish-Culturist* 31, 51–59.

Mueller-Schmid, A., Rinder, H., Lottspeich, F., Gertzen, E.M.M. and Hoffmann, W. (1992) Ependymins from the cerebrospinal fluid of salmonid fish: gene structure and molecular characterization. *Gene* 118, 189–196.

Muir, W.M. and Howard, R.D. (1999) Possible ecological risks of transgenic organism release when transgenes affect mating success: sexual selection and the Trojan gene hypothesis. *Proceedings of the National Academy of Sciences USA* 96, 13853–13856.

Muir, W.M. and Howard, R.D. (2001) Fitness components and ecological risk of transgenic release: a model using Japanese medaka (*Oryzias latipes*). *American Naturalist* 158, 1–16.

Muller, J., Skakkebaek, N.E., Jacobsen, B.B., Keller, E., Heinrich, U., Hartmann, K., Hokken-Koelega, A.C. and Delemarre van de Waal, H.A. (1999) Norditropin SimpleXx: a liquid human growth hormone formulation, a pen system and an auto-insertion device. *Hormone Research* 51 (Suppl. 3), 109–112.

Müller-Belecke, A. and Hörstgen-Schwark, G. (2000) Performance testing of clonal *Oreochromis niloticus* lines. *Aquaculture* 184, 67–76.

Mulrenin, E.M., Witkin, J.W. and Silverman, A.J. (1999) Embryonic development of the gonadotropin-releasing hormone (GnRH) system in the chick: a spatiotemporal analysis of GnRH neuronal generation, site of origin, and migration. *Endocrinology* 140, 422–433.

Munoz-Cueto, J.A., Gonzalez, D., Madigou, T., Zmora, N., Anglade, I., Saligaut, D., Mananos, E., Zanuy, S., Zohar, Y., Elizur, A. and Kah, O. (2000) The GnRH systems of perciform fish: an *in situ* hybridization and immunohistochemical study in the European sea bass, *Dicentrarchus labrax.* In: *Proceedings of the 4th International Symposium on Fish Endocrinology, 31 July–3 August 2000, Seattle, Washington.*

Muntzung, A. (1936) The evolutionary significance of autoploidy. *Hereditas* 21, 263–278.

Murakami, S., Seki, T., Wakabayashi, K., and Arai, Y. (1991) The ontogeny of luteinizing hormone-releasing hormone (LHRH) producing neurons in the chick-embryo – possible evidence for migrating LHRH neurons from the olfactory epithelium expressing a highly polysialylated neural cell-adhesion molecule. *Neuroscience Research* 12, 421–431.

Murakami, S., Seki, T., Rutishauser, U. and Arai, Y. (1998) LHRH neurons migrate into the trigeminal nerve when the developing olfactory nerve fibers are physically interrupted in chick embryos. *General and Comparative Endocrinology* 112, 312–321.

Murdoch, M.H. and Hebert, P.D.N. (1994) Mitochondrial DNA diversity of brown bullhead from contaminated and relatively pristine sites in the Great Lakes. *Environmental Toxicology and Chemistry* 13, 1281–1289.

Murphy, T.M., Cotter, D. and Wilkins, N.P. (2000) Histological studies on the gonads of triploid and diploid Atlantic salmon (*Salmo salar* L.). In: Norberg, B., Kjesbu, O.S., Taranger, G.L., Andersson, E. and Stefansson, S.O. (eds) *Reproductive Physiology of Fish 2000. 6th International Symposium on the Reproductive Physiology of Fish, Bergen, Norway, 4–9 July 1999*. Department of Fisheries and Marine Biology, University of Bergen, Bergen, Norway.

Murray, D.R. (1991) *Advanced Methods in Plant Breeding and Biotechnology*. CAB International, Wallingford, UK.

Myers, D.A., Myers, T.R., Grober, M.S. and Nathanielsz, P.W. (1993) Levels of corticotropin-releasing hormone messenger ribonucleic acid (mRNA) in the hypothalamic paraventricular nucleus and proopiomelanocortin mRNA in the anterior pituitary during late gestation in fetal sheep. *Endocrinology* 132, 2109–2116.

Myers, J.M. (1985) An assessment of spawning methodologies and induction of tetraploidy in two oreochromid species. MSc thesis, University of Washington, Seattle, Washington.

Myers, J.M. (1986) Tetraploid induction in *Oreochromis* spp. *Aquaculture* 57, 281–286.

Myers, J.M., Hershberger, W.K. and Iwamoto, R.N. (1986) The induction of tetraploidy in salmonids. *Journal of the World Aquaculture Society* 17, 1–7.

Myers, J.M., Penman, D.J., Basavaraju, Y., Powell, S.F., Baoprasertkul, P., Rana, K.J., Bromage, N. and McAndrew, B.J. (1995) Induction of diploid androgenetic and mitotic gynogenetic Nile tilapia (*Oreochromis niloticus*). *Theoretical and Applied Genetics* 90, 205–210.

Myers, J.M., Kope, R.G., Bryant, G.J., Teel, D., Lierheimer, L.J., Wainwright, T.C., Grand, W.S., Waknitz, F.W., Neely, K., Linsey, S.T. and Waples, R.S. (1998) *Status Review of Chinook Salmon from Washington, Idaho, Oregon, and California*. NOAA Technical Memorandum NMFS-NWFSC-35. NOAA, US Department of Commerce, Northwest Fisheries Science Center, National Marine Fisheries Service, Seattle, Washington.

Mylonas, C.C. and Zohar, Y. (2001) Endocrine regulation and artificial induction of oocyte maturation and spermiation in basses of the genus *Morone*. *Aquaculture* 202, 205–220.

Naevdal, G. (1983) Genetic factors in connection with age at maturation. *Aquaculture* 33, 97–106.

Nagler, L.J., Parsons. J.E. and Cloud, J.G. (2000) Single pair mating indicates maternal effects on embryo survival in rainbow trout, *Oncorhynchus mykiss*. *Aquaculture* 184, 177–183.

Nagy, A., Rajki, K., Bakos, J. and Csanyi, V. (1979) Genetic analysis in carp (*Cyprinus carpio*) using gynogenesis. *Heredity* 43, 35–40.

Nagy, A., Csanyi, V., Bakos, J. and Bercsenyl, M. (1984) Utilization of gynogenesis and sex-reversal in commercial carp breeding: growth of the first gynogenetic hybrids. *Aquacultura Hungarica (Szarvas)* 4, 7–16.

Nagy, A., Rossant, J., Nagy, R., Abramow-Newerly, W. and Roder, I.C. (1993) Derivation of completely cell culture-derived mice from early-passage embryonic stem cells. *Proceedings of the National Academy of Sciences USA* 90, 8424–8428.

Nakamura, H. and Kasahara, S. (1955) A study on the phenomenon of the Tobi Koi or shoot carp. I. On the earliest stage at which the shoot carp appear. *Bulletin of the Japanese Society of Scientific Fisheries* 21, 73–76 (in Japanese with English summary).

Nakamura, H. and Kasahara, S. (1956) A study on the phenomenon of the Tobi Koi or shoot carp. II. On the effect of particle size and quantity of food. *Bulletin of the Japanese Society of Scientific Fisheries* 21, 1022–1024 (in Japanese with English summary).

Nakamura, H. and Kasahara, S. (1957) A study on the phenomenon of the Tobi Koi or shoot carp. III. On the results of culturing the modal group and the growth of carp fry reared individually. *Bulletin of the Japanese Society of Scientific Fisheries* 22, 674–678 (in Japanese with English summary).

Nakamura, H. and Kasahara, S. (1961) A study of the phenomenon of the Tobi Koi or shoot carp. IV. Effects of adding a small number of larger individuals to the experimental batches of carp fry and of culture density upon the occurrence of shoot carp. *Bulletin of the Japanese Society of Scientific Fisheries* 22, 958–962 (in Japanese with English summary).

Nakamura, M. (1981) Feminization of masu salmon *Oncorhynchus masou* by administration of estradiol-17β. *Bulletin of the Japanese Society of Scientific Fisheries* 47, 1529.

Nakamura, M., Nahaghama, Y., Iwahashi, M. and Kojima, M. (1987) Ovarian structure and plasma steroid hormones of triploid female rainbow trout. *Bulletin of the Japanese Society of Scientific Fisheries* 53, 105.

Nakamura, S., Imai, N., Ohya, S. and Kobayashi, H. (1989) Oxygen uptake rate and hematological characteristics in triploid amago salmon, *Oncorhynchus masou macrostomus*. In: *No. 22*. Mimeo, Faculty of the Agricultural Kinki University/Kinkidai No Kiyo, pp. 31–38.

Nakanishi, S., Inoue, A., Kita, T., Nakamura, M., Chang, A.C.Y., Cohen, S.N. and Numa, S. (1979) Nucleic acid sequence of cloned cDNA for bovine corticotropin-beta-lipotropin precursor. *Nature* 278, 423–427.

Nakayama, I., Biaga, C.A., Koide, N. and Devlin, R.H. (1999) Identification of a sex-linked GH pseudogene in one of two species of Japanese salmon (*Oncorhynchus masou* and *O. rhodorus*). *Aquaculture* 173, 65–72.

Na-Nakorn, U. and Legrand, E. (1992) Induction of triploidy in *Puntius gonionotus* (Bleeker) by cold shock. *Kasetsart University Fisheries Research Bulletin* 18, 10.

Nancarrow, C.D., Marshall, J.T.A., Clarkson, J.L., Murray, J.D., Millard, R.M., Shanahan, C.M., Wynn, P.C. and Ward, K.A. (1991) Expression and physiology of performance regulating genes in transgenic sheep. *Journal of Reproduction and Fertility Supplement* 43, 277–291.

Nankervis, L., Matthews, S.J. and Appleford, P. (2000) Effect of dietary non-protein energy source on growth, nutrient retention and circulating insulin-like growth factor I and triiodothyronine levels in juvenile barramundi, *Lates calcarifer*. *Aquaculture* 191, 323–335.

Narine, K.M., Kallman, K.D., Walter, R.B. and Morizot, D.C. (1992) Linkage of two enzyme loci in the fish genus *Poecilia* (Teleostei: Poeciliidae). *Cytogenetics and Cell Genetics* 61, 75–77.

Naruse, K., Fukamachi, S., Mitani, H., Kondo, M., Matsuoka, T., Kondo, S., Hanamura, N., Morita, Y., Hasegawa, K., Nishigaki, R., Shimada, A., Wada, H., Kusakabe, T., Suzuki, N., Kinoshita, M., Kanamori, A. and Shima, A. (2000) A detailed linkage map of medaka, *Oryzias latipes*: comparative genomics and genome evolution. *Genetics* 154, 1773–1784.

Nasevicius, A. and Ekker, S.C. (2000) Effective targeted gene 'knockdown' in zebrafish. *Nature Genetics* 26, 216–220.

Navarro, A., Zamorano, M.J., Hildebrandt, S., Gines, R., Aguilera, C., and Afonso, J.M. (2009a) Estimates of heritabilities and genetic correlations for growth and carcass traits in gilthead seabream (*Sparus auratus* L.), under industrial conditions. In: *Book of Abstracts of 10th International Symposium on Genetics in Aquaculture (ISGA X) 'Roles of Aquaculture Genetics in Addressing Global Food Crisis', 22–26 June 2009, Bangkok, Thailand*. Faculty of Fisheries, Kasetsart University, Bangkok, p. 115.

Navarro, A., Zamorano, M.J., Hildebrandt, S., Gines, R., Aguilera, C., and Afonso, J.M. (2009b) Estimates of heritabilities and genetic correlations for body composition traits and G × E interactions, in gilthead seabream (*Sparus auratus* L.). In: *Book of Abstracts of 10th International Symposium on Genetics in Aquaculture (ISGA X) 'Roles of Aquaculture Genetics in Addressing Global Food Crisis', 22–26 June 2009, Bangkok, Thailand*. Faculty of Fisheries, Kasetsart University, Bangkok, p. 136.

Naylor, R.L., Goldburg, R.J., Primavera, J.H., Kautsky, N., Beveridge, M.C.M., Clay, J., Folke, C., Lubchenco, J., Mooney, J. and Troell, M. (2000) Effect of aquaculture on world fish supplies. *Nature* 405, 1017–1024.

Neira, R., Diaz, N.F., Gall, G.A.E., Gallardo, J.A., Lhorente, J.P. and Alert, A. (2006) Genetic improvement in coho salmon (*Oncorhynchus kisutch*). II: Selection response for early spawning date. *Aquaculture* 257, 1–8.

Neira, R., Garcia, X., Lhorente, J. P., Alvarado, C., Miranda, R., Fernandez, A. and Filip, M. (2009) Evaluation of cross breeding with four stocks in Nile tilapia (*Oreochromis niloticus*), results for growth and carcass quality traits from a diallel cross. In: *Book of Abstracts of 10th International Symposium on Genetics in Aquaculture (ISGA X) 'Roles of Aquaculture Genetics in Addressing Global Food Crisis', 22–26 June 2009, Bangkok, Thailand*. Faculty of Fisheries, Kasetsart University, Thailand, p. 99.

Nell, J.A. (2002) Farming triploid oysters. *Aquaculture* 210, 69–88.

Nell, J.A., McMahon, G.A. and Hand, N.S.W. (1998) *Tetraploidy Induction in Sydney Rock Oysters*. Fisheries Final Report Series No. 9. New South Wales Fisheries, Cronulla, Australia.

Nell, J.A., Smith, I.R. and Sheridan, A.K. (1999) Third generation evaluation of Sydney rock oyster *Saccostrea commercialis* (Iredale and Roughley) breeding lines. *Aquaculture* 170, 195–203.

Nelson, K. and Hedgecock, D. (1980) Enzyme polymorphism and adaptive strategy in the decapod Crustacea. *American Naturalist* 116, 238–280.

Newkirk, G.F. (1980) Review of the genetics and the potential for selective breeding of commercially important bivalves. *Aquaculture* 19, 209–228.

Newman, D. and Tallmon, D.A. (2001) Experimental evidence for beneficial fitness effects of gene flow in recently isolated populations. *Conservation Biology* 15, 1054–1063.

Newport, J.H. and Kirschner, M. (1982) A major developmental transition on early embryos: I. Characterization and timing of cellular changes at midblastula stage. *Cell* 30, 675–686.

Newton, J.R., Smith-Keune, C. and Jerry, D.R. (2010) Thermal tolerance varies in tropical and sub-tropical populations of barramundi (*Lates calcarifer*) consistent with local adaptation. *Aquaculture* 308, S128–S132.

Nichols, K.M., Bartholomew, J. and Thorgaard, G.H. (2003) Mapping multiple genetic loci associated with *Ceratomyxa shasta* resistance in *Oncorhynchus mykiss*. *Diseases of Aquatic Organisms* 56, 145–154.

Nichols, K.M., Wheeler, P.A. and Thorgaard, G.H. (2004) Quantitative trait loci analyses for meristic traits in *Oncorhynchus mykiss*. *Environmental Biology of Fishes* 69, 317–331.

Nickelsen, T.E., Solazzi, M.F. and Johnson, S.L. (1986) Use of hatchery coho salmon *Oncorhynchus kisutch* presmolts to rebuild wild populations in Oregon USA coastal streams. *Canadian Journal of Fisheries and Aquatic Sciences* 43, 2443–2449.

Nicoll, C.S., Steiny, S.S., King, D.S., Nishioka, R.S., Mayer, G.L., Eberhardt, N.L., Baxter, J.D., Yamanaka, M.K., Miller, J.A., Seilhamer, J.J. *et al.* (1987) The primary structure of coho salmon growth hormone and its cDNA. *General and Comparative Endocrinology* 68, 387–399.

Nielsen, C., Holdensgaard, G., Petersen, H.C., Bjoernsson, B.T. and Madsen, S.S. (2001) Genetic differences in physiology, growth hormone levels and migratory behaviour of Atlantic salmon smolts. *Journal of Fish Biology* 59, 28–44.

Nielsen, H.M., Odegard, J., Olesen, I., Gjerde, B., Ardo, L., Jeney, G. and Jeney, Z. (2009) Genetic parameters and hetersois for growth traits and pond survival in four Hungarian carp strains. In: *Book of Abstracts of 10th International Symposium on Genetics in Aquaculture (ISGA X) 'Roles of Aquaculture Genetics in Addressing Global Food Crisis', 22–26 June 2009, Bangkok, Thailand.* Faculty of Fisheries, Kasetsart University, Bangkok, p. 151.

Nielsen, H.M., Ødegård, J., Olesen, I., Gjerde, B., Ardo, L., Jeney, G. and Jeney, Z. (2010) Genetic analysis of common carp (*Cyprinus carpio*) strains: I: Genetic parameters and heterosis for growth traits and survival. *Aquaculture* 304, 14–21.

NIH (1990) Bovine somatotropin. NIH Technology Assessment Statement Online 1990, Dec 5–7. http://consensus.nih.gov/1990/1990BovineSomatotropinta007html.htm (accessed April 2011).

Niiler, E. (1999) Terminator technology temporarily terminated. *Nature Biotechnology* 17, 1054.

Nikolaevskaia, V.R. (1989) Characteristics of proteolysis in the gastrointestinal tract in the early postnatal period. *Biulleten' eksperimental'noi biologii i meditsiny* 107, 133–135.

Nindl, B.C., Hymer, W.C., Deaver, D.R. and Kraemer, W.J. (2001) Growth hormone pulsatility profile characteristics following acute heavy resistance exercise. *Journal of Applied Physiology* 91, 163–172.

Niu, P.-D., Perez-Sanchez, J. and Le Bail, P.-Y. (1993) Development of a protein binding assay for teleost insulin-like growth factor (IGF)-like: relationship between growth hormone (GH) and IGF-like in the blood of rainbow trout (*Oncorhynchus mykiss*). *Fish Physiology and Biochemistry* 11, 381–391.

Nixon, S.E. and Mawer, G.E. (1970a) The digestion and absorption of protein in man. 1. The site of absorption. *British Journal of Nutrition* 24, 227–240.

Nixon, S.E. and Mawer, G.E. (1970b) The digestion and absorption of protein in man. 2. The form in which digested protein is absorbed. *British Journal of Nutrition* 24, 241–258.

Nomura, T., Arai, K., Hayashi, T. and Suzuki, R. (1998) Effect of temperature on sex ratios of normal and gynogenetic diploid loach. *Fisheries Science, Tokyo* 64, 753–758.

Nordlee, J.A., Taylor, S.L., Townsend, J.A., Thomas, L.A. and Bush, R.K. (1996) Identification of a Brazil-nut allergen in transgenic soybeans. *New England Journal of Medicine* 334, 688–692.

Norgren, K.G., Dunham, R.A., Smitherman, R.O. and Reeves, W.C. (1986) Biochemical genetics of largemouth bass populations in Alabama. *Proceedings of the Annual Conference Southeastern Association of Fish and Wildlife Agencies* 40, 194–205.

Norgren, R.B. and Brackenbury, R. (1993) Cell-adhesion molecules and the migration of LHRH neurons during development. *Developmental Biology* 160, 377–387.

Norgren, R.B. and Lehman, M.N. (1991) Neurons that migrate from the olfactory epithelium in the chick express luteinizing-hormone-releasing hormone. *Endocrinology* 128, 1676–1678.

Normand, J., Ernande, B., Haure, J., McCombie, H. and Boudry, P. (2009) Reproductive effort and growth in *Crassostrea gigas*: comparison of young diploid and triploid oysters issued from natural crosses or chemical induction. *Aquatic Biology* 7, 229–241.

Norris, A.T., Bradley, D.G. and Cunningham, E.P. (1999) Microsatellite genetic variation between and within farmed and wild Atlantic salmon (*Salmo salar*) populations. *Aquaculture* 180, 247–264.

Northcutt, R.G. and Muske, L.E. (1994) Multiple embryonic origins of gonadotropin-releasing hormone (GnRH) immunoreactive neurons. *Developmental Brain Research* 78, 279–290.

Noy, R., Lavie, B. and Nevo, E. (1987) The niche-width variation hypothesis revisited: genetic diversity in the marine gastropods *Littorina punctata* (Gmelin) and *L. neritoides* (L.). *Journal of Experimental Marine Biology and Ecology* 109, 109–116.

Nukwan, S. (1987) Preliminary study on genetic parameters of Big head carp *Aristicthys nobilis* (Richardson). MSc thesis, Kasetsart University, Bangkok, Thailand.

Nwadukwe, F.O. (1995) Hatchery propagation of five hybrid groups by artifical hybridization of *Clarius gariepinus* (B) and *Heterobranchus longifilis* (Val.) (Clariidae) using dry, powdered carp pituitary hormone. *Journal of Aquaculture in the Tropics* 10, 1–11.

Obi, A. and Shelton, W.L. (1983) Androgen and estrogen sex reversal in *Tilapia hornorum*. In: Fishelson, L. and Yaron, Z. (compilers) *Proceedings of the International Symposium on Tilapia in Aquaculture*. Tel Aviv University, Tel Aviv, Israel, p. 165.

O'Brien, S. (1993) *Genetic Maps: Locus Maps of Complex Genomes*, 6th edn. Cold Spring Harbor Laboratory Press, Cold Spring Harbor, New York.

Odegard, J., Olesen, I., Gjerde, B. and Klemetsdal, G. (2007) Positive genetic correlation between resistance to bacterial (furunculosis) and viral (infectious salmon anaemia) diseases in farmed Atlantic salmon (*Salmo salar*). *Aquaculture* 271, 173–177.

Ødegård, J., Olesen, I., Dixon, P., Jeney, Z., Nielsen, H.-M., Way, K., Joiner, C., Jeney, G., Ardó, L., Rónyai, A. and Gjerde, B. (2010) Genetic analysis of common carp (*Cyprinus carpio*) strains. II: Resistance to koi herpesvirus and *Aeromonas hydrophila* and their relationship with pond survival. *Aquaculture* 304, 7–13.

OECD (1993) *Safety Evaluation of Foods Produced by Modern Biotechnology: Concepts and Principles*. Organisation for Economic Co-operation and Development, Paris,

Ohno, S. (1970) *Evolution by Gene Duplication*. Springer-Verlag, New York.

Ojanguren, A.F., Reyes-Gavilan, F.G. and Brana, F. (1996) Effects of egg size on offspring development and fitness in brown trout, *Salmo trutta* L. *Aquaculture* 147, 9–20.

Oka, Y., Rozek, L.M. and Czech, M.P. (1985) Direct demonstration of rapid insulin-like growth factor II receptor internalization and recycling in rat adipocytes. Insulin stimulates ^{125}I-insulin-like growth factor II degradation by modulating the IGF-II receptor recycling process. *Journal of Biological Chemistry* 260, 9435–9442.

Okada, H., Matumoto, H. and Yamazaki, F. (1979) Functional masculinization of genetic females in rainbow trout. *Bulletin of the Japanese Society of Scientific Fisheries* 45, 413–419.

Okamoto, H., Segawa, H. and Higashijima, S. (2002) Toward genetic dissection of motor neuron differentiation. In: Shimizu, N., Aoki, T., Hirono, I. and Takashima, F. (eds) *Aquatic Genomics: Steps Toward a Great Future*. Springer-Verlag, New York, pp. 139–152.

Okamoto, N., Tayaman, T., Kawanobe, M., Fujiki, N., Yasuda, Y. and Sano, T. (1993) Resistance of a rainbow trout strain to infectious pancreatic necrosis. *Aquaculture* 117, 71–76.

O'Keefe, R.A. and Benfey, T.J. (1999) The growth and food consumption of diploid and triploid brook trout (*Salvelinus fontinalis*) monitored by radiography. In: *Fish Performance Studies*. Department of Fisheries and Oceans, Vancouver, British Columbia, Canada and Towson University, Baltimore, Maryland, pp. 13–18.

O'Keefe, R., Stillwell, E. and Benfey, T.J. (2000) The effect of transportation and handling on the hematology of diploid and triploid Atlantic salmon (*Salmo salar*). *Bulletin of the Aquaculture Association of Canada* 4, 25–27.

Okubo, K., Yoshiura, Y., Amano, M., Suetake, H. and Aida, K. (2002) The GnRH system in teleosts. In: Shimizu, N., Aoki, T., Hirono, I. and Takashima, F. (eds) *Aquatic Genomics: Steps Toward a Great Future*. Springer-Verlag, New York, pp. 244–262.

Okutsu, T., Suzuki, K., Takeuchi, Y., Takeuchi, T. and Yoshizaki, G. (2006a) Testicular germ cell can colonize sexually undifferentiated embryonic gonad and produce functional egg in fish. *Proceedings of the National Academy of Sciences USA* 103, 2725–2729.

Okutsu, T., Yano, A., Nagasawa, K., Shikina, S., Kobayashi, T., Takeuchi, T. and Yoshizaki, G. (2006b) Manipulation of fish germ-cell: visualization, cryopreservation and transplantation. *Journal of Reproduction and Development* 52, 685–693.

Okutsu, T., Shikina, S., Kanno, M., Takeuchi, Y. and Yoshizaki, G. (2007) Production of trout offspring from triploid salmon parents. *Science* 317, 1517.

Okuzawa, K., Amano, M., Kobayashi, M., Aida, K., Hanyu, I., Hasegawa, Y. and Miyamoto, K. (1990) Differences in salmon GnRH and chicken GnRH-II contents in discrete brain areas of male and female rainbow trout according to age and stage of maturity. *General and Comparative Endocrinology* 80, 116–126.

Okuzawa, K., Granneman, J., Bogerd, J., Goos, H.J.T., Zohar, Y. and Kagawa, H. (1997) Distinct expression of GnRH genes in the red seabream brain. *Fish Physiology and Biochemstry* 17, 71–79.

Okwoche, V.O. and Lovell, R.T. (1997) Cool weather feeding influences responses of channel catfish to *Edwardsiella ictaluri* challenge. *Journal of Aquatic Animal Health* 9, 163–171.

Olesen, I., Shezhad, A., Odegard, J., Kongsro, J., Olsen, R.S. and Poppe, T. (2009) Prevalence and genetic parameters of pericarditis and cardiac abnormalities in farmed Atlantic salmon. In: *Book of Abstracts of 10th International Symposium on Genetics in Aquaculture (ISGA X) 'Roles of Aquaculture Genetics in Addressing Global Food Crisis', 22–26 June 2009, Bangkok, Thailand.* Faculty of Fisheries, Kasetsart University, Bangkok, p. 43.

Oliva-Teles, A. and Kaushik, S.J. (1987) Nitrogen and energy metabolism during the early ontogeny of diploid and triploid rainbow trout (*Salmo gairdneri* R.). *Comparative Biochemistry and Physiology, Part A: Comparative Physiology* 87A, 157–160.

Oliva-Teles, A. and Kaushik, S.J. (1990a) Growth and nutrient utilization by 0+ and 1+ triploid rainbow trout, *Oncorhynchus mykiss. Journal of Fish Biology* 37, 125–133.

Oliva-Teles, A. and Kaushik, S.J. (1990b) Effect of temperature on utilization of endogenous energy reserves during embryonic development of diploid and triploid rainbow trout (*Salmo gairdneri* R.). *Aquaculture* 84, 373–382.

Oliva-Teles, A., Kaushik, S.J., Billard, R. and de Pauw, N. (1989) The effect of the protein/energy ratio of the diets on the nitrogen and energy balance in rainbow trout (*Salmo gairdneri*). In: *Aquaculture Europe '89. Short Communications and Abstracts of Review Papers, Films/Slideshows and Poster Papers, Presented at the International Aquaculture Conference held in Bordeaux, France, 2–4 October 1989.* Special Publication 10. European Aquaculture Society, Bredene, Belgium, pp. 195–196.

Oliver, M.J., Quisenberry, J.E., Trolinder, N.L.G. and Keim, D.L. (1998) Control of Plant Gene Expression. US Patent No. 5,723,765.

Olson, E., Arnold, H., Rigby, W. and Wold, B. (1996) Know your neighbors: three phenotypes in null mutants of the myogenic bHLH gene MRF4. *Cell* 85, 1–4.

Olson, K.R. (1992) Bood and extracellular fluid volume regulation: role of the renin–angiotensin system, kallikrein–kinin system, and atrial natriuretic peptides. In: Hoar, W.S., Randell, D.J. and Farrell, A.P. (eds) *Fish Physiology*, Vol. XII, Part B, *The Cardiovasacular System.* Academic Press, New York, pp. 136–154.

O'Malley, K.G., Sakamato, T., Danzmann, R.G. and Ferguson, M.M. (2003) Quantitative trait loci for spawning date and body weight in rainbow trout: testing for conserved effects across ancestrally duplicated chromosomes. *Journal of Heredity* 94, 273–284.

Ordon, F., Schiemann, A., Pellio, B., Dauck, V., Bauer, E., Streng, S., Friedt, W. and Graner, A. (1999) Application of molecular markers in breeding for resistance to the barley yellow mosaic virus complex. *Zeitschrift für Pflanzenkrankheiten und Pflanzenschutz* 106, 256–264.

Ostenfeld, T.H., Devlin, R.H. and McLean, E. (1998) Transgenesis changes body and head shape in Pacific salmon. *Journal of Fish Biology* 52, 850–854.

OSTP (Office of Science and Technology Policy) (1985) Coordinated framework for the regulation of biotechnology: establishment of the Biotechnology Science Coordinating Committee. *Federal Register* 50, 47174–47195.

OSTP (Office of Science and Technology Policy) (1986) Coordinated framework for the regulation of biotechnology. *Federal Register* 51, 23301–23350.

Otsen, M., den Bieman, M., Kuiper, M.T., Pravenec, M., Kren, V., Kurtz, T.W., Jacob, H.J., Lankhorst, A. and van Zutphen, B.F. (1996) Use of AFLP markers for gene mapping and QTL detection in the rat. *Genomics* 37, 289–294.

Owen, J.B. and Axford, R.F.E. (1991) *Breeding for Disease Resistance in Farm Animals.* CAB International, Wallingford, UK.

Ozaki, A., Sakamoto, T., Khoo, S., Nakamura, K., Coimbra, M.R., Akutsu, T. and Okamoto, N. (2001) Quantitative trait loci (QTLs) associated with resistance/susceptibility to infectious pancreatic necrosis virus (IPNV) in rainbow trout (*Oncorhynchus mykiss*). *Molecular Genetics and Genomics* 265, 23–31.

Ozato, K., Kondoh, H., Inohara, H., Iwamatsu, T., Wakamatsu, Y. and Okada, T.S. (1986) Production of transgenic fish: introduction and expression of chicken δ-crystallin gene in medaka embryos. *Cell Differentiation and Development* 19, 237–244.

Pabo, C.O., Peisach, E. and Grant, R.A. (2001) Design and selection of novel Cys$_2$His$_2$ zinc finger proteins. *Annual Review of Biochemistry* 70, 313–340.

Padi, J.N. (1995) Response and correlated responses to four generations of selection for increased body weight in the Kansas strain channel catfish, *Ictalurus punctatus*, grown in earthen ponds. MSc thesis, Auburn University, Auburn, Alabama.

Paget, G.E., Walpole, A.L. and Richardson, D.N. (1961) Non-steroidal inhibitors of pituitary gonadotropic function. *Nature* 192, 1191–1192.

Palada-de Vera, M.S. and Eknath, A.E. (1993) Predictability of individual growth rates in tilapia. *Aquaculture* 111, 147–158.

Palmiter, R.D. and Brinster, R.L. (1986) Germ-line transformation of mice. *Annual Review of Genetics* 20, 465–499.

Palmiter, R.D., Brinster, R.L., Hammer, R.E., Trumbauer, M.E., Rosenfeld, M.G., Birnberg, N.C. and Evans, R.M. (1982) Dramatic growth of mice that develop from eggs microinjected with metallothionein–human growth hormone fusion genes. *Nature* 300, 611–615.

Palmiter, R.D., Norstedt, G., Gelinas, R.E., Hammer, R.E. and Brinster, R.L. (1983) Metallothionein–human GH fusion genes stimulate growth of mice. *Science* 222, 809–814.

Palti, Y., Li, J.J. and Thorgaard, G.H. (1997) Improved efficiency of heat and pressure shocks for producing gynogenetic rainbow trout. *Progressive Fish-Culturist* 59, 1–13.

Palti, Y., Parsons, J.E. and Thorgaard, G.H. (1999) Identification of candidate DNA markers associated with IHN virus resistance in backcrosses of rainbow and cutthroat trout. *Aquaculture* 173, 81–94.

Palti, Y., Nichols, K.M., Waller, K.I., Parsons, J.E. and Thorgaard, G.H. (2001) Association between DNA polymorphisms tightly linked to MHC class II genes and IHN virus resistance in backcrosses of rainbow and cutthroat trout. *Aquaculture* 194, 283–289.

Palti, Y., Shirak, A., Cnaani, A., Hulata, G., Avtalion, R.R. and Ron, M. (2002) Detection of genes with deleterious alleles in an inbred line of tilapia (*Oreochromis aureus*). *Aquaculture* 206, 151–164.

Pandey, S. and Leatherland, J.F. (1970) Comparison of the effects of methallibure and thiourea on the testes, thyroid and adenohypophysis of the adult and juvenile guppy *Poecilia reticulata* (Peters). *Canadian Journal of Zoology* 48, 445–450.

Pandian, T.J. and Varadaraj, K. (1988a) Sterile female triploidy in *Oreochromis mossambicus*. In: *Proceedings of Aquaculture International Congress and Exposition, Vancouver Trade and Convention Centre, Vancouver, British Columbia, Canada, 6–9 September 1988*, p. 57.

Pandian, T.J. and Varadaraj, K. (1988b) Techniques for producing all-male and all-triploid *Oreochromis mossambicus*. In: Pullin, R.S.V., Bhukasawan, T., Tonguthai, K. and Maclean, J.L. (eds) *Proceedings of the Second International Symposium on Tilapia in Aquaculture, Bangkok, Thailand, 16–20 March 1987*. ICLARM Conference Proceedings No. 15. Department of Fisheries Thailand and ICLARM, Manila, pp. 243–249.

Parhar, I.S. and Sakuma, Y. (1997) Regulation of forebrain and midbrain GnRH neurons in juvenile teleosts. *Fish Physiology and Biochemistry* 17, 81–84.

Parhar, I.S., Koibuchi, N., Sakai, M., Iwata, M. and Yamaoka, S. (1994) Gonadotropin-releasing hormone (GnRH): expression during salmon migration. *Neuroscience Letters* 172, 15–18.

Parhar, I.S., Soga, T., Ishikawa, Y., Nagahama, Y. and Sakuma, Y. (1998) Neurons synthesizing gonadotropin-releasing hormone mRNA subtypes have multiple developmental origins in the medaka. *Journal of Comparative Neurology* 401, 217–226.

Parsons, J.E. and Thorgaard, G.H. (1985) Production of androgenetic diploid rainbow trout. *Journal of Heredity* 76, 177–181.

Parsons, J.E., Busch, R.A., Thorgaard, G.H. and Scheerer, P.D. (1986) Increased resistance of triploid rainbow trout × coho salmon hybrids to infectious hematopoietic necrosis virus. *Aquaculture* 57, 337–343.

Pasdar, M., Philipp, D.P. and Whitt, G.S. (1984) Linkage relationships of nine enzyme loci in sunfishes (Lepomis: Centrarchidae). *Genetics* 107, 435–446.

Patanjali, S.R., Parimoo, S. and Weissman, S.M. (1991) Construction of a uniform-abundance (normalized) cDNA library. *Proceedings of the National Academy of Sciences USA* 88, 1943–1947.

Patterson, A.P. (2001) Translational machinery of channel catfish: complementary DNA and expression of the complete set of 47 60S ribosomal proteins. MS thesis, Auburn University, Auburn, Alabama.

Patterson, A., Karsi, A., Feng, J. and Liu, Z. (2003) Translational machinery of channel catfish: II. Complementary DNA and expression of the complete set of 47 60S ribosomal proteins. *Gene* 305, 151–160.

Paul, A.J., Paul, J.M. and Smith, R.L. (1995) Compensatory growth in Alaska yellowfin sole, *Pleuronectes asper*, following food deprivation. *Journal of Fish Biology* 46, 442–448.

Pauly, D., Christensen, V., Dalsgaard, J., Froese, R. and Torres, F. Jr (1998) Fishing down marine food webs. *Science* 279, 860–863.

Pearce, K.H. Jr, Cunningham, B.C., Fuh, G., Teeri, T. and Wells, J.A. (1999) Growth hormone binding affinity for its receptor surpasses the requirements for cellular activity. *Biochemistry* 38, 81–89.

Pearlstein, S. (1999) Seattle tries to return to normal: key negotiating points. *Washington Post*, Saturday 4 December, A15.

Peatman, E., Bao, B., Peng, X., Baoprasertkul, P., Brady, Y. and Liu, Z. (2006) Catfish CC chemokines: genomic clustering, duplications, and expression after bacterial infection with *Edwardsiella ictaluri*. *Molecular Genetics and Genomics* 275, 297–309.

Peatman, E., Baoprasertkul, P., Terhune, J., Xu, P., Nandi, S., Kucuktas, H., Li, P., Wang, S., Somridhivej, B., Dunham, R. and Liu, Z. (2007) Expression analysis of the acute phase response in channel catfish (*Ictalurus punctatus*) after infection with a Gram negative bacterium. *Developmental and Comparative Immunology* 31, 1183–1196.

Peatman, E., Terhune, J., Baoprasertkul, P., Xu, P., Nandi, S., Wang, S., Somridhivej, B., Kucuktas, H., Li, P., Dunham, R. and Liu., Z. (2008) Microarray analysis of gene expression in the blue catfish liver reveals early activation of the MHC class I pathway after infection with *Edwardsiella ictaluri*. *Molecular Immunology* 45, 553–566.

Peng, J.H., Fahima, T., Roder, M.S., Li, Y.C., Grama, A. and Nevo, E. (2000) Microsatellite high-density mapping of the stripe rust resistance gene YrH52 region on chromosome 1B and evaluation of its marker-assisted selection in the F_2 generation in wild summer wheat. *New Phytologist* 146, 141–154.

Penman, D.J., Skibinski, D.O.F. and Beardmore, J.A. (1987) Survival, growth and maturity in triploid tilapia. In: Tiews, K. (ed.) *Selection Hybridization and Genetic Engineering in Aquaculture*, Vol. 2. Heeneman Verlagsgesellschaft, Berlin, pp. 277–288.

Penman, D.J., Beeching, A.J., Penn, S. and Maclean, N. (1990) Factors affecting survival and integration following microinjection of novel DNA into rainbow trout eggs. *Aquaculture* 85, 35–50.

Penman, D.J., Beeching, A.J., Penn, S., Rhaman, A., Sulaiman, Z. and Maclean, N. (1991) Patterns of trans-gene inheritance in rainbow trout (*Oncorhynchus mykiss*). *Molecular Reproduction and Development* 30, 201–206.

Perez-Sanchez, J., Marti-Palanca, H. and Kaushik, S.J. (1995) Ration size and protein intake affect circulating growth hormone concentration, hepatic growth hormone binding and plasma insulin-like growth factor-I immunoreactivity in a marine teleost, the gilthead seabream (*Sparus aurata*). *Journal of Nutrition* 125, 546–552.

Perry, G.M.L., Danzmann, R.G., Ferguson, M.M. and Gibson, J.P. (2001) Quantitative trait loci for upper thermal tolerance in outbred strains of rainbow trout (*Oncorhynchus mykiss*). *Heredity* 86, 333–341.

Perry, G.M.L., Martyniuk, C.M., Ferguson, M.M. and Danzmann, R.G. (2005) Genetic parameters for upper thermal tolerance and growth-related traits in rainbow trout (*Oncorhynchus mykiss*). *Aquaculture* 250, 120–128.

Perry, R.P. and Meyuhas, O. (1990) Translational control of ribosomal protein production in mammalian cells. *Enzyme* 44, 83–92.

Peters, H.M. (1957) Uber die Regulation der Gelegrosusse bei Fischen. *Zeitschrift für Naturforschung/Journal of Biosciences* 12-b, 255.

Peters, T.J. (1970) Intestinal peptidases. *Gut* 11, 720–725.

Peterson, F.C. and Brooks, C.L. (2000) The species specificity of growth hormone requires the cooperative interaction of two motifs. *FEBS Letters* 472, 276–282.

Petrosky, C.E. (1985) Competitive effects from stocked catchable-size rainbow trout on wild trout population dynamics. Doctoral dissertation, University of Idaho, Moscow, Idaho.

Pfrender, M.E., Spitze, K. and Lehman, N. (2000) Multi-locus genetic evidence for rapid ecologically based speciation in *Daphnia*. *Molecular Ecology* 9, 1717–1735.

Philipp, D.P. and Whitt, G.S. (1991) Survival and growth of northern, Florida, and reciprocal F_1 hybrid largemouth bass in central Illinois. *Transactions of the American Fisheries Society* 120, 58–64.

Philipp, D.P., Childers, W.F. and Whitt, G.S. (1981) Management implications of different genetic stocks of largemouth bass (*Micropterus salmoides*) in the United States. *Canadian Journal of Fisheries and Aquatic Sciences* 38, 1715–1723.

Philipp, D.P., Childers, W.F. and Whitt, G.S. (1983a) A biochemical genetic evaluation of the northern and Florida subspecies of largemouth bass. *Transactions of the American Fisheries Society* 112, 1–20.

Philipp, D.P., Parker, H.R. and Whitt, G.S. (1983b) Evolution of gene regulation: isozymic analysis of patterns of gene expression during hybrid fish development. In: Rattazzi, M.C., Scandalios, J.G. and Whitt, G.S.

(eds) *Isozymes: Current Topics in Biological and Medical Research*. Vol. 10. *Genetics and Evolution*. Alan R. Liss, New York, p. 193.

Philipp, D.P., Childers, W.F. and Whitt, G.S. (1985) Correlations of allele frequencies with physical and environmental variables for populations of largemouth bass, *Micropterus salmoides* (Lacepede). *Journal of Fish Biology* 27, 347–365.

Philipps, A.F., Rao, R., Anderson, G.G., McCracken, D.M., Lake, M. and Koldovsky, O. (1995) Fate of insulin-like growth factors I and II administered orogastrically to suckling rats. *Pediatric Research* 37, 586–592.

Philipps, A.F., Dvorak, B., Kling, P.J., Grille, J.G. and Koldovsky, O. (2000) Absorption of milk-borne insulin-like growth factor-I into portal blood of suckling rats. *Journal of Pediatric Gastroenterology and Nutrition* 31, 128–135.

Phillips, P.C. (1989) *In vitro* fertilization and gene transfer in tilapia. Doctoral dissertation, Southern Illinois University, Carbondale, Illinois.

Phillips, P.C., Whitlock, M.C. and Fowler, K. (2001) Inbreeding changes the shape of the genetic covariance matrix in *Drosophila melanogaster*. *Genetics* 158, 1137–1145.

Phillips, R.B., Zajicek, K.D., Ihssen, P.E. and Johnson, O. (1986) Application of silver staining to the identification of triploid fish cells. *Aquaculture* 54, 313–319.

Phillips, R.B., Matsuoka, M.P. and Reed, K.M. (2000) Genome mapping in trout and zebrafish using fluorescence *in situ* hybridization. In: *Abstracts of the International Symposium 'A Step Toward the Great Future of Aquatic Genomics', 10–12 November 2000*. Tokyo University of Fisheries, Tokyo, p. 21.

Pickering, A.D., Pottinger, T.G., Sumpter, J.P., Carragher, J.F. and Le Bail, P.Y. (1991) Effects of acute and chronic stress on the levels of circulating growth hormone in the rainbow trout, *Oncorhynchus mykiss*. *General and Comparative Endocrinology* 83, 86–93.

Piferrer, F., Benfey, T.J. and Donaldson, E.M. (1994) Gonadal morphology of normal and sex-reversed triploid and gynogenetic diploid coho salmon (*Oncorhynchus kisutch*). *Journal of Fish Biology* 45, 541–553.

Piferrer, F., Cal, R.M., Alvarez-Blazquez, B., Sanchez, L. and Martinez, P. (2000) Induction of triploidy in the turbot (*Scophthalmus maximus*) I. Ploidy determination and the effects of cold shocks. *Aquaculture* 188, 79–90.

Piva, F., Maggi, R., Limonta, M. and Mantini, L. (1985) Effect of naloxone on luteinizing hormone, follicle stimulatory hormone and prolactin secreting in the different phases of the estrous cycle. *Endocrinology* 117, 766–772.

Plasterk, R.H. (1996) The *Tc1/mariner* transposon family. *Current Topics in Microbiology and Immunology* 204, 125–143.

Plosila, D.S. (1977) Relationship of strain and size at stocking to survival of lake trout in Adirondack lakes. *New York Fish and Game Journal* 24, 1–24.

Polskie, A. and Flajshans, M. (1997) Reproduction sterility caused by spontaneous triploidy in tench (*Tinca tinca*). *Hydrobiologii/Polish Archives of Hydrobiology* 44, 39–45.

Pomp, D., Nancarrow, C.D., Ward, K.A. and Murray, J.D. (1992) Growth, feed-efficiency and body-composition of transgenic mice expressing a sheep metallothionein 1a–sheep growth hormone fusion gene. *Livestock Production Science* 31, 335–350.

Pongthana, N., Penman, D.J., Karnasuta, J. and McAndrew, B.J. (1995) Induced gynogenesis in the silver barb (*Puntius gonionotus* Bleeker) and evidence for female homogamety. *Aquaculture* 135, 267–276.

Pongthana, N., Penman, D.J., Baoprasertkul, P., Hussain, M.G., Islam, M.S., Powell, S.F. and McAndrew, B.J. (1999) Monosex female production in the silver barb (*Puntius gonionotus* Bleeker). *Aquaculture* 173, 247–256.

Pontiroli, A.E. (1998) Peptide hormones: review of current and emerging uses by nasal delivery. *Advanced Drug Delivery Reviews* 29, 81–87.

Ponzoni, R., Khaw, H.L., Hamzah, A., Kamaruzzaman, N. and Nguyen, N.H. (2009) Genetic evaluation of seven generations of selection for increased harvest weight in the genetically improved farmed tilapia (GIFT strain, *Oreochromis niloticus*). In: *Book of Abstracts of 10th International Symposium on Genetics in Aquaculture (ISGA X) 'Roles of Aquaculture Genetics in Addressing Global Food Crisis', 22–26 June 2009, Bangkok, Thailand*. Faculty of Fisheries, Kasetsart University, Bangkok, p. 69.

Poompuang, S. and Hallerman, E.M. (1997) Toward detection of quantitative trait loci and genetic marker-assisted selection in fish. *Reviews in Fisheries Science* 5, 253–277.

Poon, A. and Otto, S.P. (2000) Compensating for our load of mutations: freezing the meltdown of small populations. *Evolution* 54, 1467–1479.

Popma, T.J. (1987) *Fry Production and Sex Reversal of* Tilapia nilotica *in Ponds*. Final Technical Report. Freshwater Fishculture Development Project, ESPOL, Guayaquil, Ecuador.

Postlethwait, J.H., Johnson, S.L., Midson, C.N., Talbot, W.S., Gates, M., Ballinger, E.W., Africa, D., Andrews, R., Carl, T., Eisen, J.S., Horne, S., Kimmel, C.B., Hutchinson, M., Johnson, M. and Rodriguez, A. (1994) A genetic linkage map for the zebrafish. *Science (Washington)* 264, 699–703.

Postlethwait, J.H., Woods, I.G., Ngo-Hazelett, P., Yan, Y.L., Kelly, P.D., Chu, F., Huang, H., Hill-Force, A. and Talbot W.S. (2000) Zebrafish comparative genomics and the origins of vertebrate chromosomes. *Genome Research* 10, 1890–1902.

Postlethwait, J.H., Amores, A., Yan, Y.-L. and Austin, C. (2002) Duplication of a portion of human chromosome 20q containing Topoisomerase (TopI) and Snail genes provides evidence on genome expansion and the radiation of teleost fish. In: Shimizu, N., Aoki, T., Hirono, I. and Takashima, F. (eds) *Aquatic Genomics: Steps Toward a Great Future.* Springer-Verlag, New York, pp. 20–34.

Pouyaud, L., Desmarais, E., Chenuil, A., Agnese, J.F. and Bonhomme, F. (1999) Kin cohesiveness and possible inbreeding in the mouthbrooding tilapia *Sarotherodon melanotheron* (Pisces Cichlidae). *Molecular Ecology* 5, 803–812.

Powell, W., Thomas, W.T.B., Baird, E., Lawremce, P., Booth, A., Harrower, B., McNicol, J.W. and Waugh, R. (1997) Analysis of quantitative traits in barley by the use of amplified fragment length polymorphisms. *Heredity* 79, 48–59.

Powers, D.A. (1990) The adaptive significance of allelic isozyme variation in natural populations. In: Whitmore, D.H. (ed.) *Electrophoretic and Isoelectric Focusing Techniques in Fisheries Management.* CRC Press, Boca Raton, Florida, pp. 323–340.

Powers, D.A., Cole, T., Creech, K., Chen, T.T., Lin, C.M., Kight, K. and Dunham, R. (1992) Electroporation: a method for transferring genes into the gametes of zebrafish, *Brachydanio rerio*, channel catfish, *Ictalurus punctatus*, and common carp, *Cyprinus carpio. Molecular Marine Biology and Biotechnology* 1, 301–309.

Prack, M.M., Antoine, M., Caiati, R., Roskowski, T., Treag, M.J., Vodicnick, M.J. and de Vlaming, V.L. (1980) The effects of mammal prolactin and growth hormone on goldfish *Carassius auratus* growth, plasma amino acids levels and liver amino acid uptake. *Comparative Biochemistry and Physiology, Part A: Comparative Physiology* 67A, 307–310.

Praebel, K, Odegard, J., Skagemo, V. and Nielsen, H.M. (2009) Heritability estimates and deformities in Atlantic cod (*Gadus morhua* L.). In: *Book of Abstracts of 10th International Symposium on Genetics in Aquaculture (ISGA X) 'Roles of Aquaculture Genetics in Addressing Global Food Crisis', 22–26 June 2009, Bangkok, Thailand.* Faculty of Fisheries, Kasetsart University, Bangkok, p. 26.

Pralong, F.P., Miell, J.P., Corder, R. and Gaillard, R.C. (1991) Dexamethasone treatment in man induces changes in 24-hour growth hormone (GH) secretion profile without altering total GH released. *Journal of Clinical Endocrinology and Metabolism* 73, 1191–1196.

Prarom, W. (1990) The effect of strain crossing of Gunther's walking catfish (*Clarias macrocephalus*) on growth and disease resistance. MSc thesis, Kasetsart University, Bangkok.

Prasertlux, S., Khamnamtong, B., Chumtong, P., Klinbunga, S. and Menasveta, P. (2010) Expression levels of RuvBL2 during ovarian development and association between its single nucleotide polymorphism (SNP) and growth of the giant tiger shrimp *Penaeus monodon. Aquaculture* 308, S83–S90.

Prat, F., Sumpter, J.P. and Tyler, C.R. (1996) Validation of radioimmunoassay for two salmon gonadotropins (GtH-I and GtH-II) and their plasma concentration throughout the reproductive cycle in male and female rainbow trout (*Oncorhynchus mykiss*). *Biology of Reproduction* 54, 1375–1382.

Prather, R.S., Sims, M.M. and First, N.L. (1989) Nuclear transplantation in early pig embryos. *Biology of Reproduction* 41, 414–418.

Pratt, T. (1999) International biosafety protocol negotiated in Columbia. *ISB (Information Systems for Biotechnology) News Report*, March 1999; available at www.nbiap.vt.edu.

Preston, N.P. and Crocos, P.J. (1999) Comparative growth of wild and domesticated *Penaeus japonicus* in commercial production ponds (abstract). In: *Proceedings of World Aquaculture '99, Annual International Conference and Exposition of the World Aquaculture Society, Sydney, Australia, 26 April–2 May 1999.*

Preston, N.P., Brennan, D.C. and Crocos, P.J. (1999) Comparative costs of postlarval production from wild or domesticated *Penaeus japonicus* broodstock. *Aquaculture Research* 30, 191–197.

Primmer, C.R., Aho, T., Piironen, J., Estoup, A., Cornuet, J.M. and Ranta, E. (1999) Microsatellite analysis of hatchery stocks and natural populations of Arctic charr, *Salvelinus alpinus*, from the Nordic region: implications for conservation. *Hereditas* 130, 277–289.

Proudfoot, N. (1991) Poly(A) signals. *Cell* 64, 671–674.

Purdom, C.E. (1973) Induced polyploidy in plaice (*Pleuronectes platessa*) and its hybrid with the flounder (*Platichythys flesus*). *Heredity* 29, 11–24.

Purdom, C.E. (1976) Genetic techniques in flatfish culture. *Journal of Fisheries Research Board of Canada* 33, 1088–1093.

Purdom, C.E. (1983) Genetic engineering by the manipulation of chromosomes. *Aquaculture* 33, 287–300.

Pursel, V.G., Pinkert, C.A., Miller, K.F., Bolt, D.J., Campbell, R.G., Palmiter, R.D., Brinster, R.L. and Hammer, R.E. (1989) Genetic engineering of livestock. *Science* 244, 1281–1288.

Pursel, V.G., Bolt, D.J., Miller, K.F., Pinkert, C.A., Hammer, R.E., Palmiter, R.D. and Brinster, R.L. (1990) Integration, expression and germ-line transmission of growth-related genes in pigs. *Journal of Reproduction and Fertility Supplement* 41, 77–80.

Pursov, G.M. (1975) The period of gonadal differentiation in fishes. In: *Sex Differentiation in Fishes*. University of Leningrad Publishing House, Leningrad, USSR. Tr. Fisheries Marine Science of Canada, Translation Series No. 4069.

Pycha, R.L. and King, G.R. (1967) Returns of hatchery-reared lake trout in southern Lake Superior 1955–62. *Journal of Fisheries Research Board of Canada* 24, 281–298.

Qi, R. (1988) *Diagnostics*. People's Medical Press, Beijing, pp. 419–427.

Qi, X. and Lindhout, P. (1997) Development of AFLP markers in barley. *Molecular and General Genetics* 254, 330–336.

Qin, J.G., Fast, A.W. and Ako, H. (1998) Growout performance of diploid and triploid Chinese catfish *Clarias fuscus*. *Aquaculture* 166, 247–258.

Quackenbush, J., Liang, F., Holt, I., Pertea, G. and Upton, J. (2000) The TIGR gene indices: reconstruction and representation of expressed gene sequences. *Nucleic Acids Research* 28, 141–145.

Quanbeck, C., Sherwood, N.M., Millar, R.P. and Terasawa, E. (1997) Two populations of luteinizing hormone-releasing hormone neurons in the forebrain of the rhesus macaque during embryonic development. *Journal of Comparative Neurology* 380, 293–309.

Quarrie, S.A., Stojanovic, J. and Pekic, S. (1999) Improving drought resistance in small-grained cereals: a case study, progress and prospects. *Plant Growth Regulation* 29, 1–21.

Queller, D.C., Strassmann, J.E. and Hughes, C.R. (1993) Microsatellites and kinship. *Trends in Ecology and Evolution* 8, 285–288.

Quillet, E. and Gaignon, J.L. (1990) Thermal induction of gynogenesis and triploidy in Atlantic salmon (*Salmo salar*) and their potential interest for aquaculture. *Aquaculture* 89, 351–364.

Quillet, E., Chevassus, B. and Krieg, F. (1987) Characterization of auto- and allotetraploid salmonids for rearing in seawater cages. In: Tiews, K. (ed.) *Selection, Hybridization and Genetic Engineering in Aquaculture*, Vol. 2. Heeneman Verlagsgesellschaft, Berlin, pp. 239–253.

Quillet, E., Chevassus, B., Blanc, J.M., Krieg, F. and Chourrout, D. (1988) Performances of auto and allotriploids in salmonids. 1. Survival and growth in fresh water. *Aquatic Living Resources* 1, 29–43.

Quillet, E., Le Guillou, S., Aubin, J. and Fauconneau, B. (2005) Two-way selection for muscle lipid content in pan-size rainbow trout (*Oncorhynchus mykiss*). *Aquaculture* 245, 49–61.

Quinton, C.D., Kause, A., Ruohonen, K. and Koskela, J. (2007) Genetic relationships of body composition and feed utilization traits in European whitefish (*Coregonus lavaretus* L.) and implications for selective breeding in fishmeal- and soybean meal-based diet environments. *Journal of Animal Science* 85, 3198–3208.

Raben, M.S. (1959) Human growth hormone. *Recent Progress in Hormone Research* 15, 71–114.

Radice, A.D., Bugaj, B., Fitch, D.H. and Emmons, S.W. (1994) Widespread occurrence of the *Tc1* transposon family: *Tc1*-like transposons from teleost fish. *Molecular and General Genetics* 244, 606–612.

Rahman, M.A. and Maclean, N. (1992) Production of transgenic tilapia (*Oreochromis niloticus*) by one-cellstage microinjection. *Aquaculture* 105, 219–232.

Rahman, M.A. and Maclean, N. (1999) Growth performance of transgenic tilapia containing an exogenous piscine growth hormone gene. *Aquaculture* 173, 333–346.

Rahman, M.A., Mak, R., Ayad, H., Smith, A. and Maclean, N. (1998) Expression of a novel piscine growth hormone gene results in growth enhancement in transgenic tilapia (*Oreochromis niloticus*). *Transgenic Research* 7, 357–369.

Rahman, M.A., Hwang, G.-L., Razak, S.A., Sohm, F. and Maclean, N. (2000) Copy number related trans-gene expression and mosaic somatic expression in hemizygous and homozygous transgenic tilapia (*Oreochromis niloticus*). *Transgenic Research* 9, 417–427.

Rahman, M.A., Ronyai, A., Engidaw, B.Z., Jauncey, K., Hwang, G., Smith, A., Roderick, E., Penman, D., Varadi, L. and Maclean, N. (2001) Growth performance of transgenic tilapia containing an exogenous piscine growth hormone gene. *Journal of Fish Biology* 59, 62–78.

Rainboth, W.J., Buth, D.G. and Joswiak, G.R. (1986) Electrophoretic and karyological characters of the gyrinocheilid fish, *Gyrinocheilus aymonieri*. *Biochemical Systematics and Ecology* 14, 531–537.

Rakitin, A., Ferguson, M.M. and Trippe, E.A. (2001) Male reproductive success and body size in Atlantic cod, *Gadus morhua* L. *Marine Biology* 138, 1077–1085.

Ranzani-Paiva, M.J.T., Tabata, Y.A. and Das Eiras, A.C. (1998) A comparison of the haematology of diploid and triploid rainbow trout, *Oncorhynchus mykiss* Walbaum (Pisces, Salmonidae). *Revista Brasileira de Zoologia* 15, 1093–1102.

Rasatapana, D. (1996) Estimation of heritability of disease resistance, body weight and body length of Thai walking catfish, *Clarias macrocephalus*. MSc Thesis, Kasetsart University, Bangkok, p. 67.

Rast, A.T.B. (ed.) (1975) *The Variability of Tobacco Mosaic Virus in Relation to Control of Tomato Mosaic Glasshouse Tomato Crops by Resistance Breeding and Cross Protection.* Centre for Agricultural Publishing and Documentation, Wageningen, The Netherlands.

Rathje, T.A., Rohrer, G.A. and Johnson, R.K. (1997) Evidence for quantitative trait loci affecting ovulation rate in pigs. *Journal of Animal Science* 75, 1486–1494.

Rawson, D.S. (1941) The eastern brook trout in the Maligne River system, Jasper National Park. *Transactions of the American Fisheries Society* 70, 221–235.

Razak, S.A., Hwang, G.-L., Rahman, M.A. and Maclean, N. (1999) Growth performance and gonadal development of growth enhanced transgenic tilapia, *Oreochromis niloticus* (L.) following heat-shock induced triploidy. *Marine Biotechnology* 1, 533–544.

Reagan, R.E. (1979) *Heritabilities and Genetic Correlations of Desirable Commercial Traits in Channel Catfish.* Resource Report. Mississippi Agriculture and Forestry Station, Stoneville, Mississippi.

Reagan, R.E. and Conley, C.M. (1977) Effect of egg diameter on growth of channel catfish. *Progressive Fish-Culturist* 39, 133–134.

Recoubratsky, A.V., Gomelsky, B.I., Emelyanova, O.V. and Pankratyeva, E.V. (1992) Triploid common carp produced by heat shock with industrial fish-farm technology. *Aquaculture* 108, 13–19.

Reddy, P.K. and Lam, T.J. (1992) Role of thyroid hormones in tilapia larvae (*Oreochromis mossambicus*): 1. Effects of the hormones and an antithyroid drug on yolk absorption, growth and development. *Fish Physiology and Biochemistry* 9, 473–485.

Reddy, P.V.G.K. (2000) *Genetic Resources of Indian Major Carps.* FAO Fisheries Technical Paper No. 387. Food and Agriculture Organization of the United Nations, Rome.

Reddy, P.V.G.K., Khan, H.A., Gupta, S.D., Tantia, M.S. and Kowtal, G.V. (1990) On the ploidy of three intergeneric hybrids between common carp (*Cyprinus carpio* communis L.) and Indian major carps. *Aquacultura Hungarica* 6, 5–11.

Reed, D.H. (2001) Fitness, genetic load and purging in experimental populations of the housefly. *Conservation Genetics* 2, 57–61.

Reed, D.H. and Bryant, E.H. (2001) The relative effects of mutation accumulation versus inbreeding depression on fitness in experimental populations of the housefly. *Zoo Biology* 20, 145–156.

Reed, D.H. and Frankham, R. (2001) How closely correlated are molecular and quantitative measures of genetic variation? A meta-analysis. *Evolution* 55, 1095–1103.

Reed, K.M. and Phillips, R.B. (1995) Molecular characterization and cytogenetic analysis of highly repeated DNAs of lake trout, *Salvelinus namaycush*. *Chromosoma* 104, 242–251.

Reed, K.M., Oakley, T.H. and Phillips, R.B. (1997) An AluI fragment isolated from lake trout (*Salvelinus namaycush*), maps to the intergenic spacer region of the rDNA cistron. *Gene* 186, 7–11.

Refstie, T. (1981) Tetraploid rainbow trout produced by cytochalasin B. *Aquaculture* 25, 51–58.

Refstie, T. (1984) Production of sterile (triploid) salmonids. *Norsk Fiskeoppdrett* 9, 40–41.

Reichhardt, T. (2000) Will souped up salmon sink or swim? *Nature* 406, 10–12.

Reid, D.P., Szanto, A., Glebe, B., Danzmann, R.G. and Ferguson, M.M. (2005) QTL for body weight and condition factor in Atlantic salmon (*Salmo salar*): comparative analysis with rainbow trout (*Oncorhynchus mykiss*) and Arctic charr (*Salvelinus alpinus*). *Heredity* 94, 166–172.

Reik, W., Collick, A., Norris, M.L., Bartov, S.C. and Surani, M.A. (1987) Genomic imprinting determines methylation of parental alleles in transgenic mice. *Nature* 328, 248–251.

Reilly, A., Elliott, N.G., Grewe, P.M., Clabby, C., Powell, R. and Ward, R.D. (1999) Genetic differentiation between Tasmanian cultured Atlantic salmon (*Salmo salar* L.) and their ancestral Canadian population: comparison of microsatellite DNA and allozyme and mitochondrial DNA variation. *Aquaculture* 173, 459–469.

Reimers, N. (1979) A history of stunted brook trout population in an alpine lake: a lifespan of 24 years. *California Fish and Game* 65, 196–215.

Reinhardt, U.G. (2001) Selection for surface feeding in farmed and sea-ranched masu salmon juveniles. *Transactions of the American Fisheries Society* 130, 155–158.

Reisenbichler, R.K. and McIntyre, J.D. (1977) Genetic differences in growth and survival of juvenile hatchery and wild steelhead trout, *Salmo gairdneri. Journal of Fisheries Research Board of Canada* 34, 123–128.

Rexroad, C.E. Jr, Mayo, K., Bolt, D.J., Elsasser, T.H., Miller, K.F., Behringer, R.R., Palmiter, R.D. and Brinster, R.L. (1991) Transferrin- and albumin-directed expression of growth-related peptides in transgenic sheep. *Journal of Animal Science* 69, 2995–3004.

Rexroad, C.E. III, Coleman, R.L., Gustafson, A.L., Hershberger, W.K. and Killefer, J. (2002) Development of rainbow trout microsatellite markers from repeat enriched libraries. *Marine Biotechnology* 4, 12–16.

Reynders, H., van der Ven, K., Moens, L.N., van Remortel, P., De Coen, W.M. and Blust, R. (2006) Patterns of gene expression in carp liver after exposure to a mixture of waterborne and dietary cadmium using a custom-made microarray. *Aquatic Toxicology* 80, 180–193.

Rezk, M. (1988) Efficiency of heat pressure shocks for inducing tetraploidy in interspecific hybrids of catfish, *Ictalurus* spp. MSc thesis, Auburn University, Auburn, Alabama.

Rezk, M.S. (1993) Response and correlated responses to three generations of selection for increased body weight in channel catfish, *Ictalurus punctatus*. PhD dissertation, Auburn University, Auburn, Alabama.

Rezk, M.A., Ponzoni, R.W., Khaw, H.L., Kamel, E., Dawood, T. and John, G. (2009) Selective breeding for increased body weight in a synthetic breed of Egyptian Nile tilapia, *Oreochromis niloticus*: response to selection and genetic parameters. *Aquaculture* 293, 187–194.

Ribeiro, A., Moran, P. and Caballero, A. (2008) Genetic diversity and effective size of the Atlantic salmon *Salmo salar* L. inhabiting the River Eo (Spain) following a stock collapse. *Journal of Fish Biology*. 72, 1933–1944.

Rice, V.A. (1979) *Breeding and Improvement of Farm Animals*. McGraw-Hill, New York.

Rice, W.R. and Chippendale, A.K. (2001) Sexual recombination and the power of natural selection. *Science* 294, 555–559.

Richardson, B.J., Baverstock, P.R. and Adams, M. (1986) *Allozyme Electrophoresis: a Handbook for Animal Systematics and Population Studies*. Academic Press, Sydney, Australia.

Richter, C.J.J., Henken, A.M., Eding, E.H., van Dossum, J.H. and de Boer, P. (1986) Induction of triploidy by cold shocking eggs and performance of triploids on the African catfish, *Clarias gariepinus* (Burchell, 1822). In: *Symposium E26*. European Inland Fisheries Advisory Commission, Bordeaux, France.

Richter, C.J.J., Henken, A.M., Eding, E.H., Van Doesum, J.H. and de Boer, P. (1987) Induction of triploidy by cold-shocking eggs and performance of triploids in the African catfish, *Clarius gariepinus* (Burchell, 1822). In: Tiews, K. (ed.) *Selection, Hybridization and Genetic Engineering in Aquaculture*, Vol. 2. Heeneman Verlagsgesellschaft, Berlin, pp. 225–237.

Ricker, W.E. (1975) *Computation and Interpretation of Biological Statistics of Fish Population*. Bulletin 191. Department of Environment, Fisheries and Marine Services, Ottawa.

Ricker, W.E. (1981) Changes in the average size and average age of Pacific salmon. *Canadian Journal of Fisheries and Aquatic Sciences* 38, 1636–1656.

Rico, C., Rico, I. and Hewitt, G. (1996) 470 million years of conservation of microsatellite loci among fish species. *Proceedings of the Royal Society of London, Series B: Biological Sciences* 263, 549–557.

Riedy, M.F., Hamilton, W.J. and Aquadro, C.F. (1992) Excess of non-parental bands in offspring from known primate pedigree assayed using RAPD PCR. *Nucleic Acids Research* 20, 918.

Rindi, G. (1966) Digestion and absorption of proteins. *Minerva Pediatric* 18, 604–615.

Rio, D.C., Barnes, G., Laski, F.A., Rine, J. and Rubin, G.M. (1988) Evidence for *Drosophila P* element transposase activity in mammalian cells and yeast. *Journal of Molecular Biology* 200, 411–415.

Rise, M.L., Jones, S.R., Brown, G.D., von Schalburg, K.R., Davidson, W.S. and Koop, B.F. (2004) Microarray analyses identify molecular biomarkers of Atlantic salmon macrophage and hematopoietic kidney response to *Piscirickettsia salmonis* infection. *Physiological Genomics* 20, 21–35.

Roberts, S.B. and Goetz, F.W. (2001) Differential skeletal muscle expression of myostatin across teleost species, and the isolation of multiple myostatin isoforms. *FEBS Letters* 491, 212–216.

Robertson, H.M. (1993) The *mariner* transposable element is widespread in insects. *Nature* 362, 241–245.

Robertson, O.H. (1958) *Accelerated Development of Testis after Unilateral Gonadectomy, with Observation on Normal Testis of Rainbow Trout*. Fisheries Bulletin No. 127. US Fish and Wildlife Service, Washington, DC.

Robertson, O.H. (1961) Prolongation of the life span of kokanee salmon (*Oncorhynchus nerka kennerlyi*) by castration before beginning of gonad development. *Proceedings of the National Academy of Sciences USA* 47, 609–621.

Robison, B.D., Wheeler, P.A., Sundin, K., Sikka, P., and Thorgaard, G.H. (2001) Composite interval mapping reveals a major locus influencing embryonic development rate in rainbow trout (*Oncorhynchus mykiss*). *Journal of Heredity* 92, 16–22.

Robison, O.W. and Luempert, L.G. (1984) Genetic variation in weight and survival of brook trout (*Salvelinius fontinalis*). *Aquaculture* 38, 155–170.

Rodgers, B.D., Weber, G.M., Sullivan, C.V. and Levine, M.A. (2001) Isolation and characterization of myostatin complementary deoxyribonucleic acid clones from two commercially important fish: *Oreochromis mossambicus* and *Morone chrysops*. *Endocrinology* 142, 1412–1418.

Rodriguez, M.C., Zambudio, F., Torres, J.A., Gonzalez-Ceron, L., Possani, L.D. and Rodriguez, M.H. (1995) Effect of cecropin-like synthetic peptide (Shiva-3) on the sporogonic development of *Plasmodium berghei*. *Experimental Parasitology USA* 80, 596–604.

Rodriguez, M.F., LaPatra, S., Williams, S., Famula, T and May, B. (2004) Genetic markers associated with resistance to infectious hematopoietic necrosis in rainbow and steelhead trout (*Oncorhynchus mykiss*) backcrosses. *Aquaculture* 241, 93–115.

Rodriguez-Guerrero, D. (1979) Factors influencing the androgen sex reversing of *Tilapia aurea*. MSc thesis, Auburn University, Auburn, Alabama.

Rognon, X., Andriamanga, M., McAndrew, B. and Guyomard, R. (1996) Allozyme variation in natural and cultured populations in two tilapia species: *Oreochromis niloticus* and *Tilapia zilli*. *Heredity* 76, 640–650.

Rokkones, E., Alestrom, P., Skjervold, H. and Gautvik, K.M. (1985) Development of a technique for microinjection of DNA into salmonid eggs. *Acta Physiologica Scandinavica* 124 (Suppl. 542), 417.

Rokkones, E., Alestrom, P., Skjervold, H. and Gautvik, K.M. (1989) Micro-injection and expression of a mouse metallothionein human growth hormone fusion gene in fertilized salmonid eggs. *Journal of Comparative Physiology, Part B: Biochemical, Systemic, and Environmental Physiology* 158, 751–758.

Romagosa, I., Feng, H., Ullrich, S.E., Haynes, P.M. and Wesenberg, D.M. (1999) Verification of yield QTL through realized molecular marker-assisted selection responses in a barley cross. *Molecular Breeding* 5, 143–152.

Rosenblum, P.M. and Peter, R.E. (1989) Evidence for the involvement of endogenous opioids in the regulation of gonadotropin secretion in male gold fish, *Carassius auratus*. *General and Comparative Endocrinology* 73, 21–27.

Rosenstein, S. and Hulata, G. (1993) Sex reversal in the genus *Oreochromis*: optimization of feminization protocol. *Aquaculture and Fisheries Management* 25, 329–339.

Rossiter, S.J., Jones, G., Ransome, R.D. and Barratt, E.M. (2001) Outbreeding increases offspring survival in wild greater horseshoe bats (*Rhinolophus ferrumequinum*). *Proceedings of the Royal Society of London, Series B: Biological Sciences* 268, 1055–1061.

Rothbard, S., Solnik, E., Shabbath, S., Amado, R. and Grabie, J. (1983) The technology of mass production of hormonally sex-inversed all-male tilapias. In: Fishelson, L. and Yaron, Z. (compilers) *Proceedings of the International Symposium on Tilapia in Aquaculture*. Tel Aviv University, Tel Aviv, Israel, p. 425.

Rothbard, S., Yaron, Z. and Moav, B. (1988) Field experiments on growth enhancement of tilapia (*Oreochromis niloticus* × *O. aureus* F$_1$ hybrids) using pellets containing an androgen (17α-ethynyltestosterone). In: Pullin, R.S.V., Bhukasawan, T., Tonguthai, K. and Maclean, J.L. (eds) *Proceedings of the Second International Symposium on Tilapia in Aquaculture, Bangkok, Thailand, 16–20 March 1987*. ICLARM Conference Proceedings No. 15. Department of Fisheries Thailand and ICLARM, Manila, pp. 367–375.

Royal Society (UK) (2001) The use of genetically modified animals. http://royalsociety.org/The-use-of-genetically-modified-animals-report/ (accessed April 2011).

Rueda, F.M., Martinez, F.J., Zamora, S., Kentouri, M. and Divanach, P. (1998) Effect of fasting and refeeding on growth and body composition of red porgy, *Pagrus pagrus* L. *Aquaculture Research* 29, 447–452.

Ruiz-Verdugo, C.A., Allen, S.K. and Ibarra, A.M. (2001a) Family differences in success of triploid induction and effects of triploidy on fecundity of catarina scallop (*Argopecten ventricosus*). *Aquaculture* 201, 19–33.

Ruiz-Verdugo, C.A., Racotta, I.S. and Ibarra, A.M. (2001b) Comparative biochemical composition in gonad and adductor muscle of triploid and diploid catarina scallop (*Argopecten ventricosus* Sowerby II, 1842). *Journal of Experimental Marine Biology and Ecology* 259, 155–170.

Ruzzante, D.E., Hansen, M.M. and Meldrup, D. (2001) Distribution of individual inbreeding coefficients, relatedness and influence of stocking on native anadromous brown trout (*Salmo trutta*) population structure. *Molecular Ecology* 10, 2107–2128.

Rye, M. and Eknath, A.E. (1999) Genetic improvement of tilapia through selective breeding – experience from Asia. In: *Aquaculture Europe '99 – Towards Predictable Aquaculture. Abstracts of contributions presented at the International Conference 'Aquaculture Europe '99' held in Trondheim, Norway, 7–10 August*. Special Publication No. 27. European Aqauculture Society, Oostende, Belgium, pp. 207–208.

Rye, M., Lillevik, K.M. and Gjerde, B. (1990) Survival in early life of Atlantic salmon and rainbow trout: estimates of heritabilities and genetic correlations. *Aquaculture* 89, 209–216.

Ryman, N. and Laikre, L. (1991) Effects of supportive breeding on the genetically effective population size. *Conservation Biology* 5, 325–329.

Ryman, N. and Stahl, G. (1980) Genetic changes in hatchery stocks of brown trout (*Salmo trutta*). *Canadian Journal of Fisheries and Aquatic Sciences* 32, 82–87.

Ryman, N. and Stahl, G. (1981) Genetic perspectives of the identification and conservation of Scandinavian stocks of fish. *Canadian Journal of Fisheries and Aquatic Sciences* 38, 1562–1575.

Ryman, N., Allendorf, F.W. and Stahl, G. (1979) Reproductive isolation with little genetic divergence in sympatric populations of brown trout (*Salmo trutta*). *Genetics* 92, 247–262.

Saccheri, I.J., Nichols, R.A. and Brakefield, P.M. (2001) Effects of bottlenecks on quantitative genetic variation in the butterfly *Bicyclus anynana*. *Genetical Research* 77, 167–181.

Sadler, J., Pankhurst, N.W., Pankhurst, P.M. and King, H. (2000a) Physiological stress responses to confinement in diploid and triploid Atlantic salmon. *Journal of Fish Biology* 56, 506–518.

Sadler, J., Wells, R.M.G., Pankhurst, P.M. and Pankhurst, N.W. (2000b) Blood oxygen transport, rheology and haematological responses to confinement stress in diploid and triploid Atlantic salmon, *Salmo salar*. *Aquaculture* 184, 349–361.

Sadler, J., Pankhurst, P.M. and King, H.R. (2001) High prevalence of skeletal deformity and reduced gill surface area in triploid Atlantic salmon (*Salmo salar* L.). *Aquaculture* 198, 369–386.

Saegrov, H., Hindar, K., Kalas, S. and Lura, H. (1997) Escaped farmed Atlantic salmon replace the original salmon stocks in the River Vosso, western Norway. *ICES Journal of Marine Science* 54, 1166–1172.

Sagar, A., Daemmrich, A. and Ashiya, M. (2000) The tragedy of the commoners: biotechnology and its publics. *Nature Biotechnology* 18, 2–4.

Saillant, E., Dupont-Nivet, M., Sabourault, M., Haffray, P., Laureau, S., Vidal, M.O. and Chatain, B. (2009) Genetic variation for carcass quality traits in cultured sea bass (*Dicentrarchus labrax*). *Aquatic Living Resources* 22, 105–112.

Saito, M., Kuwada, T., Arai, M., Yamashita, Y., Aoki, T. and Kunisaki, N. (1997) Changes of the proximate composition and some minor constituents of the muscular tissue from cultured diploid and triploid amago during growth. *Fisheries Science* 63, 639–643.

Saito, T., Fujimoto, T., Maegawa, S., Inoue, K., Tanaka, M., Arai, K. and Yamaha, E. (2006) Visualization of primordial germ cells *in vivo* using GFP-nos1 3′UTR mRNA. *International Journal of Developmental Biology* 50, 691–699.

Saito, T., Goto-Kazeto, R. Arai, K. and Yamaha, E. (2008) Xenogensis in teleost fish through generation of germ-line chimeras by single primordial germ cell transplantation. *Biology of Reproduction* 78, 159–166.

Saitoh, K., Takai, A. and Ojima, Y. (1984) Chromosomal study on the three local races of the striated spined loach (*Cobitis taenia striata*). *Proceedings of the Japan Academy, Series B* 60B, 187–190.

Sakamoto, T. (2000) A microsatellite linkage map of rainbow trout (*Oncorhynchus mykiss*) characterized by large sex-specific differences in recombination rates. In: *Abstracts of the International Symposium 'A Step Toward the Great Future of Aquatic Genomics', 10–12 November 2000*. Tokyo University of Fisheries, Tokyo, p. 25.

Sakamoto, T., Danzmann, R.G., Okamoto, N., Ferguson, M.M. and Ihssen, P.E. (1999) Linkage analysis of quantitative trait loci associated with spawning time in rainbow trout (*Oncorhynchus mykiss*). *Aquaculture* 173, 33–43.

Sakamoto, T., Danzmann, R.G., Gharbi, K., Howard, P., Ozaki, A., Khoo, S.K., Woram, R.A., Okamoto, N., Ferguson, M.M., Holm, L.E., Guyomard, R. and Hoyheim, B. (2000) A microsatellite linkage map of rainbow trout (*Oncorhynchus mykiss*) characterized by large sex-specific differences in recombination rates. *Genetics* 155, 1331–1345.

Salami, A.A., Fagbenro, O.A. and Sydenham, D.H.J. (1993) The production and growth of clariid catfish hybrids in concrete tanks. *Israeli Journal of Aquaculture/Bamidgeh* 45, 18–25.

Sanchez, T. and Ponce, R. (1988) Empresa Nac. de Acuicultura, Havana (Cuba).

Sanger, F., Nicklen, S. and Coulson, A.R. (1977) DNA sequencing with chain-terminating inhibitors. *Proceedings of the National Academy of Sciences USA* 74, 5463–5467.

Sanico, A.F. (1977) Effects of 17-ethynyltestosterone and estrone on sex ratio and growth of *Tilapia aurea* (Steindachner). MSc thesis, Auburn University, Auburn, Alabama.

Sarder, M.R.I., Penman, D.J., Myers, J.M. and McAndrew, B.J. (1999) Production and propagation of fully inbred clonal lines in the Nile tilapia (*Oreochromis niloticus* L.). *Journal of Experimental Zoology* 284, 675–685.

Sarder, M.R.I., Thompson, K.D., Penman, D.J. and McAndrew, B.J. (2001) Immune responses of Nile tilapia (*Oreochromis niloticus* L.) clones: I. Non-specific responses. *Developmental and Comparative Immunology* 25, 37–46.

Sari-Gorla, M., Rampoldi, L., Binelli, G., Frova, C. and Pe, M.E. (1996) Identification of genetic factors for Alachlor tolerance in maize by molecular markers analysis. *Molecular and General Genetics* 251, 551–555.

Sari-Gorla, M., Krajewski, P., Binelli, G., Frova, C., Taramino, G. and Villa, M. (1997) Genetic dissection of herbicide tolerance in maize by molecular markers. *Molecular Breeding: New Strategies in Plant Improvement* 3, 481–493.

Sarmasik, A., Chun, C.Z., Jang, I.-K., Lu, J.K. and Chen, T.T. (2001a) Production of transgenic live-bearing fish and crustaceans with replication-defective pantropic retroviral vectors. *Marine Biotechnology* 3, 177–184.

Sarmasik, A., Jang, I.-K., Chun, C.Z., Lu, J.K. and Chen, T.T. (2001b) Transgenic live-bearing fish and crustaceans produced by transforming immature gonads with replication-defective pantropic retroviral vectors. *Marine Biotechnology* 3, 470–477.

Sarmasik, A., Warr, G. and Chen, T.T. (2002) Production of transgenic medaka with increased resistance to bacterial pathogens. *Marine Biotechnology* 4, 310–322.

Sasaki, Y.F., Ayusawa, D. and Oishi, M. (1994) Construction of a normalized cDNA library by introduction of a semi-solid mRNA–cDNA hybridization system. *Nucleic Acids Research* 22, 987–992.

Sato, R. (1980) Variations in hatchability and hatching time, and estimation of the heritability for hatching time among individuals within the strain of coho salmon (*Oncorhynchus kisutch*). *Bulletin of National Research Institute of Aquaculture (Japan)* 1, 21–28.

Schaefer-Pregl, R., Ritter, E., Concilio, L., Hesselbach, J., Lovatti, L., Walkemeier, B., Thelen, H., Salmini, F. and Gebhardt, C. (1998) Analysis of quantitative trait loci (QTLs) and quantitative trait alleles (QTAs) for potato tuber yield and starch content. *Theoretical and Applied Genetics* 97, 834–846.

Schaperclaus, W. (1933) *Textbook of Pond Culture*. FL 311. Fish and Wildlife Service, US Department of the Interior, Washington, DC.

Schaperclaus, W. (1962) Bedeutung der Startkondition, insbesondere des Fettgehaltes, fur die Entstehung der infektiosen Bauchwassersucht beim Karpfen in Fruhjahr. *Deutsche Fischerei-Zeitung* 9, 321–327.

Scharf, J.M., Endrizzi, M.G., Wetter, A., Huang, S., Thompson, T.G., Zerres, K., Dietrich, W.F., Wirth, B. and Kunkel, L.M. (1998) Identification of a candidate modifying gene for spinal muscular atrophy by comparative genomics. *Nature Genetics* 20, 83–86.

Scheerer, P.D. and Thorgaard, G.H. (1983) Increased survival in salmonid hybrids by induced triploidy. *Canadian Journal of Fisheries and Aquatic Science* 40, 2040–2044.

Scheerer, P.D. and Thorgaard, G.H. (1987) Performance and developmental stability of triploid tiger trout (brown trout female × brook trout male). *Transactions of the American Fisheries Society* 116, 92–97.

Scheerer, P.D., Thorgaard, G.H., Allendorf, F.W. and Knudsen, K.L. (1986) Androgenetic rainbow trout produced from inbred and outbred sperm sources show similar survival. *Aquaculture* 57, 289–298.

Schneider, J.F., Hallerman, E.M., Yoon, S.J., Li, H., Gross, M.L., Liu, Z., Faras, A.J., Hackett, P.B., Kapuscinski, A.R. and Guise, K.S. (1989) Transfer of the bovine growth hormone gene into northern pike, *Esox lucius* (abstract). *Journal of Cell Biochemistry* 13B.

Schoonjans, L., Albright, G.M., Li, J.L., Collen, D. and Moreadith, R.W. (1996) Pluripotential rabbit embryonic stem cells are capable of forming overt coat color chimeras following injection into blastocysts. *Molecular Reproduction and Development* 45, 439–443.

Schreck, C.B. (1973) Uptake of 3-testosterone and influence of an anti androgen in tissues of rainbow trout (*Salmo gairdneri*). *General and Comparative Endocrinology* 21, 60–68.

Schroeder, D. and Palmer, C. (2003) Technology assessment and the ethical matrix. *Poiesis and Praxis* 1, 295–307.

Schrooten, C., Bovenhuis, H., Coppieters, W. and van Arendonk, J.A.M. (2000) Whole genome scan to detect quantitative trait loci for conformation and functional traits in dairy cattle. *Journal of Dairy Science* 83, 795–806.

Schuler, G.D. (1997) Pieces of the puzzle: expressed sequence tags and the catalog of human genes. *Journal of Molecular Medicine* 75, 694–698.

Schuler, G.D., Boguski, M.S., Hudson, T.J., Hui, L., Ma, J., Castle, A.B., Wu, X., Silva, J., Nusbaum, H.C., Birren, B.B., Slonim, D.K., Rozen, S., Stein, L.D., Page, D., Lander, E.S., Stewart, E.A., Aggarwal, A., Bajorek, E., Brady, S., Chu, S., Fang, N., Hadley, D., Harris, M., Hussain, S., Hudson, J.R. Jr *et al.* (1996) Genome maps 7. The human transcript map. Wall chart. *Science* 274, 547–562.

Schulte, P.M., Glémet, H.C., Fiebig, A.A. and Powers, D.A. (2000) Adaptive variation in lactate dehydrogenase-B gene expression: role of stress-responsive regulatory element. *Proceedings of the National Academy of Sciences USA* 97, 6597–6602.

Schultz, R.J. (1980) Role of polyploidy in the evolution of fishes. In: Lewis, W.H. (ed.) *Polyploidy: Biological Relevance*. Plenum Press, New York.

Schulz, W.R. and Miura, T. (2002) Spermatogenesis and its endocrine regulation. *Fish Physiology and Biochemistry* 26, 43–56.

Schwanzel-Fukuda, M. and Pfaff, D.W. (1989) Origin of luteinizing-hormone-releasing hormone neurons. *Nature* 338, 161–164.

Schwanzel-Fukuda, M., Bick, D. and Pfaff, D.W. (1989) Luteinizing-hormone-releasing hormone (LHRH)-expressing cells do not migrate normally in an inherited hypogonadal (Kallmann) syndrome. *Molecular Brain Research* 6, 311–326.

Schwanzel-Fukuda, M., Abraham, S., Crossin, K.L., Edelman, G.M. and Pfaff, D.W. (1991) Immunocytochemical demonstration of neural cell-adhesion molecule (NCAM) along the migration route of luteinizing-hormone-releasing hormone (LHRH) neurons in mice. *Journal of Comparative Neurology* 321, 1–8.

Schwanzel-Fukuda, M., Jorgenson, K.L., Bergen, H.T., Weesner, G.D. and Pfaff, D.W. (1992a) Biology of normal luteinizing hormone-releasing hormone neurons during and after their migration from olfactory placode. *Endocrine Reviews* 13, 623–634.

Schwanzel-Fukuda, M., Zheng, L.M., Bergen, H., Weesner, G. and Pfaff, D.W. (1992b) LHRH neurons: functions and development. *Progress in Brain Research* 93, 189–201.

Schwarz, F.J., Plank, J. and Kirchgessner, M. (1985) Effects of protein or energy restriction with subsequent realimentation on performance parameters of carp (*Cyprinus carpio* L.). *Aquaculture* 48, 23–33.

Scientists' Working Group on Biosafety (1998) *Manual for Assessing Ecological and Human Health Effects of Genetically Engineered Organisms. Part One: Introductory Materials and Supporting Text for Flowcharts. Part Two: Flowcharts and Worksheets*. Edmunds Institute, Edmunds, Washington; available at www.Edmonds-Institute.org

Scott, A.G., Penman, D.J., Beardmore, J.A. and Skibinski, D.O.F. (1989) The 'YY' supermale in *Oreochromis niloticus* (L.) and its potential in aquaculture. *Aquaculture* 78, 237–251.

Scott, M.P., Haymes, K.M. and Williams, S.M. (1992) Parentage analysis using RAPD PCR. *Nucleic Acids Research* 20, 5493.

Scovel, G., Ben-Meir, H., Ovadis, M., Itzhaki, H. and Vainstein, A. (1998) RAPD and RFLP markers tightly linked to the locus controlling carnation (*Dianthus caryophyllus*) flower type. *Theoretical and Applied Genetics* 96, 117–122.

Seafood Datasearch (1999) Controversy over genetically modified salmon erupts in UK. Seafood Datasearch, Lexington, Massachusetts; available at www.seafood.com.

Sealey, W.N., Davis, J.T. and Gatlin, D.M. III (1998) *Restricted Feeding Regimes Increase Production Efficiency in Channel Catfish*. Publication No. 189. Southern Regional Aquaculture Center, Stoneville, Mississippi.

Seaman, W.J., Nappier, J.L., Olsen, R.F., Charlton, M.D., Skinner, P.J., Weaver, R.J. and Hoffman, G.A. (1988) The lack of a growth-promoting effect of orally administered bovine somatotropin in the rat body-weight-gain bioassay. *Fundamental and Applied Toxicology* 10, 287–294.

Sechen, S. (1989) Review of bovine somatotropin by the Food and Drug Administration. In: *Advanced Technologies Facing the Dairy Industry: bST*. Mimeo Series No. 133. Cornell Cooperative Extension Service, Cornell University, Ithaca, New York, p. 101.

Seeb, J., Thorgaard, G., Hershberger, W.K. and Utter, F.M. (1986) Survival in diploid and triploid Pacific salmon hybrids. *Aquaculture* 57, 375.

Seeb, J.E., Seeb, L.W., Oates, D.W. and Utter, F.M. (1987) Genetic variation and postglacial dispersal of populations of northern pike (*Esox lucius*) in North America. *Canadian Journal of Fisheries and Aquatic Sciences* 44, 556–561.

Seeb, J.E., Thorgaard, G.H. and Utter, F.M. (1988) Survival and allozyme expression in diploid and triploid hybrids between chum, chinook, and coho salmon. *Aquaculture* 72, 31–48.

Seeb, J.E., Thorgaard, G.H. and Tynan, T. (1993) Triploid hybrids between chum salmon female × chinook salmon male have early sea-water tolerance. *Aquaculture* 117, 37–45.

Seelbach, P.W. and Whelan, G.E. (1988) Identification and contribution of wild and hatchery steelhead stocks in Lake Michigan tributaries Michigan USA. *Transactions of the American Fisheries Society* 117, 441–451.

Segal, D.J., Beerli, R.R., Blancafort, P., Dreier, B., Effertz, K., Huber, A., Koksch, B., Lund, C.V., Magnenat, L., Valente, D. and Barbas, C.F. III (2003) Evaluation of a modular strategy for the construction of novel polydactyl zinc finger DNA-binding proteins. *Biochemistry* 42, 2137–2148.

Sekine, S., Mizukami, T., Saito, A., Kawauchi, H. and Itoh, S. (1989) Isolation and characterization of a novel growth hormone cDNA from chum salmon (*Oncorhynchus keta*). *Biochimica et Biophysica Acta* 1009, 117–120.

Sellars, M.J., Wood, A.T., Dixon, T.J., Dierens, L.M. and Coman, G.J. (2009) A comparison of heterozygosity, sex ratio and production traits in two classes of triploid *Penaeus* (*Marsupenaeus*) *japonicus* (Kuruma shrimp): polar body I vs II triploids. *Aquaculture* 296, 207–212.

Selvamani, M.J.P., Degnan, S.M. and Degnan, B.M. (2001) Microsatellite genotyping of individual abalone larvae: parentage assignment in aquaculture. *Marine Biotechnology* 3, 478–485.

Senhorini, J.A., Figueiredo, G.M., Fontes, N.A. and Carolsfeld, J. (1988) Larval and fry culture of pacu, *Piaractus mesopotamicus*, tambaqui, *Colossoma macropomum*, and their reciprocal hybrids. *Bol. Tec. CEPTA* 1, 19–30.

Senthilkumaran, B., Okuzawa, K., Gen, K. and Kagawa, H. (2001) Effects of serotonin, GABA and neuropeptide Y on seabream gonadotropin releasing hormone release *in vitro* from preoptic-anterior hypothalamus and pituitary of red seabream, *Pagrus major*. *Journal of Neuroendocrinology* 13, 395–400.

Service, R.F. (1998) Seed-sterilizing 'terminator technology' sows discord. *Science* 28, 850–851.

Shah, M.S. (1988) Female homogamety in tilapia (*Oreochromis niloticus*) revealed by gynogenesis. In: Pullin, R.S.V., Bhukasawan, T., Tonguthai, K. and Maclean, J.L. (eds) *Proceedings of the Second International Symposium on Tilapia in Aquaculture, Bangkok, Thailand, 16–20 March 1987*. ICLARM Conference Proceedings No. 15. Department of Fisheries Thailand and ICLARM, Manila.

Shah, M.S. and Beardmore, J.A. (1986) Production of triploid tilapia (*Oreochromis niloticus*) and their comparative growth performance with normal diploids. In: *Symposium E22*. European Inland Fisheries Advisory Commission, Bordeaux, France.

Shah, M.S., Thorgaard, G.H. and Wheeler, P. (1999) Viability of diploid and triploid hybrids between a few salmon and trout species. In: Joseph, M.M., Menon, N.R. and Nair, N.U. (eds) *Proceedings of the Fourth Indian Fisheries Forum, 24–28 November 1996, Kochi, India*. Asian Fisheries Society, Indian Branch, Mangalore, India, pp. 259–262.

Shamblott, M.J. and Chen, T.T. (1992) Identification of a second insulin-like growth factor in a fish species. *Proceedings of the National Academy of Sciences USA* 89, 8913–8917.

Shapira, Y., Magen, Y., Zak, T., Kotler, M., Hulata, G and Levavi-Sivan, B. (2005) Differential resistance to koi herpes virus (KHV)/carp interstitial nephritis and gill necrosis virus (CNGV) among common carp (*Cyprinus carpio* L.) strains and crossbreds. *Aquaculture* 245, 1–11.

Sharara, F.I. and Giudice, L.C. (1997) Role of growth hormone in ovarian physiology and onset of puberty. *Journal of the Society for Gynecologic Investigation* 4, 2–7.

Sharma, M., Kambadur, R., Matthews, K.G., Somers, W.G., Devlin, G.P., Conaglen, J.V., Fowke, P.J. and Bass, J.J. (1999) Myostatin, a transforming growth factor-beta superfamily member, is expressed in heart muscle and is upregulated in cardiomyocytes after infarct. *Journal of Cell Physiology* 80, 1–9.

Shatkin, A.J. and Manley, J.L. (2000) The ends of the affair: capping and polyadenylation. *Nature Structural Biology* 7, 838–842.

Shatkin, G.M. and Allen, S.K. (1989) Recommendations for commercial production of triploid oysters. *Journal of Shellfish Research* 8, 449.

Shears, M.A., Fletcher, G.L., Hew, C.L., Gauthier, S. and Davies, P.L. (1991) Transfer, expression, and stable inheritance of antifreeze protein genes in Atlantic salmon (*Salmo salar*). *Molecular Marine Biology and Biotechnology* 1, 58–63.

Sheehan, R.J., Shasteen, S.P., Suresh, A.V., Kapuscinski, A.R. and Seeb, J.E. (1999) Better growth in all-female diploid and triploid rainbow trout. *Transactions of the American Fisheries Society* 128, 491–498.

Shelton, C.J., MacDonald, A.G. and Johnstone, R. (1986) Induction of triploidy in rainbow trout using nitrous oxide. *Aquaculture* 58, 155–159.

Shelton, W.L. (1986) Broodstock development for monosex production of grass carp. *Aquaculture* 57, 311–319.

Shelton, W.L. (1987) Genetic manipulations – sex control of exotic fish for stocking. In: Tiews, K. (ed.) *Selection, Hybridization and Genetic Engineering in Aquaculture*, Vol. 2. Heenemann Verlagsgesellschaft, Berlin, pp. 175–197.

Shelton, W.L. and Jensen, G.L. (1979) *Production of Reproductively Limited Grass Carp for Biological Control of Aquatic Weeds*. Bulletin No. 39. Water Resources Research Institute, Auburn University, Auburn, Alabama.

Shelton, W.L., Hopkins, K.D. and Jensen, G.L. (1978) Use of hormones to produce monosex tilapia for aquaculture. In: Smitherman, R.O., Shelton, W.L. and Grover, J.H. (eds) *Culture of Exotic Fishes Symposum Proceedings*. Fish Culture Section, American Fisheries Society, Auburn University, Auburn, Alabama, p. 10.

Shelton, W.L., Meriwether, F.H., Semmens, K.J. and Calhoun, W.E. (1983) Progeny sex ratios from intraspecific pair spawnings of *Tilapia aurea* and *T. nilotica*. In: Fishelson, L. and Yaron, Z. (compilers) *Proceedings of the International Symposium on Tilapia in Aquaculture*. Tel Aviv University, Tel Aviv, Israel, pp. 270–280.

Shepperd, V.D. (1984) Androgen sex inversion and subsequent growth of red tilapia and Nile tilapia. MSc thesis, Auburn University, Auburn, Alabama.

Sheridan, M.A., Plisetskaya, E.M., Bern, H.A. and Gorbman, A. (1987) Effects of somatostatin-25 and urotensin-II on lipid and carbohydrate-metabolism of coho salmon, *Oncorhynchus kisutch*. *General and Comparative Endocrinology* 66, 405–414.

Sherwood, N.M. and Wu, S. (2005) Developmental role of GnRH and PACAP in a zebrafish model. *General and Comparative Endocrinology* 142, 74–80.

Shields, B.A., Guise, K.S. and Underhill, J.C. (1990) Chromosomal and mitochondrial DNA characterization of a population of dwarf cisco (*Coregonus artedii*) in Minnesota. *Canadian Journal of Fisheries and Aquatic Sciences* 47, 1562–1569.

Shikano, T. and Fujio, Y. (1997) Successful propagation in seawater of the guppy *Poecilia reticulata* with reference to high salinity tolerance at birth. *Fisheries Science* 63, 573–575.

Shikano, T., Nakadate, M., Nakajima, M. and Fujio, Y. (1997) Heterosis and maternal effects in salinity tolerance of the guppy *Poecilia reticulata*. *Fisheries Science* 63, 893–896.

Shikano, T., Chiyokubo, T. and Taniguchi, N. (2001a) Effect of inbreeding on salinity tolerance in the guppy (*Poecilia reticulata*). *Aquaculture* 202, 45–55.

Shikano, T., Chiyokubo, T. and Taniguchi, N. (2001b) Temporal changes in allele frequency, genetic variation and inbreeding depression in small populations of the guppy, *Poecilia reticulata*. *Heredity* 86, 153–160.

Shikina, S., Ihara, S. and Yoshizaki, G. (2008) Culture conditions for maintaining the survival and mitotic activity of rainbow trout tranplantable type A spermatogonia. *Molecular Reproduction and Development* 75, 529–537.

Shim, H., Gutierrez-Adan, A., Chen, L.R., BonDurant, R.H., Behboodi, E. and Anderson, G.B. (1997) Isolation of pluripotent stem cells from cultured porcine primodial germ cells. *Biology of Reproduction* 57, 1089–1095.

Shimizu, N., Lee, J.-Y., Shimizu, Y., Ohtake, H., Sato, Y. and Asakawa, S. (2002) Genomics of the Pacific oyster *Crassostrea gigas*. In: Shimizu, N., Aoki, T., Hirono, I. and Takashima, F. (eds) *Aquatic Genomics: Steps Toward a Great Future*. Springer-Verlag, New York, pp. 128–132.

Shimomura, K., Low-Zeddies, S.S., King, D.P., Steeves, T.D.L., Whiteley, A., Kushla, J., Zemenides, P.D., Lin, A., Vitaterna, M.H., Churchill, G.A. and Takahashi, J.S. (2001) Genome-wide epistatic interaction analysis reveals complex genetic determinants of circadian behavior in mice. *Genome Research* 11, 959–980.

Shirak, A., Vartin, J., Don, J. and Avtalion, R.R. (1998) Production of viable diploid mitogynogenetic *Oreochromis aureus* using the cold shock and its optimization through definition of cleavage time. *Israeli Journal of Aquaculture/Bamidgeh* 50, 140–150.

Shirak, A., Palti, Y., Cnaani, A., Korol, A., Hulata, G., Ron, M. and Avtalion, R.R. (2002) Association between loci with deleterious alleles and distorted sex ratios in an inbred line of tilapia (*Oreochromis aureus*). *Journal of Heredity* 93, 270–276.

Shirak, A., Seroussi, E., Cnaani, A., Howe, A.E., Domokhovsky, R., Zilberman, N., Kocher, T.D., Hulata, G. and Ron, M. (2006) Amh and Dmrta2 genes map to tilapia (*Oreochromis* spp.) linkage group 23 within quantitative trait locus regions for sex determination. *Genetics* 174, 1573–1581.

Silva, A.J. and White, R. (1988) Inheritance of allelic blueprints for methylation patterns. *Cell* 54, 145–152.

Silverstein, J.T., Shearer, K.D., Dickhoff, W.W. and Plisetskaya, E.M. (1998) Effects of growth and fatness on sexual development of chinook salmon (*Oncorhynchus tshawytscha*) parr. *Canadian Journal of Fisheries and Aquatic Sciences* 55, 2376–2382.

Silverstein, J.T., Wolters, W.R., Shimizu, M. and Dickhoff, W.W. (2000) Bovine growth hormone treatment of channel catfish: strain and temperature effects on growth, plasma IGF-I levels, feed intake and efficiency and body composition. *Aquaculture* 190, 77–88.

Simpson, T.H. (1975) Endocrine aspects of salmonid culture. *Proceedings of the Royal Society of Edinburgh (B)* 17, 252.

Simpson, T.H., Johnstone, R. and Youngson, A.F. (1976) *Sex Reversal in Salmonids*. CM/E:48. International Council for the Exploration of the Sea, Copenhagen.

Sims, M. and First, N.L. (1994) Production of calves by transfer of nuclei from cultured inner cell mass cells. *Proceedings of the National Academy of Sciences USA* 91, 6143–6147.

Siraj, S.S., Smitherman, R.O., Castillo-Gallusser, S. and Dunham, R.A. (1983) Reproductive traits for three-year classes of *Tilapia nilotica* and maternal effects on their progeny. In: Fishelson, L. and Yaron, Z. (compilers) *Proceedings of the International Symposium on Tilapia in Aquaculture*. Tel Aviv University, Tel Aviv, Israel, pp. 208–216.

Slate, J., Kruuk, L.E.B., Marshall, T.C., Pemberton, J.M. and Clutton-Brock, T.H. (2000) Inbreeding depression influences lifetime breeding success in a wild population of red deer (*Cervus elaphus*). *Proceedings of the Royal Society of London, Series B: Biological Sciences* 267, 1657–1662.

Small, S.A. and Benfey, T.J. (1987) Cell size in triploid salmon. *Journal of Experimental Zoology* 241, 339–342.

Small, S.A. and Randall, D.J. (1989) Effects of triploidy on the swimming performance of coho salmon (*Oncorhynchus kisutch*). *Canadian Journal of Fisheries and Aquatic Sciences* 46, 243–245.

Smisek, J. (1979) Considerations of body conformation, heritability and biochemical characters in genetic studies of carp in Czechoslovakia. *Bulletin VURH Vodnany* 15, 3–6. In *Animal Breeding Absstracts* 1980, 48, 302.

Smit, A.F. and Riggs, A.D. (1996) Tiggers and DNA transposon fossils in the human genome. *Proceedings of the National Academy of Sciences USA* 93, 1443–1448.

Smith, D.Y. and Benfey, T.J. (1999) The reproductive physiology of triploid female brook trout (*Salvelinus fontinalis*). In: *Molecular Studies in Fish Endocrinology*. Department of Fisheries and Oceans, Vancouver, British Columbia, Canada and Towson University, Baltimore, Maryland, pp. 91–94.

Smith, D.Y. and Benfey, T.J. (2000) Effects of long term estradiol-17β treatment on the growth and physiology of female triploid brook trout (*Salvelinus fontinalis*). In: Norberg, B., Kjesbu, O.S., Taranger, G.L., Andersson, E. and Stefansson, S.O. (eds) *Reproductive Physiology of Fish 2000. 6th International Symposium on the Reproductive Physiology of Fish, Bergen, Norway, 4–9 July 1999*. Department of Fisheries and Marine Biology, University of Bergen, Bergen, Norway, p. 299.

Smith, E.J., Lyons, L.A., Cheng, H.H. and Suchyta, S.P. (1997) Comparative mapping of the chicken genome using the East Lansing reference population. *Poultry Science* 76, 743–747.

Smith, H.T. (1976) Various aspects of reproductive control in freshwater fishes. MSc thesis, Virginia Polytechnic Institute and State University, Blacksburg, Virginia.

Smith, L.C. and Wilmut, I. (1989) Influence of nuclear and cytoplasmic activity on the development *in vivo* of sheep embryos after nuclear transplantation. *Biology of Reproduction* 40, 1027–1035.

Smith, L.T. and Lemoine, H.L. (1979) Colchicine-induced polyploidy in brook trout. *Progressive Fish-Culturist* 41, 86–88.

Smith, R.R., Kincaid, H.L., Regenstein, J.M. and Rumsey, G.L. (1988) Growth, carcass composition, and taste of rainbow-trout of different strains fed diets containing primarily plant or animal protein. *Aquaculture* 70, 309–321.

Smith, S.B. (1957) Survival and growth of wild and hatchery rainbow trout (*Salmo gairdneri*) in Corbett Lake, BC. *Canadian Fisheries Research Board* 20, 7–12.

Smith, T.I.J. (1988) Aquaculture of striped bass and its hybrids in North America. *Aquaculture Magazine* 14, 40–49.

Smith, T.P., Lopez-Corrales, N.L., Kappes, S.M. and Sonstegard, T.S. (1997) Myostatin maps to the interval containing the bovine mh locus. *Mammalian Genome* 8, 742–744.

Smitherman, R.O. and Dunham, R.A. (1985) Genetics and breeding. In: Tucker, C.S. (ed.) *Channel Catfish Culture*. Elsevier Scientific Publishing, Amsterdam, pp. 283–316.

Smitherman, R.O., Dunham, R.A., Bice, T.O. and Horn, J.L. (1984) Reproductive efficiency in the reciprocal pairings between two strains of channel catfish. *Progressive Fish-Culturist* 46, 106–110.

Smitherman, R.O., Khater, A.A., Cassell, N.I. and Dunham, R.A. (1988) Reproductive performance of three strains of *Oreochromis niloticus*. *Aquaculture* 70, 29–37.

Smithies, O. (1955) Zone electrophoresis in starch gels: group variations in the serum proteins of normal human adults. *Biochemical Journal* 61, 629–641.

Sneed, K.E. (1968) Warmwater fish cultural laboratories. In: *Progress in Sport Fishery Research*. US Fish and Wildlife Service, Washington, DC, pp. 218–255.

Soether, B.-S. and Jobling, M. (1999) The effects of ration level on feed intake and growth, and compensatory growth after restricted feeding, in turbot *Scophthalmus maximus* L. *Aquaculture Research* 30, 647–653.

Sogin, M.L. (1991) Early evolution and the origin of eukaryotes. *Current Opinion in Genetics and Development* 1, 457–463.

Solar, I.I. and Donaldson, E.M. (1985) *Studies on Genetic and Hormonal Sex Control in Domesticated Rainbow Trout. 1. The Effect of Heat Shock Treatment for Induction of Triploidy in Cultured Rainbow Trout* (Salmo gairdneri *Richardson*). Canadian Technical Report Fisheries Aquatic Science No. 1379. West Vancouver Laboratory, Department of Fisheries and Oceans, West Vancouver, British Columbia, Canada.

Solar, I.I., Donaldson, E.M. and Hunter, G.A. (1984) Induction of triploidy in rainbow trout (*Salmo gairdneri* Richardson) by heat shock, and investigation of early growth. *Aquaculture* 42, 57–67.

Solemdal, P. (1997) Maternal effects – a link between the past and the future. *Journal of Sea Research* 37, 213–227.

Soller, M. (1994) Marker-assisted selection – an overview. *Animal Biotechnology* 5, 193–207.

Somorjai, I.M.L., Danzmann, R.G. and Ferguson, M.M. (2003) Distribution of temperature tolerance quantitative trait loci in arctic charr (*Salvelinus alpinus*) and inferred homologies in rainbow trout (*Oncorhynchus mykiss*). *Genetics* 165, 1443–1456.

Sonesson, A.K. and Meuwissen, T.H.E. (2001) Minimization of rate of inbreeding for small populations with overlapping generations. *Genetical Research* 77, 285–292.

Songlin, C., Yunhan, H. and Manfred, S. (2002) Development of a positive:negative selection procedure for gene targeting in fish cells. *Aquaculture* 214, 67–79.

Sonneman, E. (1971) Regeneration of internal organs in carps. *Naturwissenschaftliche Rundschau* 24, 471.

Southern, E.M. (1975) Detection of specific sequences among DNA fragments separated by gel electrophoresis. *Journal of Molecular Biology* 98, 503–517.

Souza, S.C., Frick, G.P., Wang, X., Kopchick, J.J., Lobo, R.B. and Goodman, H.M. (1995) A single arginine residue determines species specificity of the human growth hormone receptor. *Proceedings of the National Academy of Sciences USA* 92, 959–963.

Speksnijder, J., Hage, W., Lanser, P., Collodi, P. and Zivkovic, D. (1997) *In vivo* assay for the developmental competence of embryo-derived zebrafish cell lines. *Molecular Marine Biology and Biotechnology* 6, 21–32.

Spelman, R.J. and Garrick, D.J. (1998) Genetic and economic responses for within-family marker-assisted selection in dairy cattle breeding schemes. *Journal of Dairy Science* 81, 2942–2950.

Spidle, A.P., Schill, W.B., Lubinski, B.A. and King, T.L. (2001) Fine-scale population structure in Atlantic salmon from Maine's Penobscot River drainage. *Conservation Genetics* 2, 11–24.

Springate, J.R.C. and Bromage, N.R. (1985) Effects of egg size on early growth and survival in rainbow trout (*Salmo gairdneri* Richardson). *Aquaculture* 47, 163–172.

Springate, J.R.C., Bromage, N.R. and Cumaranatunga, P.R.T. (1985) The effects of different ration on fecundity and egg quality in the rainbow trout (*Salmo gairdneri*). In: Cowey, C.B., Mackie, A.M. and Bell, J.G. (eds) *International Symposium on Feeding and Nutrition in Fish, Aberdeen, UK*. pp. 371–393.

Squire, R.D. (1968) The effects of colchicine on various stages of the guppy, *Poecilia reticulatus*, and the effects of colchicine and mercaptoethanol on embryos of the Siamese fighting fish, *Betta splendens*. MSc thesis, Hofstra University, New York.

Srinivas, R.V., Birkedal, B., Owens, R.J., Anantharamaiah, G.M., Segrest, J.P. and Compans, R.W. (1990) Antiviral effects of apolipoprotein A-I and its synthetic amphipathic peptide analogs. *Virology* 176, 48–57.

Sriramula, V. (1962) Effects of colchicine on the somatic chromosomes of *Oryzias elastigma* (McClelland). *La Cellule* 63, 369.

Stahl, G. (1983) Differences in the amount and distribution of genetic variation between natural populations and hatchery stocks of Atlantic salmon. *Aquaculture* 33, 23–32.

Standal, M. and Gjerde, B. (1987) Genetic variation in survival of Atlantic salmon during the searearing period. *Aquaculture* 66, 197–207.

Stanley, J.G. (1976a) Production of hybrid, androgenetic, and gynogenetic grass carp and carp. *Transactions of the American Fisheries Society* 105, 10–16.

Stanley, J.G. (1976b) Female homogamety in grass carp (*Ctenopharyngodon idella*) determined by gynogenesis. *Journal of Fisheries Research Board of Canada* 33, 1372–1374.

Stanley, J.G. (1979) Control of sex in fishes, with special reference to the grass carp. In: Shireman, J.V. (ed.) *Proceedings Grass Carp Conference*. University of Florida Press, Gainesville, Florida, p. 21.

Stanley, J.G. and Jones, J.B. (1976) Morphology of androgenetic and gynogenetic grass carp, *Ctenopharyngodon idella* (Valenciennes). *Journal of Fish Biology* 9, 523–528.

Stanley, J.G., Allen, S.K. and Hidu, H. (1981) Polyploidy induced in the early embryo of the American oyster with cytochalasin B. *Journal of Shellfish Research* 1, 123.

Stanley, J.G., Hidu, H. and Allen, S.K. Jr (1984) Growth of American oysters increased by polyploidy induced by blocking meiosis I but not meiosis II. *Aquaculture* 37, 147–155.

Starnes, T., Rasila, K.K., Robertson, M.J., Brahmi, Z., Dahl, R., Christopherson, K. and Hromas, R. (2006) The chemokine CXCL14 (BRAK) stimulates activated NK cell migration: implications for the downregulation of CXCL14 in malignancy. *Experimental Hematology* 34, 1101–1105.

Steeb, C.B., Lamb, J., Shoubridge, C.A., Tivey, D.R., Penttila, I. and Read, L.C. (1998) Systemically but not orogastrically delivered insulin-like growth factor (IGF)-I and long [Arg3]IGF-I stimulates intestinal disaccharidase activity in two age groups of suckling rats. *Pediatric Research* 44, 663–672.

Steele, N.C. and Pursel, V.G. (1990) Nutrient partitioning by transgenic animals. *Annual Review of Nutrition* 10, 213–232.

Steffens, W. (1974) Results of crossing between wild and domestic carp (*Cyprinus carpio* L.). *Biologisches Zentralblatt* 93, 129–139.

Steffens, W., Jaehnichen, H. and Fredrich, F. (1990) Possibilities of sturgeon culture in Central Europe. *Aquaculture* 89, 101–122.

Stein, J., Phillips, R.B. and Devlin, R.H. (2001) Identification of the Y chromosome in chinook salmon (*Oncorhynchus tshawytscha*). *Cytogenetics and Cell Genetics* 92, 108–110.

Steiner, H., Hultmark, D., Engstrom, A., Bennick, H. and Boman, H.G. (1981) Sequence and specificity of two antibacterial proteins involved in antibacterial immunity. *Nature* 292, 246–248.

Stephens, L.B. and Downing, S.L. (1988) Inhibiting first polar body formation in *Crassostrea gigas* produces tetraploids, not meiotic I triploids. *Journal of Shellfish Research* 7, 550–551.

Stevens, E.D. and Devlin, R.H. (2000a) Gill morphometry in growth hormone transgenic Pacific coho salmon, *Oncorhynchus kisutch*. *Environmental Biology of Fishes* 58, 113–117.

Stevens, E.D. and Devlin, R.H. (2000b) Intestinal morphology in growth hormone transgenic Pacific coho salmon, *Oncorhynchus kisutch* Walbaum. *Journal of Fish Biology* 56, 191–195.

Stevens, E.D. and Sutterlin, A. (1999) Gill morphometry in growth hormone transgenic Atlantic salmon. *Environmental Biology of Fishes* 54, 405–411.

Stevens, E.D., Sutterlin, A. and Cook, T. (1998) Respiratory metabolism and swimming performance in growth hormone transgenic Atlantic salmon. *Canadian Journal of Fisheries and Aquatic Sciences* 55, 2028–2035.

Stice, S.L., Strelchenko, N.S., Keefer, C.L. and Matthews, L. (1996) Pluripotent bovine embryonic cell lines direct embryonic development following nuclear transfer. *Biology of Reproduction* 54, 100–110.

Stillwell, E.J. and Benfey, T.J. (1999) The blood hemoglobin concentration of triploid brook trout. In: *Fish Performance Studies*. Department of Fisheries and Oceans, Vancouver, British Columbia, Canada and Towson University, Baltimore, Maryland, pp. 1155–1159.

Stoffregen, D.A., Bowser, P.R. and Babish, J.G. (1996) Antibacterial chemotherapeutants for finfish aquaculture: a synopsis of laboratory and field efficacy and safety studies. *Journal Aquatic Animal Health* 8, 181–207.

Stoll, B.R., Leipold, H.R., Milstein, S. and Edwards, D.A. (2000) A mechanistic analysis of carrier-mediated oral delivery of protein therapeutics. *Journal of Controlled Release* 64, 217–228.

Stone, N.M. (1981) Growth of male and female *Tilapia nilotica* in ponds and cages. MSc thesis, Auburn University, Auburn, Alabama.

Stoneking, M., Wagner, D.J. and Hildebrand, A.C. (1981) Genetic evidence suggesting subspecific differences between northern and southern populations of brook trout, *Salvelinus fontinalis*. *Copeia* 1981, 810.

Storebakken, T., Hung, S.S.O., Calvert, C.C. and Plisetskaya, E.M. (1991) Nutrient partitioning in rainbow trout at different feeding rates. *Aquaculture* 96, 191–203.

Strickberger, M.W. (1985) *Genetics*. Macmillan, New York.

Stromberg, L.D., Dudley, J.W. and Rufener, G.K. (1994) Comparing conventional early generation selection with molecular marker-assisted selection in maize. *Crop Science* 34, 1221–1225.

Stuart, G.W., McMurray, J.V. and Westerfield, M. (1988) Replication, integration and stable germ-line transmission of foreign sequences injected into early zebrafish embryos. *Development* 103, 403–412.

Stuart, G.W., Vielkind, J.R., McMurray, J.V. and Westerfield, M. (1990) Stable lines of transgenic zebrafish exhibit reproducible patterns of transgenic expression. *Development* 109, 577–584.

Suarez, J.A., Gitterle, T., Angarita, M.R. and Rye, M. (1999) Genetic parameters for harvest weight and pond survival in *Litopenaeus vannamei*. In: *Aquaculture Europe '99 – Towards Predictable Aquaculture. Abstracts of contributions presented at the International Conference 'Aquaculture Europe '99' held in Trondheim, Norway, 7–10 August*. Special Publication No. 27. European Aqauculture Society, Oostende, Belgium, pp. 232–233.

Sugama, K., Taniguchi, N., Arakawa, T. and Kitajima, C. (1988) Isozyme expression of artificially induced ploidy in red sea bream, black sea bream and their hybrid. In: *Report No. 10*. US Marine Biology Institute, Kochi University/Kochidai Kaiyoseibutsu Kenpo, pp. 75–81.

Sugimoto, S., Tamura, I., Yamaguchi, K. and Yokoo, Y. (1991) Characterization of recombinant salmon growth hormone 1 expressed in *Escherichia coli*. *Bulletin of the Japanese Society of Scientific Fisheries* 57, 1759–1765.

Suhrbier, A.D., Cheney, D., Elston, R., Friedman, C., Griffin, F., Langdon, C., Davis, J. and Christy, A. (2005) Update on research concerning the summertime mortality of the Pacific oyster. *Journal of Shellfish Research* 24, 336.

Sumanas, S. and Larson, J.D. (2002) Morpholino phosphorodiamidate oligonucleotides in zebrafish: a recipe for functional genomics? *Briefings in Functional Genomics & Proteomics* 1, 239–256.

Sumanas, S., Zhang, B., Dai, R. and Lin, S. (2005) 15-zinc finger protein Bloody Fingers is required for zebrafish morphogenetic movements during neurulation. *Developmental Biology* 283, 85–96.

Summerton, J. and Weller, D. (1997) Morpholino antisense oligomers: design, preparation, and properties. *Antisense and Nucleic Acid Drug Development* 7, 187–195.

Summerton, J., Stein, D., Huang, S.B., Matthews, P., Weller, D. and Partridge, M. (1997) Morpholino and phosphorothioate antisense oligomers compared in cell-free and in-cell systems. *Antisense and Nucleic Acid Drug Development* 7, 63–70.

Sumpter, J.P., LeBail, P.Y., Pickering, A.D., Pottinger, T.G. and Carragher, J.F. (1991a) The effect of starvation on growth and plasma growth hormone concentration of rainbow trout, *Oncorhynchus mykiss*. *General and Comparative Endocrinology* 83, 84–102.

Sumpter, J.P., Lincoln, R.F., Bye, V.J., Carragher, J.F. and Le Bail, P.Y. (1991b) Plasma growth hormone levels during sexual maturation in diploid and triploid rainbow trout (*Oncorhynchus mykiss*). *General and Comparative Endocrinology* 83, 103–110.

Sun, H.S. and Kirkpatrick, B.W. (1996) Exploiting dinucleotide microsatellites conserved among mammalian species. *Mammalian Genome* 7, 128–132.

Sun, L., Bradford, S. and Barnes, D. (1995a) Feeder cell cultures for zebrafish embryonal cells *in vitro*. *Molecular Marine Biology and Biotechnology* 4, 43–50.

Sun, L., Bradford, C.S., Ghosh, C., Collodi, P. and Barnes, D.W. (1995b) ES-like cell cultures derived from early zebrafish embryos. *Molecular Marine Biology and Biotechnology* 4, 193–199.

Sun, L., Bradford, S., Ghosh, C., Collodi, P. and Barnes, D. (1995c) Cell cultures derived from early zebrafish embryos exhibit *in vitro* characteristics of pluripotent cells. *Journal of Marine Biotechnology* 3, 211–215.

Sun, X. and Liang, L. (2004) A genetic linkage map of common carp (*Cyprinus carpio* L.) and mapping of a locus associated with cold tolerance. *Aquaculture* 238, 165–172.

Sundararaj, B.I. and Naygar, S.K. (1969) Effects of castration and/or hypophysectomy on the seminal vesicles of the catfish, *Heteropneustes fossilis* (Bloch). *Journal of Experimental Zoology* 172, 369–384.

Sundström, L.F., Löhmus, M., Tymchuk, W.E. and Devlin, R.H. (2007) Gene–environment interactions influence ecological consequences of transgenic animals. *Proceedings of the National Academy of Sciences USA* 104, 3889–3894.

Sunyer, J.O., Zarkadis, I.K., Lambris, J.D. (1998) Complement diversity: a mechanism for generating immune diversity? *Immunology Today* 19, 519–523.

Supan, J.E., Wilson, C.E. and Allen, S.K. Jr (2000a) The effect of cytochalasin B dosage on the survival and ploidy of *Crassostrea virginica* (Gmelin) larvae. *Journal of Shellfish Research* 19, 125–128.

Supan, J.E., Allen, S.K. Jr and Wilson, C.A. (2000b) Tetraploid eastern oysters: an arduous effort. *Journal of Shellfish Research* 19, 655.

Supsumrarn, M. (1987) Preliminary study on genetics of *Labeo rohita* (Boche). MSc thesis, Kasetsart University, Bangkok.

Surridge, A.K., Bell, D.J., Rico, C. and Hewitt, G.M. (1997) Polymorphic microsatellite loci in the European rabbit (*Oryctolagus cuniculus*) are also amplified in other lagomorph species. *Animal Genetics* 28, 302–305.

Sutterlin, A.M., Holder, J. and Benfey, T.J. (1987) Early survival rates and subsequent morphological abnormalities in landlocked, anadromous and hybrid (landlocked anadromous) diploid and triploid Atlantic salmon. *Aquaculture* 64, 157–164.

Suzuki, R., Yamaguchi, M., Ito, T. and Toi, J. (1976) Differences in growth and survival in various races of the common carp. *Bulletin Freshwater Fisheries Research Laboratory* 26, 59–69 (Japanese with English summary).

Suzuki, R., Nakanishi, T. and Oshiro, T. (1985) Survival, growth and sterility of induced triploids in the cyprinid loach *Misgurnus anguillicaudatus*. *Bulletin of the Japanese Society of Scientific Fisheries/Nissuishi* 51, 889–894.

Suzuki, Y.A., Shin, K. and Lonnerdal, B. (2001) Molecular cloning and functional expression of a human intestinal lactoferrin receptor. *Biochemistry* 40, 15771–15779.

Svanberg, E., Frost, R.A., Lang, C.H., Isgaard, J., Jefferson, L.S., Kimball, S.R. and Vary, T.C. (1998) hGH Tissue distribution, turnover, and glycosylation of the long and short growth hormone receptor isoforms in rat tissues. *Endocrinology* 139, 2824–2830.

Svobodova, Z., Kola, J. and Flajshans, M. (1998) The first findings on the differences in complete blood count between diploid and triploid tench, *Tinca tinca* L. *Acta Veterinaria (Brno)* 67, 243–248.

Svobodova, Z., Flajshans, M., Kolarova, J., Modra, H., Svoboda, M. and Vajcova, V. (2001) Leukocyte profiles of diploid and triploid tench, *Tinca tinca* L. *Aquaculture* 198, 159–168.

Swain, D.P., Sinclair, A.F. and Hanson, J.M. (2007) Evolutionary response to size-selective mortality in an exploited fish population. *Proceedings of the Royal Society of London, Series B: Biological Sciences* 274, 1015–1022.

Swain, J.L., Stewart, T.A. and Leader, P. (1987) Parental legacy determines methylation and expression of an autosomal transgene: a molecular mechanism for paternal imprinting. *Cell* 50, 719–727.

Swanson, P. (1991) Salmon gonadotropins: reconciling old and new ideas. In: Scott, A.P., Sumpter, J.P., Kime, D.E. and Rolf, M.S. (eds) *Proceedings of the 4th International Symposium on the Reproduction Physiology of Fish*. East Anglia Printing Unit, Sheffield, pp. 2–7.

Swarup, H. (1959a) Production of triploidy in *Gasterosteus aculeatus* (L.). *Journal of Genetics* 56, 129–142.

Swarup, H. (1959b) Effect of triploidy on the body size, general organization and cellular structure in *Gasterosteus aculeatus* (L.). *Journal of Genetics* 56, 143–155.

Sweeting, R.M. and McKeown, B.A. (1987) Growth hormone and seawater adaptation in coho salmon, *Oncorhynchus kisutch*. *Comparative Biochemistry and Physiology, Part A: Comparative Physiology* 88A, 147–151.

Szarski, H. (1970) Changes in the amount of DNA in cell nuclei during vertebrate evolution. *Nature* 226, 651–652.

Szczepek, M., Brondani, V., Büchel, J., Serrano, L., Segal, D.J. and Cathomen, T. (2007) Structure-based redesign of the dimerization interface reduces the toxicity of zinc-finger nucleases. *Nature Biotechnology* 25, 786–793.

Tabarini, C.L. (1984a) Induced triploidy in the bay scallop, *Argopecten irradians*, and its effect on growth and gametogenesis. *Aquaculture* 42, 151–160.

Tabarini, C.L. (1984b) Induced triploidy in the Atlantic bay scallop *Argopecten irradians* and its effect on growth and gametogenesis. *Journal of Shellfish Research* 4, 101–102.

Tabata, Y.A., Rigolino, M.G. and Tsukamoto, R.Y. (1999) Production of all female triploid rainbow trout, *Oncorhynchus mykiss* (Pisces, Salmonidae). III. Growth up to first sexual maturation. *Boletim do Instituto de Pesca Sao Paulo* 25, 67–76.

Tacon, A.G.J. (1996) *Global Trends in Aquaculture and Aqua Feed Production*. International Milling Directory and Buyer's Guide, Turret Group. Available at: www.turret-rai.co.uk/agrifood_group/info.htm.

Takada, Y., Yoshida, M., Sakairi, M. and Koizumi, H. (1995) Detection of gamma-aminobutyric acid in a living rat brain using *in vivo* microdialysis–capillary electrophoresis/mass spectrometry. *Rapid Communications in Mass Spectrometry* 9, 895–896.

Takagi, Y., Warashina, M., Stec, W.J., Yoshinari, K. and Taira, K. (2001) Recent advances in the elucidation of the mechanisms of action of ribozymes. *Nucleic Acids Research* 29, 1815–1834.

Takahashi, Y., Kaji, H., Okimura, Y., Goji, K., Abe, H. and Chihara, K. (1996) Brief report: short stature caused by mutant growth hormone. *New England Journal of Medicine* 334, 432–436.

Takeuchi, K. and Toyohara, H. (2002) Taurine transporter: hyperosmotic stress-responsive gene. In: Shimizu, N., Aoki, T., Hirono, I. and Takashima, F. (eds) *Aquatic Genomics: Steps Toward a Great Future*. Springer-Verlag, New York, pp. 207–216.

Takeuchi, Y., Yoshizaki, G., Tominaga, H., Kobayashi, T. and Takeuchi, T. (2002) Visualization and isolation of live primordial germ cells aimed at cell-mediated gene transfer in rainbow trout. In: Shimizu, N., Aoki, T., Hirono, I. and Takashima, F. (eds) *Aquatic Genomics: Steps Toward a Great Future*. Springer-Verlag, New York, pp. 310–319.

Takeuchi, Y., Yoshizaki, G. and Takeuchi, T. (2003) Generation of live fry from intraperitoneally transplanted primordial germ cells in rainbow trout. *Biology of Reproduction* 69, 1142–1149.

Tan, G., Karsi, A., Li, P., Kim, S., Zheng, X., Kucuktas, H., Argue, B.J., Dunham, R.A. and Liu, Z.J. (1999) Polymorphic microsatellite markers in *Ictalurus punctatus* and related catfish species. *Molecular Ecology* 59, 190–194.

Tan, Y.D., Wan, C., Zhu, Y., Lu, C., Xiang, Z. and Deng, H.W. (2001) An amplified fragment length polymorphism map of the silkworm. *Genetics* 157, 1277–1284.

Tanck, M.W.T., Palstra, A.P., Van De Weerd, M., Leffering, C.P., Van Der Poel, J.J., Bovenhuis, H. and Komen, J. (2001) Segregation of microsatellite alleles and residual heterozygosity at single loci in homozygous androgenetic common carp (*Cyprinus carpio* L.). *Genome* 44, 743–751.

Tang, S.J., Sun, K.H., Sun, G.H., Lin, G., Lin, W.W. and Chuang, M.J. (1999) Cold-induced ependymin expression in zebrafish and carp brain: implication for cold acclimation. *FEBS Letters* 459, 95–99.

Tang, Y., Shepherd, B.S., Nichols, A.J., Dunham, R. and Chen, T.T. (2001) Influence of environmental salinity on messenger RNA levels of growth hormone, prolactin, and somatolactin pituitary of the channel catfish (*Ictalurus punctatus*). *Marine Biotechnology* 3, 205–217.

Tanhuanpaa, P. and Vilkki, J. (1999) Marker-assisted selection for oleic acid content in spring turnip rape. *Plant Breeding* 118, 568–570.

Taniguchi, N., Kijima, A., Tamura, T., Takegami, K. and Yamasaki, I. (1986) Color growth and maturation in ploidy-manipulated fancy carp. *Aquaculture* 57, 321–328.

Taniguchi, N., Kijima, A. and Fukai, J. (1987) High heterozygosity at Gpi-1 in gynogenetic diploids and triploids of ayu *Plecoglossus altivelis. Bulletin of the Japanese Society of Scientific Fisheries/Nissuishi* 53, 717–720.

Tashiro, F. (1972) Effects of irradiation by ^{60}Co gamma rays on the maturation of rainbow trout. *Bulletin of the Japanese Society of Scientific Fisheries/Nissuishi* 38, 793–797.

Tave, D. (1985) *Genetics for Fish Hatchery Managers*. AVI Publishing, Westport, Connecticut.

Tave, D. (1993) *Genetics for Fish Hatchery Managers*, 2nd edn. Van Nostrand Reinhold, New York.

Tave, D. (2003) Genetics and stock improvement. In: Lucas, J.S. and Southgate, P.C. (eds) *Aquaculture: Farming Aquatic Animals and Plants*. Blackwell Publishing, Oxford, UK, pp. 123–145.

Tave, D., Bartels, J.E. and Smitherman, R.O. (1983) Saddleback: a dominant, lethal gene in *Sarotherodon aureus* (Steindanchner) (*Tilapia aurea*). *Journal of Fish Diseases* 6, 59–73.

Tave, D., Smitherman, R.O. and Jayaprakas, V. (1990) Estimates of additive genetic effects, maternal genetic effects, individual heterosis, maternal heterosis, and egg cytoplasmic effects for growth in *Tilapia nilotica. Journal of the World Aquaculture Society* 21, 263–270.

Teichert-Coddington, D. and Smitherman, R.O. (1988) Lack of response by *Tilapia nilotica* to mass selection for rapid early growth. *Transactions of the American Fisheries Society* 117, 297–300.

Teplitsky, C., Mills, J.A., Yarrall, J.W. and Merilä, J. (2009) Heritability of fitness components in a wild bird population. *Evolution* 63, 716–726.

Tewari, R., Michard-Vanhée, C., Perrot, E. and Chourrout, D. (1992) Mendelian transmission, structure and expression of transgenes following their injection into the cytoplasm of trout eggs. *Transgenic Research* 1, 250–260.

Tessier, N., Bernatchez, L. and Wright, J.M. (1997) Population structure and impact of supportive breeding inferred from mitochondrial and microsatellite DNA analyses in land-locked Atlantic salmon *Salmo salar* L. *Molecular Ecology* 6, 735–750.

Thanakijkorn, J. (1997) Improvement of salinity tolerance in Nile tilapia (*Oreochromis niloticus*) by means of interspecific-hybridization with Java tilapia (*Oreochromis mossambicus*). MSc Thesis, Kasetsart University, Bangkok.

Thanh, N.M., Ponzoni, R.W., Nguyen, N.H., Vu, N.T., Barnes, A. and Mather, P.B. (2009) Evaluation of growth performance in a diallel cross of three strains of giant freshwater prawn (*Macrobrachium rosenbergii*) in Vietnam. *Aquaculture* 287, 75–83.

Thanh, N.M., Nguyen, N.H., Ponzoni, R.W., Vu, N.T., Barnes, A.C. and Mather, P.B. (2010) Estimates of strain additive and non-additive genetic effects for growth traits in a diallel cross of three strains of giant freshwater prawn (*Macrobrachium rosenbergii*) in Vietnam. *Aquaculture* 299, 30–36.

Thavornyutikarn, M. (1994) Realized heritability estimation on growth of oyster *Saccostrea cucullata*. MSc Thesis, Chulalongkorn University, Bangkok.

Thelen, G.C. and Allendorf, F.W. (2001) Heterozygosity–fitness correlations in rainbow trout: effects of allozyme loci or associative overdominance? *Evolution* 55, 1180–1187.

Thien, T.M. and Trong, T.D. (1995) Genetic resources of common carp in Vietnam. *Aquaculture* 129, 216.

Thissen, J.-P., Ketelslegers, J.-M. and Underwood, L.E. (1994) Nutritional regulation of the insulin-like growth factors. *Endocrine Reviews* 15, 80–101.

Thodesen, J. (1999) Selection for improved feed conversion in salmon. *Norsk Fiskeoppdrett* (5A), 20–21.

Thomas, D.D., Donelly, C.A., Wood, R.J. and Alphey, L.S. (2000) Insect population control using a dominant, repressible, lethal genetic system. *Science* 287, 2474–2476.

Thomas, G.L., Pauley, G.P. and Bonar, S. (1986) Estimation of triploid white amur stocking densities for aquatic plant control in the Pacific Northwest. In: *Proceedings of 6th Annual International Symposium on Lake and Reserve Management: Influences of Nonpoint Source Pollutants and Acid Precipitation, 5–8 November 1986, Portland, Oregon*. North American Lake Management Society, Madison, Wisconsin, p. 21.

Thompson, C.E., Poole, W.R., Matthews, M.A. and Ferguson, A. (1998) Comparison, using minisatellite DNA profiling, of secondary male contribution in the fertilisation of wild and ranched Atlantic salmon (*Salmo salar*) ova. *Canadian Journal of Fisheries and Aquatic Sciences* 55, 2011–2018.

Thompson, D. and Purdom, C.E. (1986) Induced diploid gynogenesis by mitotic interference in rainbow trout. *Aquaculture* 57, 376.

Thompson, S., Clarke, A., Pow, M., Hooper, M. and Melton, D.W. (1988) Germ line transmission and expression of a corrected HPRT gene produced by gene targeting in embryonic stem cells. *Cell* 56, 313–324.

Thorgaard, G.H. (1976) Robertsonian polymorphism and constitutive heterochromatin distribution in chromosomes of the rainbow trout (*Salmo gairdneri*). *Cytogenetics and Cell Genetics* 17, 174–186.

Thorgaard, G.H. (1986) Ploidy manipulation and performance. *Aquaculture* 57, 57–64.

Thorgaard, G. and Gall, G.A.E. (1979) Adult triploids in a rainbow trout family. *Genetics* 93, 961–973.

Thorgaard, G.H., Jazwin, M.E. and Stier, A.R. (1981) Polyploidy induced by heat shock in rainbow trout. *Transactions of the American Fisheries Society* 110, 546–550.

Thorgaard, G.H., Scheerer, P.D. and Parsons, J.E. (1985) Residual paternal inheritance in gynogenetic rainbow trout: implications for gene transfer. *Theoretical and Applied Genetics* 71, 119–121.

Thorpe, J.E., Morgan, R.I.G., Talbot, C. and Miles, M.S. (1983) Inheritance of developmental rates in Atlantic salmon, *Salmo salar* L. *Aquaculture* 33, 119–128.

Thresher, R.E., Hinds, L., Grewe, P., Patil, J., McGoldrick, D., Nesbitt, K., Lumb, C., Whyard, S. and Hardy, C. (2001) Repressible sterility in aquaculture species: a genetic system for preventing the escape of genetically improved stocks. In: *Aquaculture 2001: Book of Abstracts*. Elsevier, Amsterdam, p. 637.

Thresher, R., Dunham, R., Grewe, P., Whyard, S., Patil, J., Templeton, C.M., Chaimongkol, A., Hardy, C. and Hinds, L. (2009) Development of repressible sterility to prevent the establishment of feral populations of exotic and genetically modified animals. *Aquaculture* 290, 104–109.

Thune, R.L. (1993) Bacterial diseases of catfishes. In: Stoskopf, M.K. (ed.) *Fish Medicine*. W.B. Saunders, Philadelphia, Pennsylvania, pp. 511–520.

Thyagarajan, B., Guimaraes, M.J., Groth, A.C. and Calos, M.P. (2000) Mammalian genomes contain active recombinase recognition sites. *Gene* 244, 47–54.

Tian, C., Liang, Y., Wang, R., Yu, R. and Wang, Z. (1999a) Triploid of *Crassostrea gigas* induced with 6-dimethylaminopurine: blocking polar body I. *Journal of Fisheries China/Shuichan Xuebao* 23, 128–132.

Tian, C., Wang, R., Liang, Y., Wang, Z. and Yu, R. (1999b) Triploidy of *Crassostrea gigas* induced with 6-DMAP: by blocking the second polar body of the zygotes. *Journal of Fishery Sciences of China/Zhongguo Shuichan Kexue, Beijing* 6, 1–4.

Tian, C., Wang, R., Liang, Y., Yu, R. and Wang, Z. (1999c) Triploid Pacific oyster, *Crassostrea gigas*, as induced by cold shock. *Marine Sciences/Haiyang Kexue, Qingdao* 1, 53–55.

Tian, Y., Kong, J. and Yang, C. (2006) Comparative growth and viability of hybrids between two populations of Chinese shrimp (*Fennropenaeus chinensis*). *Chinese Science Bulletin* 51, 2369–2374.

Tidwell, J.H., Webster, C.D. and Clark, J.A. (1992) Growth, feed conversion, and protein utilization of female green sunfish × male bluegill hybrids fed isocaloric diets with different potein levels. *Progressive Fish-Culturist* 54, 234–239.

Tiemeier, O.W. (1957) Notes on stunting and recovery of the channel catfish. *Transactions of the Kansas Academy of Science* 60, 294–296.

Tilghman, S.M. (1996) Lessons learned, promises kept: a biologist's eye view of the genome project. *Genome Research* 6, 773–780.

Tiwary, B.K., Kirubagaran, R. and Ray, A.K. (1999) Altered body shape as a morphometric indicator of triploidy in Indian catfish *Heteropneustes fossilis* (Bloch). *Aquaculture Research* 30, 907–910.

Tiwary, B.K., Kirubagaran, R. and Ray, A.K. (2000) Gonadal development in triploid *Heteropneustes fossilis*. *Journal of Fish Biology* 57, 1343–1348.

Tobet, S.A., Chickering, T.W., King, J.C., Stopa, E.G., Kim, K., KuoLeblank, V. and Schwarting, G.A. (1996) Expression of γ-aminobutyric acid and gonadotropin-releasing hormone during neuronal migration through the olfactory system. *Endocrinology* 137, 5415–5420.

Toro, J.E., Aguila, P. and Vergara, A.M. (1996) Spatial variation in response to selection for live weight and shell length from data on individually tagged Chilian native oysters (*Ostrea chilensis* Philippi, 1845). *Aquaculture* 146, 27–36.

Townes, T.M., Chen, H.Y., Lingrel, J.B., Palmiter, R.D. and Brinster, R.L. (1985) Expression of human beta-globin genes in transgenic mice: effects of a flanking metallothionein–human growth hormone fusion gene. *Molecular and Cellular Biology* 5, 1977–1983.

Tran, M.T. and Nguyen, C.T. (1993) Selection of common carp (*Cyprinus carpio* L.) in Vietnam. *Aquaculture* 111, 301–302.

Travis, S.E., Maschinski, J. and Keim, P. (1996) An analysis of genetic variation in *Astragalus cremnophylax* var. *cremnophylax*, a critically endangered plant, using AFLP markers. *Molecular Ecology* 5, 735–745.

Trenkle, A. (1978) Relation of hormonal variations to nutritional studies and metabolism of ruminants. *Journal of Dairy Science* 61, 281–293.

Trombka, D. and Avtalion, R.R. (1993) Sex determination in tilapia – a review. *Israeli Journal of Aquaculture/ Bamidgeh* 45, 26–37.

Trudeau, V.L., Spanswick, D., Fraser, E.J., Lariviere, K., Crump, D., Chiu, S., MacMillan, M. and Schulz, R.W. (2000) The role of amino acid neurotransmitters in the regulation of pituitary gonadotropin release in fish. *Biochemistry and Cell Biology* 78, 241–259.

Tsoi, R.M. (1969) Action of nitroso-methylurea and dimethylsulfate on the sperm cells of rainbow trout and the peled. *Doklady Akademii Nauk SSSR Ser. Biol.* 189, 411.

Tsuji, S., Uehori, J., Matsumoto, M., Suzuki, Y., Matsuhisa, A., Toyoshima, K. and Seya, T. (2001) Human intelectin is a novel soluble lectin that recognizes galactofuranose in carbohydrate chains of bacterial cell wall. *Journal of Biological Chemistry* 276, 23456–23463.

Tsunoda, Y. and Kato, Y. (1993) Nuclear transplantation of embryonic stem cells in mice. *Journal of Reproduction and Fertility* 98, 537–540.

Tuan, P.A., Mair, G.C., Little, D.C. and Beardmore, J.A. (1999) Sex determination and the feasibility of genetically male tilapia production in the Thai-Chitralada strain of *Oreochromis niloticus* (L.). *Aquaculture* 173, 257–269.

Tufto, J. (2001) Effects of releasing maladapted individuals: a demographic-evolutionary model. *American Naturalist* 158, 331–340.

Tung, T.C. (1980) Nuclear transplantation in teleosts. I. Hybrid fish from the nucleus of carp and the cytoplasm of crucian. *Scientia (Peking)* 23, 517–523.

Tung, T.C. and Yan, S.Y. (1985) Nuclear transplantation in teleosts. V. Hybrid fish from the nucleus of crucian and the cytoplasm of goldfish (abstract). In: *China Japan Joint Meeting on Cell Engineering, Shanghai, China,* pp. 42–46.

Tveranger, B. (1985) Variation in growth rate, liver weight and body composition at first sexual maturation in rainbow trout. *Aquaculture* 49, 89–99.

Ueda, T., Ojima, Y., Sato, R. and Fukuda, Y. (1984) Triploid hybrids between female rainbow trout and male brook trout. *Bulletin of the Japanese Society of Scientific Fisheries/Nissuishi* 50, 1331–1336.

Ueda, T., Sato, R. and Kobayashi, J. (1988) Triploid rainbow trout induced by high-pH multiplied by high-calcium. *Nippon Suisan Gakkaishi/Bulletin of the Japanese Society of Scientific Fisheries* 54, 2045.

Ueda, T., Ojima, Y. and Kobayashi, J. (1990) Hypodiploid and hypotriploid hybrids between female Japanese char and male rainbow trout. *Kromosomo* 2, 2008–2012.

Ueno, K. (1984) Induction of triploid carp and their haematological characteristics. *Journal of Genetics* 59, 585–591.

Újvári, B., Madsen, T., Kotenko, T., Olsson, M., Shine, R. and Wittzell, H. (2002) Low genetic diversity threatens imminent extinction for the Hungarian meadow viper (*Vipera ursinii rakosiensis*). *Biological Conservation* 105, 127–130.

UNCED (United Nations Conference on Environment and Development) (1994) *Convention on Biological Diversity – Text and Annexes.* UNEP/CBD/94/1. United Nations Environment Programme, Geneva, Switzerland.

Underwood, J.L., Hestand, R.S. III and Thompson, B.Z. (1986) Gonad regeneration in grass carp following bilateral gonadectomy. *Progressive Fish-Culturist* 48, 54–56.

Ungemach, F.R. and Weber, N.E. (1998) Recombinant bovine somatotropins (addendum). Toxicological evaluation of certain veterinary drug residues in food. In: *The Fiftieth Meeting of the Joint FAO/WHO Expert Committee on Food Additives (JECFA).* WHO Food Additive Series No. 41, World Health Organization, Geneva, Switzerland, pp. 130–146.

Upton, Z., Moriyama, S., Degger, B.G., Francis, G.L. and Ballard, F.J. (1996) Evolution of insulin-like growth factor-I (IGF-I) function: *in vitro* comparison of barraundi, salmon, chicken and human IGF I. In: *Program and Abstracts, 10th International Conference on Endocrinology,* p. 65.

Upton, Z., Francis, G.L., Chan, S.J., Steiner, D.F., Wallace, J.C. and Ballard, F.J. (1997) Evolution of insulin-like growth factor (IGF) function: production and characterization of recombinant hagfish IGF. *General and Comparative Endocrinology* 105, 79–90.

Upton, Z., Yandell, C.A., Degger, B.G., Chan, S.J., Moriyama, S., Francis, G.L. and Ballard, F.J. (1998) Evolution of insulin-like growth factor-I (IGF-I) action: *in vitro* characterization of vertebrate IGF-I proteins. *Comparative Biochemistry and Physiology, Part B: Biochemistry & Molecular Biology* 121B, 35–41.

Uraiwan, S. (1990) Artificial selection on growth and age at maturation of tilapia (*Oreochromis niloticus* Linn.) in Thailand. PhD dissertation, Dalhousie University, Halifax, Nove Scotia, Canada.

Urmaza, E.B. and Aguilar, R.O. (2005) Growth performance of saline-tolerant tilapia produced from cross combinations of various tilapia species. *Journal of Aquaculture in the Tropics* 20, 11–27.

Utter, F.M., Johnson, O.W., Thorgaard, G.H. and Rabinovitch, P.S. (1983) Measurement and potential applications of induced triploidy in Pacific salmon. *Aquaculture* 35, 125–135.

Uzbekova, S., Alestrom, P., Ferriere, F., Bailhache, T., Prunet, P. and Breton, B. (2000a) Expression of recombinants GnRH-antisense RNA under the control of either specific or constitutive promoters in transgenic rainbow trout. *ISFE Abstracts* W-296.

Uzbekova, S., Chyb, J., Ferriere, F., Bailhache, T., Prunet, P., Alestrom, P. and Breton, B. (2000b) Transgenic rainbow trout expressed sGnRH-antisense RNA under the control of sGnRH promoter of Atlantic salmon. *Journal of Molecular Endocrinology* 25, 337–350.

Vadopalas, B.A. and Davis, J.P. (1998) Induction of triploidy in the geoduck clam, *Panope abrupta*. *Journal of Shellfish Research* 17, 1285.

Vaiman, D., Schibler, L., Bourgeois, F., Oustry, A., Amigues, Y. and Cribiu, E.P. (1996) A genetic linkage map of the male goat genome. *Genetics* 144, 279–305.

Valenti, R.J. (1975) Induced polyploidy in *Tilapia aurea* (Steindachner) by means of temperature shock treatment. *Journal of Fisheries Biology* 7, 519–528.

Vallin, L. and Nissling, A. (2000) Maternal effects on egg size and egg buoyancy of Baltic cod, *Gadus morhua*: implications for stock structure effects on recruitment. *Fisheries Research* 49, 21–37.

Vandeputte, M., Peignon, E., Vallod, D., Haffray, P., Komen, J. and Chevassus, B. (2002) Comparison of growth performances of three French strains of common carp (*Cyprinus carpio*) using hemi-isogenic scaly carp as internal control. *Aquaculture* 205, 19–36.

Vandeputte, M., Kocour, M., Mauger, S., Rodina, M., Launay, A., Gela, D., Dupont-Nivet, M., Hulak, M. and Linhart, O. (2008) Genetic variation for growth at one and two summers of age in the common carp (*Cyprinus carpio* L.): heritability estimates and response to selection *Aquaculture* 277, 7–13.

Vandeputte, M., Garouste, R., Dupont-Nivet,M. Haffray, P., Chavanne, H., Laureau, S., and Chatain, B. (2009) Multi-local evaluaton of the rearing performances of 5 natural populations of European sea bass (*Dicentrarchus labrax*). In: *Book of Abstracts of 10th International Symposium on Genetics in Aquaculture (ISGA X) 'Roles of Aquaculture Genetics in Addressing Global Food Crisis', 22–26 June 2009, Bangkok, Thailand*. Faculty of Fisheries, Kasetsart University, Bangkok, p. 37.

Vander, A.J., Sherman, J.H., and Luciano, D.S. (eds) (1990) *Human Physiology*, 5th edn. McGraw-Hill Publishing, New York.

Vandersteen Tymchuk, W.E., Abrahams, M.V. and Devlin, R.H. (2005) Competitive ability and mortality of growth-enhanced transgenic coho salmon fry and parr when foraging for food. *Transactions of the American Fisheries Society* 134, 381–389.

Van Eck, H.J., Vander Voort, J.R., Draaistra, J., Van Zandvoort, P., Enckevort, V., Segers, B., Peleman, J., Jacobsen, E., Helder, J. and Bakker, J. (1995) The inheritance and chromosomal localization of AFLP markers in a non-inbred potato offspring. *Molecular Breeding: New Strategies in Plant Improvement* 1, 397–410.

Van Eenennaam, L. and Olin, P.G. (2006) Careful risk assessment needed to evaluate transgenic fish. *California Agriculture* 60, 126–131.

Van Eenennaam, J.P., Stocker, R.K., Thiery, R.G., Hagstrom, N.T. and Doroshov, S.I. (1990) Egg fertility, early development and survival from crosses of diploid female triploid male grass carp (*Ctenopharyngodon idella*). *Aquaculture* 86, 111–125.

Van Velson, R.C. (1978) *The McConaughy Rainbow: Life History and a Management Plan for the North Platt River Valley*. Communication Nebraska Technical Series 2. Nebraska Game Parks Department, Lincoln, Nebraska.

Van Vugt, D.A., Badst, G., Dyrenfurth, I. and Ferrin, M. (1983) Naloxone stimulation of luteinizing hormone secretion in the female monkey: influence of endocrine and experimental conditions. *Endocrinology* 113, 1858–1864.

Varadaraj, K. and Pandian, T.J. (1989) First report on production of supermale tilapia by integrating endocrine sex reversal with gynogenetic technique. *Current Science* 58, 434–441.

Vasemägi, A., Gross, R., Paaver, T., Koljonen, M.L., Saeisae, M. and Nilsson, J. (2005) Analysis of gene associated tandem repeat markers in Atlantic salmon (*Salmo salar* L.) populations: implications for restoration and conservation in the Baltic Sea. *Conservation Genetics* 6, 385–397.

Vejaratpimol, R. and Pewnim, T. (1990) Induction of triploidy in *Clarias macrocephalus* by cold shock. In: Hirano, R. and Hanyu, I. (eds) *Proceedings of The Second Asian Fisheries Forum, Tokyo, Japan, 17–22 April 1989*. Asian Fisheries Society, Manila, pp. 531–534.

Vestling, M., Murphy, C., Fenselau, C. and Chen, T.T. (1991) Disulfide bonds in native and recombinant fish growth hormones. *Molecular Marine Biology and Biotechnology* 1, 73–77.

Veuthey, A.L. and Bittar, G. (1998) Phylogenetic relationships of fungi, plantae, and animalia inferred from homologous comparison of ribosomal proteins. *Journal of Molecular Evolution* 47, 81–92.

Vincent, R. (1960) Some influences of domestication upon three stocks of brook trout (*Salvelinus fontinalis* Mitchell). *Transactions of the American Fisheries Society* 89, 35–52.

Virtanen, E., Forsman, L. and Sundby, A. (1990) Triploidy decreases the aerobic swimming capacity of rainbow trout (*Salmo gairdneri*). *Comparative Biochemistry and Physiology, Part A: Comparative Physiology* 96A, 117–121.

Vlahakis, C. and Hazebroek, J. (2000) Phytosterol accumulation in canola, sunflower, and soybean oils: effects of genetics, planting location, and temperature. *Journal of the American Oil Chemists Society* 77, 49–53.

Volckaert, F.A., Hellemans, B.A., Galbusera, P.A., Ollevier, F., Sekkali, B. and Belayew, A. (1994) Replication, expression, and fate of foreign DNA during embryonic and larval development of the African catfish (*Clarias gariepinus*). *Molecular Marine Biology and Biotechnology* 3, 57–69.

Voorrips, R.E., Jogerius, M.C. and Kanne, H.J. (1997) Mapping of two genes for resistance to clubroot (*Plasmodiphora brassicas*) in a population of doubled haploid lines of *Brassica oleracea* by means of RFLP and AFLP markers. *Theoretical and Applied Genetics* 94, 75–82.

Vos, J.C. and Plasterk, R.H. (1994) *Tc1* transposase of *Caenorhabditis elegans* is an endonuclease with a bipartite DNA binding domain. *EMBO Journal* 13, 6125–6132.

Vos, P., Hogers, R., Bleeker, M., Reijans, M., van de Lee, T., Hornes, M., Frijters, A., Pot, J., Peleman, J., Kuiper, M. and Zabeay, M. (1995) AFLP: a new technique for DNA fingerprinting. *Nucleic Acids Research* 23, 4407–4414.

Vucetich, J.A. and Waite, T.A. (2001) Migration and inbreeding: the importance of recipient population size for genetic management. *Conservation Genetics* 2, 167–171.

Vunnam, S., Juvvadi, P. and Merrifield, R.B. (1995) Synthesis and antibacterial action of cecropin and proline–arginine-rich peptides from pig intensine. *Journal of Peptide Research* 49, 59–66.

Wachinger, M., Kleinschmidt, A., Winder, D., von Pechmann, N., Ludvigsen, A., Neumann, M., Holle, R., Salmons, B., Erfle, V. and Brack-Werner, R. (1998) Antimicrobial peptides melittin and cecropin inhibit replication of human immunodeficiency virus 1 by suppressing viral gene expression. *Journal of General Virology* 79, 731–740.

Wachirachaikarn, A., Rungsin, W., Srisapoome, P. and Na-Nakorn, U. (2009) Crossing of African catfish, *Clarias gariepinus* (Burchell, 1822), strains based on strain selection using genetic diversity data. *Aquaculture* 290, 53–60.

Wada, H., Naruse, K., Shimada, A. and Shima, A. (1995) Genetic linkage map of a fish, the Japanese medaka *Oryzias latipes*. *Molecular Marine Biology and Biotechnology* 4, 269–274.

Wada, K.T., Komaru, A. and Uchimura, Y. (1989) Triploid production in the Japanese pearl oyster, *Pinctada fucata martensii*. *Aquaculture* 76, 11–19.

Wada, K.T., Scarpa, J. and Allen, S.K. Jr (1990) Karyotype of the dwarf surf clam *Mulinia lateralis* (Say, 1822) (Mactridae, Bivalvia). *Journal of Shellfish Research* 9, 279–281.

Wahle, E. and Keller, W. (1996) The biochemistry of polyadenylation. *Trends in Biochemical Sciences* 21, 247–250.

Wakamatsu, Y., Ozato, K. and Sasado, T. (1994) Establishment of a pluripotent cell line derived from a medaka (*Oryzias latipes*) blastula embryo. *Molecular Marine Biology and Biotechnology* 3, 185–191.

Waldbieser, G.C. and Wolters, W.R. (1999) Application of polymorphic microsatellite loci in a channel catfish *Ictalurus punctatus* breeding program. *Journal of the World Aquaculture Society* 30, 256–262.

Waldbieser, G.C., Bosworth, B.G., Nonneman, D.J. and Wolters, W.R. (2001) A microsatellite-based genetic linkage map for channel catfish, *Ictalurus punctatus*. *Genetics* 158, 727–734.

Walker, D.S. (1993) Effect of electroporation and microinjection on survival of ictalurid catfish embryos. MSc thesis, Auburn University, Auburn, Alabama.

Walker, W.E., Harremoës, P., Rotmans, J. van der Sluijs, J., van Asselt, M., Janssen, P. and Krayer von Krauss, M. (2003) Defining uncertainty – a conceptual basis for uncertainty management in model-based decision support. *Integrated Assessment* 4 (1), 5–17.

Wallace, D.H. and Yan, W. (1998) *Plant Breeding and Whole-System Crop Physiology: Improving Crop Maturity, Adaption, and Yield*. CAB International, Wallingford, UK.

Wallis, A.E. and Devlin, R.H. (1993) Duplicate insulin-like growth factor I genes in salmon display alternative splicing pathways. *Molecular Endocrinology* 7, 409–422.

Wallis, M. (1975) The molecular evolution of pituitary hormones. *Biological Reviews* 50, 35–98.

Wang, I.A., Gilk, S.E., Smoker, W.W. and Gharrett, A.J. (2007) Outbreeding effect on embryo development in hybrids of allopatric pink salmon (*Oncorhynchus gorbuscha*) populations, a potential consequence of stock translocation. *Aquaculture* 272, S152–S160.

Wang, J. and Ryman, N. (2001) Genetic effects of multiple generations of supportive breeding. *Conservation Biology* 15, 1619–1631.

Wang, Y., Cui, Y., Yang, Y. and Cai, F. (2000) Compensatory growth in hybrid tilapia, *Oreochromis mossambicus* × *O. niloticus*, reared in seawater. *Aquaculture* 189, 101–108.

Wang, Z., Guo, X., Allen, S.K. and Wang, R. (2002) Heterozygosity and body size in triploid Pacific oysters, *Crassostrea gigas* Thunberg, produced from meiosis II inhibition and tetraploids. *Aquaculture* 204, 337–348.

Ward, R.D., Billington, N. and Hebert, P.D.N. (1989) Comparison of allozyme and mitochondrial DNA variation in populations of walleye, *Stizostedion vitreum*. *Canadian Journal of Fisheries and Aquatic Sciences* 46, 2074–2084.

Ward, R.D., English, L.J., McGoldrick, D.J., Maguire, G.B., Nell, J.A. and Thompson, P.A. (2000) Genetic improvement of the Pacific oyster, *Crassostrea gigas*, in Australia. *Aquaculture Research* 31, 35–44.

Warner, J.R. and Nierras, C.R. (1998) Trapping human ribosomal protein genes. *Genome Research* 8, 419–421.

Warrillow, J.A., Josephson, D.C., Youngs, W.D. and Krueger, C.C.A. (1997) Difference in sexual maturity and fall emigration between diploid and triploid brook trout (*Salvelinus fontinalis*) in an Adirondack lake. *Canadian Journal of Fisheries and Aquatic Sciences* 54, 1808–1812.

Was, A. and Wenne, R. (2002) Genetic differentiation in hatchery and wild sea trout (*Salmo trutta*) in the Southern Baltic at microsatellite loci. *Aquaculture* 204, 493–506.

Watanabe, K., Igarashi, A., Noso, T., Chen, T.T., Dunham, R.A. and Kawauchi, H. (1992) Chemical identification of catfish growth hormone and prolactin. *Molecular Marine Biology and Biotechnology* 1, 239–249.

Watanabe, T., Sugii, K., Saito, H., Okazaki, E. and Yamada, J. (1988) Properties of lipids at development of cultured triploid ayu. *Bulletin of Tokai Regional Fisheries Research Laboratory/Tokaisuikenho* 126, 49–59.

Waterhouse, P.M., Graham, M.W. and Wang, M.B. (1998) Virus resistance and gene silencing in plants can be induced by simultaneous expression of sense and antisense RNA. *Proceedings of the National Academy of Sciences USA* 95, 13959–13964.

Waters, P.J. (2001) Response and correlated response to selection for resistance to *Edwardsiella ictaluri*, *Flexibacter columnaris*, low levels of dissolved oxygen and toxic levels of nitrite in channel catfish, *Ictalurus punctatus*. MSc thesis, Auburn University, Auburn, Alabama.

Wattendorf, R.J. (1986) Rapid identification of grass carp with a Coulter counter and channelyzer. *Progressive Fish-Culturist* 48, 125–132.

Wattendorf, R.J. and Anderson, R.S. (1986) Hydrilla consumption by triploid grass carp in aquaria. *Proceedings Annual Conference Southeastern Association of Fish and Wildlife Agencies* 40.

Webb, K.E. Jr (1986) Amino acid and peptide absorption from the gastrointestinal tract. *Federation Proceedings* 45, 2268–2271.

Webster, D.A. and Flick, W.A. (1981) Performance of indigenous, exotic, and hybrid strains of brook trout (*Salvelinus fontinalis*) in waters of the Adirondack Mountains, New York. *Canadian Journal of Fisheries and Aquatic Sciences* 38, 1701–1707.

Wehrhahn, C.F. and Powell, R. (1987) Electrophoretic variation, regional differences, and gene flow in the coho salmon (*Oncorhynchus kisutch*) of southern British Columbia. *Canadian Journal of Fisheries and Aquatic Sciences* 44, 822–831.

Weissberger, A.J., Ho, K.K. and Lazarus, L. (1991) Contrasting effects of oral and transdermal routes of estrogen replacement therapy on 24-hour growth hormone (GH) secretion, insulin-like growth factor I, and GH-binding protein in postmenopausal women. *Journal of Clinical Endocrinology and Metabolism* 72, 374–381.

Welcomme, R.L. (1988) *International Introductions of Inland Aquatic Species*. Fisheries Technical Paper No. 294, Food and Agriculture Organization of the United Nations, Rome.

Wellbrock, C., Gomez, A. and Schartl, M. (2002) Melanoma development and pigment cell transformation in *Xiphophorus*. *Microscopy and Research Technique* 58, 456–463.

Weller, J.I. (ed.) (1994) *Economic Aspects of Animal Breeding*. Chapman and Hall, London.

Wells, J.A. (1995) Structural and functional epitopes in the growth hormone receptor complex. *Bio/Technology* 13, 647–651.

Wells, J.A. (1996) Binding in the growth hormone receptor complex. *Proceedings of the National Academy of Sciences USA* 93, 1–6.

Welsh, J. and McClelland, M. (1990) Fingerprinting genomes using PCR with arbitrary primers. *Nucleic Acids Research* 18, 7213–7218.

Westerfield, M., Devoto, S. and Melancon, B. (2000) Inductive interactions regulating skeletal muscle cell fates. In: *Abstracts of the International Symposium 'A Step Toward the Great Future of Aquatic Genomics', 10–12 November 2000.* Tokyo University of Fisheries, Tokyo.

Wheeler, M.B. (1994) Development and validation of swine embryonic stem cells: a review. *Reproduction, Fertility, and Development* 6, 563–568.

White, R.B. and Fernald, R.D. (1998a) Genomic structure and expression sites of three gonadotropin-releasing hormone genes in one species. *General and Comparative Endocrinology* 112, 17–25.

White, R.B. and Fernald, R.D. (1998b) Ontogeny of gonadotropin-releasing hormone (GnRH) gene expression reveals a distinct origin for GnRH-containing neurons in the midbrain. *General and Comparative Endocrinology* 112, 322–329.

Whitehead, P.K. (1994) Mapping and electroporation of mitochondrial DNA from three species of ictalurid catfish. Doctoral dissertation, Auburn University, Auburn, Alabama.

Whitmore, D.H. (1990) *Electrophoretic and Isoelectric Focusing Techniques in Fisheries Management.* CRC Press, Boca Raton, Florida.

WHO (1994) *Application of the Principles of Substantial Equivalence to the Safety Evaluation of Foods or Food Components from Plants Derived by Modern Biotechnology.* Report of a WHO Workshop, Copenhagen, Denmark, 31 October–4 November 1994. World Health Organization, Geneva, Switzerland.

Wiebe, J.P. (1968) Inhibition of pituitary gonadotropic activity in the viviparous seaperch *Cymatogaster aggregata* (Gibbons) by a dithiocarbamoylhydrazine derivative (ICI 33828). *Canadian Journal of Zoology* 46, 751–758.

Wiegertjes, G.F., Stet, R.J.M. and Van Muiswinkel, W.B. (1994) Divergent selection for antibody production in common carp (*Cyprinus carpio* L.) using gynogenesis. *Animal Genetics* 25, 251–257.

Wieghart, M., Hoover, J.L., McGrane, M.M., Hanson, R.W., Rottman, F.M., Holtzman, S.H., Wagner, T.E. and Pinkert, C.A. (1990) Production of transgenic pigs harboring a rat phosphoenolpyruvate carboxykinase bovine growth-hormone fusion gene. *Journal of Reproduction and Fertility Supplement* 41, 89–96.

Will, P.S., Paret, J.M. and Sheehan, R.J. (1994) Pressure induced triploidy in hybrid *Lepomis. Journal of the World Aquaculture Society* 25, 507–511.

Willadsen, S.M. (1986) Nuclear transplantation in sheep embryos. *Nature* 320, 6365.

Willi, Y., van Kleunen, M., Dietrich, S. and Fischer, M. (2007) Genetic rescue persists beyond first-generation outbreeding in small populations of a rare plant. *Proceedings, Biological Sciences/The Royal Society* 274, 2357–2364.

Williams, J.G.K., Kubelik, A.R., Livak, K.J., Rafalski, J.A. and Tingey, S.V. (1990) DNA polymorphisms amplified by arbitrary primers are useful as genetic markers. *Nucleic Acids Research* 18, 6531–6535.

Wills, P.S., Sheehan, R.J. and Allen, S.K. Jr (2000) Reduced reproductive capacity in diploid and triploid hybrid sunfish. *Transactions of the American Fisheries Society* 129, 30–40.

Wilmut, I., Schnieke, A.E., McWhir, J., Kind, A.J. and Campbell, K.H.S. (1997) Viable offspring derived from fetal and adult mammalian cells. *Nature* 385, 810–813.

Wilson, K., Li, Y., Whan, V., Lehnert, S., Byrne, K., Moore, S., Pongsomboon, S., Tassanakajon, A., Rosenberg, G., Ballment, E., Fayazi, Z., Swan, J., Kenway, M. and Benzie, J. (2002) Genetic mapping of the black tiger shrimp *Penaeus monodon* with amplified fragment length polymorphism. *Aquaculture* 204, 297–309.

Wilson, M.R., Marcuz, A., van Ginkel, F., Miller, N.W., Clem, L.W., Middleton, D. and Warr, G.W. (1990) The immunoglobulin M heavy chain constant region gene of the channel catfish, *Ictalurus punctatus*: an unusual mRNA splice pattern produces the membrane form of the molecule. *Nucleic Acids Research* 18, 5227–5233.

Winans, G.A. (1980) Geographic variation in the milkfish *Chanos chanos*. I. Biochemical evidence. *Evolution* 34, 558–574.

Winkelman, D.C., McKinnon, J.J. and Schmutz, S.M. (1996) *Selection of Beef Cattle for Carcass Quality using Marker Assisted Selection and Ultrasound Techniques.* Final Report 92000159. Agriculture Development Fund, Regina, Saskatchewan, Canada.

Wirgin, I.I., Proenca, R. and Grossfield, J. (1989) Mitochondrial DNA diversity among populations of striped bass in the southeastern United States. *Canadian Journal of Zoology* 67, 891–907.

Withler, R.E., Clarke, W.C., Blackburn, J. and Baker, I. (1998) Effect of triploidy on growth and survival of pre-smolt and post-smolt coho salmon (*Oncorhynchus kisutch*). *Aquaculture* 168, 413–422.

Wohlfarth, G.W. (1977) Shoot carp. *Israeli Journal of Aquaculture/Bamidgeh* 29, 35.

Wohlfarth, G.W. (1983) Genetics of fish: application to warm water fishes. *Aquaculture* 33, 373–381.

Wohlfarth, G.W. (1993) Heterosis for growth rate in common carp. *Aquaculture* 113: 31–46.

Wohlfarth, G.W. (1994) The unexploited potential of tilapia hybrids in aquaculture. *Aquaculture and Fisheries Management* 25, 781–788.

Wohlfarth, G. and Hulata, G. (1983) *Applied Genetics of Tilapias*. Studies and Reviews 6. ICLARM, Manila.

Wohlfarth, G. and Moav, R. (1970) The effect of variation in spawning time on subsequent relative growth rate and viability in carp. *Israeli Journal of Aquaculture/Bamidgeh* 22, 42–47.

Wohlfarth, G., Moav, R. and Hulata, G. (1975a) Genetic differences between the Chinese and European races of common carp. II. Multi-character variation – a response to the diverse methods of fish cultivation in Europe and China. *Heredity* 34, 341–350.

Wohlfarth, G., Moav, R., Hulata, G. and Beiles, A. (1975b) Genetic variation in seine escapability of the common carp. *Aquaculture* 5, 375–387.

Wohlfarth, G., Shroeder, G., Hulata, G., Barash, H., Lahmi, M., Moav, R., Abramson, Z. and Brody, T. (1976) Summary of the integrated research in polyculture, feeding, intensive fertilizing by manure and genetic improvement carried out in the years 1973–1974. *Fisheries and Fishbreeding Israel* 11, 11–33.

Wohlfarth, G.W., Moav, R. and Hulata, G. (1983) A genotype–environment interaction for growth rate in the common carp, growing in intensively manured ponds. *Aquaculture* 33, 187–195.

Wolfe, B.A. and Kraemer, D.C. (1992) Interspecies nuclear transplantation. In: Seidel, E. (ed.) *Symposium on Cloning Mammals by Nuclear Transplantation, Colorado State University, 17 January*. Colorado State University, Fort Collins, Colorado, pp. 43–45.

Wolfsberg, T.G. and Landsman, D. (1997) A comparison of expressed sequence tags (ESTs) to human genomic sequences. *Nucleic Acids Research* 25, 1626–1632.

Wolters, W.R. and DeMay, R. (1996) Production characteristics of striped bass × white bass and striped bass × yellow bass hybrids. *Journal of the World Aquaculture Society* 27, 202–207.

Wolters, W.R., Chrisman, C.L. and Libey, G.S. (1981a) Induction of triploidy in channel catfish. *Transactions of the American Fisheries Society* 110, 310–312.

Wolters, W., Chrisman, R.C.L. and Libey, G.S. (1981b) Lymphocyte culture for chromosomal analyses of channel catfish, *Ictalurus punctatus*. *Copeia* 1981, 503–504.

Wolters, W.R., Chrisman, C.L. and Libey, G.S. (1982a) Erythrocyte nuclear measurements of diploid and triploid channel catfish, *Ictalurus punctatus* (Rafinesque). *Journal of Fish Biology* 20, 253–258.

Wolters, W.R., Libey, G.S. and Chrisman, C.L. (1982b) Effect of triploidy on growth and gonad development of channel catfish. *Transactions of the American Fisheries Society* 111, 102–105.

Wolters, W.R., Lilyestrom, C.G. and Craig, R.J. (1991) Growth, yield, and dress-out percentage of diploid and triploid channel catfish in earthen ponds. *Progressive Fish-Culturist* 53, 33–36.

Wong, A.C. and Van Eenennaam, A.L. (2008) Transgenic approaches for the reproductive containment of genetically engineered fish. *Aquaculture* 275, 1–12.

Woodward, C.C. and Strange, R.J. (1987) Physiological stress responses in wild and hatchery-reared rainbow trout. *Transactions of the American Fisheries Society* 116, 574–579.

Wool, I.G. (1979) The structure and function of eukaryotic ribosomes. *Annual Review of Biochemistry* 48, 719–754.

Wool, I.G. (1986) Studies of the structure of eukaryotic (mammalian) ribosomes. In: Hardesty, B. and Kramer, G. (eds) *Structure, Function, and Genetics of Ribosomes*. Springer-Verlag, New York, pp. 391–411.

Wool, I.G., Chan, Y.L. and Gluck, A. (1995) Structure and evolution of mammalian ribosomal proteins. *Biochemistry and Cell Biology* 73, 933–947.

Wool, I.G., Chan, Y.L. and Gluck, A. (1996) Mammalian ribosomes: the structure and the evolution of the proteins. In: Hershey, J.W.B., Matthews, M.B. and Sonenberg, N. (eds) *Translational Control*. Cold Spring Harbor Laboratory Press, Cold Spring Harbor, New York, pp. 685–732.

Wootton, A.N., Herring, C., Spry, J.A., Wiseman, A., Livingstone, D.R. and Goldfarb, P.S. (1995) Evidence for the existence of cytochrome P450 gene families (CYP1A, 3A, 4A, 11A) and modulation of gene expression (CYP1A) in the mussel *Mytilus* spp. *Marine Environmental Research* 39, 21–26.

Wray, S., Grant, P. and Gainer, H. (1989) Evidence that cells expressing luteinizing-hormone releasing hormone messenger-RNA in the mouse are derived from progenitor cells in the olfactory placode. *Proceedings of the National Academy of Sciences USA* 86, 8132–8136.

Wray, S., Key, S., Qualls, R. and Fueshko, S.M. (1994) A subset of peripherin positive olfactory axons delineates the luteinizing hormone releasing hormone neuronal migratory pathway in developing mouse. *Developmental Biology* 166, 349–354.

Wray, S., Fueshko, S.M., Kusano, K. and Gainer, H. (1996) GABAergic neurons in the embryonic olfactory pit/vomeronasal organ: maintenance of functional GABAergic synapses in olfactory explants. *Developmental Biology* 180, 631–645.

Wright, J., Krueger, C.C., Brussard, P.F. and Hall, M.C. (1985) Sea lamprey (*Petromyzon marinus*) populations in northeastern North America, genetic differentiation and affinities. *Canadian Journal of Fisheries and Aquatic Sciences* 42, 776.

Wright, J.E., Johnson, K., Hollister, A., May, B. and Delaney, M. (1982) Derivation and relative growth of triploids in a rainbow trout family. In: *Northeast Fish and Wildlife Conference, 13–15 April 1982, Cherry Hill, New Jersey*. Northeastern Association of Fish and Wildlife Agencies.

Wright, S. (1943) Isolation by distance. *Genetics* 28, 114.

Write, J.M. and Cook, D. (1995) DNA fingerprint based analysis of paternal and maternal effect on offspring growth and survival in communally reared rainbow trout. *Aquaculture* 137, 245–256.

Wroblewski, V.J., Masnyk, M. and Becker, G.W. (1991) Proteolytic cleavage of human growth hormone (hGH) by rat tissues *in vitro*: influence on the kinetics of exogeously administered hGH. *Endocrinology* 129, 465–474.

Wu, G., Sun, Y. and Zhu, Z. (2003) Growth hormone gene transfer in common carp. *Aquatic Living Resources* 16, 416–420.

Wu, P. and Childs, G.V. (1991) Changes in rat pituitary POMC mRNA after exposure to cold or a novel environment, detected by *in situ* hybridization. *Journal of Histochemistry and Cytochemistry* 39, 843–852.

Wu, Q., Fu, H. and Ye, Y. (1997) Effect of enzymic gene dosage on the distant hybridization of fishes. *Acta Hydrobiologica Sinica/Shuisheng Shengwu Xuebao* 21, 143–151.

Wu, T., Yang, H., Dong, Z., Xia, D., Shi, Y., Ji, X., Shen, Y. and Sun, W. (1994) The integration and expression of human growth gene in blunt snout bream and common carp. *Journal of Fisheries China/Shuichan Xuebao* 18, 284–289.

Xian, C.J., Shoubridge, C.A. and Read, L.C. (1995) Degradation of IGF-I in the adult rat gastrointestinal tract is limited by a specific antiserum or the dietary protein casein. *Journal of Endocrinology* 146, 215–225.

Xiang, J., Li, F., Zhang, C., Zhang, X., Yu, K., Zhou, L. and Wu, C. (2006) Evaluation of induced triploid shrimp *Penaeus (Fenneropenaeus) chinensis* cultured under laboratory conditions. *Aquaculture* 259, 108–115.

Xie, Y., Chen, X. and Wagner, T.E. (1997) A ribozyme-mediated, gene 'knockdown' strategy for the identification of gene function in zebrafish. *Proceedings of the National Academy of Sciences USA* 95, 13777–13781.

Xu, Z., Primavera, J.H., Pena, L.D. de la, Pettit, P., Belak, J. and Alcivar-Warren, A. (2001) Genetic diversity of wild and cultured black tiger shrimp (*Penaeus monodon*) in the Philippines using microsatellites. *Aquaculture* 199, 13–40.

Yamaha, E. and Onozato, H. (1985) Histological investigation on the gonads of artificially induced triploid crucian carp, *Carassius auratus*. *Bulletin of the Faculty of Fisheries, Hokkaido University* 36, 170–176.

Yamaki, M., Satou, H., Taniura, K. and Arai, K. (1999) Progeny of the diploid–tetraploid mosaic amago salmon. *Bulletin of the Japanese Society of Scientific Fisheries* 65, 1084–1089.

Yamamoto, E. (1999) Studies on sex-manipulation and production of cloned populations in hirame, *Paralichthys olivaceus* (Temminck et Schlegel). *Aquaculture* 173, 235–246.

Yamamoto, S. and Sugawara, Y. (1988) Induced triploidy in the mussel, *Mytilus edulis*, by temperature shock. *Aquaculture* 72, 21–29.

Yamamoto, T. (1975) A YY male goldfish from mating estrone-induced XY female and normal male. *Journal of Heredity* 66, 2–4.

Yamano, K., Yamaha, E. and Yamazaki, F. (1988) Increased viability of allotriploid pink salmon × Japanese char hybrids. *Bulletin of the Japanese Society of Scientific Fisheries* 54, 1477–1481.

Yamazaki, F. (1976) Application of hormones in fish culture. *Journal of Fisheries Research Board of Canada* 33, 948–958.

Yamazaki, F. (1983) Sex control and manipulation in fish. *Aquaculture* 33, 329–354.

Yan, S. (1998) *Cloning in Fish – Nucleocytoplasmic Hybrids*. IUBS Educational and Cultural Press, Hong Kong.

Yan, S.Y., Lu, D.Y., Du, M., Li, G.S., Lin, L.T., Jin, G.Q., Wang, H., Yang, Y.Q., Xia, D.Q., Liu, A.Z., Zhu, Z.Y., Yi, Y.L. and Chen, H.X. (1984a) Nuclear transplantation in teleosts. II. Hybrid fish from the nucleus of crucian and the cytoplasm of carp. *Scientia Sinica (B)* 27, 1029–1034.

Yan, S.Y., Wu, N.H., Yan, J.Z., Xue, G.Z. and Li, G.S. (1984b) Some evidence of cytoplasmic influences on the gene expression of hybrid fish CtMe obtained from the combination of nucleus and cytoplasm from two subfamilies of fresh water teleosts, *Ctenopharyngoden idellus* (Ct) and *Megalobrama amblycaephala* (Me) (abstract). *Journal of Embryology and Experimental Morphology* 82 (Suppl.), 105.

Yan, S.Y., Lu, D.Y., Du, M., Li, G.S., Han, Z., Yang, H.Y. and Wu, Z.A. (1985) Nuclear transplantation in teleosts: nuclear transplantation between different subfamilies – hybrid fish from the nucleus of grass carp (*Ctenopharyngoden idellus*) and the cytoplasm of blunt-snout bream (*Megalobrama amblycaephala*). *Chinese Journal of Biotechnology* 1, 1526.

Yan, S.Y., Tu, M., Yang, H.Y., Mao, Z.R., Zhao, Z.Y., Fu, L.J., Li, G.S., Huang, G.P., Li, S.H., Guo, L., Jin, G.Q., Ma, L., Lin, L.D., Xia, S.L. and Yu, L.N. (1990) Developmental incompatibility between cell nucleus and cytoplasm as revealed by nuclear transplantation experiments in teleost of different families and orders. *International Journal of Developmental Biology* 34, 255–265.

Yan, S.Y., Mao, Z.R., Yang, H.Y., Tu, M., Li, S.H., Huang, G.P., Li, G.S., Guo, L., Jin, G.Q., He, R.F., Lin, L.D., Xia, S.L., Yu, L.N., Xia, D.Q. and Zhang, N.C. (1991) Further investigation on nuclear transplantation in different orders of teleost: the combination of the nucleus of tilapia (*Oreochromis nilotica*) and the cytoplasm of loach (*Paramisgurnus dabryanus*). *International Journal of Developmental Biology* 35, 429–435.

Yan, S.Y., Mao, Z.R., Li, S.H., Huang, G.P., Chen, H.P., Jin, G.Q., Lin, L.D., Xia, S.L., Yu, L.N. and Xia, D.Q. (1993) Nuclear transplantation in different families of teleost: combination of the nucleus of zebrafish (*Brachydanio rerio*) and the cytoplasm of loach (*Paramisgurnus dabryartus*). *Journal of the Islamic Academy of Sciences* 6, 82–87.

Yang, A., Wang, Q., Kong, J., Liu, P., Liu, Z., Sun, H., Li, F., Wang, R. and Jiang, M. (1999) Triploid induction in *Chlamys farreri* by application of 6-dimethylaminopurine. *Journal of Fisheries China/Shuichan Xuebao* 23, 241–247.

Yang, D., Kusser, I., Kopke, A.K., Koop, B.F. and Matheson, A.T. (1999) The structure and evolution of the ribosomal proteins encoded in the spc operon of the archaeon (Crenarchaeota) *Sulfolobus acidocaldarius*. *Molecular Phylogenetics and Evolution* 12, 177–185.

Yang, H., Wang, R., Guo, X. and Yu, Z. (1997) Tetraploid induction by blocking polar body 1 and mitosis 1 in fertilized eggs of the scallop *Chlamys* (*Azumapecten*) *farreri* with cytochalasin B. *Journal of Ocean University Qingdao/Qingdao Haiyang Daxue Xuebao* 27, 166–172.

Yang, H., Guo, X. and Zhang, F. (1999) Tetraploid zhikong scallop (*Chlamys farreri*) produced by inhibiting polar body I. *Journal of Shellfish Research* 18, 294.

Yang, H.-S., Ting, Y.-Y., Chen and Hon-Cheng (1997) Effect of cold shock on the production of triploid zygotes and the embryonic development of small abalone, *Haliotis diversicolor supertexta* Lischke. *Acta Zooligica Taiwanica* 8, 67–78.

Yang, K., Challis, J.R., Han, V.K. and Hammond, G.L. (1991) Pro-opiomelanocortin messenger RNA levels increase in the fetal sheep pituitary during late gestation. *Journal of Endocrinology* 131, 483–489.

Yang, Y.P. and Womack, J.E. (1997) Construction of a bovine chromosome 19 linkage map with an inter-species hybrid backcross. *Mammalian Genome* 8, 262–266.

Yant, R., Smitherman, R.O. and Green, O.L. (1975) Production of hybrid (blue × channel) catfish and channel catfish in ponds. *Proceedings of the Annual Conference of the Southeastern Association of Game and Fish Commissioners* 26, 86–91.

Yee, J.K., Miyanohara, A., LaPorte, P., Bouic, K., Burns, J.C. and Friedmann, T. (1994) A general method for the generation of high-titer, pantropic retroviral vectors: highly efficient infection of primary hepatocytes. *Proceedings of the National Academy of Sciences USA* 91, 9564–9568.

Yin-Xiong, L., Farell, M.J., Lie, R., Mohanty, N. and Kirby, M.L. (2000) Double-stranded RNA injection produces null phenotypes in zebrafish. *Developmental Biology* 217, 394–405.

Yonash, N., Heller, E.D., Hillel, J. and Cahaner, A. (2000) Detection of RFLP markers associated with antibody response in meat-type chickens: haplotype/genotype, single-band, and multiband analyses of RFLP in the major histocompatibility complex. *Journal of Heredity* 91, 24–30.

Yoon, S.J., Hallerman, E.M., Gross, M.L., Liu, Z., Schneider, J.F., Faras, A.J., Hackett, P.B., Kapuscinski, A.R. and Guise, K.S. (1990) Transfer of the gene for neomycin resistance into goldfish, *Carassius auratus*. *Aquaculture* 85, 21–33.

Yoshizaki, G., Takeuchi, Y., Sakatani, S. and Takeuchi, T. (2000) Germ cell-specific expression of green fluorescent protein in transgenic rainbow trout under control of the rainbow trout vasa-like gene promoter. *International Journal of Developmental Biology* 44, 323–326.

Young, W.P., Wheeler, P.A. and Thorgaard, G.H. (1995) Asymmetry and variability of meristic characters and spotting in isogenic lines of rainbow trout. *Aquaculture* 137, 67–76.

Young, W.P., Wheeler, P.A., Coryell, V.H., Keim, P. and Thorgaard, G.H. (1998) A detailed linkage map of rainbow trout produced using doubled haploids. *Abstracts in Plant and Animal Genome* 6, 165.

Yuan, L.X., Fu, J.H., Warburton, M., Li, X.H., Zhang, S.H., Khairallah, M., Liu, X.Z., Peng, Z.B. and Li, L.C. (2000) Comparison of genetic diversity among maize inbred lines based on RFLPs, SSRs, AFLPs and RAPDs. *Yi Chuan Xue Bao/Acta Genetica Sinica* 27, 725–733.

Zapata, A.G., Chiba, A. and Varas, A. (1996) Cells and tissues of the immune system of fish. In: Iwama, G. and Nakanishi, T. (eds) *The Fish Immune System: Organism, Pathogen and Environment.* Academic Press, San Diego, California, pp. 1–62.

Zardoya, R., Vollmer, D.M., Craddock, C., Streelman, J.T., Karl, S. and Meyer, A. (1996) Evolutionary conservation of microsatellite flanking regions and their use in resolving the phylogeny of cichlid fishes (Pisces: Perciformes). *Proceedings of the Royal Society of London, Series B: Biological Sciences* 263, 1589–1598.

Zeng, Z., Lin, Q., Wu, J. and Chen, M. (1999a) Method for distinguishing oyster as diploid or triploid by outward appearance. *Marine Sciences/Haiyang Kexue Qingdao* 1, 51–53.

Zeng, Z., Lin, Q., Wu, J., Chen, P. and Chen, M. (1999b) Comparison of growth between diploids and triploids oyster, *Crassostrea gigas. Journal of Shanghai Fisheries University/Shanghai Shuichan Daxue Xuebao* 8, 119–123.

Zeng, Z., Lin, Q., Wu, J., Chen, P. and Chen, M. (1999c) The ultrastructure of development in oocytes of diploid and triploid oyster, *Crassostrea gigas. Journal of Fisheries China/Shuichan Xuebao* 23, 109–114.

Zhang, F., Wang, Y., Hu, W., Cui, Z. and Zhu, Z. (2000) Physiological and pathological analysis of the mice fed 'all-fish' gene transferred Yellow River carp. *High Technology Letters* 7, 17–19.

Zhang, G., Wang, Z., Chang, Y., Song, J., Ding, J., Wang, Y. and Wang, R. (1998) Triploid induction in Pacific abalone *Haliotis discus hannai* Ino by 6-dimethylaminopurine and the performance of triploid juveniles. *Journal of Shellfish Research* 17, 783–788.

Zhang, G., Chang, Y., Song, J., Ding, J. and Wang, Z. (2000a) Comprehensive comparison between triploid oysters induced with CB and 6-DMAP. *Journal of Fisheries China/Shuichan Xuebao* 24, 324–328.

Zhang, G., Wang, Z., Chang, Y., Song, J., Ding, J., Zhao, S. and Guo, X. (2000b) Tetraploid induction in the Pacific abalone *Haliotis discus hannai* Ino with 6-DMAP and CB. *Journal of Shellfish Research* 19, 540–541.

Zhang, J., Nguyen, H.T. and Blum, A. (1999) Genetic analysis of osmotic adjustment in crop plants. *Journal of Experimental Botany* 50, 291–302.

Zhang, P.J., Hayat, M., Joyce, C., Gonzalez, V.L., Lin, C.M., Dunham, R.A., Chen, T.T. and Powers, D.A. (1990) Gene transfer, expression and inheritance of pRSV-rainbow trout-GH cDNA in the common carp, *Cyprinus carpio* (Linnaeus). *Molecular Reproduction and Development* 25, 3–13.

Zhang, Q., Tiersch, T.R. and Cooper, R.K. (1998) Inducible expression of green fluorescent protein within channel catfish cells by a cecropin gene promoter. *Gene* 216, 207–213.

Zhang, Q., Nakayama, I., Fujiwara, A., Kobayashi, T., Oohara, I., Masaoka, T., Kitamura, S. and Devlin, R. (2001) Sex identification by male-specific growth hormone pseudogene (GH-psi) in *Oncorhynchus masou* complex and a related hybrid. *Genetica* 111, 111–118.

Zhao, X., Wong, T.K. and Batten, B. (1989) Electric field mediated protein transfer into mouse oocytes. *Techniques* 1, 37–42.

Zhao, X., Zhang, P.J. and Wong, T.K. (1993) Application of baekonization: a new approach to produce transgenic fish. *Molecular Marine Biology and Biotechnology* 2, 63–69.

Zhou, Y., Song, J., Li, X., Li, H. and Zhang, G. (2000) Comparison of bioenergetics between triploid *Crassostrea gigas* and sibling diploid. *Journal of Fisheries China/Shuichan Xuebao* 24, 504–509.

Zhu, H., Briceno, G., Dovel, R., Hayes, P.M., Liu, B.H., Liu, C.T. and Ullrich, S.E. (1999) Molecular breeding for grain yield in barley: an evaluation of QTL effects in a spring barley cross. *Theoretical and Applied Genetics* 98, 772–779.

Zhu, Z. (1992) Generation of fast growing transgenic fish: methods and mechanisms. In: Hew, C.L. and Fletcher, G.L. (eds) *Transgenic Fish.* World Scientific Publishing, Singapore, pp. 92–119.

Zhu, Z., Li, G., He, L. and Chen, S. (1985) Novel gene transfer into the fertilized eggs of goldfish (*Carassius auratus* 1758). *Journal of Applied Ichthyology* 1, 31–33.

Zhu, Z., Xu, K., Li, G., Xie, Y. and He, L. (1986) Biological effects of human growth hormone gene microinjected into the fertilized eggs of loach, *Misgurnus anguillicaudatus. Kexue Tongbao Academia Sinica* 31, 988–990.

Zhu, Z., Xie, Y., Xu, K., He, L. and Li, G. (1988) The production of transgenic mirror carp and its hereditability (abstract). In: Gjedrem, T. (ed.) *Genetics in Aquaculture III. Proceedings of the Third International Symposium on Genetics in Aquaculture, Trondheim, Norway, 20–24 June 1988.* International Association for Genetics in Aquaculture.

Zimmerman, A.M., Evenhuis, J.P., Thorgaard, G.H. and Ristow, S.S. (2004) A single major chromosomal region controls natural killer cell-like activity in rainbow trout. *Immunogenetics* 55, 825–835.

Zimmerman, A.M., Wheeler, P.A., Ristow, S.S. and Thorgaard, G.H. (2005) Composite interval mapping reveals three QTL associated with pyloric caeca number in rainbow trout, *Oncorhynchus mykiss*. *Aquaculture* 247, 85–95.

Ziv, E. and Bendayan, M. (2000) Intestinal absorption of peptides through the enterocytes. *Microscopy Research and Technique* 49, 346–352.

Ziv, E., Lior, O. and Kidron, M. (1987) Absorption of protein via the intestinal wall: a quantitative model. *Biochemical Pharmacology* 36, 1035–1039.

Zohar, Y. (1996) New approaches for the manipulation of ovulation and spawning in farmed fish. *Bulletin of the National Research Institute of Aquaculture Supplement* 2, 43–48.

Zohar, Y. and Mylonas, C.C. (2001) Endocrine manipulations of spawning in cultured fish: from hormones to genes. *Aquaculture* 197, 99–136.

Zubay, G. (1983) *Biochemistry*. Addison-Wesley, Reading, Massachusetts.

Zuerner, R.L. (1994) Nucleotide sequence analysis of IS1533 from *Leptospira borgpetersenii*: identification and expression of two IS-encoded proteins. *Plasmid* 31, 1–11.

Index